The Bacteria

VOLUME III: BIOSYNTHESIS

THE BACTERIA

A TREATISE

Volume I: Structure
Volume II: Metabolism
Volume III: Biosynthesis
Volume IV: Growth
Volume V: Heredity

The Bacteria

A TREATISE ON STRUCTURE AND FUNCTION

edited by

I. C. Gunsalus
Department of Chemistry
University of Illinois
Urbana, Illinois

Roger Y. Stanier
Department of Bacteriology
University of California
Berkeley, California

VOLUME III: BIOSYNTHESIS

1962

ACADEMIC PRESS • NEW YORK AND LONDON

ACADEMIC PRESS INC.
111 Fifth Avenue
New York 3, N. Y.

United Kingdom Edition
Published by
ACADEMIC PRESS INC. (London) Ltd.
Berkeley Square House, London W. 1

Library of Congress Catalog Card Number 59-13831

PRINTED IN THE UNITED STATES OF AMERICA

CONTRIBUTORS TO VOLUME III

Bernard D. Davis, *Department of Bacteriology and Immunology, Harvard Medical School, Boston, Massachusetts*

S. R. Elsden, *Department of Microbiology, The University, Sheffield, England*

Ernest F. Gale, *Medical Research Council Unit for Chemical Microbiology, Department of Biochemistry, University of Cambridge, England*

Shlomo Hestrin,* *Department of Biological Chemistry, The Hebrew University, Jerusalem, Israel*

Riley D. Housewright, *Fort Detrick, Frederick, Maryland*

June Lascelles, *Microbiology Unit, Department of Biochemistry, University of Oxford, England*

Boris Magasanik, *Massachusetts Institute of Technology, Cambridge, Massachusetts*

J. G. Morris,† *Microbiology Unit, Department of Biochemistry, University of Oxford, England*

Leonard E. Mortenson,¶ *Central Research Department, E. I. du Pont de Nemours and Company, Wilmington, Delaware*

Arthur B. Pardee,‡ *Departments of Biochemistry and Virology, University of California, Berkeley, California*

Rune L. Stjernholm, *Department of Biochemistry, Western Reserve University School of Medicine, Cleveland, Ohio*

Jack L. Strominger, *Department of Pharmacology, Washington University School of Medicine, St. Louis, Missouri*

Edwin Umbarger, *The Biological Laboratory, Long Island Biological Association, Cold Spring Harbor, Long Island, New York*

Harland G. Wood, *Department of Biochemistry, Western Reserve University School of Medicine, Cleveland, Ohio*

* Deceased

† Present address: Department of Biochemistry, University of Leicester, England

¶ Present address: Department of Biological Sciences, Purdue University, Lafayette, Indiana

‡ Present address: Biology Department, Princeton University, Princeton, New Jersey

PREFACE

Our basic concepts concerning the mechanisms of energy-yielding metabolism were developed largely through studies on mammalian tissues and on yeast. The subsequent exploration of energy-yielding metabolism of bacteria did not really lead to the discovery of new principles, but served chiefly to reveal the diversity of the biochemical pathways which cells can use to satisfy their energetic needs. With respect to the biosynthetic aspects of metabolism, the development of knowledge has followed a different path. When the techniques and materials required for the analysis of biosynthetic pathways became available, the advantages of performing biochemical studies with unicellular organisms were fully appreciated, and bacteria and other microorganisms thus became from the start the experimental objects of choice. Thus, in large part, our present understanding of biosynthesis has grown out of studies on the bacteria; studies on biosynthetic pathways in higher organisms have been in the main secondary. The general biochemical picture which emerges is more uniform than that which emerged from the exploration of energy-yielding metabolism. Certain small portions of biosynthesis can be pointed out as more or less specific to bacteria, for example, the synthesis of the unique structural heteropolymers of the cell wall; but in its totality, the process of biosynthesis in the bacterial cell does not seem to differ markedly from that in other types of cells.

The editors wish to thank the contributors to the present volume for their cooperation and patience, particularly in view of the delay in the appearance of this volume. We also wish to thank the publishers and the members of their staff for their encouragement and expert help in the preparation of this, as of the previous volumes of "The Bacteria."

While this volume was in the final stages of its preparation, we learned to our profound regret of the untimely death of Professor S. Hestrin, whose great contributions to knowledge of bacterial carbohydrate metabolism are amply evidenced in the chapter which he prepared for the present volume shortly before his death.

I. C. Gunsalus
R. Y. Stanier

June 1962

CONTENTS OF VOLUME III

CONTRIBUTORS TO VOLUME III v
PREFACE vii
CONTENTS OF VOLUME I xiii
CONTENTS OF VOLUME II xiv
CONTENTS OF VOLUME IV xv

1. Photosynthesis and Lithotrophic Carbon Dioxide Fixation 1

S. R. ELSDEN

I. Introduction 1
II. The Photolithotrophic Bacteria 5
III. The Autotrophic Mechanism 11
IV. The Chemolithotrophic Bacteria 19
V. The Mechanism of Carbon Dioxide Fixation in Photosynthetic Bacteria 26
VI. Energetics 29
References 37

2. Assimilation of Carbon Dioxide by Heterotrophic Organisms 41

HARLAND G. WOOD AND RUNE L. STJERNHOLM

I. Introduction 42
II. Original Proof of Assimilation of CO_2 by Heterotrophic Bacteria 44
III. Early Studies on the Mechanism of CO_2 Fixation by Heterotrophs 46
IV. Primary Reactions of CO_2 Assimilation 51
V. Methylmalonyl-Oxalacetic Transcarboxylase and the Formation of Propionate 81
VI. Reversal of α-Decarboxylation as a Mechanism of CO_2 Fixation 86
VII. The Function of Biotin in the Fixation of CO_2 89
VIII. Total Synthesis of Acetate from CO_2 98
IX. Conversion of CO_2 to Methane 108
X. Concluding Comments 109
References 112

3. Inorganic Nitrogen Assimilation and Ammonia Incorporation 119

L. E. MORTENSON

I. Present Status of Inorganic Nitrogen Metabolism by Microorganisms .. 119
II. Nitrogen Fixation 121
III. Incorporation of Ammonia into Organic Compounds 152
References 161

4. Pathways of Amino Acid Biosynthesis 167

EDWIN UMBARGER AND BERNARD D. DAVIS

I. Introduction 168
II. The Formation of Glutamic Acid; Its Role in Ammonia Assimilation and Transamination 176
III. The Conversion of Glutamic Acid to Glutamine, Proline, and Arginine 182
IV. Aspartic Acid, Asparagine, and the Key Intermediate: β-Aspartic Semialdehyde 192
V. The Conversion of Aspartic Semialdehyde to Threonine and Methionine 194
VI. The Conversion of Aspartic Semialdehyde to Diaminopimelate and Lysine 198
VII. The Formation of Alanine, Serine, Glycine, and Cysteine 202
VIII. The Formation of Isoleucine, Valine, and Leucine 208
IX. Histidine 216
X. The Aromatic Amino Acids: Tyrosine, Phenylalanine, and Tryptophan 222
XI. The Fomation of D-Amino Acids 236
XII. The Formation of Amino Acids as Major Excretion Products 237
XIII. General Considerations 238
References 243

5. The Synthesis of Vitamins and Coenzymes 253

J. G. MORRIS

I. General Introduction 253
II. Experimental Methods of Approach 254
III. Individual Vitamins and Coenzymes—Synthetic Pathways 259
References 287

6. Biosynthesis of Purine and Pyrimidine Nucleotides 295

BORIS MAGASANIK

I. Introduction 295
II. Formation of Ribose Phosphates 296
III. Biosynthesis of Purine Nucleotides 298
IV. Biosynthesis of Pyrimidine Nucleotides 320
V. Conclusion 330
References 331

7. Tetrapyrrole Synthesis in Microorganisms 335

JUNE LASCELLES

I. Introduction 335
II. Tetrapyrroles as Growth Factors 339
III. Excretion of Porphyrins by Cultures 342
IV. The Path of Tetrapyrrole Synthesis 347
V. Synthesis of Heme and Hemoproteins 358
VI. Synthesis of the Chlorophylls 362
VII. Prodigiosin 365

VIII. The Regulation of Synthesis of Tetrapyrroles 366
References 368

8. Synthesis of Polymeric Homosaccharides 373

Shlomo Hestrin

I. Historical Retrospect 373
II. Dextrans 374
III. Amyloses and Glycogen 378
IV. Cellulose 380
V. Levan 382
References 386

9. The Biosynthesis of Homopolymeric Peptides 389

Riley D. Housewright

I. Introduction 389
II. Morphological Evidence of Biosynthesis 391
III. Nutritional Requirements for Polyglutamic Acid Biosynthesis 394
IV. Enzyme Systems for Polyglutamate Biosynthesis 397
V. Hydrolysis of Homopolymers 400
VI. Composition and Structural Properties of the Peptides 402
VII. Biological Activity 407
VIII. Biosynthetic Homopolymers and the Genetic Code 409
References 409

10. Biosynthesis of Bacterial Cell Walls 413

Jack L. Strominger

I. Introduction 413
II. Structure of Bacterial Cell Walls 415
III. Isolation of Intermediates in Bacterial Cell Wall Synthesis 426
IV. Enzymic Synthesis of the Nucleotide Cell Wall Precursors 444
V. Biosynthesis of the Cell Wall 457
VI. Some Additional Problems: The Role of Thymidine and Guanosine Nucleotides in Cell Wall Synthesis 461
References 465

11. The Synthesis of Proteins and Nucleic Acids 471

Ernest F. Gale

I. The Chemical Nature of Proteins 472
II. The Chemical Nature of Nucleic Acids 479
III. Evidence for a Relationship between the Synthesis of Proteins and the Presence of Nucleic Acids 485
IV. Practical Considerations 489
V. The Site of Protein Synthesis in the Cell 499
VI. Components of the Protein-Synthesizing Mechanism 502
VII. Protein Synthesis 519

VIII. Peptides as Intermediates in Protein Synthesis 531
IX. The Synthesis of Specific Microbial Peptides 537
X. Ribonucleic Acid Synthesis 540
XI. Deoxyribonucleic Acid Synthesis 549
XII. Chloramphenicol 554
XIII. Discussion (1959) 557
Discussion (1961) 559
Addendum (1962) 562
References 565

12. The Synthesis of Enzymes 577

ARTHUR B. PARDEE

I. The Problems of Enzyme Formation 577
II. The Kinds of Enzymes Synthesized by Bacteria 578
III. The Quantities of Enzymes Synthesized by a Bacterium 592
IV. Metabolic Control and the Regulation of Enzyme Synthesis 615
V. Summary and Current Problems 618
References 621

AUTHOR INDEX 631
SUBJECT INDEX 669

The Bacteria

A TREATISE ON STRUCTURE AND FUNCTION

VOLUME I: STRUCTURE

The Bacterial Protoplasm: Composition and Organization
S. E. Luria

The Internal Structure of the Cell
R. G. E. Murray

Surface Layers of the Bacterial Cell
M. R. J. Salton

Movement
Claes Weibull

Morphology of Bacterial Spores, Their Development and Germination
C. F. Robinow

Bacterial Protoplasts
Kenneth McQuillen

L-Forms of Bacteria
E. Klieneberger-Nobel

Bacterial Viruses—Structure and Function
Thomas F. Anderson

Antigenic Analysis of Cell Structure
E. S. Lennox

Localization of Enzymes in Bacteria
Allen G. Marr

AUTHOR INDEX—SUBJECT INDEX

VOLUME II: METABOLISM

Energy-Yielding Metabolism in Bacteria
I. C. Gunsalus and C. W. Shuster

Fermentation of Carbohydrates and Related Compounds
W. A. Wood

Fermentations of Nitrogenous Organic Compounds
H. A. Barker

Cyclic Mechanisms of Terminal Oxidation
L. O. Krampitz

The Dissimilation of High Molecular Weight Substances
H. J. Rogers

Survey of Microbial Electron Transport Mechanisms
M. I. Dolin

Cytochrome Systems in Aerobic Electron Transport
Lucile Smith

Cytochrome Systems in Anaerobic Electron Transport
Jack W. Newton and Martin D. Kamen

Cytochrome-Independent Electron Transport Enzymes of Bacteria
M. I. Dolin

Bacterial Photosynthesis
David M. Geller

Bacterial Luminescence
W. D. McElroy

AUTHOR INDEX—SUBJECT INDEX

VOLUME IV: GROWTH

Synchronous Growth
O. Maaløe

Nutritional Requirements of Microorganisms
B. M. Guirard and E. E. Snell

Ecology of Bacteria
R. E. Hungate

Exoenzymes
M. R. Pollock

Permeation
G. N. Cohen and A. Kepes

Physiology of Sporulation
Harlyn O. Halvorson

Temperature Relationships
John Ingraham

Halophilism
Helge Larsen

Antimicrobial Agents
Bernard D. Davis and David S. Feingold

AUTHOR INDEX—SUBJECT INDEX

CHAPTER 1

Photosynthesis and Lithotrophic Carbon Dioxide Fixation

S. R. ELSDEN

I. Introduction . . . 1
II. The Photolithotrophic Bacteria . . . 5
III. The Autotrophic Mechanism . . . 11
IV. The Chemolithotrophic Bacteria . . . 19
V. The Mechanism of Carbon Dioxide Fixation in Photosynthetic Bacteria . . . 26
VI. Energetics . . . 29
References . . . 37

I. Introduction*

Biochemical reaction sequences that result in the fixation of carbon dioxide are conveniently separated into two types, which will be referred to as heterotrophic and autotrophic, respectively. The former, which occur in all cells and which will be discussed by H. G. Wood in Chapter 2 of this volume, involve the addition of carbon dioxide to organic acceptors but do not result in the total synthesis of the acceptor from carbon dioxide. In general the heterotrophic reactions are steps in the synthesis of specific compounds. The autotrophic reaction involves the addition of carbon dioxide to a specific acceptor and the over-all process is so constituted that the acceptor is regenerated and is ultimately synthesized entirely from carbon dioxide; the process is in fact cyclic. Organisms capable of synthesizing all their organic matter from carbon dioxide possess both the heterotrophic and the autotrophic enzyme systems, whereas those that grow only upon organic compounds contain, with certain exceptions which will be discussed later, only the heterotrophic systems.

This chapter is concerned with the nature of the autotrophic reaction which enables bacteria to use carbon dioxide as sole source of carbon for growth. The organisms that fix carbon dioxide in this way fall into two distinct groups depending on the nature of their primary energy source. The first group, for which the name chemolithotrophic bacteria has been suggested[1] obtain their energy by the oxidation of inorganic compounds. The second group, the photolithotrophic bacteria, use light as their energy

* The following abbreviations are used: DPN and DPNH, oxidized and reduced diphosphopyridine nucleotide; TPN and TPNH, oxidized and reduced triphosphopyridine nucleotide; ATP, adenosine triphosphate; ADP, adenosine diphosphate.

source; the photometabolic process of these latter organisms has already been discussed by D. M. Geller in Volume II (Chapter 10) of this treatise.

We owe the discovery of the chemolithotrophic bacteria to Winogradsky, who summarized the results of his studies of the colorless sulfur bacteria[2] as follows:

"En résumé, les Sulfobactéries présentent un type physiologique nouveau et inattendu, dont l'énergétique est différente du type dominant. Leur processus vital se joue d'après un schéma beaucoup plus simple, empruntant toute l'énergie nécessaire à une réaction chimique inorganique, l'oxidation du soufre."

A year later, in 1888, Winogradsky[3] showed that the iron bacteria obtained their energy for growth by the oxidation of ferrous salts to ferric oxide. The iron bacteria, like the sulfur bacteria, grew in the complete absence of organic carbon and Winogradsky concluded that both types of organism were able to use carbon dioxide as their source of carbon.

Winogradsky next turned his attention to the nitrifying bacteria and in his second paper of the series[4] he showed that coincident with the oxidation of ammonia the organic carbon of the medium, determined chemically by a wet combustion process, increased. His conclusions merit repetition in full:

"En résumant, l'agent de la nitrification nous apparaît comme doué de propriétés marquantes, qui en font *un type physiologique nouveau dans la Science.* Ces caractères se résument comme suit:

1° *Développement dans un milieu purement minéral pourvu de substance inorganique oxydable.*

2° *Processus vital étroitement lié à la présence de cette substance, qui est l'ammoniac dans le cas de la nitrification.*

3° *Oxydation de cette substance, comme seule source d'énergie*

4° *Aucun besoin en aliment organique, ni en qualité de materiel plastique, ni comme source d'énergie.*

5° *Incapacité de décomposer les substances organique, dont la présence ne fait qu'entraver le dévelopment.*

6° *Assimilation de l'acide carbonique—seul source de carbone—par chimiosynthèse.*

Table I, taken from his third paper,[5] shows the relationship he found between the amount of ammonia oxidized and the amount of organic carbon formed. Subsequent work showed that the oxidation of ammonia to nitrate proceeded in two stages: first, the oxidation of ammonia to nitrite; second, the oxidation of nitrite to nitrate and culminating with the isolation of the two organisms concerned, *Nitrosomonas* and *Nitrobacter*, respectively.

This work, justly called classic, set the pattern of research for the next fifty years. During this period attention was focused, first, on the isolation of other types of chemolithotrophic organisms, and second, on establishing the dimensions of the ratio between the amount of carbon dioxide assimilated and the amount of the energy source oxidized. From this ratio the thermodynamic efficiency of the process could be calculated. Table II, taken from Baas-Becking and Parks[6] gives some indication of the results obtained. It will be seen that the thermodynamic efficiency of the process is low. Hofmann and Lees[13] have pointed out that these calculations apply only to the stage of growth the culture had reached at the time the analyses

TABLE I

RELATIONSHIP BETWEEN AMMONIA OXIDIZED AND CARBON DIOXIDE ASSIMILATED[5]

	Culture number			
	11	12	26	30
Ammonia N oxidized (mg.)	722	506.1	928.3	815.4
Carbon assimilated (mg.)	19.7	15.2	26.4	22.4
Ammonia N oxidized (carbon assimilated)	36.6	33.3	35.2	36.4

TABLE II

THERMODYNAMIC EFFICIENCIES[a]

Reaction	Free energy efficiency	Reference
$H_2 + 0.5O_2 = H_2O$	26.4	7
$CH_4 + 2O_2 = CO_2 + 2H_2O$	0.6–29.6	8
$NH_4^+ + 1.5O_2 = NO_2^- + H_2O + 2H^+$	7.9	9
$NO_2 + 0.5O_2 = NO_3^-$	5.9	9
$S + 1.5O_2 + H_2O = H_2SO_4$	8.3	10
$6KNO_3 + 5S + 2CaCO_3 = 3K_2SO_4 + 2CaSO_4 + 2CO_2 + 2\ N_2$	5.0	11
$8KNO_3 + 5Na_2S_2O_3 + 2NaHCO_3 = 6Na_2SO_4 + 4\ K_2SO_4 + 4N_2 + 2CO_2 + H_2O$	9.0	12

[a] Calculated by Baas-Becking and Parks.[6]

were made and that, if old cultures are examined, then the values obtained may be low. Certainly, in the case of *Nitrosomonas*, which Hofmann and Lees examined, the efficiency of young cultures appears to be of the order of 40% whereas that of old cultures was 7%.

The amount of oxygen consumed may also be taken as an index of the amount of energy made available by an oxidative process; the data for a number of species showing the relationship between carbon dioxide fixed and oxygen consumed are given in Table III. It will be seen that the values obtained for this ratio, using both sulfur and iron-oxidizing bacteria, and measured in both short-term *in vitro* experiments and in long-term growth experiments agree surprisingly well with one exception, namely, the results of Vogler and Umbreit.[16, 17] Interesting as the calculations of the thermodynamic efficiency may be, such calculations have given no indication whatsoever of the mechanism by which carbon dioxide is fixed.

The first attempt to understand the nature of the processes involved came from the work of Vogler[16] and Umbreit.[17] These authors measured the amount of oxygen consumed and the amount of carbon dioxide assimilated during the oxidation of elementary sulfur by washed suspensions of *Thiobacillus thiooxidans*.

TABLE III

RELATIONSHIP BETWEEN OXYGEN CONSUMED AND CARBON DIOXIDE ASSIMILATED[a]

Organism	Substrate	Oxidant	O_2/CO_2	Reference
Thiobacillus denitrificans[b]	$S_2O_3^{--}$	NO_3^-[d]	9	12
Thiobacillus thiooxidans[b]	S	O_2	18	14
Thiobacillus thioparus[b]	$S_2O_3^{--}$	O_2	19	15
Thiobacillus thiooxidans[c]	S	O_2	2.9	16
Thiobacillus thiooxidans[c]	S	O_2	1.5	17
Thiobacillus thiooxidans[c]	$S_2O_3^{--}$	O_2	9–26	18, 23
Thiobacillus thioparus[c]	$S_2O_3^{--}$	O_2	9–26	
Thiobacillus denitrificans[c]	$S_2O_3^{--}$	NO_3^-[d]	4.6–11	18, 23
Iron-oxidizing bacterium[c]	Fe^{++}	O_2	37	19
Iron-oxidizing bacterium[c]	S	O_2	32	19
Hydrogenomonas facilis[c]	H_2	O_2	2.0	20
Hydrogenomonas ruhlandii[c]	H_2	O_2	2.7	21

[a] Data of this table are taken in part from reference 25.
[b] Experiments made with growing cultures.
[c] Experiments made with washed cell suspensions.
[d] Value for oxygen is calculated according to the equation

$$2HNO_3 \rightarrow H_2O + N_2 + 2.5O_2$$

bacillus thiooxidans. Vogler[16] observed that washed suspensions which had been allowed to oxidize sulfur in the absence of carbon dioxide acquired the ability to take up measurable amounts of carbon dioxide when placed in an oxygen-free atmosphere, i.e., under conditions where no further oxidation of sulfur could occur. These experiments were considered to show that the oxidation of sulfur provides the organism with a store of energy which could subsequently be used to fix carbon dioxide. In other words, the fixation of carbon dioxide and the energy supply could be separated in time. It was then shown[17] that if sulfur was oxidized in the absence of carbon dioxide, inorganic phosphate disappeared from the medium and reappeared when carbon dioxide was admitted to the system. It was concluded from these results that during the oxidation of sulfur *Thiobacillus thiooxidans* synthesized adenosine triphosphate (ATP) and that this compound provided the energy for the fixation of carbon dioxide. The results of these experiments have been subjected to a devastating analysis by Baalsrud and Baalsrud,[23] Baalsrud,[22] Vishniac and Santer.[24] The essential feature of their criticisms was that the amount of carbon dioxide fixed by the cells under anaerobic conditions was very much greater than could have reasonably been accounted for by the inorganic phosphate taken up—the observed ratio of carbon dioxide fixed: inorganic phosphate taken up was 47:1. It is difficult to see how 1 mole of ATP could supply the energy for the fixation of 47 moles of carbon dioxide; indeed, according to the current hypothesis the CO_2:ATP ratio should be 1:3. Attempts to repeat these experiments, while confirming the uptake of inorganic phosphate during the oxidation of sulfur, have failed to confirm the anaerobic fixation of carbon dioxide in the amounts claimed by Umbreit and Vogler.[16,17] Although, in my opinion, these criticisms were justified, recent work on the mechanism of carbon dioxide fixation has shown that ATP does indeed play an intimate part in the fixation of carbon dioxide. It thus has transpired that although the evidence provided by Vogler and Umbreit was inadequate to substantiate their main conclusion, their concept that the oxidation of sulfur to sulfate is coupled to the synthesis of ATP, which is then used for the fixation of carbon dioxide, is probably correct.

II. The Photolithotrophic Bacteria

Pure cultures of photosynthetic bacteria were first isolated and described by van Niel.[26] The organisms, which were strict anaerobes, were of two main types: the green sulfur bacteria or Chlorobacteriaceae, the type species of which is *Chlorobium limicola*; and the purple sulfur bacteria or Thiorhodaceae, of which representatives of two genera *Chromatium* and *Thiocystis* were obtained in pure culture. The isolates of *Chlorobium limicola* converted carbon dioxide to cell material and oxidized hydrogen sulfide to elementary sulfur anaerobically in the light; the sulfur thus formed was deposited outside of the cell. Analysis of the culture fluid at the end of the growth period showed that the chemical changes which occurred fitted equation (1)

$$2H_2S + CO_2 \rightarrow [CH_2O] + 2S \qquad (1)$$

A more detailed study of the green bacteria was published by Larsen[27,28] some twenty years later. In addition to strains of *Chlorobium limicola*, Larsen, with the aid of the enrichment culture technique, isolated a new species, *Chlorobium thiosulfatophilum*, which was capable of using thiosulfate as well as hydrogen sulfide. In contrast to the organisms isolated by van Niel, Larsen's two species oxidized hydrogen sulfide to a mixture of sulfur and sulfate. The elementary sulfur, which accumulated in the medium in the early stages of the growth, was subsequently oxidized to sulfate, so that the latter became the major end product. Calculation of the amount of carbon dioxide assimilated from the amounts of sulfur and sulfate formed according to equations (1) and (2) gave a value which agreed with that observed.

$$H_2S + 2CO_2 + 2H_2O \rightarrow 2[CH_2O] + H_2SO_4 \quad (2)$$

Manometric experiments with washed cell suspensions of both species of green sulfur bacteria showed that, in the light, sulfide was oxidized quantitatively to sulfate and the amount of carbon dioxide assimilated was 90% of that predicted by equation (2). Although *Chlorobium thiosulfatophilum* would not grow photosynthetically upon tetrathionate, illuminated washed cell suspensions, in the presence of carbon dioxide, oxidized both this compound and thiosulfate with the assimilation of carbon dioxide according to equations (3) and (4).

$$2CO_2 + Na_2S_2O_3 \rightarrow 2[CH_2O] + Na_2SO_4 + H_2SO_4 \quad (3)$$

$$7CO_2 + Na_2S_4O_6 \rightarrow 7[CH_2O] + Na_2SO_4 + 4H_2SO_4 \quad (4)$$

Washed cell suspensions of both species reduced carbon dioxide with hydrogen in the light according to equation (5).[29, 30]

$$2H_2 + CO_2 \rightarrow [CH_2O] + H_2O \quad (5)$$

In addition, Larsen demonstrated that *Chlorobium thiosulfatophilum* was able to grow upon a mixture of hydrogen and carbon dioxide. The ability of *Chlorobium limicola* to do likewise was not tested.

The Thiorhodaceae like the green sulfur bacteria, metabolize sulfur compounds in the light with the fixation of an amount of carbon dioxide equivalent to the amount of sulfur compound oxidized.[26] During the early stages of growth with hydrogen sulfide droplets of elementary sulfur accumulated within the cells but, when all the sulfide had disappeared from the medium, the intracellular sulfur was oxidized quantitatively to sulfate and the over-all reaction observed was identical with that found in the green bacteria isolated by Larsen. These organisms also resembled the green bacteria in their ability to use thiosulfate; Roelofsen[29] showed that they used hydrogen to reduce carbon dioxide.

In the reactions so far discussed there is an almost stoichiometric relationship between the amount of carbon dioxide assimilated and the amount of the electron donor oxidized. The ability of these photosynthetic bacteria to use radiant energy for growth with carbon dioxide as the carbon source, coupled with the fact that the pigments responsible for the light reaction are related to chlorophyll a, suggests that there is a close relationship between green plant photosynthesis on the one hand and bacterial photosynthesis on the other. This led van Niel[31-34] to develop his general hypothesis to cover all types of both photosynthesis, both that of bacteria and that of the green plant. According to van Niel, light is used to split water and the over-all chemistry of the process is usually expressed by equation (6)

$$2H_2A + CO_2 \rightarrow [CH_2O] + H_2O + 2A \tag{6}$$

The general implications of the equation have been discussed by Geller (Volume II, Chapter 10) and will not be enlarged upon further here. What is significant in the present context is the implication that the mechanism of carbon dioxide fixation is the same both in bacteria and in the green plant.

The discovery of Müller[35] that the Thiorhodaceae will grow anaerobically upon organic compounds if the cultures are illuminated complicated matters. Analysis of the culture medium at the end of growth showed that most of the organic compound supplied was assimilated and depending on the oxidation level of the substrate, carbon dioxide was either produced or assimilated. While there was no doubt that growth under these conditions was photosynthetic in the sense that radiant energy was essential, in the case of those substrates in which there was a net output of carbon dioxide it was not established that fixation of this gas was involved; in contrast to growth in the presence of inorganic hydrogen donors there appeared to be no stoichiometric relationship between the amount of carbon dioxide fixed and the amount of organic substrate assimilated.

The third group of bacteria which grow at the expense of radiant energy is the Athiorhodaceae.[36] These organisms, in the main, use organic compounds as hydrogen donors but many strains will, in addition, use hydrogen to reduce carbon dioxide in the light[30] in a manner similar to the Chlorobacteriacae[27] and the Thiorhodaceae.[29] Gaffron[37] introduced the use of the manometric method to study the photometabolism of organic compounds by members of the Athiorhodaceae, and examined in particular the photometabolism of fatty acids from acetic to nonanoic. He observed that, with acetate, there was a net output of carbon dioxide, whereas with the higher fatty acids carbon dioxide was assimilated. The amount of carbon dioxide fixed was proportional to, but not equivalent to, the chain length. Results similar to those of Gaffron was subsequently obtained by van Niel[32] using *Rhodospirillum rubrum*. Gaffron[37] considered that radiant energy was used for the assimilation of the fatty acids by these organisms and claimed to have isolated the assimilation product formed from acetate. This view was not generally accepted but it has recently received support from the work of Stanier and his colleagues,[38] who have isolated and identified the assimilation product as poly-β-hydroxybutyric acid.

There is one recorded exception to the general rule that during the photometabolism of organic compounds the carbon of the substrate is converted to cell material. Foster,[39] using the enrichment culture technique with isopropanol as the substrate, isolated an organism which in the light oxidized isopropanol to acetone and assimilated carbon dioxide. The acetone so produced accumulated in the medium and did not appear to be further metabolized. Analysis of the culture medium showed that over-all changes which accompanied growth could be approximated to equation (7)

$$2 \begin{array}{l} CH_3 \diagdown \\ \qquad CHOH \\ CH_3 \diagup \end{array} + CO_2 \rightarrow 2 \begin{array}{l} CH_3 \diagdown \\ \qquad C{:}O \\ CH_3 \diagup \end{array} + [CH_2O] + H_2O \qquad (7)$$

These observations of Foster's were used by van Niel[32-34] to support his view that the photometabolism of organic compounds involved reactions similar to those found in the Chlorobacteriacae and the Thiorhodaceae, namely, that the primary light reaction is the photolysis of water followed by the oxidation of the H-donor and the reduction of carbon dioxide. Unfortunately the organism isolated by Foster was lost. Subsequent attempts by Siegel and Kamen[40] to isolate new strains capable of growing upon an isopropanol in the light in a manner similer to that of Foster's organism failed. They did, however, obtain a pure culture of an organism identified as *Rhodopseudomonas gelatinosa* which would use isopropanol in the light; but during the growth of this organism, the isopropanol was assimilated rather than converted quantitatively to acetone.

The Athiorhodaceae share with the Thiorhodaceae and the Chlorobacteriacae the property of being able to reduce carbon dioxide with hydrogen on illumination and the fact that the chlorophyll of the Athiorhodaceae is identical with that of the Thiorhodaceae suggests that the metabolic processes of all three have much in common. But, at the same time, the facts thus far discussed do not permit us to draw the conclusion that, during the photometabolism of organic compounds, such carbon dioxide as is fixed is assimilated by the autotrophic process.

During the photometabolism of acetate carbon dioxide is produced, approximately 0.2 moles of carbon dioxide per mole of acetate metabolized.[32, 37] Cutinelli and his colleagues[41-43] investigated in detail the metabolism of acetate by *R. rubrum*. They showed that during this process some carbon dioxide was fixed and at least some of the acetate was assimilated without rupture of the carbon chain. To establish this point they used, first, unlabeled carbon dioxide and acetate labeled as follows: $C^{13}H_3C^{14}OOH$, and second, $C^{14}O_2$ and unlabeled acetate. In this way they could follow the fate not only of the individual carbon atoms involved but also of the intact acetate molecule. At the end of the experiment the cells were harvested, the protein extracted, hydrolyzed, and the amino acids separated and degraded.

Analysis in this way of the alanine and aspartate suggested very strongly that a primary step in the assimilation of acetate involved the addition of carbon dioxide to the carboxyl group of acetate. Whether acetate itself was involved or a derivative was not and is not clear.

These experiments suggested that during the photometabolism of acetate some carbon dioxide is fixed despite the fact that there is a net output of carbon dioxide during the metabolism of acetate in the light. A more detailed study of the metabolism of carbon dioxide when organic substrates are used in the light was made by Ormerod.[44] In these experiments washed suspensions of *R. rubrum* were incubated with a variety of organic substrates in the light in the presence of $C^{14}O_2$. The net output or uptake of carbon dioxide was measured manometrically and the amount of $C^{14}O_2$

TABLE IV

Carbon Dioxide Fixation by *Rhodospirillum rubrum*[a]

Substrate	$C^{14}O_2$ fixed (c.p.m.)		$C^{14}O_2$ (μmoles)	Manometric CO_2 (μmoles)
	Endogenous	+ Substrate		
Acetate (20 μmoles)	10,715	5,970	—	+4.0
Propionate (10 μmoles)	24,550	62,075	−5.67	−2.3
Butyrate (8.7 μmoles)	10,900	32,300	−3.24	−3.2
Pyruvate (18.5 μmoles)	22,500	29,200	−1.03	+11.5
DL-Lactate (14 μmoles)	17,250	52,550	−5.34	+0.8
Succinate (10 μmoles)	15,750	48,800	−5.09	+7.4
Fumarate (10 μmoles)	15,800	34,000	−2.81	+11.0
L-Malate (10 μmoles)	20,600	38,900	−2.82	+11.8

[a] Data taken from Ormerod.[44]

fixed by the suspension was obtained by assaying the radioactivity of the suspension after removal of the dissolved $C^{14}O_2$. The results he obtained are summarized in Table IV. Only two of the substrates examined, namely, propionate and *n*-butyrate, were metabolized with a net uptake of carbon dioxide. In the case of propionate the amount of $C^{14}O_2$ assimilated was more than double the net uptake measured by the manometric method. On the other hand, when *n*-butyrate was the substrate the amount of carbon dioxide taken up, measured gasometrically, agreed with the value obtained by the tracer method after correcting the latter value for the fixation observed in the absence of substrate—there was invariably no pressure change in the manometers when the cells were incubated in the light in the absence of substrate. The photometabolism of pyruvate, lactate, succinate, fumarate, and malate was accompanied by an output of carbon dioxide but despite this, the amount of $C^{14}O_2$ assimilated was greater in the presence of these substrates than that observed when the cells were incubated in the absence of substrate. In contrast to these seven substrates acetate behaved quite differently: when suspensions of *R. rubrum* were incubated with acetate in the presence of $C^{14}O_2$ the cells were less radioactive than when they had been incubated with $C^{14}O_2$ alone.[45] This inhibition of carbon dioxide fixation by acetate suggests that carbon dioxide plays little part in the photometabolism of acetate. A similar conclusion was reached by Stanier *et al.*[38] in their study of the metabolism of organic acids by both *Rhodospirillum rubrum* and by *Rhodopseudomonas spheroides*. On the other hand, these these results do seem to conflict with those of Cutinelli *et al.*[41-43]

The results of Ormerod's experiments show that, with the exception of acetate, the photometabolism of organic compounds by washed suspensions of *R. rubrum* is associated with the assimilation of carbon dioxide; at the same time, however, they provide no indication of the mechanism of carbon dioxide fixation associated with the metabolism of organic compounds. As far as these experiments go the $C^{14}O_2$ incorporated might either have been fixed by conventional heterotrophic reactions (see Chapter 2) or by the autotrophic pathway. The fact that cells incubated in the light in the absence of substrate fixed considerable quantities of $C^{14}O_2$ without any detectable pressure change in the manometers suggests that part at least of the fixation observed was the result of exchange reactions. *Rhodospirillum rubrum* was shown by Vernon and Kamen[46] to contain the malic enzyme and it is reasonable to suggest that in the presence of this enzyme the intracellular malate will become labeled. Propionate metabolism was also peculiar, in the sense that the amount of $C^{14}O_2$ fixed greatly exceeded the uptake of carbon dioxide measured manometrically; the suggestion was made that the photometabolism of propionate is associated with a carboxylation reaction.[44]

At high light intensity, photosynthesis by green plants is inhibited by cyanide. It has been concluded from this and other evidence[47, 48] that cyanide inhibits some step in the assimilation of carbon dioxide. This inhibitor is thus a possible means of testing whether the autotrophic mechanism of carbon dioxide fixation plays a part in the photometabolism of organic compounds. The effect of cyanide (10^{-3} *M*) on the photometabolism of *Rhodospirillum rubrum* and *Rhodopseudomonas spheroides* has been studied by Elsden.[49] The results of his experiments are given in Table V. It will be seen that in the case of *R. rubrum* the photometabolism of succinate, propionate, and butyrate is some 80% inhibited; that of lactate, malate and fumarate some 60%; and that of acetate only 20%. The pattern with *R. spheroides* is similar save that, in the case of acetate, cyanide was slightly stimulatory. Comparing these results with those of Ormerod[44] (Table IV) it would appear that the greater the amount of carbon dioxide assimilated per mole of substrate the greater the sensitivity to cyanide. The anomalous results with acetate again suggest that fixation of carbon dioxide is not essential for the photometabolism of this compound. If cyanide does in fact inhibit a specific step in the autotrophic fixation of carbon dioxide by green plants (and recent *in vitro* experiments by Trebst *et al.*[50] suggest that it is carboxydismutase which is inhibited), then these results with cyanide are consistent with the view that the autotrophic mechanism of carbon dioxide fixation plays a part in the photometabolism of most but not all organic compounds by these two members of the Athiorhodaceae.

TABLE V

Action of Cyanide (10^{-3} *M*) on the Rate of Light Metabolism of Organic Compounds

Organism	Substrate	Q_{CO_2} − Cyanide	Q_{CO_2} + Cyanide	Inhibition (%)
Rhodospirillum rubrum	Succinate	19.5	1.67	91.5
	Fumarate	14.0	5.7	65
	L-Malate	13.4	5.06	62
	n-Butyrate	18.8	2.89	84
	Acetate	52.3	43.7	16.5
	Propionate	20.2	2.76	87
	DL-Lactate	6.12	2.07	65
Rhodopseudomonas spheroides	Succinate	21.4	0	100
	Fumarate	19.1	1.92	90
	L-Malate	17.5	1.68	90.5
	n-Butyrate	10.2	0	100
	Acetate	6.46	10.7	−65
	DL-Lactate	17.9	0.66	96

III. The Autotrophic Mechanism

No further progress in the study of the mechanism of carbon dioxide fixation in either the photolithotrophic or the chemolithotrophic bacteria was made until the biochemistry of the process occurring in green plants had been worked out. Once this was achieved there became available, first, a working hypothesis on which to base experiments; second, an experimental approach for investigating the process in whole cells; third, assay methods for the various enzymes participating in the reaction. As will be seen later, in those cases where these procedures have been applied to the bacteria, it has been shown that the mechanism of carbon dioxide fixation in the bacteria is the same as that found in the green plant.

In order to understand the significance of the experiments that have been performed with the various bacterial systems it is necessary to describe the biochemistry of the process found in green plants and to discuss the evi-

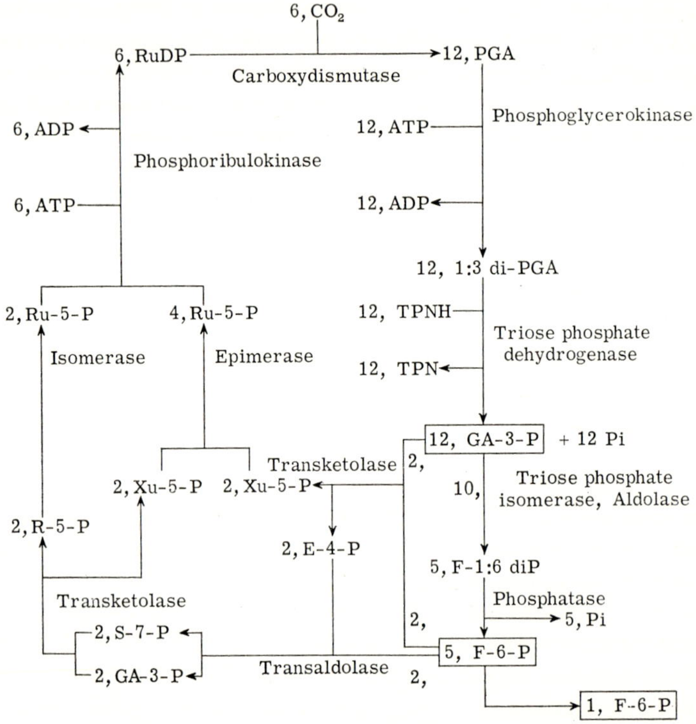

SUM: $6\ CO_2 + 18\ ATP + 12\ TPNH \longrightarrow 1\ F\text{-}6\text{-}P + 17\ ADP + 12\ TPN$

FIG. 1. The autotrophic mechanism of carbon dioxide fixation. PGA = 3-phosphoglyceric acid; 1:3 di-PGA = 1,3-diphosphoglyceric acid; GA-3-P = glyceraldehyde-3-phosphate; E-4-P = erythrose-4-phosphate; R-5-P = ribose-5-phosphate; Ru-5-P = ribulose-5-phosphate; RuDP = ribulose-1,5-diphosphate; Xu-5-P = xylulose-5-phosphate; F-6-P = fructose-6-phosphate; F-1:6 diP = fructose-1,6-diphosphate; S-7-P = sedoheptulose-7-phosphate; ATP = adenosine triphosphate; ADP = adenosine diphosphate; TPN = triphosphopyridine nucleotide; TPNH = reduced triphosphopyridine nucleotide; Pi = inorganic orthophosphate.

dence upon which the current theory is based. Figure 1 is a scheme of the events leading to the formation of one molecule of fructose-6-phosphate from carbon dioxide. It will be seen that carbon dioxide is involved in one reaction only; namely, the formation of 3-phosphoglyceric acid from ribulose-1,5-diphosphate. Only two hitherto unknown and probably unique enzymes are involved: namely, phosphoribulokinase and carboxydismutase. The former, analogous in its action to but different in specificity from the phosphofructokinase of the Embden-Meyerhof system, phosphorylates ribulose-5-phosphate with ATP, giving ribulose-1,5-diphosphate and ADP. Carboxydismutase catalyzes a reaction between carbon dioxide and ribulose-1,5-diphosphate to produce two molecules of 3-phosphoglyceric acid. According to this scheme the formation of 3-phosphoglyceric acid is the first step in the process leading to the formation of carbohydrate from carbon dioxide. ATP plays a part in one other step, namely, the formation of 1,3-diphosphoglyceric acid which is subsequently reduced to triosephosphate. This is the only reduction step in the process; in the green plants reduced triphosphopyridine nucleotide (TPNH) appears to be the pyridine nucleotide concerned. The remaining enzymes participating in this scheme are not unique to photosynthetic systems and are found in most, if not all cells.

The unraveling of this complex pathway is due largely to the efforts of Calvin and his colleagues, who identified the various intermediates and followed the changes in the amounts of the intermediates when the external conditions were altered.

The key procedures employed in this work involved the use of $C^{14}O_2$ to label the intermediates and of two-dimensional paper chromatography for the separation of the labeled intermediates, which were located on the chromatograms by radioautography. A full description of the procedures employed is given in refs. 51 and 52. The first experiments were designed to isolate and identify the first stable product of carbon dioxide fixation. This entailed exposing illuminated suspensions of *Chlorella* to $C^{14}O_2$ for very short intervals of time. Analysis of extracts prepared after 60 seconds' exposure to $C^{14}O_2$ showed the presence of a large number of labeled compounds, indicating that carbon dioxide very rapidly incorporated into cell constituents. By decreasing the exposure time the number of radioactive areas on the chromatograms was progressively reduced so that samples taken 2–5 seconds after addition of $C^{14}O_2$ contained one major radioactive area which was subsequently identified as carboxyl-labeled 3-phosphoglyceric acid. Triosephosphate and the sugar phosphate areas also became labeled very early. The separation of the sugar phosphates with the solvent systems employed (phenol:water and butanol:propionic acid:water mixtures) was not adequate for the rigid identification of the various sugars. The following procedure was therefore adopted: each radioactive area was eluted from the chromatogram, hydrolyzed with a phosphatase preparation, and the free sugars separated chromatographically. In this way the important intermediates ribulose-1,5-diphosphate and sedoheptulose-7-phosphate were identified.[53] Ribulose diphosphate was a major constituent of the diphosphate area. These experiments were not carried out under steady state conditions and in consequence did not permit of a kinetic analysis of the fixation process.

Gaffron *et al.*[54] confirmed that carboxyl-labeled phosphoglyceric acid was the first stable product of carbon dioxide assimilation; in addition they made the very interesting observation that, under steady state conditions, carbon dioxide continued to be fixed for a short time after switching off the light and that the carbon dioxide so fixed appeared almost exclusively in the carboxyl group of phosphoglyceric acid. They concluded from these experiments that photosynthesizing cells contain a compound which is converted by carboxylation to phosphoglyceric acid by a dark reaction.

Similar experiments were performed by Calvin and Massini.[55] Under steady state conditions with all intermediates saturated with the tracer, it is possible to use the radioactivity of an intermediate as an index of the amount of that intermediate present in the cells. Calvin and Massini[55] used this procedure to study the effect of stopping illumination on the amounts of the various intermediates which had been previously identified. They observed that turning off the light had no effect on the radioactivity (and hence the amount) of the hexose monophosphates present. On the other hand, the radioactivity of phosphoglyceric acid continued to increase, which agreed with the observations of Gaffron's group,[54] while that of the diphosphate area decreased by an amount roughly equivalent to the increased observed phosphoglyceric acid. Since ribulose diphosphate is the main constituent of the diphosphate area, it was suggested that this compound, in the presence of carbon dioxide, is converted to phosphoglyceric acid.

The effect of altering the concentration of carbon dioxide on the amounts of the various intermediates was examined.[56] It was found that, under steady state conditions, reducing the pressure of carbon dioxide from 1 to 0.003% caused a precipitous fall in the amount of phosphoglyceric acid present in the cells and at the same time the amount of ribulose diphosphate increased. No significant changes were found in the amounts of the other intermediates. This experiment clearly complements the one previously described and reinforces the conclusion that ribulose diphosphate is a precursor of phosphoglyceric acid.

In addition, Bassham *et al.*[56] studied the time course of carbon dioxide fixation under steady-state conditions. These experiments confirmed that phosphoglyceric acid was the earliest stable product; triose, pentose, and sedoheptulose phosphates next became labeled, at approximately equal rates. Degradation of these intermediates isolated from cells that had been exposed for a very short time to $C^{14}O_2$ revealed a labeling pattern which was characteristic for each intermediate and which is shown in Fig. 2 and Table

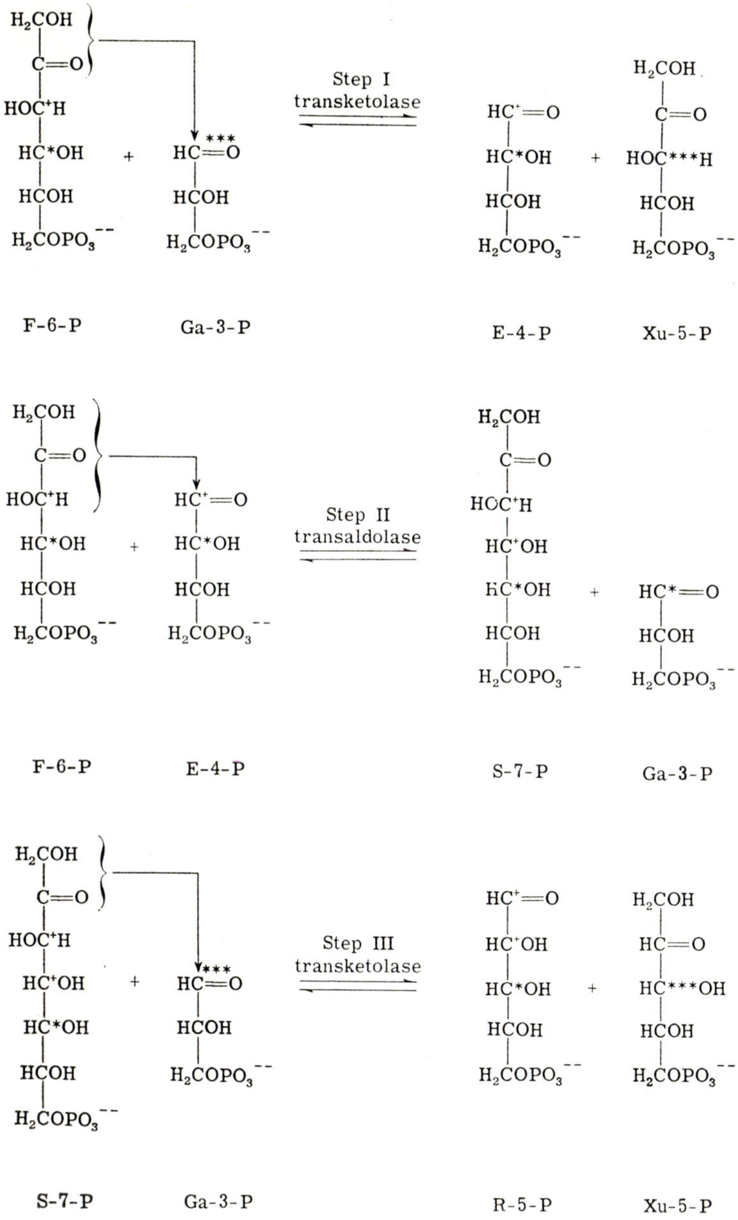

FIG. 2. Isotope distribution in pentoses and heptuloses formed from 3,4-labeled fructose-6-phosphate and 1-labeled glyceraldehyde-3-phosphate. F-6-P = fructose-6-phosphate; GA-3-P = glyceraldehyde-3-phosphate; E-4-P = erythrose-4-phosphate; R-5-P = ribose-5-phosphate; Xu-5-P = xylulose-5-phosphate; S-7-P = sedoheptulose-7-phosphate (Taken from Vishniac *et al.*[68]).

VI. It will be seen that phosphoglyceric acid was predominantly carboxyl labeled, with small but equal amounts of radioactivity in carbons 2 and 3. Fructose was equally labeled on carbons 3 and 4, suggesting that it was formed from 2 molecules of triosephosphate which could be derived from the carboxyl labeled phosphoglyceric acid by reduction. In sedoheptulose, the bulk of the tracer was approximately evenly distributed between carbons 3, 4, and 5, whereas ribulose was labeled predominantly on carbon 3 with smaller but equal amounts of the isotope on carbons 1 and 2, respectively.

These labeling patterns, taken in conjunction with the reciprocal behavior of phosphoglyceric acid and ribulose diphosphate and the recently

TABLE VI

DISTRIBUTION OF RADIOACTIVITY AS PER CENT TOTAL ACTIVITY IN PHOTOSYNTHETIC INTERMEDIATES PRODUCED BY ALGAE (A)[a] AND BY *Thiobacillus denitrificans* (B)[b]

	Glyceric acid			Fructose			Sedoheptulose			Ribulose	
	A	B		A	B		A	B		A	B
						C_1	2	0			
			C_1	3	1	C_2	2	0			
			C_2	3	1	C_3	28	32	C_1	11	10
			C_3	43	49	C_4	24	29	C_2	10	10
C_1	82	94	C_4	42	47	C_5	27	39	C_3	69	80
C_2	6	4	C_5	3	1	C_6	2	0	C_4	5	0
C_3	6	2	C_6	3	1	C_7	2	0	C_5	3	0

[a] Algae data are taken from Bassham *et al.*[56]

[b] *Thiobacillus denitrificans* data taken from Aubert *et al.*[77]

discovered pentose cycle transformations mediated by the enzymes transaldolase, transketolase, and epimerase, led Bassham *et al.*[56] to put forward a mechanism for the autotrophic fixation process similar to that set out in Fig. 1. It is important to realize that this mechanism was arrived at solely on the basis of studies of the nature and behavior of intermediates within the photosynthesizing cell, i.e., without resort to cell-free extracts. Their work has accordingly provided biochemists with a new and potent technique.

The mechanism was confirmed by the demonstration of all the postulated reactions in cell-free extracts of photosynthetic organisms and the purification of the various enzymes concerned.

Fager[57] was the first to demonstrate the synthesis of carboxyl-labeled phosphoglyceric acid in a cell-free system. The components of this system were a chloroplast preparation, a heat-stable extract of either spinach leaves or algae, and carbon dioxide. The active component of the leaf extract was shown to be a phosphorylated compound but was not further identified. Subsequently it was shown[58, 59] that either ribulose diphosphate or ribulose-5-phosphate plus ATP would replace the heat-stable preparation used by Fager. Ribose-5-phosphate was shown to be converted to ribulose diphosphate by a combination of phosphoribose isomerase and a new enzyme, phosphoribulokinase.[60] Once the second reactant of the carboxylating system was identified, purification of the enzyme responsible, carboxydismutase, was undertaken. Preparations purified some 10- to 20-fold have been obtained.[61, 62]

The reaction catalyzed by carboxydismutase is complex. Calvin[56, 63] has suggested that the first step involves the addition of carbon dioxide to the enediol form of ribulose diphosphate; the hypothetical carboxylic acid so formed is then cleaved, giving 2 molecules of phosphoglyceric acid. The further transformation of phosphoglyceric acid proceeds by a series of reactions which, with one exception, are not peculiar to the autotrophic process. The exception is the reduction reaction which in green plant systems appears to be catalyzed by a TPN-specific triosephosphate dehydrogenase.[64, 65] In this context it is significant that the light reaction catalyzed by chloroplast preparations reduces TPN but not DPN.[66, 67] The triose phosphate so produced is converted to hexose diphosphate by reactions involving, first, the triose phosphate isomerase and, second, aldolase. Hydrolysis of the hexose diphosphate by a specific phosphatase gives fructose-6-phosphate which, in the presence of transketolase and transaldolase, yields a mixture of ribose-5-phosphate and xylulose-5-phosphate. These, in the presence of pentose phosphate isomerase and epimerase, respectively, give rise to ribulose-5-phosphate, which is then phosphorylated by phosphoribulokinase in the presence of ATP to give back the ribulose diphosphate. All of these enzymes have been demonstrated in the green leaf. For a more detailed statement of the evidence the reader is referred to the excellent review by Vishniac *et al.*[68]

Although the labeling pattern of the various intermediates (Table VI and Fig. 2) appears to conform precisely with that required by the autotrophic mechanism, it is important to note that the procedure used to degrade hexose does not estimate separately the radioactivity of each carbon atom of the molecule and that, in particular, it is not possible to distinguish between carbon atoms 3 and 4. Gibbs and Kandler,[69] using a degradation method which permits the separate assay of the activity of each carbon atom, have observed that hexose is in fact asymmetrically labeled with carbon 4 containing significantly more C^{14} than carbon 3. The significance of this observation for the autotrophic mechanism as outlined in Fig. 1 is not as yet clear.

Considering this reaction sequence as a whole the first point which strikes one is that, in contradistinction to all other reactions involving the fixation of carbon dioxide, this autotrophic process results in the total synthesis of hexose from carbon dioxide. By so doing it provides the plant with the raw materials for growth from which all the other constituents of the plant are synthesized. As is well known, some of these syntheses involve the fixation of carbon dioxide by the heterotrophic reactions. Their secondary nature, insofar as the green plant is concerned, is obvious.

Although it is convenient and conventional to regard hexose as the end product of the autotrophic mechanism, it is nonetheless misleading to do so. Studies of biosynthetic mechanisms over the past few years have made it abundantly clear that, apart from structural polysaccharides such as cellulose, and reserve polysaccharides such as starch and sucrose, hexoses are not immediate precursors of cell constitutents although they can be transformed into such precursors. These precursors are in fact provided by the autotrophic mechanism without first proceeding to hexose. Thus phosphoglyceric acid, the product of the carboxydismutase reaction, is converted, through the intervention of the appropriate enzymes, to pyruvate, and thence to the Krebs' cycle intermediates from which are produced so many of the monomers required for cell synthesis.[70] The autotrophic mechanism also supplies erythrose-4-phosphate, an intermediary in the synthesis of the aromatic amino acids,[71] and ribose-5-phosphate, required for the synthesis of nucleotides[72] and such amino acids as histidine[73] and tryptophan.[74] Thus it is more precise to regard the autotrophic mechanism as a synthetic process which can be tapped at a number of points. These drain-off points are phosphoglyceric acid, erythrose-4-phosphate, ribose-5-phosphate, and hexose phosphate and for the process to continue it is sufficient to ensure that the amount of any intermediate drained off does not exceed the amount necessary to maintain the reaction sequence.

The second striking feature is that the process is driven by ATP and TPNH, with no other high energy compounds involved. It is accordingly clear that if the photolithotrophic and chemolithotrophic bacteria utilize carbon dioxide by the same process, their energy-yielding reactions must supply, first, ATP and, second, reduced pyridine nucleotide, possibly TPNH. The over-all equation shows that 18 molecules of ATP are required per molecule of hexose synthesized. As Vishniac[75] has pointed out, the possibility exists that 5 molecules of ATP would be saved per hexose unit synthesized were there a transphosphorylation reaction between hexose diphosphate and ribulose-5-phosphate. However, the occurrence of such a reaction has still to be proved; were it of significance, the role of phosphoribulokinase would be somewhat obscure.

In order to stress what is entailed in the demonstration that this autotrophic pathway operates in the bacteria, it is useful to summarize the evidence which has led to the general acceptance of the scheme outlined in Fig. 1 as the mechanism by which the green plant fixes carbon dioxide. It is first necessary to demonstrate that carboxyl-labeled phosphoglyceric acid accounts for most of the carbon dioxide fixed during the first few seconds of the exposure to $C^{14}O_2$. Second, it is necessary to demonstrate that the hexose phosphates, the pentose phosphates, and the sedoheptulose phosphates are also rapidly labeled and, under steady-state conditions, have a labeling pattern similar to that predicted by the scheme. Third, the pool sizes of ribulose diphosphate and phosphoglyceric acid must be shown to vary in a reciprocal fashion; thus, under conditions of low CO_2 tension, the amount of ribulose diphosphate should increase and the amount of phosphoglyceric acid decrease; conversely, when the reducing agent is removed from the system, the phosphoglyceric acid must increase and the ribulose diphosphate decrease. Finally, it is essential to demonstrate that the organism contains all the enzymes concerned; in addition, it is desirable to demonstrate that the rates of the various reactions catalyzed by these enzymes are adequate to account for the fixation of carbon dioxide observed in the whole cell. As has been emphasized earlier, the two enzymes peculiar to the process are carboxydismutase and phosphoribulokinase. Fortunately there are good methods for the assay of both of these enzymes.[60, 61]

IV. The Chemolithotrophic Bacteria

The current state of affairs with these organisms is that the autotrophic mechanism outlined above has been demonstrated with varying degrees of certainty in all organisms so far examined. The evidence in support of this assertion will now be examined.

The evidence is most complete in the case of *Thiobacillus denitrificans*. Following the work of Baalsrud and Baalsrud,[18] the isolation and mass culture of this organism presents no difficulty. Exploiting this fact, Trudinger[76] made a study of the fixation of carbon dioxide by cell-free extracts; as will be seen below he was able to demonstrate the presence of various enzymes either directly or indirectly in these extracts. Trudinger's work was complemented by that of Aubert and his colleagues[77] using the techniques developed by Calvin.

Thiobacillus denitrificans obtains its energy for growth by an oxidoreduction action in which either thiosulfate or sulfur is oxidized to sulfate and nitrate reduced to gaseous nitrogen. Aubert *et al.*[77, 78] showed that this organism in the presence of its energy-yielding substrates rapidly fixed $C^{14}O_2$; ethanolic extracts, prepared from cells exposed for 10 seconds to the tracer, contained five labeled compounds: 3-phosphoglyceric acid, ribulose diphosphate, hexose phosphates, sedoheptulose phosphate, and aspartic acid. The phosphoglyceric acid at this stage accounted for 75% of the tracer present in the extract. Increasing the time of the exposure resulted in a progressive fall in the percentage of the total $C^{14}O_2$ fixed found in phosphoglyceric acid, as is required by theory. The phosphorylated compounds were isolated and degraded; the results are shown in Table VI. The pattern of labeling found agrees closely with that required by theory (see Fig. 2). Finally, using cells in which the pools of phosphoglyceric acid and ribulose diphosphate were fully labeled, they found that reduction in the concentration of carbon dioxide in the system resulted in a marked rise in the amount of ribulose diphosphate and a fall in the amount of phosphoglyceric acid. On the other hand, when the reducing agent, thiosulfate, was removed from the system the phosphoglyceric acid increased as predicted, and at the same time the ribulose diphosphate decreased. These experiments provide strong support for the view that the autotrophic mechanism functions in this organism. The comparatively rapid labeling of aspartic acid observed is not inconsistent with this view; similar observations have been made by Calvin and his colleagues using algae.

Trudinger[76] prepared extracts of *Thiobacillus denitrificans* with the Hughes Press[79] and found that such extracts fixed $C^{14}O_2$ in the presence of magnesium ions and ATP; supplementation of the reaction mixture with ribose-5-phosphate increased the amount of $C^{14}O_2$ fixed. The crude cell-free extracts produced some eight radioactive compounds. One, which accounted for about 50% of the total radioactivity, was identified as phosphoglyceric acid; a second compound was identified as phosphoenolpyruvic acid. Partial purification of the extracts by ammonium sulfate fractionation gave preparations which, in the presence of ribose-5-phosphate, ATP, and magnesium ions produced, mainly, carboxyl-labeled phosphoglyceric acid. Finally these partially purified extracts were shown to carboxylate ribulose diphosphate with the formation of phosphoglyceric acid. These experiments provide a direct demonstration of the presence of carboxydismutase. From the fact that phosphoglyceric acid was formed when extracts were reinforced with ribose-5-phosphate and ATP it is evident that the extracts also contained phosphopentose isomerase and phosphoribulokinase. Trudinger also demonstrated the presence of transketolase, transaldolase, and triosephosphate dehydrogenase. The triosephosphate dehydrogenase, which required cobaltus or ferrous ions for maximum activity, reduced DPN but not TPN.

The evidence outlined above shows that in *Thiobacillus denitrificans* the autotrophic pathway completely accounts for the conversion of carbon dioxide to carbohydrate; there is no evidence at all for an alternative mechanism. Santer and Vishniac[80] have shown that extracts prepared from

Thiobacillus thioparus produce carboxyl-labeled phosphoglyceric acid from added ribulose diphosphate and $C^{14}O_2$, thus demonstrating the presence of carboxydismutase in this organism. A somewhat more detailed account of the carbon dioxide metabolism of a third sulfur-oxidizing organism, *Thiobacillus thiooxidans*, has been published by Suzuki and Werkman.[81, 82] Extracts prepared from *Thiobacillus thiooxidans*, when incubated with ribose-5-phosphate, ATP, magnesium ions, and $C^{14}O_2$ produced, among other compounds, carboxyl-labeled phosphoglyceric acid. The formation of phosphoglyceric acid from ribose-5-phosphate and carbon dioxide is consistent with the view that the extracts prepared contained phosphoribose isomerase, phosphoribulokinase, and carboxydismutase. A more detailed study of the products formed from ribose-5-phosphate confirmed this and at the same time provided evidence that the extracts of *Thiobacillus thiooxidans* contained all the enzymes participating in the autotrophic mechanism. Thus, incubation of the extracts with ribose-5-phosphate, ATP, magnesium ions, and $C^{14}O_2$ gave, in addition to the phosphoglyceric acid, sedoheptulose-7-phosphate and sedoheptulose diphosphate. When carbon dioxide was omitted from the reaction mixture, no phosphoglyceric acid was formed; the products were then the two sedoheptulose phosphates already mentioned above, together with ribulose diphosphate and ribulose-5-phosphate. If both the ATP and the carbon dioxide were omitted, only ribulose-5-phosphate and sedoheptulose-7-phosphate were formed in short-term experiments; when the incubation period was extended to 1 hour, some fructose-6-phosphate was also produced. The fact that the hexose and heptulose phosphates were formed from ribose phosphate shows that the extracts contained, in addition to the phosphoriboisomerase, the epimerase, the transketolase, and transaldolase. The extracts also contained aldolase, phosphoglycerokinase, and triose phosphate dehydrogenase which, like that found by Trudinger[76] in extracts of *T. denitrificans*, reduced DPN but not TPN. To conclude this part of the discussion of the sulfur bacteria it would seem that two of the three species of sulfur bacteria, namely, *T. denitrificans* and *T. thiooxidans* contain all the enzymes required for the operation of the autotrophic mechanism. Furthermore, in the case of *T. denitrificans* there is clear evidence of the operation of this mechanism within the cell. So far only carboxydismutase has been demonstrated in *T. thioparus*. However, it seems reasonable to assume that in all three *Thiobacillus* spp. carbon dioxide is converted to carbohydrate by the same mechanism as that demonstrated to occur in the green plant.

The hydrogen bacteria differ from the sulfur bacteria in that they are able to use either carbon dioxide or organic compounds as their source of carbon for growth. When carbon dioxide is used, the energy required for the fixation process is made available by the oxidation of hydrogen by molecular oxygen or, in the case of *Micrococcus denitrificans*, by nitrate. On the other hand, when these organisms grow on organic compounds they appear to do so by means of conventional heterotrophic reactions. This group thus forms a bridge between with the autotrophic mode of life on the one hand and the heterotrophic on the other. Vishniac and Santer[24, 68, 80] showed the presence of carboxydismutase in *Hydrogenomonas ruhlandii* and other hydrogen-oxidizing bacteria, and made the interesting observation that when these organisms were grown heterotrophically upon lactate the amount of carboxydismutase in the cells was markedly reduced (*vide infra*). Thus these organisms contain one of the key enzymes of the autotrophic mechanism for carbon dioxide fixation only when they are grown autotrophically. These enzymic studies were complemented by those of Bergmann *et al.*,[83] who studied the fixation of $C^{14}O_2$ by *Hydrogenomonas facilis*. Growing cultures were exposed to a hydrogen-oxygen-carbon dioxide mixture (39:10:1) for 30 minutes and $C^{14}O_2$ as sodium bicarbonate then added. At various intervals of time after addition of the tracer, the cells were killed and the soluble fraction analyzed by ion exchange chromatography using a column of the formate form of Dowex-1. Phosphoglyceric acid was the main labeled compound in the early sample; after 5 seconds it accounted for 25% of the total $C^{14}O_2$ fixed. After 45 seconds, this value had fallen to 10%. The mixture of sugar phosphates, which could not be resolved by ion exchange chromatography accounted for some 30% of the total tracer fixed after 5 seconds' exposure. In addition to these compounds malic acid was rapidly labeled; the sample isolated from the 5-second sample of cells accounted for some 8% of the total $C^{14}O_2$ assimilated. In the later samples both glutamic and aspartic acid also became labeled. In contrast to phosphoglyceric acid, the sugar phosphates, and malic acid, these two amino acids became progressively more and more radioactive with time. It had been previously reported that this organism produced both labeled acetate and labeled formate,[84] but Bergmann and his colleagues[83] were unable to detect these compounds under the conditions of their experiments. Extracts of *Hydrogenomonas facilis* were prepared and shown to fix carbon dioxide. The amount fixed was increased by the addition of ribose-5-phosphate, and a further increase was obtained by the addition of ATP at low concentrations. On the other hand, 0.01 *M* ATP appeared to inhibit carbon dioxide fixation.

Rhodopseudomonas capsulata, a member of the Athiorhodaceae, will, under certain conditions, oxidize hydrogen in the dark with molecular oxygen and at the same time fix carbon dioxide. Stoppani *et al.*[85] studied this process, using the techniques worked out in Calvin's laboratory. They observed that fixation of carbon dioxide in the dark was contingent upon the presence of both hydrogen and oxygen in the system. A kinetic study of the process showed that in the earlier samples both phosphoglyceric acid and malic acid were labeled, and that the amount of the tracer present in each of these two compounds, when expressed as a percentage of the total fixed, decreased with time, suggesting that both compounds are important intermediates in the assimilation of carbon dioxide. The results obtained resembled those obtained with *Thiobacillus denitrificans*[77, 78] and *Hydrogenomonas facilis*.[83] As was found with *H. facilis*, both glutamic acid and aspartic acid were also rapidly labeled.

The observation of Vishniac and Santer[24, 68, 80] that species of hydrogen bacteria grown heterotrophically contain no carboxydismutase has been confirmed and extended by Kornberg using *Micrococcus denitrificans*[86] When this organism was grown autotrophically it contained both carboxydismutase and phosphoribulokinase, the two unique enzymes of the autotrophic mechanism for carbon dioxide fixation. Additional support for the view that these enzymes play a part in carbon dioxide assimilation was obtained by using Calvin's techniques. Thirty seconds after exposure to $C^{14}O_2$, 70% of the total isotope fixed was present in phosphoglyceric acid. This compound behaved as an intermediate, for with the passage of time, the proportion of the total $C^{14}O_2$ assimilated that was present in phosphoglyceric acid sharply decreased. When, however, *Micrococcus denitrificans* was grown upon acetate, formation of both carboxydismutase and phosphoribulokinase was repressed; the cells now contained isocitratase, a key enzyme of the glyoxalate cycle, the mechanism which Kornberg[87] has shown to play a key part in the utilization of acetate carbon for growth.

The hydrogen bacteria, when grown autotrophically, resemble the sulfur bacteria in that they appear to fix carbon dioxide by the autotrophic mechanism. However, it must be emphasized that the evidence so far produced is not complete. Thus, although both carboxydismutase and phosphoribulokinase have been found in these organisms and although phosphoglyceric acid has been shown to become very rapidly labeled when suspensions are incubated in the presence of $C^{14}O_2$, it has still to be shown that these organisms contain the other enzymes which play a part in the autotrophic process. Second, it remains to be shown that the fixation process is in fact cyclic. In two organisms, namely, *Hydrogenomonas facilis* and *Rhodopseudomonas capsulata*, malic, glutamic, and aspartic acids are also rapidly labeled when cell suspensions are incubated with $C^{14}O_2$. So far there has been no detailed study of the mechanism by which this labeling occurs; nor, for that matter, has the distribution of the tracer in these three compounds been determined. Presumably conventional heterotrophic reactions are involved, but this has still to be demonstrated.

Bacteria are known which will grow upon methane,[8, 88, 89] methanol,[90-92] methylamine,[91] formate,[90, 92-94] and carbon monoxide.[90, 95] These compounds serve as both energy and carbon sources; because they contain only one carbon atom, growth upon them presents the cell with a biosynthetic task analogous to that which faces the chemolithotrophic and photolithotrophic bacteria, i.e., the synthesis of all cell material from a one-carbon compound. This raises the question of whether they should be considered as (*1*) chemolithotrophic organisms, or (*2*) as a separate physiological group. The answer to this question hinges upon what is meant by a chemolithotrophic organism. If the definition given in Ref. 1 is rigidly adhered to, then only those organisms which use carbon monoxide qualify, since the other one-carbon compounds—formate, methanol, methylamine, and methane—are generally considered to be organic and not inorganic! It would therefore appear that the definition of the term chemolithotrophy depends on how one defines the terms organic and inorganic. However the chemolithotrophs are remarkable not only for the materials which they use as sources of energy, but also for the fact that they build their substance from carbon dioxide. In this context Bhat and Barker[96] have pointed out that organisms which grow upon compounds containing one carbon atom may, like the chemolithotrophs, use their substrates exclusively as sources of energy and synthesize their cell material from carbon dioxide, the end product of the oxidation of the substrate; see also.[97] Evidence in keeping with this point of view has been provided by Mevius[92] and by Näveke and Engel[98] who showed that *Hyphomicrobium vulgare* assimilates carbon dioxide during the metabolism of methanol and formate and that, in addition, carbon dioxide is required for the growth of this organism upon formate. Likewise, Leadbetter, and Foster[89] observed that strains of *Pseudomonas methanica* assimilate considerable amounts of carbon dioxide during growth. The hypothesis of Bhat and Barker requires that the substrate carbon be assimilated by the same route as carbon dioxide; hence the demonstration that, during growth, an organism assimilates carbon dioxide is consistent with the hypothesis. Direct evidence has now been obtained by Quayle and Keech[99, 100] in the course of their studies of the metabolism of formate by *Pseudomonas oxalaticus*.[94]

The technique employed was essentially that developed by Calvin for the study of carbon dioxide fixation by algae. To cultures growing upon formate either $C^{14}O_2$ or $HC^{14}OOH$ was added, and the subsequent distribution of the isotope in the cells was determined by paper chromatography and radioautography. In the case of the labeled formate, samples taken within 90 seconds of adding the tracer showed that some 70% of the isotope was to be found in the phosphate ester fraction. When $C^{14}O_2$ was added to the cultures growing upon unlabeled formate the distribution of the labeling pattern was indistinguishable from that obtained with C^{14}-formate. This was not only consistent with the view that formate carbon is assimilated as carbon dioxide but also suggests that the autotrophic pathway is involved. This view was further strengthened by the demonstration that, when either $C^{14}O_2$ or C^{14}-formate was used, phosphoglyceric acid was the earliest compound to become labeled. Hydrolysis of the phosphate ester fraction with phosphatase followed by chromatography of the free sugars showed that, in addition to phosphoglyceric acid, glucose, fructose, and sedoheptulose were rapidly labeled. The evidence thus obtained with whole cells, while not as complete as that adduced for the sulfur bacteria—for example, the distribution of the C^{14} in both the hexoses and in sedoheptulose has not as yet been determined—nonetheless gives strong support to the idea that formate is assimilated via carbon dioxide.

Extracts prepared from *Pseudomonas oxalaticus* grown upon formate formed labeled phosphoglyceric acid when incubated with ribulose diphosphate and $C^{14}O_2$. Replacement of ribulose diphosphate by a mixture of ribose-5-phosphate and ATP also resulted in the fixation of $C^{14}O_2$. These experiments showed the presence of carboxydismutase, phosphoribulokinase, and phosphoribose isomerase. These extracts, when incubated with C^{14}-formate and ribulose diphosphate, also produced labeled phosphoglyceric acid. The assimilation of formate carbon depended on the presence of the particulate fraction of the extracts; if this was removed by high speed centrifugation no assimilation occurred. The particulate fraction was shown to contain a formic dehydrogenase, which presumably oxidizes formate to carbon dioxide. The carbon dioxide is then used in the carboxydismutase reaction.

A completely different picture was obtained when *Pseudomonas oxalaticus* was grown upon oxalate.[101] Cultures grown under these conditions fixed carbon dioxide much more slowly than they fixed oxalate carbon. Further, extracts prepared from oxalate-grown cells contained neither carboxydismutase nor phosphoribulokinase. It is quite clear from this and other evidence that the assimilation of oxalate carbon proceeds by an entirely different mechanism from that concerned with the assimilation of formate carbon. The demonstration that cells grown on formate contain the unique enzymes of the autotrophic mechanism, whereas cells grown on oxalate do not, is important because it provides additional evidence for the idea that formate carbon is assimilated as carbon dioxide by the autotrophic mechanism.

The results of Quayle and Keech make it essential to study the anabolic processes of the other organisms which grow upon one-carbon compounds using the experimental approach of these authors. If these organisms resemble *Pseudomonas oxalaticus* growing upon formate, insofar as they assimilate their carbon in the form of carbon dioxide by means of the autotrophic mechanism, there will be a strong case for reconsidering the terminology of nutritional categories. There are two aspects to the problem: first, the form in which carbon enters the anabolic stream, and second, the nature of the energy source. Thus a chemolithotroph is defined as an organism that obtains its carbon in the form of carbon dioxide and its energy by the oxidation of inorganic compound. The procedures used by Quayle and Keech[99, 100] make it easy to determine the form in which carbon is assimilated; we are thus faced with the problem of defining what we mean by the term "inorganic compound." In practice this amounts to deciding whether the line should be drawn between carbon monoxide and formate, or between formate and methanol, or between methanol and methane. Clearly, whatever decision is reached this decision will be arbitrary. In contrast to this the form in which carbon is assimilated can be determined experimentally. Because of the difficulty of defining an inorganic compound one is driven to ask whether the ability to use inorganic electron donors is of fundamental biological importance. In this context it is worth noting that no particular importance is attached to the ability of many microorganisms to use inorganic electron acceptors as part of their energy-yielding system. In the definition of nutritional categories, there would appear to be a *prima facie* case for emphasizing the ability of an organism to synthesize cell material from carbon dioxide, and reducing the importance attributed to the ability to use inorganic compounds as energy sources. The profound significance previously attached to this ability seems to be a heritage from the time when organisms so endowed were believed to represent the most primitive form of living things; the fallacy in this hypothesis has been clearly stated by Oparin.[102]

The fundamental characteristic of the nutrition of the so-called chemolithotrophic organisms is, in my opinion, their ability to use carbon dioxide as their sole source of carbon, the process being driven by chemical energy as opposed to the radiant energy used by photosynthetic organisms. If this opinion is acceptable then the term "chemolithotrophy" should be abandoned, and the older term "chemoautotrophy" reinstated. Chemoautotrophy should be defined as the ability of an organism to synthesize its cell material from carbon dioxide using chemical as opposed to radiant energy, the origin of the carbon dioxide so used being of little significance.

V. The Mechanism of Carbon Dioxide Fixation in Photosynthetic Bacteria

Glover *et al.*[45] studied the photometabolism of carbon dioxide and acetate by *Rhodospirillum rubrum* using techniques similar to those developed by Calvin and his associates. When the hydrogen donor was gaseous hydrogen, $C^{14}O_2$ was rapidly converted to water-soluble organic compounds; for the first 24 seconds following the addition of the tracer phosphoglyceric acid accounted for most of the $C^{14}O_2$ fixed. Like Ormerod,[44] they observed that acetate had an inhibitory affect on the fixation of $C^{14}O_2$, but at the same time they noted that even in the presence of this hydrogen donor such $C^{14}O_2$ as was assimilated occurred mainly in the phosphoglyceric acid, at least in the early samples.

Some three years later a more detailed investigation of the photometabolism of carbon dioxide in the presence of hydrogen was published by Stoppani *et al.*[85] *Rhodopseudomonas capsulata* was used in these experiments. The pattern of labeled compounds was similar to that obtained with green plants, phosphoglyceric acid, phosphate esters of glucose, fructose, ribose, and ribulose being prominent on the chromatograms; malic acid and glutamic acid were also labeled. A kinetic analysis of the fixation process showed that in the first sample phosphoglyceric acid accounted for somewhat more than 50% of the total $C^{14}O_2$ fixed and malic acid about 18%. In succeeding samples the proportion of the total tracer assimilated found in these two compounds decreased steadily; this decrease coincided with an increase in the proportion of the tracer found in the sugar phosphates and in glutamic acid. These results suggested that two parallel fixation processes were occurring, one which involved the autotrophic mechanism, and the other a heterotrophic fixation reaction that yields malic acid. The effect of cutting off the light after 20 minutes was also studied. Just as in the experiments with algae[55] this resulted in a sharp decrease in the amount of sugar phosphate and an increase in the amount of phosphoglyceric acid. There was also an initial drop in the proportion of the tracer found in malic acid, but as the dark period proceeded this increased steadily. There is thus some evidence that in two members of the Athiorhodaceae the autotrophic mechanism plays a part in the assimilation of carbon dioxide by whole cells.

Enzymic studies using cell-free extracts have shown the presence of carboxydismutase in *Rhodospirillum rubrum* and *Chromatium*[103] and in *Rhodopseudomonas spheroides* and *Rhodopseudomonas palustris*.[104] The adaptive nature of carboxydismutase in certain of the chemoautotrophic bacteria has already been discussed and similar studies on the formation of this enzyme in photosynthetic bacteria have now been made both by Vishniac[105] and by Lascelles.[104] Vishniac observed that when *Chromatium* was grown in the light with either hydrogen or a reduced sulfur compound as the hydrogen donor the cells contained carboxydismutase. On the other hand, when acetate was the hydrogen donor the amount of carboxydismutase was reduced. Unfortunately no other organic hydrogen donors were studied so it is impossible to say whether the absence of the enzyme was a specific response to acetate or whether this was a general response to growth on all organic compounds. The experiments of Lascelles show that when the organisms she studied were grown upon a malate-glutamate mixture under appropriate conditions they produced carboxydismutase.

Lascelles[104] investigated the effect of aeration and illumination on the formation of carboxydismutase by *R. spheroides* and *R. palustris.* The former organism, when grown aerobically in the dark, contained a negligible amount of carboxydismutase. Transfer of the culture to anaerobic light conditions resulted in a high differential rate of synthesis of the enzyme provided that the cells contained a small but definite amount of bacteriochlorophyll. This experiment does not enable one to decide whether formation of this enzyme is light dependent or whether the formation is inhibited by oxygen. The latter explanation was subsequently shown to be correct, for when air was bubbled through illuminated cultures the formation of carboxydismutase was completely inhibited. Light intensity also had an effect on the synthesis of carboxydismutase. Thus, at high light intensities the differential rate of synthesis of the enzyme is slower than at low light intensities. In contrast to this, neither high light intensity nor oxygen had any effect upon the synthesis of either phosphoribose isomerase, phosphoglycerokinase, or triosephosphate dehydrogenase. It would have been of interest to study the effect of light intensity and oxygen on the other enzyme peculiar to the autotrophic mechanism—phosphoribulokinase.

Rhodopseudomonas palustris also contained carboxydismutase when grown anaerobically in the light. However, it differed from *R. spheroides* in that when it was grown aerobically in the dark it still contained a small but significant amount of this enzyme.

Despite the fragmentary nature of the evidence—it should be noted that no experiments have so far been carried out with the green bacteria—it seems reasonable to conclude that the autotrophic mechanism participates in the assimilatory processes of the photosynthetic bacteria. What remains to be decided is the extent to which it participates. It would seem reasonable to assume that, when carbon dioxide is the sole source of carbon, the autotrophic mechanism plays quantitatively the same part in the assimilation of carbon dioxide in the photosynthetic bacteria as it does in the green plant. The difficulties arise in those cases when organic hydrogen donors are used. Since a considerable proportion of the organic substrate is assimilated it is obvious that only part of the cell carbon is derived from carbon dioxide and the actual amount will depend on the oxidation level of the substrate.[32, 35, 37] In the case of those substrates more oxidized than cell material there is a net output of carbon dioxide and, as Ormerod[44] showed (Table IV), even using $C^{14}O_2$ it is impossible to get a meaningful estimate of carbon dioxide assimilated when such substrates are photometabolized, at least by washed suspensions. If cyanide specifically inhibits carboxydismutase[50] then the observed inhibition of the photometabolism of all the substrates tested save acetate (Table V) indicates that the autotrophic mechanism plays an essential part in the photometabolism regardless of whether there is a net uptake or output of carbon dioxide.

VI. Energetics

So far only the mechanism of carbon dioxide fixation has been considered. The available evidence, admittedly incomplete, leads to the conclusion that the reaction sequence outlined in Fig. 1, supplemented by heterotrophic fixation reactions, will account for the assimilation of carbon dioxide by both chemoautotrophic bacteria and photosynthetic bacteria growing under conditions where carbon dioxide is the sole source of carbon for growth. Examination of this scheme shows that, in addition to carbon dioxide, two other reactants, ATP and a reduced pyridine nucleotide (TPNH in the green plants), are involved. These two compounds provide the energy required; for the total synthesis of 1 mole of hexose, 12 moles of reduced pyridine nucleotide and 18 moles of ATP are used. In the green plant both are synthesized at the expense of radiant energy absorbed by chlorophyll. Arnon and his colleagues[106-109] have shown that two types of reaction occur. In one, called cyclic photophosphorylation, radiant energy is used to generate ATP from ADP and inorganic phosphate—equation (8)

$$\text{ADP} + \text{HPO}_4^{--} \xrightarrow{\text{light}} \text{ATP} \tag{8}$$

The second, called noncyclic photophosphorylation, yields ATP, TPNH, and molecular oxygen in stoichiometric amounts—equation (9)

$$2\text{ADP} + 2\text{HPO}_4^{--} + 2\text{TPN} + 2\text{H}_2\text{O} \xrightarrow{\text{light}} 2\text{ATP} + 2\text{TPNH} + \text{O}_2 \tag{9}$$

In order to supply the ATP and TPNH required it is clear that in the green plant both cyclic and noncyclic photophosphorylation must take place. Since the photosynthetic bacteria do not produce oxygen it seems unlikely that noncyclic photophosphorylation occurs in these organisms, at least in the form depicted by equation (9).

ATP plays a part in two reactions of the autotrophic mechanism. It is involved, first, in the conversion of 3-phosphoglyceric acid to 1,3-diphosphoglyceric acid catalyzed by the enzyme phosphoglycerokinase, and second, in the conversion of ribulose-5-phosphate to ribulose-1,5-diphosphate catalyzed by phosphoribulokinase. TPNH is involved in only one reaction, namely, the reduction of 1,3-diphosphoglyceric acid to triosephosphate. This must not be taken to imply that in the synthesis of cell material from carbon dioxide reduced pyridine nucleotide and ATP are only required for these three reactions. Clearly, the subsequent conversion of the intermediates provided by the autotrophic mechanism to cell material will require both ATP, or its equivalent, and reduced pyridine nucleotide. In addition to this, a supply of ATP will be required for the performance of osmotic and other forms of work.

Since the autotrophic mechanism is believed to play an essential part in the assimilation of carbon dioxide by the bacteria it follows that the energy-yielding systems of these organisms must supply both the ATP and the reduced pyridine nucleotide. The evidence at present available suggests that, in the bacteria, the pyridine nucleotide concerned is DPN and not, as in the green plant, TPN. Thus Trudinger[76] demonstrated a DPN-specific triosephosphate dehydrogenase in *Thiobacillus denitrificans*. His attempts to show the presence of a TPN-specific enzyme in his extracts of this organism were uniformly negative; the triosephosphate dehydrogenase of *Thiobacillus thiooxidans* is also DPN-specific.[82] Likewise, Lascelles[104] found that her extracts of *Rhodopseudomonas spheroides* contained a DPN-specific but no TPN-specific triosephosphate dehydrogenase. While these findings may reflect more the method of preparation of the extracts than the enzymic constitution of the cells, it seems reasonable at this stage to accept them at their face value and hence to draw the conclusion that in these two organisms the reducing enzyme of the autotrophic mechanism is linked with DPN, and not with TPN.

Photophosphorylation has been demonstrated in all three groups of the photosynthetic bacteria; a detailed account of this reaction in these bacteria has been given by Geller (Volume II, Chapter 10). The process, which resembles the cyclic photophosphorylation found in the chloroplasts of green plants, is catalyzed by the chlorophyll-containing chromatophores. Most of the work on bacterial photophosphorylation has been done with chromatophores prepared from *Rhodospirillum rubrum*. In addition to inorganic phosphate and ADP, a catalytic amount of either succinate, lactate, or DPNH is essential for the photophosphorylation of ADP;[110, 111] it is important to note that the oxidized forms of these compounds, namely, fumarate, pyruvate, and DPN, are inactive. While there is no evidence for the occurrence in these preparations of a light reaction which yields equivalent amounts of ATP and reduced pyridine nucleotide, Frenkel has shown that in the presence of succinate, illuminated chromatophore preparations reduce DPN to DPNH and the amount formed is equivalent to the amount of succinate added.[111] In addition, Vernon[112] has shown that chromatophores of *R. rubrum* reduce TPN in the light provided the plant TPN reductase is added.[66, 67]

In contrast, the origin of both the ATP and the reduced pyridine nucleotide required by the chemoautotrophs for the assimilation of carbon dioxide is, for the most part, a matter of conjecture. Cytochromes have been observed in some of these organisms. Trudinger[113] has demonstrated that addition of thiosulfate either to suspensions of a *Thiobacillus* species or to cell-free extracts causes a reduction of cytochrome. From these extracts Trudinger isolated three basic pigments with α-band maxima at 550, 553.5, and 557 mμ, respectively. The extracts also contained two acidic cytochromes with α-band maxima at 550 and 557 mμ, respectively. An enzyme was separated from the extracts that oxidized thiosulfate with either ferricyanide or the 553.5 cytochrome. Cytochromes have also been shown to be present in *Thiobacillus denitrificans*.[17, 114] More recently the partial purification of a cytochrome c from an iron-oxidizing organism has been reported.[19, 115] *Nitrobacter* has been shown by Lees and Simpson[116] to contain two cytochromes, a c-type with an absorption maximum at 551 mμ and an a-type with a maximum at 589 mμ. These hemoproteins appear to be concerned with the oxidation of nitrite, for, on addition of nitrate to a suspension of *Nitrobacter*, these two bands appeared along with a broader band with an absorption maximum between 520 and 525 mμ; presumably this last represents the β-bands of the two cytochromes.

The presence of a cytochrome system is presumptive evidence for the occurrence of oxidative phosphorylation and it is therefore of great interest that Aleem and Nason[117] have prepared cytochrome-bearing particles from *Nitrobacter agilis* which carry out oxidative phosphorylation when supplied with nitrite. The maximum P/O ratio obtained was 0.14. Compared with the P/O ratios obtained with mitochondrial preparations from animal tissues this is very low. Phosphorylating particles have been prepared from a number of heterotrophic bacteria;[118] in all cases so far studied the P/O ratios are low compared with the P/O ratios obtained with preparations from animal tissues. It may well be the case that the methods used for breaking the bacterial cells partially inactivate the enzyme systems concerned with oxidative phosphorylation.

The cytochrome c prepared from the iron-oxidizing bacteria had an E_0' of 0.31 at pH 7.0.[115] This organism grows at pH 2.9 at which pH the E_0' of this cytochrome is 0.38 and that of the Fe^{++}:Fe^{+++} system has an E_0' of 0.77. Vernon and his colleagues point out that if this cytochrome system is concerned with oxidation of ferrous ions to ferric ions the state of the substrate within the cell must be quite different from that of ferrous ions at pH 2.9.

There is thus evidence that the chemoautotrophic bacteria obtain at least part of the ATP required for assimilation of carbon dioxide by oxidative phosphorylation. The question which now has to be answered is the origin of the reduced pyridine nucleotide in these organisms. The demonstration that the hydrogenase of *Hydrogenomonas ruhlandii* reduces DPN[119] suggests that, at least in the hydrogen bacteria, reduced pyridine nucleotide is generated by a direct reaction involving the substrate. If this is the case, and assuming a P/O ratio of 3 for the oxidation of reduced pyridine nucleotide, it is possible to draw up a set of reactions which equate the synthesis of hexose from carbon dioxide with the oxidation of hydrogen to water by molecular oxygen; these reactions are set out in equations (10–13).

$$18\ H_2 + 18\ DPN \rightarrow 18\ DPNH \quad (10)$$

$$6DPNH + 3O_2 + 18\ HPO_4^{--} + 18\ ADP \rightarrow 6\ DPN + 18\ ATP + 6\ H_2O \quad (11)$$

$$18\ ATP + 12\ DPNH + 6\ CO_2 \rightarrow 18\ ADP + C_6H_{12}O_6 + 18\ HPO_4^{--} \quad (12)$$

$$18\ H_2 + 3\ O_2 + 6\ CO_2 \rightarrow C_6H_{12}O_6 + 12\ H_2O \quad (13)$$

It will be seen that the calculated CO_2:O_2:H_2 ratios are 1.0:0.5:3.0. The ratios reported in the literature are about 1.0:2.0:6.2 for *Hydrogenomonas facilis*[20] and 1.0:2.7:6.6 for *Hydrogenomonas ruhlandii*.[21] If the assumed P/O ratio is correct, this means that in the whole cells the ATP/CO_2 ratio is 12:1 as opposed to the value of 3:1 required by theory. It must be remembered that these calculations refer to the synthesis of hexose. Since additional ATP, or its equivalent, is required both for the synthesis of the monomers out of which the cell is built and for the polymerization of these monomers, and since ATP or its equivalent is probably required for the performance of osmotic and similar forms of work, the high value of 12 for the ATP/CO_2 ratio may not be as extravagant as it appears as first sight.

The enzymic steps involved in the oxidation of sulfur compounds to sulfate, ammonia to nitrite, and nitrite to nitrate are largely unknown, and it is accordingly impossible to say whether or not pyridine nucleotides participate in any of these oxidation reactions. In one instance, however, namely that of the iron-oxidizing bacteria, the potential of the system is such that the direct formation of reduced pyridine nucleotide is highly improbable. This means that other, and perhaps novel, mechanisms for the generation of reduced pyridine nucleotides have to be considered. One such mechanism has been discussed by Davies and Krebs.[120] They suggested that reduced pyridine nucleotide might be formed by a process which amounts to a reversal of oxidative phosphorylation, i.e., a reaction in which the energy made available by the breakdown of ATP is used to drive an oxidoreduction reaction of the type depicted in equation (14).

$$XH_2 + ATP + DPN \rightarrow X + ADP + DPNH + HPO_4^{--} \quad (14)$$

where XH_2 is either the substrate itself or a reduced compound formed by the interaction of the substrate with a suitable H-acceptor, for instance, a cytochrome or a flavoprotein. In this example only one ATP is used per

DPNH produced but it is possible that up to three such steps, each requiring one ATP, might be needed to produce one DPNH. An extra push to such reactions would be given if the ATP underwent pyrophosphorolysis yielding AMP and pyrophosphate, the latter being subsequently hydrolyzed to phosphate.

Very recently Klingenberg *et al.*,[121] and Chance and Hollunger[122] have demonstrated that reactions of the type postulated by Davies and Krebs[120] occur in mitochondria prepared from animal tissues. Such preparations reduced the bound, intramitochondrial pyridine nucleotide while oxidizing either α-glycerophosphate or succinate. Since the potentials of the dehydrogenases concerned are so much more positive than that of the pyridine nucleotide system, direct reduction of pyridine nucleotide to the extent observed is impossible. Both groups have therefore suggested that the energy made available by the oxidation of succinate and α-glycerophosphate is used for the reduction reaction. Chance and Hollunger[122] found that addition of ATP to their preparations did not increase the rate of reduction of pyridine nucleotide and the possibility therefore exists that in these systems ATP itself is not involved.

If ATP or its equivalent is used to generate reduced pyridine nucleotide, in bacteria where this occurs the amount of ATP required for the synthesis of hexose from carbon dioxide will be greater than that predicted by the autotrophic mechanism. For example, if an organism makes reduced pyridine nucleotide by a mechanism which required one ATP per pyridine nucleotide reduced, then the total ATP requirement for the synthesis of hexose will be 30 moles, i.e., 18 + 12 moles; or, putting it another way, 5 moles ATP per mole of CO_2 assimilated. Further, if the ATP so used is generated by oxidative phosphorylation it follows that the amount of oxygen used per mole of CO_2 fixed will also be increased. It is perhaps significant that, with the exception of the hydrogen bacteria, the O_2/CO_2 ratios which have been measured (cf. Table II) are very high and the calculated thermodynamic efficiencies are correspondingly low (cf. Table I). It may be argued that these low efficiencies merely imply that the over-all P/O ratio for the catabolic reaction is low. This is certainly possible; but it still leaves the origin of the reduced pyridine nucleotide to be accounted for.

The object of these speculations is not only to suggest a plausible origin of the reducing power in certain of the chemoautotrophic bacteria, but also to emphasize both the gaps in our knowledge and the pressing need for more work on the electron transport systems of these organisms. It is true that, in most cases, these bacteria are not so easy to work with as, for example, *Escherichia coli*; a major difficulty is the low yield of cells. On the other hand, there is a very real possibility that the knowledge gained from such investigations will be of far-reaching importance, not only for an understanding of the metabolism of the chemoautotrophic bacteria but for cell physiology and biochemistry as a whole.

Two hypotheses have been advanced to account for the reducing power used by the photosynthetic bacteria for the reduction of carbon dioxide. The first of these was developed by van Niel[31-34] as part of his general hypothesis of photosynthesis. According to van Niel the light reaction brings about the cleavage of water yielding a reduced component [H] and an oxidized component [OH]. In the photosynthetic bacteria [H] is used to reduce carbon dioxide and the oxidized component is disposed of by an oxidoreduction reaction with the H-donor present in the medium. This fruitful hypothesis was advanced before the details of the assimilatory

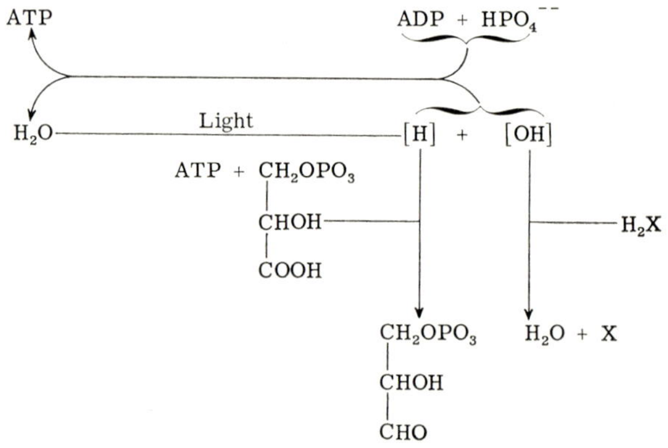

FIG. 3. The van Niel scheme for bacterial photosynthesis.

process had been worked out and before the nature of the reactants which had to be provided, i.e., the reduced pyridine nucleotide and ATP, was known. We now know that phosphoglyceric acid is the form in which carbon dioxide is reduced, but substitution of phosphoglyceric acid for carbon dioxide in van Niel's scheme does not alter its basic principles. A version of this scheme which also suggests a source of ATP is set out in Fig. 3.

The second hypothesis was proposed by Arnon *et al.*[123, 124] In this scheme the photolysis of water is no longer considered to be the primary light reaction; instead, Arnon suggests that radiant energy is used solely for the synthesis of ATP by cyclic photophosphorylation and that the substrate acts as the electron donor in the assimilatory processes. This is in marked contrast to the van Niel scheme in which the electron donor is produced by the photolysis of water. The essential features of Arnon's hypothesis as applied to the photosynthetic bacteria are set out in Fig. 4.

If radiant energy is used solely for the synthesis of ATP, and in the bacteria the available evidence supports this view,[123, 124] then it follows that given a supply of ATP, fixation of carbon dioxide and oxidation of the substrate should proceed anaerobically in the dark. This has in effect been demonstrated by Arnon *et al.*[124] who have prepared extracts of *Chromatium* which, in the presence of ATP converted acetate to much the same products as those made by whole cells when incubated with acetate in the light. Arnon[123] also reports that chromatophores prepared from *Chromatium* contain a powerful hydrogenase which reduces pyridine nucleotides in the dark (see Packer and Vishniac[119]).

Stanier *et al.*[38] have also come to the same conclusion, namely, that the light reaction serves as a source of energy for the assimilation of organic substrates by the Athiorhodaceae. They studied the photometabolism of acetate by *Rhodospirillum rubrum* and found that this process, which does

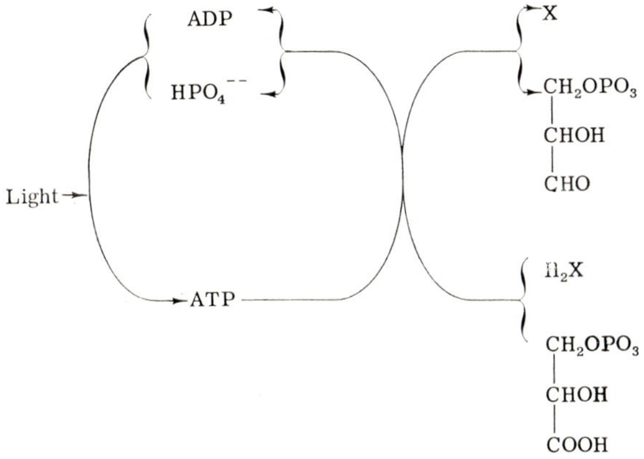

FIG. 4. The Arnon scheme for bacterial photosynthesis.

not require carbon dioxide, results in the conversion of acetate to poly-β-hydroxybutyric acid.

Carbon dioxide is produced when acetate is photometabolized; yields of 0.2 CO_2 per mole acetate[32] and 0.25 CO_2 per mole acetate[44] have been reported. Stanier *et al.*[38] also confirmed Gaffron's[37] earlier finding that when acetate is photometabolized under an atmosphere of hydrogen, there is a rapid uptake of this gas; they showed that here too acetate is reduced to poly-β-hydroxybutyrate. They suggested that when acetate is photometabolized under an inert gas phase part of the acetate is oxidized by the tricarboxylic acid cycle[125] and the remainder reduced by the electrons thus made available, according to the following equations:

$$CH_3COOH + 2\ H_2O \rightarrow 2\ CO_2 + 8\ [H] \quad (15)$$

$$8\ CH_3COOH + 8\ [H] \rightarrow (CH_3CHOH\ CH_2COOH)_4 + 2\ CO_2 \quad (16)$$

$$9\ CH_3COOH \rightarrow (CH_3CHOHCH_2COOH)_4 + 2\ CO_2 + 2\ H_2O \quad (17)$$

According to this mechanism, 0.22 moles CO_2 should be produced per mole of acetate, a value which agrees well with the observed values reported above. There is only one slight difficulty, which does not materially affect the correctness of this interpretation. These reactions suggest that one mole of acetate is completely oxidized to carbon dioxide whereas tracer experiments have established that the carbon dioxide produced during the metabolism of acetate is derived exclusively from the carboxyl group.[126] These experiments[38] give powerful support to the Arnon hypothesis. The fact that carbon dioxide is not essential for the photometabolism of acetate is consistent with Ormerod's[44] observation that fixation of $C^{14}O_2$ is actually inhibited by acetate, and with the fact that cyanide only slightly inhibits the photometabolism of this compound (see Table V).

However, the photometabolism of acetate is an exceptional process because the conversion of acetate to the precursor of poly-β-hydroxybutyrate, presumably acetoacetyl coenzyme A, requires only ATP and coenzyme A. The reduction of this precursor is then brought about either by molecular hydrogen or by acetate, serving (via the tricarboxylic acid cycle) as the H-donor. In the case of the other organic substrates utilized by the photosynthetic bacteria, the intermediates in the assimilation process are produced by oxidation of the substrate which, in contrast to acetate, does not at the same time act as the H-acceptor. For example, butyrate, like acetate, is converted to poly-β-hydroxybutyric acid[38] by *R. rubrum*, a process which necessitates the oxidation of butyrate. Succinate, on the other hand, is converted mainly to polysaccharide; this process involves the oxidation of succinate presumably as far as oxaloacetate. van Niel suggests that the H-acceptor in these reactions is the oxidized component [OH] produced by the light reaction. Arnon, on the other hand, proposes that carbon dioxide, in the form of phosphoglyceric acid, is the H-acceptor. Arnon's scheme thus entails an oxidoreduction reaction between, for example, succinate and phosphoglyceric acid, driven by ATP and catalyzed by the appropriate dehydrogenases; see equation (18).

$$\text{Succinic acid} + \text{phosphoglyceric acid} + \text{ATP} \rightleftharpoons \text{fumaric acid} + \text{triose phosphate} + \text{ADP} + \text{phosphate} \quad (18)$$

Taken in isolation this reaction will proceed from right to left; but if the ADP produced is rapidly converted back to ATP by cyclic photophosphorylation, and if at the same time the triose phosphate is removed by aldolase, the reaction will in fact proceed from left to right.[127] Thus, if Arnon's scheme is correct, the function of the autotrophic mechanism in the photometabolism of organic compounds is, not so much to provide a source of cell carbon, as to provide the H-acceptor in assimilatory reactions which in all other respects save the source of ATP resemble those found in heterotrophic organisms.

References

[1] See "Autotrophic Micro-organisms." *Symposium Soc. Gen Microbiol* **4**(1954).
[2] S. Winogradsky, *in* "Microbiologie du Sol," p. 24. Masson, Paris, 1949.
[3] S. Winogradsky, *in* "Microbiologie du Sol," p. 52. Masson, Paris, 1949.
[4] S. Winogradsky, *in* "Microbiologie du Sol," p. 159. Masson, Paris, 1949.
[5] S. Winogradsky, *in* "Microbiologie du Sol," p. 179. Masson, Paris, 1949.
[6] L. G. M. Baas-Becking and G. S. Parks, *Physiol. Revs.* **7,** 85 (1927).
[7] W. Ruhland, *Jahrb. wiss. Botan.* **63,** 321 (1924).
[8] K. Söhngen, *Zentr. Bakteriol. Parasitenk. Abt. II* **15,** 513 (1906).
[9] O. Meyerhof, *Pflügers Arch. ges. Physiol.* **164,** 353 (1916)
[10] R. L. Starkey, *J. Bacteriol.* **10,** 165 (1925).
[11] M. W. Beijerinck, *Proc. Koninkl. Akad. Wetenschap. Amsterdam* **22,** 899 (1920).
[12] R. Lieske, *Sitzber. heidelberg. Akad. Wiss. Math. naturw. Kl. Abhandl. (B)* p. 6 (1912).
[13] T. Hofmann and H. Lees, *Biochem. J.* **52,** 140 (1952).
[14] S. Waksman and R. L. Starkey, *J. Gen. Physiol.* **5,** 285 (1922).
[15] R. L. Starkey, *J. Gen. Physiol.* **18,** 325 (1935).
[16] K. G. Vogler, *J. Gen. Physiol.* **26,** 103 (1942).
[17] K. G. Vogler and W. W. Umbreit, *J. Gen. Physiol.* **26,** 157.
[18] K. Baalsrud and K. S. Baalsrud, *Arch. Mikrobiol.* **20,** 34 (1954).
[19] Jay V. Beck, *J. Bacteriol.* **79,** 502 (1960).
[20] A. Schatz, *J. Gen. Microbiol.* **6,** 329 (1952).
[21] L. Packer and W. Vishniac, *J. Bacteriol.* **70,** 216 (1955).
[22] K. Baalsrud, *in* "Autotrophic Micro-organisms." *Symposium Soc. Gen. Microbiol.* **4,** 54 (1954).
[23] K. Baalsrud and K. S. Baalsrud, *in* "Phosphorus Metabolism," (W. McElroy and B. Glass, eds.) Vol. 2, p. 544. Johns Hopkins Press, Baltimore, Maryland, 1952.
[24] W. Vishniac and M. Santer, *Bacteriol. Revs.* **21,** 195 (1957).
[25] R. W. Newburgh, *J. Bacteriol.* **68,** 93 (1954).
[26] C. B. van Niel, *Arch. Mikrobiol.* **3,** 1 (1931).
[27] H. Larsen, *K. Norske Videnskab. Selskabs, Skrifter* **1,** 1 (1952).
[28] H. Larsen, *J. Bacteriol.* **64,** 187 (1952).
[29] P. A. Roelofsen, *Proc. Koninkl Acad. Wetenschap. Amsterdam* **37,** 660 (1934).
[30] H. Gaffron, *Biochem. Z.* **275,** 301 (1935).
[31] C. B. van Niel, *Cold Spring Harbor Symposium Quant. Biol.* **3,** 138 (1935).
[32] C. B. van Niel, *Advances in Enzymol.* **1,** 263 (1941).
[33] C. B. van Niel, *in* "Photosynthesis in Plants," (James Franck and W. E. Loomis, eds.) p. 473. Iowa State College Press, Ames, Iowa, 1949.
[34] C. B. van Niel, *in* "The Enzymes" (J. B. Sumner and K. Myrbäck, eds.), Vol. 2, Part 2, p. 1074. Academic Press, New York, 1952.
[35] F. M. Müller, *Arch. Mikrobiol.* **4,** 131 (1933).
[36] C. B. van Niel, *Bacteriol. Revs.* **8,** 1 (1944).
[37] H. Gaffron, *Biochem. Z.* **260,** 1 (1933).
[38] R Y. Stanier, M. Doudoroff, R. Kunisawa, and R. Contopoulos, *Proc. Natl. Acad. Sci. U.S.* **45,** 1246 (1959).
[39] J. W. Foster, *J. Gen. Physiol.* **24,** 123 (1940).
[40] J. M. Siegel and M. D. Kamen, *J. Bacteriol.* **59,** 693 (1950).

[41] C. Cutinelli, G. Ehrensvärd, and L. Reid, *Arkiv. Kemi* **2,** 357 (1950).
[42] C. Cutinelli, G. Ehrensvärd, L. Reid, E. Saluste, and R. Stjernholm, *Arkiv Kemi* **3,** 315 (1951).
[43] C. Cutinelli, G. Ehrensvärd, G. Hösgström, L. Reid, E. Saluste, and R. Stjernholm, *Arkiv Kemi* **3,** 501 (1951).
[44] J. G. Ormerod, *Biochem. J.* **64,** 373 (1956).
[45] J. Glover, M. D. Kamen, and H. van Genderen, *Arch. Biochem. Biophys.* **35,** 384 (1952).
[46] L. P. Vernon and M. D. Kamen, *Arch. Biochem. Biophys.* **44,** 298 (1953).
[47] O. Warburg, *Biochem. Z.* **103,** 188 (1920).
[48] C. P. Whittingham, *Nature* **169,** 1017 (1952).
[49] S. R. Elsden, unpublished data.
[50] A. V. Trebst, M. Losada, and D. I. Arnon, *J. Biol. Chem.* **235,** 840 (1960).
[51] J. A. Bassham and M. Calvin, *in* "The Path of Carbon in Photosynthesis," 1st ed. Prentice-Hall, Englewood Cliffs, New Jersey, 1957.
[52] A. A. Benson, J. A. Bassham, M. Calvin, T. C. Goodale, V. A. Haas, and W. Stepka, *J. Am. Chem. Soc.* **72,** 1710 (1950).
[53] A. A. Benson, J. A. Bassham, M. Calvin, A. G. Hill, H. E. Hirsch, S. Kawaguchi, V. Lynch, and N. E. Tolbert, *J. Biol. Chem.* **196,** 703 (1952).
[54] H. Gaffron, E. W. Fager, and J. L. Rosenberg, *in* "Carbon Dioxide Fixation and Photosynthesis." *Symposium Soc. Exptl. Biol.* **5,** 262 (1951).
[55] M. Calvin and P. Massini, *Experientia* **8,** 445 (1952).
[56] J. A. Bassham A. A. Benson, L. D. Kay, A. Z. Harris, A. T. Wilson, and M. Calvin, *J. Am. Chem. Soc.* **76,** 1760 (1954).
[57] E. W. Fager, *Biochem. J.* **57,** 264 (1954).
[58] J. R. Quayle, R. C. Fuller, A. A. Benson, and M. Calvin, *J. Am. Chem. Soc.* **76,** 3610 (1954).
[59] A. Weissbach, P. Z. Smyrniotis, and B. L. Horecker, *J. Am. Chem. Soc.* **76,** 3611 (1954).
[60] J. Hurwitz, A. Weissbach, B. L. Horecker, and P. Z. Smyrniotis, *J. Biol. Chem.* **218,** 769 (1956).
[61] A. Weissbach, B. L. Horecker, and J. Hurwitz, *J. Biol. Chem.* **218,** 795 (1956).
[62] W. Jakoby, D. O. Brummond, and S. Ochoa, *J. Biol. Chem.* **218,** 811 (1956).
[63] M. Calvin and Ning G. Pon, *J. Cellular Comp. Physiol.* **54,** Suppl. 1, 51 (1959).
[64] M Gibbs, *Nature* **170,** 164 (1952).
[65] D. I. Arnon, *Science* **116,** 635 (1952).
[66] A. San Pietro and H. M. Lang, *J. Biol. Chem.* **231,** 211 (1958).
[67] A. San Pietro, *in* "The Photochemical Apparatus its Structure and Function." *Brookhaven Symposia in Biol. No.* **11,** 262 (1959).
[68] W. Vishniac, B. L. Horecker, and S. Ochoa, *Advances in Enzymol.* **19,** 1 (1957).
[69] M. Gibbs and O. Kandler, *Proc. Natl. Acad. Sci. U.S.* **43,** 447 (1957).
[70] R. B. Roberts, D. B. Cowie, P. H. Abelson, E. T. Bolton, and R. J. Britten, "Studies of Biosynthesis in *Escherichia coli.*" *Carnegie Inst. Wash. Publ. No.* **607** (1955).
[71] P. R. Srinivasan, M. Katagiri, and D. B. Sprinson, *J. Biol. Chem.* **234,** 713 (1959).
[72] J. M. Buchanan and S. C. Hartman, *Advances in Enzymol.* **21,** 199 (1959).
[73] H. S. Moyed and B. Magasanik, *J. Biol. Chem.* **235,** 149 (1960).
[74] C. Yanofsky, *J. Biol. Chem.* **223,** 171 (1956).
[75] W. Vishniac, *Ann. Rev. Plant Physiol.* **6,** 115 (1955).
[76] P. A. Trudinger, *Biochem. J.* **64,** 274 (1956).

[77] J. P. Aubert, G. Milhaud, and J. Millet, *Ann. inst. Pasteur* **92,** 515 (1957).
[78] J. P. Aubert, G. Milhaud, and J. Millet, *Ann. inst. Pasteur* **92,** 679 (1957).
[79] D. E. Hughes, *Brit. J. Exptl. Pathol.* **32,** 97 (1951).
[80] M. Santer and W. Vishniac, *Biochim et Biophys. Acta.* **18,** 157 (1955).
[81] I. Suzuki and C. H. Werkman, *Arch. Biochem. Biophys.* **76,** 103 (1958).
[82] I. Suzuki and C. H. Werkman, *Arch. Biochem. Biophys.* **77,** 112 (1958).
[83] F. H. Bergmann, J. C. Towne, and R. H. Burris, *J. Biol. Chem.* **230,** 13 (1958).
[84] G. Orgel, N. E. Dewar, and H. Koffler, *Biochim. et Biophys. Acta* **21,** 409 (1956).
[85] A. O. M. Stoppani R. C. Fuller and M. Calvin, *J. Bacteriol.* **69,** 491 (1951).
[86] H. L. Kornberg, J. F. Collins, and D. Bigley, *Biochim. et Biophys. Acta* **39,** 9 (1960).
[87] H. L. Kornberg, *Ann. Rev. Microbiol.* **13,** 49 (1959).
[88] M. Dworkin and J. W. Foster, *J. Bacteriol.* **72,** 646 (1956).
[89] E. R. Leadbetter and J. W. Foster, *Arch. Mikrobiol.* **30,** 91 (1958).
[90] G. T. P. Schnellen, Doctoral Thesis, University of Delft, Netherlands, 1947.
[91] L. E. den Dooren de Jong, Doctoral Thesis, University of Delft, Netherlands, 1926.
[92] W. Mevius, *Arch. Mikrobiol.* **19,** 1 (1953).
[93] T. C. Stadtman and H. A. Barker, *J. Bacteriol.* **62,** 269 (1951).
[94] S. R. Khambata and J. V. Bhat, *J. Bacteriol.* **66,** 505 (1953).
[95] A. J. Kluyver and G. T. P. Schnellen, *Arch. Biochem.* **14,** 57 (1947).
[96] J. V. Bhat and H. A. Barker, *J. Bacteriol.* **55,** 359 (1948).
[97] C. B. van Niel, *Ann. Rev. Microbiol.* **8,** 105 (1954).
[98] R. Näveke and H. Engel, *Arch. Mikrobiol.* **21,** 371 (1954).
[99] J. R. Quayle and D. B. Keech, *Biochem. J.* **72,** 623 (1959).
[100] J. R. Quayle and D. B. Keech, *Biochem. J.* **72,** 631 (1959).
[101] J. R. Quayle and D. B. Keech, *Biochem. J.* **75,** 515 (1960).
[102] A. I. Oparin, "The Origin of Life on Earth," 3rd ed. (Translated by Academic Press, New York; Ann Synge.) 1957.
[103] R. C. Fuller and M. Gibbs, *Plant Physiol.* **34,** 324 (1959).
[104] J. Lascelles, *J. Gen. Microbiol.* **23,** 499 (1960).
[105] W. Vishniac, *Abstr. Proc. Intern. Congr. Microbiol., 7th Congr., Stockholm* p. 79. (1958).
[106] D. I. Arnon, F. R. Whatley, and M. B. Allen, *Science* **127,** 1026 (1958).
[107] D. I. Arnon, F. R. Whatley, and M. B. Allen, *Biochim. et Biophys. Acta* **32,** 32 (1959).
[108] F. R. Whatley M. B. Allen, and D. I. Arnon, *Biochim et Biophys. Acta* **32,** 47 (1959).
[109] D. I. Arnon, *in* "The Photochemical Apparatus: Its Structure and Function." *Brookhaven Symposia in Biol. No.* **11,** 181 (1959).
[110] A. W. Frenkel, *J. Biol. Chem.* **222,** 823 (1956).
[111] A. W. Frenkel, *in* "The Photochemical Apparatus: Its Structure and Function." *Brookhaven Symposia in Biol. No.* **11,** 276 (1959).
[112] L. P. Vernon, *J. Biol. Chem.* **233,** 212 (1958).
[113] P. A. Trudinger, *Biochim. et Biophys. Acta* **30,** 211 (1958).
[114] J. P. Aubert, G. Milhaud, C. Moncel, and J. Millet. *Compt. rend. acad. sci.* **246,** 1616 (1958).
[115] L. P. Vernon, J. H. Mangum, and Jay V. Beck, *Arch. Biochem. Biophys.* **88,** 227 (1960).
[116] H. Lees and J. R. Simpson, *Biochem. J.* **65,** 297 (1957).
[117] M. I. H. Aleem and A. Nason, *Proc. Natl. Acad. Sci. U.S.* **46,** 763 (1960).
[118] S. R. Elsden and J. L. Peel, *Ann. Rev. Microbiol.* **12,** 145 (1958).
[119] L. Packer and W. Vishniac, *Biochim. et Biophys. Acta* **17,** 153 (1955).

[120] R. E. Davies and H. A. Krebs, *Biochem Soc. Symposia* (*Cambridge, Engl.*) **8,** 77 (1952).
[121] M. Klingenberg, W. Slenczka, and E. Ritt, *Biochem. Z.* **332,** 47 (1959).
[122] B. Chance and G. Hollunger, *Nature* **185,** 666 (1960).
[123] D. I. Arnon, *Nature* **184,** 10 (1959).
[124] M. Lasada, A. V. Trebst, S. Ogata, and D. I. Arnon, *Nature* **186,** 753 (1960).
[125] S. R. Elsden and J. G. Ormerod, *Biochem. J.* **64,** 691 (1956).
[126] M. D. Kamen, S. J. Ajl, S. L. Ranson, and J. M. Siegel, *Science* **113,** 302 (1951).
[127] H. Gest and M. D. Kamen, *in* "Handbuch der Pflamzenphysiologie" (W. Ruhland, ed.) Vol. 5, Chapter 4, p. 568. Springer, Berlin (1960).

CHAPTER 2

Assimilation of Carbon Dioxide by Heterotrophic Organisms

HARLAND G. WOOD AND RUNE L. STJERNHOLM

I. Introduction 42
II. Original Proof of Assimilation of CO_2 by Heterotrophic Bacteria 44
III. Early Studies on the Mechanism of CO_2 Fixation by Heterotrophs 46
A. $C_3 + C_1$ Addition 46
B. CO_2 Incorporation into Oxalacetate by an Exchange Reaction 47
C. Attempts to Obtain Net CO_2 Fixation with Oxalacetic Decarboxylase 48
IV. Primary Reactions of CO_2 Assimilation 51
A. Reactions with Reduced Pyridine Nucleotides as an Energy Source 52
1. Malic Enzyme 53
2. Isocitric Dehydrogenase 57
3. Phosphogluconic Dehydrogenase 59
B. Reactions without an Apparent Extra Energy Source 61
1. Phosphoenolpyruvic Carboxylase 61
2. Phosphoenolpyruvic Carboxykinase 64
3. Phosphoenolpyruvic Carboxytransphosphorylase 69
4. Ribulose Diphosphate Carboxylase 69
5. Aminoimidazoleribonucleotide Carboxylase 70
C. Reactions with ATP as an Energy Source 71
1. Propionyl Carboxylase 71
2. β-Methylcrotonyl Carboxylase 75
3. Acetyl Carboxylase 77
4. Pyruvic Carboxylase 79
V. Methylmalonyl-Oxalacetic Transcarboxylase and the Formation of Propionate 81
VI. Reversal of α-Decarboxylation as a Mechanism of CO_2 Fixation 86
VII. The Function of Biotin in the Fixation of CO_2 89
VIII. Total Synthesis of Acetate from CO_2 98
A. Introduction 98
B. Proof of the Total Synthesis of Acetate from CO_2 by *Clostridium thermoaceticum* 100
C. Experiments with Labeled Glucose, CO_2, and Formate with *Clostridium thermoaceticum* 102
D. Tracer Experiments on the Mechanism of Acetate Formation from CO_2 by *Butyribacterium rettgeri* 103
E. Fixation of CO_2 by *Diplococcus glycinophilus* 104
F. Fixation of CO_2 by *Clostridium acidi-urici* and *C. cylindrosporum* 105
G. Summary 107
IX. Conversion of CO_2 to Methane 108
X. Concluding Comments 109
References 112

I. Introduction*

In 1890 the brilliant work of Winogradsky[1, 2] showed that certain bacteria are able to grow and reproduce in an entirely inorganic medium in the absence of light, thus laying the foundation for the classification of bacteria into two major categories: (*1*) the *autotrophs*, which synthesize all cell constituents from inorganic materials, and (*2*) the *heterotrophs*, which require some form of organic carbon for their nutrition, and which were considered unable to utilize CO_2. The utilization of carbon dioxide by heterotrophic bacteria was not discovered until 1935.[3] Until that time the idea had been firmly fixed in the minds of scientists that the assimilation of carbon dioxide is a process uniquely limited to the autotrophic forms of life. This view seems to have arisen from Winogradsky's work, which showed not only that the nitrifying and sulfur bacteria are obligate autotrophs, but also that the presence of organic compounds actually inhibits the growth of these bacteria. The inability of these microorganisms to grow on anything but inorganic compounds was, and still is, a very strange phenomenon. It encouraged the view that autotrophs differ from all other forms of life; this difference was associated with a unique ability to utilize CO_2.

It is difficult for the student of microbiology today to realize how firmly these views on CO_2 utilization were held.** All mechanisms of carbohydrate

* Abbreviations used in this chapter: DPN, DPNH—diphosphopyridine nucleotide and its reduced form; TPN, TPNH—triphosphopyridine nucleotide and its reduced form; ADP, ATP—adenosine diphosphate, adenosine triphosphate; GDP, GTP—guanosine diphosphate, guanosine triphosphate; IDP, ITP—inosine diphosphate, inosine triphosphate; P_i—inorganic orthophosphate; THFA—tetrahydrofolic acid PEP—Phosphoenolpyruvate.

** The only documented exception to this view is that advanced by the Russian scientist, A. F. Lebedev.[4] Lebedev is best known for his introduction of a special type of yeast maceration juice (Leberdeusaft). However, in 1921 he published a remarkably penetrating discussion of the relationship between photosynthesis, chemosynthesis, and heterotrophic synthesis. He stated, "The aggregate of these facts, together with methodological considerations, leads us to conclude that: (1) the accepted division of organisms into two groups, based on their ability to assimilate carbon from carbon dioxide, cannot be considered demonstrated; and (2) already established facts permit us to suppose that even under so-called heterotrophic conditions the assimilation of carbon is accomplished as under autotrophic conditions.

And though this new hypothesis cannot be considered proved by the analysis of already existing knowledge, in view of the total lack of foundation of the accepted view, it deserves deep attention and study."

Although Lebedev obtained no experimental results which can be considered direct evidence supporting his view that heterotrophs are able to utilize CO_2, there can be no doubt that he was "ahead of the times" and the first to visualize a broad function of CO_2 in the synthetic processes in all forms of life, be they autotrophs or heterotrophs. He also clearly applied the principle of comparative biochemistry to aid in his analysis of the metabolism of CO_2.

metabolism were constructed on the premise that CO_2 is completely inert in heterotrophic organisms, in fact, some pathways of metabolism were specifically proposed to meet this contingency. For example, at that time succinate was thought to be formed by a methyl to methyl condensation of two molecules of acetate.

$$COOH—CH_3 + CH_3—COOH \rightarrow COOH—CH_2—CH_2—COOH + 2\ H$$

Acetate was believed to arise via the decarboxylation of a C_3 compound, and accordingly, for each acetate molecule formed, there should be produced one molecule of CO_2 or some other C_1 compound, such as formate. Likewise, for each molecule of succinate produced, two molecules of either CO_2 or some other C_1 compound were expected. Virtanen[5] and Virtanen and Karström[6] found, however, that succinate and acetate were the only products formed when glucose was fermented by a suspension of propionic acid bacteria in the presence of toluene. Virtanen was thus forced to consider other possible mechanisms of glucose breakdown, and he proposed a cleavage of glucose to 4 and 2 carbon compounds to account for the formation of succinate and acetate unaccompanied by a C_1 compound.[5, 6] The idea of a C_4–C_2 cleavage also was applied to the *Escherichia coli* fermentation by Virtanen[7] and by Kluyver[8] because it offered an explanation for the lack of C_1 compounds among products in this fermentation.

The same consideration arose relative to the formation of citrate by molds. It was argued that if citrate were formed by condensation of three molecules of acetate, the maximum possible yield of citrate would be 66.6%, since one third of the six carbons of glucose would be converted to CO_2. When yields of citrate greater than 66.6% were obtained, this pathway was considered eliminated as a possibility.[9, 10] One of the authors (HGW) recalls a discussion of this problem during a seminar at the University of Wisconsin in 1936. It was proposed that one explanation might be that the molds assimilate CO_2 and convert it in part to citrate. This idea was promptly dismissed with the reply, "You can explain anything if you assume CO_2 utilization by molds." In effect, to assume CO_2 utilization was to destroy one of the basic assumptions on which the schemes of fermentation were constructed.

It therefore is understandable that the evidence of CO_2 utilization by heterotrophic bacteria was viewed with much skepticism. In fact, acceptance of this concept would have been delayed much longer were it not for the production of radioactive and stable carbon isotopes in 1940. The advent of these tracers made it possible for a number of scientists to demonstrate by simple and direct procedures that heterotrophic bacteria,[11, 12] molds,[13] and protozoa,[14] and even animals[15-17] utilized carbon dioxide.

II. Original Proof of Assimilation of CO_2 by Heterotrophic Bacteria

The original observation of heterotrophic CO_2 assimilation was obtained with propionic acid bacteria.[3, 18, 19] The proof was by a quantitative determination of all the products of a glycerol fermentation, thus permitting the calculation of the carbon and oxidation-reduction balances. The oxidation-reduction balance[20] is seldom used in present-day studies, which most frequently employ tracers and purified enzymes, but it is still good logic to recognize that for every oxidative event taking place there must be a concomitant reduction. In large part, it was a consideration of the oxidation-reduction balance which first led to the conclusion that CO_2 is utilized by the propionic acid bacteria. Although CO_2 is utilized by practically all, if not all, heterotrophic bacteria, in most cases more CO_2 is produced than is utilized, so the CO_2 uptake is concealed. The fermentation of glycerol by the propionic acid bacteria proved to be an exceptional case because balance studies showed a net uptake of CO_2. Previous balance studies by numerous investigators had in all instances except with autotrophic bacteria shown a net output of CO_2. Another exception occurs in the dissimilation of glucose by trypanosomes.[21] In the propionic acid fermentations, $CaCO_3$ was included in the medium to neutralize the organic acids formed from the glycerol. The CO_2 liberated from the $CaCO_3$ during the fermentation was collected and, at the conclusion of the fermentation, acid was added to convert the remaining carbonate to CO_2. The total CO_2 collected was less than the amount added originally in the form of $CaCO_3$; thus a net decrease of CO_2 had occurred in the system. It is of interest that this observation, made in 1934, was at first considered to be due to some unaccountable error made either in weighing out the $CaCO_3$ or in determining the CO_2 evolved. Another disturbing fact became apparent, however, because besides the CO_2 deficiency, there seemed to be a discrepancy in the oxidation-reduction balance. A substantial amount of succinate was produced in the fermentation, and the only other product present in significant quantity was propionate. Also, carbon of the succinate and propionate, when added together, accounted for somewhat more carbon than was contained in the glycerol fermented. Since propionic acid has the same oxidation-reduction value as glycerol, the data did not appear to provide an oxidation reduction balance. The basis for these calculations was proposed by Johnson *et al.*[20]

		Oxidation-reduction value
$CH_2OH—CHOH—CH_2OH$	(8H and 3O, balance 2H)	1 red.
CH_3CH_2COOH	(6H and 2O, balance 2H)	1 red.
$COOHCH_2CH_2COOH$	(6H and 4O, balance 1O)	1 oxid.

In this notation, 1 atom of oxygen is assigned an oxidation value of 1 while 2 hydrogen atoms are represented by a reduction value of 1. Thus the net oxidation-reduction value of both glycerol and propionate is "1 reduced," while that of succinate is "1 oxidized." Obviously, no net oxidation or reduction is involved in the conversion of glycerol to propionate.

$$CH_2OH—CHOH—CH_2OH \rightarrow CH_3CH_2COOH + H_2O$$

Glycerol Propionic acid

Since succinate is more oxidized than glycerol there had to be a reduced compound to balance this oxidized compound. Since all the carbon of the

TABLE I

CARBON AND OXIDATION-REDUCTION BALANCES CALCULATED ON THE BASIS OF ASSIMILATION OF CO_2 AND NONASSIMILATION OF CO_2[a]

Culture No.	Products per 100 mmoles of fermented glycerol (mmoles)			CO_2 per 100 mmoles glycerol (mmoles)	Carbon balance (%)		O/R[b]	
	Propionate	Succinate	Acetate		Glycerol + CO_2	Glycerol only	Glycerol + CO_2	Glycerol only
34W	59.3	34.5	2.0	−43.2	93	107	0.93	2.27
49W	55.8	42.1	2.9	−37.7	101	114	1.08	2.55

[a] Medium: glycerol 2%, $CaCO_3$ 1.14% and yeast extract 0.4%. The propionic acid fermentation was conducted anaerobically in an atmosphere of N_2 at 30°C. for 35 days.

[b] If the substrate is a compound at 0 level of oxidation and reduction, such as glucose (12 H, 6 O) the O/R balance equals 1. If the substrate is at a reduced level, one adjusts for the reduced state of the substrate, glycerol (reduction value of 1), by considering its conversion to the 0 level as an oxidation; the glycerol fermented therefore is given an oxidation value of 1. Likewise, utilization of CO_2 (an oxidized compound, oxidation valve, 2) represents a reduction; CO_2 utilized thus is assigned a reduction value of 2. For the fermentation with 34W the calculation is therefore as follows: $O/R = (100 \times 1) + (34.5 \times 1)/(59.3 \times 1) + (43.2 \times 2) = 0.93$.

fermented glycerol was recovered in the propionate and succinate, and no hydrogen gas was observed, one was at a loss to see a possibility of a reduced product. The answer to the riddle is, of course, obvious on the basis of our present-day knowledge. The missing CO_2, which was thought to be the consequence of an error, was in fact the carbon compound being reduced.

Results from the original experiments[18] are shown in Table I. The carbon and oxidation-reduction balances have been calculated with and without inclusion of CO_2 as one of the substrates. When CO_2 utilization is not included, the O/R balances are 2.27 and 2.55, whereas they should be 1.0, i.e., oxidized products equal reduced products. When the utilization of CO_2 is taken into consideration, the ratios are 0.93 and 1.08. Also, unless CO_2 utilization is included, the carbon recovered in the products is more than 100%. This is to be expected, since part of the carbon of the products is from the assimilated CO_2.

III. Early Studies on the Mechanism of CO_2 Fixation by Heterotrophs

A. $C_3 + C_1$ Addition

The first indication of the mechanism of CO_2 utilization by heterotrophs came from a study of the relationship between the amount of CO_2 fixed and

TABLE II

The Relationship between CO_2 Utilization and Succinate Formation by *Propionibacterium pentosaceum* (49W)[a, b]

Interval (days)	Glycerol fermented per liter (mmoles)	Products per 100 mmoles of fermented glycerol		
		CO_2 (mmoles)	Succinate (mmoles)	Propionate (mmoles)
0–4	46.6	−14.6	13.6	68.3
6–9	48.1	−27.8	26.7	62.2
12.5–19.5	30.6	−38.5	34.7	59.4
28.5–46.5	52.9	−62.9	59.0	42.1

[a] Values presented refer to the change during each interval.

[b] Medium: glycerol 2.5%, yeast extract 0.4%, and $NaHCO_2$ 1.5%. Incubation: anaerobic under CO_2 at 30°C. Small amounts of acetate and propyl alcohol were formed also; the values are given in the original paper.[19]

the yield of succinate found (see Table II).[19] These data were taken from an experiment in which serial analyses were made of the products of a glycerol fermentation and the values are presented on the basis of the amount of fermentation which occurred during each interval. It is seen that with time, an increased amount of CO_2 was utilized per mole of glycerol fermented, with concurrent increase in the formation of succinate. Furthermore, the amounts of CO_2 utilized and succinate produced were equivalent, which indicated that the CO_2 was fixed by the condensation of CO_2 with a C_3 compound arising from the glycerol. This hypothesis was further supported by the observation that when the fermentation occurred in phosphate buffer in the absence of CO_2, little if any succinate was formed.[19] It therefore appeared that CO_2 played a direct role in succinate formation. In addition, NaF was found to inhibit the fixation of CO_2 and to cause simultaneously an equivalent decrease in the yield of succinate.[22]

Pyruvate had been isolated previously from the propionic acid fermentation by the use of bisulfite as a trapping agent;[23] it was proposed, therefore, that CO_2 might combine with the pyruvate[19, 24] to form oxalacetate.

$$CO_2 + CH_3—CO—COOH \rightleftharpoons COOH—CH_2—CO—COOH$$

The reaction is frequently called the Wood and Werkman reaction. Barron[25] was the first to use this designation. When isotopes of carbon became available, it was possible to test this proposal in a more direct manner. Fermentations were carried out with propionic acid bacteria and with *E. coli* in the presence of $C^{13}O_2$, and the succinate produced was isolated and degraded. If CO_2 were fixed by the above reaction and succinate formed by its reduction, the C^{13} would be expected to occur exclusively in the carboxyl groups of the succinate.

$$C^{13}O_2 + CH_3—CO—COOH \rightleftharpoons C^{13}OOH—CH_2—CO—COOH$$

$$C^{13}OOH—CH_2—CO—COOH + 2\ H \rightleftharpoons C^{13}OOH—CH_2—CHOH—COOH$$

$$C^{13}OOH—CH_2—CHOH—COOH \rightleftharpoons H_2O + C^{13}OOH—CH{=}CH—COOH$$

$$C^{13}OOH—CH{=}CH—COOH + 2\ H \rightarrow C^{13}OOH—CH_2—CH_2—COOH$$

Carboxyl-labeled succinate was found as a product with both the propionic acid bacteria and *E. coli*.[24] In addition, the tracer was also found in propionate and there, too, exclusively in the carboxyl.[26, 27] The latter finding will be discussed in greater detail in Section V.

Another important clue indicating CO_2 fixation into oxalacetate came from the studies of Evans and Slotin[15] with pigeon liver. The $C^{11}O_2$ used was fixed exclusively into the α-carboxyl of α-ketoglutarate.[17, 28] Furthermore, Wood *et al.*[17] showed the distribution of C^{13} in the other compounds of the tricarboxylic acid cycle such as malate, fumarate, and succinate to be consistent with the idea of fixation in oxalacetate, i.e., the CO_2 carbon was in the carboxyl groups in all cases. See Krampitz, Vol. II, Chapter 4, p. 214 for further discussion of these data in relation to oxalacetate and the tricarboxylic acid cycle.

B. CO_2 Incorporation in Oxalacetate by an Exchange Reaction

Although oxalacetate seemed a logical candidate as the product of CO_2 fixation, there was no direct evidence for the reaction at the time it was proposed. Evidence for such a reaction was, however, not long in appearing. Krampitz and Werkman[29] found in *Micrococcus lysodeikticus*, a heat-labile enzyme which catalyzes the decarboxylation of oxalacetate to pyruvate and CO_2. When the reaction was tested in reverse, i.e., for CO_2 fixation, no oxalacetate could be shown to accumulate from pyruvate and CO_2. Krampitz *et al.*[30] were able, nevertheless, to present evidence for the partial reversal of the reaction. Testing the reaction by the exchange method, they demonstrated the entrance of $C^{13}O_2$ into the β-carboxyl of oxalacetate.

Because this procedure set the pattern for other studies based on exchange reactions, it is necessary to consider it here in some detail. Oxalacetate, labeled bicarbonate, and the enzyme preparation were incubated until about 50% of the oxalacetate was broken down. Then the residual oxalacetate was degraded to determine the location of any tracer in the molecule. This approach was based on the following reasoning. Even though the equilibrium is far toward pyruvate and CO_2, if the reaction were reversible, one would expect synthesis as well as breakdown. In this way, labeled CO_2 would enter the oxalacetate. By terminating the reaction before all of the oxalacetate was decarboxylated, there would be sufficient substrate left for the degradation. To localize the tracer carbon, the oxalacetate was decarboxylated by heating the acid in solution to liberate the β-carboxyl as CO_2. The resulting pyruvate was then treated with ceric sulfate to obtain the α-carboxyl. Typical results, values in per cent excess C^{13}, were as follows:

$$\begin{array}{ccc} 0.14 & & 0.02 \\ \text{HOOC—CH}_2\text{—CO—} & & \text{COOH} \\ \beta & & \alpha \end{array}$$

The original bicarbonate contained 7.1% excess C^{13}. Thus there was some incorporation of CO_2 into the β-carboxyl group of oxalacetate and little or none into the α-carboxyl. No $C^{13}O_2$ exchange was observed during the oxidative decarboxylation of either pyruvate or α-ketoglutarate with *M. lysodeikticus* nor during the nonenzymic decarboxylation of oxalacetate.

C. Attempts to Obtain Net CO_2 Fixation with Oxalacetic Decarboxylase

At first sight, the results of Krampitz *et al.*[30] seemed to provide fairly direct evidence for a fixation reaction by the proposed Wood and Werkman reaction. To date, however, no one has been able to demonstrate a net fixation of CO_2 with oxalacetic decarboxylase. Herbert[31] has done the most complete studies with the *M. lysodeikticus* enzyme which, although isolated in crystalline form, was not homogeneous when subjected to ultracentrifugal studies. The enzyme was found to be specific for oxalacetate; it did not decarboxylate pyruvate, α-ketoglutarate, acetoacetate, β-ketoglutarate, oxalosuccinate, dihydroxymaleic acid, or dihydroxytartaric acid.[32] The purified enzyme, however, was found by Utter and Wood (see p. 57 of ref. 33) to promote a very rapid exchange of CO_2 with oxalacetate.

It seemed possible that the decarboxylase could be coupled with malic dehydrogenase to pull the reaction and thus to obtain a net fixation of CO_2, i.e.:

$$\text{pyruvate} + CO_2 \rightleftharpoons \text{oxalacetate} + H^+$$

$$\text{oxalacetate} + \text{DPNH} + H^+ \rightleftharpoons \text{malate} + \text{DPN}^+$$

$$\text{Sum: pyruvate} + CO_2 + \text{DPNH} \rightleftharpoons \text{malate} + \text{DPN}^+$$

The "malic" enzyme (see Section IV, A, 1) catalyzes the same over-all reaction which is shown as the sum above except the nucleotide is TPN^+. Harary *et al.*[34] have measured the equilibrium of the malic enzyme at pH 7.4 and 22 to 25°C., using the enzyme obtained from wheat germ. They found:

$$K = \frac{(\text{L-malate}^{2-})(\text{TPN}^+)}{(\text{pyruvate}^-)(\text{CO}_2)(\text{TPNH})} = 19.6 \text{ (liters} \times \text{mole}^{-1})$$

Krebs and Roughton[35] presented evidence that CO_2, not HCO_3^-, is the primary product of pyruvate decarboxylation by yeast carboxylase; Harary *et al.*[34] assumed this to be true also for the malic enzyme. As may be noted from the above equations, the protons cancel out; thus the hydrogen ion concentration does not enter into the calculation of the equilibrium constant. This is not the case if bicarbonate rather than CO_2 is the reactant.

The $\Delta F°$ of the "malic" reaction was calculated to be -1.8×10^3 cal.

$$\Delta F° = -RT\ 2.303 \log K$$

$$\Delta F° = -1.987 \times 296 \times 2.303 \log 19.6$$

$$\Delta F° = -1.8 \times 10^3 \text{ cal.}$$

Krebs and Kornberg,[36] from other data, have calculated a free energy of -0.3×10^3 cal. for the same reaction. Thus the over-all reaction appears to favor malate synthesis.*

In spite of the fact that a coupled oxalacetic decarboxylase and a malic dehydrogenase would have the same thermodynamic feasibility as the malic enzyme reaction, since they would yield the same products, numerous investigators, Herbert,[31] Mehler *et al.*,[41] and Ochoa *et al.*[42] have shown that such coupling does not take place. Failure to couple occurs even though the oxalacetic decarboxylase of Herbert does catalyze an exchange of CO_2

* The equilibrium of the oxalacetate decarboxylase reaction theoretically can be calculated from the $\Delta F°$ of the malic reaction and the malic dehydrogenase reaction. Burton and Wilson[37] have calculated the free energy toward oxalacetate formation in the latter reaction to be 16.7×10^3 cal. If one assumes that the DPN and TPN systems have the same free energy, i.e., DPN^+ cancels a TPN^+, etc., one may calculate the free energy of the oxalacetate decarboxylase reaction.

	$\Delta F°$ *cal.*
$\text{malate}^{2-} + \text{DPN}^+ \rightleftharpoons \text{oxalacetate}^{2-} + \text{DPNH} + \text{H}^+$	16.7×10^3
$\text{pyruvate}^- + \text{CO}_2 + \text{TPNH} + \text{H}^+ \rightleftharpoons \text{malate}^{2-} + \text{TPN}^+ + \text{H}^+$	-1.8×10^3
Sum: $\text{pyruvate}^- + \text{CO}_2 \rightleftharpoons \text{oxalacetate}^{2-} + \text{H}^+$	14.9×10^3

The value, 14.9×10^3 cal., written for the carboxylation of pyruvate by CO_2, is much greater than the values calculated if bicarbonate rather than CO_2 is the assumed substrate for the carboxylation, For bicarbonate. Evans *et al*,[38] calculated the $\Delta F°$ as 5.25×10^3 cal. and Burton and Krebs,[39] 6.39×10^3. Johnson[40] gives a value of -17×10^3 cal. for the $\Delta F°$ of the decarboxylation of oxalacetate to pyruvate and CO_2.

with the β-carboxyl of the oxalacetate.[33] It must be realized that the thermodynamic data show only what the ultimate equilibrium will be—they do not indicate the time that is required for the equilibrium to be reached. The malic enzyme may promote a rapid synthesis from pyruvate and CO_2 which can be measured, while the coupled reaction of the decarboxylase and malic dehydrogenase may be so slow as to be not measurable. It is important to note in this connection that Salles *et al.*[43] have not found free oxalacetate as a product of the malic enzyme reaction. When malate was synthesized from pyruvate, $C^{14}O_2$, and TPNH by the malic enzyme in the presence of unlabeled oxalacetate, a rapid incorporation of $C^{14}O_2$ into the β-carboxyl group of malate occurred, but oxalacetate remained unlabeled.[43] Thus, the site of the condensation between pyruvate and CO_2 and the site for the reduction appear to be such that the oxalacetate never leaves the enzyme surface. The malic enzyme therefore may provide a very favorable situation for a rapid synthesis.

The exchange of CO_2, which is catalyzed by oxalacetic decarboxylase, could occur as shown in Scheme I:

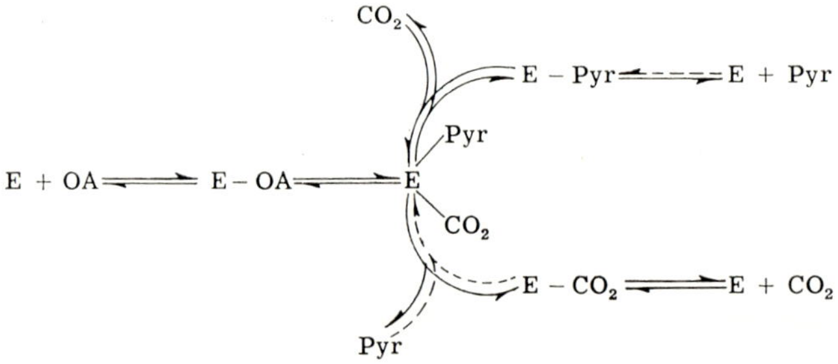

SCHEME I

In Scheme I the dotted arrows indicate a slow reaction. It is seen that one would expect CO_2 to exchange with oxalacetate, but the synthesis of oxalacetate from pyruvate and CO_2 would be much slower. In this connection it is of interest that Steinberger and Westheimer[44] have presented evidence that the nonenzymic decarboxylation of oxalacetate involves the formation of a chelate of a metal ion with the α-COOH group and the keto group of oxalacetate. Utter[45] has noted with the *M. lysodeikticus* enzyme, that oxalacetate is decarboxylated more rapidly with Mn^{++} than with Mg^{++}. Since in the exchange reaction, however, the reverse situation applies, i.e., Mg^{++} is more effective than Mn^{++}, Utter believes the nature of the metal chelate may be of considerable importance in directing the kinetics of the decarboxylation reaction. Plaut and Lardy[46] have purified oxalacetate decarboxylase from *Azotobacter vinelandii* and found the enzyme from this source does not promote a CO_2 exchange with oxalacetate. Presumably, the extent of the exchange reaction may depend on the stability of the metal chelate, and this may differ in the *M. lysodeikticus* and *Azotobacter* enzymes.

Although the malic enzyme from animal tissue catalyzes the decarboxylation of oxalacetate, it does not promote the exchange reaction with CO_2.[43] McManus[47] has shown with *M. lysodeikticus* that the exchange reaction does not occur via the malic enzyme nor is it stimulated by ATP. The last observation indicates that the exchange probably does not proceed via the phosphoenolpyruvic carboxykinase reaction, which will be described below. Decarboxylation of oxalacetic has been observed with enzyme preparations isolated from *E. coli*[48] and from *Lactobacillus arabinosus*.[49] These reactions may occur via oxalacetic decarboxylase or possibly by the malic enzyme or by other enzymes. Kaltenbach and Kalnitsky[50] have observed a net formation of oxalacetate from pyruvate and CO_2 using extracts of *E. coli* and *Proteus morganii*, but the mechanism of this synthesis is not known.

We have seen that oxalacetic decarboxylase catalyzes the breakdown of oxalacetate and that the *M. lysodeikticus* enzyme also catalyzes the exchange with CO_2. However, it has not yet been possible to show that this enzyme will function as a component of a reaction sequence that yields a net fixation of CO_2. It was necessary to consider the exchange reaction at some length because this procedure has frequently been used to study CO_2-fixation reactions. It is evident that the exchange reaction as such cannot be used to determine whether an enzyme will function in a net fixation of CO_2. Our attention will now be directed to those reactions in which a net fixation of CO_2 clearly occurs.

IV. Primary Reactions of CO_2 Assimilation

A primary CO_2 fixation is defined here as a reaction by which CO_2 is combined with some compound to form a new carbon-carbon bond and results in a net fixation of CO_2. The primary reaction is to be contrasted with secondary reactions whereby a great many compounds are derived from the primary product. For 10 years following the discovery of the heterotrophic assimilation of CO_2, no clearly defined primary fixation reaction was found. However, in 1945, Ochoa[51] described such a reaction involving the synthesis of isocitrate from α-ketoglutarate, CO_2, and TPNH by the "isocitric" enzyme. In 1948, another primary fixation was discovered:[42] the "malic" enzyme. Since 1948, numerous additions have been made to the list of primary fixation reactions. These may be considered conveniently in the three categories described by Calvin and Pon.[52]

A. Reactions requiring reduced pyridine nucleotides as a source of energy.

1. Malic enzyme

$$\text{pyruvate} + CO_2 + \underset{(\text{DPNH})}{\text{TPNH}} \rightleftharpoons \text{malate} + \underset{(\text{DPN}^+)}{\text{TPN}^+}$$

2. Isocitric dehydrogenase

$$\alpha\text{-ketoglutarate} + CO_2 + \underset{(DPNH)}{TPNH} \rightleftharpoons \text{isocitrate} + \underset{(DPN^+)}{TPN^+}$$

3. Phosphogluconic dehydrogenase

$$\text{ribulose-5-P} + CO_2 + TPNH \rightleftharpoons \text{6-phosphogluconate} + TPN^+$$

B. Reactions requiring no apparent extra energy source.

1. Phosphoenolpyruvic carboxylase

$$\text{phosphoenolpyruvate} + CO_2 \rightarrow \text{oxalacetate} + P_i$$

2. Phosphoenolpyruvic carboxykinase

$$\text{phosphoenolpyruvate} + CO_2 + \underset{(ADP)}{GDP} \rightleftharpoons \text{oxalacetate} + \underset{(ATP)}{GTP}$$

3. Phosphoenolpyruvic carboxytransphosphorylase

$$\text{phosphoenolpyruvate} + CO_2 + P_i \rightleftharpoons \text{oxalacetate} + P\text{—}P$$

4. Ribulose diphosphate carboxylase

$$\text{ribulose-1,5-di-P} + CO_2 \rightarrow \text{two 3-phosphoglycerate}$$

5. Aminoimidazoleribonucleotide carboxylase

$$\text{5-aminoimidazoleribonucleotide} + CO_2 \rightarrow \text{5-amino-4-carboxyimidazoleribonucleotide}$$

C. Reactions requiring ATP as a source of energy.

1. Propionyl carboxylase

$$\text{propionyl CoA} + CO_2 + ATP \rightleftharpoons \text{methylmalonyl CoA} + ADP + P_i$$

2. β-methylcrotonyl carboxylase

$$\beta\text{-methylcrotonyl CoA} + CO_2 + ATP \rightleftharpoons \beta\text{-methylglutaconyl CoA} + ADP + P_i$$

3. Acetyl carboxylase

$$\text{acetyl CoA} + CO_2 + ATP \rightleftharpoons \text{malonyl CoA} + ADP + P_i$$

4. Pyruvic carboxylase

$$\text{pyruvate} + CO_2 + ATP \rightleftharpoons \text{oxalacetate} + ADP + P_i$$

A. Reactions with Reduced Pyridine Nucleotide as an Energy Source

These reactions, in general, are carboxylations α to a carbonyl group but the β-keto acid does not occur as a *free* intermediate. Presumably there is a simultaneous reduction of the enzyme-bound keto acid to a hydroxy acid.

1. Malic Enzyme

Although the fixation of CO_2 via isocitric dehydrogenase[51] was observed prior to fixation by the "malic" enzyme,[42] we shall consider the malic enzyme first and in greater detail because the synthesis of dicarboxylic acids appears to be of greater importance. Oxalacetate is required in the tricarboxylic acid cycle, which serves not only as a pathway of oxidation to supply energy but also as the pathway for synthesis of glutamate and its derivatives such as ornithine, proline, etc. For each such synthesis, oxalacetate and acetyl CoA are the starting materials. Furthermore, the synthesis of phosphoenolpyruvate probably occurs by way of oxalacetate.[36, 53-55] Thus carbons flow into amino acids and carbohydrates via the dicarboxylic acids, so the synthesis of dicarboxylic acids from pyruvate occupies a pivotal role in metabolism.

The malic enzyme has a wide distribution in animals[56, 57] and plants;[58, 59] it is present in insect blood[60] and in the intestinal round worm, *Ascaris*.[61] It can be adaptively formed in *Lactobacillus arabinosus*[62, 63] and probably occurs in other microorganisms. As we shall see, the bacterial malic enzyme may differ from the animal and plant enzymes, e.g., it may actually be a decarboxylase yielding lactate and CO_2 from malate and may not reverse the reaction to yield a net fixation of CO_2. Ochoa[64] should be consulted for additional information and references on the malic enzyme.

Malic enzyme usually has been found to catalyze two reactions:

$$CH_3COCOO^- + CO_2 + TPNH \xrightleftharpoons{Mn^{++}} COO^-CH_2CHOHCOO^- + TPN^+ \quad (1)$$

$$COO^-CH_2COCOO^- + H^+ \xrightarrow{Mn^{++}} CH_3COCOO^- + CO_2 \quad (2)$$

During purification, these two activities parallel each other.[56] This was the case with the pigeon liver enzyme which Rutter and Lardy[57] purified more than 1000-fold. On the other hand, Saz and Hubbard[61] seem to have obtained from *Ascaris* a malic enzyme which lacks oxalacetic decarboxylase activity.

With the pigeon liver enzyme, reaction 1 has been shown to be readily reversible.[42] The reaction can be followed in either direction spectrophotometrically, as illustrated in Fig. 1. Curve 1 shows the course of the oxidation of malate. The cuvette contained the liver enzyme, TPN, buffer, and $MnCl_2$; at time 0, L-malate was added, and rapid reduction of the TPN occurred as the malate was converted to pyruvate and CO_2 . Curve 2 shows the course of the reaction when CO_2 is being fixed. The cuvette contained the enzyme, TPNH, buffer, and $MnCl_2$. At time 0, pyruvate was added and a slow oxidation of the TPNH started; this was due to the presence of lactic dehydrogenase in the malic enzyme preparation. At the time indicated by the arrow, $NaHCO_3$ saturated with CO_2 was added; this initiated an immediate oxidation of the TPNH as the pyruvate and CO_2 were converted to malate. Curve 3 shows the reaction when both pyruvate and CO_2 were added at 0 time.

The decarboxylation of oxalacetate by malic enzyme (reaction 2) is markedly stimulated by either TPN or TPNH. The optimum pH is 4.5, whereas the optimum for reaction 1 is pH 7.5. There is no exchange of $C^{14}O_2$ with the β-carboxyl of oxalacetate during the decarboxylation, nor is oxalacetate reduced by TPNH or DPNH (no malic dehydrogenase activity). Because of this independent dual activity, i.e., nonoxidative decarboxylation of oxalacetate and oxidative decarboxylation of malate to pyruvate and CO_2, the malic enzyme has been spoken of as a "double-headed" enzyme. We have noted that even when $C^{14}O_2$ is fixed in malate there is no incorporation of C^{14} in the added oxalacetate;[43] therefore, free oxalacetate is not an intermediate of the malic reaction. If oxalacetate is produced, it is reduced to malate without leaving the enzyme surface; thus the reaction appears to be a "simultaneous process with no detectable intermediates."[57] With regard to this view, Boyer[65] presented Scheme II and stated, "This is both logical and attractive and, taken together with the demonstrated role of metals in the non-enzymic decarboxylation of α-keto

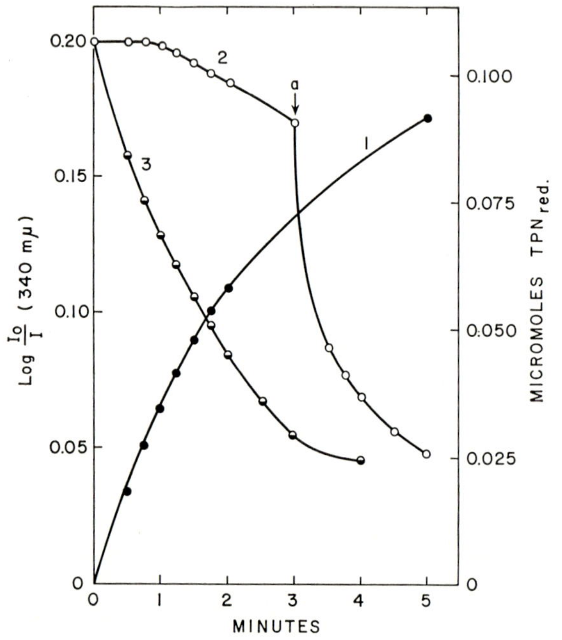

FIG. 1. Optical demonstration of the reversibility of the fixation of CO_2 by the malic enzyme.

dicarboxylic acids . . . and the demonstration of direct hydrogen transfer in pyridine nucleotide reactions, suggests to the reviewer a mechanism . . . as depicted . . . Decarboxylation is facilitated by electron withdrawal from the β-carbon atom and concomitant withdrawal from the carboxyl group, which yields CO_2. Both transfer of a hydride ion to the pyridine nucleotide and chelation of the metal would promote electron deficiency on the α-carbon atom; thus, the dehydrogenation and decarboxylation processes can logically proceed simultaneously. The metal chelate of the enol form of pyruvate would be released to solution as the free pyruvate and metal . . ."

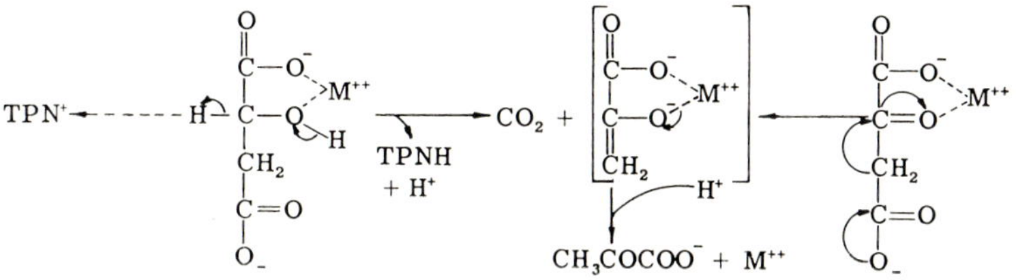

SCHEME II

It was formerly considered that the equilibrium of the malic reaction was far toward decarboxylation, but we have seen (p. 49) that the equilibrium is actually toward synthesis of malate.[34] The former view probably arose because the breakdown reaction was measured without the addition of CO_2, the affinity of the malic enzyme for CO_2 is low,[54] and the CO_2 produced is lost in the gas phase. All these factors favor decarboxylation.

The synthesis of malate has been accomplished[42] by linking the malic enzyme with glucose-6-phosphate dehydrogenase, which serves to regenerate the TPNH.

$$\text{glucose-6-P} + \text{TPN}^+ \rightleftharpoons \text{6-P-gluconate} + \text{TPNH} + \text{H}^+$$

$$\text{pyruvate} + CO_2 + \text{TPNH} + \text{H}^+ \rightleftharpoons \text{malate} + \text{TPN}^+$$

$$\text{Sum: glucose-6-P} + \text{pyruvate} + CO_2 \rightleftharpoons \text{6-P-gluconate} + \text{malate}$$

We may now turn to a consideration of the malic enzyme of *L. arabinosus*. The reaction with *L. arabinosus* has been pictured as follows:

$$\text{malate} + \text{DPN}^+ \rightleftharpoons \text{pyruvate} + CO_2 + \text{DPNH} + \text{H}^+$$

$$\text{pyruvate} + \text{DPNH} + \text{H}^+ \rightleftharpoons \text{lactate} + \text{DPN}^+$$

$$\text{Sum: malate} \rightleftharpoons \text{lactate} + CO_2 \quad (3)$$

The conversion is observed as the decarboxylation of malate to lactate and CO_2; DPN^+ is required. The analogy to the malic enzyme from animal sources is by assumption rather than by direct evidence. The bacterial enzyme[62, 63] has the following properties: (*1*) rapidly decarboxylates malate in the presence of DPN to yield lactate and CO_2, (*2*) decarboxylates oxalacetate but does not reduce oxalacetate to malate with DPNH, (*3*) in an exchange experiment[62] with a mixture of malate and pyruvate, $C^{14}O_2$ is incorporated into the β-carboxyl of malate, (*4*) does not incorporate $C^{14}O_2$ into oxalacetate by an exchange reaction during the decarboxylation of oxalacetate, (*5*) when malate labeled in both carboxyls is decarboxylated in the presence of pyruvate, the pyruvate remains unlabeled.[63] The last fact indicates that *free* pyruvate is not an intermediate in the over-all reaction and requires the assumption that malic enzyme and lactic dehydrogenase are bound in an enzyme complex in such a manner that the pyruvate produced remains enzyme-bound and is reduced to lactate rather than equilibrated with pyruvate in solution.

It has not been possible with the bacterial enzyme to obtain evidence for CO_2 fixation as presented in Fig. 1 for the mammalian malic enzyme. The reduction of DPN (curve 1) is not observed during malate decarboxylation, presumably because the DPNH which is formed is immediately reoxidized by the conversion of pyruvate to lactate, due to the presence of lactic dehydrogenase. The CO_2 fixation reaction cannot be shown (curve 2) because the pyruvate is reduced by the lactic dehydrogenase even without the addition of CO_2.

Therefore, direct evidence for the existence of the malic enzyme in heterotrophic bacteria is still lacking. Until the enzyme is obtained free of lactic dehydrogenase and a primary CO_2-fixation is demonstrated, there will remain doubt whether this enzyme should be called a "malic" enzyme. Furthermore, if the bacterial enzyme is a "triple-headed" enzyme possessing lactic dehydrogenase activity, the competing reactions would be:

	ΔF cal. $\times 10^{-3}$
$\text{pyruvate}^- + CO_2 + \text{DPNH} + H^+ \rightleftharpoons \text{malate}^{--} + \text{DPN}^+$	−1.5
$\text{pyruvate}^- + \text{DPNH} + H^+ \rightleftharpoons \text{lactate}^- + \text{DPN}^+$	−15.5[37]

The lactic dehydrogenase reaction is much more exergonic than the malic reaction; thus pyruvate would be converted to lactate instead of malate. The fixation of CO_2 is the most distinctive property of the malic enzyme, and the bacterial enzyme would lack this property.

As we have noted, the malic enzyme from *Ascaris* lacks oxalacetic decarboxylase activity and thus differs from the other malic enzymes. When the incorporation of $C^{14}O_2$ was tested in an exchange experiment with this enzyme, some fixation occurred, but the amount was small.[61] Unfortunately, the spectrophotometric test for the carboxylation of pyruvate, as illustrated in curve 2 of Fig. 1, was not performed. One would like to have direct proof of a net fixation of CO_2 with the enzyme before concluding that a malic enzyme may be obtained free of oxalacetate decarboxylase activity. The

Ascaris enzyme was more active with DPN than with TPN. We shall see, however, there are two types of isocitric dehydrogenase, one using TPN, which is reversible, and one using DPN, which is irreversible. The DPN-dependent isocitric dehydrogenase catalyzes the oxidative decarboxylation reaction but not the reverse, i.e., the reductive fixation of CO_2. The *Ascaris* enzyme could be similar to the latter.

2. Isocitric Dehydrogenase

As noted above, two types of isocitric dehydrogenase are known; one, a reversible TPN^+-requiring enzyme, which catalyzes both the reductive carboxylation of α-ketoglutarate and the oxidative decarboxylation of isocitric acid, and a second, a DPN^+-requiring enzyme, which catalyzes the irreversible oxidative decarboxylation of isocitrate to form CO_2 and α-ketoglutarate. Both the DPN^+ and TPN^+ enzymes have been found in yeast by Kornberg and Pricer.[66] A DPN^+-requiring enzyme similar to the yeast enzyme has been found to occur in mitochondria prepared from animal tissue[67] while the TPN^+-requiring enzyme is found largely in the soluble fraction of animal tissue. CO_2-fixation by isocitric dehydrogenase isolated from either bacteria or molds has not been reported, but the reaction probably is present.

Ochoa,[51] using an enzyme isolated from pig heart, was the first to show the reductive carboxylation of α-ketoglutarate to isocitrate.

$$\underset{\alpha\text{-Ketoglutarate}}{COO^-CH_2CH_2COCOO^-} + CO_2 + TPNH \xrightleftharpoons{Mn^{++}} \underset{\text{Isocitrate}}{COO^-CH_2\underset{\displaystyle COO^-}{\underset{|}{C}}HCHOHCOO^-} + TPN^+ \qquad (4)$$

Associated with this protein are two other enzymic activities, viz., the decarboxylation of oxalosuccinate to form α-ketoglutarate,

$$\underset{\text{Oxalosuccinate}}{COO^-CH_2\underset{\displaystyle COO^-}{\underset{|}{C}}HCOCOO^-} + H^+ \xrightarrow{Mn^{++}} CO_2 + \underset{\alpha\text{-Ketoglutarate}}{COO^-CH_2CH_2COCOO^-} \qquad (5)$$

and the reduction of oxalosuccinate to isocitrate.

$$\underset{\text{Oxalosuccinate}}{COO^-CH_2\underset{\displaystyle COO^-}{\underset{|}{C}}HCOCOO^-} + TPNH + H^+ \rightarrow \underset{\text{Isocitrate}}{COO^-CH_2\underset{\displaystyle COO^-}{\underset{|}{C}}HCHOHCOO^-} + TPN^+ \qquad (6)$$

At first it was thought that the fixation of CO_2 involved two separate enzymes, one catalyzing reaction 5 from right to left and the second enzyme catalyzing reaction 6. It now appears that all three reactions are catalyzed by one and the same enzyme.[51, 68, 69] The enzyme from heart has been purified extensively[68, 69] and all three activities remain associated. It should be noted that the isocitric enzyme differs from the malic enzyme in that the former reduces the corresponding keto acid to isocitrate while the m lic enzyme, it will be recalled, does not reduce oxalacetate to malate.

The CO_2 fixation can be followed by optical methods just as with the malic enzyme (Fig. 1). Also, by linking the reaction with glucose-6-phosphate dehydrogenase to regenerate the TPNH, an increased conversion of α-ketoglutarate to isocitrate is obtained. The net reaction can also be shifted toward synthesis by linkage to aconitase, which converts the isocitrate to aconitate and citrate.[51]

An equilibrium constant of 1.3 liters × $mole^{-1}$ at pH 7.0 and 22°C. has been calculated from the data of Ochoa;[70] Ceithaml and Vennesland[70] obtained a value of 2.0 with the enzyme isolated from parsley root.

$$K = \frac{(\text{L-isocitrate}^{3-})(\text{TPN}^+)}{(\alpha\text{-ketogutarate}^{2-})(\text{CO}_2)(\text{TPNH})} = 1.3 \text{ liter} \times \text{mole}^{-1}$$

The equilibrium thus is not as favorable for carboxylation as it is with the malic enzyme, e.g., $K = 19.6$ liters × $mole^{-1}$ (see page 49).

Free oxalosuccinate cannot be detected when α-ketoglutarate and CO_2 are converted to isocitrate (reaction 4 from left to right) or when isocitrate is dehydrogenated (reaction 4 from right to left). For example, when Siebert *et al.*[69] converted $C^{14}O_2$ and α-ketoglutarate to isocitrate in the presence of oxalosuccinate, there was no C^{14} found in the oxalosuccinate at the end of the reaction. Likewise, when C^{14}-labeled isocitrate was converted to α-ketoglutarate and CO_2 in the presence of oxalosuccinate, no C^{14} entered the oxalosuccinate. From these experiments, it is apparent that free oxalosuccinate does not occur as an intermediate; hence, oxalosuccinate cannot be considered as a true intermediate of the tricarboxylic acid cycle.

Thus far, no important physiological function has been found for the synthesis of isocitrate from α-ketoglutarate and CO_2. We know that α-ketoglutarate is synthesized via the tricarboxylic acid cycle, and the fixation reaction would only serve to reverse the isocitric dehydrogenase step. If another pathway of α-ketoglutarate synthesis from C_3 compounds were found, this CO_2 fixation might assume greater importance. The α-ketoglutarate formed could be converted to isocitrate and give rise to glyoxalate and succinate via the isocitrase reaction or to oxalacetate and acetate via the citrase reaction. In this way, C_4 compounds could be synthesized by an organism lacking the ability to form oxalacetate by CO_2 fixation.

3. Phosphogluconic Dehydrogenase

Fixation of CO_2 by phosphogluconic dehydrogenase resembles the fixation catalyzed by both the malic enzyme and the isocitric dehydrogenase in that a reduction by TPNH occurs and a keto acid might be expected as an intermediate. The reaction is as follows:

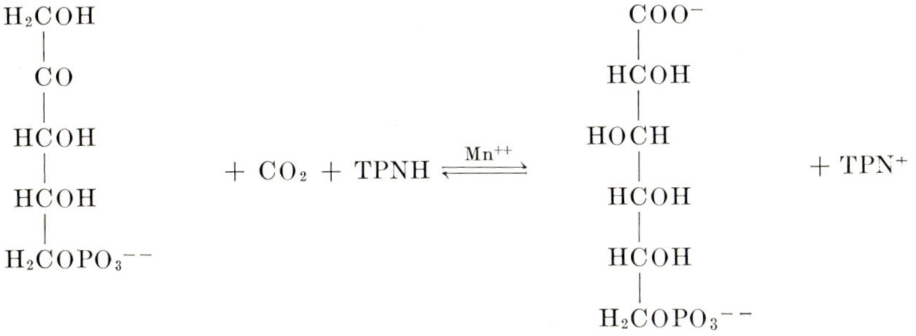

$$\begin{array}{c} H_2COH \\ | \\ CO \\ | \\ HCOH \\ | \\ HCOH \\ | \\ H_2COPO_3^{--} \end{array} + CO_2 + TPNH \underset{}{\overset{Mn^{++}}{\rightleftarrows}} \begin{array}{c} COO^- \\ | \\ HCOH \\ | \\ HOCH \\ | \\ HCOH \\ | \\ HCOH \\ | \\ H_2COPO_3^{--} \end{array} + TPN^+ \quad (7)$$

Ribulose-5-P   6-Phosphogluconate

Although no evidence of its occurrence has been obtained, 3-keto-6-phosphogluconate would be the expected intermediate. Apparently 3-keto-6-phosphogluconate has not been tested directly with the dehydrogenase to determine if it is reduced or decarboxylated. If these reactions occurred, however, by analogy with the malic enzyme and the isocitric dehydrogenase, the keto acid probably would not be a free intermediate in reaction 7 but would occur only as a transient compound bound to the enzyme (see the illustration with malic enzyme p. 50).

Horecker and Smyrniotis[71] first demonstrated the reversibility of the phosphogluconic dehydrogenase reaction and proved it to be a primary fixation reaction. Using a purified enzyme from yeast, they showed: (a) $C^{14}O_2$ is fixed in 6-phosphogluconate during the course of a partial oxidative decarboxylation of the gluconate, (b) TPNH is oxidized in the presence of ribulose-5-phosphate and CO_2, and (c) a net synthesis of 6-phosphogluconate occurs from ribulose-5-phosphate and CO_2. The energy for this synthesis was furnished by coupling the reaction with glucose-6-phosphate dehydrogenase to regenerate the TPNH.

$$\text{glucose-6-P} + TPN^+ + H_2O \rightarrow \text{6-P-gluconate} + TPNH + H^+$$

$$\text{ribulose-5-P} + CO_2 + TPNH + H^+ \rightarrow \text{6-P-gluconate} + TPN^+$$

$$\text{glucose-6-P} + \text{ribulose-5-P} + CO_2 + H_2O \rightarrow 2 \text{ 6-P-gluconate} \quad (8)$$

For each mole of glucose-6-phosphate oxidized, 2 moles of 6-phosphogluconate should be formed, one of which would arise from the reductive carboxylation of ribulose-5-phosphate. The results from an experiment[71] are shown in Table III: in the complete system, 1.13 μmoles of glucose-6-

phosphate were utilized, and 2.13 μmoles of 6-phosphogluconate were formed. On omission of ribulose-5-phosphate, very little phosphogluconate was formed because ribulose-5-phosphate and CO_2 are essential as hydrogen acceptor to promote the reaction.

The equilibrium constant at 22°C. and pH 7.4 for reaction 7 was calculated to be 1.9 liters/mole, which is very close to the value of 1.3 for isocitric dehydrogenase.

Phosphogluconic dehydrogenase has a wide distribution in nature. It is present in animal tissues,[72] in many bacteria (*Bacillus subtilis* and *B. megaterium*,[73] *Pseudomonas fluorescens*,[74] *Leuconostoc mesenteroides*,[75] *E. coli*[76]) in yeast,[71] and in plants.[77] The enzyme from *L. mesenteroides* catalyzes

TABLE III

FORMATION OF 6-PHOSPHOGLUCONATE FROM RIBULOSE-5-PHOSPHATE AND CO_2 IN A COUPLED REACTION WITH GLUCOSE-6-PHOSPHATE DEHYDROGENASE (SEE REACTION 8)[a]

System	Products	
	Glucose-6-P (μmoles)	6-P-gluconate (μmoles)
Complete	−1.13	2.13
Ribulose-5-P omitted	−0.08	0.03
Glucose-6-P omitted	0	−0.04

[a] Reaction mixture contained in 1 ml: 0.13 μmoles TPN, 2.25 μmoles $NaHCO_3$, 1.1 mg. purified yeast phosphogluconic dehydrogenase, 3.5 μmoles ribulose-5-phosphate and 2.2 μmoles glucose-6-phosphate (the phosphogluconic dehydrogenase contained glucose-6-phosphate dehydrogenase). Incubation 15 min; 25°C.

the reaction more effectively in the presence of DPN^+ than in the presence of TPN^+.[75] Generally, it is assumed that phosphogluconic dehydrogenase from any source will catalyze the carboxylation of ribulose-5-phosphate, but in most instances, this has not been demonstrated.

The conversion of pentose to hexose via phosphogluconic dehydrogenase and glucose-6-phosphate dehydrogenase is not a very probable reaction sequence, since the equilibrium of the hydrolysis of the 6-phosphoglucono-lactone is far toward 6-phosphogluconate. Hence the rate of 6-phosphogluconolactone formation from the 6-phosphogluconate would be very slow. However, Moses *et al.*[78] have found that the mold *Zygorrhynchus moelleri* fixed CO_2 rapidly into phosphogluconate. The label appeared more slowly in hexosemonophosphate and even more slowly in phosphoglyceric acid. None entered the hexosediphosphate. The authors concluded that the $C^{14}O_2$ was incorporated into hexoses by carboxylation of pentose phosphate to form phosphogluconate followed by reduction to glucose-6-phosphate. It is unfortunate that the hexosemonophosphate was not degraded to establish the location of C^{14}, which should be in carbon 1. In animal tissue, no evidence has been obtained to suggest that hexoses are synthesized via fixation of CO_2 into 6-phosphogluconate.

No important physiological role has been described for the fixation of CO_2 by phosphogluconic dehydrogenase. 6-Phosphogluconate can be metabolized by the Entner and Doudoroff pathway or by the heterolactic type of cleavage of *L. mesenteroides* and possibly by other pathways; see Gunsalus *et al.*[79] for a review of pathways. Conceivably, some organisms may metabolize pentose by conversion to 6-phosphogluconate and fermentation of the latter.

B. Reactions without an Apparent Extra Energy Source

In these reactions, the source of energy is present in the substrate itself. Calvin and Pon[52] state, "The substrate is already in an 'active' form in the sense that it is unstable with respect to the more stable isomers. For example in phosphoenolpyruvate, the energy is stored in the form of enol phosphate; in the imidazole, there is the carboxylation of an ene-amine; and in the ribulose diphosphate, there is presumably the carboxylation of the non-cyclic form of the ribose, which is constrained to go through an ene-diol since cyclic acetal formation is prohibited by small ring size."

1. Phosphoenolpyruvic Carboxylase

This enzyme, discovered in spinach leaves by Bandurski and Greiner,[80] catalyzes the following irreversible reaction:

$$\begin{array}{l} \quad\;\; PO_3^{2-} \\ \quad\;\; | \\ \quad\;\; O \\ \quad\;\; | \\ CH_2{=}C{-}COO^- + CO_2 + H_2O \rightarrow COO^-CH_2COCOO^- + HPO_4^{2-} + H^+ \end{array} \qquad (9)$$

Phosphoenolpyruvate Oxalacetate

or

$$\begin{array}{l} \quad\;\; PO_3^{2-} \\ \quad\;\; | \\ \quad\;\; O \\ \quad\;\; | \\ CH_2{=}C{-}COO^- + HCO_3^- \rightarrow COO^-CH_2COCOO^- + HPO_4^{2-} \end{array} \qquad (9a)$$

Two equations are shown because it is not certain whether free CO_2 or the bicarbonate ion combines with the phosphoenolpyruvate.*

* Tchen *et al.*[81] have calculated the free energy of the phosphoenolpyruvic carboxylase reaction. Similar calculations using values quoted by Johnson[40] are also shown as well as $\Delta F^{\circ\prime}$ values which will be discussed on bottom of p. 62.

Phosphoenolpyruvic carboxylase occurs in plants,[81-84] in wheat germ,[85] and in autotrophic bacteria.[86]

The carboxylase has a very high affinity for both CO_2 and phosphoenolpyruvate (see Table IV). For this reason, the carboxylase is assigned a prominent role in dicarboxylic acid formation. For example, crassulacean plants in the dark are capable of removing practically all the CO_2 from the atmosphere. The enzyme responsible for this CO_2 fixation must have a low K_m value (i.e., a high affinity) for CO_2. The K_m value of the phosphopyru-

	$\Delta F°$ kcal. Tchen *et al.*	$\Delta F°$ kcal. Johnson	$\Delta F°'$ kcal. pH 7 Johnson
(a) Phosphoenolpyruvate-2-PO_3^{3-} + ADP^{3-} + $H^+ \rightarrow$ pyruvate$^-$ + ATP^{4-}	−14.5	−15.8	−5.72
(b) $ATP^{4-} + H_2O \rightarrow ADP^{3-} + HPO_4^{2-} + H^+$	+0.9	+1.18	−8.14
(c) Pyruvate$^-$ + HCO_3^- $\rightarrow$ oxalacetate^{2-} + H_2O	+6.39	+6.39[a]	6.08[a]
(d) Phosphoenolpyruvate-2-PO_3^{3-} + HCO_3^- $\rightarrow$ oxalacetate^{2-} + HPO_4^{2-}	−7.21	−8.23	−7.77

[a] Calculated from reactions 37 and 38 of Table IV of Johnson.[40]

Tchen *et al.* also made a calculation by a second procedure which gave a value of $\Delta F° = -7.95$ kcal, The equilibrium constant calculated from the latter value was $K = 6.6 \times 10^5$. This values is sufficiently large to account for the apparent irreversibility of the carboxylase reaction.

If free CO_2 is the reactant in this reaction the equations become:

	$\Delta F°$ kcal. Johnson	$\Delta F°'$ kcal pH 7 Johnson
(a) Phosphoenolpyruvate-2-PO_3^{3-} + ADP^{3-} + $H^+ \rightarrow$ pyruvate$^-$ + ATP^{4-}	−15.8	−5.72
(b) $ATP^{4-} + H_2O \rightarrow ADP^{3-} + HPO_4^{2-} + H^+$	1.18	−8.14
(e) Pyruvate$^-$ + $CO_2 \rightarrow$ oxalacetate^{2-} + H^+	17.05	+7.06
(f) Phosphoenolpyruvate-2-PO_3^{3-} + $CO_2 + H_2O \rightarrow$ oxalacetate^{2-} + H^+ + HPO_4^{2-}	+2.43	−6.80

It is seen with CO_2 the calculated $\Delta F°$ actually favors formation of phosphoenoll pyruvate. However, the $\Delta F°$ values are the standard free energy change when all reactants are present at unit concentration. In this case the compounds would be un-ionized, pH = 0. When the reaction occurs at pH 7 and substrates and products are ionized, the free energy change ($\Delta F°'$) is quite different from the $\Delta F°$ at standard conditions (see Johnson[40] and Rutman and George[82]). The values used are those presented by Johnson.[40] In this case $\Delta F°'$ of reactions (d) and (f) are almost the same ($\Delta F° = -7.77$ with bicarbonate and -6.80 with CO_2). Thus at pH 7 the energetics are approximately the same for both reactions, the equilibria are far toward oxalacetate, and the reactions are essentially irreversible. These calculations serve to illustrate the danger of using $\Delta F°$ per se for judging an equilibrium under physiological conditions.

vic carboxylase for CO_2 is shown in Table IV to be of a different order of magnitude than other enzymes listed. Walker and Brown[87] and Walker[59] have compared the reaction rates of phosphopyruvic carboxylase and malic enzyme as a function of the concentration of CO_2. The maximum rate for the carboxylase[87] was reached at 0.25% CO_2, and the rate was almost as high at 0.05%, whereas with the malic enzyme,[59] the highest rate was not reached until the CO_2 concentration was 30% of the atmosphere.

Suzuki and Werkman[86] observed a similar relationship with *Thiobacillus thiooxidans*. Phosphoenolpyruvic carboxylase has a lower K_m value for CO_2 than ribulose diphosphate carboxylase; therefore, other factors being equal, the phosphopyruvic carboxylase should be more active at low CO_2 concentrations.

TABLE IV

COMPARISON OF SUBSTRATE K_m VALUES OF SOME CO_2-FIXING ENZYMES

Enzyme	Phosphoenol-pyruvate (K_m)	CO_2 (K_m)	Pyruvate (K_m)	Ref.
Phosphoenolpyruvic carboxylase	1.5 to 1.9 × 10^{-4}	2.2 × 10^{-4}	—	84
Phosphoenolpyruvic carboxykinase	2 × 10^{-4}	2 × 10^{-3}	—	88
Malic enzyme	—	6 × 10^{-3}	3 × 10^{-3}	54
Ribulose diphosphate carboxylase	—	1.1 × 10^{-2}	—	89

The mechanism of the phosphoenolpyruvic carboxylase reaction is of great theoretical interest and has received considerable attention. Bandurski and Greiner[80] suggested that phosphoenoloxalacetate might be an intermediate product which is subsequently cleaved to inorganic phosphate and enoloxalacetate. Tchen *et al.*[81] showed however, that the enoloxalacetate does not occur in the reaction. They coupled the carboxylase reaction with malic dehydrogenase in a medium containing D_2O: if the enol form of oxalacetate were an essential component, one atom of deuterium would be introduced into each molecule of malate formed, as illustrated in Fig. 2, pathway B. If the reaction proceeded via pathway A, some deuterium would be introduced by the tautomerization of oxalacetate (broken arrows); the amount would be less than 1 atom, providing the tautomerization were slower than the reduction of oxalacetate by malic dehydrogenase. Only 0.05 and 0.10 atoms of deuterium were found per molecule of malate synthesized in two trials. Since these values are far below the minimal figure of 1.0 deuterium atom required for the enol mechanism, the synthesis must proceed by the keto form of oxalacetate. The mechanism of this reaction will be considered again under the heading, Phosphoenolpyruvic Carboxykinase, because the two reactions appear to be quite similar.

We have seen that the phosphoenolpyruvic carboxylase is an enzyme of great interest; it has a very strong affinity for CO_2 and phosphoenolpyruvate and therefore should be very effective in fixing CO_2 in oxalacetate. It is not certain, however, whether CO_2 or bicarbonate is the reactant. The keto form of oxalacetate is the product of the reaction.

2. Phosphoenolpyruvic Carboxykinase

This enzyme was discovered in avian liver by Utter and Kurahashi[90] at approximately the same time that Bandurski and Greiner[80] described

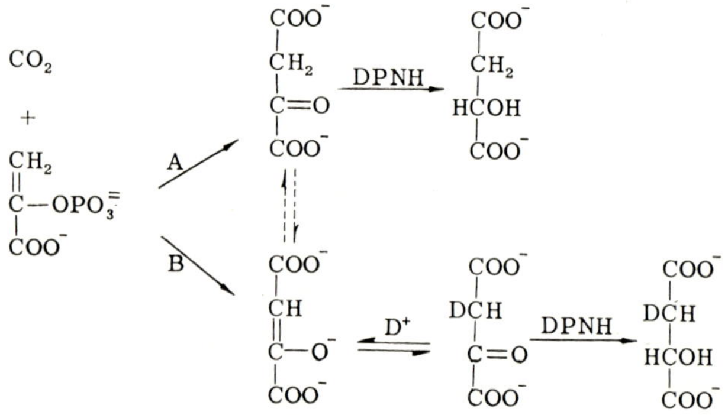

Fig. 2. The use of D_2O to investigate the mechanism of formation of malate from phosphoenolpyruvate and CO_2 by phosphoenolpyruvic carboxylase and malic dehydrogenase.

phosphoenolpyruvic carboxylase from spinach leaves. It catalyzes the following reaction:

$$\begin{array}{l} \quad\;\; PO_3^{2-} \\ \quad\;\;\;\; | \\ \quad\;\;\;\; O \\ \quad\;\;\;\; | \\ CH_2{=}C{-}COO^- + GDP^{3-} + CO_2 \rightleftharpoons COO^-CH_2COCOO^- + GTP^{4-} \end{array} \tag{10}$$

Phosphoenolpyruvate Oxalacetate

The first observations relating to this enzyme were made in 1943 by Evans *et al.*[38] They found, with fumarate and pyruvate as substrates, that a cell-free extract of an acetone powder of pigeon liver fixed a substantial amount of $C^{11}O_2$. The isotope was next shown[91] to occur exclusively in the carboxyl groups of pyruvate, lactate, fumarate, and malate. The preparation, however, did not catalyze an exchange reaction of CO_2 with the β-carboxyl of oxalacetate and therefore seemed to lack the ability to fix CO_2 in oxalacetate. This dilemma was resolved later by Utter and Wood,[92] who found the exchange reaction did occur when ATP was added to the pigeon liver extract. When the malic enzyme was discovered in the liver extract, some doubt[43] arose as to the existence of a separate enzyme that fixed CO_2 in oxalacetate directly. Utter[93] then established the distinct and separate reactions catalyzed by malic enzyme and phosphopyruvic carboxykinase by showing that the addition of ATP stimulated CO_2 fixation in oxalacetate but not in malate, while TPN stimulated fixation in malate but not in oxalacetate. At that time, phosphopyruvic carboxykinase was called oxalacetate carboxylase; it was given its present name by Graves *et al.*[94] in 1956. Utter and his co-workers later showed that IDP, ITP,[95] and GDP, GTP[96] are the active nucleotides for the avian liver enzyme and not ADP, ATP as originally believed. The ATP used in the earlier experiments contained trace amounts of ITP, which was continuously regenerated by the ATP through the action of nucleoside diphosphokinase, a contaminant of the avian enzyme preparation. It has recently been found,[97] however, that in the case of the yeast enzyme, the only nucleotides that are active are the adenosine derivatives. The carboxykinase also occurs in wheat germ[85] and in the bacteria *Thiobacillus thiooxidans*[86] and *Pseudomonas oxalaticus*.[98] Its occurrence in bacteria is probably more widespread than presently recognized.

The apparent equilibrium of the carboxykinase reaction, using the avian liver enzyme, has been calculated in two ways by Utter and Kurahashi,[53] one with CO_2 as the active form and the other with HCO_3^-.

	K_{eq}	$\Delta F°$ cal.
phosphoenolpyruvate^{3-} + IDP^{3-} + $CO_2 \rightleftharpoons$ oxalacetate^{2-} + ITP^{4-}	2.69	−589
phosphoenolpyruvate^{3-} + IDP^{3-} + H^+ + $HCO_3 \rightleftharpoons$ oxalacetate^{2-} + ITP^{4-} + H_2O	5.86×10^6	−9270

Note the necessity of introducing a H^+ ion into the equation if HCO_3^- is considered to be the active form and the consequently very large K_{eq} however, see footnote on p. 62.

The carboxykinase reaction involves the keto form of oxalacetate just as does the carboxylase. Graves *et al.*[94] employed D_2O in an experiment similar to the one of Tchen *et al.*[81] with phosphoenolpyruvic carboxylase, as illustrated in Fig. 2. The results were the same as those obtained with the carboxylase. Only 0.05 and 0.10 atoms of deuterium were introduced per molecule of malate in two trials, thus providing evidence that the enol form of oxalacetate is not formed in the reaction.

Tchen *et al.*[81] advanced an interesting proposal relative to the mechanism of the two reactions; they suggested the transfer of phosphate of the phosphopyruvate to one of the two carboxyl groups of oxalacetate during the fixation of the CO_2, thus forming oxalacetylphosphate. The conversion is illustrated below:

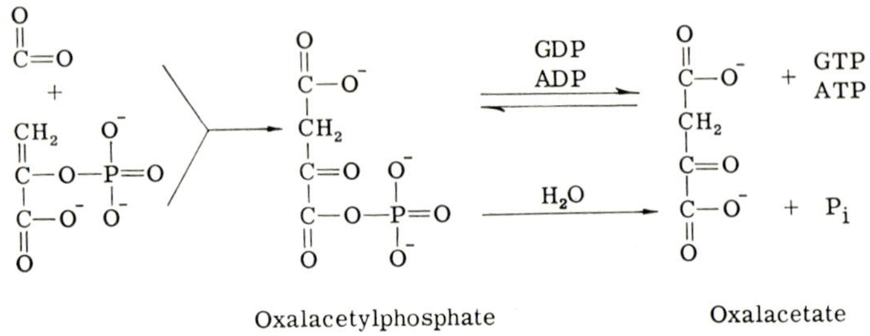

As shown in the illustration, the carboxylase reaction would involve the irreversible hydrolysis of the acyl phosphate, while the carboxykinase would involve the reversible transfer of the phosphate to GDP or ATP. Neither reaction would require the enol form of oxalacetate. This mechanism is of interest in relation to Utter's[93] studies of the ATP-stimulated exchange of $C^{14}O_2$ with the β-carboxyl group of oxalacetate, using the pigeon liver enzyme. After incubation, the mixture was deproteinized and the labeled oxalacetate allowed to decarboxylate spontaneously. The resulting CO_2 was collected during the first, second, and third hours, and each sample was analyzed for C^{14}. At the end of the third hour, the oxalacetate remaining was degraded by decarboxylation with Al^{3+}. The results are shown in Table V. As can be seen, the specific activity of the CO_2 increased markedly with time, and Al^{3+} decarboxylation of the oxalacetate remaining after 3 hours yielded CO_2 with the highest activity. Utter has concluded that the "oxalacetate" formed in the exchange reaction is not homogeneous. However, after chromatography, the "oxalacetate" attains homogeneity, suggesting it is converted to oxalacetate by mild procedures. Tchen *et al.*[81] have suggested that oxalacetylphosphate probably would be converted to oxaloacetate under these mild conditions and that this compound may be the source of the high activity oxalacetate. This observation of inhomogeneity of "oxalacetate" by Utter and the mechanism proposed by Tchen *et al.* are of great interest, but there have been no further developments in this area.

There has been much speculation[36, 53-55] concerning the sequential action of the malic enzyme and phosphoenolpyruvic carboxykinase as a means of circumventing the energy barrier encountered at phosphopyruvate in the synthesis of carbohydrate from pyruvate. The $\Delta F°$ of the phosphopyruvokinase reaction

$$\text{phosphopyruvate}^{3-} + ADP^{3-} + H^+ \rightarrow \text{pyruvate}^- + ATP^{4-}$$

is -6.1 kcal., and the equilibrium is far to the right. When it was shown by Utter and Kurahashi[53, 90] that phosphoenolpyruvate could be synthesized from oxalacetate by the carboxykinase reaction, it was obvious that

TABLE V

INHOMOGENEITY OF OXALACETATE AS DEMONSTRATED BY CHANGING SPECIFIC ACTIVITY OF CO_2[a]

Type of decarboxylation	Time of treatment (min.)	CO_2 from oxalacetate (c.p.m. per mg. C)
Spontaneous	0–60	2390
Spontaneous	60–120	3730
Spontaneous	120–180	4450
Al^{3+}	Residual at 180	5490
	Average, calc.	4350
Al^{3+}	Zero time	4690

[a] $C^{14}OOH\text{-}CH_2\text{-}CO\text{-}COOH$ was prepared by incubating a pigeon liver extract with oxalacetate, Mn^{2+}, ATP, and $C^{14}O_2$ for 4 minutes. The mixture was deproteinized by addition of sulfuric acid. After centrifuging, the supernatant was gassed with CO_2 and CO_2-free air successively. An aliquot was removed for degradation by the Al^{3+} treatment at 0 time. The remainder of the supernatant was placed in a Warburg vessel containing NaOH in the center well and was permitted to decarboxylate at 38°C. while shaking. At the specified intervals, the alkali was rinsed out of the center well and the absorbed CO_2 plated and counted as $BaCO_3$.

this enzyme offered a new possibility of bypassing this energy barrier. This pathway is illustrated below.

	ΔF° kcal.
pyruvate + CO_2 + TPNH $\rightleftharpoons$ malate + TPN^+ Malic enzyme	-1.8
malate + DPN^+ $\rightleftharpoons$ oxalacetate + DPNH + H^+ Malic dehydrogenase	16.5
oxalacetate + GTP $\rightleftharpoons$ phosphoenolpyruvate + CO_2 + GDP Phosphoenolpyruvic carboxykinase	-0.5
DPNH + H^+ + $\frac{1}{2}$ O_2 $\rightarrow$ H_2O + DPN^+	-52.4

The oxidation of the DPNH would serve to pull the reaction toward phosphoenolpyruvate formation. There are, however, some difficulties with this proposal because of the intracellular location of the enzymes involved. The phosphoenolpyruvic carboxykinase is located in the mitochondria whereas the malic enzyme is found in the cytoplasm. Therefore the pyruvate would need to be converted to malate in the cytoplasm; the malate would then either have to pass into or onto the mitochondria to be oxidized by the malic dehydrogenase to oxalacetate, which in turn would be decarboxylated to phosphoenolpyruvate by the carboxykinase. Since the glycolytic enzymes are located in the cytoplasm, the phosphoenolpyruvate would then have to reenter the cytoplasm to be converted to glucose. These relationships are illustrated in Fig. 3. A further obstacle to this dicarboxylic acid pathway is the relatively low affinity of the malic enzyme for CO_2 ($K_m = 6 \times 10^{-3}\ M$).

Utter and Keech[99] recently have discovered in liver mitochondria a new CO_2-fixation reaction that yields oxalacetate directly from pyruvate, CO_2, and ATP (see Section IV, C, 4). This enzyme has a high affinity for CO_2,

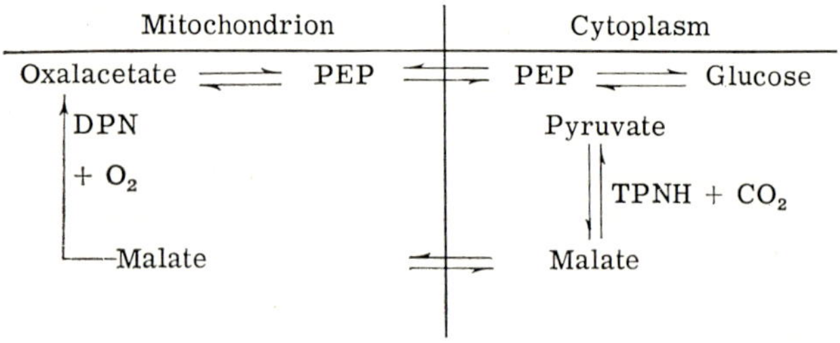

FIG. 3. Relationship of enzymes and substrates in cytoplasm and mitochondrion.

and it seems likely that this enzyme rather than the malic enzyme may be involved in the conversion of pyruvate to phosphoenolpyruvate.

So far, we have considered only the thermodynamics of the reactions of the dicarboxylic acid pathway of phosphopyruvate synthesis, but the kinetic considerations are of equal or greater importance. The amounts and properties of the enzymes and the level of the metabolites in the tissue play an important role in the evaluation of the kinetic data. See Utter[54] for a consideration of this problem.

We have seen that phosphoenolpyruvic carboxykinase catalyzes a reversible primary reaction of CO_2 fixation in which a high energy phosphate is involved but is preserved as an enol phosphate of pyruvate or anhydride phosphate of GTP or ATP. The reaction has a great potential as a means of circumventing the energy hurdle involved in the formation of phosphopyruvate from pyruvate. This CO_2-fixation reaction has received only limited attention from investigators using bacterial sources of the enzyme, but it seems likely that this mechanism, or a modification thereof, may be widespread in bacteria.

3. Phosphoenolpyruvic Carboxytransphosphorylase

Siu *et al.*[100] have recently discovered in propionibacteria a new primary reaction of CO_2 assimilation which occurs as illustrated below:

$$\underset{\text{Phosphoenol pyruvate}}{CH_2{=}\overset{\overset{PO_2^{3-}}{|}}{\underset{}{\overset{O}{\overset{|}{C}}}}{-}COO^-} + CO_2 + HPO_2^{3-} \rightleftharpoons \underset{\text{Oxalacetate}}{COO^-{-}CH_2COCOO^-} + \underset{\text{Pyrophosphate}}{HP_2O_7^{3-}} \qquad (11)$$

Phosphoenolpyruvate and CO_2 are converted to oxalacetate. Inorganic phosphate is required and acts as an acceptor of the phosphate from the phosphoenol pyruvate, yielding pyrophosphate. Addition of pyrophosphatase stimulates the conversion of phosphopyruvate to oxalacetate. The reaction is reversible, i.e., phosphoenolpyruvate is formed from oxalacetate and pyrophosphate. It is of interest that Tchen and Vennesland[85] have shown that the rate of fixation of CO_2 by wheat germ phosphoenolpyruvic carboxylase is stimulated by the addition of inorganic phosphate. They could find no evidence, however, that pyrophosphate was involved in the reaction when they used P_i^{32} and a pool of pyrophosphate to trap P^{32}-labeled pyrophosphate which might be formed in the reaction. Furthermore, exchange of P_i^{32} with the phosphate of phosphoenolpyruvate did not occur. The latter would be expected if phosphoenolpyruvic carboxylase were reversible. We shall give additional consideration to the fixation of CO_2 by propionic acid bacteria under the heading, Methylmalonyl-Oxalacetic Transcarboxylase and the Formation of Propionate (Section V).

4. Ribulose Diphosphate Carboxylase

This enzyme is generally thought to catalyze the reaction through which CO_2 is assimilated by photosynthetic and chemosynthetic organisms. The reaction is discussed in detail by Elsden Chapter 1 of this volume and also is reviewed by Vishniac *et al.*[101] The reaction is believed to proceed as follows:

$$\text{ribulose-1,5-diphosphate}^{4-} + CO_2 \rightarrow \text{two 3-phosphoglycerate}^{3-} + 2H^+ \qquad (12)$$

We shall confine our attention to its occurrence in heterotrophic bacteria. Fuller[102] has stated that cell-free extracts of *E. coli* which are grown on xylose or arabinose fix a substantial amount of $C^{14}O_2$ when incubated with ribulose diphosphate, and the C^{14} occurs in phosphoglycerate, pyruvate, and lactate. This report is in the form of an abstract; a full account of the work has not appeared. Two publications by Quayle and Keech[98, 103] show quite conclusively that *Pseudomonas oxalaticus*, when grown on formate, utilizes a metabolic pathway similar to that of the autotrophs. One set of experiments[103] dealt with the distribution of C^{14} in the nonvolatile constituents of the cells after 12 seconds and to 5 minutes' exposures to formate-C^{14} or bicarbonate-C^{14}. In 12 seconds, over 80% of the C^{14} was in phosphorylated compounds, and 80% of this C^{14} was in 3-phosphoglyceric acid. These workers estimated that at least 94% of the formate passed through the stage of CO_2 before being utilized. The fixation of CO_2 and formate by cell-free extracts was also investigated.[98] The extracts catalyzed the incorporation of C^{14}-bicarbonate into phosphoglyceric acid with ribulose-1,5-diphosphate or ribulose-5-phosphate and ATP as substrates. This organism, when grown on formate, appears to live like an autotrophic bacterium except that it obtains its energy by oxidizing formate rather than oxidizing an inorganic compound.

It is of considerable interest that Kaneda and Roxburgh[104] have performed similar studies with a pseudomonad grown on a methanol medium containing biotin. They found no indication of the conversion of $C^{14}O_2$ or methanol-C^{14} to phosphoglyceric acid in 15- to 80-second incubations. The first product to be labeled was serine. They suggest that this organism may utilize a C_1 compound via serine hydroxymethylase rather than by way of CO_2 and ribulose diphosphate carboxylase.

5. Aminoimidazoleribonucleotide Carboxylase

This enzyme is responsible for the incorporation of CO_2 into the 6 position of the purine ring in the *de novo* synthesis of purine nucleotides. It has been purified about 20-fold from chicken livers by Lukens and Buchanan,[105] but the enzyme has not been studied in microorganisms or plants. The reaction is as follows:

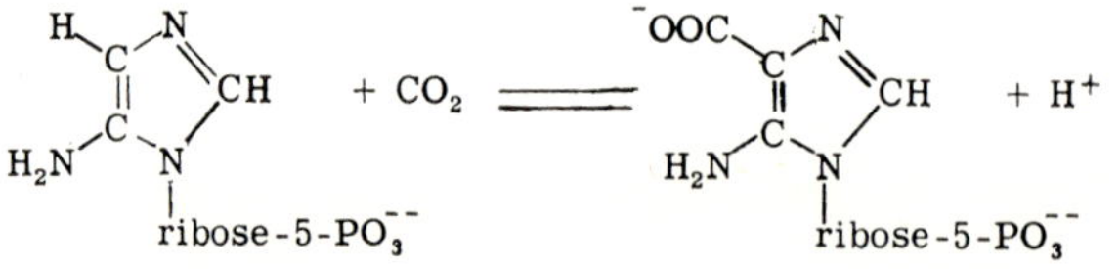

(13)

5-Aminoimidazole ribonucleotide

5-Amino-4-imidazole-carboxylic acid ribonucleotide

There is no requirement for ATP or any other nucleotide. The equilibrium of the reaction favors decarboxylation rather than synthesis. In the presence of 0.3 *M* bicarbonate, however, the yield of the carboxylated compound was 50% on the basis of the aminoimidazole ribonucleotide.

The carboxylated compound was identified by comparison with 5-amino-

4-imidazolecarboxylic acid. The two compounds were found to decarboxylate spontaneously at very similar rates and their absorption spectra were similar. 5-Amino-4-imidazolecarboxylic acid nucleotide was shown to be converted enzymatically in the presence of ATP and aspartic acid to 5-amino-4-imidazole-*N*-succinocarbamide ribonucleotide, which is the next intermediate in the sequence of reactions in the *de novo* synthesis of the purine ring.

C. Reactions with ATP as an Energy Source

Recently, exciting discoveries have been made relative to the fixation of CO_2 in the reactions requiring ATP. All reactions of this group appear to involve biotin-containing enzymes and usually acyl CoA compounds. It has been known for a long time that biotin has some role in CO_2 fixation, but how it functions has been a mystery. For a review of the early studies on biotin as related to CO_2 fixation, see Utter and Wood.[33] It now appears that ATP is utilized in the conversion of CO_2 to an "active" form through some reaction involving biotin, which is a component of these enzymes. There is a simultaneous splitting of the ATP to ADP and P_i. The enzyme-biotin-CO_2 complex is then believed to react with the acyl CoA compound to yield the carboxylated CoA derivative. The carboxylation occurs either on the α-carbon of the acyl CoA compound or on a carbon adjacent to a conjugate system of a CoA derivative.* Pyruvic carboxylase may prove to be an exception, but the mechanism of this reaction is unknown. We shall review the individual reactions of this group and then return to a consideration of the mechanism of the reaction of biotin and CO_2.

1. Propionyl Carboxylase

In 1955 Flavin *et al.*[107] discovered the fixation of CO_2 with propionyl CoA yielding methylmalonyl CoA. A dialyzed extract of pig heart was incubated with propionate, $KHC^{14}O_3$, ATP, CoA, and $Mg.^{++}$ The conversion of propionate to methylmalonate was observed independently by Katz and Chaikoff[108] in rat liver slices. Flavin and Ochoa,[109] Tietz and

* The methyl of acetyl CoA and the methylene of propionyl CoA are activated groups by virtue of their position, alpha to the thioacyl group. It is a well-known principle of organic chemistry that the activating effect of such functional groups as nitriles, carbonyls, and esters, which are ordinarily exerted on an α-methyl or an α-methylene group, may be observed at a remote point in a molecule if the activating group is part of a conjugated system of double bonds.[106] Thus the activation by the CoA ester is qualitatively unchanged by the interposition of a vinyl group and the carboxylation of the methyl group of methylcrotonyl CoA is in complete agreement with a recognized principle of organic chemistry.

Ochoa,[110] and Kaziro *et al.*[111, 112] have established that the fixation of CO_2 occurs by the reversible reaction illustrated below.

$$\underset{\text{propionyl CoA}}{CH_3CH_2COSCoA} + CO_2 + ATP^{4-} + H_2O \rightleftharpoons \underset{\text{Methylmalonyl CoA}}{CH_3\overset{\overset{\textstyle COO^-}{|}}{C}HCOSCoA} + ADP^{3-} + HPO_4^{2-} + 2H^+ \quad (14)$$

or

$$\underset{\text{Propionyl CoA}}{CH_3CH_2COSCoA} + HCO_3^- + ATP^{4-} \rightleftharpoons \underset{\text{Methylmalonyl CoA}}{CH_3\overset{\overset{\textstyle COO^-}{|}}{C}HCOSCoA} + ADP^{3-} + HPO_4^{2-} + H^+ \quad (14a)$$

Table VI shows some results from their earlier experiments. It can be seen that ATP, Mg, and propionyl CoA are essential for the fixation of CO_2. The enzyme activity can be determined spectrophotometrically by coupling it with pyruvic kinase and lactic dehydrogenase as illustrated below:

$$\text{propionyl CoA} + CO_2 + \text{ATP} \rightleftharpoons \text{methylmalonyl CoA} + \text{ADP} + P_i$$

$$\text{ADP} + \text{phosphoenolpyruvate} \rightleftharpoons \text{ATP} + \text{pyruvate}$$

$$\text{pyruvate} + \text{DPNH} + H^+ \rightleftharpoons \text{lactate} + \text{DPN}^+$$

Beck *et al.*,[113] Beck and Ochoa,[114] and Lengyel *et al.*[115] have shown that the methylmalonyl CoA is converted to succinyl CoA by another enzyme, methylmalonyl isomerase. This conversion is illustrated in reaction 15.

$$\underset{\text{Methylmalonyl CoA}}{CH_3\overset{\overset{\textstyle COO^-}{|}}{C}HCOSCoA} \rightleftharpoons \underset{\text{Succinyl CoA}}{COO^-CH_2CH_2COSCoA} \quad (15)$$

The isomerase was not observed in the pig heart extracts but was demonstrated in extracts prepared from liver, kidney, or rat heart. These extracts, in contrast to the extract from pig heart, catalyzed the conversion of propionyl CoA to succinyl CoA by the combined action of the enzymes catalyzing reactions 14 and 15. Lardy and Peanasky[116] and Lardy and Adler[117] had shown earlier that extracts of liver mitochondria fix CO_2 in an ATP-dependent reaction in which propionyl CoA is converted to succinyl CoA.

This conversion no doubt occurred via the above sequence, although at first it was considered to be a direct carboxylation of propionyl CoA to succinyl CoA.

There has been only one brief report of this reaction occurring in microorganisms, and this was in an autotroph. Gibson and Knight[118] have prepared cell-free extracts of *Rhodospirillum rubrum* and have shown that $C^{14}O_2$ is fixed in the presence of propionyl CoA, Mg^{++}, and ATP. Methylmalonate and succinate have been identified as reaction products. The fixation was completely inhibited by avidin. The authors in unpublished experiments have observed that this fixation occurs in *Mycobacterium smegmatis*. It probably will be found to occur in propionibacteria and *Micrococcus*. The reverse reaction, i.e., the decarboxylation of succinyl CoA to propionyl CoA and CO_2 by these bacteria has received attention because it has been considered that propionate is formed by this mechanism. Recently,[119] however, it has been found with propionibacteria that propionate is formed by a new type of biochemical reaction, a transcarboxylation rather than a decarboxylation to CO_2. We shall consider this mechanism of formation of propionate under the heading, Methylmalonyl-Oxalacetic Transcarboxylase and the Formation of Propionate (Section V).

TABLE VI

FIXATION OF CO_2 BY PARTIALLY PURIFIED PROPIONYL CARBOXYLASE FROM PIG HEART[a]

System	$C^{14}O_2$-fixed (μmole)
Complete	0.600
Complete + 1 μmole CoA	0.600
ATP omitted	0
Mg^{++} omitted	0.003
Propionyl CoA omitted	0.002
Enzyme omitted	0

[a] The complete system contained (in micromoles/ml.): Tris buffer, pH 7.0, 50; $MgCl_2$, 2; GSH, 5; ATP, 2; $KHC^{14}O_3$, 10; propionyl CoA, 1; and pig heart enzyme, 5 mg. protein. Incubation, 60 min. at 25°C.

The early studies by Ochoa and his co-workers using propionyl carboxylase presented an exciting possibility that carbonyl phosphate was an active form of CO_2 and was formed from CO_2 and ATP.[120] With crude preparations of the propionyl carboxylase, and in the absence of propionyl CoA but in the presence of fluoride, a CO_2-dependent reaction between ATP and fluoride yielded ADP and monofluorophosphate. This was called the fluorokinase reaction.[109, 120] It was considered that fluorokinase might actually be carbonokinase yielding carbonyl phosphate as the initial product. If the acceptor propionyl CoA were present, methylmalonyl CoA would be formed, but if fluoride were present, the carbonyl phosphate would react with fluoride to give fluorophosphate. When fluorokinase[121] and the propropionyl carboxylase[110] were highly purified, it was found that the two reactions were completely independent. Fluorokinase was shown to be pyruvic kinase,[121] and furthermore, there was no resolution of the carboxylase[110] into separate enzymes for CO_2 activation and propionyl CoA carboxylation.

Lardy and co-workers[116, 117] had observed earlier that extracts from mitochondria of livers of biotin-deficient rats carboxylated propionate at a greatly reduced rate as compared to the extracts from livers of normal animals. When Tietz and Ochoa assayed their purified carboxylase for biotin, none was found present.[110] However, Halenz and Lane,[122] using a partially purified propionyl carboxylase from mitochondria of bovine liver, found that avidin inhibited the carboxylase reaction just as Wakil *et al.*[123] had observed with acetyl carboxylase. Furthermore, Lynen *et al.*[124] found that there was a linear relationship between the biotin content and the purity of β-methylcrotonyl carboxylase. We shall see that these enzymes are very similar to propionyl carboxylase. Therefore, Kaziro *et al.*[111] reinvestigated propionyl carboxylase and found that the enzyme did contain biotin. *L. arabinosus* was used in the earlier assay and yeast in the latter assay. *L. arabinosus* is sensitive to both pH and salt concentration, which may account for the previous failure to detect the biotin. Kaziro *et al.*[112] found one mole of biotin per 175,000 g. of propionyl carboxylase. The molecular weight is stated to be 700,000 therefore, one molecule of the protein contains 4 molecules of biotin. The activity of the enzyme is inhibited by equivalent amounts of avidin. Propionyl carboxylase is an —SH enzyme and is highly sensitive to sulfhydryl binding reagents. The enzyme has been crystallized.[112]

The reversibility of the carboxylase reaction has been shown[111] by linking it with hexokinase and glucose-6-phosphate dehydrogenase.

$$\text{methylmalonyl CoA} + P_i + \text{ADP} \rightleftharpoons \text{propionyl CoA} + CO_2 + \text{ATP}$$

$$\text{ATP} + \text{glucose} \rightleftharpoons \text{ADP} + \text{glucose-6-phosphate}$$

$$\text{glucose-6-phosphate} + \text{TPN}^+ \rightarrow \text{6-phosphogluconate} + \text{TPNH} + H^+$$

The net formation of ATP from P_i and ADP by the reverse reaction was measured by determining the increase in the light absorption at 340 mμ.

The K_m values for the substrates of the forward reaction were found[111] to be: propionyl CoA 2.7×10^{-4} *M*, CO_2 2.5×10^{-3} *M*, and ATP 8×10^{-5} *M*. Thus the affinity of the enzyme for CO_2 is quite low.

Halenz and Lane,[122] Lane *et al.*,[125] and Kaziro *et al.*[111] have investigated the exchange of P_i^{32} and ADP-C^{14} with ATP as catalyzed by propionyl carboxylase. These studies relate directly to the proposed[124] mechanism of the activation of CO_2 by biotin and will be considered under the heading of Function of Biotin in the Fixation of CO_2 (Section VII).

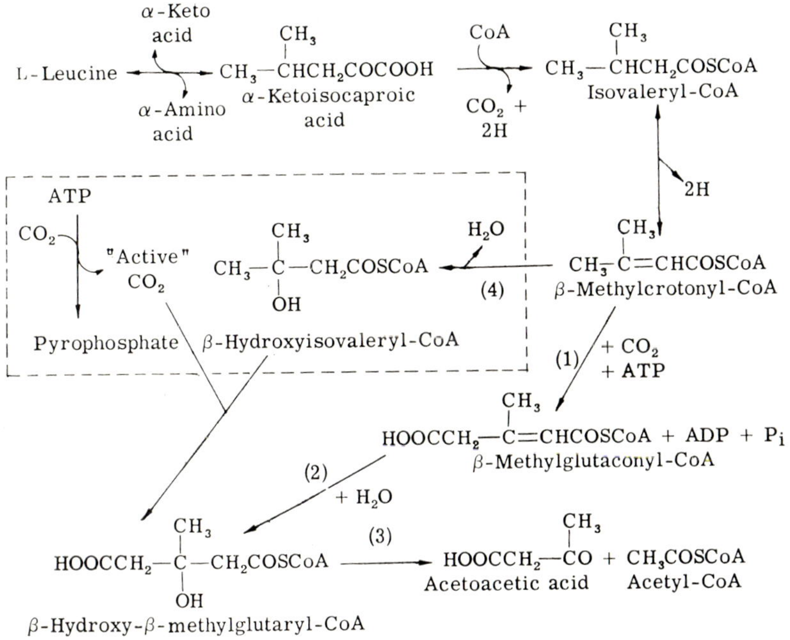

FIG. 4. The degradation of leucine in animals. The portion of the scheme that is set off by dashed lines has been found to be incorrect. Reaction (1) is catalyzed by β-methylcrotonyl carboxylase, reaction (2) by methylglutaconase, reaction (3) by cleavage enzyme, and reaction (4) by enoyl hydrase.

2. β-METHYLCROTONYL CARBOXYLASE

The studies of this fixation reaction were initiated by Coon and co-workers[126] during their investigations of the degradation of L-leucine in animal tissue. The reaction sequence that they and others postulated is shown in Fig. 4. Bachhawat *et al.*[127] observed that there was fixation of CO_2 during this degradation, and they considered that this fixation occurred by the mechanism which is set off by dashed lines in Fig. 4. It was subsequently shown by Lynen and Knappe[128] that the fixation does not occur with β-hydroxyisovaleryl CoA but rather with β-methylcrotonyl CoA. Lynen and Knappe[128] used a mycobacterium species as a source of the enzyme. The bacterium was isolated by enrichment culture technique using isovaleric acid as a source of carbon. The fixation reaction is illustrated below.

$$\underset{\beta\text{-Methylcrotonyl CoA}}{CH_3\overset{\overset{\displaystyle CH_3}{|}}{C}{=}CHCOSCoA} + ATP^{4-} + CO_2 + H_2O \rightleftharpoons$$

$$\underset{\beta\text{-Methylglutaconyl CoA}}{CH_3\overset{\overset{\displaystyle CH_2COO^-}{|}}{C}{=}CHCOSCoA} + ADP^{3-} + HPO_4^{2-} + 2H^+ \qquad (16)$$

or

$$\underset{\beta\text{-Methylcrotonyl CoA}}{CH_3\overset{\overset{\displaystyle CH_3}{|}}{C}{=}CHCOSCoA} + ATP^{4-} + HCO_3^- \rightleftharpoons$$

$$\underset{\beta\text{-Methylglutaconyl CoA}}{CH_3\overset{\overset{\displaystyle CH_2COO^-}{|}}{C}{=}CHCOSCoA} + ADP^{3-} + HPO_4^{2-} + H^+ \qquad (16a)$$

Coon and co-workers[129] subsequently confirmed this mechanism using enzyme preparations isolated from both chicken and ox liver. Reaction 4 of Fig. 4 is catalyzed by enoyl hydrase, which presumably was present in the carboxylase used by Bachhawat *et al.*[127] when they observed CO_2 fixation starting with β-hydroxyisovaleryl CoA.

β-Methylcrotonyl carboxylase has been purified about 175-fold by Lynen *et al.*;[124] see also Lynen.[130] The enzyme is inhibited by avidin; the purified enzyme contained 1 mole of biotin per 344,000 g. of protein. Woessner *et al.*[131] had previously shown that extracts from the livers of biotin-deficient rats lacked this carboxylase. They could not restore the activity with biotin or boiled juice prepared from livers of normal rats.

β-Methylcrotonyl carboxylase can be assayed by coupling it with pyruvic kinase and lactic dehydrogenase, just as Tietz and Ochoa[110] did with the propionyl carboxylase. Reaction 16 can be reversed[130] by linking it with hexokinase to trap the ATP, and during this decarboxylation there is generation of ATP from ADP and P_i. The same type of conversion occurs with propionyl carboxylase.[111]

Interest was aroused by Bachhawat and Coon[132] when they described what appeared to be an enzymic activation of CO_2 by the following mechanism.

$$CO_2 + ATP \rightleftharpoons AMP\text{-}CO_2 + PP$$

This proposal was based on the observation of a CO_2 and hydroxylamine-dependent liberation of pyrophosphate from ATP. The hydroxylamine was presumed to react with the AMP-CO_2 and thus pull the reaction toward ATP breakdown. In the absence of hydroxylamine and in the presence of the natural CO_2 acceptor, the AMP-CO_2 was thought to bring about carboxylation. The CO_2 and hydroxylamine dependent reaction was carried out with a crystalline enzyme obtained from pig heart.[132] Kupiecki and Coon[133] later reexamined this reaction and found that the products were not AMP and pyrophosphate but were ADP and a mononucleotide phosphate compound that was neither phosphate nor pyrophosphate. The phosphate compound does not contain CO_2, although the cleavage of ATP has a specific requirement for CO_2 and hydroxylamine. It is now clear that this enzyme plays no part in the carboxylation reaction. We have seen that two "active" forms of CO_2, i.e., carbonyl phosphate (p. 73) and adenyl-CO_2 have been eliminated from schemes of CO_2 fixation. There remains the biotin CO_2 complex, which holds great promise; see Function of Biotin in the Fixation of CO_2 (Section VII).

Rilling and Coon[129] have recently found that β-methylcrotonyl carboxylase is present in *Pseudomonas oleovorans*. This bacterium was isolated by enrichment culture technique using hexane as a substrate.

3. Acetyl Carboxylase

This enzyme catalyzes the carboxylation of acetyl CoA, yielding malonyl-CoA as illustrated below:

$$\underset{\text{Acetyl CoA}}{CH_3COSCoA} + CO_2 + ATP^{4-} + H_2O \rightleftharpoons \underset{\text{Malonyl CoA}}{{}^{-}OOCCH_2COSCoA} + ADP^{3-} + HPO_4^{2-} + 2\,H^+ \quad (17)$$

or

$$CH_3COSCoA + HCO_3^- + ATP^{4-} \rightleftharpoons {}^{-}OOCCH_2COSCoA + ADP^{3-} + HPO_4^{2-} + H^+ \quad (17a)$$

Great interest was aroused in this reaction when it was demonstrated that CO_2 is required for fatty acid synthesis. As early as 1950, Brady and Gurin[134] recognized that synthesis of fatty acids occurred in bicarbonate buffer but not in phosphate buffer. In 1957, Klein[135] observed with a yeast homogenate that acetate was incorporated into a nonsaponifiable fraction in the absence of CO_2, but there was a several-fold stimulation of fatty acid synthesis on the addition of CO_2 to the system. The significance of these observations was not realized until 1958, when Gibson *et al.*[136, 137] and Wakil *et al.*[123] reported that two protein fractions from chicken liver R_{1g} and R_{2g} catalyzed the synthesis of palmitic acid from acetyl CoA in the presence of Mn^{2+}, ATP, TPNH, and HCO_3^-. However, $C^{14}O_2$ was not incorporated into the long-chain fatty acid, and therefore it was assumed that the bicarbonate played only a catalytic role. No apparent intermediates could be demonstrated with these two enzyme systems. After passage of the protein through ion exchange resins, Wakil[138] obtained a purified enzyme system designated R_{1gc} which, with acetyl CoA in the presence of Mn^{2+}, ATP, and HCO_3^-, formed a compound that was converted to palmitic acid in the presence of TPNH and a second purified enzyme system R_{2gc}. The compound produced by the R_{1gc} fraction contained equal amounts of acetyl CoA and CO_2, as shown by incorporation of C^{14} from either component. On hydrolysis, an acid was isolated containing all the original C^{14}. This acid proved by several criteria to be identical with malonic acid.

Brady[139] independently described the formation of malonyl CoA as the first step in the synthesis of long-chain fatty acids. Formica and Brady[140] later obtained an ammonium sulfate fraction of extracts of pigeon liver and pig heart capable of carboxylating acetyl-CoA. The requirement of CO_2 for fatty acid synthesis now has been demonstrated in enzyme systems from avocado,[141] yeast,[130] and *Mycobacterium avium*,[142] and the participation of malonyl CoA in fatty acid synthesis has been confirmed by Wakil and Ganguly[143] and by Lynen.[130] An interesting proposal, based on his work with yeast enzymes, was made by Lynen.[130] He concluded that ATP is required for the carboxylation step only and that malonyl CoA, once formed, condenses with itself according to the equation illustrated below:

$$\text{acetyl CoA} + CO_2 + \text{ATP} \rightleftharpoons \text{malonyl CoA} + \text{ADP} + P_i$$

$$8\ \text{malonyl CoA} + 14\ \text{TPNH} \rightarrow \text{palmityl CoA} + 8\ CO_2 + 7\ \text{CoA} + 14\ \text{TPN} + 7\,H_2O$$

Assuming this sequence of events to be correct, 8 moles of ATP would be required for the formation of one mole of palmitic acid from acetyl CoA.

A very interesting observation was made by Vagelos,[144] who reported that an enzyme preparation obtained from *Clostridium kluyveri* was capable of catalyzing an exchange reaction between malonyl CoA and bicarbonate-C^{14} in the presence of caproic, or valeric, or butyric acid and acetyl CoA. Since malonyl CoA is not decarboxylated by this enzyme and C^{14}-acetyl CoA is not incorporated into the malonyl CoA, the acids seemed to induce the exchange of CO_2 with the carboxyl group of malonyl CoA. Furthermore, caproyl-CoA replaced the requirement for a free fatty acid and acetyl CoA (indicating the presence of a CoA transferase enzyme). Vagelos therefore proposed that the observed exchange reaction is consistent with a reaction involving a reversible condensation between malonyl CoA and an acyl

CoA coupled with a decarboxylation as illustrated below:

$$\underset{\text{Malonyl CoA}}{HOOC—CH_2—COSCoA} + \underset{\text{Caproyl CoA}}{C_5H_{11}COSCoA} \rightleftharpoons C_5H_{11}COCH_2COSCoA + CO_2 + CoASH$$

then

$$C_5H_{11}COCH_2COSCoA + C^{14}O_2 + CoASH \rightleftharpoons \underset{\text{Malonyl CoA}}{C^{14}OOH—CH_2—COSCoA} + \underset{\text{Caproyl CoA}}{C_5H_{11}COSCoA}$$

This reaction may indicate that an acyl CoA can be lengthened by two carbon atoms by reaction with malonyl CoA. This concept of fatty acid synthesis is in good agreement with the findings of Brady *et al.*[145, 146] using purified preparations from rat liver and rat brain. These workers showed that one mole of acetyl CoA, 7 moles of malonyl CoA, and 14 moles of reduced TPN are required for the biosynthesis of one mole of palmitic acid. The addition of one mole of malonyl CoA to one mole of acyl CoA is assumed to cause simultaneous decarboxylation of the unesterified carboxyl group of malonyl CoA. The β-ketoacyl-CoA thus formed is then reduced with two equivalents of TPNH.*

The enzymic carboxylations of acetyl-CoA, propionyl-CoA, and butyryl-CoA all occur on the α-carbon and require one mole of ATP per mole of acyl CoA. These enzymes apparently contain biotin, since the carboxylation reactions are completely inhibited by avidin. Lane *et al.*[125] report that mitochondrial propionyl CoA carboxylase is slightly active toward acetyl CoA and butyryl CoA. Stern *et al.*[147] observed in their study of the carboxylation of butyryl CoA that the relative rates of carboxylation of acetyl CoA, propionyl CoA, and butyryl CoA remained essentially the same during the purification process, suggesting that only one enzyme is involved. It would be of interest to determine if the "fatty acid" enzyme system capable of forming malonyl CoA from acetyl CoA also is capable of producing methylmalonyl CoA from propionyl CoA. Since acetyl CoA carboxylase has not been purified or investigated sufficiently, one cannot be completely certain this enzyme is different from propionyl carboxylase.

4. Pyruvic Carboxylase

Recently, Utter and Keech[99] have discovered a very interesting enzyme in liver mitochondria which catalyzes the following reaction:

$$CH_3COCOO^- + CO_2 + ATP^{4-} + H_2O \xrightarrow[\text{acetyl CoA}]{Mg^{++}} COO^-CH_2COCOO^- + ADP^{3-} + HPO_4^{2-} + 2H^+ \quad (18)$$

or

$$CH_3COCOO^- + HCO_3^- + ATP^{4-} \xrightarrow[\text{acetyl CoA}]{Mg^{++}} COO^-CH_2COCOO^- + ADP^{3-} + HPO_4^{2-} + H^+ \quad (18a)$$

* See F. Lynen. Fed. Proc. **20,** 941 (1961) for a recent review.

The enzyme forms oxalacetate from pyruvate per se, CO_2, and ATP and requires the presence of acetyl CoA and Mg^{++}. Since this enzyme utilizes free pyruvate, it catalyzes a reaction similar to the original reaction proposed by Wood and Werkman. Using $C^{14}O_2$, or pyruvate-1-C^{14} or -2-C^{14}, or acetyl-1-C^{14}-CoA, it was shown[99] that $C^{14}O_2$ labels only the β-carboxyl group of oxalacetate, pyruvate-1-C^{14} labels the α-carboxyl group, and pyruvate-2-C^{14} the carbonyl carbon, while acetyl-1-C^{14} CoA does not contribute any C^{14} to the oxalacetate. There is little change in the amount of acetyl CoA during the reaction. Propionyl CoA is as effective as acetyl CoA in catalyzing the reaction, but methylmalonyl CoA and malonyl CoA will not replace their function. The reaction is inhibited by avidin and reagents which inactivate thiols. Little information is available concerning the details of the reaction mechanism. An obvious possibility is that the acetyl CoA acts as the initial CO_2 acceptor forming malonyl CoA, which then transcarboxylates to pyruvate to yield oxalacetate and regenerate the acetyl CoA (see Methylmalonyl-Oxalacetic Transcarboxylase, Section V). However, the enzyme does not catalyze the transcarboxylase reaction shown below.

$$\text{malonyl CoA} + \text{pyruvate} \rightleftharpoons \text{acetyl CoA} + \text{oxalacetate}$$

The participation of ATP and, apparently, of biotin in this reaction suggests a general similarity to the acyl CoA carboxylases that have been discussed above in which an "activated" CO_2 is involved in the form of an enzyme-biotin-CO_2 complex.

The apparent K_m values[45] are CO_2 $4.8 \times 10^{-5}\ M$, pyruvate $4.4 \times 10^{-4}\ M$, ATP $3.7 \times 10^{-5}\ M$, Mg^{++} $1.3 \times 10^{-3}\ M$, acetyl CoA $1.9 \times 10^{-5}\ M$. In view of these high affinities for the substrates and the close association of this enzyme with phosphopyruvic carboxykinase in mitochondria, it is an attractive proposal[99] that phosphoenolpyruvate may be synthesized from pyruvate by the sequential action of these two enzymes.

$$\text{pyruvate} + CO_2 + \text{ATP} \rightarrow \text{oxalacetate} + \text{ADP} + P_i$$

$$\text{oxalacetate} + \text{GTP} \rightleftharpoons \text{phosphoenolpyruvate} + CO_2 + \text{GDP} + P_i$$

$$\overline{\text{pyruvate} + \text{ATP} + \text{GTP} \rightarrow \text{phosphoenolpyruvate} + \text{ADP} + \text{GDP} + 2\ P_i}$$

See Malic Enzyme (Section IV, A, 1) for a further discussion of the synthesis of phosphopyruvate.

Recently Seubert and Remberger[147a] have purified and described a pyruvic carboxylase from *Pseudomonas citronellolis* which catalyzes the carboxylation of pyruvate with CO_2 using ATP just as does the enzyme described by Utter and Keech.[99] The bacterial enzyme differs from the liver enzyme in that it does not require acetyl CoA or other CoA esters as a cofactor. The enzyme is inhibited by avidin and it contains biotin. It was purified 500-fold from an extract obtained by breaking the cells with glass beads in a shaker. The purification included protamine sulfate treatment, ammonium sulfate fractionation (38 to 44% fraction), and chromatography on diethylaminoethylcellulose. The enzyme may be assayed spectrophotometrically by coupling the reaction with malic dehydrogenase.

$$\text{pyruvate} + CO_2 + \text{ATP} \rightleftharpoons \text{oxalacetate} + \text{ADP} + P_i$$

$$\text{oxalacetate} + \text{DPNH} + H^+ \rightleftharpoons \text{malate} + \text{DPN}^+$$

It also may be assayed spectrophotometrically with hexokinase and glucose-6-phosphate dehydrogenase

$$\text{oxalacetate} + \text{ADP} + P_i \rightleftharpoons \text{pyruvate} + CO_2 + \text{ATP}$$
$$\text{ATP} + \text{glucose} \rightarrow \text{glucose-6-P} + \text{ADP}$$
$$\text{glucose-6-P} + \text{TPN}^+ \rightleftharpoons \text{6-phosphogluconate} + \text{TPNH} + H^+$$

$$\text{Sum: oxalacetate} + P_i + \text{TPN}^+ \rightarrow \text{pyruvate} + CO_2 + \text{6 phosphogluconate} + \text{TPNH} + H^+$$

The mechanism of the reaction was studied by exchange reactions and will be considered in Section VI. The K_m for ATP is 1.1×10^{-4} M and for CO_2-$KHCO_3$ 2.2×10^{-2} M.

V. Methylmalonyl-Oxalacetic Transcarboxylase and the Formation of Propionate

Swick and Wood[119] recently have demonstrated a new type of biochemica reaction in which one compound, a carboxyl donor, is decarboxylated and a second compound, a carboxyl acceptor, is carboxylated. Thus it is possible to accomplish a direct carboxylation without the intervention of CO_2 or the expenditure of energy to activate the CO_2. Although this reaction does not involve fixation of CO_2, it is considered here because of its similarity to CO_2 fixation and because it may be coupled with the fixation of CO_2.

The conversion is illustrated in reaction 19.

$$\underset{\text{Methylmalonyl CoA}}{CH_3\overset{\overset{\displaystyle COO^-}{|}}{C}HCOSCoA} + \underset{\text{Pyruvate}}{CH_3COCOO^-} \rightleftharpoons \underset{\text{Propionyl CoA}}{CH_3CH_2COSCoA} + \underset{\text{Oxalacetate}}{COO^-CH_2COCOO^-} \quad (19)$$

It should be emphasized that neither ATP nor Mg^{++} is required for this transfer and that free CO_2 is not involved. The reaction is inhibited by avidin,[119] and an enzyme biotin-C_1 complex probably is formed as part of the mechanism.

The discovery of this enzyme was a consequence of studies which showed that the formation of propionate did not involve the expected turnover of CO_2. It had generally been considered that the dissimilation of glucose, glycerol, and lactate by propionibacteria occurred by conversion to pyruvate which, through fixation of CO_2, was converted to oxalacetate. It was thought that the oxalacetate was then reduced to succinate, which in turn was esterified with CoA and decarboxylated to yield propionyl CoA and CO_2. Therefore, the reduction of one mole of pyruvate to propionate would be expected to involve the fixation and release of one mole of CO_2. Because the succinate is symmetrical, half of the CO_2 released would be derived from the original CO_2 fixation and the other half from pyruvate. By $C^{14}O_2$ dilution experiments, Wood and Leaver[148] found that the turnover of CO_2 was insufficient to account for the amount of propionate formed, and it was concluded that a C_1 unit other than free CO_2 was a product of the cleavage of the C_4-dicarboxylic acid. Numerous other studies likewise indicated that a C_1 other than CO_2 was formed during the decarboxylation of succinate.[149, 150]

The elucidation of the transcarboxylase reaction provides an explanation for the observed small turnover of CO_2. It now is apparent that the formation of propionate involves two linked cycles, one a C_1 cycle involving methylmalonyl isomerase and methylmalonyl-oxalacetic transcarboxylase and the other a CoA cycle involving propionyl CoA transferase. The mechnism is outlined in Fig. 5. It is seen that oxalacetate is formed by two reactions; one is by the CO_2 fixation catalyzed by phosphoenolpyruvic carboxytransphosphorylase described by Siu *et al.*[100] and the other is by the transcarboxylase reaction. Let us assume for the moment that oxalacetate is formed by the latter mechanism, i.e., by the reaction of a catalytic amount of methylmalonyl CoA with pyruvate, simultaneously yielding propionyl CoA. The oxalacetate then is reduced to succinate via malate and fumarate with accompanying oxidation of pyruvate to CO_2 and acetate. The succinate, in turn, is converted to succinyl CoA by reaction with the propionyl CoA simultaneously yielding propionate, the reaction being catalyzed by CoA transferase.[119] The methylmalonyl CoA is regenerated by conversion of the succinyl CoA to methylmalonyl CoA by the enzyme methylmalonyl isomerase. This latter enzyme has been demonstrated in propionic acid bacteria by Swick and Wood,[119] Stadtman *et al.*,[151] and Phares *et al.*[152] It is seen that the unesterified carboxyl group of the C_4-dicarboxylic acids is not released but is transferred to pyruvate. This mechanism avoids the formation of free CO_2 and the subsequent necessity of activating the CO_2 for carboxylation of the pyruvate. Thus, only catalytic amounts of oxalacetate need be produced by fixation of CO_2. Likewise, CoA is recycled, and only catalytic amounts of the acyl derivative need to be formed by the *de novo* synthesis using ATP. Therefore, this sequence of events provides a mechanism for the reduction of pyruvate to propionate with a minimum expenditure of energy and without fixation of CO_2. For simplicity, the arrows of Fig. 5 are shown in the direction of propionate formation, although a number of the reactions are reversible.

It is noted that the mechanism of Fig. 5 accounts for the occurrence of fixed CO_2 in propionate. The fixed CO_2 passes through succinate, which is a symmetrical molecule, and therefore the CO_2 carbon is randomized before it reaches succinyl CoA. When the succinyl CoA is converted to

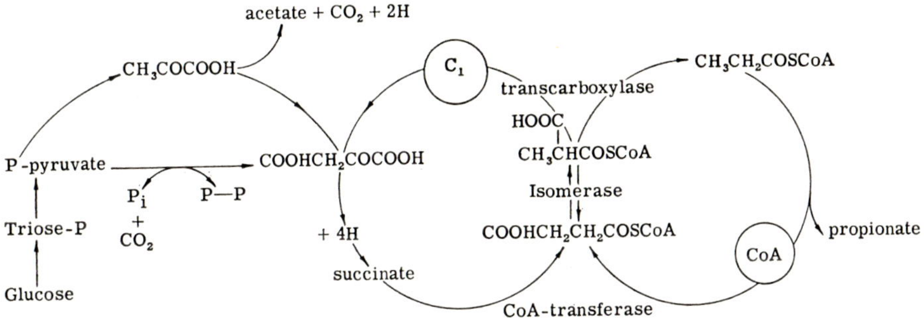

FIG. 5. Formation of propionate by interlocked reactions involving a C_1 transcarboxylation cycle and a CoA transfer cycle. The reactions of the two cycles may be summarized by the following equations:

pyruvate + methylmalonyl CoA ⇌ oxalacetate + propionyl CoA

oxalacetate + 4 H → succinate

succinate + propionyl CoA ⇌ succinyl CoA + propionate

succinyl CoA ⇌ methylmalonyl CoA

Sum: pyruvate + 4 H → propionate

propionate, the carboxyl group contains CO_2 carbon equivalent to the average of the two succinate carboxyls.

Wood and Stjernholm[153] and Stjernholm and Wood[154] have purified and described the properties of the methylmalonyl-oxalacetic transcarboxylase and of methylmalonyl isomerase. The transcarboxylase can be assayed spectrophotometrically in either direction.[153] In the one case, it is coupled with malic dehydrogenase.

$$\text{methylmalonyl CoA} + \text{pyruvate} \rightleftharpoons \text{oxalacetate} + \text{propionyl CoA}$$

$$\text{oxalacetate} + \text{DPNH} + H^+ \rightleftharpoons \text{malate} + \text{DPN}^+$$

In the other case, it is coupled with lactic dehydrogenase:

$$\text{oxalacetate} + \text{propionyl CoA} \rightleftharpoons \text{pyruvate} + \text{methylmalonyl CoA}$$

$$\text{pyruvate} + \text{DPNH} + H^+ \rightleftharpoons \text{lactate} + \text{DPN}^+$$

The equilibrium constant

$$K = \frac{(\text{pyruvate})(\text{methylmalonate})}{(\text{oxalacetate})(\text{propionyl CoA})} = 1.9$$

at 30°C. and pH 6.5. The $\Delta F°$ of the reaction, as calculated from this constant, is -3.9×10^2 calories. The enzyme has a broad specificity for the CoA component. With oxalacetate as the carboxyl donor, propionyl CoA, acetyl CoA, butyryl CoA, or acetoacetyl CoA will serve as acceptors. Acetyl CoA is ½ as effective as propionyl CoA, butyryl CoA 1/10, and acetoacetyl CoA 1/40. In contrast, the specificity for the keto acid is narrow. Pyruvate was the only keto acid which was found to serve as a carboxyl acceptor from methylmalonyl CoA.

It is of considerable interest that the reaction catalyzed by pyruvic carboxylase can be duplicated by coupling propionyl carboxylase from pig heart[111] with the transcarboxylase of propionibacteria.[155]

$$\text{ATP} + CO_2 + \text{propionyl CoA} \rightleftharpoons \text{methylmalonyl CoA} + \text{ADP} + P_i$$

$$\text{methylmalonyl CoA} + \text{pyruvate} \rightleftharpoons \text{oxalacetate} + \text{propionyl CoA}$$

$$\text{Sum: ATP} + CO_2 + \text{pyruvate} \rightleftharpoons \text{oxalacetate} + \text{ADP} + P_i$$

It is seen that the propionyl CoA is regenerated in the above reactions, and it thus acts catalytically. With a concentration of 10^{-5} *M* propionyl CoA, a rapid reaction occurs. At 10^{-6} *M* the reaction still occurs, but there is no reaction in the absence of propionyl CoA. If propionyl carboxylase is present in propionic acid bacteria, and preliminary evidence[119] indicates that it is, oxalacetate could be formed by this mechanism as well as by the two reactions shown in Fig. 5. It is seen that the transcarboxylase promotes the synthesis of oxalacetate by shifting the carboxyl from one reaction sequence to a second. This may prove to be but one example of this type of synthesis.

It is noteworthy that the propionic acid fermentation of glucose by a cell-free extract of *P. shermanii* can be converted to a succinate fermentation by addition of avidin.[156] The avidin inhibits the transcarboxylase reaction, thus preventing the formation of propionate. The fermentation therefore occurs with a net fixation of CO_2 which is catalyzed by phosphoenolpyruvic transphosphorylase and is not inhibited by avidin (cf. Fig. 5). The following over-all conversion would be expected:

$$3\ C_6H_{12}O_6 + 2\ CO_2 \rightarrow 4\ COOH{-}CH_2CH_2COOH + 2\ CH_3COOH + 2\ H_2O$$

This conversion is remindful of the results of Virtanen,[5] who obtained succinate and acetate as the only products of a propionic acid fermentation and postulated the C_4 and C_2 cleavage of hexose (see p. 43).

It is apparent from Fig. 5 that the only way the net fixation of CO_2 can occur when the complete enzyme system is functioning is for either the isomerase, transcarboxylase, or CoA transferase to become rate-limiting so as to cause the accumulation of succinate. Apparently, this must occur when the propionic acid bacteria are grown on glycerol. Thus it was possible to observe net fixation of CO_2.

The methylmalonyl isomerase may be assayed[154] spectrophotometrically by coupling it with transcarboxylase and malic dehydrogenase as shown below:

$$\text{succinyl CoA} \rightleftharpoons \text{methylmalonyl CoA}$$

$$\text{methylmalonyl CoA} + \text{pyruvate} \rightleftharpoons \text{propionyl CoA} + \text{oxalacetate}$$

$$\text{oxalacetate} + \text{DPNH} + \text{H}^+ \rightleftharpoons \text{malate} + \text{DPN}^+$$

The isomerase equilibrium at pH 7.0 and 30° is in favor of succinyl CoA.[154]

$$K = \frac{(\text{succinyl CoA})}{(\text{methylmalonyl CoA})} = 10$$

The $\Delta F°$ as calculated from this equilibrium constant is -1.4×10^3 cal.

The cobamide coenzymes (vitamin B_{12} or pseudovitamin B_{12} derivatives) which were discovered through the brillliant work of Barker and co-workers[157] during their studies of the glutamate-β-methylaspartate isomerization are now known to be essential for methylmalonyl isomerase activity.[115, 151, 152, 154, 158, 159] The carboxyl esterified with CoA has also been shown to be the carbonyl shifted during the isomerization of the succinyl CoA[152, 160, 160a] and not the free carboxyl, as was originally suggested by Beck and Ochoa.[114]

Several bacterial systems, including the propionic acid bacteria[161, 162] and *Micrococcus lactilyticus*,[163] have been shown to decarboxylate succinate to propionate and CO_2. The rate of this decarboxylation is much slower with propionibacteria than with *M. lactilyticus.* The mechanism of the decarboxylation has been studied extensively by Whiteley[164] and by Carson and co-workers.[149, 152, 165] In their most recent work, Carson and co-workers have used a mixture of two enzymic preparations: one from propionibacteria, which contains methylmalonyl isomerase, and one from *M. lactilyticus*, which contains a decarboxylase for methylmalonate. It now appears that the decarboxylation of succinate occurs via methylmalonyl CoA. It is not clear from the brief abstracts, however, whether methylmalonyl decarboxylase is identical with propionyl carboxylase, but catalyzing the reaction toward breakdown rather than fixation of CO_2, or whether the former is a separate enzyme which only catalyzes the breakdown. It is to be noted that succinate may be decarboxylated through the coupled action of methylmalonyl isomerase, methylmalonyl oxalacetic transcarboxylase, and oxalacetic decarboxylase, as illustrated below:

$$\text{succinyl CoA} \rightleftharpoons \text{methylmalonyl CoA}$$

$$\text{methylmalonyl CoA} + \text{pyruvate} \rightleftharpoons \text{oxalacetate} + \text{propionyl CoA}$$

$$\text{oxalacetate} \rightarrow \text{pyruvate} + CO_2$$

$$\text{Sum: succinyl CoA} \rightarrow \text{propionyl CoA} + CO_2$$

The pyruvate would be regenerated and could act catalytically in this sequence.

The importance of transcarboxylase as a mechanism of readily reversible transfer of carboxyls is clearly evident from the above discussion. It is interesting to consider the possibility that other decarboxylations and carboxylations may be found to be coupled reactions and that methylmalonyl-oxalacetic transcarboxylase is but one example of a group of transcarboxylases. The synthesis of fatty acids involves the formation of malonyl CoA from acetyl CoA. This synthesis is usually considered to occur via CO_2 fixation but might very well occur via a transcarboxylation. At one time, the only type of heterotrophic CO_2 fixation which was known was that of the propionic acid bacteria. Now many types of CO_2 fixation are known. The same development seems possible for transcarboxylation.

VI. Reversal of α-Decarboxylation as a Mechanism of CO_2 Fixation

The discovery in 1945 by Utter *et al.*[166] that a soluble enzyme system from *E. coli* catalyzes a very rapid exchange of formate-C^{13} with the carboxyl group of pyruvate caused high hopes that it would be possible to demonstrate a net synthesis of pyruvate by the carboxylation of a C_2 compound with either CO_2 or formate. This exchange reaction occurs by the phosphoroclastic reaction, so named because it was thought that the pyruvate is cleaved by orthophosphate to yield acetyl phosphate[167]* and formate.

The original work on the phosphoroclastic type of cleavage was performed by Koepsell and Johnson[168] using extracts prepared from *C. butylicum*. In this case, hydrogen and CO_2 were products of the reaction, rather than formate. In 1948, Wilson *et al.*[169] showed that this enzyme catalyzed an exchange of CO_2 with the carboxyl group of pyruvate, but no exchange was observed with formate-C^{13}. However, Novelli[170] has found that formate does exchange with the carboxyl group of pyruvate at pH 8.0 with extracts of *C. butylicum* prepared by grinding the cells with alumina.

* Authors' note in reference list.

It is now known that phosphate is not involved in the initial cleavage of pyruvate, and therefore the enzyme is not properly named. Mortlock *et al.*[171] found, when a substrate amount of CoA was used, that cleavage to acetyl CoA occurred with a *C. butyricum* enzyme in the absence of orthophosphate. The reaction is illustrated below:

$$CH_3{-}CO{-}COOH \xrightarrow[HCOOH]{CoASH} CH_3{-}CO{-}S{-}CoA \underset{CoASH}{\overset{H_3PO_4}{\rightleftharpoons}} CH_3{-}CO{-}O{-}P{\overset{OH}{\underset{OH}{=}}}O \qquad (20)$$

Acetyl phosphate

The enzyme that transfers the acetyl group from acetyl CoA to orthophosphate is called phosphotransacetylase, transphosphorylase, or transacetylase. This enzyme has not been found in animal tissues. The formate which is formed may be oxidized either by formic dehydrogenase to CO_2 in the presence of a hydrogen carrier or may be decomposed in the presence of the enzyme hydrogenlyase to CO_2 and H_2.

When labeled acetate was used,[166, 169] the rate of exchange with pyruvate was found to be extremely slow compared to the rate observed with formate or CO_2. Furthermore, when labeled acetyl phosphate was used as a substrate for the *E. coli* system, the exchange rate was also slow,[172] in spite of the fact that the enzyme preparation probably contained phosphotransacetylase. More recently, Mortlock *et al.*[171] using an enzyme from *C. butyricum*, have tested the exchange reaction using acetyl-1-C^{14} CoA, and again the exchange has been found to be very slow.

The phosphoroclastic reaction is no doubt more complicated than shown in reaction 20. Strecker[173] showed that thiamine pyrophosphate, coenzyme A, orthophosphate, and possibly manganese are required for the reaction to occur with *E. coli* extracts. Likewise, the exchange reaction catalyzed by extracts of *C. butyricum* has similar cofactor requirements,[174] while ferrous iron was found to stimulate the over-all forward reaction. However, the more recent experiments by Mortlock *et al.*[171] have shown that the presence of orthophosphate is not necessary for the exchange reaction to occur, a finding which is consistent with the view that the formation of acetyl phosphate occurs by a secondary reaction catalyzed by phosphotransacetylase (see reaction 20).

Mortlock and Wolfe[175] observed a small but significant formation of pyruvate from acetyl-1-C^{14} phosphate, CO_2, and hydrogen. In 30 minutes, they found that about 1% of the total counts added as acetylphosphate was incorporated into pyruvate. These experiments were performed using a crude enzyme preparation from *C. butyricum*, and no evidence was presented to establish that the C^{14} resided exclusively in the carbonyl group of the pyruvate, as is required by the reversal of reaction 20. Strecker[173] found, using cell suspensions of *E. coli*, that acetate-1-C^{14} was incorporated into pyruvate, but much of the C^{14} was in the carboxyl group of the pyruvate. At best, the reversal of the phosphoroclastic reaction is very slow under the experimental conditions which have thus far been devised.

Evidence for the reversal of the oxidative decarboxylation of pyruvate is meager. Goldberg and Sanadi[176] observed a slow exchange of CO_2 with pyruvate using pyruvic oxidase. Similarly, the exchange reaction using $C^{14}O_2$, α-ketoglutarate, and α-ketoglutaric oxidase was slow. We have discussed previously (p. 47) the significance of such exchange reactions and have shown that the exchange per se is not considered to be evidence for the existence of a primary fixation reaction, i.e., a reaction which yields net synthesis from CO_2. For further references to the early work on the reversal of α-decarboxylation and for additional discussion of its significance, see Utter and Wood,[33] also Korkes.[177]

The more recent experiments by Nutting and Carson[178] should be mentioned because they have been considered strong evidence for a C_2 plus C_1 condensation. They reported that *E. coli* ferments xylose at a low pH with the formation of 1.3 to 1.4 moles of lactate per mole of fermented xylose and that a net utilization of CO_2 occurs concurrent with the high yield of lactate. The high yields of lactate suggested that lactate was being formed by C_2 and C_1 addition. At that time, pentose was assumed to cleave to a triose and a C_2 compound with the triose converted to lactate while some of the C_2 compound combined with CO_2 to form additional lactate. We now know that pentose may be fermented by *E. coli* by the transketolase and transaldolase reactions, yielding fructose-6-phosphate, which is then fermented. In this way, more than one mole of lactate can be formed from xylose without using a C_2 and C_1 addition mechanism; see Gunsalus *et al.*[79]

Nutting and Carson offered isotopic evidence for the C_2 plus C_1 condensation by fermenting xylose in the presence of $C^{14}H_3COONa$ and $C^{13}O_2$.

The distribution of the tracers in the lactate formed was as follows:

specific activity of C^{14}	2.19	0.00	0.19
	CH_3—	CHOH—	COOH
% excess C^{13}	0.00	0.00	0.287

The acetate also was isolated from the fermentation mixture and the methyl group had a specific activity of 147, but there was little or no C^{14} in the carboxyl carbon. Accordingly, a maximum of 1.5% [(2.19 × 100)/147] of the lactate methyl group originated from the methyl of acetate. The C^{13} concentration in the final CO_2 was not reported, but it probably was not diluted much, since 15.2 mmoles of CO_2 were present and only 0.55 mmole of xylose was fermented. The original CO_2 contained 12% C^{13}. If the final CO_2 contained 10% excess C^{13}, then a maximum 2.87% of the carboxyl carbon of lactate originated from CO_2. These figures indicate that there may have been some synthesis of lactate via a C_2 plus C_1 addition, but the amount was small.

Cutinelli *et al.*[179] have postulated a C_2 plus C_1 addition to account for the results obtained when *Rhodospirillum rubrum* was grown photosynthetically on $C^{13}H_3$—$C^{14}OOH$ and CO_2 and alternatively on unlabeled acetate and $C^{14}O_2$. The alanine isolated from the bacteria was degraded and found to contain $C^{14}O_2$ in the carboxyl group; the carboxyl group of acetate became the α-carbon, while the methyl group of acetate provided the β-carbon. Thus, the photoassimilation of acetate appears to take place by a C_2 plus C_1 addition reaction. Similar experiments with *E. coli*[180] and yeast[181], however, provided no evidence for the carboxylation of either acetate or a C_2 derivative of acetate.

Thus, we conclude that the utilization of CO_2 by the reversal of α-decarboxylation is of little practical significance in the heterotrophic assimilation of CO_2, at least in the organisms so far studied. The mechanism of the rapid exchange of formate or CO_2 with the carboxyl group of pyruvate by the phosphoroclastic reaction is an intriguing problem. The exchange reaction may occur at the level of the hydroxyethylthiamine diphosphate. Both Breslow[182] and Krampitz *et al.*[183, 184] believe the decarboxylation of pyruvate by enzymes utilizing thiamine pyrophosphate as a coenzyme occurs as illustrated in Fig. 6. There will be further consideration of this interesting subject in the section on the Function of Biotin in the Fixation of CO_2 (Section VII).

VII. The Function of Biotin in the Fixation of CO_2

Currently one of the most exciting biochemical problems under investigation is the elucidation of the role of biotin in the fixation of CO_2. Since 1947, biotin has been known to have some function in CO_2 utilization. Aspartate was found to partially replace the requirement for biotin in the growth medium of *Torula cremoris*. Next, stimulation of aspartate synthesis was observed in the presence of CO_2. It therefore seemed plausible that biotin was acting by stimulating the utilization of CO_2, thus yielding oxalacetate which in turn is converted to aspartate. It then was found that CO_2 could replace aspartate in stimulating the growth of *Lactobacillus* and *Streptococcus* on a biotin-deficient medium and, as a final step, Lardy *et al.*[185] showed that the amount of assimilation of $NaHC^{14}O_3$ into cellular aspartate by *L. arabinosus* was related to the biotin concentration in the growth medium. With a very low concentration of biotin, very little $C^{14}O_2$ was converted to aspartate. In addition, biotin analogs inhibited the fixation of CO_2. A more complete discussion of these early developments with references may be found in the review by Utter and Wood.[33] In the intervening years, a large number of reports have been published relating biotin to a host of biochemical reactions; however, many of the observed effects may be of a secondary nature, because a biotin deficiency is thought to have a deleterious effect on enzyme synthesis.

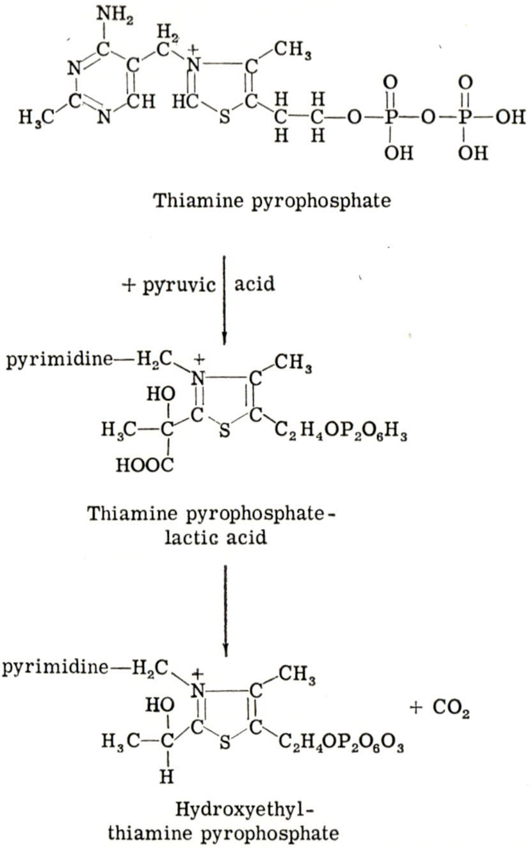

FIG. 6. Mechanism of thiamine pyrophosphate action.

We shall confine our attention to the function of biotin in the fixation of CO_2. The fixation reactions in which biotin has been implicated are those in which ATP is required as a source of energy and include the enzymes propionyl carboxylase, β-methylcrotonyl carboxylase, acetyl carboxylase, and pyruvic carboxylase. All are inhibited by avidin, and the first three enzymes have been shown to contain biotin by direct assay of the protein. In the case of propionyl carboxylase, the enzyme was of very high purity.[111] In addition, methylmalonyl-oxalacetic transcarboxylase is a biotin enzyme, since it is inhibited by avidin and has been shown to contain biotin by direct assay.[156]

There is no evidence as yet that the other primary CO_2-fixation reactions involve biotin. There was a report[186] that phosphoenolpyruvic carboxykinase contained biotin, but recalculation[187] has shown that the biotin content is very low, about 1 mole per 10^8 g. of protein. The activity of this enzyme is not inhibited by avidin.[45, 188] In addition, Semenza *et al.*[188] have shown that phosphoenolpyruvic carboxykinase and the protein-bound biotin do not fractionate alike on avidin-azocellulose. Likewise, biotin is not a part of the "malic" enzyme. It was not found in the purified enzyme[189] and the reaction is not inhibited by avidin.[111] The amount of malic enzyme in livers from biotin-deficient turkeys was less, however, than that in livers of normal turkeys, although several other dehydrogenases were present at normal levels in the deficient livers.[189] Biotin may possibly be involved in the synthesis of the malic enzyme rather than being concerned directly with its enzymic action. Phosphoenolpyruvic carboxytransphosphorylase activity is not inhibited by avidin and probably is not a biotin enzyme.[100]

Shuster and Lynen[190] have recently shown that avidin inhibits a CO_2-pyruvate exchange reaction which is catalyzed by an enzyme from *C. kluyveri*. This exchange differs from the phosphoroclastic exchange reaction in that CoA and inorganic phosphate are not required. The reaction requires adenosine triphosphate, thiamine pyrophosphate (TPP), and magnesium ion. Shuster and Lynen propose an exchange by the following reactions:

a. pyruvate + TPP $\rightleftharpoons$ α-TPP-lactate

b. α-TPP-lactate + biotin-enzyme $\rightleftharpoons$ α-hydroxyethyl-TPP + CO_2-biotin-enzyme

c. CO_2-biotin-enzyme + ADP + P_i $\rightleftharpoons$ CO_2 + biotin-enzyme + ATP

---

d. pyruvate + TPP + ADP + P_i $\rightleftharpoons$ α-hydroxyethyl-TPP + CO_2 + ATP

The α-TPP-lactate and α-hydroxyethyl-TPP are the compounds shown in Fig. 6. This preliminary report is very interesting because it suggests a role of biotin in a reaction which does not involve CoA but does involve thiamine pyrophosphate and an α-carboxyl of pyruvate. Perhaps the biotin-containing enzymes are far more numerous than is presently known. A more detailed report on the purity of the enzyme and on the identification of the pyruvate is needed to establish this reaction more firmly. Of course, the existence of an exchange reaction does not imply that a net fixation of CO_2 can be enacted by the enzyme. We have noted in a previous section that tracer studies in animals and heterotrophic microorganisms have provided no evidence of a C_2 plus C_1 addition reaction which yields pyruvate in amounts which are of major importance in the over-all metabolism of the organism.

The first direct experimental results concerning the mechanism of action of biotin in CO_2 fixation reactions were obtained by Lynen *et al.*,[124, 130, 192a] when they showed that free biotin in substrate amounts may be substituted for β-methylcrotonyl CoA as a CO_2 acceptor in a reaction catalyzed by β-methylcrotonyl carboxylase. The reaction is illustrated below.

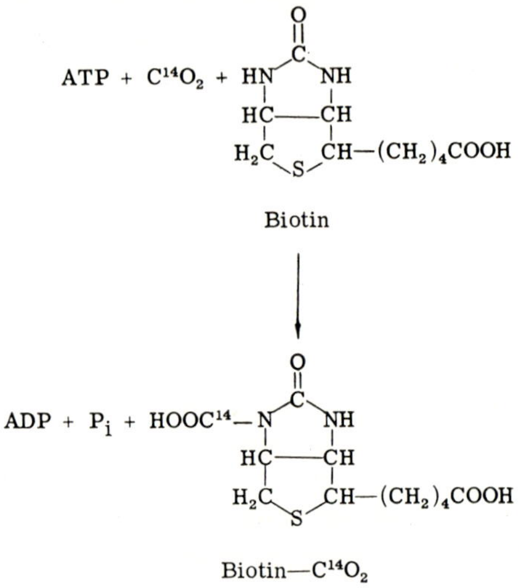

(21)

Reaction 21 was followed spectrophotometrically by determining the rate of ADP formation using phosphoenolpyruvate, pyruvic kinase, DPNH, and lactic dehydrogenase (Fig. 7). The enzyme activates only the D- form of biotin; L-biotin and homobiotin were inactive. The reaction was slower, however, with D-biotin than with β-methylcrotonyl CoA, even though the concentration of biotin was 300 times greater. The $C^{14}O_2$-biotin product was very labile. At pH 2 it decomposed almost instantly but was more stable at a neutral or slightly alkaline pH (lifetime ~20 min. at 0°C.). In 0.03 *N* KOH at 0°C., only 7% was destroyed in 25 minutes. To obtain a stabilized derivative, the crude reaction product was dried directly and the dimethylester prepared using diazomethane. This product was shown by carrier dilution and paper chromatographic techniques to be identical with chemically synthesized *N*-carbomethoxy-biotin-methylester.

Lynen *et al.*[124, 130, 192a] have proposed that the same reaction occurs when biotin is bound to the enzyme. The enzyme-biotin-CO_2 complex is the "active" CO_2 which promotes carboxylation reactions. The biotin may be bound to the enzyme by a peptide linkage involving the ε-amino group of a lysine in a manner similar to that found in biocytin.

Lynen *et al.* point out the acidic properties of the ureido nitrogen of biotin and that the CO_2 addition product would have the properties of an acid anhydride. Lynen states[130] "Thus its propensity for condensation to the α-carbon of acyl-CoA becomes understandable. The C—N bond is polarized in the direction of the nitrogen so that the CO_2 group can act as an acylating agent with the carbanion of the acyl-CoA." Clearly, this is a very attractive mechanism. It remains possible, of course, that free biotin when acting as a substrate for carboxylation does not react with CO_2 in the same manner that it does when it is enzyme-bound and when it is serving to activate CO_2.

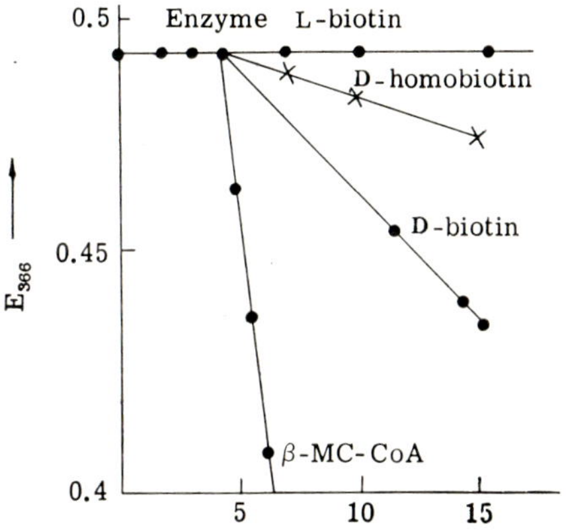

FIG. 7. The enzymic carboxylation of biotin. The reaction mixture contained 280 μmoles triethanolamine-HCl buffer pH 7.4, 160 μmoles KCl, 40 μmoles $MgSO_4$, 10 μmoles $KHCO_3$, 1.3 μmoles K phosphoenolpyruvate, 0.3 μmoles DPNH and 1 mg. serum albumin, 10 μg. pyruvic kinase, 25 μg. lactic dehydrogenase, 200 μg. β-methylcrotonyl carboxylase and as substrates 0.1 μmole β-methylcrotonyl CoA (β-MC-CoA) or 30 μmoles of the different forms of biotin. Total volume = 1.8 ml., T = 37°C.

Lynen *et al.*[124] have suggested the following mechanism for the combination of CO_2 with the biotin of the enzyme and for the carboxylation reaction.

a. ATP + biotin-enzyme $\xrightleftharpoons{Mg^{++}}$ ADP ~ biotin-enzyme + P_i

b. ADP ~ biotin-enzyme + CO_2 $\xrightleftharpoons{Mg^{++}}$ CO_2 ~ biotin-enzyme + ADP

c. CO_2 ~ biotin-enzyme + β-methylcrotonyl CoA ⇄ biotin-enzyme +
or
free biotin

β-methylglutaconyl CoA
or
free CO_2-biotin

This mechanism was proposed on the basis of the following observations:

1. When the carboxylase was incubated with ATP, P_i^{32} and Mg^{++} with

no other substrates, ATP^{32} was formed. This exchange reaction was inhibited by avidin. The exchange was proposed to occur as follows:

$$ATP + \text{biotin-E} \underset{}{\overset{Mg^{++}}{\rightleftarrows}} ADP \sim \text{biotin-E} + P_i$$

$$ADP \sim \text{biotin-E} + P_i^{32} \underset{}{\overset{Mg^{++}}{\rightleftarrows}} ATP^{32} + \text{biotin-E}$$

2. The presence of $KHCO_3$ inhibited the above exchange reaction. A concentration of 2.5×10^{-3} M $KHCO_3$ in the reaction mixture resulted in a 65% inhibition of the exchange, while 10^{-2} M brought about an 80% reduction in activity.

The observed reduction in the exchange reaction was believed to be caused by the following competing reaction:

$$ADP \sim \text{biotin-E} + CO_2 \underset{}{\overset{Mg^{++}}{\rightleftarrows}} CO_2 \sim \text{biotin-E} + ADP$$

3. When 1,3,5-C^{14}-β-methylglutaconyl CoA was incubated with the carboxylase and nonlabeled β-methylcrotonyl CoA, the β-methylcrotonyl CoA became labeled. Neither Mg^{++} nor P_i was required for this exchange to occur. The postulated mechanism is:

$$\underset{\text{β-Methylglutaconyl CoA}}{\overset{\overset{CH_2-C^{14}OOH}{|}}{CH_3C^{14}=CHC^{14}OSCoA}} + \text{biotin-E} \rightleftarrows \underset{\text{β-Methylcrotonyl CoA}}{\overset{\overset{CH_3}{|}}{CH_3C^{14}=CHC^{14}OSCoA}} + C^{14}O_2 \sim \text{biotin-E}$$

$$\overset{\overset{CH_3}{|}}{CH_3C=CHCOSCoA} + C^{14}O_2 \sim \text{biotin-E} \rightleftarrows \overset{\overset{CH_2C^{14}OOH}{|}}{CH_3C=CHCOSCoA} + \text{biotin-E}$$

Sum:

$$\overset{\overset{CH_2C^{14}OOH}{|}}{CH_3C^{14}=CHC^{14}OSCoA} + \overset{\overset{CH_3}{|}}{CH_3C=CHCOSCoA} \rightleftarrows$$

$$\overset{\overset{CH_3}{|}}{CH_3C^{14}=CHC^{14}OSCoA} + \overset{\overset{CH_2C^{14}OOH}{|}}{CH_3C=CHCOSCoA}$$

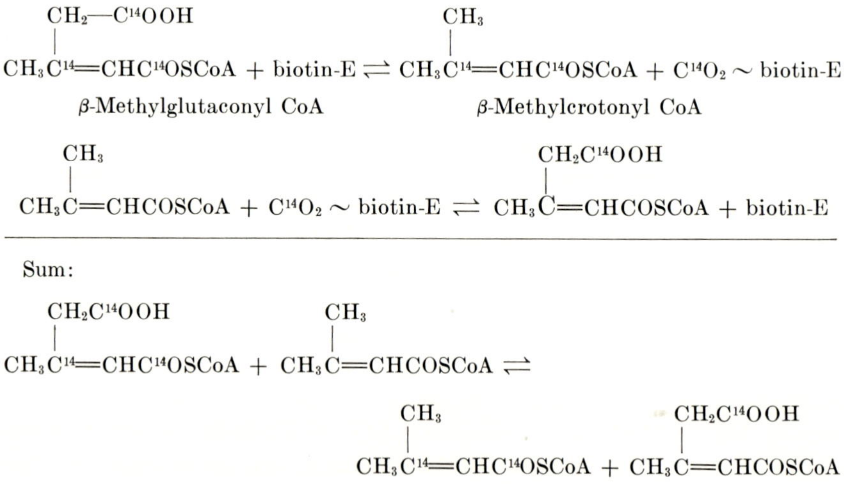

4. When β-methylglutaconyl CoA, β-methylcrotonyl CoA, Mg^{++}, $C^{14}O_2$, and ADP were incubated with the carboxylase, fixation of $C^{14}O_2$ occurred. P_i did not replace the requirement for ADP. The exchange was thought to occur as follows:

β-methylglutaconyl CoA + biotin-E $\rightleftarrows$ β-methylcrotonyl CoA + $CO_2 \sim$ biotin-E

$$CO_2 \sim \text{biotin-E} + ADP \underset{}{\overset{Mg^{++}}{\rightleftarrows}} ADP \sim \text{biotin-E} + CO_2$$

Then

$$ADP \sim \text{biotin} + C^{14}O_2 \xrightleftharpoons{Mg^{++}} C^{14}O_2 \sim \text{biotin-E} + ADP$$

$$C^{14}O_2 \sim \text{biotin-E} + \beta\text{-methylcrotonyl CoA} \rightleftharpoons CH_3-\overset{\overset{CH_2C^{14}OOH}{|}}{C}=CH-COSCoA + \text{biotin-E}$$

No experimental data were presented in support of the exchange mechanism described under 3 and 4 above. Lynen *et al.* proposed that these reactions would probably apply to acetyl CoA carboxylase and propionyl carboxylase. However, Halenz and Lane,[122] Lane *et al.*,[125] and Kaziro *et al.*[111] did not obtain similar results with propionyl carboxylase. Halenz and Lane[122] and Kaziro *et al.*[191] found that propionyl carboxylase catalyzed an exchange of ADP-C^{14} with ATP and no exchange occurred between ATP and P_i^{32}. These findings contrasted with the results of Lynen *et al.*,[124] and Halenz and Lane[122] proposed the following reactions:

$$ATP + \text{biotin-E} \rightleftharpoons \text{P-biotin-E} + ADP$$

$$\text{P-biotin-E} + CO_2 \rightleftharpoons CO_2\text{-biotin-E} + P_i$$

$$CO_2\text{-biotin-E} + \text{propionyl-CoA} \rightleftharpoons \text{biotin-E} + \text{methylmalonyl CoA}$$

Subsequently, Lane *et al.*[125] found that both ADP and P_i were required for the incorporation of $C^{14}O_2$ into the carboxyl group of methylmalonyl CoA (no propionyl CoA was added), whereas by the last two equations above, CO_2 should equilibrate with methylmalonyl CoA in the absence of ADP. Likewise, Kaziro *et al.*[111] have found that both ADP and P_i are required for the exchange reaction between ADP and ATP, whereas only ADP should be required according to the first reaction of the above mechanism. The previous findings were caused by using an ADP solution which was contaminated with P_i. Kaziro *et al.*[111] have pointed out that both ATP and ADP decompose on storage (even in the cold and in solid form) to ADP plus P_i and AMP plus P_i, respectively. They also found that in addition to Mg^{++}, ADP, and P_i there is a CO_2 requirement for the ATP exchange reaction. However the ATP-P_i exchange was slower than the rate of synthesis of ATP from ADP, P_i and methylmalonyl CoA, i.e., reaction 24 from right to left, see below. In addition they failed to observe an exchange between propionyl-C^{14}-CoA and methylmalonyl CoA in the presence or absence of Mg^{++} (reaction 23) and therefore proposed[111] a scheme based on these observations which has subsequently been discarded. Halenz and Lane [191a] and Friedman and Stern[191b] then demonstrated that exchange between propionyl CoA and methylmalonyl CoA does occur with the propionyl carboxylase from liver and that no cofactors or metals are required. In addition Friedman and Stern[191b] tested a sample of propionyl carboxylase which was supplied by Dr. S. Ochoa and found that the exchange occurred. The exchange was more rapid than the over-all back reaction (reaction 24, from right to left) and thus was rapid enough to be a step in the over-all process.[191b] Kaziro *et al.*[112] subsequently confirmed the occurrence of the exchange reaction. The previous failure to observe the exchange is ascribed to impurities in the methylmalonyl CoA-C^{14} (private communication from S. Ochoa).

Although there still remains some uncertainty it now seems probable that the over-all mechanism of the carboxylation can be broken down to two partial reactions:

$$\text{ATP} + \text{CO}_2 + \text{biotin-E} \xrightleftharpoons{\text{Mg}^{++}} \text{ADP} + \text{P}_i + \text{CO}_2 \sim \text{biotin-E} \quad (22)$$

$$\text{CO}_2 \sim \text{biotin-E} + \text{propionyl CoA} \rightleftharpoons \text{biotin-E} + \text{methylmalonyl CoA} \quad (23)$$

$$\text{Sum: ATP} + \text{CO}_2 + \text{propionyl CoA} \xrightleftharpoons{\text{Mg}^{++}} \text{ADP} + \text{P}_i + \text{methylmalonyl CoA} \quad (24)$$

Reaction 22 possibly is a concerted reaction since CO_2, ADP, P_i, and Mg^{++} are all required for the exchange reaction with ATP. It is not entirely clear why the exchange is slower than the over-all back reaction but Ochoa (private communication) has suggested that this may be attributed to competition between ATP and ADP since ADP inhibits the over-all forward reaction and in addition at concentrations above optimum also inhibits the P_i^{32}-ATP exchange.

Direct evidence of partial reaction 23 has recently been accomplished by carboxylation of propionyl carboxylase with $C^{14}O_2$.[111, 191c] This was done by incubating crystalline propionyl carboxylase with Mg^{++}, ATP, and $C^{14}O_2$. The enzyme-$C^{14}O_2$ complex then was obtained free of nucleotides by passing it through a Dowex-1 column. It was found that the radioactivity of the complex was transferred quantitatively to propionyl CoA in the presence of a large excess of unlabeled bicarbonate. The first report[111] indicated that ATP was required for this transfer but subsequent study[191c] showed that ATP was not necessary if the reaction was conducted at 3°C. and for a short time. It has been found that the $CO_2 \sim$ biotin-enzyme complex is unstable. The effect of ATP was indirect and the following explanation has been offered by Kaziro and Ochoa.[191c] At 30°C. in the presence of propionyl CoA the $C^{14}O_2$-enzyme complex is subject to two competing reactions:

a. $C^{14}O_2 \sim \text{biotin-E} + \text{propionyl CoA} \rightleftharpoons \text{methylmalonyl CoA-}C^{14} + \text{biotin-E}$

b. $C^{14}O_2 \sim \text{biotin-E} \rightarrow C^{14}O_2 + \text{biotin-E}$

Reaction a is reversible and $C^{14}O_2 \sim$ biotin-E is in part converted to methylmalonyl CoA-C^{14}, but reaction b is irreversible and yields $C^{14}O_2$. However, since reaction a is reversible, the C^{14} of the methylmalonyl CoA-C^{14} is subsequently drained off through reaction b. When ATP is present, however, methylmalonyl CoA is formed not only by reaction a but also by the over-all reaction 24 from propionyl CoA, ATP, and $NaHC^{12}O_3$. The latter forms a pool of methylmalonyl CoA-C^{12} in which the methylmalonyl CoA-C^{14} is trapped so that reversal of reaction a and followed by reaction b does not cause a rapid loss of the C^{14} from the methylmalonyl CoA. Thus ATP functions indirectly by providing a "protective pool" for the C^{14} of the methylmalonyl CoA.

It is of considerable interest that the stability of the $CO_2 \sim$ biotin-enzyme complex was found to be of the same order of magnitude as that reported by Lynen *et al.*[124] for the free $CO_2 \sim$ biotin compound.

Lynen *et al.*[192a] also have shown recently that ADP, P_i, and CO_2 are necessary for the ATP-P_i^{32} exchange reaction which is catalyzed by β-methylcrotonyl carboxylase. Similar results have been obtained by Seubert and Remberger.[147a] They also investigated the following partial reaction by determining the incorporation of pyruvate-C^{14} into oxalacetate.

$$CO_2 \sim \text{biotin-E} + \text{pyruvate} \rightleftharpoons \text{oxalacetate} + \text{biotin-E}$$

They found there was a rapid incorporation of C^{14} into oxalacetate with no metal or cofactor requirement. Thus the results with the three enzymes are in agreement. This is to be expected if the underlying mechanism of the reactions involves a $CO_2 \sim$ biotin complex and transfer of CO_2 to the acceptor molecule.

Methylmalonyl-oxalacetic transcarboxylase also is a biotin enzyme and it seems likely that the over-all reaction involves the following partial reactions:

$$\text{oxalacetate} + \text{biotin-enzyme} \rightleftharpoons \text{pyruvate} + CO_2 \sim \text{biotin-enzyme} \qquad (25)$$

$$CO_2 \sim \text{biotin-enzyme} + \text{propionyl CoA} \rightleftharpoons \text{biotin-enzyme} + \text{methylmalonyl CoA} \qquad (26)$$

$$\text{Sum: oxalacetate} + \text{propionyl CoA} \rightleftharpoons \text{pyruvate} + \text{methylmalonyl CoA} \qquad (27)$$

Accordingly the enzyme should catalyze each of the partial reactions. It has been shown by incubating pyruvate-C^{14} and oxalacetate or propionyl-C^{14}-CoA and methylmalonyl CoA with methylmalonyl-oxalacetic transcarboxylase that exchange occurs by partial reactions 25 and 26,[156] yielding, respectively, oxalacetate-C^{14} and methylmalonyl CoA-C^{14}.

Propionyl carboxylase can also function as transcarboxylase. It catalyzes the formation of ethylmalonyl CoA from methylmalonyl CoA and butyryl CoA.[192] However, the unique feature of methylmalonyl-oxalacetic transcarboxylase is that it catalyzes the transfer between different types of compounds, an α-keto acid and CoA ester of a fatty acid. Thus, it may shuttle carboxyl groups generated in the Krebs cycle to the pathways of fatty acid synthesis. It therefore differs from the transcarboxylation of carboxyl groups between homologous types of CoA esters such as is catalyzed by propionyl carboxylase or β-methylcrotonyl carboxylase or between keto acids as by pyruvic carboxylase. Methylmalonyl-oxalacetic transcarboxylase has the ability to combine both types of transcarboxylation.

A rather complete account has been given of the studies of the mechanism of the biotin enzymes even though some of the observations have been in error. The change in concepts of the mechanism with each new bit of information illustrates how the mechanisms are investigated and derived. The above discussion also illustrates the precautions which must be taken concerning the purity of the compounds used in exchange studies. Clearly, the understanding of the mechanism of the action of biotin in the fixation of CO_2 is still in the developmental stage. Before present concepts can be accepted, it will be necessary to prove that the $C^{14}O_2$ is linked to the biotin of the enzyme complex at the same position which Lynen *et al.*[124] established for free biotin. To accomplish this, it will be necessary to cleave the biotin-$C^{14}O_2$ from the $C^{14}O_2 \sim$ biotin-enzyme complex. Obviously there is much exciting work still to be done on the function of biotin.

VIII. Total Synthesis of Acetate from CO_2

A. Introduction

The first example of synthesis of acetic acid from CO_2 by a heterotrophic bacterium was observed by Wieringa in 1936.[193] He isolated a microorganism, *Clostridium aceticum*, which was capable of reducing CO_2 to acetate in the presence of hydrogen. At first this microorganism was considered to be a chemoautotroph, but a subsequent study by Karlsson *et al.*[194] has shown that it requires glucose, glutamic acid, biotin, pyridoxamine, and pantothenic acid for growth. In 1940, Barker *et al.*[11] reported that another clostridium, *C. acidi-urici*, fermented hypoxanthine and produced 1.25 moles of acetate per mole of purine utilized.

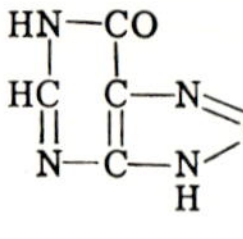

$$+ 6.5\ H_2O \longrightarrow 1.25\ CH_3COOH + 4\ NH_3 + 2.5\ CO_2$$

Since there is only one C—C chain in the hypoxanthine molecule, the acetate in excess of 1.0 mole must have been formed by a condensation of C_1 units (CO_2). When Barker *et al.*[11] studied this fermentation in the presence of radioactive CO_2, they observed CO_2 fixation equally into both the methyl and the carboxyl group of acetate and there was a definite possibility that acetate was being synthesized entirely from CO_2.

In 1942, another microorganism, *C. thermoaceticum*, was isolated and described by Fontaine *et al.*[195] This thermophilic anaerobe produces almost three moles of acetic acid per mole of fermented glucose. This yield is most striking and cannot be accounted for by the conventional glycolytic pathway. No volatile compounds other than acetate were found, and acetate accounted for 85% of the glucose carbon utilized. Although there was no net fixation of CO_2, these workers suggested that either C_1 compounds were produced and then reutilized or that there was a primary cleavage of glucose other than the traditional C_3–C_3 breakdown. Barker and Kamen[196] studied this fermentation in the presence of $C^{14}O_2$ and confirmed that acetate was the only end product and showed in addition that the $C^{14}O_2$ was incorporated into both carbon atoms of the acetate. Another heterotroph, *Butyribacterium rettgeri*,[197, 198] likewise fixes CO_2 in both carbons of acetate. It ferments either glucose or lactate to yield principally acetate and butyrate along with some CO_2. When lactate was fermented in the presence of $C^{14}O_2$, a comparison of the specific activities of the initial CO_2 and the acetate produced indicated that at least 25% of the acetate formed was derived from CO_2.[198] There was a net production of 0.4 mole of CO_2 per mole of fermented lactate, but the total CO_2 formed was estimated by the isotope dilution method to be 0.8 mole per mole of lactate utilized. This difference between the total amount of CO_2 formed and that remaining at the end of the fermentation apparently represents the CO_2 used in the synthesis of acetate.

Two other microorganisms have been shown to be capable of fixing CO_2 into both positions of acetate.[199] One, *C. cylindrosporum*, is very similar to *C. acidi-urici* and ferments uric acid anaerobically, forming CO_2, ammonia, acetate, and glycine. It fixed more CO_2 in the methyl group than in the carboxyl group of acetate[200] and thus differed from *C. thermoaceticum* and *B. rettgeri*, which yield either an equal or higher rate of fixation into the carboxyl group. Carbon dioxide was fixed in glycine as well, but the labeled carbon was present almost exclusively in the carboxyl group. The other microorganism, *Diplococcus glycinophilus*,[201] is noteworthy in that it possesses a remarkable substrate specificity for glycine. Under strictly anaerobic conditions, it yields the following products:

$$4 \text{ glycine} + 2\,H_2O \rightarrow 4 \text{ ammonia} + 3 \text{ acetic acid} + 2\,CO_2$$

When glycine was fermented in the presence of $C^{14}O_2$, more radioactivity was found in the carboxyl group of the acetate than in the methyl group.[202]

From the above discussion, it is apparent that a number of microorganisms are capable of fixing CO_2 in both the methyl and carboxyl groups of acetate, and it has been suggested that some of these bacteria may accomplish a total synthesis of acetate from CO_2. A mixture of unlabeled ($C^{12}H_3—C^{12}OOH$) and two singly labeled molecules ($C^{14}H_3C^{12}OOH$ and $C^{12}H_3C^{14}OOH$) could give the same isotope distribution as a mixture of unlabeled and double-labeled molecules. The mode of biosynthesis of two singly labeled molecules might well be greatly different from that of a doubly labeled molecule, so it became essential to establish the actual constitution of the acetate produced in the fermentation.

B. Proof of the Total Synthesis of Acetate from CO_2 by *Clostridium thermoaceticum*

Wood[203] has pointed out that on a physical basis (mass analysis) it should be possible to differentiate among various types of acetate synthesized from $C^{13}O_2$. Accordingly, *C. thermoaceticum* was studied with $C^{13}O_2$ in a culture fermenting glucose. Since the presence of oxygen isotopes would complicate the mass analysis, the acetate obtained from the fermentation was converted to ethylene. Singly, doubly, and unlabeled acetate molecules yield ethylene molecules possessing different masses:

$$C^{12}H_3—C^{12}OOH \rightarrow C^{12}H_2{=}C^{12}H_2 = \text{mass } 28$$

$$C^{13}H_2—C^{12}OOH \rightarrow C^{13}H_2{=}C^{12}H_2 = \text{mass } 29$$

$$C^{12}H_3—C^{13}OOH \rightarrow C^{12}H_2{=}C^{13}H_2 = \text{mass } 29$$

$$C^{13}H_3—C^{13}OOH \rightarrow C^{13}H_2{=}C^{13}H_2 = \text{mass } 30$$

If the fermentation formed only $C^{13}H_3—COOH$ and $CH_3—C^{13}OOH$, the number of molecules with mass 30 would be almost nil (there would be some molecules of mass 30, since normal carbon contains about 1% C^{13}). The reliability of the method of mass analysis was tested with known mixtures of different types of C^{13}-ethylene. One type of C^{13}-ethylene (singly labeled) was made from $C^{12}H_3C^{13}OOH$ synthesized from unlabeled methyl iodide and C^{13}-cyanide, and another type of C^{13}-ethylene (doubly labeled) was made from $C^{13}H_3—C^{13}OOH$ synthesized from barium carbide (BaC_2^{13}). From these preliminary experiments, it was apparent that the composition of C^{13}-ethylene mixtures could be determined with precision. For a complete discussion of methods, assumptions, and calculations, see Wood.[203]

Results obtained from the mass analysis of the acetate produced during the fermentation of glucose in the presence of $C^{13}O_2$ are shown in Table VII. As shown, approximately 1/3 of the acetate was doubly labeled, 1/3 was singly labeled, and 1/3 was unlabeled. An aliquot of the acetate was degraded in order to determine the C^{13} distribution in the individual carbon atoms; these results also are recorded in Table VII. If calculations are made using the average per cent C^{13} of the $C^{13}O_2$ in the gas phase, it is found that about

TABLE VII

TYPES OF ACETATE AND DISTRIBUTION OF C^{13} IN ACETATE FORMED FROM $C^{13}O_2$ BY *Clostridium thermoaceticum* DURING GLUCOSE FERMENTATION[a]

Exp. no.	Types of acetate recovered (determined on ethylene gas)			Distribution of C^{13} (values are excess C^{13})		
	Doubly labeled (%)	Singly labeled (%)	Unlabeled (%)	CH_3— (%)	—COOH (%)	CO_2 (av.) (%)
1	32	33	35	9.4	16.5	25.4
2	29	31	40	7.1	12.3	21.4

[a] The medium contained 0.056 *M* glucose, 0.5% yeast extract, 0.5% tryptone, 0.05% $(NH_4)_2SO_4$, 0.05% sodium thioglycolate, 0.4% $MgSO_4 \cdot 7H_2O$, 0.08 *M* $NaHC^{13}O_3$, 0.04 *M* K_2HPO_4, 0.04 *M* KH_2PO_4, and $C^{13}O_2$ at 60 cm. Hg, and was in a 3-liter flask. Exp. no. 1, 150 ml. of medium and fermented 6 days; No. 2, 300 ml. of medium and fermented 4 days. Incubation was at 55°C. with continuous shaking.

65% of the carboxyl groups (16.5/25.4 = 0.65) and approximately 35% of the methyl groups (9.4/25.45 = 0.37) contained C^{13}. Since the mass analysis shows that ⅓ of the acetate was formed by a double labeling process and ⅓ by a single labeling mechanism, it follows that the acetate obtained from the fermentation had approximately the composition given in the tabulation below:

	Per cent labeled with $C^{13}O_2$	
	CH_3	COOH
$C^{13}H^3$-$C^{13}OOH$	33	33
$C^{12}H_3$-$C^{13}OOH$	0	33
$C^{12}H_3$-$C^{12}OOH$	0	0
	33	66

Barker and Kamen[198] postulated that the fermentation by *C. thermoaceticum* occurs as follows:

$$C_6H_{12}O_6 + 2\ H_2O \rightarrow 2\ CH_3COOH + 8\ H + 2\ CO_2 \quad (28)$$

$$8\ H + 2\ CO_2 \rightarrow CH_3COOH + 2\ H_2O \quad (29)$$

Accordingly, one would have expected ⅔ of the acetate molecules to have been unlabeled and ⅓ to have been doubly labeled. The presence of carboxyl-labeled acetate indicated that there was a single labeling process occurring in the fermentation. This apparently occurs by an exchange reaction between CO_2 and the carboxyl group of acetate. Evidence of such exchange was obtained[203] when glucose was fermented together with methyl- or carboxyl-labeled acetate and in the presence of a large pool of unlabeled bicarbonate-CO_2. It was observed that over 50% of the C^{14} of the carboxyl-labeled acetate was converted to CO_2, in contrast to only 15% of the methyl group. This occurred without a net destruction of an equivalent amount of acetate.

Although the mass analysis and degradation data show that molecules are formed with fixed CO_2 in adjacent carbon atoms, they provide no information on the actual mechanism of formation of doubly labeled acetate.

C. Experiments with Labeled Glucose, CO_2, and Formate with *Clostridium thermoaceticum*

If *C. thermoaceticum* ferments glucose by the Embden-Meyerhof cleavage, according to equations 28 and 29, the 1,2 and 5,6 positions of glucose would give rise to acetate and the 3,4 positions to CO_2, which would subsequently be converted to acetate. If, however, the fermentations were conducted in the presence of a very large pool of unlabeled $NaHCO_3$ and CO_2, most of the $C^{14}O_2$ formed from the glucose-3,4-C^{14} would be trapped in this pool and the acetate would be unlabeled. Wood[204] and Lentz[205] have conducted such fermentations, and the over-all results were in accord with the Embden-Meyerhof cleavage. As expected from this cleavage, glucose-1-C^{14} and -6-C^{14} gave rise to acetate which was largely methyl-labeled, glucose-2-C^{14} yielded acetate with the highest C^{14} in the carboxyl position, and glucose-3,4-C^{14} to acetate with low activity.

To date, the only pathway known for the formation of carbon-carbon bonds with both carbon atoms arising from CO_2 is by the autotrophic carbon cycle. This cycle involves ribulose diphosphate carboxylase as the primary fixation reaction, then synthesis of fructose-6-phosphate from the resulting phosphoglycerate, then formation of pentose-phosphate through transketolase-transaldolase reactions (see Elsden, Chapter 1). The pentoses are important intermediates in this scheme, and therefore Ljungdahl[206] has investigated the distribution of C^{14} in ribose-5-P formed by *C. thermoaceticum* during the fermentation of xylose in the presence of $C^{14}O_2$. The ribose contained more than 95% of the C^{14} in position 3. The acetate was also degraded, and contained a high concentration of C^{14} in both the methyl and carboxyl carbons. Thus, no evidence was obtained that the totally labeled acetate was formed by *C. thermoaceticum* via a modification of the autotrophic carbon cycle.

In summary, it appears that the glucose is cleaved in the fermentation to two C_3 compounds which in turn are converted to two acetate molecules and two C_1 compounds that are either CO_2 or are equilibrated with CO_2. The two C_1 compounds are then converted to acetate, but as yet the mechanism of the synthesis of acetate from CO_2 is unknown. One possibility is that formate or an "activated" derivative of formate may condense with CO_2 to give a C_2 compound which is converted to acetate. Lentz and Wood[207] have observed that formate-C^{14} is a better precursor of the methyl group of acetate than it is of the carboxyl group. Thus formate differs from CO_2 since the latter is a better precursor of the carboxyl group of the acetate. It is apparent therefore that the formate is not entering the acetate exclusively via CO_2.

D. Tracer Experiments on the Mechanism of Acetate Formation from CO_2 by *Butyribacterium rettgeri*

Extensive studies of the fixation of $C^{14}O_2$ by this organism have been carried out by Pine and Barker.[208] Of all the products of the fermentation of glucose, the carboxyl group of lactate had the highest radioactivity. The C^{14} in this position was equal to or slightly higher than that of the final CO_2 and 3 to 4 times more active than the carboxyl of acetate. Carbons 1 and 2 of lactate had a much lower activity than either the CO_2 or acetate. The ratio of C^{14} in the methyl to carboxyl groups of acetate varied from 0.72 to 0.8. When either carboxyl- or methyl-labeled acetates were added to fermentations of glucose, very little label appeared in the CO_2; 0.8% of acetate-1-C^{14} and 0.1% of acetate-2-C^{14}. However, acetate-1-C^{14} and -2-C^{14} were found to enter the 2 and 3 carbons, respectively, of lactate. Furthermore, experiments with lactate-2-C^{14} and lactate-3-C^{14} showed that these positions gave rise to acetate. This organism thus appears to differ from most bacteria, since it appears to convert acetate to a C_3 by C_2 plus C_1 addition. Although some organisms utilize acetate, they appear to do so by the malate synthetase-glyoxalate pathway (see Krampitz, Vol. II, Chapter 4). In this case, the carboxyl group of acetate does not become carbon 2 of lactate. For a discussion of C_2 plus C_1 addition see Section VI.

Three other labeled compounds were tested, glycine-2-C^{14} and glycolate-1- and -2-C^{14}, but were not utilized to a significant extent during glucose fermentations.

Pine and Barker also performed experiments in which they attempted to identify the intermediate compounds by use of brief exposures to $C^{14}O_2$ in the presence of lactate. After 5 minutes' exposure, the most highly labeled compound was lactate, and it was almost entirely carboxyl-labeled. Acetate was found to be methyl-labeled as well as carboxyl-labeled, the ratio of specific activities of the methyl to carboxyl being 0.82. By the radioautogram technique, it was shown that the glycine and serine were unlabeled and other amino acids had very low C^{14} levels. When glucose was fermented in the presence of formate-C^{14}, the CO_2 was found to be highly labeled, and the C^{14} patterns in the acetate and lactate were the same as those following experiments with $C^{14}O_2$. It has been shown that glucose-adapted cells are able to decompose formate to CO_2 and H_2.[208] The failure to observe preferential incorporation of formate into the methyl carbon of acetate such as found by Lentz and Wood[207] with *C. thermoaceticum* may be due to a very rapid conversion of formate to CO_2, and therefore the two compounds become equivalent in tracer experiments. In order to accentuate a preferential incorporation into the methyl carbon of acetate, it probably is necessary to use a large pool of unlabeled CO_2 to trap any C^{14} of the formate which is converted to CO_2.

When formaldehyde-C^{14} was fermented together with glucose by *B. rettgeri*, C^{14} was recovered in appreciable amounts in carbon dioxide, acetate, and lactate. The distribution of C^{14} in the latter compounds was significantly different from that obtained with $C^{14}O_2$. The ratio of methyl to carboxyl carbon of acetate was 1.76 compared to values of 0.82 to 0.67 with $C^{14}O_2$. The lactate became preferentially labeled in carbons 1 and 3 with predominant labeling occurring in carbon 3. Furthermore, carbon 3 of the lactate was considerably more active than the methyl carbon of acetate. In addition, it was found that unlabeled formaldehyde acquired C^{14} activity when fermented together with glucose and $C^{14}O_2$.

A comparison between *B. rettgeri* and *C. thermoaceticum* indicates a similarity in the mechanisms of CO_2 utilization by the two organisms. Both incorporate CO_2 into methyl and carboxyl carbons of acetate. An exchange between CO_2 and the carboxyl of acetate has been observed with both organisms, although the rate of the exchange is slower in *B. rettgeri*. Formaldehyde can be utilized to a limited extent by both organisms and labels the methyl carbon of acetate preferentially. Formate can be utilized by both organisms, although *B. rettgeri* apparently possesses hydrogenlyase, since H_2 is released from formate. *C. thermoaceticum* probably possesses formic dehydrogenase, since no hydrogen production has been observed from formate. The experimental data can be accounted for by a mechanism in which CO_2 and formaldehyde (or activated forms of these C_1 compounds) yield doubly labeled acetate. Thus far no mechanism for such conversion is known.

E. Fixation of CO_2 by *Diplococcus Glycinophilus*

It may be recalled that the glycine fermentation by *D. glycinophilus* occurs according to the reaction illustrated below:[201]

$$4 \text{ glycine} + 2\ H_2O \rightarrow 4\ NH_3 + 2\ CO_2 + 3 \text{ acetate}$$

A small amount of hydrogen also is produced, but the evolution of this gas ceases when the medium becomes saturated with H_2. From the above equation, it might be expected that the glycine fermentation occurs by a complete oxidation of 1 mole of glycine to CO_2, coupled with reduction of

3 moles of glycine to acetate. This possibility was tested by Barker *et al.*[202] by fermenting glycine-1-C^{14}, glycine-2-C^{14}, and unlabeled glycine in the presence of C^{14} bicarbonate (see Table VIII). The results show that CO_2 is derived almost entirely from the carboxyl group of glycine (Exp. 1 and 2). Both carbon atoms of the acetate originate partly from the methylene carbon of glycine (Exp. 1) but also to some extent from CO_2 (Exp. 3). These results exclude a direct reduction of glycine to acetate. The data in Experiments 4 and 5 indicate that acetate, once formed, is not metabolized further; they exclude a possible redistribution of C^{14} in the acetate by secondary

TABLE VIII
GLYCINE FERMENTATIONS WITH C^{14}-LABELED SUBSTRATES BY *Diplococcus glycinophilus*

Exp. no.	Labeled compound		Final products			Recovery (%)
	Compound	c.p.m. per mmole	CO_2 (c.p.m. per mmole)	CH_3— (c.p.m. per mmole)	—COOH (c.p.m. per mmole)	
1	2-C^{14}-Glycine + $NaHCO_3$	12,600	160	9,400	6,800	91
2	1-C^{14}-Glycine	17,200	15,500	2,450	8,200	96
3	Glycine + $NaHC^{14}O_3$	37,600	14,500	1,260	7,460	99
4	Glycine + acetate-1-C^{14}	1.5×10^6	180	150	18,200	116
5	Glycine + acetate-2-C^{14}	4.5×10^6	85	12,100	160	70

Washed cell suspensions of *D. glycinophilus* incubated anaerobically for 16 to 20 hours at 37°C. in a medium containing approximately 0.04 *M* glycine, 0.02 *M* phosphate buffer (pH 7.0), and 0.02% $Na_2S \cdot 9H_2O$. When bicarbonate was added, the initial concentration was from 0.11 to 0.15 *M*. Acetate was added to experiments 4 and 5 in tracer amounts (5×10^{-4} *M*).

reactions. We shall consider a possible mechanism for these conversions in the summary of this section.

F. FIXATION OF CO_2 BY *Clostridium acidi-urici* AND *C. cylindrosporum*.

We noted previously that the fermentation of purines by *C. acidi-urici* occurs with the fixation of CO_2 in both carbon atoms of acetic acid. The isolation of the two anaerobic bacteria, *C. acidi-urici* and *C. cylindrosporum*, was described by Barker and Beck.[209] These bacteria ferment xanthine, uric acid, and guanine, and also hypoxanthine, guanosine, and inosine after adaptation.[199] Ammonia, carbon dioxide, and acetic acid were identified as the products of the fermentation by *C. acidi-urici*, and glycine was identified as an additional product with *C. cylindrosporum*. Considerable evidence was obtained which implicated glycine as an intermediate in the fermentation of purines by *C. acidi-urici*. It therefore is apparent that a close physiological relationship exists between the two organisms. When xanthine is fermented by *C. acidi-urici*, the reaction can be represented as follows:

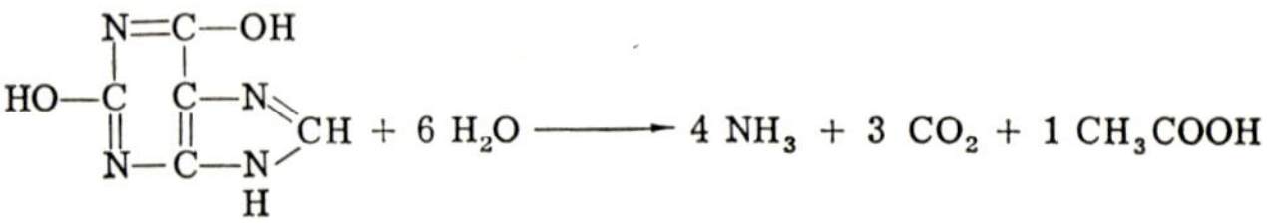

Karlsson and Barker[210] prepared C^{14}-labeled samples of uric acid by feeding formate-C^{14}, or bicarbonate-C^{14}, or glycine-C^{14} to pigeons, and the uric acid obtained from the excreta was purified and used to determine the origin of the CO_2 and acetic acid in the *C. acidi-urici* fermentations.[211] The calculations were complicated, however, because the biosynthetic samples

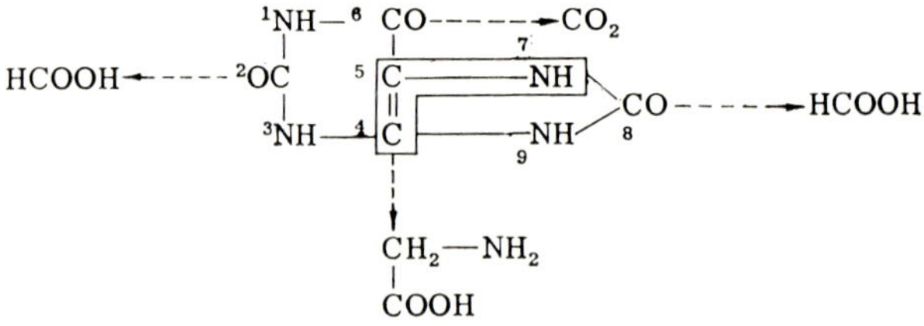

FIG. 8. Breakdown of purine by *Clostridium cylindrosporum*.

of purines were not specifically labeled. These experiments have been discussed in some detail by Utter and Wood.[33] Rabinowitz and Barker[212] have reinvestigated this problem. The results are summarized in Fig. 8. In many respects the breakdown seems to resemble the reverse of the synthesis that has been studied in mammalian systems in which glycine is a precursor of positions 4, 5, and 7, CO_2 position 6, and formate of 2 and 8. In addition, it was observed that CO_2 could be converted to formic acid and that the methyl group of acetic acid was derived mainly from carbon 5 of the purine ring, with significant contributions coming from CO_2 and formic acid. The carboxyl group of acetate also was derived from carbon 5 of the purine. In an excellent series of investigations, Rabinowitz and Barker,[213] Rabinowitz,[214] Rabinowitz and Pricer,[215] Sagers *et al.*[216] Sagers and Beck,[217] and Beck *et al.*[218] have determined the individual enzymic steps in the conversion of xanthine to glycine. This sequence of catabolic reactions involves three imidazole derivatives as illustrated in Fig. 9. Tetrahydrofolic acid (THFA) plays an integral part in the formation of glycine and formate from formiminoglycine and a pathway for the formation of acetate is illustrated below:

a. formiminoglycine + THFA → glycine + formimino-THFA

b. formimino-THFA $\xrightarrow{-NH_3}$ 5,10-anhydro-formyl-THFA

c. 5,10-anhydroformyl-THFA → 10-formyl-THFA

d. glycine + 10-formyl-THFA → serine + THFA

e. serine $\xrightarrow{-NH_3}$ pyruvate → acetate + CO_2

Sagers and Beck[217] and Rabinowitz and Pricer[219] have shown by a number of tracer experiments and enzymic reactions that this series of reactions does occur.

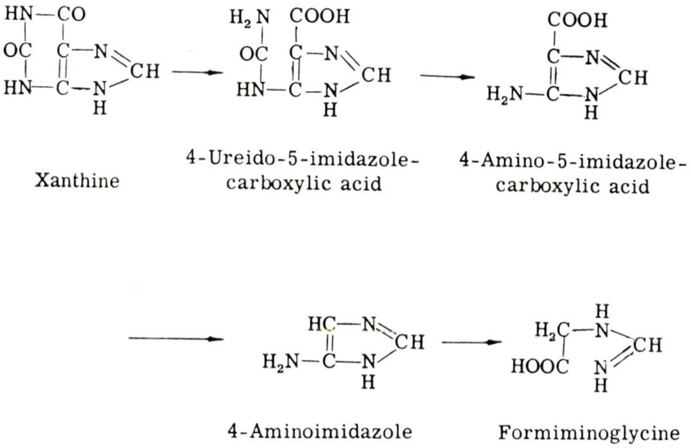

FIG. 9. Conversion of xanthine to formiminoglycine.

G. SUMMARY

There are a number of heterotrophic bacteria capable of fixing CO_2 into both carbon atoms of acetate. Only with *C. thermoaceticum* has it been proved by mass spectrometric analysis of the acetate that both carbon atoms within an individual molecule are derived from CO_2, but it seems quite certain that other organisms of the above group carry out a total synthesis of acetate. The mechanism of the synthesis of acetate from CO_2 is not known. The glycine-serine interconversion outlined in Section VIII, F (reactions d and e), can account for parts of the labelling data; however, it should be noted that neither glycine-1-C^{14} nor $C^{14}O_2$ can produce carboxyl-labeled acetate by this route. Nevertheless, these conversions do occur with *C. acidi-urici* and *D. glycinophilus* (see Table VIII). The scheme below is based on known reactions which have been found in microorganisms. It is presented to illustrate that reactions a, b, and c do not account for the formation of doubly labeled acetate from CO_2 and that an extension of this mechanism could account for such synthesis.

a. C^*O_2 + 2 H → HC*OOH → HC*HO

b. $HC^*HO + NH_2CH_2COOH \rightleftharpoons HOC^*H_2CH(NH_2)COOH$

c. $HOC^*H_2CH(NH_2)COOH \rightleftharpoons C^*H_3COCOOH + NH_3 \rightarrow C^*H_3COOH + CO_2$

d. $C^*O_2 + C^*H_3COCOOH + ATP \rightleftharpoons HOOC^*C^*H_2COCOOH + ADP + P_i$

e. $HOOC^*C^*H_2COCOOH + H_2O \rightarrow HOOC^*C^*H_3 + HOOCCOOH$

f. $HOOCCOOH + ATP + CoASH \rightleftharpoons HOOCCOSCoA + ADP + P_i$

g. $HOOCCOSCoA + DPNH + H^+ \rightleftharpoons HOOCCHO + DPN + CoASH$

h. $CHOCOOH + NH_3 \rightleftharpoons NH_2CH_2COOH$

In reaction a, the formation of formaldehyde from CO_2 via formate is shown. Formaldehyde then condenses with glycine to yield serine in reaction b. The latter is converted to pyruvate, which can be oxidatively decarboxylated to acetate reaction c. These three reactions would account for the formation of methyl labeled acetate but do not account for the formation of doubly labeled acetate. Fixation of CO_2 with pyruvate to yield oxalacetate is shown in reaction d. This is followed by a hydrolytic cleavage resulting in doubly labeled acetate and oxalate (reaction e). The cleavage of oxalacetate to acetate and oxalate has been observed by Hayaishi *et al.*[220] to occur in *Aspergillus niger*. In reaction f, oxalate is converted to the CoA ester with ATP and CoASH before being reduced to glyoxylate (reaction g). Reactions f and g have been suggested by Quayle and Keech[221] to occur in *Pseudomonas oxalaticus*. Glycine is then regenerated by amination of the glyoxylate (reaction h), and may then be recycled via reaction b. This sequence of reactions could account for the metabolic events observed with *D. glycinophilus*, *C. cylindrosporum*, and *C. acidi-urici*. However, Lentz[205] and Pine and Barker[208] concluded that a mechanism involving some of the above features could not be operating in either *C. thermoaceticum* or *B. rettgeri*. With $C^{14}O_2$, carbon 3 of serine and lactate had low C^{14} activities, whereas it would have been expected to have a high activity if the $C^{14}O_2$ were reduced to formaldehyde and then utilized in reaction b.

IX. Conversion of CO_2 to Methane

The anaerobic decomposition of organic materials with the concomitant production of methane has been called the "methane fermentation." All methane-producing bacteria are strictly anaerobic heterotrophs. In most cases, the only end products found are CH_4 and CO_2. The ratio of CO_2 to CH_4 varies for different substrates that are fermented, being higher for more oxidized compounds than for more reduced ones. The proof that at least some "methane bacteria" reduce CO_2 to methane was obtained by Barker in studies with *Methanobacterium omelianskii* fermenting ethanol and *M. suboxidans* and *M. propionicum* fermenting butyrate and propionate, respectively.

M. omelianskii converts ethyl alcohol almost quantitatively to acetate according to the following equation:[222]

$$2\ CH_3CH_2OH + CO_2 \rightarrow 2\ CH_3COOH + CH_4$$

This oxidation is dependent on the presence of CO_2. Direct evidence for the conversion of CO_2 to methane was obtained in experiments where unlabeled ethyl alcohol was incubated with $C^{14}O_2$.[223] The specific activity of the methane from this fermentation was almost the same as that of the CO_2, thus proving that the methane was derived entirely from CO_2. *M. suboxidans* ferments 2 moles of butyrate to 4 moles of acetate with concomitant reduction of 1 mole of CO_2 to methane according to the equation below:[226]

$$2\ CH_3CH_2CH_2COOH + 2\ H_2O + CO_2 \rightarrow 4\ CH_3COOH + CH_4$$

Tracer experiments with this bacterium indicated that almost all of the methane produced is derived from CO_2.

The fermentation of propionate by *M. propionicum* is more complicated, since CO_2 is both formed and utilized. The reactions as outlined by Stadtman and Barker[224] are pictured below:

$$4\ CH_3CH_2COOH + 8\ H_2O \rightarrow 4\ CH_3COOH + 4\ CO_2 + 24\ H$$

$$3\ CO_2 + 24\ H \rightarrow 3\ CH_4 + 6\ H_2O$$

$$\text{Sum:}\quad 4\ CH_3CH_2COOH + 2\ H_2O \rightarrow 4\ CH_3COOH + CO_2 + 3\ CH_4$$

Although only 1 mole of CO_2 was found per 4 moles of propionate, it was assumed that the oxidation of 4 moles of propionate resulted in 4 moles of acetate, 4 moles of CO_2 and 24 "hydrogens." Three moles of CO_2 were then reduced to yield methane.

The observations described above represent anaerobic fermentations in which CO_2 serves as an electron acceptor. For the sake of completeness, however, it should be mentioned here that at least two other compounds, methanol and acetate, can be converted to methane by catabolic reactions not involving CO_2.[225, 226] Pine and Barker,[227] using enrichment cultures of acetate-fermenting and methane-producing bacteria, have demonstrated in experiments with deuterium-labeled acetate that the methyl group of the substrate is transferred intact into methane.

X. Concluding Comments

The major part of this chapter has dealt with primary reactions of CO_2 assimilation which have been defined as reactions by which CO_2 combines with an acceptor molecule in a carbon-to-carbon bond yielding a net fixation of CO_2. We have seen that there are twelve primary reactions of CO_2 assimilation known to occur in heterotrophic organisms. Three of the primary reactions have not been demonstrated in bacteria, and there is some uncertainty concerning a fourth, the "malic" enzyme. Ten of the twelve reactions have been described within the last seven years, and it is almost certain that others will be discovered in the near future. The energy supply for three of the reactions is from TPNH or DPNH, for four it is from ATP, and for five no apparent external energy source is required; the substrate is already in an "active" form in the sense that it is unstable with respect to its more stable isomers.

Sufficient knowledge now has been gained to demonstrate the importance of these primary fixation reactions in the over-all metabolism of the cell. They are key reactions in such diverse metabolic pathways as fatty acid synthesis, the reversal of glycolysis starting from pyruvate, the *de novo* synthesis of purines, and the maintenance of the supply of the C_4 dicarboxylic acids for all synthesis that occurs via the tricarboxylic acid cycle. It should be pointed out that enzymes such as the "malic" enzyme and the TPN-dependent isocitric enzyme may be very important to the cell when they operate in the direction of decarboxylation. In this manner, they provide the cell with TPNH, which is a very valuable commodity in reductive synthetic reactions.

We have seen that CO_2 exchange with a carboxyl group does not provide a reliable indication of a primary fixation reaction, since exchange may occur in reactions which will not yield a net fixation of CO_2. Even if the reversal is thermodynamically feasible, the kinetics of the reaction may be such as to prohibit a significant or useful net fixation of CO_2. We have observed that it is not known whether free CO_2 or bicarbonate ion is the reactive species in these primary fixation reactions.

Perhaps the most exciting current development is the "breakthrough" relative to the mechanism of action of biotin in fixation reactions. For the past twelve years, biotin has been known to play a role in CO_2 fixation, but until recently, its function remained obscure. The discovery by Lynen's group that biotin can be carboxylated enzymically has provided great impetus to this field. Of additional interest is the role of CO_2 and biotin in fatty acid synthesis in which fixation of CO_2 serves to convert acetyl CoA to an "active" form (malonyl CoA) which may undergo condensation to thus lengthen the fatty acid chain.

Carboxylation of a C_2 compound by CO_2 to yield pyruvate seemed within grasp when the rapid exchange of formate and CO_2 with the carboxyl of pyruvate was discovered some fifteen years ago. Nevertheless, the net synthesis of pyruvate by α-carboxylation is still to be demonstrated in an amount which is significant in terms of the over-all metabolism of the cell. Tracer studies in animals and heterotrophic microorganisms have not provided clear indication that such a conversion occurs from acetate or any of its derivatives; thus one wonders if there is a primary fixation of CO_2 by a reaction of this type in heterotrophic organisms.

We have seen that the reaction catalyzed by methylmalonyl-oxalacetic transcarboxylase is a new type of biochemical reaction that permits the movement of carboxyl groups from compounds involved in one metabolic pathway to compounds involved in another pathway without the intervention of free CO_2, i.e., between the CoA esters of fatty acid metabolism and the keto acids of carbohydrate metabolism. One wonders if the transcarboxylase of the propionic acid bacteria is but one example of a group of transcarboxylases. From the view of economy and control of reactions, it would seem advantageous for the cell to be able to transfer carboxyl groups just as anhydride phosphates are transferred. Carboxyl groups, like phosphate anhydrides, may be utilized in coupled reactions instead of being generated each time they are needed.

One of the most intriguing problems of heterotrophic fixation of CO_2 which remains unsolved is the total synthesis of acetate from CO_2. It seems likely that some new and novel biochemical reactions lie hidden here.

Finally, a comment should be made about the autotrophs, which are not a subject of the chapter. One wonders if, just as some heterotrophs utilize CO_2 via ribulose diphosphate carboxylase, some photosynthetic organisms may not utilize some of the ATP and TPNH generated in the light reaction to fix CO_2 by a mechanism that does not involve the autotrophic carbon cycle. It would be strange if the ATP or reduced nucleotides arising in the light reaction or in the oxidation of inorganic material were used exclusively in the autotrophic carbon cycle. This would imply that in order for an autotroph to utilize CO_2 in a heterotrophic type of CO_2 fixation, it first would be necessary to make the ATP and TPNH autotrophically and then convert the CO_2 to a hexose phosphate, then the hexose would need to be oxidized again to regenerate ATP for the heterotrophic type of CO_2 fixation. This seems to be a very cumbersome and circuitous route. One is led to wonder if all primary CO_2-fixation reactions may not be driven directly by ATP and TPNH generated by the light reaction or oxidation of inorganic compounds. In this case, ribulose diphosphate carboxylase would catalyze the major pathway of carbon incorporation in photosynthesis only because it is required to provide the units essential as building blocks for the other types of heterotrophic fixation reactions. Viewed in this light, one is led to consider the possibility that there may be more than one way of synthesizing the building blocks. For instance, may not some autotrophs make acetate directly from CO_2 as do the heterotrophs? There are known reactions by which acetate may be utilized as a sole carbon source. Thus, once acetate was synthesized, known routes could be followed to form all cellular components. There are often several pathways of metabolism in heterotrophic bacteria. It seems unlikely that the evolutionary process would lead to one and only one process of photosynthesis or chemosynthesis.

ACKNOWLEDGMENTS

The authors wish to thank Dr. D. B. Keech and Dr. Henry Z. Sable for their careful review of the manuscript and for many helpful suggestions during discussions of CO_2 fixation. They also wish to acknolwedge the expert and careful assistance of Adalyn B. Sakami (Mrs. W.) in the preparation of the manuscript.

REFERENCES

[1] S. Winogradsky, *Ann. inst. Pasteur* **4,** 213 (1890).
[2] S. Winogradsky, *Ann. inst. Pasteur* **5,** 92, 577 (1891).
[3] H. G. Wood and C. H. Werkman, *J. Bacteriol.* **30,** 332 (1935).
[4] A. F. Lebedev, *Izvest. Donskovo Gosudarstvenno Universiteta* **3,** 25 (1921). Eng. translation, *Am. Rev. Soviet Med.* **5,** 15 (1947–1948).
[5] A. I. Virtanen, *Soc. Sci. Fennica Commentationes Phys.-Math.* **2,** 1 (1925).
[6] A. I. Virtanen and H. Karström, *Acta Chem. Fennica Ser.* **B7,** 17 (1931).
[7] A. I. Virtanen, *Soumalaisen Tiedeakatemian Toimituksia* **29,** 26 (1928).
[8] A. J. Kluyver, "The Chemical Activities of Micro-Organisms." London Univ. Press, London, 1931.
[9] V. S. Butkevich and M. S. Gaevskaya, *Compt. rend. acad. sci. U.R.S.S.* **3,** 405 (1935).
[10] P. A. Wells, A. J. Moyer, and O. E. May, *J. Am. Chem. Soc.* **58,** 555 (1936).
[11] H. A. Barker, S. Ruben, and J. V. Beck, *Proc. Natl. Acad. Sci. U.S.* **26,** 477 (1940).
[12] H. D. Slade, H. G. Wood, A. O. Nier, A. Hemingway, and C. H. Werkman, *J. Biol. Chem.* **143,** 133 (1942).
[13] J. W. Foster, S. F. Carson, S. Ruben, and M. D. Kamen, *Proc. Natl. Acad. Sci. U.S.* **27,** 590 (1941).
[14] C. B. van Niel, J. O. Thomas, S. Ruben, and M. D. Kamen, *Proc. Natl. Acad. Sci. U.S.* **28,** 157 (1942).
[15] E. A. Evans, Jr. and L. Slotin, *J. Biol. Chem.* **136,** 301 (1940).
[16] A. K. Solomon, B. Vennesland, F. W. Klemperer, J. M. Buchanan, and A. B. Hastings, *J. Biol. Chem.* **140,** 171 (1941).
[17] H. G. Wood, C. H. Werkman, A. Hemingway, and A. O. Nier, *J. Biol. Chem.* **139,** 483 (1941); **142,** 31 (1942).
[18] H. G. Wood and C. H. Werkman, *Biochem. J.* **30,** 48 (1936).
[19] H. G. Wood and C. H. Werkman, *Biochem. J.* **32,** 1262 (1938).
[20] M. J. Johnson, W. H. Peterson, and E. B. Fred, *J. Biol. Chem.* **91,** 569 (1931).
[21] D. S. Searle and L. Reiner, *J. Biol. Chem.* **141,** 563 (1941).
[22] H. G. Wood and C. H. Werkman, *Biochem. J.* **34,** 129 (1940).
[23] H. G. Wood, R. W. Stone, and C. H. Werkman, *Biochem. J.* **31,** 349 (1937).
[24] H. G. Wood, C. H. Werkman, A. Hemingway, and A. O. Nier, *J. Biol. Chem.* **135,** 789 (1940); **139,** 377 (1941).
[25] E. S. G. Barron, *Advances in Enzymol.* **3,** 162 (1943).
[26] H. G. Wood, C. H. Werkman, A. Hemingway, and A. O. Nier, *Proc. Soc. Exptl. Biol. Med.* **46,** 313 (1941).
[27] S. F. Carson, J. W. Foster, S. Ruben, and H. A. Barker, *Proc. Natl. Acad. Sci. U.S.* **27,** 229 (1941).
[28] E. A. Evans, Jr. and L. Slotin, *J. Biol. Chem.* **141,** 439 (1941).

[29] L. O. Krampitz and C. H. Werkman, *Biochem. J.* **35,** 595 (1941).
[30] L. O. Krampitz, H. G. Wood, and C. H. Werkman, *J. Biol. Chem.* **147,** 243 (1943).
[31] D. Herbert, *Symposia Soc. Exptl. Biol.* **5,** 52 (1951).
[32] D. Herbert, *in* "Methods in Enzymology" (S. P. Colowick and N. O. Kaplan, eds.), Vol. I, p. 753. Academic Press, New York, 1955.
[33] M. F. Utter and H. G. Wood, *Advances in Enzymol.* **12,** 41 (1951).
[34] I. Harary, S. R. Korey, and S. Ochoa, *J. Biol. Chem.* **203,** 595 (1953).
[35] H. A. Krebs and F. J. W. Roughton, *Biochem. J.* **43,** 550 (1948).
[36] H. A. Krebs and H. L. Kornberg, "Energy Transformation in Living Matter." Springer, Berlin, 1957.
[37] K. Burton and T. H. Wilson, *Biochem. J.* **54,** 86 (1953).
[38] E. A. Evans, Jr., B. Vennesland, and L. Slotin, *J. Biol. Chem.* **147,** 771 (1943).
[39] K. Burton and H. A. Krebs, *Biochem. J.* **54,** 94 (1953).
[40] M. J. Johnson. *in* "The Enzymes" (P. D. Boyer, H. A. Lardy, and K. Myrbäck, eds.), Vol. III, p. 407. Academic Press, New York, 1960.
[41] A. H. Mehler, A. Kornberg, S. Grisolia, and S. Ochoa, *J. Biol. Chem.* **174,** 961 (1948).
[42] S. Ochoa, A. H. Mehler, and A. Kornberg, *J. Biol. Chem.* **174,** 979 (1948).
[43] J. B. V. Salles, I. Harary, R. F. Banfi, and S. Ochoa, *Nature* **165,** 675 (1950).
[44] R. Steinberger and F. H. Westheimer, *J. Am. Chem. Soc.* **73,** 429 (1951).
[45] M. F. Utter, *in* "The Enzymes" (P. D. Boyer, H. A. Lardy, and K. Myrbäck, eds.), Vol. V, pp. 319–340. Academic Press, New York, 1961.
[46] G. W. E. Plaut and H. A. Lardy, *J. Biol. Chem.* **180,** 13 (1949).
[47] I. R. McManus, *J. Biol. Chem.* **188,** 729 (1951).
[48] G. Kalnitsky and C. H. Werkman, *Arch. Biochem. Biophys.* **4,** 25 (1944).
[49] S. Korkes and S. Ochoa, *J. Biol. Chem.* **176,** 463 (1948).
[50] J. P. Kaltenbach and G. Kalnitsky, *J. Biol. Chem.* **192,** 629, 641 (1951).
[51] S. Ochoa, *J. Biol. Chem.* **159,** 243 (1945); **174,** 133 (1948); A. L. Grafflin and S. Ochoa, *Biochim. et Biophys. Acta* **4,** 205 (1950).
[52] M. Calvin and N. G. Pon, *J. Cellular Comp. Physiol.* **54,** 51 (1959).
[53] M. F. Utter and K. Kurahashi, *J. Biol. Chem.* **207,** 821 (1954).
[54] M. F. Utter, *Ann. N. Y. Acad. Sci.* **72,** 451 (1959).
[55] H. A. Krebs, *Bull. Johns Hopkins Hosp.* **95,** 19 (1954).
[56] J. B. Viellas-Salles and S. Ochoa, *J. Biol. Chem.* **187,** 849 (1950).
[57] W. J. Rutter and H. A. Lardy, *J. Biol. Chem.* **233,** 374 (1958).
[58] B. Vennesland, M. C. Gollub, and J. F. Speck, *J. Biol. Chem.* **178,** 301 (1949).
[59] D. A. Walker, *Biochem. J.* **74,** 216 (1960).
[60] P. Faulkner, *Biochem. J.* **64,** 430 (1956).
[61] H. J. Saz and J. A. Hubbard, *J. Biol. Chem.* **225,** 921 (1957).
[62] S. Korkes, A. del Campillo, and S. Ochoa, *J. Biol. Chem.* **187,** 891 (1950).
[63] S. Kaufman, S. Korkes, and A. del Campillo, *J. Biol. Chem.* **192,** 301 (1951).
[64] S. Ochoa, *Symposia Soc. Exptl. Biol.* **5,** 29 (1951); *Physiol. Revs.* **31,** 56 (1951).
[65] P. D. Boyer, *Ann. Rev. Biochem.* **29,** 15 (1960).
[66] A. Kornberg and W. E. Pricer, Jr., *J. Biol. Chem.* **189,** 123 (1951).
[67] G. W. E. Plaut and S. C. Sung, *J. Biol. Chem.* **207,** 305 (1954).
[68] J. Moyle and M. Dixon, *Biochem. J.* **63,** 548 (1956); J. Moyle, *Biochem. J.* **63,** 552 (1956).
[69] G. Siebert, J. Dubuc, R. C. Warner, and G. W. E. Plaut, *J. Biol. Chem.* **226,** 965 (1957); G. Siebert, M. Carsiotis, and G. W. E. Plaut, *J. Biol. Chem.* **226,** 977 (1957).
[70] J. Ceithaml and B. Vennesland, *J. Biol. Chem.* **178,** 133 (1949).
[71] B. L. Horecker and P. Z. Smyrniotis, *J. Biol. Chem.* **196,** 135 (1952).

[72] F. Dickens and G. E. Glock, *Biochem. J.* **50,** 81 (1951).
[73] R. Dedonder and C. Noblesse, *Ann. inst. Pasteur* **85,** 71 (1953).
[74] W. A. Wood and R. F. Schwerdt, *J. Biol. Chem.* **206,** 625 (1954).
[75] R. D. DeMoss and M. Gibbs, *J. Bacteriol.* **70,** 730 (1955).
[76] D. B. M. Scott and S. S. Cohen, *Biochem. J.* **65,** 686 (1957).
[77] B. Axelrod, R. S. Bandurski, C. M. Greiner, and R. Jang, *J. Biol. Chem.* **202,** 619 (1953).
[78] V. Moses, O. Holm-Hansen, and M. Calvin, *J. Bacteriol.* **77,** 70 (1959).
[79] I. C. Gunsalus, B. L. Horecker, and W. A. Wood, *Bacteriol. Revs.* **19,** 79 (1955).
[80] R. S. Bandurski and C. M. Greiner, *J. Biol. Chem.* **204,** 781 (1953).
[81] T. T. Tchen, F. A. Loewus, and B. Vennesland, *J. Biol. Chem.* **213,** 547 (1955).
[82] R. J. Rutman and P. George, *Proc. Natl. Acad. Sci. U.S.* **47,** 1094 (1961).
[83] R. S. Bandurski, *J. Biol. Chem.* **217,** 137 (1955).
[84] D. A. Walker, *Biochem. J.* **67,** 73 (1957); D. A. Walker and S. L. Ranson, *Plant Physiol.* **33,** 226 (1958).
[85] T. T. Tchen and B. Vennesland, *J. Biol. Chem.* **213,** 533 (1955).
[86] I. Suzuki and C. H. Werkman, *Arch. Biochem. Biophys.* **76,** 103 (1958).
[87] D. A. Walker and J. M. A. Brown, *Biochem. J.* **67,** 79 (1957).
[88] R. G. Stickland, *Biochem. J.* **73,** 660 (1959).
[89] A. Weissbach, B. L. Horecker, and J. Hurwitz, *J. Biol. Chem.* **218,** 795 (1956).
[90] M. F. Utter and K. Kurahashi, *J. Am. Chem. Soc.* **75,** 758 (1953); *J. Biol. Chem.* **207,** 787 (1954).
[91] H. G. Wood, B. Vennesland, and E. A. Evans, Jr., *J. Biol. Chem.*, **159,** 153 (1945).
[92] M. F. Utter and H. G. Wood, *J. Biol. Chem.* **164,** 455 (1946).
[93] M. F. Utter, *J. Biol. Chem.* **188,** 847 (1951).
[94] J. L. Graves, B. Vennesland, M. F. Utter, and R. J. Pennington, *J. Biol. Chem.* **223,** 551 (1956).
[95] M. F. Utter, K. Kurahashi, and I. A. Rose, *J. Biol. Chem.* **207,** 803 (1954).
[96] K. Kurahashi, R. J. Pennington, and M. F. Utter, *J. Biol. Chem.* **226,** 1059 (1957).
[97] J. Cannata and A. O. M. Stoppani, *Biochim. et Biophys. Acta* **32,** 284 (1959).
[98] J. R. Quayle and D. B. Keech, *Biochem. J.* **72,** 631 (1959).
[99] M. F. Utter and D. B. Keech, *J. Biol. Chem.* **235,** 17P (1960).
[100] P. M. L. Siu, H. G. Wood and R. L. Stjernholm, *J. Biol. Chem.*, **236,** PC21 (1961).
[101] W. Vishniac, B. L. Horecker, and S. Ochoa, *Advances in Enzymol.* **19,** 1 (1957).
[102] R. C. Fuller, *Bacteriol.Proc.* (*Soc. Am. Bacteriologists*) **56,** 112 (1956).
[103] J. R. Quayle and D. B. Keech, *Biochem. J.* **72,** 623 (1959).
[104] T. Kaneda and J. M. Roxburgh, *Biochim. et Biophys. Acta* **33,** 106 (1959).
[105] L. N. Lukens and J. M. Buchanan, *J. Biol. Chem.* **234,** 1799 (1959).
[106] R. C. Fuson, *Chem. Revs.* **16,** 1 (1935).
[107] M. Flavin, P. J. Ortiz, and S. Ochoa, *Nature* **176,** 832 (1955).
[108] J. Katz and I. L. Chaikoff, *J. Am. Chem. Soc.* **77,** 2659 (1955).
[109] M. Flavin and S. Ochoa, *J. Biol. Chem.* **229,** 965 (1957); M. Flavin, H. Castro-Mendoza, and S. Ochoa, *J. Biol. Chem.* **229,** 981 (1957).
[110] A. Tietz and S. Ochoa, *J. Biol. Chem.* **234,** 1394 (1959).
[111] Y. Kaziro, E. Leone, and S. Ochoa, *Proc. Natl. Acad. Sci. U.S.* **46,** 1319 (1960).
[112] Y. Kaziro, S. Ochoa R. C. Warner, and J. Chen, *J. Biol. Chem.*, **236,** 1917 (1961).
[113] W. S. Beck, M. Flavin, and S. Ochoa, *J. Biol. Chem.* **229,** 997 (1957).
[114] W. S. Beck and S. Ochoa, *J. Biol. Chem.* **232,** 931 (1958).
[115] P. Lengyel, R. Mazumder, and S. Ochoa, *Proc. Natl. Acad. Sci. U.S.* **46,** 1312 (1960).
[116] H. A. Lardy and R. Peanasky, *Physiol. Revs.* **33,** 560 (1953).
[117] H. A. Lardy and J. Adler, *J. Biol. Chem.* **219,** 933 (1956).

[118] J. Gibson and M. Knight, *Biochem. J.*, **78,** 8B (1961).
[119] R. W. Swick and H. G. Wood, *Proc. Natl. Acad. Sci. U.S.* **46,** 28 (1960).
[120] M. Flavin, H. Castro-Mendoza, and S. Ochoa, *Biochim. et Biophys. Acta* **20,** 591 (1956).
[121] A. Tietz and S. Ochoa, *Arch. Biochem. Biophys.* **78,** 477 (1958).
[122] D. R. Halenz and M. D. Lane, *J. Biol. Chem.* **235,** 878 (1960).
[123] S. J. Wakil, E. B. Titchener, and D. M. Gibson, *Biochim. et Biophys. Acta* **29,** 225 (1958); S. J. Wakil and D. M. Gibson, *Biochim. et Biophys. Acta* **41,** 122 (1960).
[124] F. Lynen, J. Knappe, E. Lorch, G. Jütting, and E. Ringelmann, *Angew. Chem.* **71,** 481 (1959).
[125] M. D. Lane, D. R. Halenz, D. P. Kosow, and C. S. Hegre, *J. Biol. Chem.* **235,** 3082 (1960).
[126] B. K. Bachhawat, W. G. Robinson, and M. J. Coon, *J. Am. Chem. Soc.* **76,** 3098 (1954).
[127] B. K. Bachhawat, W. G. Robinson, and M. J. Coon, *J. Biol. Chem.* **216,** 727 (1955); **219,** 539 (1956).
[128] F. Lynen, *Proc. Intern. Symposium Enzyme Chem., Tokyo and Kyoto, 1957* pp. 57–63; J. Knappe and F. Lynen, *4th Intern. Congr. Biochem. Abstr. Communs., Vienna, 1958* p. 49.
[129] A. del Campillo-Campbell, E. E. Dekker, and M. J. Coon, *Biochim. et Biophys. Acta* **31,** 290 (1959); H. C. Rilling and M. J. Coon, *J. Biol. Chem.* **235,** 3087 (1960).
[130] F. Lynen, *J. Cellular Comp. Physiol.* **54,** 33 (1959).
[131] J. F. Woessner, Jr., B. K. Bachhawat, and M. J. Coon, *J. Biol. Chem.* **233,** 520 (1958).
[132] B. K. Bachhawat and M. J. Coon, *J. Biol. Chem.* **231,** 625 (1958).
[133] F. P. Kupiecki and M. J. Coon, *J. Biol. Chem.* **234,** 2428 (1959).
[134] R. O. Brady and S. Gurin, *J. Biol. Chem.* **186,** 461 (1950).
[135] H. P. Klein, *J. Bacteriol.* **73,** 530 (1957).
[136] D. M. Gibson, E. B. Titchener, and S. J. Wakil, *J. Am. Chem. Soc.* **80,** 2908 (1958)
[137] D. M. Gibson, E. B. Titchener, and S. J. Wakil, *Biochim. et Biophys. Acta* **30,** 376 (1958).
[138] S. J. Wakil, *J. Am. Chem. Soc.* **80,** 6465 (1958).
[139] R. O. Brady, *Natl. Acad. Sci. U.S.* **44,** 993 (1958).
[140] J. V. Formica and R. O. Brady, *J. Am. Chem. Soc.* **81,** 752 (1959).
[141] C. Squires and P. K. Stumpf, *Federation Proc.* **18,** 329 (1959).
[142] M. Kusunose, E. Kusunose, and Y. Yamamura, *J. Biochem. (Tokyo)* **46,** 525 (1959).
[143] S. J. Wakil and J. Ganguly, *J. Am. Chem. Soc.* **81,** 2597 (1959).
[144] P. R. Vagelos, *J. Am. Chem. Soc.* **81,** 4119 (1959).
[145] R. O. Brady, R. M. Bradley, and E. G. Trams, *J. Biol. Chem.* **235,** 3093 (1960).
[146] R. O. Brady, *J. Biol. Chem.* **235,** 3099 (1960).
[147] J. R. Stern, D. L. Friedman, and G. K. K. Menon, *Biochim. et Biophys. Acta* **36,** 299 (1959); *Abstr 137th Meeting Am. Chem. Soc., Cleveland,* p. 47c, (1960); *Arch. Biochem. Biophys.* **92,** 280 (1961).
[147a] W. Seubert and V. Remberger, *Biochem. Zeit.* **334,** 401 (1961).
[148] H. G. Wood and F. W. Leaver, *Biochim. et Biophys. Acta* **12,** 207 (1953).
[149] E. F. Phares, E. A. Delwiche, and S. F. Carson, *J. Bacteriol.* **71,** 604 (1956).
[150] H. G. Wood, R. Stjernholm, and F. W. Leaver, *J. Bacteriol.* **72,** 142 (1956); S. H. Pomerantz, *J. Biol. Chem.* **231,** 505 (1958).
[151] E. R. Stadtman, P. Overath, H. Eggerer, and F. Lynen, *Biochem. Biophys. Research Comm.* **2,** 1 (1960).

[152] E. F. Phares, M. V. Long, and S. F. Carson, *Abstr. 138th Meeting. Am. Chem. Soc., New York*, p. 21c (1960).
[153] H. G. Wood and R. Stjernholm, *Proc. Natl. Acad. Sci. U.S.* **47,** 289 (1961).
[154] R. Stjernholm and H. G. Wood, *Proc. Natl. Acad. Sci. U.S.* **47,** 303 (1961).
[155] R. Stjernholm and H. G. Wood, *Federation Proc.* **20,** 235 (1961).
[156] R. Stjernholm and H. G. Wood, unpublished data.
[157] H. A. Barker, H. Weissbach, and R. D. Smyth, *Proc. Natl. Acad. Sci. U.S.* **44,** 1093 (1958); H. A. Barker, R. D. Smyth, H. Weissbach, J. I. Toohey, J. N. Ladd, and B. E. Volcani, *J. Biol. Chem.* **235,** 480 (1960).
[158] S. Gurnani, S. P. Mistry, and B. C. Johnson, *Biochim. et Biophys. Acta* **38,** 187 (1960).
[159] R. M. Smith and K. J. Monty, *Biochem. Biophys. Research Communs.* **1,** 105 (1959).
[160] H. Eggerer, P. Overath, and F. Lynen, *J. Am. Chem. Soc.* **82,** 2643 (1960); H. Eggerer, E. R. Stadtman, P. Overath, and F. Lynen, *Biochem. Zeit.* **333,** 1 (1960).
[160a] R. W. Swick, private communication (1960).
[161] E. A. Delwiche, *J. Bacteriol.* **56,** 811 (1948).
[162] A. T. Johns, *J. Gen. Microbiol.* **5,** 337 (1951).
[163] A. T. Johns, *J. Gen. Physiol.* **5,** 326 (1951).
[164] H. R. Whiteley, *Proc. Natl. Acad. Sci. U.S.* **39,** 772, 779 (1953).
[165] E. F. Phares and S. F. Carson, *Bacteriol. Proc.* p. 154 (1960).
[166] M. F. Utter, F. Lipmann, and C. H. Werkman, *J. Biol. Chem.* **158,** 521 (1945).
[167] F. Lipmann [*Advances in Enzymol.* **1,** 99 (1941); **6,** 231 (1946)] was the first to identify acetyl phosphate as a product of intermediary metabolism. It was isolated as a product of pyruvate oxidation by *Lactobacillus delbruckii.*
[168] H. J. Koepsell and M. J. Johnson, *J. Biol. Chem.* **145,** 379 (1942); H. J. Koepsell, M. J. Johnson, and J. S. Meek, *J. Biol. Chem.* **154,** 535 (1944).
[169] J. Wilson, L. O. Krampitz, and C. H. Werkman, *Biochem. J.* **42,** 598 (1948).
[170] G. D. Novelli, *Biochim. et Biophys. Acta* **18,** 594 (1955).
[171] R. P. Mortlock, R. C. Valentine, and R. S. Wolfe, *J. Biol. Chem.* **234,** 1653 (1959).
[172] H. J. Strecker, H. G. Wood, and L. O. Krampitz, *J. Biol. Chem.* **182,** 525 (1950).
[173] H. J. Strecker, *J. Biol. Chem.* **189,** 815 (1951).
[174] R. S. Wolfe and D. J. O'Kane, *J. Biol. Chem.* **205,** 755 (1953); **215,** 637 (1955).
[175] R. P. Mortlock and R. S. Wolfe, *J. Biol. Chem.* **234,** 1657 (1959).
[176] M. Goldberg and D. R. Sanadi, *J. Am. Chem. Soc.* **74,** 4972 (1952).
[177] S. Korkes, *Brookhaven Symposia in Biol.* No. **5,** 192 (1953).
[178] L. A. Nutting and S. F. Carson, *J. Bacteriol.* **63,** 575, 581 (1952).
[179] C. Cutinelli, G. Ehrensvärd, L. Reio, E. Saluste, and R. Stjernholm, *Arkiv Kemi* **3,** 315 (1951).
[180] C. Cutinelli, G. Ehrensvärd, L. Reio, E. Saluste, and R. Stjernholm, *Acta Chem. Scand.* **5,** 353 (1951).
[181] G. Ehrensvärd, L. Reio, E. Saluste, and R. Stjernholm, *J. Biol. Chem.* **189,** 91 (1951).
[182] R. Breslow, *J. Am. Chem. Soc.* **79,** 1762 (1957).
[183] L. O. Krampitz, G. Greull, C. S. Miller, J. B. Bicking, H. R. Skeggs, and J. M. Sprague, *J. Am. Chem. Soc.* **80,** 5893 (1958).
[184] L. O. Krampitz, I. Suzuki, and G. Greull, *Fed. Proc.* **20,** 971 (1961).
[185] H. A. Lardy, R. L. Potter, and R. H. Burris, *J. Biol. Chem.* **179,** 721 (1949).
[186] H. C. Lichstein, *J. Biol. Chem.* **212,** 217 (1955).
[187] H. C. Lichstein, *Arch. Biochem. Biophys.* **71,** 276 (1957).
[188] G. Semenza, L. S. Prestige, D. Menard-Jeker, and M. Bettex-Galland, *Helv. Chim. Acta* **42,** 669 (1959).

[189] S. Ochoa, A. Mehler, M. L. Blanchard, T. H. Jukes, C. E. Hoffmann, and M. Regan, *J. Biol. Chem.* **170,** 413 (1947).
[190] C. W. Shuster and F. Lynen, *Biochem. Biophys. Research Comm.* **3,** 350 (1960).
[191] Y. Kaziro, E. Leone, and S. Ochoa, *Abstr. 137th Meeting. Am. Chem. Soc., Cleveland,* p. 39C (1960).
[191a] D. R. Halenz and M. D. Lane, *Biochim. et Biophys. Acta* **48,** 425 (1961).
[191b] D. L. Friedman and J. R. Stern, *Biochem. Biophys. Research Comm.* **4,** 266 (1961).
[191c] Y. Kaziro and S. Ochoa, *J. Biol. Chem.* **236,** 3131 (1961).
[192] D. R. Halenz and M. D. Lane, *Biochem. Biophys. Research Comm.* **5,** 27 (1961).
[192a] F. Lynen, J. Knappe, and E. Lorch, *Proc. 5th Intern. Congr. Biochem.* Moscow, USSR. (1961).
[193] K. T. Wieringa, *Antonie van Leeuwenhoek, J. Microbiol. Serol.* **3,** 263 (1936).
[194] J. L. Karlsson, B. E. Volcani, and H. A. Barker, *J. Bacteriol.* **56,** 781 (1948).
[195] F. E. Fontaine, W. H. Peterson, E. McCoy, M. J. Johnson, and G. J. Ritter, *J. Bacteriol.* **43,** 701 (1942).
[196] H. A. Barker and M. D. Kamen, *Proc. Natl. Acad. Sci. U.S.* **31,** 219 (1945).
[197] H. A. Barker and V. Haas, *J. Bacteriol.* **47,** 301 (1944).
[198] H. A. Barker, M. D. Kamen, and V. Haas, *Proc. Natl. Acad. Sci. U.S.* **31,** 355 (1945).
[199] H. A. Barker and J. V. Beck, *J. Biol. Chem.* **141,** 3 (1941).
[200] H. A. Barker and S. R. Elsden, *J. Biol. Chem.* **167,** 619 (1947).
[201] B. P. Cardon and H. A. Barker, *Arch. Biochem.* **12,** 165 (1947).
[202] H. A. Barker, B. E. Volcani, and B. P. Cardon, *J. Biol. Chem.* **173,** 803 (1948).
[203] H. G. Wood, *J. Biol. Chem.* **194,** 905 (1952).
[204] H. G. Wood, *J. Biol. Chem.* **199,** 579 (1952).
[205] K. E. Lentz, Ph.D. Thesis. Western Reserve University, Cleveland, Ohio, 1956.
[206] L. Ljungdahl, private communication (1961).
[207] K. E. Lentz and H. G. Wood, *J. Biol. Chem.* **215,** 645 (1955).
[208] L. Pine and H. A. Barker, *J. Bacteriol.* **68,** 216 (1954).
[209] H. A. Barker and J. V. Beck, *J. Bacteriol.* **43,** 291 (1942).
[210] J. L.. Karlsson and H. A. Barker, *J. Biol. Chem.* **177,** 597 (1949).
[211] J. L Karlsson and H. A. Barker, *J. Biol. Chem.* **178,** 891 (1949).
[212] J. C. Rabinowitz and H. A. Barker, *J. Biol. Chem.* **218,** 147 (1956).
[213] J. C. Rabinowitz and H. A. Barker, *J. Biol. Chem.* **218,** 189 (1956).
[214] J. C. Rabinowitz, *J. Biol. Chem.* **218,** 175 (1956).
[215] J. C. Rabinowitz and W. E. Pricer, Jr., *J. Biol. Chem.* **218,** 189 (1956); **222,** 537 (1956); *J. Am. Chem. Soc.* **78,** 5702 (1956); **78,** 1513 (1956); **78,** 4176 (1956).
[216] R. D. Sagers, J. V. Beck, W. Gruber, and I. C. Gunsalus, *J. Am. Chem. Soc.* **78,** 694 (1956).
[217] R. D. Sagers and J. V. Beck, *J. Bacteriol.* **72,** 199 (1956).
[218] J. V. Beck, R. D. Sagers, and L. R. Morris, *J. Bacteriol.* **73,** 465 (1957).
[219] J. C. Rabinowitz and W. E. Pricer, Jr., *Federation Proc.* **15,** 332 (1956).
[220] O. Hayaishi, H. Shimazono, M. Katagiri, and Y. Saito, *J. Am. Chem. Soc.* **78,** 5126 (1956).
[221] J. R. Quayle and D. B. Keech, *Biochem. J.* **75,** 515 (1959).
[222] H. A. Barker, *J. Biol. Chem.* **137,** 153 (1941).
[223] T. C. Stadtman and H. A. Barker, *Arch. Biochem.* **21,** 256 (1949).
[224] T. C. Stadtman and H. A. Barker, *J. Bacteriol.* **61,** 67 (1951).
[225] T. C. Stadtman and H. A. Barker, *J. Bacteriol.* **61,** 81 (1951).
[226] A. M. Buswell and F. W. Sollo, Jr., *J. Am. Chem. Soc.* **70,** 1778 (1948).
[227] M. J. Pine and H. A. Barker, *J. Bacteriol.* **71,** 644 (1956).

CHAPTER 3

Inorganic Nitrogen Assimilation and Ammonia Incorporation

L. E. MORTENSON

I. Present Status of Inorganic Nitrogen Metabolism by Microorganisms.... 119
II. Nitrogen Fixation 121
A. Scope 121
B. Biological Agents That Fix Nitrogen 124
C. Requirements for Fixation 125
D. Soluble Nitrogen-Fixing Systems 127
E. Reaction Steps 135
F. Role of Hydrogenase 144
G. Proposed Mechanisms 147
H. Existing Problems 151
III. Incorporation of Ammonia into Organic Compounds 152
A. Dehydrogenases 153
B. Saturation Amination 155
C. Amide Synthesis 157
D. Conclusions 161
References 161

I. Present Status of Inorganic Nitrogen Metabolism By Microorganisms*

The importance of reactions of inorganic nitrogen to bacteria has been recognized for over a century, but most information concerning the mechanism of these reactions has been obtained only in the last decade. Pathways for the metabolism of nitrate and nitrite were suggested from known chemical reactions of inorganic nitrogen compounds and from studies with whole cells; more recently soluble enzymes have been prepared from a variety of microorganisms which have provided evidence for or against the early proposals and also suggested new routes. Studies of the fixation of molecular nitrogen have followed the same route, but reduction to enzymic studies

* Abbreviations used in this chapter: DPN, DPNH—oxidized and reduced diphosphopyridine nucleotide; TPN, TPNH—oxidized and reduced triphosphopyridine nucleotide; ADP, ATP—adenosine diphosphate, adenosine triphosphate; FAD, FMN—flavin adenine nucleotide, flavin mononucleotide; AMP, GMP, XMP—adenine-5′-phosphate, guanosine-5′-phosphate, xanthosine-5′-phosphate; CTP, UTP—cytidine triphosphate, uridine triphosphate; TPP—thiamine pyrophosphate; AGA—acetyl-L-glutamic acid; Pi—inorganic orthophosphate; UV—ultraviolet; O.D.—optical density.

with cell extracts has been more delayed and has been accomplished only in the last two years.

Today most workers agree that the assimilation of inorganic nitrogen ends with the incorporation of ammonia into organic compounds. Biological agents that assimilate inorganic nitrogen with valence more positive than -3 must have either the necessary enzymes for converting these compounds to ammonia or they must grow in symbiosis, or synergism, with organisms that do. Nitrate, the most highly oxidized form of nitrogen metabolized by

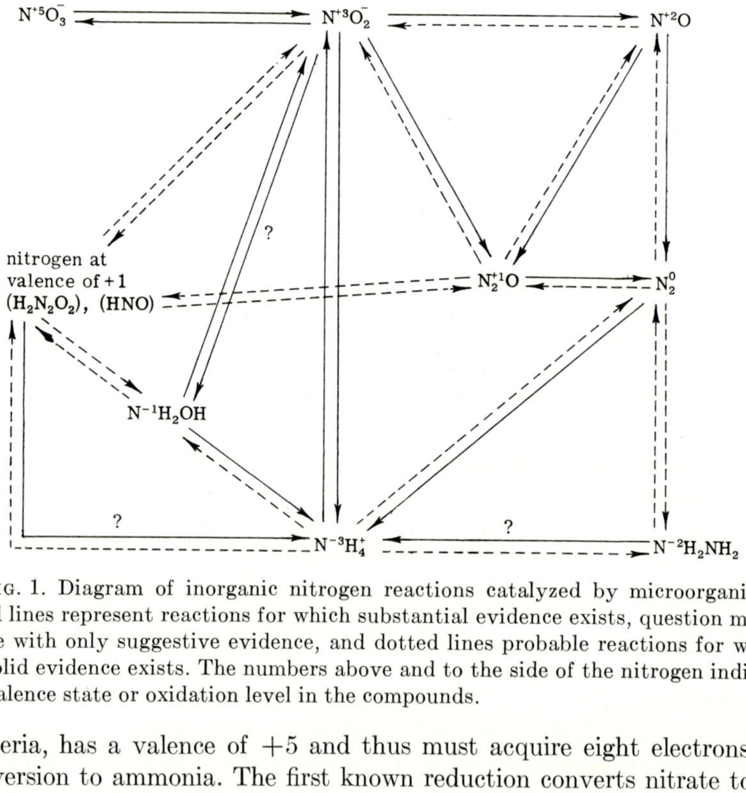

FIG. 1. Diagram of inorganic nitrogen reactions catalyzed by microorganisms. Solid lines represent reactions for which substantial evidence exists, question marks those with only suggestive evidence, and dotted lines probable reactions for which no solid evidence exists. The numbers above and to the side of the nitrogen indicate its valence state or oxidation level in the compounds.

bacteria, has a valence of $+5$ and thus must acquire eight electrons for conversion to ammonia. The first known reduction converts nitrate to nitrite, i.e., to $N^{+3}O_2$. Nitrite, in turn, is reduced by microorganisms to a variety of products, some of which are further converted to ammonia and others not. For example, *Pseudomonas stutzeri* converts NO_2^- to N_2 but cannot further metabolize N_2. A diagram of the inorganic nitrogen conversions catalyzed by microorganisms is presented in Fig. 1. Some of these reactions are thought to constitute intermediate steps of the more general pathways. For example, hyponitrite may be an intermediate in the reduction of nitrite to ammonia. As evidence accumulates, the diagram will probably be simplified; perhaps other conversions may be found and added.

Inorganic nitrogen assimilation infers actual incorporation into cellular components. Many of the reactions performed by microorganisms do not conform to this classification. For example, some organisms merely use nitrate as a terminal electron acceptor, replacing O_2, and do not use it as a source of nitrogen for growth. The same holds true for further denitrification; the nitrite formed serves as electron acceptor forming NO, N_2O, and N_2, each of which is in some organisms a side product from the cells' energy-metabolism and is wasted so far as the use for cell synthesis is concerned. In addition, the nitrifiers, *Nitrosomonas* and *Nitrobacter*, do not fit the definition of nitrogen assimilators since, rather than reduce oxidized nitrogen compounds to ammonia, they oxidize ammonia and nitrite to nitrite and nitrate, respectively, and are able to employ the resulting energy for cellular syntheses. All organisms in such groups are dependent on reactions of other organisms for their source of ammonia and/or other assimilable nitrogen for cell formation and growth.

Extensive reviews of recent date cover the information available on nitrate, nitrite, hyponitrite, and hydroxylamine reductases, as well as on denitrification and nitrification. All of these systems have been obtained in a cell-free state, and the systems either wholly or partially characterized as to steps, enzymes, and cofactor requirements. Table I presents a summary of the known biological inorganic nitrogen reductions. The table includes representative organisms that perform the reactions, the known and questionable requirements of each reaction, and reference to one or more recent reviews or papers covering the subject. For further details of these subjects, the reviews and original papers should be consulted.

Nitrogen fixation in cell-free extracts has only recently been obtained; information on the requirements and properties of the enzyme systems is now available. Since the data are of recent origin, have not been presented in general form, and reviews covering these enzymic data are not yet available, this subject will be treated in some detail. Also the status of ammonia incorporation in bacteria has not been reviewed recently and new data are available; therefore details on this subject will be included.

II. Nitrogen Fixation

A. Scope

Many reviews have been written concerning the progress of research on biological nitrogen fixation using whole cells.[12-15] Most of the original papers, and consequently the reviews, concluded that the nitrogen-fixing system must be prepared in extracts and the components studied before the mechanism of nitrogen fixation could be understood. Within the last two years, nitrogen fixation by enzyme systems in the absence of whole cells has been accomplished—first in *Clostridium pasteurianum*, by Carnahan *et al.*[16, 17]

TABLE I
KNOWN BIOLOGICAL REACTIONS OF INORGANIC NITROGEN

Reaction	Electrons required	Representative organisms performing reaction	Requirements (comments)	References
$NO_3^- \rightarrow NO_2^-$ Nitrate reductase	2	*Escherichia coli*, *Micrococcus denitrificans*, *Bacillus subtilis*, *Haemophilus influenzae*, *Pseudomonas aeruginosa*, *Neurospora crassa*, *Achromobacter fischeri*	Varies among organisms—DPN, FAD (FMN), Mo, cytochromes c, b, TPN	1, 2, 4a, 4b
$NO_2^- \rightarrow NH_4^+$ Nitrite assimilation	6	*Bacillus pyocyaneus*, *Desulfovibrio desulfuricans*, *Bacillus pumilus*, *Neurospora crassa*, *Clostridium pasteurianum*	TPNH (DPNH), FAD, Cu, Fe	1, 2, 86
NO_3^- or $NO_2^- \rightarrow NO$, N_2O, and N_2; denitrification	Varies from 1–10	*Micrococcus denitrificans*, *Denitrobacillus*, *Pseudomonas* spp., *Spirillum itersonii*, *Bacillus licheniformis*, *Achromobacter* spp., *Thiobacillus denitrificans* and *thioparus*	See requirements for individual reactions of this process	2, 3, 4a, 4b
$NO_2^- \rightarrow NO$ Nitrite reduction	1	*Pseudomonas stutzeri*, *Bacillus subtilis*, *Pseudomonas aeruginosa*	TPN (DPN), FMN (FAD), Fe^{++}	3
$NO \rightarrow N_2$ or N_2O Nitric oxide reductase	2 or 4	*Pseudomonas stutzeri*, *Bacillus subtilis*, *Clostridium pasteurianum*, *Pseudomonas aeruginosa*	TPN (DPN), FAD (FMN), (Fe^{++})	3, 3a
$N_2O \rightarrow N_2$ Nitrous oxide reductase	2	*Pseudomonas aeruginosa*, nodule preparations	?	3, 4, 4a, 5
$H_2N_2O_2 \rightarrow 2NH_4^+$ Hyponitrite assimilation	8	*Neurospora crassa*, *Escherichia coli* (*Bn*)	DPNH, Fe^{++}	6, 6a

$N_2 \rightarrow 2NH_4^+$ Nitrogen fixation	6	Aerobic			Anaerobic		Molybdenum Iron Calcium (?) Biotin (?)	See Section II
		Symbiotic Leguminoseae *Alnus* *Elaeagnus* *Hippophaë* *Shepherdia* *Casuarina* *Myrica* *Coriaria* *Ceanothus*	Heterotrophic *Azotobacter* spp. *Pseudomonas* spp. *Nocardia* *Pullalaria*	Photoautotrophic *Nostoc* *Calothrix*	Heterotrophic *Clostridium* spp. (saccharolytic) *Aerobacter aerogenes* *Bacillus polymyxa* *Desulfovibrio desulfuricans* *Achromobacter* spp.	Photoautotrophic *Chromatium* *Rhodospirillum rubrum* *Rhodomicrobium vannielii* *Rhodopseudomonas* spp. *Chlorobium* *Methanobacterium* (autotrophic)		
$NH_2OH \rightarrow NH_4^+$ Hydroxylamine reductase	2	*Neurospora crassa*, *Azotobacter vinelandii* O, *Desulfovibrio desulfuricans*, *Bacillus pumilis*, halotolerant bacterium, *Clostridium pasteurianum*, *Pseudomonas aeruginosa*					Varies—DPNH, FAD, Mn^{++}	2, 2a, 86
$NH_2NH_2 \overset{?}{\rightarrow} 2NH_4^+$ Hydrazine reductase	2	*Micrococcus lactilyticus*					Measured hydrogen uptake by whole cells with hydrazine added	7
$NH_4^+ \rightarrow$ organic compounds	In some cases 2	Large group (see Section III)					Varies (see Section III)	See Section III
$NH_4^+ \rightarrow NO_2^-$ or $NH_2OH \rightarrow NO_2^-$ Nitrification	−6 or −4	*Nitrosomonas* spp.					Acceptor—cytochrome c or phenazine methosulfate	8-10
$NO_2^- \rightarrow NO_3^-$ Nitrite oxidase	−2	*Nitrobacter* spp.					cytochrome, Fe^{+++}	10, 11

More recently, using their procedures as well as others, nitrogen-fixing preparations have been obtained from a variety of known N_2 fixing organisms.

The present manuscript will deal principally with the stoichiometry, reaction route, chemical and enzymic requirements, and mechanism of fixation by the *C. pasteurianum* system. Since the accomplishment of nitrogen fixation by enzymic systems, many of the observations and hypotheses from whole cell experiments have been confirmed while others have been altered and extended.[17-19]

B. Biological Agents That Fix Nitrogen

1. Introduction

A list of the nitrogen-fixing microorganisms is included with the inorganic nitrogen reactions in Table I. Several of these are recent additions. Two reasons for this extension of the list of known nitrogen fixers are the use of the N^{15} technique to detect fixation[20] and an added effort by some laboratories to study this property among a wider variety of organisms. Most of the nitrogen-fixing organisms could have been detected by growth measurements in nitrogen-free media had the right conditions for growth and fixation been known. The conditions required for growth on N_2 in some organisms varied considerably from those required for optimum growth on fixed nitrogen. For example, *Aerobacter aerogenes*, a facultative organism that grows more abundantly in the presence of oxygen with fixed nitrogen, fixes only small amounts of nitrogen when grown aerobically, whereas anaerobically it fixes N_2 almost as well as the best of previously recognized fixers.[21] If the N^{15} technique had not been used to detect the nitrogen fixed aerobically, this organism would still be listed among those not capable of fixing nitrogen. The trace of aerobic nitrogen fixation prompted researchers to examine further the fixation under anaerobic conditions.

No attempt will be made to review exhaustively the research on each nitrogen-fixing organism, for there are reviews which are explicit and up-to-date in this regard.[15] Some of the more recently discovered fixers will be discussed as representatives of the types of organisms that can fix nitrogen.

2. Variety of Organisms

The organisms that fix nitrogen have been divided into the heterotrophic bacteria, both aerobic, such as *Azotobacter vinelandii*, and anaerobic, such as *Clostridium pasteurianum*; photoautotrophic bacteria, such as the photosynthetic bacteria and various algae[13,15]; and the symbiotic system consisting of rhizobia in root nodules of legumes (Table I).

Recently several organisms not previously recognized as nitrogen fixers have been recorded. Proctor and Wilson,[22] using N_2^{15} uptake, detected fixation in six strains of *Pseudomonas* and eight of *Achromobacter*. Two species of *Nocardia*[23] and one of *Pullaria*[24] have been found capable of fixation by measuring growth increase on a nitrogen-free medium. *Nocardia cellulans* fixed as much as 12 mg. N/g. of decomposed cellulose, a value comparable with the best nitrogen fixer, *Azotobacter vinelandii* (15 to 20 mg. N/g. carbohydrate). Finally a new nitrogen-fixing microorganism producing a red pigment was reported.[25] This organism required the presence of some fixed nitrogen and therefore should be checked using N_2^{15}.

Symbiotic nitrogen fixation, which in the not too distant past was thought to be confined to the nodules of leguminous plants, has now been extended to nonlegumes. Gardner[26] found nitrogen fixation in *Elaeagnus* root nodules, and Bond and Gardner[27] in several other nonlegume root nodules. One recent report[28] presented evidence for nitrogen fixation in several grasses in the absence of nitrogen-fixing bacteria and nodules.

One of the most significant findings in the research on symbiotic nitrogen fixation was the recent report of Raggio *et al.*[29] that nodules formed on isolated bean roots fixed nitrogen. These workers grew roots of *Phaseolus vulgaris* L. in Petri dishes containing sand moistened with a medium containing salts, vitamins, nitrate, and sucrose. The roots were inoculated with rhizobia, and nodules formed. To determine if the nodules were effective, they were exposed to N_2^{15} and analyzed for N^{15} excess. Within 3 hr. fixation of 0.035 to 0.190 atom % excess N^{15} was obtained in the nodules whereas the roots themselves did not fix nitrogen. Since nitrogen fixation occurred in nodules on isolated roots with sucrose as energy source, a direct photosynthetic product is not needed for nitrogen fixation as previously suspected.[30]

C. Requirements for Fixation

1. Inorganic and Organic

The requirements for nitrogen fixation by whole cells for the most part have been determined by comparing the growth on ammonia with that on N_2 with and without the compound in question. For example, molybdenum is required in much higher concentrations when cells are growing on N_2 than when growing on ammonia.[31-35] Likewise, iron is required in extra amounts when cells of *Azotobacter vinelandii*,[34] *Clostridium pasteurianum*,[36] *Aerobacter aerogenes*,[35] and various Gram-negative bacteria[37] are fixing nitrogen. These studies are interpreted as indicative of a role of molybdenum and iron in the nitrogen fixation process. The functions could be in nitrogen adsorption, in the metabolic steps which transfer electrons for N_2 reduction, or even in associated carbohydrate metabolism. The danger of concluding too much from whole cell studies (growth, etc.) is illustrated by recent studies of a calcium requirement for *Azotobacter vinelandii*. By techniques similar to those above, calcium was shown to be required for nitrogen fixation[34]; later[38] the calcium requirement was found not to be specific for fixation for the requirement could be decreased by adding acetate.[39] (The reason for the calcium-sparing action of acetate is unknown.) Knowledge of the exact role of these metals in nitrogen fixation probably will have to come from studies with cell-free enzymes, perhaps only with highly purified components.

One technique employed to uncover further requirements for nitrogen fixation was a search for compounds inhibitory to nitrogen fixation but not to ammonia utilization.[40] If the inhibition of nitrogen fixation could be reversed by the addition of an extract of cells of a nitrogen-fixing organism, one might use this as an assay for the isolation of a necessary component of the nitrogen-fixing system. For example, at a given concentration, trichloromethylsulfenyl benzoate decreased the growth rate of *Clostridium pasteurianum* much more when growing on N_2 than when growing on ammonia. Subsequently, sodium molybdate was found to reverse this inhibition and to stimulate nitrogen fixation. Hydrogenase, an enzyme thought to be associated with nitrogen fixation, was also strongly inhibited by this compound; and its inhibition also was reversed by sodium molybdate. Thus, one could tentatively conclude that molybdenum is required for nitrogen fixation via a role in hydrogenase action. Since molybdenum and possibly hydrogenase[41] are required for nitrogen fixation as shown by other experiments,[31-35] the inhibition technique appears to be valid for finding new requirements. Inhibitors of nitrogen fixation also were found which could be counteracted by biotin. Inhibitors which are counteracted by other components of the nitrogen-fixing system are not presently known.

2. Culture

For growth with N_2 as nitrogen source, organisms require a medium with an *oxidizable substrate*, e.g., sucrose for the heterotrophs, a *source of carbon*, such as sucrose for the heterotrophic organisms and carbon dioxide for the photoautotrophs, the *minerals* Fe, Mo, Mn, Mg, Ca, sulfate, phosphate, and possibly other inorganic and organic compounds in trace amounts, i.e., Co, Zn, Cu, or biotin. Other inorganic components generally added to the medium of most nitrogen-fixing organisms, such as Na, K, and Cl, are assumed to be necessary on the basis of studies with other biological agents, but no study of their role has been made. Most of these components are required whether the cells are growing on ammonia or N_2 although, as already discussed, lesser amounts of some are required for growth on ammonia. In addition, anaerobic conditions are required for the growth of the photosynthetic organisms, the clostridia, *B. polymyxa*, *A. aerogenes*, and *Achromobacter* on N_2.

As mentioned earlier,[21] organisms have been discovered which do not fix nitrogen well under conditions that favor optimal growth on NH_3 or organic nitrogen compounds. A strain of *Bacillus polymyxa* and *Achromobacter* also fixes nitrogen best under anaerobic conditions; the fixation in the presence of 10% oxygen is only 0.3% of the anaerobic rate.[42] The growth rate on ammonia was approximately the same for both aerobic and anaerobic growth. With nitrate as the sole nitrogen source, growth was extremely slow unless oxygen was supplied; no explanation of this unusual result can be presented on the basis of present data.

3. Significance of Recent Findings

The recent discovery of additional nitrogen-fixing organisms and the restricted cultural conditions necessary for fixation by some of them leaves one quite certain that the list of nitrogen fixers will be further extended. One recent lesson is the apparent lack of nitrogen fixation by some facultative organisms if oxygen is present. Thus, though anaerobic growth seems to be more restrictive, one should test new organisms for nitrogen-fixing potential at low O_2 tension. It also seems likely that some organisms may possess certain parts of the nitrogen-fixing apparatus; and, if supplied with the remainder, they too would fix nitrogen. For example, one can visualize such a situation with the rhizobia. Thus, symbioses with systems other than plants could well exist, for example, with other microorganisms.

Previous calculations of the significance of nitrogen added to the soil by nitrogen-fixing microorganisms will require reevaluation in view of the discovery of nitrogen fixation by so many new organisms, and the recent addition of several symbiotic nitrogen-fixing systems besides the legume family.

D. Soluble Nitrogen-Fixing Systems

Many attempts have been made to isolate a cell-free preparation that would fix nitrogen; of the few that were successful, none were reproducible.[43-45] Recently, the first reproducible methods for obtaining cell-free nitrogen fixation have been reported.[16, 17] The results of these experiments were confirmed,[46] and one of the methods was applied successfully to *Rhodospirillum rubrum*. In addition, cell-free preparations of algae (in particular, *Mastigocladus laminoseus*) were obtained which, although fixing only small quantities of nitrogen compared to *Clostridium* and *Rhodospirillum*, did fix consistently. Nitrogen fixation with cell-free preparations of *Azotobacter vinelandii*,[47,48] *Chromatium*,[49] and *Bacillus polymyxa*[50] also have been reported recently. In this section, the methods used to prepare the active extracts, the conditions required to obtain fixation, and the properties of the enzyme system(s) will be discussed.

1. Method of Solubilization

Two methods have been used successfully to obtain cell-free preparations of *Clostridium pasteurianum*[16, 17]—the Hughes press[51] and vacuum drying plus autolysis.[16, 17, 52] For the Hughes press procedure, the assembled press was cooled to −35°C. and a suspension containing the cells from 500 ml. of a culture of *C. pasteurianum* in the late log phase (10^9 cells/ml.) in 8 ml. was poured into the chamber and broken. The crushed cells were removed from the press, thawed, and centrifuged, and the supernatant solution used directly as the nitrogen-fixing preparation.

The second method, autolysis of dried cells, was similar to a procedure used previously to extract other enzymes from this organism.[52] The cells were dried rapidly under anaerobic conditions *in vacuo*. The dried cells were suspended in phosphate buffer (0.05 *M*, pH 6.8) at a ratio of 1 g. of cells per 10 ml. of buffer, evacuated, put under H_2 to produce anaerobic conditions, rotated gently for 1 hr. at 30°C., and then centrifuged. The supernatant solution contained the solubilized nitrogen-fixing enzyme system.

Other techniques, including grinding with alumina an l ultrasonic treatment, have yielded other enzymes from clostridia and some nitrogenase (nitrogen-fixing activity), but in far smaller yield than by the autolysis of dried cells. The dried cell procedure affords the additional advantage; the cells can be stored for months without loss of activity.

The methods used successfully for preparing active extracts of *Azotobacter vinelandii* were cell rupture with lysozyme at pH 8.0, followed by adjustment of the pH to 7.2 and treatment with an ultrasonic probe (Mullard). The disrupted cell preparations were then centrifuged at 25,000 *g* for 30 min. and the supernatant solution collected. When centrifuged at higher speeds, significant but decreased activity was obtained.[47, 48]

2. Requirements, Properties, and Measurement of Fixation

Cell-free preparations of *C. pasteurianum*, obtained by Hughes press or autolysis, fix nitrogen (N_2) actively only when sodium pyruvate is added. At present only potassium α-ketobutyrate has been found to replace pyruvate but, on a molar basis, the total fixation with α-ketobutyrate is only 50% of that with pyruvate.[17] The data in Table II are typical for fixation by a cell-free preparation, supplied with pyruvate or α-ketobutyrate, and 60 atom % N_2^{15}.

The fixation of nitrogen appears today as a complex process. For example, since a product (or products) of pyruvate metabolism is necessary for nitrogen fixation, many of the enzymes and coenzymes required for the operation of the "clostridial pyruvate clastic" system appear as necessary for nitrogen to be fixed. The cofactors so far found to be indispensable for this clastic reaction are Fe^{++}, Mg^{++} (or Mn^{++}), thiamine pyrophosphate, orthophosphate, and coenzyme A. It has also been reported that flavin adenine dinucleotide, factor B (a vitamin B_{12} derivative), and possibly biotin may be necessary. Of these, only CoA has been shown to increase the rate of nitrogen fixation consistently, whereas the others are apparently present in nonlimiting amounts, and their addition either has no effect or depresses fixation slightly, especially at high concentrations. This could be a result of favoring side reactions not coupled to N_2 fixation or to competition for sites. Other components required for nitrogen fixation by whole cells, such as iron, molybdenum, and calcium, also have no effect when added to cell-free preparations. Many of these components probably are

TABLE II

ORGANIC SUBSTRATE REQUIREMENT FOR NITROGEN FIXATION BY *Clostridium pasteurianum* EXTRACTS

Organic substrate[a]	N fixed[b] (atom % excess N^{15}) Exp. 1	Exp. 2
Sodium pyruvate	0.64	0.39
Potassium α-ketobutyrate	0.25	0.13
None	0.00	0.00

[a] The main compartment of each 25-ml. reaction flask (with a side arm) contained 400 μmoles potassium phosphate buffer, pH 6.5; 15 mg. protein nitrogen as cell extract; and 7.5 ml. H_2O. The substrate (0.5 ml.), 100 mg. sodium or potassium salt, was in the side arm. The gas phase was 0.6 atm. of 61.5 atom % N_2^{15}. After equilibration, the substrate was tipped and the flask incubated at 30°C. for 60 min. on a Brunswick rotary shaker.

[b] The N^{15} content was determined on the whole hydrolyzate.[20]

present in the preparations in optimal amounts so that further additions would not be expected to exert an effect. For example, even after dialysis against versene, Mo, Fe, and Ca (as oxides by ashing) were found associated with the cell proteins as measured by spectrographic analysis.

If the protein concentration of a cell-free preparation in a nitrogen-fixing reaction system was less than 0.2%, little, if any, fixation occurred. The effect of varying the protein N content on nitrogen fixation is shown in Fig. 2. In this crude system, the optimum rate of fixation was found with a ratio of pyruvate to protein of 5:3, with the protein content at about 0.6%. With higher protein concentrations, more pyruvate is required to maintain fixation at the maximum rate for 60 min.; as a result, the atom % excess of fixed nitrogen was less at the higher protein concentrations. The standard assay system used for these nitrogen fixation experiments consisted of approximately 10 mg. of protein N (60 mg. protein as enzyme preparation) and 100 mg. of sodium pyruvate in 8 ml. of 0.02 *M* potassium phosphate buffer, pH 6.4 to 6.5. The flask containing these ingredients was freed completely of oxygen by evacuation and N_2^{15} (60 atom %) added to 0.5 atm. pressure. The reaction time was standardized at 1 hr. After incubation, the nitrogen in the total reaction was analyzed for N^{15} excess.[20]

Recently, a new method for measuring nitrogen fixation in extracts by direct analysis of ammonia formed has been reported by Mortenson.[53] This procedure, sensitive to 3 to 5 μg. N fixed, is simple, rapid, and much less laborious than the N^{15} method. A much more sensitive technique also has

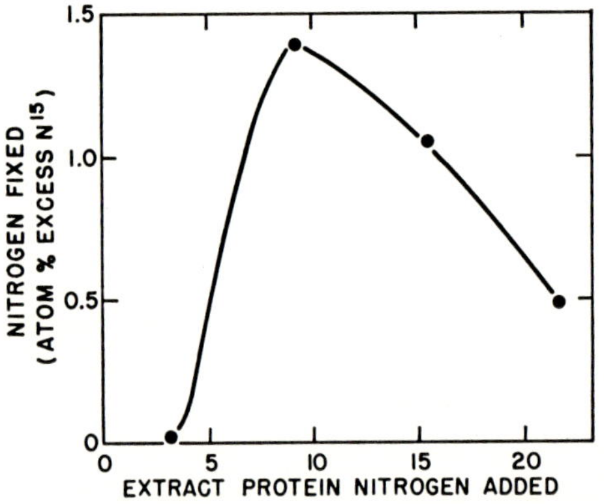

FIG. 2. Effect of extract concentration on nitrogen fixation. Reaction mixture in a 25-ml. flask consisted of: 400 μmoles potassium phosphate, pH 6.5; Extract (Hughes Press supernatant centrifuged at 144,000 g) as indicated; 100 mg. sodium pyruvate (0.5 ml.) in side arm; H_2O to 8 ml. total volume. The gas phase was 0.6 atm. of 61.5 atom % N_2^{15}. After equilibration, sodium pyruvate was tipped in from side arm. Incubation was at 30°C. for 60 min. Fixed nitrogen was determined on whole digest.[20]

been developed recently by Nicholas *et al.*[54] using N^{13}, but special equipment and handling procedures are necessary.

The pH for maximum nitrogen fixation rate with these clostridial extracts is between 6.30 and 6.60.[17] Very little fixation occurred below pH 6.3; above pH 6.6, the rate decreased, but even at 6.8 it was still about 50% of maximum. A preparation stored at a pH below 6.3 still fixed nitrogen actively when readjusted to the optimum range; but storage below pH 6.0 for 8 to 24 hr. destroyed the fixation activity completely. It must be remembered also that this is the optimum pH of the whole nitrogen-fixing system (N_2 to NH_3), which includes not only "nitrogenase", the nitrogen-activating system, but also the reducing system—hydrogen-donating enzymes. Thus, the pH for optimum nitrogen fixation would be influenced most by the pH optimum of the first limiting reaction in the over-all system.

Prolonged exposure of the nitrogen-fixing preparations to air (oxygen) resulted in inactivation. Brief exposure (5 to 10 min.) to oxygen was not detrimental provided the temperature was kept below 5°C. The oxygen sensitivity of the nitrogen-fixing activity closely parallels that of hydrogenase,[52, 55] an enzyme thought to be involved in the nitrogen-fixing process.[15] Both hydrogenase and nitrogen-fixing activity can be partially protected from oxygen inactivation by sulfhydryl compounds.

The nitrogen-fixing system of *C. pasteurianum* is soluble, as shown by the retention of the activity in the supernatant solution after centrifugation for 4 hr. at 144×10^3 *g*. This treatment rules out the possibility of whole cells or large subcellular particles being responsible for fixation activity.[17] The nitrogen-fixing activity can be recovered in the pellet by centrifugation at 144×10^3 *g* for 16 to 24 hr.[17a] Many of the other proteins in the crude solution also are removed under these conditions, as witnessed by the absence of the pyruvate-metabolizing system in the supernatant solution.

3. Kinetics of Fixation

The fixation of nitrogen by cell-free preparations of *Clostridium pasteurianum* is coupled to the metabolism of pyruvate.[17, 18] The amount of pyruvate required for optimal activity for 1 hr. is approximately 0.1 *M* for a preparation 0.6% with respect to protein. No fixation occurred in the absence of pyruvate; and, at concentrations of 0.2 *M* or higher, nitrogen fixation occurred at a greatly reduced rate.

Pyruvate is actively metabolized by cell-free preparations of *C. pasteurianum* by the "phosphoroclastic" system[52, 55] disclosed by studies with other clostridial extracts.[55-62c] A possible mechanism based on these studies is:

$$CH_3\overset{O}{\overset{\|}{C}}-\overset{O}{\overset{\|}{C}}-OH \xrightarrow[]{\text{thiamine pyrophosphate}} [\alpha\text{-lactyl}\cdot\text{TPP}]$$

$$\xrightarrow[\text{biotin?}]{\text{CoA?}} [\alpha\text{-hydroxyethyl}\cdot\text{TPP}] + CO_2$$

$$[\alpha\text{-hydroxyethyl}\cdot\text{TPP}] \xrightarrow[Fe^{++},X]{\text{CoA}} \text{acetyl} \sim \text{CoA} + \text{"reduced X"}$$

$$\text{acetyl} \sim \text{CoA} \xrightarrow[HPO_4^{--}]{\text{phosphotransacetylase}} \text{acetyl} \sim PO_4 + \text{CoA}$$

$$\text{acetyl} \sim PO_4 + \text{ADP} \xrightarrow{\text{acetokinase}} \text{acetate} + \text{ATP}$$

$$\text{"reduced X"} \xrightarrow{\text{hydrogenase}} H_2$$

Ferrous iron, CoA, and thiamine pyrophosphate were shown to be required for the *C. pasteurianum* system, and acetyl phosphate accumulated.[17, 63] High concentrations of the products, acetyl phosphate and ATP, for reasons presently unknown, are inhibitory to nitrogen fixation.[17] These preparations probably contain enzymes capable of further metabolizing acetylphosphate[17]; its level in normal preparations is below inhibitory concentrations.

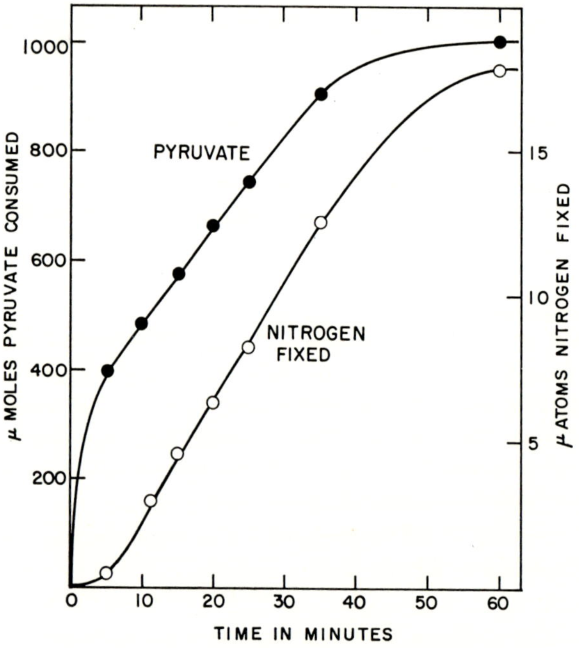

FIG. 3. Time course of pyruvate consumption and nitrogen fixation. For conditions, see Fig. 2. Protein N (14 mg.) as cell extract (dry cell extract) was used. Incubation was for periods indicated. Each flask was analyzed for excess N^{15} and residual pyruvate.

A time-course study showed that fixation did not begin until 2 to 4 min. after the addition of sodium pyruvate. Once fixation began, however, it proceeded linearly for 40 min. or more (Fig. 3). The metabolism of pyruvate, measured by its disappearance, was initially rapid—as much as 40% of the added pyruvate disappeared during the first 10 min.—and then the rate became steady at one-fifth the initial value. The initial rapid pyruvate metabolism that occurred before nitrogen fixation was detected, suggested that some product of pyruvate metabolism was essential for fixation or possibly that sites on the nitrogen-activating enzyme must be reduced or activated before nitrogen could be adsorbed.

Recent experiments[86] have shown that the initial lag in nitrogen fixation was a result of the high phosphate concentration (0.05 *M*). With lower phosphate (0.01 to 0.02 *M*) not only is nitrogen fixation linear but pyruvate utilization is also linear. Time course studies show that with high phosphate present an initial rapid acetyl phosphate synthesis resulted. As the rate of acetyl phosphate synthesis decreased, the rate of nitrogen fixation increased. This suggests that there is a control mechanism present that regulates the amount of pyruvate metabolism that is coupled to nitrogen fixation. The control mechanism could be the concentration of a precursor to acetyl CoA and "reduced X," α-hydroxyethyl·TPP (see page 131). Details of these results will be published soon.

Pyruvate must serve as the electron source for the reduction of nitrogen since no compound other than pyruvate has been added, and without pyruvate no fixation occurred. There is also evidence that a "reduced coenzyme" produced during pyruvate metabolism is rapidly reoxidized by hydrogenase with the formation of one mole of hydrogen for each mole of pyruvate metabolized.[63] If hydrogen evolution is measured under nitrogen-fixing conditions, less H_2 is evolved than under nonfixing conditions, i.e., under an atmosphere of argon.[86] This difference, however, is small when compared to the total H_2 evolved. The evidence indicates a competition between the nitrogen-fixing system and the hydrogen-evolving system for the reductant, with hydrogen evolution being greatly favored. Such large amounts of pyruvate could be required for nitrogen fixation in these preparations because the available pool of reduced coenzymes would be readily exhausted with lower concentrations. This also might explain why it has not been possible to obtain nitrogen fixation with hydrogen gas as an electron donor, i.e., by a reversal of the hydrogenase reaction. In this regard attempts were made without success to aid hydrogen reduction of nitrogen by adding artificial acceptors such as methyl viologen (E_0' at pH 7 = −0.46). Most of these acceptors inhibited nitrogen fixation.[86] Attempts also were made to obtain fixation with limiting amounts of pyruvate added together with hydrogen (0.7 atm.) and nitrogen (0.3 atm.), but no increase in nitrogen fixation above a control under 0.3 atm. N_2 and 0.7 atm. argon occurred. Thus, even though one might expect hydrogen to serve as the reductant for nitrogen since ample hydrogenase is present in these preparations and compounds such as methylene blue are readily reduced by these preparations on exposure to H_2, only a "coenzyme" reduced during pyruvate metabolism appears to function in the nitrogen-fixing system.

4. Components of the Clostridial System

Recently two fractions were obtained from the cell-free system of *C. pasteurianum*, both of which were required for nitrogen fixation.[64, 65] One of these, obtained by a selective heat denaturation (60° C. for 10 min.), contained the enzymes necessary for the conversion of pyruvate to acetyl $\sim PO_4$, CO_2, and H_2 and therefore was designated the *hydrogen-donating system*. The other fraction, obtained from the original preparation by protamine sulfate and calcium phosphate gel treatment, did not degrade pyruvate but contained a component tentatively assumed to be nitrogenase. Therefore, it was referred to as the *nitrogen-activating system*. Neither fraction fixed nitrogen with sodium pyruvate as reducing and energy source, but together the two fractions fixed nitrogen as well as, if not better than, the parent preparation (original cell extract).

Evidence for the "nitrogenase" activity in the protamine sulfate and calcium phosphate gel-treated fraction was obtained by fractionating cells grown with ammonium sulfate as nitrogen source. Cells grown on ammonia are known not to fix nitrogen presumably because ammonia prevents the synthesis of the "nitrogenase" enzyme. Thus, extracts from NH_3-grown cells fail to fix nitrogen but possess the system to convert pyruvate to CO_2, H_2, and acetyl $\sim PO_4$. A heat-treated preparation (60° C. for 10 min.) from NH_3-grown cells degraded pyruvate 50% as actively as the heated preparations from N_2-grown cells. Furthermore, when added to the "nitrogenase fraction" from N_2-grown cells, the heat-treated preparation (hydrogen-donating system) from NH_3-grown cells activated nitrogen fixation 50% as well as the hydrogen-donating system from N_2-grown cells. In contrast, the protamine-phosphate gel fraction from NH_3-grown cells was not activated to fix nitrogen by hydrogen-donating systems from either NH_3- or N_2-grown cells. Thus, the protamine-phosphate gel fraction from N_2-grown cells contains a component that is either not synthesized by cells grown on NH_3 (ammonium sulfate) or is in some way repressed.

Further characterization of the two fractions, "hydrogen-donating" and "nitrogen-activating" systems, was made on the basis of their individual reactions with carbon monoxide.[65, 66] When the "nitrogen-activating" system (protamine-phosphate gel fraction) was exposed for varying periods of time to CO and tested in the recombined nitrogen-fixing system, an irreversible inhibition of one of its components occurred to the extent of 50% within the first 15 min. of CO exposure. Further exposure to CO, for 16 hr., resulted in complete loss of activity. The hydrogen-donating system was also inhibited by CO, but the inhibition was reversible, full activity being restored after 15 min. in the presence of H_2, N_2, and pyruvate. Hydrogenase, a component of the hydrogen-donating system, is inhibited reversibly also by CO. Hydrogenase activity is expressed immediately on removal of CO and thus can be distinguished from the over-all hydrogen-donating activity by the lack of lag time in removal of inhibition.

Based on these data, the following scheme is postulated as a working hypothesis:

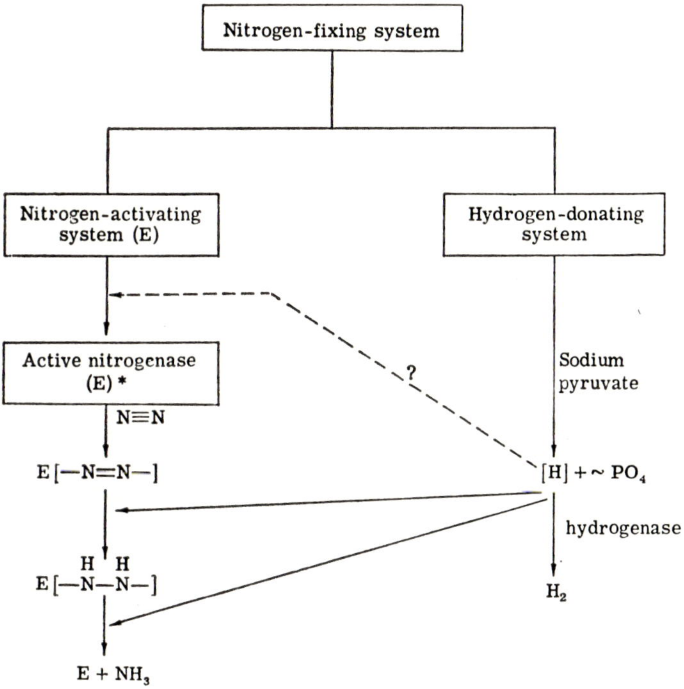

Whether nitrogenase (E) is reduced and then simultaneously adsorbs and reduces nitrogen, or whether nitrogen is adsorbed and then reduced remains to be determined. The initial lag in nitrogen fixation suggests the former. The need for high energy phosphate in the nitrogen-fixing reaction has been suggested on the basis of arsenate inhibition.[67] This point remains to be clarified, however, since the high-energy phosphate, acetyl $\sim PO_4$ and ATP, synthesized during the metabolism of pyruvate by these cell-free preparations, depending on concentration, is either inhibitory or has no effect on nitrogen fixation. For further discussion of postulated reaction steps in nitrogen fixation see Section II, G.

E. Reaction Steps

1. Evidence for Ammonia as the Product of Nitrogen Fixation

Attempts to approach the mechanism of biological nitrogen fixation by identifying the first "stable intermediate" from N_2 were greatly aided by the development of the N^{15} tracer technique.[20] By following the progress of nitrogen fixation in whole cells with this tracer the following general reactions were deduced to occur:

$$N_2^{15} \longrightarrow N^{15}H_3 \xrightarrow{\alpha\text{-ketoglutarate}} COOHCH_2CH_2CHN^{15}H_2COOH$$

The evidence can be enumerated as follows:

(*a*) Exposure of growing cultures (*Azotobacter vinelandii*,[68] *Clostridium pasteurianum*,[69] photosynthetic bacteria,[70] algae,[71] or soybean nodules[72]) for a short time to N_2^{15} or $N^{15}H_4^+$ resulted in the highest specific activity of N^{15} in glutamic acid. The concentration of N^{15} in glutamic acid was usually two or more times that of the next highest compound. Since both N^{15} or $N^{15}H_4^+$ appeared in the same compounds and in the same proportions, it was concluded that N_2 is converted first to ammonia, then to glutamic acid.

(*b*) A culture of *Clostridium pasteurianum* grown under N_2 with little agitation and then exposed with mixing to N_2^{15} excretes a considerable amount of N^{15}-labeled nitrogen,[73] presumably because the cell is unable to provide acceptors for the increase in fixed nitrogen.[52] Ammonia was one of the compounds excreted; it contained about one-half the N^{15} excess of the N_2^{15} added originally. N^{15}-labeled glutamine and asparagine were also isolated and found to contain much less label than the ammonia. Hydroxylamine and other postulated partial reduction products of nitrogen fixation were not found.

(*c*) Addition of ammonia to cultures actively fixing N_2^{15} results in its immediate and preferential use. The two sources of nitrogen can be used simultaneously by *C. pasteurianum*[74] and *A. vinelandii*[75] but ammonia supplants most of the utilization of N_2. This behavior would be expected (but not necessary) if ammonia were an intermediate in nitrogen fixation.[15]

(*d*) The first demonstrable product from short-time exposure of growing cultures of *A. vinelandii* to N_2^{15} was ammonia and amides.[76] The latter equilibrate rapidly with ammonia; on longer exposure the ammonia and amide N^{15} decreased and the α-amino N^{15} of glutamic acid increased. Thus, the label must proceed from N_2 to ammonia, then to glutamic acid.

(*e*) Exposure of "resting" cells of *A. vinelandii* or *C. pasteurianum* to N_2^{15} in the presence of an $N^{14}H_4^+$ pool results in labeling of the ammonia pool with N^{15} before any cellular material is labeled.[74, 75]

With a cell-free preparation of *C. pasteurianum*, N^{15} was found first in the alkali-distillable nitrogen compounds (ammonia and amide) and only after some time in the non-distillable nitrogen-containing compounds, such as amino acids, purines, or pyrimidines.[17, 18] For example, after incubation of 10 to 12 min., all the N_2^{15} fixed was in the ammonia-amide fraction (Fig 4); by 30 minutes' incubation the total amount of N^{15} fixed increased, but the amount of label present in the ammonia-amide fraction had decreased to only 4% of the total N^{15} fixed. If a pool of ammonia or glutamine was added to the reaction mixture before the fixation occurred to trap any am-

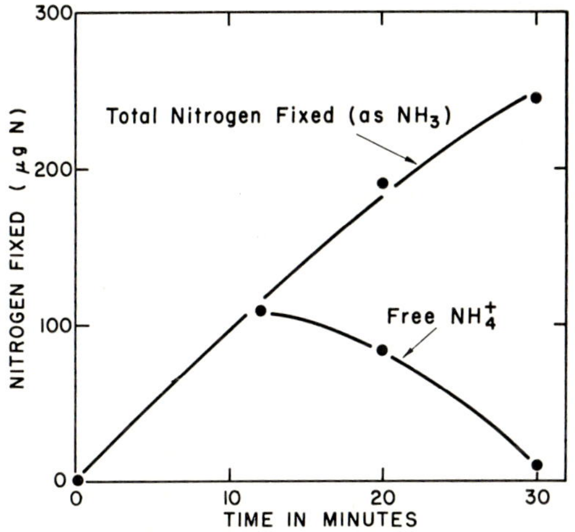

FIG. 4. Form of nitrogen fixed by *Clostridium pasteurianum* extracts as a function of time. For conditions, see Fig. 2. Protein N (14 mg.) as cell extract (Hughes Press) was used. Contents of flasks were analyzed for total nitrogen fixed and nitrogen fixed as ammonia.

TABLE III

NH_3 AND AMIDE RETENTION OF N_2^{15} FIXED[17]

Additions[a]	Amount added (μmoles/ml.)	N_2^{15} fixed (μg. $N^{15}H_3$-N per 60 min.) Total	NH₃
None	—	244	10
NH_4^+	15	214	180
Glutamine	35	240	250
None	—	157	2
NH_4^+	35	78	51
Glutamine	35	138	118

[a] For other contents of reaction mixture, see Table II. Sodium pyruvate was the organic substrate.

monia or amide that nitrogen fixation produced, most of the N_2^{15} fixed was in the alkali-distillable fraction even after 60 min. (Table III), thus indicating the product of fixation is ammonia, an amide, or both. The addition of potassium α-ketoglutarate to the cell-free preparations greatly decreased the amount of N^{15} found in the alkali-distillable nitrogen pool, probably by formation of glutamic acid by addition of ammonia and reduction by glutamic acid dehydrogenase. To distinguish between ammonia and amide as the product of nitrogen fixation, the ammonia produced during fixation was selectively distilled from possible amides and the amount of N^{15} compared with the N^{15} of the total alkaline distillate and the total N^{15} determined after Kjeldahl hydrolysis.[20] The results (Table IV) show 96% of the N_2^{15} fixed to be ammonia.

In some experiments, when no ammonia pool was added to trap the fixed nitrogen, the N^{15} was incorporated into nondistillable compounds.[17, 18] Evidence from whole cell experiments with N^{15} showed that glutamic acid

TABLE IV

COMPOUNDS INTO WHICH NITROGEN IS FIXED BY *Clostridium pasteurianum* EXTRACTS

Form	Nitrogen present[a] (μg. as NH_3)		
	Zero time	60 Min.	Fixed
NH_3[b]	22	258	236
Amide[c]	437	442	5
Nonalkali labile[d]	—	—	6
Total[d]	—	247	247

[a] See footnote of table II for reaction mixture and conditions. Organic substrate was 100 mg. sodium pyruvate; 16.8 mg. protein N as cell extract was used.

[b] Distilled from K_2CO_3 by micro-Conway technique.[53]

[c] Distilled from NaOH by micro-Conway technique.[53]

[d] Total nitrogen fixed determined from N^{15}/N^{14} ratio and total nitrogen present. Total-(NH_3 + amide) = nonalkali labile

became labeled after ammonia. Similarly, with a cell-free preparation glutamic acid contained the highest content of N^{15} of the organic compounds present and was most abundant. For example, the reaction mixture from cell-free fixation in the presence of 60 atom % N_2^{15} contained glutamic acid at 21 atom % excess.[17, 18] The next highest N_2^{15} content (5.5 atom % excess) was found in an unidentified amino acid fraction. In conclusion, it appears that cell-free preparations, as well as whole cells, fix nitrogen rapidly to ammonia; the ammonia is next incorporated into glutamic acid.

2. INTERMEDIATES BETWEEN N_2 AND AMMONIA

Most workers in the field of nitrogen fixation consider ammonia to be the final inorganic product of nitrogen fixation. It is, however, assumed that several intermediates between nitrogen and ammonia must exist, but no experimental evidence of their structure is available at the present time. Hydrazine and hydroxylamine, both strongly suspected intermediates because of their reduction states, are inhibitory, even at low concentrations, to both cell growth and to nitrogen fixation by cell-free preparations; thus their role remains unclear.

The reduction of N_2 to NH_3 requires a valence change in each nitrogen atom from 0 to —3; therefore, six electrons are required to reduce one mole of N_2 to two of NH_3. The conversion could take place in six 1-electron transfer steps, three 2-electron transfer steps, or by any combination of these. Electron transfer above 2 per step has been suggested in the electron transfer oxidases[76a] and could function here as well. Any information obtained on the mechanism of electron transfer to nitrogen would obviously aid in the prediction of intermediates, reduction states, and the clarification of the mechanism of nitrogen fixation.

We shall tabulate the approaches made toward identification of nitrogen fixation intermediates and the information gained from these studies.

a. Inhibition Studies. Nitrous oxide was found by Molnar *et al.*[77] to be a specific inhibitor for nitrogen fixation. Since one proposed pathway of nitrogen fixation is via oxidation, Repaske and Wilson[78] considered in more detail the effect of nitrous oxide on fixation. They confirmed the observation of a competitive inhibitor effect on nitrogen fixation and reported further that ammonia utilization was not affected. Thus, they concluded that the inhibition occurred in one of the reaction steps between N_2 and ammonia. Wilson and Roberts[79] made similar observations. Mozen and Burris[80] showed that $N_2^{15}O$ is utilized slowly by cultures of *Azotobacter vinelandii* and, as expected, this utilization of N_2O is inhibited by N_2.[81, 82] This evidence could suggest the possibility of N_2O as an intermediate in nitrogen fixation were it not for the rate, which is too slow to be compatible with the over-all reaction rate. One could explain the slow utilization of N_2O by assuming it is first reduced to N_2. It would thus be interesting to see if any N_2^{15} is produced from $N_2^{15}O$. This could be done by trapping the N_2^{15} in a pool of N_2^{14} supplied to cells at a concentration which does not prevent fixation of $N_2^{15}O$. If N_2^{15} did not appear in the N_2^{14} pool, one could conclude either that N_2O fixation did not proceed through N_2 or that the N_2 produced from N_2O was further converted to ammonia without release from the enzyme surface. Evidence for the conversion of N_2O to N_2 has been reported in recent studies with soybean root nodules.[5]

After ammonia was found to inhibit nitrogen fixation by cell-free preparations, it was hoped that added ammonia might inhibit only the conversions of precursors to ammonia and not the initial steps in fixation. If so, addition of ammonia to a nitrogen-fixing reaction would allow intermediates to accumulate. No indication of intermediates such as hydrazine, hydroxylamine, oximes, or amines has so far been obtained either by this technique or by direct analysis of noninhibited reaction mixtures.

b. Utilization of Possible Intermediates. Hydroxylamine has been suggested as an intermediate in nitrogen fixation for two reasons: (*1*) its oxidation-reduction state is between that of N_2 and ammonia; (*2*) root nodules fixing N_2 excrete both hydroxylamine and its derivatives, the oximes.[14] If hydroxylamine is a fixation intermediate, it should precede ammonia (follow N_2) and therefore it should support cell growth in preference to N_2 and its use be supplanted by ammonia. Unfortunately, hydroxylamine is so toxic to nitrogen-fixing cells that concentrations above 3 to 5 p.p.m. inhibit cell growth either on ammonia or N_2. Even at nontoxic concentration, neither *Azotobacter vinelandii* nor *Clostridium pasteurianum* cells have been shown to use the compound.[83, 84] Derivatives of hydroxylamine, the oximes of pyruvic, oxalacetic acid, and α-ketoglutaric acids have also been added as nitrogen sources to cultures of these organisms but, like hydroxylamine and hydrazine, little, if any, utilization occurred.[83, 84]

From the evidence now at hand, one is tempted to conclude that neither hydroxylamine nor its oximes are intermediates in nitrogen fixation. Two possible deterrents to this conclusion are that the cells might not be readily permeable to these compounds, and the recent finding of hydroxylamine reductase in *Azotobacter vinelandii*[85] and *Clostridium pasteurianum*.[86] The latter strongly suggests that, if hydroxylamine entered the cell, it should be converted readily to ammonia.

Another reason the nonutilization of these compounds by cells does not eliminate them as intermediates is that they may require activation for attachment to an enzyme site. Further, a relatively high concentration might be required for attachment to the active site, and this concentration may inhibit some other step in the sequence (e.g., hydroxylamine inhibits the clostridial pyruvate clastic system[86]).

The situation pertaining to hydroxylamine is also true of hydrazine. This compound, too, inhibits the growth of nitrogen-fixing cells at levels of 3 to 5 p.p.m., and all attempts to grow cells with hydrazine as the sole nitrogen source have failed.[84] In one experiment N^{15}-labeled hydrazine was fed to heavy suspensions of washed cells of *Azotobacter vinelandii* and N^{15} incorporation into organic azines was found, probably by a nonenzymic formation of 3,4-dihydropyridazinone-5-carboxylic acid from hydrazine and α-ketoglutarate.[87] Some N^{15}-labeled azine was apparently isolated from *Azotobacter* cells and from nodules of field-grown soybeans using N_2^{15} as nitrogen source. Fractions identified as azines were isolated by chromatography and found to contain higher relative and absolute atom % N^{15} excess than isolated ammonia. A recent report of the uptake of hydrogen when hydrazine was incubated with H_2 and the hydrogenase of whole cells of *Micrococcus lactilyticus* lends some support to the possibility of hydrazine as a fixation intermediate.[7] If hydrazine were an intermediate, the enzyme, hydrazine reductase, would be required to complete the conversion of N_2 to ammonia just as would the hydroxylamine reductase described above.

Nitramide and hyponitrite, possible intermediates in an oxidative pathway of fixation, also have been added to cells as possible replacement for N_2.[81, 82] No utilization of either of these compounds was shown. These compounds, however, are extremely unstable in aqueous solutions; for example, nitramide decomposed completely to N_2O and water in 5 min. (for a more detailed discussion of these compounds, see Roberts[88] and Delwiche[89]).

c. Mutant Studies. If a mutant of one of the nitrogen-fixing microorganisms could be found that was impaired in the later steps of its nitrogen-fixing pathway so that the cells would grow on ammonia but not on N_2, an intermediate between N_2 and ammonia might accumulate because of the blocked pathway. A mutant of *Azotobacter vinelandii*, produced through UV irradiation, was found to fix nitrogen only when grown with either pyruvate or lactate as carbon source. This mutant would grow using glucose or sucrose as carbon and energy sources only when ammonia was supplied.[90] The blocked reaction, or missing step, appeared to be in the conversion of D-glyceraldehyde-3-phosphate to pyruvate. Since the cells grew well on ammonia with glucose or sucrose as carbon source, the failure to grow on N_2 could not be a lack of ammonia acceptors but could be a result of the lack of sufficient concentration of pyruvate. This reminds one of the high pyruvate concentration required for nitrogen fixation by cell-free preparations of *C. pasteurianum*.[17]

The blocked reaction step apparently was not complete for some fixation occurred, although not of sufficient amount to maintain growth. No intermediates of nitrogen fixation were found when this mutant was exposed to N_2^{15} under conditions where N^{15} was incorporated but no growth occurred.

d. Spectral Differences on Contact of Cell-Free Preparations with N_2. Until recently only one method was available for measuring nitrogen fixation, i.e., analysis for an increase in total fixed nitrogen. This technique requires the conversion of cellular material to ammonia by the Kjeldahl method followed by distillation and titration. If N^{15} incorporation is to be measured, further laborious steps must be performed.[20] Recently, a simple and rapid distillation-titration technique was developed to measure ammonia, the product of nitrogen fixation by cell-free preparations.[53] This technique eliminated many of the objections of the earlier methods and allowed the direct analysis of a product of nitrogen fixation. In order to study the mechanism of nitrogen fixation, however, a technique is needed to measure the primary reaction of N_2.

The report of a spectrophotometric measurement of nitrogen adsorption and/or reduction in the absence of total fixation to ammonia appeared to offer a new approach to the study of fixation, as well as to information concerning intermediates.

In 1956, Shug *et al.* reported optical density changes when crude cell-free *C. pasteurianum* preparations were exposed first to hydrogen and then to nitrogen. The maxima, which were 450, 425, 405, and 390 mμ,[91] led them to suggest that nitrogen had caused the oxidation of a flavin, previously shown to be reversibly reduced by molecular hydrogen.[41] The maxima at 405 and 425 mμ appeared to be specific for nitrogen whereas those at 450 and 390 mμ occurred on exposure to NO as well as nitrogen. Exposure of similar preparations to O_2 was accompanied by formation of maxima at 450 and 418 mμ. Since the maxima with O_2 differed from those under N_2, the possibility of oxygen-contaminated nitrogen was eliminated.

Preparations of *Azotobacter vinelandii* and of Lincoln variety soybean nodules were assayed for these spectral shifts with N_2.[92] With *Azotobacter* preparations, N_2 appeared to oxidize the flavins and the cytochrome b, i.e., optical density increases occurred at about 430, 530, and 555 mμ. With nodule extracts, N_2 appeared to oxidize hemoglobin (legume) to hemiglobin, and H_2 reversed the process, i.e., reduced hemiglobin to hemoglobin.

In these experiments the question of the possible presence of O_2, as a contaminant of the gases or from small air leaks, was not ruled out for the same spectral shifts occurred by oxidation with O_2. Bergersen and Wilson[93] showed that the hemoglobin of nodule preparations exposed 10 min. to 8 p.p.m. O_2, underwent oxygenation but not oxidation whereas, during the same period, exposure to N_2 containing less than 8 p.p.m. O_2, caused the oxidation of hemoglobin to hemiglobin.

Nitrogen adsorption by leghemoglobin is further substantiated by recent work[94] which showed that nitrogen is adsorbed preferentially on hemoglobin and leghemoglobin; the experiments were performed by adsorbing nitrogen on hemoglobin and then measuring the amount that could be desorbed.

Cell-free preparations of *Clostridium pasteurianum* fix nitrogen only when supplied with pyruvate or α-ketobutyrate. Consequently, the preparations were first incubated with sodium pyruvate under H_2 and then exposed to N_2, O_2, or CO to see what spectra would appear. On exposure to N_2, a sharp increase in O.D. at 365 mμ resulted[17, 19] (Fig. 5). If H_2 were removed from a cuvette containing a similar reaction mixture and either O_2 or argon admitted, a decrease in optical density at 365 mμ occurred. These results are consistent with the interpretation that H_2 is adsorbed on the same site as N_2, but less strongly; when desorbed (pumped off) a negative adsorption at 365 mμ results. If the preparation was allowed to age 16 hr. before the experiment was repeated, the optical density change on exposure to O_2 was identical to the change on exposure to N_2, and the difference spectrum under argon was equivalent to that under H_2. Apparently, the protein component that adsorbed N_2 changed on aging so that oxygen could attack a site previously inaccessible and hydrogen no longer could be desorbed.

The spectral difference at 365 mμ could result from the oxidation of a

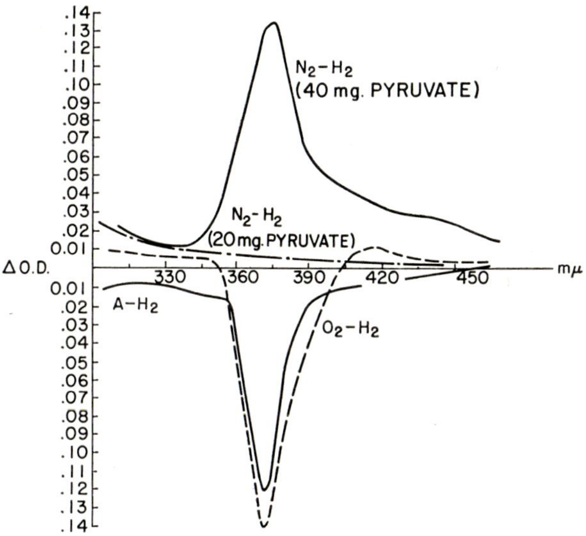

FIG. 5. Difference spectra of pyruvate-conditioned extracts exposed to N_2, A. and O_2 relative to H_2. The reactions were run in specially designed 25-ml. flasks with 1-cm. cuvettes attached.[17] The reaction mixture for each curve consisted of 1.1 mg. protein N as cell-free extract, 150 μmoles potassium phosphate at pH 6.8, and sodium pyruvate (as indicated) in a total volume of 3 ml. The gas phase during the 15-min. incubation with pyruvate at 25°C. was H_2. After incubation, the H_2 was removed, the designated gases introduced, and the spectra determined in a water-cooled Beckman DU spectrophotometer against a control under H_2. Reproduced from Carnahan *et al.*, *Biochim. et Biophys. Acta* **44,** 520 (1960).

reduced coenzyme by the N_2 adsorbed on nitrogenase. An alternative explanation is that the adsorption of N_2 on a protein-metal site caused perturbation of a metal atom that resulted in an increased adsorption. Proof of such conclusions may well require a physical and chemical study of the pure components.

3. SUMMARY

At the present time the identity of the intermediates in the pathway of nitrogen fixation remains unknown. It is doubtful that the oximes of pyruvic, α-ketoglutaric, or oxalacetic acid are intermediates, although it is possible that an oxime derivative of these keto acids could be present as an intermediate attached to the enzyme. As such, the oxime might not dissociate and the nitrogen could be reduced *in situ* to an amino group which then hydrolzed to free amino. Hydroxylamine and hydrazine are not utilized by cells of nitrogen-fixing cultures nor are they found as intermediates in the cell-free fixation of nitrogen by *C. pasteurianum* preparations (the assays used in the latter studies were capable of detecting these compounds in concentrations of 0.1 μg./ml.). Therefore, if these compounds are intermediates, they probably are never free in solution. This would mean that nitrogen once adsorbed never leaves the enzyme surface until it has been reduced to ammonia. Such a conclusion, if true, can only be substantiated by further experimentation, perhaps only by experiments with the isolated enzymes of the nitrogen-fixing system.

F. Role of Hydrogenase

When hydrogen was found to act as a competitive inhibitor of nitrogen fixation by the symbiotic system of red clover,[95] a general role was suggested for hydrogenase in the scheme of nitrogen fixation. This possibility was strengthened when *Azotobacter* was found to possess a powerful hydrogenase,[96] the concentration of which was greatest in cells actively fixing nitrogen.[97] In a more recent study[98] the levels of hydrogenase measurable in *Azotobacter* were found to be reduced 67 to 94% when cells of this organism were grown on fixed nitrogen such as NH_4Cl or KNO_3 instead of N_2. It appears that either a low level of ammonia or the presence of N_2 stimulates the production of hydrogenase, as is also true of a component of the "nitrogenase" system (see Section II, D, 4).

To explore further the connection between hydrogenase and nitrogen fixation, *Azotobacter* mutants that failed to fix molecular nitrogen were selected and examined for hydrogenase activity.[99] If there were a connection between hydrogenase and nitrogenase, one might expect an absence or diminished level of hydrogenase in nitrogen-fixing mutants. The mutants selected did contain a greatly decreased concentration of hydrogenase which agreed with the results obtained when the wild-type cells were grown on fixed nitrogen.

Further evidence that hydrogenase is part of the nitrogen-fixing system came from the finding that photoproduction of hydrogen by organisms of the Athiorhodaceae and Thiorhodaceae was inhibited by N_2 or ammonia.[100-103] When this inhibition was discovered, the organisms were tested for and found to possess the ability to fix nitrogen. Since ammonia also inhibited hydrogen evolution, it appeared that reductive amination and/or the synthesis of carbon acceptors for ammonia occurred at the expense of hydrogen evolution. If this were true, one would expect more inhibition with one mole of N_2 than one mole of ammonia since, in addition to the moles of H_2 required to "fix" ammonia into organic compounds, three moles would be required for the fixation of one mole of N_2 to two of ammonia. Bregoff and Kamen[104] have analyzed the photogas-evolving system in *Rhodospirillum rubrum* and have found that in the absence of N_2 or NH_4Cl one mole of CO_2 and one mole of H_2 each were evolved per mole of malate (the substrate for photogas evolution) dissilimated in the first 3 hours of metabolism. The hydrogen evolution stopped in the presence of N_2 or NH_4Cl but the evolution of CO_2 and dissimilation of malate continued. Variable amounts of ammonia were required to inhibit the evolution of hydrogen, thus tending to eliminate the possibility that a stoichiometric relation existed between the amount of hydrogen expected from malate consumption and the amount of ammonia assimilated. Since the inhibition of H_2 evolution cannot be explained solely on the basis of the utilization of an intermediate, H_2X, other systems in the cell under the influence of N_2 or ammonia must also be using H_2X as a hydrogen source. Knowledge of the fate of this hydrogen is needed.

For many years one of the inconsistencies in the hypothesis of nitrogenase being coupled to hydrogenase was the lack of any measurable hydrogenase in the nitrogen-fixing system of root nodules.[105] On further experimentation[106] it was found that preparations from soybean nodules showed a spectral shift on exposure to hydrogen which indicated that some function for hydrogen existed in this system. It was not until Hoch *et al.*[107] discovered that hydrogen was evolved from soybean nodules that the first real evidence for hydrogenase in the symbiotic system was presented. These workers were attempting to determine whether D_2 would exchange with combined hydrogen (H) to yield HD. In these measurements evidence for the exchange was found, and in addition and less expected was the finding of an increase in the mass 2 peak indicating the liberation of H_2. In a detailed analysis of this H_2 evolution, a dependence on oxygen was observed; sliced nodules were less active and ground nodules were inactive. The results with hydrogen evolution were identical to those found with nitrogen fixation, i.e., only whole nodules fixed nitrogen to any extent. In addition, the ability of these nodules to evolve H_2 coincided with nitrogen fixation activity.[108, 109]

Since N_2, N_2O, and CO were shown to affect hydrogenase and nitrogen fixation in other systems, the effects of these gases on hydrogen evolution by soybean root nodules was studied.[109] N_2 was found to inhibit H_2 evolution, a result consistent with that found with the photosynthetic microorganisms.[101, 102] H_2 evolution was also inhibited by N_2O, a known competitive inhibitor of nitrogen fixation,[78] and in some experiments was inhibited by CO. More significant was the finding that N_2 was required for the formation of HD from D_2 and either a nodular hydrogen donor or a hydrogen donor present in the medium. N_2O, which inhibited hydrogen evolution and nitrogen fixation, did not promote the deuterium-hydrogen exchange reaction, and carbon monoxide completely inhibited it at levels of 0.02 and 0.03 atm. Thus, as these authors have suggested, it appears that nitrogen is adsorbed on the enzyme surface and then hydrogenated; and it is at this site of hydrogenation that the deuterium exchange occurs. Carbon monoxide inhibited this hydrogenation (and thus the exchange) presumably by attaching itself to and blocking the site.

The discussion until now has concerned those nitrogen-fixing organisms readily inhibited by H_2. On the other hand, nitrogen fixation by *Clostridium pasteurianum*, an anaerobic organism, is not readily affected by hydrogen.[110] Recently[111] hydrogen was discovered to act as a competitive inhibitor of nitrogen fixation by this organism but with a K_i, 0.5 ± 0.05 atm., i.e., five times the value reported for *A. vinelandii*. As these authors have suggested, the organism probably has developed a system for protecting its nitrogen-fixing system from one of its major metabolic products, H_2. Unlike the photosynthetic organisms and nodules, N_2 does not inhibit H_2 evolution by this organism (H_2 is evolved as a product of pyruvate metabolism by the clostridial pyruvate clastic system).[112] Again, this would follow since, in this organism, the H_2-evolving and N_2-fixing systems apparently are protected from one another; this must also be true of the cell-free preparations of this organism, since 0.7 atm. of H_2, at 0.3 atm. N_2, only slightly inhibited nitrogen fixation.

Cell-free preparations of *C. pasteurianum* contain a potent hydrogenase which in the presence of H_2 can readily reduce dyes such as methylene blue (E_0', pH 7.0 = +0.011); any hydrogen acceptor with an E_0' at pH 7.0 equal to or higher than the hydrogenase system (−0.42) could presumably be reduced by hydrogenase provided the appropriate coupling components were present. From this reasoning one would expect H_2 to reduce nitrogen in cell-free preparations, and therefore to replace pyruvate. Since H_2 has so far proved inactive, one must conclude either that the equilibrium of the reaction is unfavorable or a hydrogen acceptor link is lacking. Conversely, since pyruvate and α-ketobutyrate can support fixation, a coenzyme must be reduced during their metabolism, which has a standard electrode potential more negative than is possible by reduction with hydrogenase and H_2, or the rate of reduction of a required coenzyme is sufficient to allow fixation to occur. Whether or not the conditions are optimal is still unknown. At least one step in the reduction of N_2 to ammonia, perhaps to M=N—N=M or M—N=N—M,[113] must require such a reduced coenzyme. The other reduction steps might be mediated by hydrogenase or other reduced coenzyme systems.

G. Proposed Mechanisms

The earlier mechanisms proposed for nitrogen fixation were based mainly on the necessary oxidation-reduction states through which the nitrogen must pass for reduction to ammonia. Since then mechanisms proposed have included the available experimental data as argument for or against certain possible steps. Thus, for example, Shug *et al.*[55] incorporated into their scheme the idea of hydrogenase as a flavoprotein coupled to the reduction of chemisorbed N_2 (Fig. 6).

Two metals (M), Fe and Mo, are known to be required for fixation; and on theoretical grounds one of them, Fe, has been associated with the adsorption of N_2.[113] The evidence for a flavoprotein nature of hydrogenase consists of the presence of flavin in the fiftyfold-purified enzyme and its

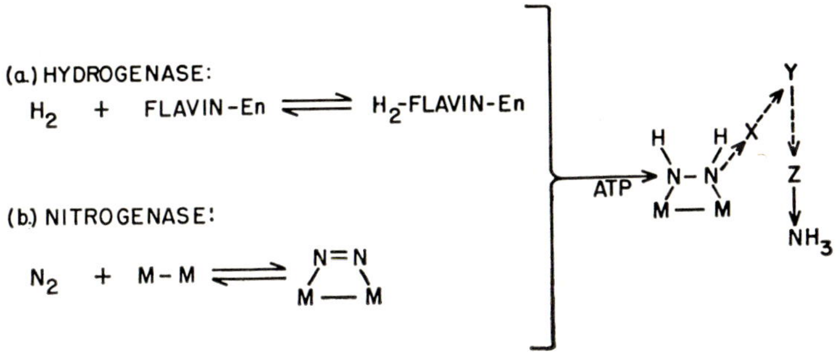

Fig. 6. Suggested model for interaction of hydrogenase and nitrogenase. From Shug *et al.*[55]

reduction on exposure of the enzyme to hydrogen. The coupling of hydrogenase and nitrogenase was suggested by spectral differences observed in a crude enzyme preparation of *C. pasteurianum* in contact with N_2 and H_2.[55] If the preparation was incubated with H_2 at 0.8 atm. and then N_2 admitted to atmospheric pressure, no optical density change was noted. On the other hand, of N_2 was admitted first to 0.8 atm. and then H_2 admitted, or if a low pH_2 was used, an optical density change occurred with maxima at 450 and 390 mμ. These data suggust that hydrogen and nitrogen act at the same site.

Winfield[113] has presented arguments, on theoretical as well as some experimental grounds, that nitrogenase could be depicted as a "multifunctional enzyme" which will react with H_2, O_2, and CO as well as N_2. In his view nitrogenase consists of two mononuclear hydrogenase prosthetic groups set face to face and held at a distance of about 3 to 4 A. by the protein of nitrogenase. This structure is represented as

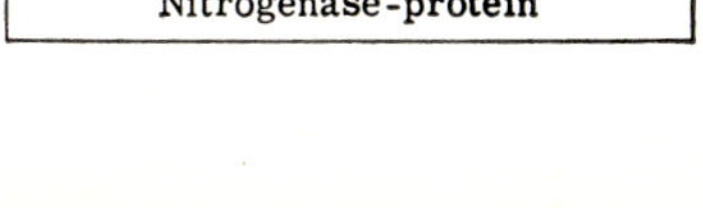

in which C and Fe atoms are in planar prosthetic groups. The prosthetic groups are vizualized as parallel to each other and at right angles to the plane of the paper. H_2, O_2, or CO is assumed to react with this enzyme at either of the prosthetic groups (Fig. 7). N_2, however, must react with both

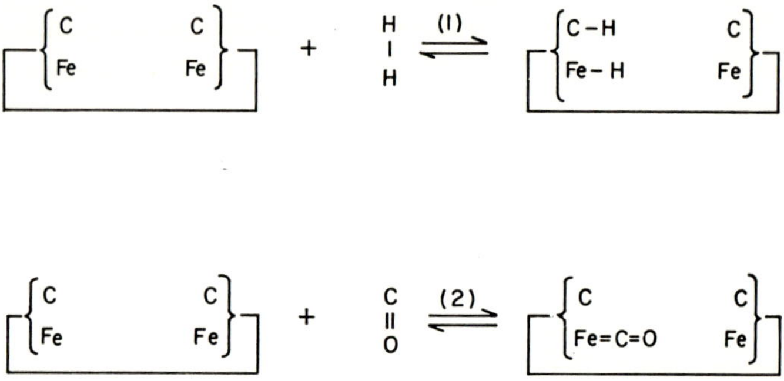

FIG. 7. Model for reaction of nitrogenase sites with hydrogen and carbon monoxide. For function as nitrogenase, see Fig. 8. Adapted from Winfield.[113]

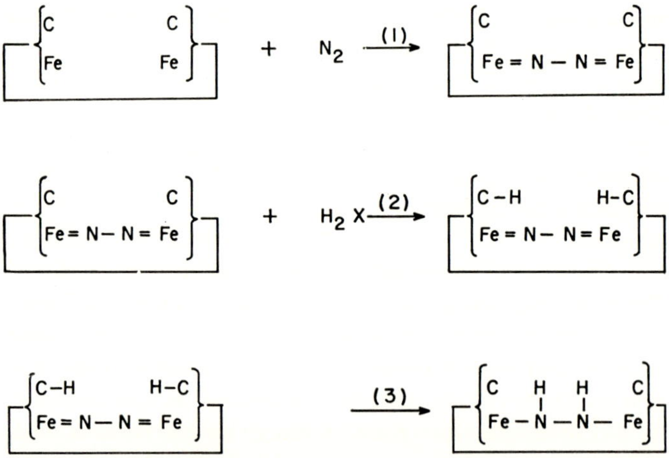

FIG. 8. Model for action of nitrogenase as a nitrogen chemisorbent and hydrogen as a reductant for the chemisorbed nitrogen. Adapted from Winfield.[113]

prosthetic groups in order to avoid a charge on the adsorbed atoms and to assume a positive heat of reaction.

$$(\text{Fe}::\overset{+}{\underset{..}{\text{N}}}::\overset{-}{\text{N}}: \text{ versus } \text{Fe}::\underset{..}{\text{N}}:\underset{..}{\text{N}}::\text{Fe})$$

Nitrogen fixation would proceed as shown in Fig. 8. Reaction 1 represents the chemisorption of N_2 by the enzyme nitrogenase (for more detail on this complex, see Winfield[113]). The next step (Reaction 2) requires the reduction of C by a hydrogen-donating system. Since the H atoms attached to this complex when it functions as hydrogenase can be removed as H_2 gas (see Fig. 7, Reaction 1), the H atoms donated to the complex must be at an energy level close to that of gaseous hydrogen, i.e., the halfcell (E_0', pH 7.0 = −0.42). In a manner similar to the hydrogenase system (Fig. 7, Reaction 1), the H atoms become attached to nitrogen (Fig. 8, Reaction 3). Further reduction could occur by repetition of Reaction 3 (Fig. 8) or a related reaction.

This mechanism is proposed to clarify many unexplained observations. For example, hydrogen inhibition of nitrogen fixation would be expected because of competition of H_2 for the same adsorption sites as nitrogen. The inhibition of H_2 evolution by N_2 could result because nitrogen occupies the sites that are required for the system to function as hydrogenase. This proposed mechanism would also explain why H_2 gas will not reduce nitrogen in cell-free preparations of *C. pasteurianum* since the reduction of coenzymes by hydrogenase would be mediated through the same enzyme sites as nitrogen adsorption. One might, however, be able to reduce the hydrogen acceptors by hydrogenase, remove the hydrogen to prevent inhibition of nitrogen fixation, and then introduce N_2, and measure nitrogen fixation as a function of reduction by coenzymes reduced by hydrogenase. The amount of fixation would depend, of course, on the concentration of reduced coenzymes in the preparation but this limitation might be overcome by rapidly alternating the gas phase from H_2 to N_2. This scheme does not consider the inhibition of H_2 evolution by NH_3 in the photosynthetic organisms.

Recently, Hoch *et al.*[5] proposed a scheme similar to Winfield's, based on results obtained while examining the exchange between [H] and D_2 (Fig. 9). They found N_2 inhibition of H_2 evolution by soybean root nodules—an indication of N_2 and H_2 competing for the same enzyme sites. Also the exchange reaction between D_2 and [H] depended on the presence of N_2, indicative of a hydrogen-deuterium exchange on the chemisorbed nitrogen molecule. These results are consistent with the mechanism suggested by Winfield (Fig. 8) in which a reduced coenzyme would transfer its hydrogen to chemisorbed nitrogen.

The second step in the reduction of nitrogen should yield an intermediate at the reduction level of hydrazine or hydrazine-enzyme complex. All attempts to identify or accumulate such an intermediate, free or combined, in cell-free preparations have so far failed.[17, 18] Thus either hydrazine or an azine is not an intermediate, the levels present cannot be detected with present methods, or the intermediates are either transferred directly to another enzyme or never leave the original enzyme surface and cannot be removed by the chemical methods used in attempts to identify. Small amounts of a trapped derivative of this intermediate may have been detected in *Azotobacter* during fixation of N_2^{15}.[87] Proof that a hydrazine-enzyme complex (azide) is involved may require experiments with high concentrations of purified "nitrogenase."

The terminal step in conversion of N_2 to ammonia would require the reduction of hydrazine (or a compound at this level of reduction) to ammonia. The only evidence for such a biological system is the Woolfolk *et al.*[7] observation that *Micrococcus lactilyticus* cells would reduce hydrazine with

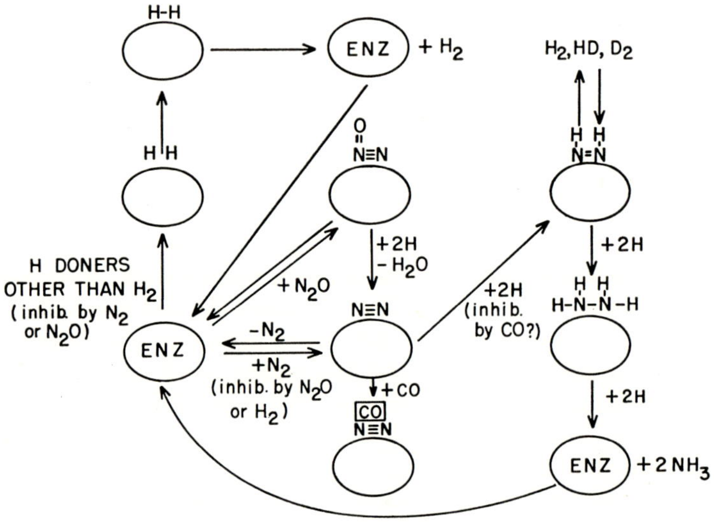

FIG. 9. Schematic presentation of working hypothesis for the interaction of N_2, N_2O, H_2, and CO in leguminous root nodules and other nitrogen-fixing agents. From Hoch *et al.*, *Biochim. et Biophys. Acta* **37,** 273 (1960).

H_2 gas as reductant. The product could be ammonia but no evidence exists as yet. Under appropriate conditions, hydrogenase reduces many biological H-carriers, thus one cannot deduce whether hydrazine was reduced directly by hydrogenase or through intermediary carriers.

One recent proposal compared the fixation of nitrogen to nitride formation of certain metals i.e., Li.[36] The active site of nitrogenase could be a chelate of a variable-valent metal ion which transfers its electrons to N_2 by a reaction similar to nitride formation. For example, Fe^{++} chelated to nitrogenase might form a nitride with N_2 that on hydrolysis would yield ammonia and Fe^{+++}. The Fe^{++} would then be regenerated by reduction of Fe^{+++} by hydrogenase or other cellular-reducing mechanisms. Evidence for the formation of a nitride is so far lacking. Several facts seem inconsistent with this hypothesis; for example, N_2 is required for the hydrogen-deuterium exchange reaction in nodule preparations, hydrogen inhibits nitrogen fixation, and hydrogen will not act as the sole hydrogen donor for nitrogen fixation by cell-free preparations of *C. pasteurianum*. Such a mechanism, however, would explain why no intermediates have been found in the nitrogen fixation reaction since a nitride once formed would rapidly hydrolyze to ammonia.

Present data are most consistent with the nitrogenase model of Winfield[113] and the very similar one of Hoch *et al.*[5] None of these proposals explains if and where molybdenum fits into the scheme. It has been suggested[41, 55] that molybdenum is required for hydrogenase to function, and perhaps both Fe and Mo are part of the same enzyme. These mechanisms also do not explain why nitrogen fixation by cell-free preparations of *C. pasteurianum* was only slightly inhibited by H_2 at 0.7 atm. One might postulate that the Fe site has a greater affinity for N_2 than H_2, but then one would expect N_2 to inhibit hydrogenase activity. Perhaps the simplest explanation is that nitrogenase has hydrogenase activity but that there are two distinct hydrogenases. For example, in clostridia the hydrogenase, not a part of nitrogenase, would be available to act as a hydrogen liberator (electron-balancing system) or even as an electron carrier that transfers hydrogen to nitrogenase. This mononuclear hydrogenase might itself protect nitrogenase by keeping the sites free of H_2. In *Azotobacter*, which does not evolve H_2 in its metabolism, only one hydrogenase, the hydrogenase present as a part of nitrogenase, might be present and thus nitrogen fixation would be strongly inhibited by hydrogen.

One still cannot rule out the possibility of the first step in nitrogen fixation being oxidative. For example, N_2 could be converted from a valence state of 0 to +1, the valence state of the first hypothetical intermediate in nitrite reduction. If this oxidation did occur, the resulting intermediate might be reduced to ammonia by the nitrite reductase system. There is no evidence for such an oxidation at the present time. However, the finding of nitric oxide reductase in high concentration in cells of *C. pasteurianum* grown on N_2 and in low concentration in cells grown on ammonia suggests that an oxidized intermediate might be involved.[4] The recent discovery of a nitrite reductase (NO_2^- to NH_4^+) in *C. pasteurianum*[86] also is consistent with such a proposal.

H. Existing Problems

With the advent of cell-free nitrogen-fixing preparations, one major problem has been solved. This accomplishment can be considered a major breakthrough for it provided workers in this field the mechanisms to study the intervening steps and reactions in N_2 fixation. First of all, the conversion of N_2 to ammonia can be divided into two major categories, which now seem warranted as working assumptions.

1. How is nitrogen chemisorbed?
2. How is chemisorbed nitrogen reduced?

An approach to the first problem may be made possibly through a study of the spectral differences under conditions where the enzyme apparently is not converting the chemisorbed nitrogen to ammonia. Certainly, the value of this type of experimentation will now be quickly determined. The validity of such results must be established by purification and analysis of the responsible components. The first problem can be subdivided further into several other problems, such as what metal or metals are involved and what is the nature of the metal-binding sites (ligands). Considerable information on the action of the metals might be obtained from a study of the effects of metal-binding agents but again the final information can only come from analyses of the purified components.

The second problem (how is chemisorbed nitrogen reduced to ammonia?) may differ from organism to organism. In the clostridia, nitrogen is reduced at the expense of "reduced coenzymes" produced during the metabolism of pyruvate.[18] In *Azotobacter* a reduced coenzyme produced during the functioning of the tricarboxylic acid cycle may be required. Thus, cell-free preparations of *C. pasteurianum* only fix nitrogen under strictly anaerobic conditions while, in contrast, nitrogen fixation by *A. vinelandii* preparations apparently requires oxygen.[47] This presents an added problem, however, since O_2 is a competitive inhibitor of nitrogen fixation by whole cells of *Azotobacter*.[114] One then wonders why nitrogen fixation in cell-free preparations of aerobes is not also inhibited by oxygen. Needless to say, such surface inconsistencies will be eliminated only with further experimentation.

The identification of the reduced coenzymes necessary for the reduction of nitrogen also would greatly aid in the isolation of the fixing system; for example, it may become possible to add these directly instead of depending on their reduction by added substrates. The mechanism by which nitrogen is reduced by the "reduced coenzymes" (the coupling mechanism) is another problem that will have to be solved.

In conclusion, it now appears possible to isolate and identify the enzyme and coenzyme fractions which in concert accomplish the biological conversion of N_2 to NH_3. Although much work may be required, the promise of new cofactors and clarification of nitrogen and hydrogen pathways with resultant new biological and chemical understanding seem warranted.

III. Incorporation of Ammonia into Organic Compounds

The major pathway through which biological agents incorporate inorganic nitrogen into organic compounds appears to be by reactions of the ammonium ion. As early as 1926, Quastel and Woolf[115] reported the formation of aspartic acid from fumaric acid and ammonia by cells of *Escherichia coli*. Since then, several additional systems for ammonia incorporation have been demonstrated. The present discussion will outline the known pathways and, in so far as clarified, their mechanism.

A. Dehydrogenases

Until recently, glutamic acid dehydrogenase was thought to be the primary and, perhaps the sole, route for incorporation of ammonia into amino acids. In 1955, however, Wiame and Pierard[116] discovered alanine dehydrogenase in a mutant of *Bacillus subtilis*. This observation showed that the list was incomplete and signifies that other amino acid dehydrogenases may well be found.

1. Glutamic Acid Dehydrogenase

The pioneer work of von Euler *et al.*[117] demonstrated in yeast preparations the reaction:

$$\begin{array}{c} \text{glutamic acid} + \underset{\text{(TPN)}}{\text{codehydrase II}} \\ \downarrow \\ \alpha\text{-ketoglutaric acid} + \underset{\text{(TPNH)}}{\text{codehydrase II}} + NH_3 \end{array}$$

Further studies with cell-free preparations of *E. coli*[118, 119] and of yeast[120] have shown the reaction to be TPN-specific and to proceed in two steps.

$$\text{L-(+)-glutamic acid} + \text{TPN} \rightleftharpoons \text{iminoglutaric acid} + \text{TPNH} \qquad (1)$$

$$\text{iminoglutaric acid} + H_2O \rightleftharpoons \alpha\text{-ketoglutaric acid} + NH_3 \qquad (2)$$

The enzyme is specific for L-glutamate; D-glutamic acid and other amino acids do not react. More significant to cellular biosynthesis, the reaction was found to be reversible, i.e., glutamic acid is synthesized from TPNH, NH_3, and α-ketoglutaric acid in the presence of the enzyme.

A TPN-specific glutamic acid dehydrogenase has been found also in extracts of *Neurospora crassa*.[121] The reaction was followed by measuring ammonia disappearance in the presence of α-ketoglutaric acid and TPNH and by identification of glutamic acid chromatographically. Alternatively, TPNH formed during glutamic acid oxidation was followed. Mutants of the organisms required α-amino nitrogen for growth and did not possess glutamic acid dehydrogenase.[122]

Glutamic acid dehydrogenase has been demonstrated in several mammalian tissues. The beef liver[123] enzyme was crystallized[124, 125] and the pyridine nucleotide specificity and equilibrium measured. This enzyme is active with DPN, desamino-DPN, and TPN; the activity ratios, at optimum pH for enzyme activity, are 100:60:35, respectively. The maximal specific activity (change in O.D., ΔE, per mg. of enzyme) of the crystalline enzyme was 90 for glutamate oxidation and 950 for glutamate formation. A unit is that amount of enzyme which produced an increase in O.D. of 1 in a 1-cm. cuvette at 340 mμ. The equilibrium constant

$$K_{eq} = \frac{(\alpha\text{-ketoglutarate})\ (NH_4^+)\ (DPNH)}{(\text{glutamate})\ (H_2O)\ (DPN)}$$

is approximately 2.7×10^{-15} moles per liter, which corresponds to a standard free energy at pH 6.5 of +17.6 kcal. at 27°C. The equilibrium, as written in equations (1) and (2), favors glutamic acid synthesis.

The enzyme isolated from yeast and bacteria[117-120] is TPN-specific, from plants is DPN-specific,[119, 120] and from animals is more active with DPN but can function with desamino-DPN or TPN.

2. Alanine Dehydrogenase

This enzyme was first described by Wiame and Pierard[116] in a mutant of *Bacillus subtilis* which did not possess a glutamic acid dehydrogenase and yet oxidized glutamic acid. They assumed that the amino group of glutamic acid was being transaminated to another α-keto acid to form an amino acid which in turn was oxidized by an unknown enzyme. Examination of cell extracts revealed an alanine dehydrogenase specific for L(+)-alanine and DPN. Other amino acids, including L(+)-glutamic and L(+)-aspartic acids, were not oxidized; TPN did not serve as hydrogen acceptor. The oxidation product of L(+)-alanine was pyruvate; in the presence of DPNH, pyruvate, and ammonium ions, alanine was synthesized by the reverse reaction (reductive amination).

Alanine synthesis by this dehydrogenase has been studied by several workers[126-128]; they have reported its presence in other bacilli and in numerous other microorganisms.[126-128] Many bacilli studied lacked glutamic dehydrogenase. In the absence of the ability to form glutamate by reductive amination of α-ketoglutarate, the alanine dehydrogenase assumes more significance as an ammonia incorporation reaction.

Goldman[129] found an alanine dehydrogenase in *Mycobacterium tuberculosis* var. *hominis* and purified the enzyme about twenty-fold. From the kinetic data, Goldman[129] suggested an enzyme-bound iminopropionate rather than a free, spontaneously formed intermediate as assumed in the glutamate sequence.[118, 119]

The following reaction sequence best fits his data.

1. $E + NH_4^+ \rightleftharpoons [E{-}NH_4^+]$

or

1a. $E + CH_3COCOO^- \rightleftharpoons [E{-}CH_3COCOO^-]$

2. $[E{-}NH_4^+] + CH_3COCOO^- \rightleftharpoons [E{-}CH_3C(NH)COO^-] + H_2O + H^+$

3. $[E{-}CH_3C(NH)COO^-] + DPNH + H^+ \rightleftharpoons [E{-}CH_3CH(NH_2)COO^-] + DPN^+$

4. $[E{-}CH_3CH(NH_2)COO^-] \rightleftharpoons E + CH_3CH(NH_2)COO^-$

Reactions 1 and 1a can be written, since it was shown that at a given pyruvate concentration increased NH_4^+ does not affect reductive amination but, at a given NH_4^+ concentration, increased pyruvate acts as a potent inhibitor.

Reaction 2 represents the formation of enzymically (not spontaneously) formed iminopropionate. Further evidence is needed before a final mechanism can be written.

3. Reductive Amination of Glyoxylate

A single report exists on the formation of glycine from glyoxylate, ammonia, and DPNH.[130] The enzyme isolated from *Mycobacterium tuberculosis* var. *hominis* catalyzed the reaction:

$$HCOCOO^- + NH_4^+ + DPNH \longrightarrow CH_2(NH_2)COO^- + DPN^+ + H_2O$$

The free energy of glycine formation (ΔF°_{298}) for this reaction was calculated as -11 kcal./mole.

B. Saturation Amination

1. Aspartase

The catalysis of the reversible formation of L(+)-aspartic acid from fumaric acid and ammonia was first demonstrated by Quastel and Woolf[115] using whole cells of *Bacillus coli* (*Escherichia coli*) and later[131] in several other faculative anaerobes. These workers isolated the aspartic acid formed from the forward reaction and fumaric acid and ammonia from the reverse reaction and established the equilibrium:

$$HOOCCH{=}CHCOOH + NH_3 \longrightarrow HOOCCH_2CH(NH_2)COOH$$

$$K_{eq} = \frac{[\text{aspartic acid}]}{[\text{fumaric acid}]\,[\text{ammonia}]} = 20$$

With whole cells, the equilibrium was found to favor aspartate, with a K_{eq} of about 20. With whole cells and aspartate as substrate, succinate accumulated by the reduction of the fumarate unless one of several inhibitors for succinic dehydrogenase (e.g., toluene) was added.[132]

Virtanen and Tarnanen[133] extracted aspartase from *Bacillus fluorescens liquefaciens* (*Pseudomonas fluorescens*) and demonstrated with the cell-free enzyme the reversible breakdown of aspartic acid as previously suggested from studies with whole cells.[131, 132] An increase in the ammonia concentration favored aspartic acid synthesis. The enzyme was fumarate-specific. For example, the addition of methyl-fumarate or *trans*-aconitate with ammonia did not yield an amino acid.[134] Their preparations also contained succinic dehydrogenase which could be inactivated by trypsin so that succinic acid would not function as a source of fumarate for the reaction. The enzyme also was found in some higher plants and in all groups of bacteria tested. It was not found, however, in yeast or animal tissues.

The *P. fluorescens* and *E. coli* aspartase was obtained in extracts by autolysis and purified slightly by isoelectric precipitation and ammonium sulfate fractionation.[134-136] Gale[135] observed two forms of aspartase in *E. coli* extracts. Both forms deaminated L(+)-aspartic acid anaerobically. One form was inhibited by toluene treatment and required adenosine for action; the other did not have these characteristics. Ichihara *et al.*[137] further studied the enzyme system of *E. coli* and reported, after a twentyfold purification, a requirement for folic acid, cobalt ion, and reduced glutathione. This report, which remains to be confirmed or extended, likens aspartase to histidine deaminase.[138]

2. β-Methylaspartase

Clostridium tetanomorphum ferments glutamic acid readily in the presence of α,α'-dipyridyl to yield mesaconate and ammonia.[139, 140] Treatment of a crude extract with charcoal removed cofactor(s) required for glutamate decomposition, but the synthesis of an amino acid from mesaconic acid and ammonium ion still occurred. The amino acid was isolated and identified as β-methylaspartate.[141] The reaction appeared to occur as follows:

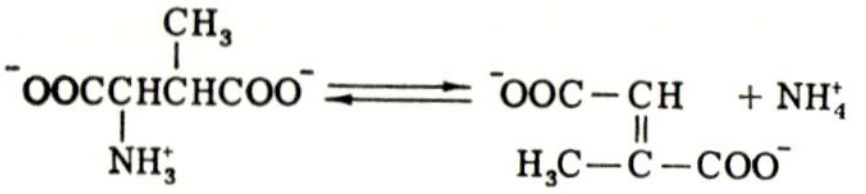

The β-methylaspartate appeared to be the L-*threo* isomer. Magnesium and a monovalent cation were required for maximum activity. L-Aspartate and L-*erythro*-β-methylaspartate are also decomposed by the purified enzyme but the rates are much slower than with the L-*threo* isomer of β-methylaspartate. Consequently, the enzyme appears to possess aspartase activity; but, since purified aspartase from other organisms did not react with mesaconate, the enzyme is not aspartase.

3. β-Alanyl-CoA Synthetase

In studies on the metabolism of propionic acid, Stadtman and Vagelos[142] found the conversion of propionyl-CoA to acrylyl-CoA (see Chapter 3 in Volume II). All attempts to convert acrylyl-CoA to lactyl-CoA failed, but it was discovered that acrylyl-CoA did react readily with ammonia to form β-alanyl-CoA.[142-144]

$$CH_2 = CHCOSCoA + NH_3 \rightleftharpoons CH_2NH_2CH_2COSCoA$$

The evidence favors free ammonia and not the ammonium ion as substrate since high ammonia concentration and high pH are required for optimal reaction. The significance of this reaction to the cell and its ubiquitousness remains to be determined.

C. Amide Synthesis

The synthesis of amides from ammonia and either glutamic or aspartic acid is a major ammonia incorporation reaction of plant tissue.[145] In some plants as much as one-half of the nonprotein nitrogen is present as glutamine nitrogen[145]; also, in spent growth media of some bacteria large amounts of glutamine and asparagine have been found.[73] The literature on these systems has been reviewed comprehensively by Meister[146] and Webster.[147] Therefore, only the pertinent data covering the reactions and the evidence for the presence of these enzymes in bacteria will be presented here.

1. Glutamine Synthetase

Although the presence of glutamine in biological materials had been known for some time, the first demonstration of glutamine synthesis appeared in 1935.[148] The oxidation of glucose was shown to serve as the energy source for the synthesis of glutamine from glutamic acid and ammonia. In 1949, Speck[149-151] and Elliot[152, 153] showed the reaction sequence to occur as follows:

$$\text{L-glutamic acid} + NH_3 + \text{ATP} \overset{Mg^{++}}{\rightleftharpoons} \text{glutamine} + \text{ADP} + \text{Pi}$$

The enzyme has been purified[153-155] and found to catalyze, in addition to glutamine formation from glutamic acid and ammonia, the formation of γ-glutamyl hydroxamate, γ-glutamyl hydrazide, and glutamyl γ-methylamide from glutamic acid and hydroxylamine, hydrazine, and methylamine, respectively. The enzyme activity has been assayed by formation of γ-glutamyl hydroxamate since a simple quantitative colorimetric method exists for measuring hydroxamic acids of this type.[156]

Cell-free preparations of *Staphylococcus aureus*[157] and *Proteus vulgaris*[158] have been shown to synthesize glutamine or glutamylhydroxamic acid from glutamic acid and ammonia or hydroxylamine. Adenosine triphosphate and magnesium are absolute requirements, a result identical to that found with animal and plant preparations.

Studies of the metal requirements for this enzyme system have shown that either magnesium[150, 152] or cobalt[159] is effective, with manganese about one-third as active. For optimal activity 0.005 M Co^{++} or 0.05 M Mg^{++} is required. With Co^{++} ions the reaction is specific for L(+)-glutamic acid but with Mg^{++} either L(+)- or D(−)-glutamic acid will react. No D(−)-glutamyl hydroxamate was formed from D(−)-glutamic acid when cobalt was added in preference to magnesium. With Co^{++} as activator, D(−)-glutamic acid partially inhibited the formation of the hydroxamic acid from L(+)-glutamic acid. The metal ion apparently determines the stereospecificity.

Mechanism studies with glutamic acid labeled in the carboxyl with O^{18} have demonstrated that the carboxyl oxygen of glutamate was transferred to the inorganic phosphate released during glutamine synthesis.[160, 161] Thus, the synthesis of glutamine was suggested to proceed by the following reaction sequence[147]:

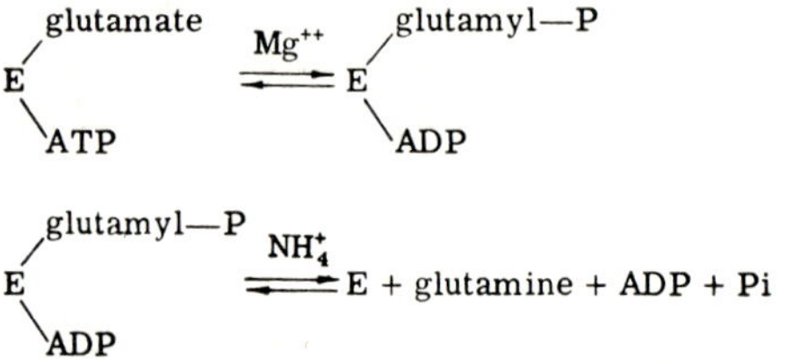

(1)

(2)

Recently, Krishnaswamy *et al.*[161a] presented evidence that the binding of glutamate to the enzyme required ATP and Mg^{++}. The binding is associated with cleavage of ATP to ADP and inorganic phosphate and the formation of γ-carboxyl-activated glutamate. The evidence for this resides in the finding that pyrrolidone carboxylate could be readily formed when glutamine synthetase was incubated with ATP, Mg^{++}, and glutamate (in the absence of NH_4^+) and the resulting product heated to 100°C. for 5 min. Only small amounts of the cyclic pyrrolidone carboxylate was formed when heat-denatured enzyme was used, ATP or Mg^{++} omitted, or when ATP was replaced by AMP, ADP, and other triphosphate derivatives. Only traces of glutamine were formed. If, however, NH_4^+ was added to the system, glutamine was formed and pyrrolidone was reduced to the blank level. These results suggest that ATP and Mg^{++} are bound to the enzyme before glutamate and that glutamine is formed by a series of reactions on the enzyme with amide formation occurring last. Enzyme-bound γ-glutamyl phosphate could be an intermediate in the synthesis, as suggested above, but no proof exists yet.

2. Asparagine Synthetase

Little knowledge concerning the reaction sequence in asparagine synthesis is available, and nothing is known of the mechanism of asparagine synthesis in bacteria. Webster and Varner[162, 163] have demonstrated a conversion of aspartate-C^{14} to asparagine-C^{14} in cell-free extracts of lupine seedlings and of wheat germ; the stoichiometry is consistent with the reaction:

$$\text{aspartate} + NH_3 + ATP \rightleftharpoons \text{asparagine} + ADP + Pi$$

This reaction appears to be patterned after the known reactions in glutamine synthesis. The enzymes have not been separated from the crude mixtures, so conclusive data are lacking. Levintow[164] has reported, in a mammalian tissue culture preparation, the transfer of the amide group of glutamine to aspartic acid to form asparagine. Ammonia was not incorporated into asparagine; thus, a transamidation was suggested. Possibly the plant system of Webster and Varner contains glutamic acid which is aminated to glutamine; the asparagine could then be formed by transamidation of the glutamine amide group to aspartic acid, as reported for the mammalian enzyme.

3. Carbamyl Phosphate Synthetase

In 1955, Jones *et al.* reported that a cell-free preparation of *Streptococcus faecalis* synthesized carbamyl phosphate from ammonium carbonate and ATP:

$$CO_2 + NH_3 + ATP \rightleftharpoons H_2N{-}\overset{\displaystyle O}{\overset{\|}{C}}{-}OPO_3^{--} + ADP$$

The equilibrium of this reaction is toward the left, and continuous regeneration of ATP was required to drive the reaction to the right. The carbamyl phosphate then reacted with ornithine to form citrulline.[165, 166] Later, an enzyme system was isolated from mammalian liver which appeared to perform the same reaction but, unlike the bacterial system, required two moles of ATP per mole of carbamylphosphate formed and acetyl-L-glutamic acid (AGA) acted as cofactor. The following reactions are consistent with their findings[167-170]:

$$CO_2 + ATP \xrightleftharpoons{\text{acetyl glutamate}} ADP + \text{“active” } CO_2 + Pi \quad (1)$$

$$\text{“active” } CO_2 + ATP + NH_4^+ \xrightleftharpoons{\text{acetyl glutamate}} H_2N{-}\overset{\overset{\displaystyle O}{\|}}{C}{-}OPO_3^{--} \quad (2)$$

The acetyl glutamate appeared to be required for both reactions; the formation of a high-energy phosphate derivative of acetyl glutamate as an intermediate has been excluded.[169]

Recently, Jones and Spector,[171] using C^{14}-bicarbonate, obtained suggestive evidence for an active carbonic acid acetyl-L-glutamic acid derivative such as

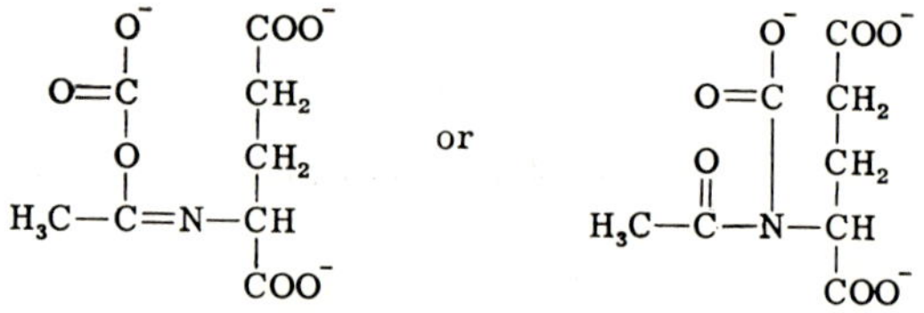

If true, then the reaction could proceed according to the mechanism they suggested:

$$ATP + HOCO_2^- + AGA \longrightarrow ADP + Pi + AGA{-}CO_2^- \quad (1)$$

$$AGA{-}CO_2^- + ATP + NH_3 \rightleftharpoons NH_2\overset{\overset{\displaystyle O}{\|}}{C}OPO_3^{--} + ADP + AGA \quad (2)$$

Further evidence is needed to firmly establish $AGA\text{-}CO_2^-$ as an intermediate, but these studies do suggest a role for AGA in the reaction sequence. For further details of these experiments, see Jones and Spector.[171]

4. Purine and Pyrimidine Synthesis

The biosynthesis of the purine and pyrimidine rings could be considered as ammonia incorporation reactions of the amide type; the synthesis of purine requires glutamine and the synthesis of pyrimidine requires carbamyl phosphate. However, since the biosyntheses of these amides has already been discussed (see Chapter 6 by Magasanik, this volume) and excellent reviews on purine and pyrimidine biosynthesis have recently been published (see Buchanan and Hartman[172] and Reichard[173]), no details will be presented here.

The synthesis of CTP (cytidine triphosphate) from UTP (uridine triphosphate) by cytidine triphosphate aminase, and GMP (guanosine-5′-phosphate) from XMP (xanthosine-5′-phosphate) by xanthosine-5′-phosphate aminase have been demonstrated respectively by Lieberman[174, 175] with an enzyme from *E. coli* and by Moyed and Magasanik[176] with a 300-fold purified enzyme from *Aerobacter aerogenes.* The reactions proceed as follows:

$$UTP + NH_3 + ATP \longrightarrow CTP + ADP + Pi \quad (1)$$

$$XMP + NH_3 + ATP \xrightarrow{Mg^{++}} GMP + AMP + \text{pyrophosphate} \quad (2)$$

In reaction (1) neither UMP, uridine, nor uracil would replace UTP whereas UDP (uridine diphosphate) was only half as effective. Ammonia was the specific amino donor; glutamine was inactive. In reaction (2) the enzyme is specific for NH_3, xanthosine-5′-phosphate, and ATP. Mg^{++} was required for the reaction.

D. Conclusions

Of the ammonia-incorporation reactions now known to occur in bacteria, aspartase was the first recognized but it has not been subjected to rigorous study with modern methods of enzymology; the mechanism of its action is still unknown. Glutamic acid dehydrogenase, for years considered to be the main ammonia pickup system, does exist in bacteria—in some cases in abundant amount. It does, however, seem to be absent from other microorganisms, and the increasing list of cells with this omission forces a reevaluation of its role in microorganisms. The widespread occurrence of glutamic-based transaminases in animal and microbial systems seemed to secure the rationalization of glutamic acid as an intermediate in nitrogen flow from ammonia to other organic nitrogen compounds. More recently, in *E. coli,* several alanine-based transamination systems have been demonstrated. This observation may only serve to accentuate the multiplicity of routes or may again lead to overrationalization based on the recent and expanding evidence of an alanine dehydrogenase in microbial species. Especially, attention is focused on alanine as an initial amination product by the apparent absence of glutamic dehydrogenase from numerous bacilli and other cells.

References

[1] Section II, *in* "Inorganic Nitrogen Metabolism" (W. D. McElroy and B. Glass, eds.). Johns Hopkins Press, Baltimore, Maryland, 1956.

[2] A. Nason and H. Takahashi, *Ann. Rev. Microbiol.* **12,** 203 (1958).

[2a] G. C. Walker and D. J. D. Nicholas, *Biochim. et Biophys. Acta* **49,** 361 (1961).

[3] Section III, *in* "Inorganic Nitrogen Metabolism" (W. D. McElroy and B. Glass, eds.). Johns Hopkins Press, Baltimore, Maryland, 1956.

[3a] C. A. Fewson and D. J. D. Nicholas, *Proc. Biochem. Soc. Biochem. J.* **78** 9P (1961).

[4] C. A. Fewson and D. J. D. Nicholas, *Nature* **188,** 794 (1960).

[4a] C. A. Fewson and D. J. D. Nicholas, *Nature* **190,** 2 (1961).

[4b] D. J. D. Nicholas, *Ann. Rev. Plant Physiol.* **12,** 63 (1961).

[5] G. E. Hoch, K. C. Schneider, and R. H. Burris, *Biochim. et Biophys. Acta* **37,** 273 (1960).
[6] A. Medina and D. J. D. Nicholas, *Nature* **179,** 533 (1957).
[6a] E. G. McNall and D. E. Atkinson, *J. Bacteriol.* **74,** 60 (1957).
[7] C. A. Woolfolk, E. J. Ordal, and H. R. Whiteley, *Soc. Am. Bacteriologists Proc.* **P15,** 152 (1960).
[8] D. J. D. Nicholas and O. T. G. Jones, *Nature* **195,** 512 (1960).
[9] M. S. Engel and M. Alexander, *J. Bacteriol.* **76,** 217 (1958).
[10] H. Lees, *Ann. Rev. Microbiol.* **14,** 83 (1960).
[11] M. I. H. Aleem and M. Alexander, *J. Bacteriol.* **76,** 510 (1958).
[12] P. W. Wilson, *in* "Bacterial Physiology" (C. W. Werkman and P. W. Wilson, eds.), p. 467. Academic Press, New York, 1951.
[13] P. W. Wilson and R. H. Burris, *Ann. Rev. Microbiol.* **7,** 415 (1953).
[14] A. I. Virtanen, *Angew. Chem.* **65,** 1 (1953).
[15] P. W. Wilson, *in* "Handbuch der Pflanzenphysiologie" (W. Ruhland, ed.), Vol. VIII, p. 9. Springer, Berlin, 1958.
[16] J. E. Carnahan, L. E. Mortenson, H. F. Mower, and J. E. Castle, *Biochim. et Biophys. Acta* **38,** 188 (1960).
[17] J. E. Carnahan, L. E. Mortenson, H. F. Mower, and J. E. Castle, *Biochim. et Biophys. Acta* **44,** 520 (1960).
[17a] H. F. Mower, unpublished result (1961).
[18] L. E. Mortenson, *Federation Proc.* **19,** 241 (1960).
[19] H. F. Mower, *Federation Proc.* **19,** 241 (1960).
[20] R. H. Burris and P. W. Wilson, *in* "Methods in Enzymology" (S. P. Colowick and N. O. Kaplan, eds.), Vol IV, p. 355. Academic Press, New York, 1957.
[21] P. B. Hamilton and P. W. Wilson, *in* "Biochemistry of Nitrogen" (N. J. Toivonen, E. Tomnila, J. Erkama, P. Roine, and J. K. Miettines, eds.), p. 139. Suomalainen Tiedeakatemia, Helsinki, Sweden, 1955.
[22] M. H. Proctor and P. W. Wilson, *Nature* **182,** 891 (1958).
[23] G. Metcalfe and M. E. Brown, *J. Gen. Microbiol.* **17,** 565 (1957).
[24] M. E. Brown and G. Metcalfe, *Nature* **180,** 282 (1957).
[25] G. Nemeth, *Nature* **183,** 1460 (1959).
[26] I. C. Gardner, *Nature* **181,** 717 (1958).
[27] G. Bond and I. C. Gardner, *Nature* **179,** 680 (1957).
[28] G. Stevenson, *Ann. Botany (London)* **23,** 622 (1959).
[29] N. Raggio, M. Raggio, and R. H. Burris, *Biochim. et Biophys. Acta* **32,** 274 (1959).
[30] E. S. Lindstrom, J. W. Newton, and P. W. Wilson, *Proc. Natl. Acad. Sci. U.S.* **38,** 392 (1952).
[31] H. L. Jensen, *Proc. Linnean Soc. N.S. Wales* **70,** 203 (1946).
[32] H. L. Jensen and D. Spenser, *Proc. Linnean Soc. N.S. Wales* **72,** 73 (1947).
[33] H. L. Jensen, *Proc. Linnean Soc. N.S. Wales* **72,** 299 (1948).
[34] R. G. Esposito and P. W. Wilson, *Proc. Soc. Exptl. Biol. Med.* **93,** 564 (1956).
[35] R. M. Pengra and P. W. Wilson, *Proc. Soc. Exptl. Biol. Med.* **100,** 436 (1959).
[36] J. E. Carnahan and J. E. Castle, *J. Bacteriol.* **75,** 121 (1958).
[37] M. H. Proctor and P. W. Wilson, *Nature* **182,** 891 (1958).
[38] J. R. Norris and H. L. Jensen, *Nature* **180,** 1493 (1957).
[39] R. G. Esposito and P. W. Wilson, *Proc. Natl. Acad. Sci. U. S.* **44,** 472 (1958).
[40] J. E. Carnahan, L. E. Mortenson, and J. E. Castle, *J. Bacteriol.* **80,** 311 (1960).
[41] A. L. Shug, P. W. Wilson, D. E. Green, and H. R. Mahler, *J. Am. Chem. Soc.* **76,** 3355 (1954).
[42] S. Hino and P. W. Wilson, *J. Bacteriol.* **75,** 403 (1958).

[43] R. H. Burris, F. J. Eppling, H. B. Wahlin, and P. W. Wilson, *J. Biol. Chem.* **148,** 349 (1943).
[44] P. G. Hamilton, W. E. Magee, and L. E. Mortenson, *Soc. Am. Bacteriologists Proc.* **A29,** 82 (1953).
[45] G. E. Hoch and D. W. S. Westlake, *Federation Proc.* **17,** 243 (1958).
[46] K. C. Schneider, C. Bradbeer, R. N. Singh, L. C. Wang, P. W. Wilson, and R. H. Burris, *Proc. Natl. Acad. Sci. U. S.* **46,** 727 (1960).
[47] D. J. D. Nicholas and D. J. Fisher, *Nature* **186,** 735 (1960).
[48] D. J. D. Nicholas and D. J. Fisher, *J. Sci. Food Agri.* **11,** 603 (1960).
[49] D. I. Arnon, M. Losada, and K. Nozaki, *Biochem. J.* **77,** 23P (1960).
[50] F. H. Grau and P. W. Wilson, *Bacteriol. Proc.* p. 193 (1961).
[51] D. E. Hughes, *Brit. J. Exptl. Pathol.* **32,** 97 (1951).
[52] L. E. Mortenson, M. S. Thesis, University of Wisconsin Press, Madison, 1951.
[53] L. E. Mortenson, *Anal. Biochem.* **2,** 216 (1961).
[54] D. J. D. Nicholas, D. J. Silvester, and J. F. Fowler, *Nature* **189,** 634 (1961).
[55] A. L. Shug, P. B. Hamilton, and P. W. Wilson, *in* "Inorganic Nitrogen Metabolism" (W. D. McElroy and B. Glass, eds.), p. 344. Johns Hopkins Press, Baltimore, Maryland, 1956.
[56] H. J. Koepsell and M. J. Johnson, *J. Biol. Chem.* **145,** 379 (1942).
[57] F. Lipman and L. C. Tuttle, *J. Biol. Chem.* **158,** 505 (1945).
[58] R. S. Wolfe and D. J. O'Kane, *J. Biol. Chem.* **205,** 755 (1953).
[59] R. P. Mortlock, R. C. Valentine, and R. S. Wolfe, *J. Biol. Chem.* **234,** 1653 (1959).
[60] R. P. Mortlock and R. S. Wolfe, *J. Biol. Chem.* **234,** 1657 (1959).
[61] J. C. Rabinowitz, *J. Biol. Chem.* **235,** PC50 (1960).
[62a] R. Breslow, *J. Am. Chem. Soc.* **80,** 3719 (1958).
[62b] H. Holzer and K. Beaucamp, *Biochim. et Biophys. Acta* **46,** 226 (1961).
[62c] C. Delavier-Klutchko, *Compt. rend. acad. sci.* **252,** 1681 (1961).
[63] A. L. Shug and P. W. Wilson, *Federation Proc.* **15,** 355 (1956).
[64] L. E. Mortenson, *Federation Proc.* **20,** 234 (1961).
[65] L. E. Mortenson, H. F. Mower, and J. E. Carnahan, *Bacteriol. Revs.* In press (1962).
[66] H. F. Mower, *Federation Proc.* **20,** 349 (1961).
[67] J. McNary, K. C. Schneider, and R. H. Burris, *Federation Proc.* **20,** 350 (1961).
[68] R. H. Burris, *J. Biol. Chem.* **143,** 509 (1942).
[69] I. Zelitch, E. D. Rosenblum, R. H. Burris, and P. W. Wilson, *J. Bacteriol.* **63,** 563 (1952).
[70] J. S. Wall, A. C. Wagenknecht, J. W. Newton, and R. H. Burris, *J. Bacteriol.* **63,** 563 (1952).
[71] A. E. Williams and R. H. Burris, *Am. J. Botany* **39,** 340 (1952).
[72] I. Zelitch, P. W. Wilson, and R. H. Burris, *Plant Physiol.* **27,** 1 (1952).
[73] I. Zelitch, E. D. Rosenblum, R. H. Burris, and P. W. Wilson, *J. Biol. Chem.* **191,** 295 (1951).
[74] I. Zelitch, *Proc. Natl. Acad. Sci. U.S.* **37,** 559 (1951).
[75] J. W. Newton, P. W. Wilson, and R. H. Burris, *J. Biol. Chem.* **204,** 445 (1953).
[76] R. M. Allison and R. H. Burris, *J. Biol. Chem.* **224,** 351 (1957).
[76a] H. S. Mason, *Advances in Enzymol.* **19,** 79 (1957).
[77] D. M. Molnar, R. H. Burris, and P. W. Wilson, *J. Am. Chem. Soc.* **70,** 1713 (1948).
[78] R. Repaske and P. W. Wilson, *J. Am. Chem. Soc.* **74,** 3101 (1952).
[79] T. G. G. Wilson and E. R. Roberts, *Biochim. et Biophys. Acta* **15,** 568 (1954).
[80] M. M. Mozen and R. H. Burris, *Biochim. et Biophys. Acta* **14,** 574 (1954).
[81] M. M. Mozen, Ph.D. Thesis, University of Wisconsin, Madison, 1955.

[82] R. H. Burris, *in* "Inorganic Nitrogen Metabolism" (W. D. McElroy and B. Glass, eds.), p. 316. Johns Hopkins Press, Baltimore, Maryland, 1956.
[83] R. Novak and P. W. Wilson, *J. Bacteriol.* **55,** 517 (1948).
[84] E. D. Rosenblum and P. W. Wilson, *J. Bacteriol.* **61,** 475 (1951).
[85] D. Spencer, H. Takahashi, and A. Nason, *J. Bacteriol.* **73,** 553 (1957).
[86] L. E. Mortenson, unpublished result (1961).
[87] M. K. Bach, *Biochim. et Biophys. Acta* **26,** 104 (1957).
[88] E. R. Roberts, *in* "Utilization of Nitrogen and Its Compounds by Plants." *Symposia Soc. Exptl. Biol.* **13,** 24 (1959).
[89] C. C. Delwiche, *in* "Inorganic Nitrogen Metabolism" (W. D. McElroy and B. Glass, eds.), p. 233. Johns Hopkins Press, Baltimore, Maryland, 1956.
[90] F. E. Mumford, J. E. Carnahan, and J. E. Castle, *J. Bacteriol.* **77,** 86 (1959).
[91] A. L. Shug, P. B. Hamilton, and P. W. Wilson, *in* "Inorganic Nitrogen Metabolism" (W. D. McElroy and B. Glass, eds.), p. 344. Johns Hopkins Press, Baltimore, Maryland, 1956.
[92] P. B. Hamilton, A. L. Shug, and P. W. Wilson, *Proc. Natl. Acad. Sci. U.S.* **43,** 297 (1957).
[93] F. J. Bergersen and P. W. Wilson, *Proc. Natl. Acad. Sci. U.S.* **45,** 1641 (1959).
[94] N. Bauer and R. G. Mortimer, *Biochim. et Biophys. Acta* **40,** 170 (1960).
[95] P. W. Wilson, S. B. Lee, and O. Wyss, *J. Biol. Chem.* **139,** 81 (1941).
[96] A. S. Phelps and P. W. Wilson, *Proc. Soc. Exptl. Biol. Med.* **47,** 473 (1941).
[97] S. B. Lee and P. W. Wilson, *J. Biol. Chem.* **151,** 377 (1943).
[98] M. Green and P. W. Wilson, *J. Bacteriol.* **65,** 511 (1953).
[99] M. Green, M. Alexander, and P. W. Wilson, *Proc. Soc. Exptl. Biol. Med.* **82,** 361 (1953).
[100] M. D. Kamen and H. Gest, *Science* **109,** 560 (1949).
[101] H. Gest, M. D. Kamen, and H. M. Bregoff, *J. Biol. Chem.* **182,** 153 (1950).
[102] J. W. Newton and P. W. Wilson, *Antonie van Leeuwenhoek, J. Microbiol. Serol.* **19,** 71 (1953).
[103] J. G. Ormerod and H. Gest, *Soc. Am. Bacteriologists Proc.* **P14,** 151 (1960).
[104] H. M. Bregoff and M. D. Kamen, *Arch. Biochem. Biophys.* **36,** 202 (1952).
[105] P. W. Wilson, R. H. Burris, and W. B. Coffee, *J. Biol. Chem.* **147,** 475 (1943).
[106] P. W. Wilson, *Science* **123,** 676 (1956).
[107] G. E. Hoch, H. N. Little, and R. H. Burris, *Nature* **179,** 430 (1957).
[108] M. H. Aprison and R. H. Burris, *Science* **115,** 264 (1952).
[109] G. E. Hoch, K. C. Schneider, and R. H. Burris, *Biochim. et Biophys. Acta* **37,** 273 (1960).
[110] E. D. Rosenblum and P. W. Wilson, *J. Bacteriol.* **59,** 83 (1950).
[111] D. W. S. Westlake and P. W. Wilson, *Can. J. Microbiol.* **5,** 617 (1959).
[112] L. E. Mortenson and P. W. Wilson, *J. Bacteriol.* **62,** 513 (1951).
[113] M. E. Winfield, *Revs. Pure Appl. Chem.* **5,** 217 (1955).
[114] C. A. Parker and P. B. Scutt, *Biochim. et Biophys. Acta* **29,** 662 (1960).
[115] J. H. Quastel and B. Woolf, *Biochem. J.* **20,** 545 (1926).
[116] J. M. Wiame and A. Piérard, *Nature* **176,** 1073 (1955).
[117] H. von Euler, E. Adler, and T. Steenholf-Eriksen, *Z. physiol. Chem. Hoppe-Seyler's* **248,** 227 (1937).
[118] E. Adler, V. Hellstrom, G. Gunther, and H. von Euler, *Z. physiol. Chem. Hoppe-Seyler's* **255,** 14 (1938).
[119] E. Adler, N. B. Das, H. von Euler, and U. Heyman, *Compt. rend. trav. lab. Carlsberg* **22,** 15 (1938).

[120] E. Adler, G. Gunther, and J. E. Everett, *Z. physiol. Chem. Hoppe-Seyler's* **255,** 27 (1938).
[121] J. R. S. Fincham, *J. Gen. Microbiol.* **5,** 793 (1951).
[122] J. R. S. Fincham, *J. Gen. Microbiol.* **11,** 236 (1954).
[123] J. A. Olson, *Federation Proc.* **11,** 266 (1952).
[124] J. A. Olson and C. B. Anfinsen, *J. Biol. Chem.* **197,** 67 (1952).
[125] J. A. Olson and C. B. Anfinsen, *J. Biol. Chem.* **202,** 841 (1953).
[126] A. S. Fairhurst, H. K. King, and C. E. Sewell, *J. Gen. Microbiol.* **15,** 106 (1956).
[127] M. M. Hong, S. C. Schen, and A. E. Braunstein, *Biochim. et Biophys. Acta* **36,** 288 (1959).
[128] M. M. Hong, S. C. Schen, and A. E. Braunstein, *Biochim. et Biophys. Acta* **36,** 290 (1959).
[129] D. S. Goldman, *Biochim. et Biophys. Acta* **34,** 527 (1959).
[130] P. P. Cohen and G. W. Brown, Jr., *in* "Comparative Biochemistry" (M. Florkin and H. S. Mason, eds.), Chapter 4, p. 161. Academic Press, New York, 1960.
[131] R. P. Cook and B. Woolf, *Biochem. J.* **22,** 474 (1928).
[132] B. Woolf, *Biochem. J.* **23,** 472 (1929).
[133] A. I. Virtanen and J. Tarnanen, *Biochem. Z.* **250,** 193 (1932).
[134] N. Ellfolk, *Acta Chem. Scand.* **8,** 151 (1954).
[135] E. F. Gale, *Biochem. J.* **32,** 1583 (1938).
[136] A. I. Virtanen and N. Ellfolk, *in* "Methods in Enzymology" (S. P. Colowick and N. O. Kaplan, eds.), Vol. II, p. 386. Academic Press, New York, 1955.
[137] K. Ichihara, H. Kanagawa, and M. Uchida, *J. Biochem.* (*Tokyo*) **42,** 439 (1955).
[138] K. Ichihara, M. Uchida, N. Matsuda, and H. Kikuoka, *Z. physiol. Chem. Hoppe-Seyler's* **295,** 220 (1953).
[139] J. T. Wachsman, *J. Biol. Chem.* **223,** 19 (1956).
[140] H. A. Barker, R. M. Wilson, and A. Munch-Peterson, *Federation Proc.* **16,** 151 (1957).
[141] H. A. Barker, R. D. Smyth, R. M. Wilson, and H. Weissbach, *J. Biol. Chem.* **234,** 320 (1959).
[142] E. R. Stadtman and P. R. Vagelos, *Proc. Intern. Symposium Enzyme Chem., Tokyo and Kyoto, 1957* p. 86 (1958).
[143] E. R. Stadtman, *J. Am. Chem. Soc.* **77,** 5765 (1955).
[144] P. R. Vagelos, J. M. Earl, and E. R. Stadtman, *J. Biol. Chem.* **234,** 490 (1959).
[145] F. C. Steward and J. F. Thompson, *in* "The Proteins" (H. Neurath and K. Bailey, eds.), Vol. II, Part A. Academic Press, New York, 1954.
[146] A. Meister, *Physiol. Revs.* **36,** 103 (1956).
[147] G. C. Webster, *in* "Utilization of Nitrogen and Its Compounds by Plants" *Symposia Soc. Exptl. Biol.* **13,** 330 (1959).
[148] H. A. Krebs, *Biochem. J.* **29,** 1951 (1935).
[149] J. F. Speck, *J. Biol. Chem.* **168,** 403 (1947).
[150] J. F. Speck, *J. Biol. Chem.* **179,** 1387 (1949).
[151] J. F. Speck, *J. Biol. Chem.* **179,** 1405 (1949).
[152] W. H. Elliot, *Biochem. J.* **49,** 106 (1951).
[153] W. H. Elliot, *J. Biol. Chem.* **201,** 661 (1953).
[154] J. E. Varner and G. C. Webster, *Plant Physiol.* **30,** 393 (1955).
[155] J. E. Varner, D. H. Slocum, and G. C. Webster, *Arch. Biochem. Biophys.* **73,** 508 (1958).
[156] F. Lipmann and L. C. Tuttle, *J. Biol. Chem.* **159,** 21 (1945).
[157] W. H. Elliot and E. F. Gale, *Nature* **161,** 129 (1948).

[158] N. Grossowicz, E. Wainfan, E. Borek, and H. Waelsch, *J. Biol. Chem.* **187,** 111 (1950).
[159] G. Denes, *Biochim. et Biophys. Acta* **15,** 296 (1954).
[160] P. D. Boyer, O. J. Koeppe, and W. W. Luchsinger, *J. Am. Chem. Soc.* **78,** 356 (1956).
[161] A. Kowalsky, C. Wyttenback, L. Langer, and D. E. Koshland, Jr., *J. Biol. Chem.* **219,** 719 (1956).
[161a] P. R. Krishnaswamy, V. Pamiljans, and A. Meister, *J. Biol. Chem.* **235,** PC39 (1960).
[162] G. C. Webster and J. E. Varner, *J. Biol. Chem.* **215,** 91 (1955).
[163] G. C. Webster and J. E. Varner, *Federation Proc.* **14,** 301 (1955).
[164] L. Levintow, *Federation Proc.* **16,** 211 (1957).
[165] M. E. Jones, L. Spector, and F. Lipmann, *J. Am. Chem. Soc.* **77,** 819 (1955).
[166] M. E. Jones, L. Spector, and F. Lipmann, *Proc. Intern. Congr. Biochem. 3rd Congr., Brussels 1955* p. 278 (1956).
[167] R. L. Metzenberg, L. M. Hall, and P. P. Cohen, *Federation Proc.* **16,** 221 (1957).
[168] R. L. Metzenberg, L. M. Hall, and P. P. Cohen, *J. Biol. Chem.* **229,** 1019 (1957).
[169] L. M. Hall, R. L. Metzenberg, and P. P. Cohen, *J. Biol. Chem.* **230,** 1013 (1958).
[170] M. Marshall, R. L. Metzenberg, and P. P. Cohen, *J. Biol. Chem.* **233,** 102 (1958).
[171] M. E. Jones and L. Spector, *J. Biol. Chem.* **235,** 2897 (1960).
[172] J. M. Buchanan and S. C. Hartman, *Advances in Enzymol.* **21,** 199 (1959).
[173] P. Reichard, *Advances in Enzymol.* **21,** 263 (1959).
[174] I. Lieberman, *J. Am. Chem. Soc.* **77,** 2661 (1955).
[175] I. Lieberman, *J. Biol. Chem.* **222,** 765 (1956).
[176] H. S. Moyed and B. Magasanik, *J. Biol. Chem.* **226,** 351 (1957).

CHAPTER 4

Pathways of Amino Acid Biosynthesis

EDWIN UMBARGER AND BERNARD D. DAVIS

I. Introduction 168
A. Demonstration of Biosynthetic Pathways 168
B. The Unity of Biosynthetic Patterns 175
II. The Formation of Glutamic Acid; Its Role in Ammonia Assimilation and Transamination 176
A. Formation of Glutamic Acid 176
B. The Route of Ammonia Assimilation 178
C. Transamination 180
III. The Conversion of Glutamic Acid to Glutamine, Proline, and Arginine 182
A. Formation of Glutamine 182
B. Formation of Proline 185
C. Formation of Arginine 186
IV. Aspartic Acid, Asparagine, and the Key Intermediate: β-Aspartic Semialdehyde 192
A. Formation of Aspartate 192
B. Formation of Asparagine 193
C. Aspartic Semialdehyde as an Intermediate 193
V. The Conversion of Aspartic Semialdehyde to Threonine and Methionine 194
A. The Formation of Homoserine 195
B. The Conversion of Homoserine to Threonine 196
C. The Conversion of Homoserine to Methionine 196
VI. The Conversion of Aspartic Semialdehyde to Diaminopimelate and Lysine 198
A. The Formation of L-Lysine from *meso*-Diaminopimelic Acid 200
B. The Formation of Diaminopimelic Acid 200
VII. The Formation of Alanine, Serine, Glycine, and Cysteine 202
A. Alanine 202
B. Serine and Glycine 203
C. Cysteine 206
VIII. The Formation of Isoleucine, Valine, and Leucine 208
A. The Origin of the Carbon Chain of Valine 208
B. The Origin of the Carbon Chain of Isoleucine 212
C. Common Steps in Valine and Isoleucine Biosynthesis 214
D. The Formation of Leucine 215
IX. Histidine 216
A. Compounds Accumulated by Histidine Auxotrophs 217
B. Enzymic Conversions of Imidazole Derivatives 217
C. Origin of the Imidazole Ring 219
X. The Aromatic Amino Acids: Tyrosine, Phenylalanine, and Tryptophan 222
A. The Common Pathway 222
B. The Formation of Tyrosine and Phenylalanine 227
C. The Formation of Tryptophan 229

XI. The Formation of D-Amino Acids 236
XII. The Formation of Amino Acids as Major Excretion Products 237
XIII. General Considerations 238
A. Comparative Biochemistry 238
B. General Features of Biosynthesis 239
C. Special Features of Amino Acid Biosynthesis 240
D. Biosynthetic versus Degradative Enzymes; Control Mechanisms 242
References 243

I. Introduction

In the last dozen years research in many areas of general biochemistry has come to depend more and more on the microbial kingdom for experimental material. This development is probably nowhere more evident than in the study of the pathways of amino acid biosynthesis. One reason has been the broad synthetic powers of many bacteria: for mammals can synthesize only half the natural amino acids. An even more important factor has been the availability of auxotrophic mutants of molds and bacteria. Such mutants offer several advantages, including the frequent accumulation of intermediates in substantial quantities; the identification of these accumulated compounds has provided the key to a number of pathways.

As will be demonstrated in this chapter, most of the steps in amino acid biosynthesis have by now been defined, although there are still significant gaps in our knowledge. It would appear that the further study of the enzymes of these pathways will soon be limited to elucidating the detailed mechanisms of their reactions, the factors that control their activity, and their arrangement in the cell.

A. Demonstration of Biosynthetic Pathways

1. Catabolic versus Biosynthetic Pathways

It is customary to classify the metabolic reactions occurring in a cell into three groups. The degradative or catabolic reactions bring about the dissimilation of the carbon source to its fermentative or respiratory end products, with the attendant supply of chemical energy; the biosynthetic or anabolic reactions lead to the formation of the building blocks of the cell; and the polymerizing reactions convert the building blocks to the macromolecular proteins, nucleic acids, lipids, and polysaccharides.

With the recent detailed analysis of biosynthetic pathways it has become clear that the so-called catabolic pathways supply to biosynthesis not only energy, but also a variety of required precursors. Thus pyruvate, a key compound in a variety of dissimilatory pathways, is also the branch point for starting the specific pathways to alanine, to valine and beyond it to part of leucine, and to part of diaminopimelic acid and lysine. Similar branch points for other pathways to amino acids are provided by phosphoenolpyruvate, acetyl CoA, oxalacetate, α-ketoglutarate, succinyl CoA, erythrose-4-phosphate, and ribose-5-phosphate (see Table VI, at end of chapter). In the formation of valine, for example, the compounds between glucose and pyruvate are just as true intermediates as those between pyruvate and valine. However, in this chapter a biosynthetic pathway will be considered to start with the first reaction that leads *specifically* to one or more biosynthetic products, and not to a dissimilatory product.*

While we shall thus disregard pathways that are generally capable of performing both a catabolic and a biosynthetic function, it should be noted that a shift in organism or in experimental conditions may give such a pathway a purely biosynthetic role. Thus under aerobic conditions the tricarboxylic acid cycle serves in many organisms both as a source of energy and as a source of the α-ketoglutarate from which glutamate, proline, and arginine are derived. Under anaerobic conditions, in contrast, the same pathway no longer functions as a complete cycle, but its biosynthetic contribution to amino acid formation continues undiminished. Similarly, when an organism grows on glucose as sole carbon source the glycolytic pathway is the classical example of catabolism; but when the carbon source is a member of the tricarboxylic acid cycle (e.g., succinate) such biosynthetic intermediates as ribose phosphate must be provided by an apparent reversal of glycolysis. This is achieved by using energy-linked reactions to circumvent those steps of glycolysis that are nearly irreversible.[1]

In this connection attention might be called to an aspect of the tricarboxylic acid cycle that was revealed by a mutant of *Escherichia coli* blocked between α-ketoglutarate and succinate. Under aerobic conditions this mutant requires succinate, which can be replaced by a mixture of methionine, threonine, and diaminopimelate or lysine.[1a] (Succinate is required as a cofactor in the formation of diaminopimelate and lysine (p. 200); its relation to methionine and threonine is not clear.) Under anaerobic conditions, in contrast, the mutant, like the wild type, grows on glucose as sole carbon source and forms succinate as a fermentation product. Evidently *E. coli* can form succinate by reduction of fumarate under anaerobic but not under aerobic conditions.

2. Criteria for Establishing a Precursor as an Obligatory Intermediate

Since this chapter will attempt to describe for each amino acid the intermediates of the "normal" biosynthetic pathway and the nature of the

* Acetolactate illustrates the difficulty encountered in attempting to distinguish sharply between catabolic and biosynthetic reactions (see Section VIII,A,2). This compound is an intermediate in valine biosynthesis; but in certain organisms it is also an intermediate in the formation of acetoin, an end product of glucose dissimilation.

enzymic steps involved, some additional definitions will be required. In the main, the definitions to be used here are those employed in an earlier review.[2] A *precursor* is a compound, either formed within the cell or supplied to it from without, that can be converted by enzymes within the cell to some product. An *intermediate* is a compound that is both formed and further converted as part of the sequence of reactions leading from a precursor to some product. An intermediate may be termed *obligatory* if it is a member of a pathway that is the only one by which a given product can be formed from a given source material at a rate adequate for growth.

The nature of the source material is sometimes important in defining a biosynthetic pathway. Thus with glucose as sole carbon source *Aerobacter aerogenes* forms glutamic acid only via the tricarboxylic acid cycle, since a mutant lacking the citrate-condensing enzyme requires glutamate or α-ketoglutarate for growth.[3] However, exogenous L-histidine is also active in this mutant because it is degraded by a route in which glutamate is an intermediate.[4] Nevertheless, endogenous histidine fails to induce formation of this degradative pathway and so the pathway to histidine does not continue into a theoretically possible alternative pathway to glutamate.

It would require little indulgence on the part of the reader to concede that this formation of glutamate from exogenous histidine can scarcely be considered a "normal" pathway. However, the example does emphasize the necessity of agreeing on what will be considered a normal medium. In this chapter, a *normal intermediate* will be defined as an obligatory intermediate in cells growing in minimal medium, i.e., one containing only a general carbon source (usually glucose), any growth factors required by the wild-type organism, and inorganic salts.

The demonstration that a compound *can serve as a precursor* of a cell constituent is easier than the proof that *it is a normal, obligatory intermediate* in the formation of that constituent; but the distinction is an important one for a cell physiologist. The screening of possible precursors (whether in nutritional, isotopic, or enzymic experiments) is a usual preliminary step in elucidating the pathway of biosynthesis of a given compound. Whether or not a given precursor is also an intermediate must usually be decided on the basis of several types of evidence. The most convincing pattern of evidence that compound A (see diagram below) is an intermediate would probably include all the following. (*a*) An intact cell or an extract, capable of forming cell constituent X from a given source material O, can also form it from A and B. (*b*) The extract contains a single enzyme that converts A to B.* (*c*) Loss of the activity of this enzyme (by mutation, by

* In order to avoid including as intermediates compounds that occur in the cell only in the enzyme-bound state [such as indole (p. 233) or histidinal (p. 219)], it should also be specified that the enzyme that converts A to B is distinct from the one that forms A and the one that utilizes B.

action of a specific inhibitor, or by omission in a reconstructed system) is invariably accompanied by loss of the ability to convert O to X, and A to X, but not B to X.

$$\begin{array}{ccccccc} O & \rightarrow\rightarrow & A & \rightarrow & B & \rightarrow\rightarrow & X \\ & & \downarrow\uparrow & & \downarrow\uparrow & & \\ & & A' & & B' & & \end{array}$$

In many instances the identity of A has been revealed by its appearance in culture filtrates of mutants blocked in the conversion of A to B. However, it is also possible that the accumulated compound is not A itself but a closely related compound, A′, enzymically interconvertible with A. Likewise, the product obtained from the enzymic conversion of A or A′, and thought to be B, may be an enzymic derivative, B′. The conversion of A′ to B′, like that of A to B, would satisfy the condition that it be effected by an extract of the wild type but not by that of a mutant blocked between A and B. The distinction between the true pair of intermediates A and B, and their derivatives A′ and B′, would then depend, as emphasized by Adelberg,[5] on whether the observed conversion involved a single enzyme or three enzymes. Hence the above criteria for an intermediate are rigorous only to the extent that the reaction can be proved to be mediated by a single enzyme rather than by a complex enzyme system.

As will be pointed out below, the above criteria have been fulfilled in the past several years for a large number of intermediates in amino acid biosynthesis. This development increases our confidence that many of the other compounds described here, which may be provisionally accepted on a less rigorous basis, are also true intermediates. At the same time it should be pointed out that these criteria have made it possible to exclude as intermediates certain active precursors: e.g., quinic acid[6] and the α-keto acids corresponding to tryptophan, histidine, and methionine.

3. Methods of Analyzing Biosynthetic Pathways

The analysis of biosynthetic pathways has profitably utilized several experimental approaches. These include the use of (*a*) auxotrophic (growth factor-requiring) mutants of bacteria which differ from the wild-type parents by single enzyme deficiencies; (*b*) isotopically labeled compounds; (*c*) extracted and purified enzymes; and (*d*) growth-inhibiting analogs of metabolites.

a. Use of Auxotrophic Mutants. The introduction of auxotrophic mutants of *Neurospora crassa* by Beadle and Tatum[7] provided the breakthrough that initiated systematic efforts to define biosynthetic pathways, and this development was soon extended to bacteria. Two basic techniques were involved. One consisted of screening available compounds that might be substituted for the required building block. This procedure was only satisfactory, of course, for identifying precursors to which the cell was permeable. The second technique was based on the fact that many auxotrophs accumulate the compound preceding the blocked reaction.*

In some cases the accumulated compound had been known previously to be active as a precursor, and the recognition of its accumulation served merely to strengthen the idea that it was on the pathway. An example would be the accumulation of the corresponding α-keto acids by certain mutants requiring valine and isoleucine.[8] In other cases the identification of the accumulated compound provided the first indication of its biological role. An example would be the α,β-dihydroxy acids that precede the α-keto acids in the biosynthesis of isoleucine and valine.[9, 10] Indeed, for a number of amino acids (tyrosine and phenylalanine, histidine, proline, lysine) the identification of an accumulated precursor provided the clue that led to the eventual elucidation of the entire biosynthetic sequence.

Many of the accumulated compounds turned out to be not only intermediates but also growth factors for mutants with an earlier block in the same pathway. Such a happenstance permitted ready recognition of the accumulation by tests for cross-feeding (syntrophism), and provided a convenient bioassay during the isolation of the compound. However, extension of the pathways by enzymic methods has revealed many intermediates that are not growth factors because of exclusion by a permeability barrier. Sometimes these compounds (e.g., shikimic acid 5-phosphate,[11, 12] prephenic acid[13]) are nevertheless heavily excreted by bacterial mutants blocked after them. Other impermeable intermediates have been found to be accumulated only within the cell, while their more permeable dephosphorylated derivatives (which are not intermediates) are excreted. Examples (in *Neurospora*) are the several phosphorylated imidazole compounds that serve as intermediates in histidine biosynthesis.[14]

b. Use of Isotopes. In their simplest form, isotopic studies with bacteria were patterned after the classical incorporation experiments with animals, viz., the addition of a labeled compound to the growth medium of the normal organism, followed by isolation and perhaps selective degradation of the cell constituents under investigation. The technique of isotope incorporation has also been applied to intermediates accumulated by auxotrophic mutants. The intermediate may be easier to isolate from the fluid than the related constituent from the cell, and may be better suited for selective degradation. In tyrosine and phenylalanine, for example, the two halves of the aromatic ring are indistinguishable, whereas in the precursor shikimic acid the isotope content of each carbon atom may be determined.[15]

* It was early observed that the greatest accumulation of various precursors occurred after the required growth factor had been exhausted. This fact was later shown to be due to the same regulatory mechanisms that prevented overproduction of metabolites by wild-type organisms.

Isotopically labeled compounds have generally been employed in three ways. The carbon source, such as glucose, lactate, or acetate, may be labeled in a specific carbon atom. Another method employs an unlabeled general carbon source such as glucose and a small addition of a heavily labeled compound (e.g., acetate-1-C^{14}) which is related to glucose by known pathways. Finally, a suspected intermediate may itself be prepared in the labeled form and added to an adequate medium. This technique is more sensitive than a growth-factor test for detecting precursor activity of a compound that penetrates the cell only poorly.

In a much less sensitive but much more convenient variant of the last method, introduced by the Carnegie group,[16] the suspected precursor in unlabeled form is tested in competition with a commercially available labeled general source of carbon (or other element); thus, in effect C^{12} becomes the label against a background of C^{14}. In this way a variety of compounds can be tested without being synthesized with C^{14}. It should be emphasized that this technique is effective only when a large proportion of the cell constituent is formed from the added precursor (at least 20%). Such extensive incorporation of an exogenous precursor has often been achieved; the mechanism, as we now know, involves not only dilution of the endogenous supply but also interference, by normal feedback control mechanisms, with its formation. Because this approach came on the scene late it revealed only a few new steps in amino acid biosynthesis; but since many known steps were confirmed in a single organism, the results obtained by this group have been of considerable interest. As with other techniques involving intact cells, poorly penetrating intermediates (e.g., cystathionine, diaminopimelate) gave misleading results.

The extension of the method of isotope incorporation to bacteria has led to an unexpected limitation. With animals it has usually been possible to introduce labeled material in an amount small enough not to alter significantly the concentration of the compound in the body fluids. However, when an isotopically labeled compound is added to a minimal medium for bacteria, in no matter how small an amount, the possibility arises that its presence may induce formation of a pathway that does not appear during growth in minimal medium. Thus, while there is no method more rigorous than the use of isotopes for demonstrating that a given compound or portion of a compound can serve as a precursor of a cell constituent, there always remains the task of proving that the compound actually does so in a cell growing in minimal medium. For example, Roberts and co-workers[16] observed that threonine, added even in very small amounts to minimal medium, was significantly converted by *E. coli* to glycine. However, these workers did not postulate a normal pathway to glycine that included threonine as an intermediate, since they had earlier observed that threonine and glycine are endogenously formed by independent pathways.

c. The Use of Extracted Enzymes. The introduction of enzymic techniques tended to come late in this field, since the earlier studies concentrated on identifying intermediates that were readily recognized by their growth-factor activity. However, it was later realized that not all intermediates could be discovered in this way, since the implicit assumption of free penetration of all intermediates was incorrect. Indeed, many of the intermediates that will be described in this chapter could be identified only after the pathway in question had been studied with extracted enzymes. Furthermore, rigorous establishment of the biosynthetic function of a reaction now requires the demonstration that absence of its enzyme is associated with auxotrophy (see Section I, A, 2).

Most biosynthetic reactions were long inaccessible to enzymologists for several reasons: their largely endergonic nature, the fact that each biosynthetic pathway is quantitatively small compared with the major catabolic pathways, and the lack of suitable assays for a bewildering array of possible intermediates. Today, however, it is possible to approach unknown biosynthetic reactions with confidence. The reasons include the availability of sensitive and specific tests (mutant bioassay, chromatography combined with radioactive isotopes; ultraviolet spectrophotometry and spectrophotofluorometry); the commercial production of such cofactors as adenosine triphosphate (ATP), diphosphopyridine nucleotide (DPN), and triphosphopyridine nucleotide (TPN), as well as a host of other biochemicals; improved techniques and equipment for the disruption of bacteria and the purification of enzymes; and the use of mutants, as well as enzyme fractionation, to produce interruptions in the sequence.

In the absence of an appropriate mutant it is possible to provide another kind of biological evidence that a given precursor is an obligatory intermediate. This technique takes advantage of the fact that the formation of the enzymes in a biosynthetic sequence is often specifically repressed by the end product.[17, 18, 19] In addition, the action of the earliest enzyme in a biosynthetic sequence is almost invariably inhibited by the end product.[20-23] Because this repression and inhibition are so specifically related to the biosynthetic function of the enzyme in known cases, their presence can be considered evidence for this function in a doubtful case. For example, in a recent study on valine biosynthesis in *E. coli*[24] no mutants were available that were blocked in acetolactate formation, the postulated initial step in this pathway. However, valine specifically inhibited this enzyme and also repressed its formation. It was therefore possible with confidence to relate this enzyme and its product to valine biosynthesis.

d. The Use of Metabolite Analogs. Another tool for the analysis of biosynthetic pathways is "inhibition analysis," based on the use of growth-inhibiting analogs of metabolites and the assumption that they act by competing with the metabolite at an enzyme surface. It would follow that the action of a growth inhibitor may be overcome competitively by the substrate, or noncompetitively by the product, of the inhibited reaction. A number of significant interrelationships have been revealed by this technique (e.g., the relation of *p*-aminobenzoate to 1-carbon transfer); but it is the least rigorous of the techniques described here, and its use has occasionally led to conclusions that are in conflict with data obtained by the other methods. It now appears that at least some of the difficulty is based on failure to consider the possibilities that the analog (*a*) might competitively interfere with penetration of the metabolite, or (*b*) might itself be competitively excluded by the metabolite.

Further complications in the interpretation of analog action have emerged with the recognition that not only the end product of a biosynthetic sequence, but also its analogs, can exert feedback control over the action of the initial enzyme of the sequence. Moyed and Friedman[25, 26] have observed that inhibition by certain tryptophan analogs could be noncompetitively reversed by tryptophan and also by intermediates in its formation. There was no evidence for the classical action of the analog: competition with tryptophan at its site of incorporation. Moyed[26] has recently analyzed the action of several analogs of histidine and obtained direct evidence that they, in fact, block the histidine biosynthetic pathway in the same way as histidine itself does. Such mimicry of the end product in inhibiting the action of the initial enzyme has been called "false feedback."

It will become apparent to the reader that there is no one method for the elucidation of a biosynthetic pathway. This chapter will attempt to outline not only the present state of our knowledge of the various pathways, but also the methods by which each intermediate was discovered.

B. The Unity of Biosynthetic Patterns

The concept of the "unity of biochemistry" has undergone a considerable evolution since the phrase was introduced by Kluyver and Donker in 1926.[27] They focused on the observation that the predominant chemical reactions in the cell were essentially hydrogen transfer reactions, and could be expressed in the general equation:

$$AH_2 + B \rightarrow A + BH_2 \qquad (1)$$

In later years, because of the similarity noted in protoplasmic building blocks from one living material to another, and because of the similarity of the major reaction sequences, both catabolic and biosynthetic, in various organisms, the unity of biochemistry came to refer to a sort of "master plan" of reactions to which all organisms adhered within certain limits. In comparison with the similarities, the dissimilarities were minor. For example, in the study of amino acid biosynthetis only two pathways have been reported to differ in *Neurospora crassa* and *E. coli*: those to ornithine[28] and to lysine.[29, 30] Consequently, while the sequences reported in this chapter have been worked out largely in these two species, it is highly probable that most of them are used by all organisms that synthesize the amino acids in question.

II. The Formation of Glutamic Acid; Its Role In Ammonia Assimilation and Transamination

A. Formation of Glutamic Acid

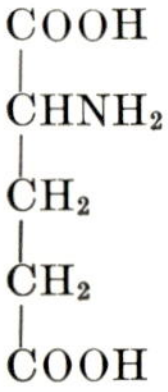

Glutamic acid

As is well known, glutamate is formed by the amination of α-ketoglutarate, which is produced in the tricarboxylic acid cycle by the oxidative decarboxylation of isocitrate. This is the only significant route for *E. coli* or *A. aerogenes* growing on glucose as carbon source, since mutants lacking the citrate-condensing enzyme[3] have an absolute requirement for glutamate or α-ketoglutarate.* The inactivity of succinate is readily understood, since its formation from α-ketoglutarate is irreversible.[33]

This utilization of α-ketoglutarate is an example of the biosynthetic function of the tricarboxylic acid cycle, which competes with its energy-producing function by preventing completion of the cycle and regeneration of oxalacetate.[34] Hence in growth on glucose one molecule of dicarboxylic acid must be formed via CO_2 fixation for every molecule of α-ketoglutarate (or other tricarboxylic acid cycle intermediate) assimilated.

The pathway to glutamate via the tricarboxylic acid cycle is, of course, the "normal" one by the definition outlined earlier in this chapter. This

* Under aerobic conditions, citrate, aconitate, and isocitrate are inactive for the *E. coli* mutant because of a permeability barrier. However, during anaerobic growth on glucose, this mutant is induced by citrate to develop a transport system that permits this compound to satisfy the glutamate requirement.[31] In *A. aerogenes*, in contrast, under anaerobic as well as aerobic conditions, glucose represses the induction of the transport system for citrate; but in the presence of certain other carbon sources (e.g., lactate) this induction occurs.[32] The effects of glucose and aerobiosis in repressing the induction of the transport system thus differ in the two species for reasons that are not clear.

does not imply that glutamate cannot be formed by other mechanisms under special circumstances. For example, its formation from arginine in *Bacillus subtilis* has been observed by Wiame.[35] Although the over-all route is the reverse of the biosynthetic pathway from glutamate to arginine, it is of interest to note that the enzymic steps are quite different. Just as in the conversion of histidine to glutamate by *A. aerogenes*[4] (cited on page 170), the reactions involved are induced only by an exogenous supply of the source material.

There are two known kinds of enzyme that can aminate α-ketoglutarate: glutamic dehydrogenase and various transaminases. Both have been found in most animal, plant, and microbial sources examined. Glutamic dehydrogenase catalyzes the reaction as shown in Reaction (I).

$$\begin{array}{c} COOH \\ | \\ C=O \\ | \\ CH_2 \\ | \\ CH_2 \\ | \\ COOH \end{array} + NH_4^+ + TPNH \rightleftharpoons \begin{array}{c} COOH \\ | \\ CHNH_2 \\ | \\ CH_2 \\ | \\ CH_2 \\ | \\ COOH \end{array} + TPN^+ + H_2O$$

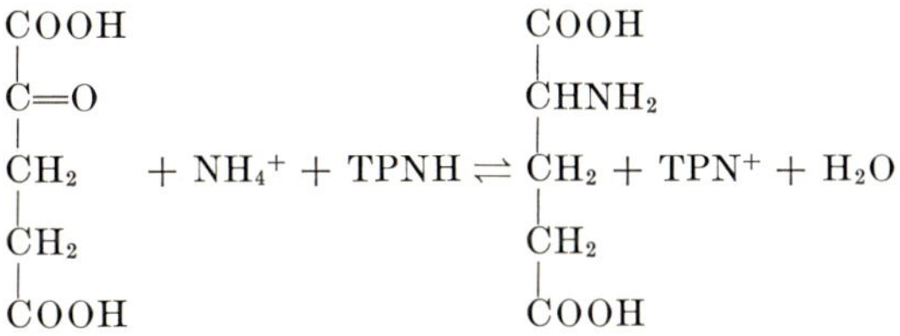

α-Ketoglutaric Acid   Glutamic acid

(I)

In *E. coli* and most other organisms the reaction is TPN-linked. However, DPN-linked glutamic dehydrogenases have been found in organisms in which the deamination appears to serve a catabolic rather than a biosynthetic function. For example, in *Clostridium sporogenes* a DPN-linked glutamic dehydrogenase, along with a glutamate-tryptophan transaminase, was shown to play an essential role in the degradation of tryptophan.[36]

As will be shown in Section II, C, 1, several transaminases catalyze the amination of α-ketoglutarate in the presence of various α-amino acids as amino donors. Such reactions have the over-all general form shown in Reaction (II).

$$\begin{array}{c} COOH \\ | \\ C=O \\ | \\ CH_2 \\ | \\ CH_2 \\ | \\ COOH \end{array} + \begin{array}{c} R-CHCOOH \\ | \\ NH_2 \end{array} \rightleftharpoons \begin{array}{c} COOH \\ | \\ CHNH_2 \\ | \\ CH_2 \\ | \\ CH_2 \\ | \\ COOH \end{array} + \begin{array}{c} R-C-COOH \\ \| \\ O \end{array}$$

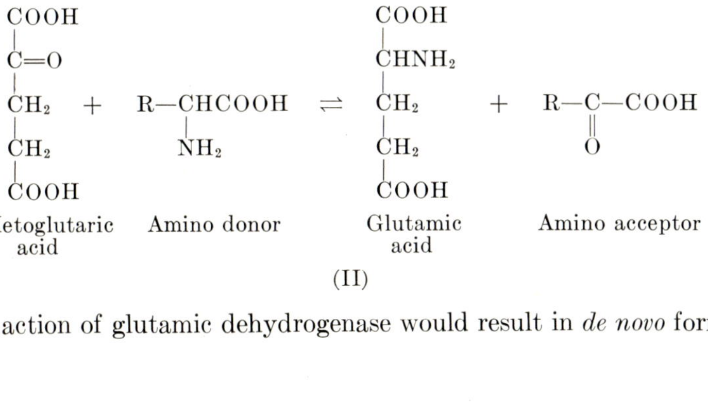

α-Ketoglutaric   Amino donor   Glutamic   Amino acceptor
acid   acid

(II)

The action of glutamic dehydrogenase would result in *de novo* formation of amino groups, which could then be distributed via transamination to the carbon chains of other amino acids. However, it is also conceivable that in some organisms the normal mechanism for the amination of α-ketoglutarate is transamination, in which case the direct amination of some other carbon chain would be required. The origin of amino groups is therefore quite pertinent to the mechanism of α-ketoglutarate amination.

B. The Route of Ammonia Assimilation

In addition to glutamic dehydrogenase, the enzyme aspartase[37, 38, 39] has also been proposed as an agent for the formation of α-amino groups from ammonia.[40] This enzyme reversibly converts aspartate to fumarate plus NH_3 (Reaction III). It was one of the earliest enzymes to be studied in bacteria.[38]

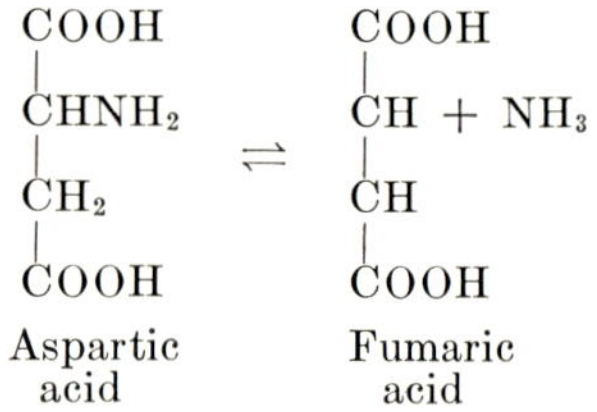

(III)

While the action of aspartase was established by these early studies, its physiological role was not, particularly because of the unfortunate use of broth-grown cells of *E. coli*.[41] (The later experiments of Warner[42] showed that in a medium containing preformed amino acids the *de novo* formation of amino groups from exogenous ammonia by *E. coli* is almost completely suppressed.) A more direct test of the mechanism of amination would be possible with a mutant that had lost glutamic dehydrogenase, but unfortunately no such mutants have been reported in bacteria. Fincham[43, 44] has described such a mutant of *Neurospora crassa*, which grows well in a minimal medium only if glutamate or any one of a number of other amino acids is supplied. It is thus clear that in *Neurospora* glutamic dehydrogenase is a nearly* obligatory enzyme for amino group formation.

To study this problem further in bacteria, aspartase and glutamic dehydrogenase have been measured[45] in extracts of *E. coli* grown on various carbon sources. After growth on glucose only the values for glutamic dehydrogenase were sufficiently high to account for the required rate of conversion of ammonia to amino nitrogen; and after growth on an amino acid mixture, which would require much amino acid degradation and little

* In this mutant the loss of glutamic dehydrogenases did not result in an absolute requirement, i.e., slow growth occurred in minimal medium.

synthesis, aspartase activity was high and glutamic dehydrogenase low. These results indicate a degradative role for aspartase. On the other hand, after growth on succinate the aspartase activities were about equal to those of glutamic dehydrogenase, and so a biosynthetic role for aspartase under these circumstances cannot be excluded.

Since glutamic dehydrogenase clearly has an important biosynthetic role, it seems unfortunate that it was named after the reverse of its biosynthetic reaction. With a high ammonia concentration (as in the usual growth medium) the reaction proceeds essentially quantitatively in the direction of glutamate formation. In fact, the transfer of hydrogen in the opposite direction, from glutamate to TPN, can be achieved *in vitro* only by using a high pH (9.0) and a high glutamate and low ammonia concentration; even then the initial rate decreases rapidly. Furthermore, function in the reductive direction is also promoted by the fact that TPN is maintained in the cell (at least in mammals) in a more reduced state than DPN.[46]

Another reversible enzyme capable of fixing ammonia is alanine dehydrogenase (Reaction IV), a DPN-linked enzyme found in *B. subtilis* by Wiame and Piérard.[47, 48] Recently Shen *et al.*[49-52] have reported that most strains of the genus *Bacillus* have this enzyme but lack glutamic dehydrogenase. These workers therefore concluded that the main pathway of ammonia assimilation in this genus is via alanine.

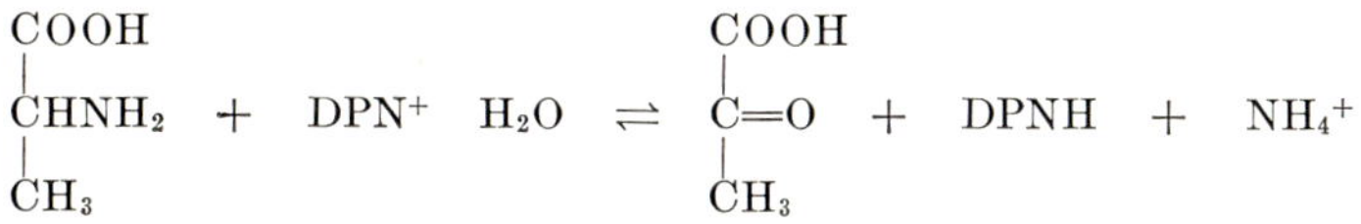

Alanine Pyruvate

(IV)

In this connection it should be noted that Wiame *et al.*[53] found alanine dehydrogenase in a strain of *B. subtilis* that was unable to assimilate ammonia. When a mutant was selected from this strain for ability to grow on ammonia it was found to contain a TPN-linked glutamic dehydrogenase. While these observations, and the linkage with DPN rather than TPN, raise some question concerning the physiological role of alanine dehydrogenase, a biosynthetic function is not excluded.

Although glutamic dehydrogenase has been favored by many as *the* mechanism for net amino group formation, it seems probable that the normal pathway of ammonia assimilation in bacteria varies with different organisms. Because the question has been investigated only for *E. coli* and members of the genus *Bacillus*, it is not yet possible to state whether the glutamic or the alanine dehydrogenase pathway is the more widespread in bacteria.

C. Transamination

Whether the initial uptake of NH_3 yields glutamate, alanine, or aspartate, the resulting amino group is then distributed to the other amino acids. The term "transamination" is restricted to an enzymic reaction in which amino nitrogen is transferred from one compound to another without the intermediate formation of free ammonia.

1. Distribution and Specificity of the Transaminases

Although transamination was described by Braunstein and Kritzmann in 1937,[54] it was not until 1950 that Feldman and Gunsalus,[55] using dried cells of *E. coli*, *Pseudomonas fluorescens*, and *Bacillus subtilis*, first demonstrated the process in a bacterial system. These experiments were especially significant because they verified the disputed claim that a wide variety of amino acids were capable of undergoing transamination. Indeed, even abnormal amino acids, such as norleucine, can be transaminated by bacterial enzymes.

TABLE I

α-Ketoglutarate-Amino Acid Transamination in *Escherichia coli*[a]

Amino acid	Parent strain extract	Fraction A	Fraction B	Isoleucine auxotroph extract
Isoleucine	40	0	84	0
Valine	37	0	58	0
Leucine	41	3	(100)	3
Methionine	31	10	32	19
Phenylalanine	37	44	26	29
Tyrosine	25	25	7	18
Tryptophan	73	59	0	72
Aspartate	(100)	(100)	0	(100)

[a] Transaminase activity is expressed in terms of the reaction of α-ketoglutarate with the most active substrate, which is arbitrarily assigned a value of 100. (Adapted from Meister).[57]

Transamination in *E. coli* has been shown by Rudman and Meister[56] to involve at least three distinct enzymes. One, designated transaminase A, transferred amino groups between any amino and any keto acid of the group glutamate, aspartate, the aromatic amino acids, methionine, and, to a limited extent, leucine. Transaminase B acted on glutamate, isoleucine, valine, and leucine. The relative activities of these fractions are presented in Table I. A third transaminase (not named) reacted with valine, α-aminobutyrate, and alanine. Loss of transaminase B by mutation resulted in an absolute requirement only for isoleucine; apparently valine and leucine receive amino groups adequately from the other transaminases.

Transaminases in other organisms may have somewhat different, but equally broad, specificities. For example, *Brucella abortus* contains a transaminase reacting leucine, isoleucine, and phenylalanine with pyruvate, as well as other(s) reacting these amino acids with α-ketoglutarate.[58]

More recently additional transaminases have been discovered in *E. coli* and other microorganisms which have a greater substrate specificity than those noted above. These enzymes appear to have a specific biosynthetic function (histidinol; *N*-succinyl diaminopimelate; possibly ornithine), and they will be discussed under the appropriate headings.

The transamination reactions discussed above are all concerned with α-amino nitrogen. As has been discussed by Meister,[59] in animal tissues several transaminases have been found which react with ω-amino acids and aldehydes. The only ω-amino acids with known roles in biosynthesis (except for glycine) are ornithine and lysine. As will be discussed in the sections devoted to these amino acids, the ω-amino group of ornithine, but not of lysine, arises in bacteria via a specific ω-transaminase.

Transamination involving D-amino acids will be discussed in a later section (p. 236), as will the irreversible transfer of the amino group from glutamine (p. 183).

2. Mechanism of Transamination[59]

Like many other enzymes reacting with amino acids, transaminases were early discovered to require a derivative of pyridoxine as a cofactor. Subsequent studies with model systems[60, 61] and with purified enzymes and crystalline pyridoxal and pyridoxamine phosphates[62] have favored the participation of a Schiff's base formed from the coenzyme and the α-keto or α-amino acid. Metzler *et al.*[60, 61] have proposed mechanisms for all pyridoxal-catalyzed amino acid reactions which include a Schiff's base-chelate with pyridoxal (not pyridoxal phosphate), an amino acid, and a trivalent metal ion (Al^{+++}). While such a complex, by appropriate electron shifts, can undergo most of the reactions that enzymes are known to catalyze for amino acids, there is no evidence that enzymic transamination requires a metal ion. However, its role in the model reaction could conceivably be performed by the protein in the enzymic reaction.

The interconversions catalyzed by the transaminases are reversible. Because the amino group is deposited on the prosthetic group of the transaminase during the reaction, any amino acid capable of reacting with a transaminase in the pyridoxal state should yield a keto acid that can react with the transaminase in the pyridoxamine state. It should therefore not be necessary, contrary to earlier views, for either the amino donor or the amino acceptor to be a dicarboxylic acid. This principle is supported by the observation[56] that transaminase A of *E. coli* catalyzed transamination between any one of its amino acid substrates and any one of their α-keto analogs.

3. Significance of Transamination in Biosynthesis

The transaminases provide the cell with a mechanism for distributing amino groups to a wide variety of carbon chains. The source of the amino groups might be glutamate (or possibly aspartate or alanine) formed from ammonia and the corresponding α-keto acid, or it might be an exogenous amino acid.

As will be shown below, the final step in the biosynthesis of alanine, phenylalanine, tyrosine, isoleucine, valine, leucine, and probably aspartate involves the corresponding α-keto acid. With other amino acids (tryptophan, histidine, lysine, arginine, threonine, methionine, and cysteine) the known pathways of biosynthesis exclude transamination of the corresponding α-keto as an important normal reaction, even though some of these α-keto acids can serve as growth factors or can be acted on by transaminases. The amino groups of these amino acids nevertheless also originate from glutamate, being introduced (usually via transamination) at an earlier stage in the biosynthetic pathway.

III. The Conversion of Glutamic Acid to Glutamine, Proline, and Arginine

The previous section emphasized the central position of glutamate in the formation and distribution of amino groups. Glutamate is also the source of the carbon and nitrogen of proline and much of arginine.* While a metabolic relationship between these three amino acids was suggested early on the basis of isotopic studies in animals,[63] and the enzymic interconversions were also first studied with mammalian material, the pathway of normal biosynthesis was revealed by studies with microorganisms.

A large proportion of the glutamate in protein exists in the form of another derivative, glutamine, which is also important as a participant in several biosynthetic reactions.

A. Formation of Glutamine

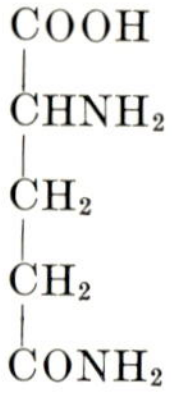

Glutamine

* Glutamic acid is also the source of hydroxyproline in higher animals, but hydroxyproline does not appear to be a constituent of bacteria.

1. The Amidation Reaction

Glutamine is formed from glutamate plus ammonia by an enzymic reaction requiring ATP (see Reaction V).

$$\begin{array}{c} \text{COOH} \\ | \\ \text{CHNH}_2 \\ | \\ \text{CH}_2 \\ | \\ \text{CH}_2 \\ | \\ \text{COOH} \\ \text{Glutamic acid} \end{array} + \text{NH}_3 + \text{ATP} \rightarrow \begin{array}{c} \text{COOH} \\ | \\ \text{CHNH}_2 \\ | \\ \text{CH}_2 \\ | \\ \text{CH}_2 \\ | \\ \text{CONH}_2 \\ \text{Glutamine} \end{array} + \text{ADP} + \text{Pi}$$

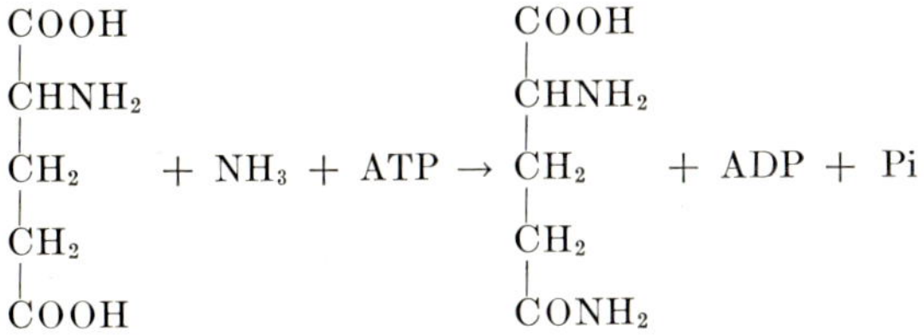

(V)

Although most of our knowledge concerning the mechanism of this reaction has been obtained using animal enzymes,[64,65] a very similar enzyme has been found in *Proteus vulgaris*.[66] In addition to Reaction (V), both the animal and the bacterial enzyme catalyze the transfer of the γ-glutamyl group to hydroxylamine, hydrazine, or ammonia. The enzyme is thus both a synthetase (Reaction V) and a transferase (Reaction VI).

$$\begin{array}{c} \text{COOH} \\ | \\ \text{CHNH}_2 \\ | \\ \text{CH}_2 \\ | \\ \text{CH}_2 \\ | \\ \text{CONH}_2 \end{array} + \text{NH}_2\text{OH} \rightarrow \begin{array}{c} \text{COOH} \\ | \\ \text{CHNH}_2 \\ | \\ \text{CH}_2 \\ | \\ \text{CH}_2 \\ | \\ \text{CONHOH} \end{array} + \text{NH}_3$$

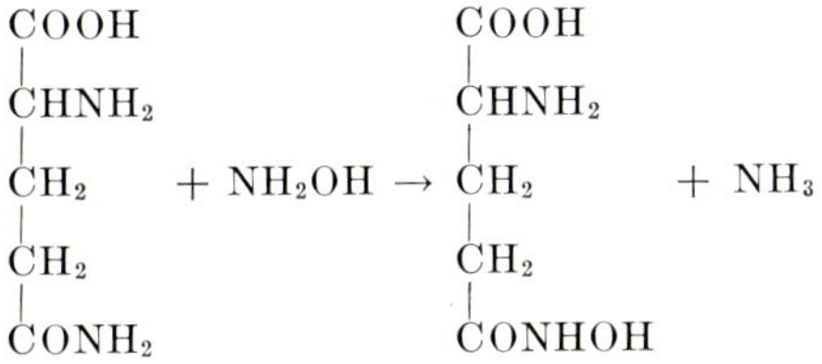

(VI)

A mutant of *A. aerogenes* requiring glutamine has been isolated.[67] It also responds to peptides of glutamine, which are present in enzymic hydrolysates of protein and which (in contrast to glutamine) survive autoclaving. Extracts of this mutant lack glutamine synthetase activity (Reaction V).[68] The occurrence of such a mutant implies that this reaction is essential and that the amide group of glutamine residues in protein is formed *before* incorporation into the polypeptide chain rather than *after*.

Glutamine is a fairly common growth factor for lactic acid bacteria.[69,70] In *Clostridium tetani* good growth was obtained without glutamine, but for satisfactory toxin production this compound was required.[71] The mechanisms underlying these requirements are not clear, although in some strains of *Lactobacillus arabinosus* glutamine appears to be superior to glutamate because of its greater effectiveness in penetrating the cell.[72,73]

2. Transfer Reactions of Glutamine

a. Transfer of the Amide Group. An important metabolic role for glutamine involves the transfer of nitrogen in the biosynthesis of purines and amino sugars. Since these reactions will be considered elsewhere in this volume, they are merely listed in Table II. While enzymic reactions (*a*) and (*b*) have not been demonstrated in bacterial systems, there is no reason to doubt their occurrence there.

Also listed in the table is the formation of imidazoleglycerol phosphate. As will be discussed in detail in the section devoted to histidine, isotope labeling experiments using intact *E. coli*,[77] and early enzyme experiments,[78] indicated that nitrogen atom 1 of the imidazole ring is derived from the amide nitrogen of glutamine in preference to ammonia. However, the preference may possibly depend on superior penetration; for in more exhaustive experiments with enzymes prepared from *A. aerogenes* nitrogen 1 was shown to be derived from ammonia.[79]

TABLE II

BIOSYNTHETIC REACTIONS INVOLVING THE AMIDE GROUP OF GLUTAMINE

Reaction	Reference
(*a*) 5-Phosphoribosyl-1-pyrophosphate + glutamine → 5-phosphoribosylamine + glutamic acid + pyrophosphate	74
(*b*) α-*N*-Formylglycinamide ribotide + glutamine → α-*N*-formylglycinamidine ribotide + glutamic acid	75
(*c*) Glucose-6-phosphate + glutamine → glucosamine-6-phosphate + glutamic acid	76
(*d*) Adenosine-5′-phosphate + 5-phosphoribosylpyrophosphate → "Compound III" + pyrophosphate; Compound III + glutamine → imidazole glycerol phosphate + 4-aminoimidazole-5-carboxamide riboside-5′-phosphate + glutamic acid (Fig. 10)	77

The action of glutamine synthetase, followed by the transfer of amide nitrogen to other carbon chains, provides a mechanism for coupling phosphate bond energy to addition reactions involving ammonia.

b. Transfer of the γ-Glutamyl Group. The energy of the amide bond is also preserved in γ-glutamyl transfer reactions (VI). One such system was studied in *B. subtilis*,[80-82] which contains a capsule composed largely of D-glutamyl residues in γ-peptide linkage. The enzyme catalyzing this reaction was isolated as a peptidase which split the capsular polypeptide. However, in the presence of glutamine γ-glutamyl peptides (up to a chain length of at least 5 or 6) are formed by a series of successive transamidation reactions. Whether this is the mechanism for the synthesis of the capsular polypeptides has not been determined.

Another γ-glutamyl peptide is glutathione (γ-L-glutamyl-L-cysteinylglycine). While animal enzymes have been described that catalyze glutamyl transfer from glutathione, it seems unlikely that reversal of this process would account for glutathione formation in living cells. Glutathione synthesis from glutamic acid, glycine, cysteine, and an ATP-generating system has been demonstrated in extracts of *E. coli*.[83] Although the individual steps of the over-all reaction were not studied it seems likely that they would be similar to those found in liver.[84, 85]

There has as yet been no evidence that γ-glutamyl transfer plays an obligatory role in any biosynthetic sequence, although it is possible that such a role will eventually be discovered. The transamidation reactions that have been observed may conceivably be adventitious actions of an enzyme whose function is hydrolytic.

c. *Transfer of the α-Amino Group*. In animal tissues Meister *et al.*[86, 87, 88] have described systems which transfer the α-amino groups of glutamine and asparagine to a large number of α-keto acids. The resulting compounds, α-ketoglutaramic and α-ketosuccinamic acids, are in turn attacked by specific ω-amidases. Because of the irreversibility of these reactions they are potentially important mechanisms by which a transamination with an otherwise unfavorable equilibrium might be promoted, indirectly using the energy in ATP (which enters into glutamine synthesis). While these reactions have not as yet been observed in any bacterial systems, they should be kept in mind as potential mechanisms for exergonic amino group transfer.

B. Formation of Proline

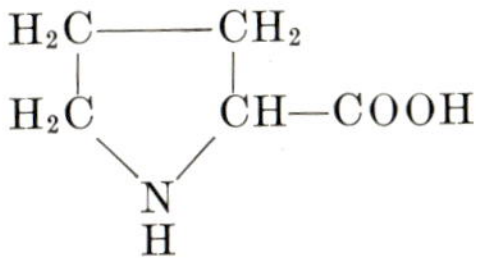

Proline

The pathway from glutamate to proline was made accessible by the isolation of two kinds of specific proline auxotrophs of *E. coli*; one accumulated a compound that could replace proline as a growth factor for the other.[89] The accumulated compound was identified[90] as glutamic γ-semialdehyde, which had been proposed earlier as a metabolic intermediate[91] in order to account for the interconversion of proline and glutamate indicated by isotopic experiments in animals.

The reduction of glutamate to glutamic γ-semialdehyde has been studied in resting cells of an *E. coli* mutant.[22] The conversion has not been reported in extracts or broken cell preparations. However, from the analogy offered by the conversion of aspartate to aspartic-β-semialdehyde (p. 194), it seems likely that at least two enzymic steps (Reactions VII and VIII, Fig. 1) are involved, with γ-glutamyl phosphate as an intermediate.

In aqueous solution glutamic γ-semialdehyde is in tautomeric equilibrium

(Reaction IX) with its cyclization product, Δ^1-pyrroline-5-carboxylic acid;[90] hence the cyclization reaction in this pathway does not require, and presumably does not utilize, an enzyme.

The second reductive step in proline biosynthesis, the conversion of Δ^1-pyrroline-5-carboxylic acid to proline (Reaction X), has been observed in extracts of *A. aerogenes* and *E. coli.*[92] Although DPNH was employed as the hydrogen donor, it was later observed that TPNH was more active.[93] TPNH is also more active than DPNH for the pyrroline-5-carboxylate reductase of *Neurospora.*[94]

The over-all formation of proline from glutamate is readily reversed, as evidenced by the fact that glutamate auxotrophs of *E. coli* will also respond to proline.[89] Such mutants require about ten times as much proline as do those blocked between glutamate and proline, indicating that proline is serving as a source of glutamate. Whether this conversion of proline to glutamate proceeds by precise reversal of the biosynthetic pathway has not been established.

C. Formation of Arginine

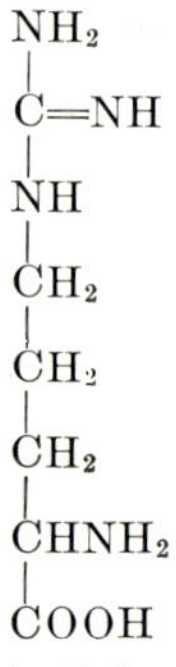

Arginine

The formation of arginine from ornithine via citrulline was one of the first biosynthetic pathways to be discovered, and one of the few to be revealed initially by enzymic analysis. This metabolic sequence was originally elucidated as part of the Krebs-Henseleit cycle for urea formation in liver.[95] The reactions leading from ornithine to arginine have been studied most extensively in animal tissues,[96] while the biosynthesis of ornithine itself has been analyzed with the aid of microbial mutants. Ornithine is also the source of the putrescine moiety of the important bases, spermine and spermidine.

1. Glutamic Acid to Ornithine

The pathway of ornithine biosynthesis (Fig. 2) has been analyzed largely by Vogel, following the earlier observation[89] that one ornithine auxotroph of

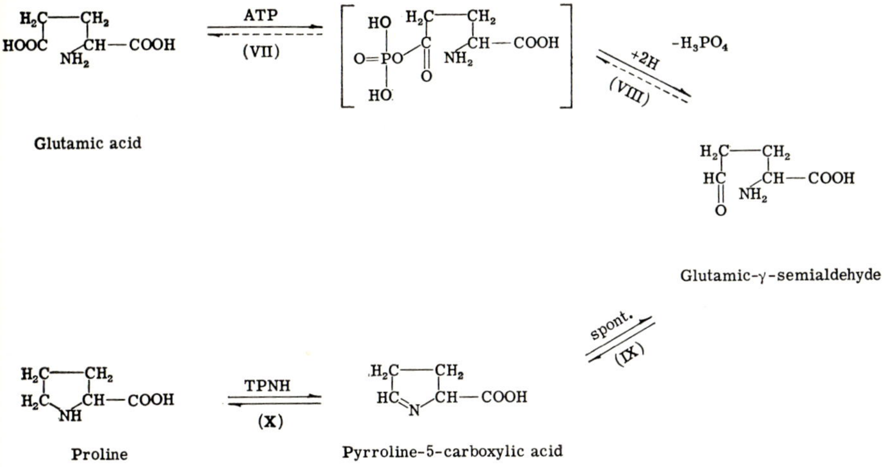

FIG. 1. The biosynthesis of proline.

E. coli accumulated a precursor to which another could respond. The accumulated compound is N^{α}-acetylornithine,[97, 98, 99] and the same strain was also found to accumulate *N*-acetylglutamic semialdehyde.[99] Structural considerations suggested *N*-acetylglutamate as an earlier intermediate.

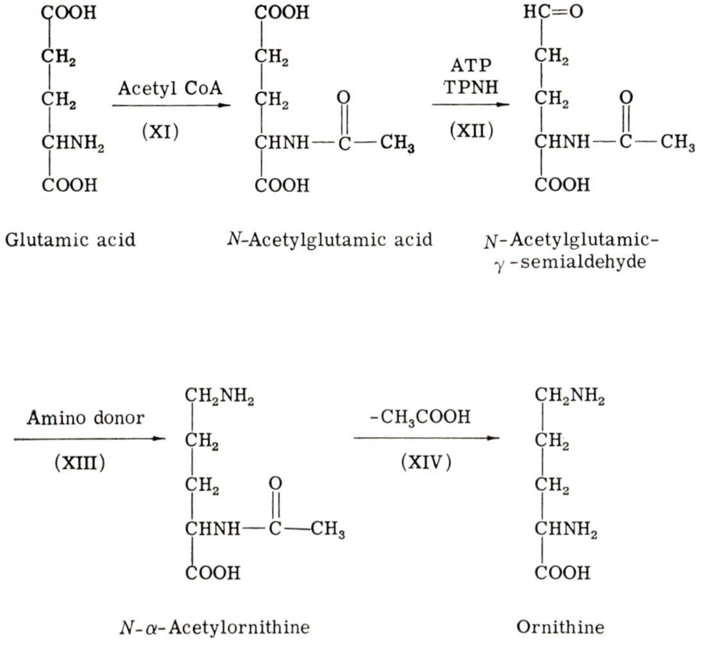

FIG. 2. The biosynthesis of ornithine.

This compound was shown to be heavily incorporated into arginine in isotopic experiments with the wild-type organism,[100] and to serve as a growth factor for a mutant with an early block in ornithine synthesis.[101] These three acetylated compounds therefore seemed likely intermediates in the conversion of glutamate to ornithine in *E. coli*.

This view has been strengthened by the results of enzymic studies. *N*-Acetylglutamate has been formed by bacterial extracts from glutamate and acetyl CoA (Reaction XI).[102] The formation of *N*-acetylglutamic-γ-semialdehyde (Reaction XII) by extracts of *Micrococcus glutamicus* has been reported by Udaka and Kinoshita.[103] These workers found that the forward reaction required ATP as well as TPNH, and a hydroxamate could be formed; while in the reverse direction inorganic phosphate was required for the reduction of TPN in the presence of *N*-acetylglutamic-γ-semialdehyde. A sequence was therefore proposed (Reaction XII a and b) involving an acyl phosphate (*N*-acetyl-γ-glutamylphosphate), as in the similar conversion of aspartate to its semialdehyde (p. 194).

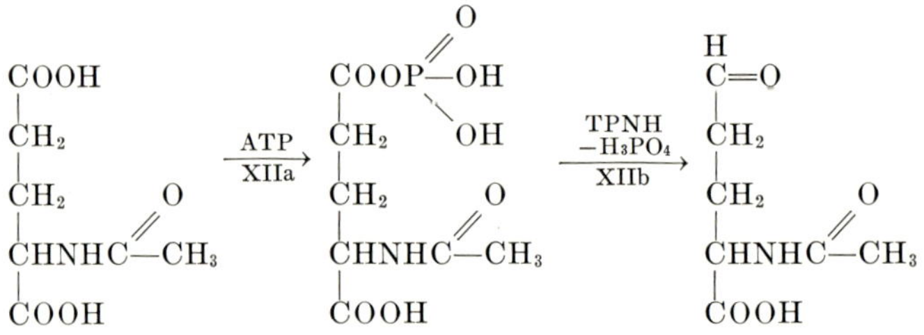

(XII)

Reaction XII has not been demonstrated in *E. coli*, but this organism does yield a transaminase that converts *N*-acetylglutamic-γ-semialdehyde to N^{α}-acetylornithine[99] (Reaction XIII, Fig. 2). Glutamate is the amino donor in this reaction. N^{α}-Acetylornithine, in turn, is hydrolyzed (Reaction XIV) by an enzyme, acetylornithinase, that is demonstrable in wild-type *E. coli* but not in a mutant that accumulates N^{α}-acetylornithine.[99, 104] The enzyme is activated by Co^{++}.*

The N^{α}-acetyl substituent undoubtedly serves in the pathway to ornithine as a blocking group, such as an organic chemist might introduce, to prevent the spontaneous cyclization that would otherwise occur at the semialdehyde stage. This cyclization is useful in the formation of proline (see above); but its potential interference in the formation of ornithine is reflected in the fact that the equilibrium of ornithine-glutamate transamina-

* It is interesting that in *M. glutamicus* acetylornithinase is absent; in its place there is a transacetylase which transfers the acetyl group to glutamate.[103] Thus in this organism the energy of the acetyl group is apparently preserved, and it serves in a cyclic manner in the conversion of glutamate to ornithine.

tion is far in the direction of the formation of glutamic semialdehyde and its cyclic tautomer, Δ^1-pyrroline-5-carboxylic acid.[105, 106]

A similar sequence is probably present in *A. aerogenes*, even though this organism differs from *E. coli* in yielding mutants that respond to either ornithine or proline.[31] This observation can be explained by the fact that *A. aerogenes* can form a transaminase that converts ornithine to glutamic γ-semialdehyde.[107] Hence mutants of *A. aerogenes* blocked in proline synthesis, between glutamate and glutamic γ-semialdehyde, respond to exogenous ornithine, just as glutamate auxotrophs of the same species can respond to exogenous histidine (p. 170).

This pathway to ornithine does not appear to be universal. Acetylornithine was not active for any mutants of *Neurospora*, and the enzyme acetylornithinase could not be detected in extracts of this organism[194] or various Gram-positive Bacillaceae. These organisms therefore appear to have a different pathway from that observed in *E. coli* and other Gram-negative bacteria. *Neurospora* and Gram-positive Bacillaceae were also found, unlike Gram-negative bacteria, to possess a constitutive ornithine δ-transaminase.[108]

Vogel and Bonner[109] have proposed for *Neurospora* a pathway to ornithine involving transamination of glutamic γ-semialdehyde. To explain why exogenous glutamic γ-semialdehyde is a precursor of proline but not of ornithine "channeling" was invoked. This pathway would have the disadvantage of the unfavorable equilibrium for the transamination noted above (unless the "channeling" should involve enzyme-bound glutamic semialdehyde, which might conceivably avoid cyclization and hence have a more favorable equilibrium). However, the available observations are equally compatible with an unknown sequence of precursors, paralleling those of *E. coli* but with a different acyl group. One possibility is suggested by the recent isolation of *N*-succinylglutamic acid from one of the Bacillaceae, *B. megaterium*, during sporulation.[110] Alternatively, the data would also be compatible with the slight modification of the *E. coli* pathway observed in *Micrococcus glutamicus*, which lacks the hydrolytic enzyme acetylornithinase but possesses instead an enzyme that transfers the acetyl group from acetylornithine to glutamate.[103]

This unresolved problem illustrates the importance of enzymic studies with mutants in firmly establishing a pathway. The proposed biosynthetic role for ornithine transaminase could be established by demonstrating loss of this enzyme in an ornithine auxotroph. However, investigation of one such mutant of *Neurospora* has shown no deficiency of this enzyme.[105]

2. Ornithine to Citrulline

Citrulline is formed by the synthesis of carbamyl phosphate and the transfer of the carbamyl group to ornithine (Fig. 3). The details of the reaction in liver[111] differ appreciably from those observed thus far in microorganisms.[112] The liver system involves the catalytic participation of a glutamate derivative (e.g., *N*-acetylglutamate) in the formation of

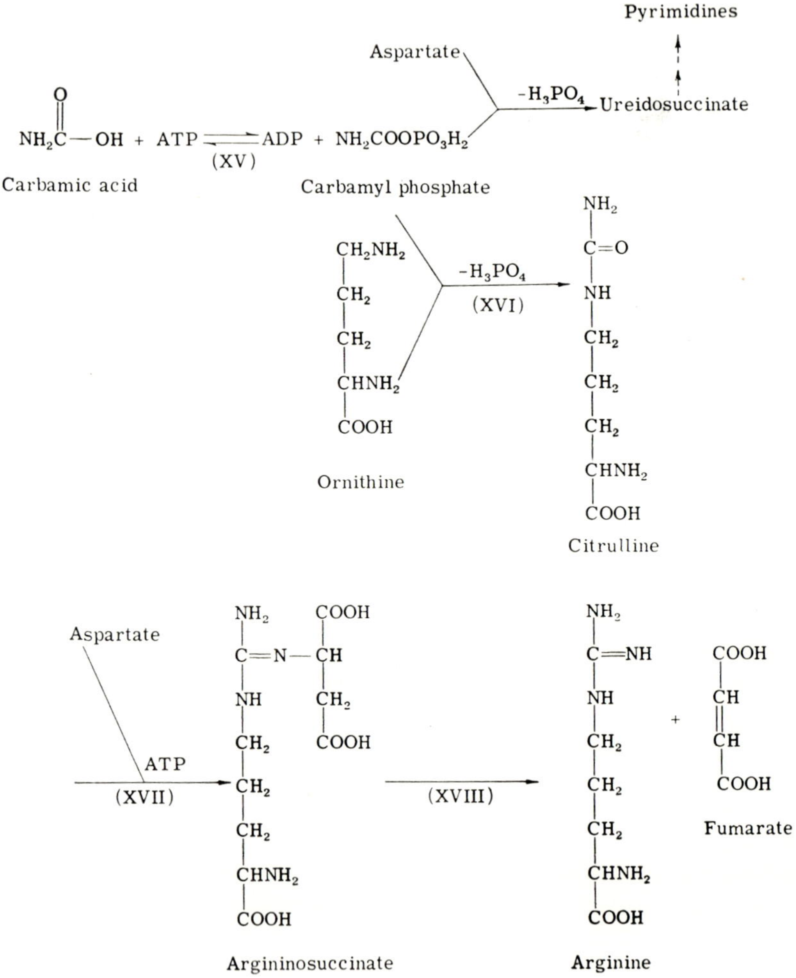

FIG. 3. The conversion of ornithine to arginine, and its relation to pyrimidine biosynthesis.

carbamyl phosphate, while no such requirement has been detected in the bacterial systems that have been studied. In addition, the irreversible mammalian system consumes two molecules of ATP per molecule of carbamyl phosphate formed, whereas the bacterial system consumes only one.

a. Formation of Carbamyl Phosphate in Bacteria. Carbamyl phosphate was discovered by Jones *et al.*[112, 113] in a study of ATP formation coupled to the conversion of citrulline to ornithine in extracts of *Streptococcus faecalis.* The reactions turned out to be reversible, and hence to provide a possible biosynthetic sequence. In this sequence (Reaction XV, Fig. 3) a specific enzyme, carbamate phosphokinase, forms carbamyl phosphate from carbamic acid and ATP. Carbamic acid, in turn, is in equilibrium with dissolved NH_3 and CO_2. Whether or not there is an enzyme to speed this equilibration remains to be established. While it may be tentatively assumed that carbamyl phosphate is provided by this reversible mechanism in bacteria, the possibility has not been eliminated that an irreversible sequence similar to that found in liver also occurs in bacteria. In this connection it might be noted that the NH_3 concentration in the usual bacterial culture medium is much higher than that in mammalian cells.

As will be noted elsewhere in this volume, carbamyl phosphate is also an obligatory intermediate in pyrimidine biosynthesis (Fig. 3). The existence of such a common precursor of citrulline and the pyrimidines was first indicated by the repeated isolation of one-step mutants of *E. coli* with a double requirement for these compounds.[31, 114]

Such organisms would presumably be blocked in the formation of carbamyl phosphate (Reaction XV). However, examination of an extract of one such mutant (in which the block was incomplete, i.e., a "leaky" mutant) revealed a capacity to catalyze Reaction (XV) (measured in the reverse direction) equal to that of the wild type.[115] Other such mutants, with an absolute requirement for citrulline plus a pyrimidine, have shown a fractional or a complete loss of carbamate phosphokinase activity, depending on the conditions of assay.[116] A more exhaustive study of this class of mutants would perhaps yield a definitive appraisal of the role of Reaction (XV).

b. The Transfer of the Carbamyl Group. In both liver and bacteria the carbamylation of ornithine is catalyzed by the enzyme ornithine transcarbamylase (Reaction XVI). This enzyme has been extensively studied in *E. coli*[19] as an example of the repression of enzyme synthesis by an end product.

3. Citrulline to Arginine

This conversion consists of two enzymic steps, which were first elucidated in liver by Ratner and her colleagues.[117] Both enzymes have been demonstrated in *E. coli*.[101]

The first step is the condensation of citrulline and aspartate in the presence of ATP to yield argininosuccinate, AMP, and inorganic pyrophosphate (Reaction XVII). The second step, the cleavage of argininosuccinic acid (Reaction XVIII), has been shown in liver to be an elimination reaction, yielding fumarate. To exclude the alternative possiblity of a hydrolysis that would yield malate, it was necessary to employ a purified enzyme preparation that lacked fumarase.[118] Although the corresponding reaction in bacteria has not been so thoroughly studied, it seems likely that the same mechanism is involved.

4. Control of Arginine Synthesis

The studies of Vogel,[17] Gorini and Maas,[19] and Gorini[119, 120] have demonstrated that in certain (but not all) strains of *E. coli* arginine represses the formation of enzymes in its biosynthetic sequence. Recent observations[121] indicate that arginine also controls the action of an early enzyme, probably involving the acetylation of glutamate. However, in *M. glutamicus* arginine inhibits the formation of a hydroxamate from *N*-acetylglutamate and ATP.[122] Hence in this organism Reaction (XIIa) may be the site of feedback inhibition.

IV. Aspartic Acid, Asparagine, and the Key Intermediate: β-Aspartic Semialdehye

Aspartate, like glutamate, is important not only as a major constituent of proteins but also as an intermediate in several other biosynthetic pathways. Isotopic studies with various organisms have shown that the 4-carbon chain of aspartate is incorporated into the pyrimidines and the amino acids methionine, threonine, isoleucine, and, in bacteria, lysine and diaminopimelate. Its amino group contributes directly to all these amino acids except isoleucine, to the guanido group of arginine, and to the purines and pyrimidines. Methionine in turn contributes the aminopropane moiety of spermine and spermidine.[123]

A. Formation of Aspartate

COOH
|
$CHNH_2$
|
CH_2
|
COOH

Aspartate

The formation of asparate from oxalacetate by transamination is well known. The most thoroughly studied example is the reaction catalyzed by transaminase A. in *E. coli*,[56] which was discussed in Section II, C, 1. Another enzyme concerned with aspartate metabolism is aspartase, which catalyzes the reversible addition of NH_3 to fumarate (Reaction III, Section

II, B). The physiological significance of this enzyme is still in doubt; mutants lacking it have not been studied. The observation discussed in Section II, B, that the aspartase level in *E. coli* grown in a glucose minimal medium is very low,[45] suggests that it has little or no biosynthetic function.

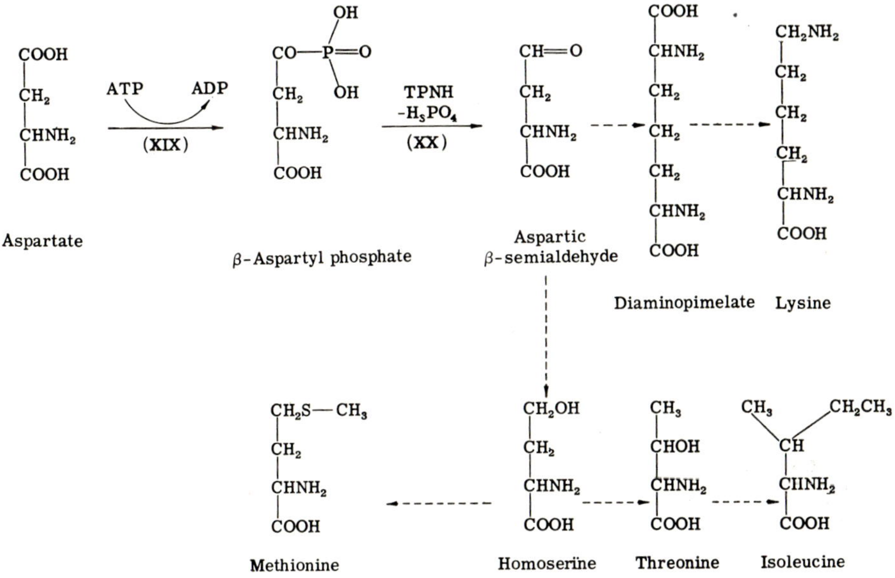

FIG. 4. The biosynthetic derivatives of aspartate.

B. Formation of Asparagine

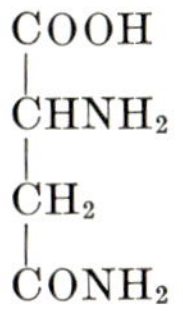

Aspargine

The biosynthetic pathway to asparagine has not been demonstrated in bacteria. In plants asparagine is a major component of the pool of soluble nitrogenous compounds. The evidence for its formation from aspartic acid, ammonia, and ATP in plant tissues has been reviewed by Meister.[59]

C. Aspartic Semialdehyde as in Intermediate

As noted above, aspartate serves as a source of several other amino acids. Figure 4 shows the over-all relationship between these compounds and aspartate. In this scheme homoserine was recognized early, through its growth factor activity for certain mutants and its accumulation by others, as an intermediate in the formation of methionine and threonine; and enzymic observations showed that the conversion of aspartate to homoserine proceeded via aspartic semialdehyde. These studies will be reviewed below. Only recently, aspartic semialdehyde has been established as a branch point by the isolation of a mutant of *E. coli* requiring methionine, threonine, and diaminopimelate, and by the demonstration that this mutant lacks aspartic semialdehyde dehydrogenase.

A possible novel cofactor in this sequence is suggested by the observation that the requirement of certain mutants for *p*-hydroxybenzoate can be satisfied by methionine plus lysine.[125]

1. The Formation of Aspartic Semialdehyde

The first observed step in the conversion of aspartate to its semialdehyde is the formation of β-aspartyl phosphate by the action of an enzyme, aspartokinase, which has been obtained from yeast (Reaction XIX).[126] In *E. coli* extracts coenzyme A stimulates the over-all reaction, suggesting the intermediate formation of aspartyl CoA;[127] but the two-step nature of the reaction has not been established. β Aspartyl phosphate is then reduced to the semialdehyde by means of aspartic-β-semialdehyde dehydrogenase, in a TPN-linked reaction (Reaction XX).[128]

The importance of feedback control mechanisms has been particularly clearly illustrated with aspartokinase, the initial enzyme of a branched pathway. *E. coli* forms at least two aspartokinases: one is both inhibited in its action and repressed in its formation by lysine, while the other is inhibited by threonine.[128a] In this way, the cell can achieve economy of biosynthesis without interference, by added lysine or threonine, with each other's synthesis.

V. The Conversion of Aspartic Semialdehyde to Threonine and Methionine

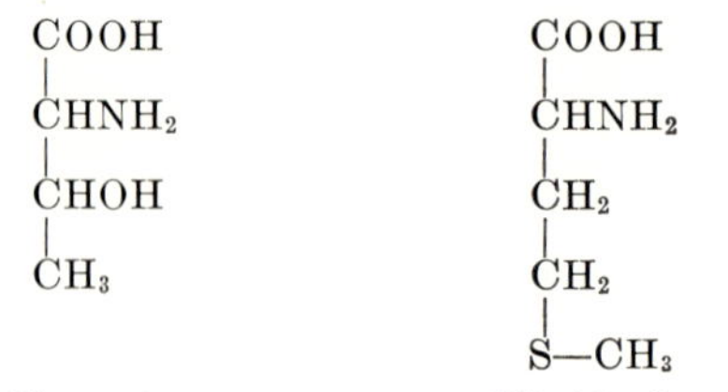

Threonine Methionine

In Section IV it was noted that threonine and methionine were both derived from aspartate through the common intermediate, homoserine.

Homoserine was first recognized as a precursor of both methionine and threonine when it was observed[129] to satisfy the double requirement for these compounds exhibited by a one-step mutant of *Neurospora*. In addition, a methionine auxotroph accumulated homoserine[130] (and smaller amounts of

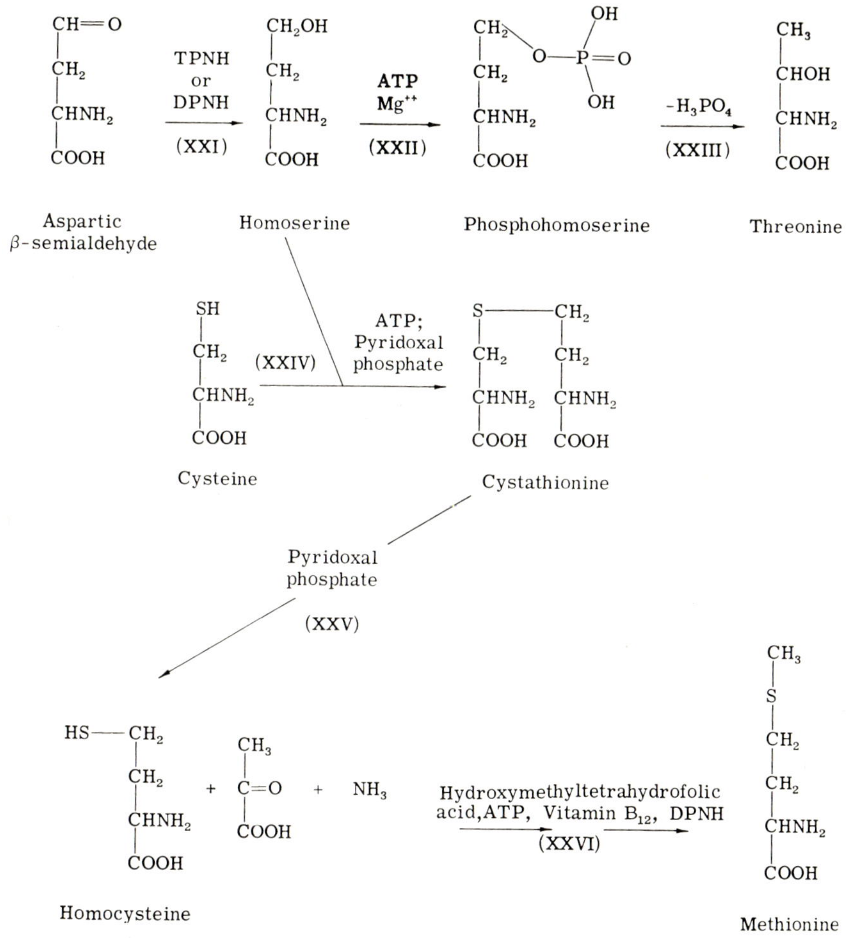

FIG. 5. The biosynthesis of threonine and methionine.

threonine) in its mycelium. The same precursor activity of homoserine was demonstrated in bacteria by nutritional observations with *B. subtilis*[131] and by isotope competition experiments with *E. coli*[132, 133]

A. The Formation of Homoserine

Homoserine is formed in yeast by the reduction of aspartic semialdehyde[134] (Reaction XXI, Fig. 5). Either DPNH or TPNH can serve as the hydrogen donor, although DPNH was more effective. Evidence has been presented for a similar reaction in extracts of *E. coli*.[135]

B. The Conversion of Homoserine to Threonine

The conversion of homoserine to threonine consists, in effect, of the migration of an hydroxyl group from the γ- to the β-carbon. The conversion has been studied in extracts of *E. coli*[135, 136, 137] and yeast.[138] In yeast two separate enzymic steps have been demonstrated.[139]

The first step is catalyzed by homoserine kinase in a reaction requiring ATP and Mg^{++} (Reaction XXII). The resulting compound, *O*-phosphohomoserine, is converted to threonine by a second enzyme (Reaction XXIII). Studies on this enzyme prepared from yeast indicate that the phosphate group provides a mechanism for driving the reaction toward threonine biosynthesis.[140]

It is of interest that in *E. coli* the action of homoserine kinase in inhibited by threonine, the end product of the sequence.[137] This feedback inhibitory effect may be important in regulating the synthesis of threonine. For further discussion of this and other feedback controls, see p. 242.

C. The Conversion of Homoserine to Methionine

In the sequence of reactions leading from homoserine to methionine (Fig. 5) the hydroxyl group of homoserine is replaced by a thiol group to yield homocysteine, and a methyl group is added from a folic acid derivative. The sulfur comes from inorganic sulfate via cysteine and cystathionine. The latter compound was disovered first as an intermediate[141] in the reverse process in the mammal: the transfer of sulfur from methionine (via homocysteine) to cysteine.

The essential features of the pathway in microorganisms were revealed with the aid of mutants. A methionine auxotroph of *Neurospora* was found[142] to respond to homocysteine and to accumulate cystathionine. A mutant blocked before cystathionine responded to this compound, though only slowly, and accumulated homoserine. Similar results have been obtained with *E. coli* mutants[31] and, except for the failure to incorporate exogenous cystathionine, isotopic experiments in bacteria[132, 133] have provided confirmation of this pathway. A difficulty in penetration presumably accounts for the failure of cystathionine to be incorporated in the isotope competition experiments and for the slow growth response to cystathionine.

1. The Formation and Cleavage of Cystathionine

The enzyme system that forms cystathionine from cysteine and homoserine has thus far been demonstrated only in *Neurospora*.[143] Reaction (XXIV) requires ATP and pyridoxal phosphate, but little is known regarding its mechanism. The cleavage, to yield pyruvate and ammonia as well as homocysteine (Reaction XXV), has been observed in extracts of both *Neurospora*[143] and *E. coli*.[144] Pyridoxal phosphate is a cofactor also in the cleavage reaction.

It is of interest to compare the biochemistry of cystathionine cleavage in bacteria, which form methionine, and in mammals, for which methionine is a dietary essential and serves as a source of cysteine. If the cleavage reaction in *E. coli* were a true reversal of mammalian cystathionine synthetase the expected products (in addition to homocysteine) would be serine, but instead ammonia and pyruvate appear. This result could not be due to the presence of serine deaminase since the same products were found with enzyme preparations that could not deaminate serine.[144] The cleavage in *E. coli* thus appears to be an example of β-elimination of a Schiff's base complex with pyridoxal phosphate, a reaction seen in several pyridoxal phosphate-linked reactions.[61]

The mammalian cleavage enzyme also produces a sulfur-containing amino acid and a sulfur-free α-keto acid. In this case the keto acid is α-ketobutyrate, whose formation must involve γ-elimination in a Schiff's base complex between pyridoxal phosphate and the homocysteine end of the molecule. Such an enzyme has been isolated in crystalline form from liver;[145] it catalyzes a γ-elimination (yielding α-ketobutyrate) from cystathionine, homocysteine, or homoserine.

While the cleavage of cystathionine in the mammalian and that in the *E. coli* system appear essentially irreversible, the cleavage in *Neurospora* appears to be a reversible hydrolysis.[143] Mutants of *Neurospora* blocked before cysteine can utilize methionine readily,[142] in contrast to such mutants of *E. coli*, which utilize methionine very slowly.[31] Whether this apparent conversion of methionine to cysteine in *E. coli* is due to a slight reversibility of the cystathionine-cleaving enzyme or to an alternative, although minor, reaction is not known.

2. The Methylation of Homocysteine

The final steps in the formation of methionine are still under active investigation. Although homocysteine can be converted to methionine by extracts and partially purified preparations from *E. coli*, the steps involved have not been clearly defined. However, because of the complexity of this pathway, the progress made has been remarkable.

The earliest inference regarding the role of *p*-aminobenzoate (as a folic acid derivative) in the formation of the methyl group arose from the observation that methionine, together with other products of 1-carbon metabolism, could noncompetitively antagonize sulfonamide inhibition[146, 147, 148] or could replace *p*-aminobenzoate as a growth factor. It was later inferred that this biosynthesis also involved vitamin B_{12}, since this compound could replace methionine as a growth factor for certain *E. coli* mutants blocked between homocysteine and methionine.[149]

The conversion of homocysteine to methionine (Reaction XXVI, Fig. 5) has recently been achieved with extracts of *E. coli*.[150, 151, 152] It requires hydroxymethyltetrahydrofolic acid, DPNH, ATP, and a mixture of at least three enzymes, one of which contains B_{12} in bound form. Recently tetrahydropteroyltriglutamic acid, or its hydroxymethyl derivative, has been found to be more active than the monoglutamate (hydroxymethyl-tetrahydrofolic acid).[152a] Tetrahydrofolic acid was reported to be inactive.[153]

The vitamin B_{12}-containing enzyme has been obtained[150] as the apoenzyme which can be activated with free vitamin B_{12}.* In this system the adenine-containing derivative of vitamin B_{12}, isolated by Barker,[154] is also active. It is not known whether this enzyme is required for the condensation or for the reduction of the hydroxymethyl group. ATP supplies the energy for the condensation, probably in a phosphorylation reaction, while DPNH is the hydrogen donor for the reduction. More recently a requirement for a flavin compound [flavinadenine nucleotide (FAD) or flavin mononucleotide (FMN)] has been reported.[155] DPNH could be replaced by either reduced flavin, but not vice versa. Thus DPNH may serve to reduce the flavin component of the system. In these studies serine aldolase has been used to regenerate the hydroxymethyl derivative of tetrahydrofolic acid.

The methyl group of methionine, via *S*-adenosylmethionine, can contribute one-carbon units to a variety of products.

VI. The Conversion of Aspartic Semialdehyde to Diaminopimelate and Lysine

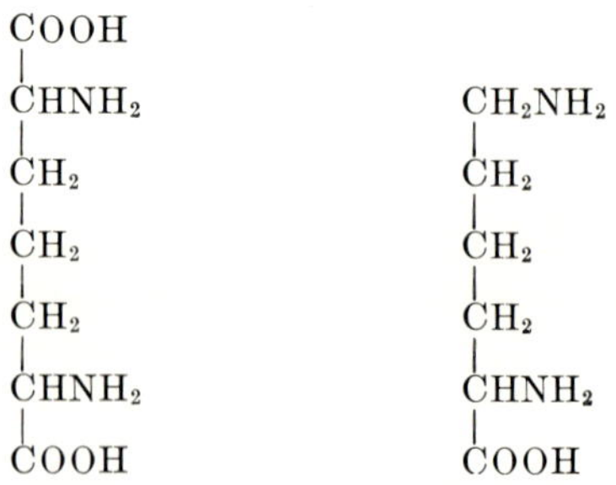

α,α'-Diaminopimelic acid   Lysine

While all the other amino acids discussed in this chapter are distributed throughout the biological kingdom, diaminopimelic acid (DAP), first

* For this purpose the methionine/B_{12} auxotroph described above was particularly valuable since it could be grown in the absence of vitamin B_{12} and thus yield a completely resolved apoenzyme.

identified in bacterial hydrolyzates in 1950 by Work,[156, 157] is found only in certain bacteria and blue-green algae. This unique distribution was clarified by the later finding that DAP is not a constituent of the intracellular proteins but is located in the cell-wall mucopeptide characteristic of these organisms.[158, 159, 160] Because of this restricted location, specific DAP starvation of a DAP auxotroph has much the same effect as penicillin treatment of a growing culture: that is, lysis in ordinary media and formation of spheroplasts in hypertonic media.[161-164]

DAP turned out to have an additional role. A DAP-requiring mutant of *E. coli* was isolated and was found to show an accelerating and sparing effect of lysine; and lysine auxotrophs were found to accumulate DAP.[30] Moreover, a decarboxylase that converted DAP to lysine was present in extracts of the wild-type organism and absent from extracts of the lysine auxotroph.[165] It was concluded that DAP is an obligatory intermediate in the biosynthesis of lysine in *E. coli.** The accumulation of DAP by a lysine auxotroph and its decarboxylation by another organism have provided the basis of a commercial process for the production of lysine.[167]

While the DAP isolated from hydrolyzates of corynebacteria and *E. coli* was found to have the *meso* (optically inactive) configuration, that accumulated in filtrates of an *E. coli* mutant was found to be a mixture of *meso*- and L-DAP.[168] Subsequently some bacterial species were found to contain the L- rather than the *meso*-isomer in their walls. (Still other species, mostly Gram-positive cocci, contain no DAP in the wall mucopeptide but instead incorporate L-lysine.) This confusing picture was resolved by the demonstration (see below) that both isomers are on the pathway to lysine.

The formation of lysine from the 7-carbon *meso*-DAP in bacteria stands in marked contrast to its as yet unanalyzed pathway of formation in *Neurospora* via a 6-carbon intermediate, α-aminoadipic acid.[29, 169] It seems evident that the evolution of the shorter pathway in bacteria has accompanied the evolution of the synthesis of DAP for wall formation. Studies on the distribution of enzymes of the DAP pathway[166] and on incorporation of a labeled precursor[170] have provided evidence for the DAP pathway to lysine even in bacteria that do not have DAP in their cell walls.

* Doubts concerning this role of DAP, based on the poor utilization of exogenous DAP in nutritional and isotopic experiments, could be explained by the assumption of slow penetration. However, Meadow and Work[166] have shown that in an incompletely blocked DAP auxotroph exogenous DAP is incorporated into cellular DAP but not into lysine. They therefore postulated an alternative route to lysine, bypassing DAP. Since biosynthesis in general involves unique rather than alternative pathways, and since the location of DAP residues in the wall suggests that exogenous DAP might reach this site more easily than the site of lysine formation, this problem merits further analysis.

DAP, or one of its precursors, may possibly provide the origin for the biosynthesis of dipicolinic acid (pyridine-2,6-dicarboxylic acid), a major constitutent of bacterial spores. This possibility, though not proved, is suggested by the structural resemblance of the compounds, by the slight incorporation of labeled exogenous DAP into dipicolinic acid,[171] and by isotopic evidence[172] for formation of dipicolinic acid from the same precursors as DAP (aspartate and pyruvate: see below) or from closely related compounds. In this connection it is of interest that in *Bacillus sphaericus* DAP is found in hydrolyzates of spores but not of vegetative cells.[173]

A. The Formation of L-Lysine from *meso*-Diaminopimelic Acid

DAP decarboxylase, in contrast to other amino acid decarboxylases, is constitutive rather than inducible in *E. coli*.[165] It has been found in extracts of a wide variety of bacteria, including some micrococci which do not contain DAP as a building block;[174] and it was absent from extracts of a DAP-accumulating lysine auxotroph of *E. coli*.[165] The enzyme is highly specific for *meso*-DAP.[175] Extracts required pyridoxal phosphate for maximal activity,[175] as did dried cells of a pyridoxal-limited mutant of *E. coli*;[176] and dried cells of a culture of *B. sphaericus* showed an absolute requirement for this cofactor.[177] (See Reaction XXXII, Fig. 6.)

B. The Formation of Diaminopimelic Acid

Isotopic studies[133] indicated that in *E. coli* aspartate, in competition with glucose, furnishes four carbons of DAP and of lysine.* Detailed analysis of the pathway was initiated by the finding of Gilvarg[124, 178] that extracts of a mutant blocked after DAP could form this compound from L-aspartate and pyruvate. The over-all reaction was promoted by Mg^{++}, ATP, TPN and DPN, and glutamate. Rhuland and Soda[179, 180] showed that succinate was also stimulatory. Analysis of the pathway was further promoted by the isolation of a DAP auxotroph that accumulated two precursors, identified as *N*-succinyl-L-DAP and its keto analog.[181]

With this information it became possible to fractionate the enzyme system and develop the sequence represented in Fig. 6. In this sequence the earlier steps, involving condensation of the 3- and the 4-carbon fragment, succinylation, and reduction, are still unknown. The *N*-succinyl group, like the *N*-acetyl group in ornithine biosynthesis (Section III, C, 1), evidently functions to prevent the spontaneous cyclization that would otherwise occur in a compound containing an amino and a carbonyl group separated by an aliphatic chain of 3 or 4 carbon atoms.

In Fig. 6 the synthesis of DAP is depicted as starting from aspartic β-

* In contrast, aspartate is not a direct precursor in the α-aminoadipic pathway found in yeast and molds.

semialdehyde, which is known also to be a precursor, via homoserine, of methionine and threonine. The formation of the aldehyde from aspartate, via β-aspartyl phosphate, has already been described (Section IV).

1. The Formation of *N*-Succinyl-α-Amino-ε-Ketopimelate

The reactions following the first known intermediate in this sequence (succinylamino-ketopimelate) consist only of transamination and deacyla-

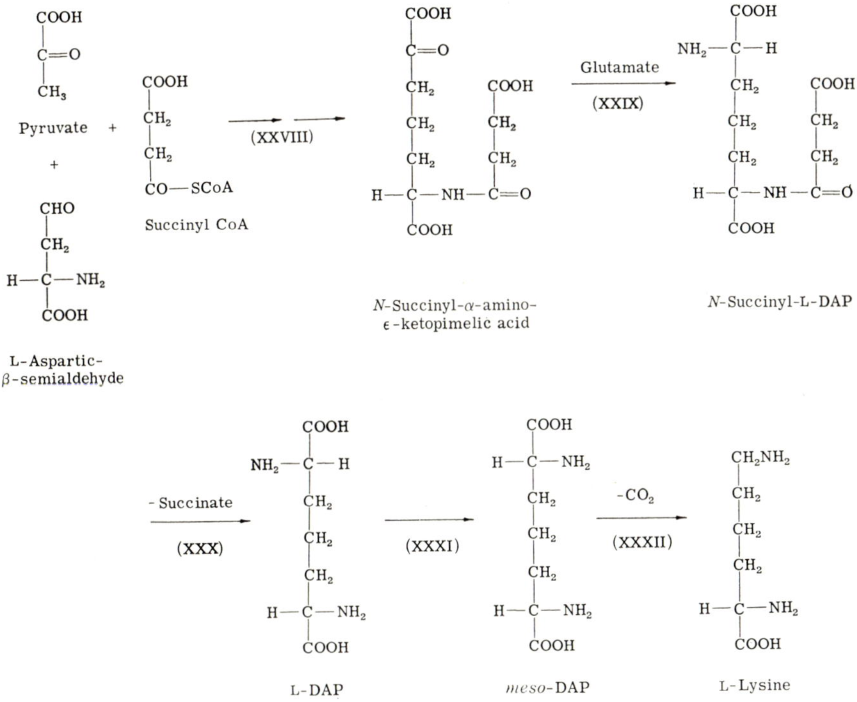

FIG. 6. The biosynthesis of diaminopimelic acid and lysine.

tion (Reaction XXVIII, Fig. 6). Since the ATP and pyridine nucleotide required for DAP formation (see above) are unlikely to participate in these reactions, they presumably function in the unanalyzed earlier reactions.[178] It seems reasonable to expect also that succinate is first activated by means of CoA, but no CoA requirement has been demonstrated even with a partly purified system.

2. The Conversion of *N*-Succinyl-α-Amino-ε-Ketopimelate to *N*-Succinyl-L-DAP

The transaminase that forms *N*-succinyl-L-DAP is specific for this compound (Reaction XXIX) and for glutamate as amino donor.[182] The purified enzyme is stimulated by pyridoxal phosphate.[124] This biosynthetic enzyme is distinct from the more general transaminases, for some of which DAP itself can serve as a substrate. The enzyme has been found in a variety of bacteria.

3. The Conversion of *N*-Succinyl-L-DAP to L-DAP

The deacylase that forms L-DAP from its succinyl precursor (Reaction XXX) has been studied in extracts of wild-type *E. coli*, and is absent from the DAP auxotroph that accumulates this precursor.[183] This enzyme is activated by Co^{++}.

4. The Conversion of L-DAP to *meso*-DAP

While acetone-dried bacteria could form lysine by decarboxylating either *meso*- or L-DAP (Reaction XXXI), the activity of the latter was found to depend on sequential action of a racemase[184] and the *meso*-DAP-specific decarboxylase. The racemase interconverts *meso*- and L-DAP, and has no action on D-DAP. The racemase proved too unstable for purification, but a mutant blocked between DAP and lysine provided a convenient source of the enzyme free of the decarboxylase. No requirement for pyridoxal phosphate has been demonstrated.

The enzyme has a broad distribution among bacteria, including many species that lack DAP in the cell wall. No striking difference in content was noted between bacteria that have *meso*- and those that have L-DAP in the wall. This finding is consistent with the view that the quantitatively major function of the racemase concerns the biosynthesis of lysine; for in hydrolyzates of *E. coli* the amount of lysine is about ten times that of DAP.[16]

VII. The Formation of Alanine, Serine, Glycine, and Cysteine

A. Alanine

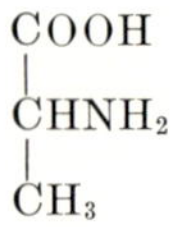

Alanine

The formation of alanine from pyruvate might appear to require little discussion. However, it should be pointed out that at the present time the enzymic pathway of pyruvate amination is uncertain. Of the three transaminases of relatively broad specificity described in *E. coli* only one, the valine-alanine-α-aminobutyrate transaminase, is known to react with pyruvate.[56] Valine could thus act as an amino group carrier, releasing its amino group to pyruvate via this transaminase and then regaining it from glutamate via transaminase B; but this mechanism could not account for alanine formation in those isoleucine auxotrophs of *E. coli*, *A. aerogenes*, and *S. typhimurium* that lack transaminase B and yet grow, though slowly, without exogenous valine, alanine, or α-aminobutyrate.[55] There must therefore be some as yet unknown mechanism that results in the amination of at least one of these three amino acids.

Table III lists several reactions that have been reported to be concerned with pyruvate amination. There is no evidence for their role in biosynthesis; but they are noted because they are reactions for which a more thorough search in *E. coli* and related organisms might be profitable.

TABLE III

KNOWN REACTIONS THAT CAN YIELD L-ALANINE

Amino donor	Organism	Reference
A. Transamination with pyruvate as amino acceptor		
1. L-Valine or L-α-aminobutyrate	*Escherichia coli*	56
2. L-Leucine, L-isoleucine, L-phenylalanine	*Brucella abortus*	58
3. L-Glutamate	Pig heart	185
4. L-Glutamine	Rat liver	186
B. Direct amination		
1. Pyruvate + NH_3 + TPNH ⇌ L-alanine + TPN	*Bacillus subtilis*	48
C. Decarboxylation		
1. Aspartate → L-alanine + CO_2	*Clostridium welchii*	187
D. Racemization		
1. D-Alanine ⇌ L-alanine	*Bacillus subtilis*	188

B. Serine and Glycine

Serine	Glycine
COOH	COOH
|	|
$CHNH_2$	CH_2NH_2
|	
CH_2OH	

Serine and glycine have been considered closely related in their biosynthesis since early studies revealed a frequent class of mutants of molds and bacteria that could respond to either compound.[89, 189, 190] Furthermore, a number of investigators have also demonstrated the interconversion of labeled serine and glycine in microorganisms and in mammals.

Serine and glycine have been further shown (mostly in mammalian systems) to participate in an extraordinarily wide variety of reactions.

Even though these offer a number of potential mechanisms for the formation of serine or glycine, the existence of serine/glycine auxotrophs suggests the existence of a single major pathway for the biosynthesis of these compounds from glucose. This pathway has not been defined. Indeed, while serine and glycine can be interconverted in a variety of organisms, their biosynthetic relationship is quite uncertain.

Glycine is an intermediate in the synthesis of the nuclei of both purines and porphyrins.

1. The Interconversion of Serine and Glycine

This interconversion (Reaction XXXIII) has been studied in extracts not only of mammals but also of *Clostridium* HF,[191, 192] an organism selected for growth on glycine as a sole carbon source. The enzyme has been called serine aldolase[193] and serine hydroxymethylase.[150] It requires pyridoxal phosphate, a conjugate of tetrahydrofolic acid, and Mn^{++}.

$$\underset{\text{Serine}}{\begin{array}{l} COOH \\ | \\ CHNH_2 \\ | \\ CH_2OH \end{array}} + \begin{array}{c}\text{tetrahydrofolic} \\ \text{acid conjugate}\end{array} \underset{\begin{array}{c}\text{Pyridoxal} \\ \text{phosphate}\end{array}}{\overset{Mn^{++}}{\rightleftharpoons}} \underset{\text{Glycine}}{\begin{array}{l} COOH \\ | \\ CH_2NH_2 \end{array}} + \begin{array}{c} N^{10}\text{-hydroxymethyl-} \\ \text{tetrahydrofolic acid} \\ \text{conjugate} \end{array}$$

(XXXIII)

This reaction has also been studied in bacterial extracts as a mechanism for the generation of the hydroxymethyltetrahydrofolic acid required for the conversion of homocysteine to methionine.[150, 151, 152] The reaction is reversible if a source of hydroxymethyl groups is available; and effective sources include formate,[194] purines,[195] and glycine.[196] Hence not only can serine serve as a source of glycine, but glycine can serve as a source of all the carbon atoms of serine.

2. The Formation of Serine and Glycine

Several indirect lines of evidence have suggested that glycine might be a biosynthetic precursor of serine. First, as noted above, folic acid derivatives are required for both the conversion of serine to glycine[150, 151, 152] and the reverse process.[193] Hence the fact that serine is one of the several compounds that spare a *p*-aminobenzoate (or folic acid) requirement in bacteria,[31, 147, 148, 197] whereas glycine is inactive, suggested that serine is the biosynthetic product of this reaction sequence (Reaction XXXIV).

$$\text{Glucose} \rightarrow\rightarrow \text{glycine} \rightarrow\rightarrow \text{serine}$$

(XXXIV)

However, the sparing effect of serine could conceivably be due to the fact that it provides 1-carbon units in the form of hydroxymethyl groups rather than formate, and hence might promote more efficient use of enzymes involved in 1-carbon transfers.*

This pathway was also supported by the occasional isolation of mutants responding only to serine. However, this evidence is also inconclusive since mutants of this class have been shown to differ from serine/glycine auxotrophs in being less permeable to glycine.[199]

Finally, the widespread occurrence of glycolate, glyoxylate, and glycoaldehyde in bacterial enzyme systems had led to consideration of these 2-carbon compounds as potential precursors of glycine. Indeed, glycolate or glyoxylate can satisfy the growth requirement of a serine/glycine auxotroph of *Neurospora* at low pH (presumably required for permeation).[200] The corresponding *E. coli* mutants do not respond to these compounds, and retain the requirement even when grown on acetate as the sole energy source (at which time glyoxylate is formed endogenously by isocitratase[201]); but these mutants might be blocked between glyoxylate and glycine.

A different pathway is supported by the results of isotopic experiments. Abelson demonstrated that *E. coli* in the presence of glucose incorporated added serine into both serine and glycine residues, whereas added glycine was incorporated only into glycine.[202] This result excludes free glycine as a normal precursor of serine in *E. coli* and favors serine as a normal precursor of glycine. However, the possibility is not excluded that both amino acids normally arise from a common precursor. The isotopic data imply that its formation from glucose would be suppressed by serine but not by glycine.

Abelson[202] has also observed that in *E. coli* pyruvate is a poor precursor of serine (in competition with glucose). This finding is in agreement with isotopic observations on serine formation from glucose-1-C^{14} and pyruvate-3-C^{14} in the rat.[203, 204] These findings indicate that serine arises from a glycolytic fragment preceding pyruvate; but decisive evidence for such an origin has not been obtained in bacteria.

Figure 7 presents a series of reactions which would provide several potential pathways from glucose to serine. This scheme has been modified from that proposed by Sallach[205] to account for serine formation in animal tissues. The key reaction would be the formation of hydroxypyruvate or phosphohydroxypyruvate from 3-phosphoglycerate.

Of this scheme Reactions (XLI), (XXXVII), and (XXXVIII) have been demonstrated in *E. coli* extracts.[206] In pea seedling extracts Reactions (XXXVI) and (XXXVII) have been observed,[207] phosphoserine being formed in the presence of D-3-phosphoglyceric acid, DPN, pyridoxal phos-

* While biosynthetic products generally exert a direct sparing effect on the utilization of their precursors (e.g., the effect of isoleucine on threonine utilization[20]), sparing effects can also be indirect (e.g., the effect of glucose on the histidine requirement of histidine auxotrophs of *A. aerogenes*[198]). Sparing effects are therefore rather unreliable guides to pathways.

phate, and glutamate. These two reactions have not yet been separated from each other. Finally, phosphoserine phosphatase, catalyzing Reaction (XXXVIII), was observed in *E. coli*[206] and has also been found and rather intensively studied in vertebrate liver.[208, 209] It appears to be specific for phosphoserine (either D or L).

On the basis of the data available it seems probable that serine arises from phosphoglyceric acid, although this pathway has not been unequivocally established. Glycine would then be expected to arise from serine. However, it might be worth while to consider the possibility that the ready interconversion of serine and glycine has been misleading, and that each has a distinct biosynthetic pathway with a separate branch point from carbohydrate metabolism. The ability of exogenous serine to satisfy the

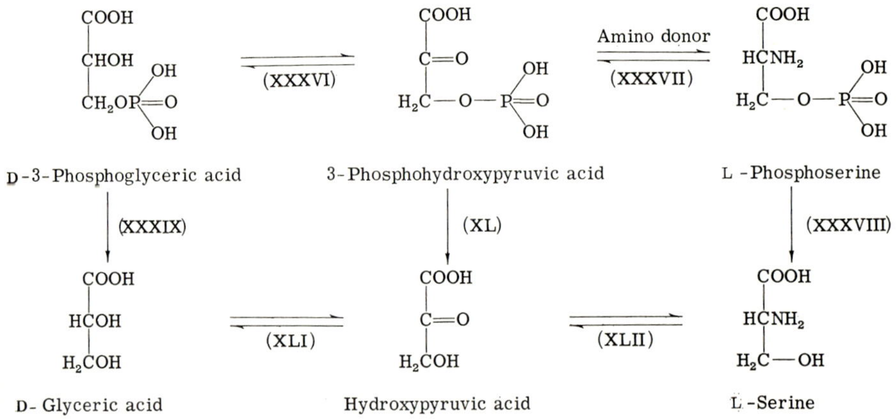

FIG. 7. Possible reactions in the biosynthesis of serine from glucose.

requirement of a true glycine auxotroph (or vice versa) would then be analogous to the ability of exogenous ornithine to give rise to proline (see p. 189) or exogenous histidine to give rise to glutamate (see p. 170) in *Aerobacter*.

C. CYSTEINE

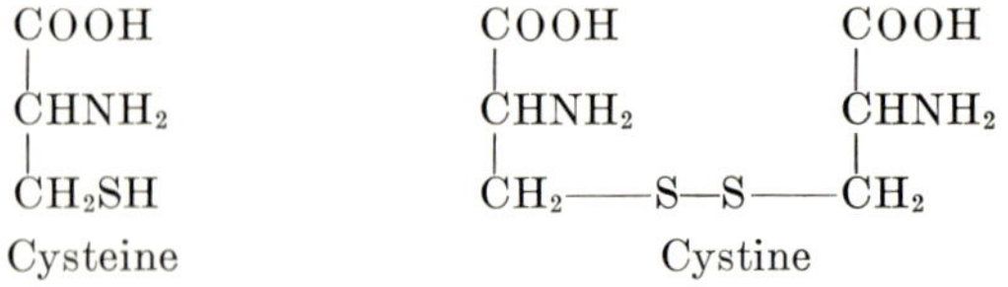

Cysteine Cystine

1. THE CARBON CHAIN OF CYSTEINE

Cystine and cysteine are readily interconvertible, cysteine being the form that participates in the known metabolic reactions of this amino acid (protein synthesis; formation of methionine). The possibility of an essential metabolic role also for cystine, however, is suggested by the observation that this compound overcomes the long lag in growth exhibited by *E. coli* under strictly anaerobic conditions.[210]

Isotope competition experiments have shown that cysteine incorporates exogenous serine extensively, but pyruvate only slightly.[202] This observation would eliminate the possibility that cysteine biosynthesis occurred via a reversal of the degradation of cysteine (catalyzed by cysteine desulfhydrase) to H_2S, ammonia, and pyruvate. Indeed, this degradative reaction can proceed in cysteine auxotrophs of *E. coli* blocked between sulfide and cysteine.[211]

The enzymic synthesis of cysteine has been achieved with extracts from yeast.[212] In this reaction (XLIII) serine and H_2S are converted to cysteine by an enzyme system requiring pyridoxal phosphate and ATP. The enzyme has been called serine sulfhydrase. It seems likely that a similar system accounts for cysteine synthesis in bacteria.

$$CH_2\text{—}CH\text{—}COOH + H_2S + ATP \xrightarrow[\text{phosphate}]{\text{pyridoxal}} CH_2\text{—}CH\text{—}COOH + ADP + Pi$$
$$\underset{OH\quad NH_2}{} \qquad\qquad\qquad\qquad\qquad \underset{SH\quad NH_2}{}$$
Serine　　　　　　　　　　　　　　　　Cysteine

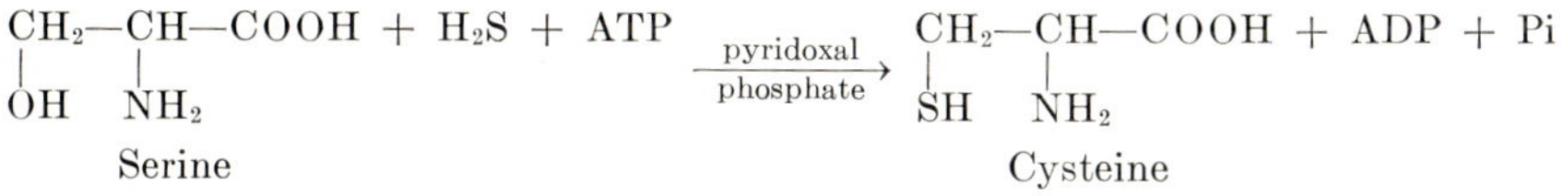

(XLIII)

2. Sulfur Assimilation

Sulfate is the source of sulfur in most minimal media. The general pathway of sulfate assimilation, which has been revealed by the nutritional responses of various *E. coli* mutants, proceeds successively via sulfite, either thiosulfate or sulfide, and cysteine.[211]

$$SO_4^= \to SO_3^= \to S^=$$

Blocks at various stages in this pathway have also been observed in many naturally occurring microorganisms. This pathway has been confirmed in *E. coli* by isotope competition experiments.[213]

Studies with extracts of various organisms have shown that sulfate can be activated by a two-step conversion to 3'-phosphoadenosine-5'-phosphosulfate (Reaction XLIII a and b).[213a, 213b]

$$ATP + SO_4^= \to \text{adenosine-5'-OP}(O_2H)OSO_3H + PP$$

(XLIIIa)

$$\text{Adenosine-5'-OP}(O_2H)OSO_3H + ATP \to \text{3'-}H_2PO_3\text{-O-adenosine-5'-OP}(O_2H)OSO_3H$$

(XLIIIb)

The role of this activation in biosynthetic sulfate reduction is strongly supported by the finding that a *Neurospora* mutant blocked in this reduction lacks the enzyme (ATP-sulfurylase) of Reaction (XLIIIa).[213c]

As noted above, cysteine provides the sulfur of methionine; it probably also provides the sulfur of other metabolites such as thiamine and biotin.

VIII. The Formation of Isoleucine, Valine, and Leucine

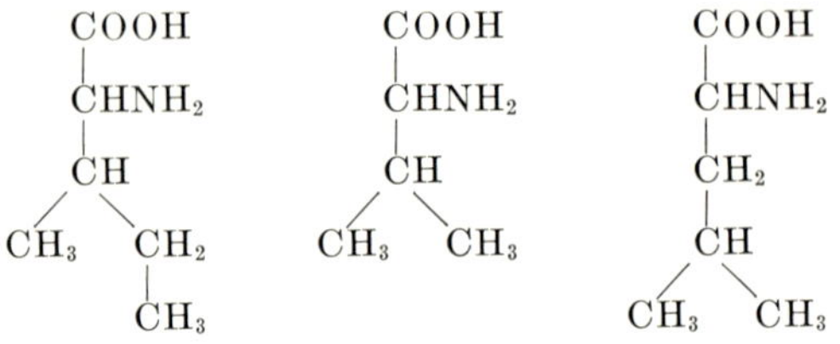

Isoleucine Valine Leucine

Valine and isoleucine illustrate well the way in which information obtained with mutants, isotopes, and enzymic studies can be combined to give a fairly complete picture of a biosynthetic pathway. It will be necessary to draw heavily on experimental results that have been obtained with other than bacterial systems, but the pathways appear to be the same in all the organisms studied. The general pattern that has emerged is that the sequences of isoleucine and valine run parallel and are mediated by a common set of enzymes.

For leucine no enzymic or mutant studies are available, but isotopic evidence indicates that this pathway branches off from the α-keto acid that immediately precedes valine (see Fig. 8). This compound (α-ketoisovaleric acid) is also an intermediate in the formation of the pantoyl moiety of pantothenate.[214]

A. The Origin of the Carbon Chain of Valine

1. Isotope Incorporation Studies

In early studies with bacteria as well as yeast on the distribution of acetate carbons in valine[215] acetate was unfortunately the sole carbon source; hence interpretation of the data was limited. However, it is of interest that the carboxyl carbon of acetate was incorporated into valine only in the carboxyl group, just as in alanine, an amino acid derived directly from pyruvate. The origin from pyruvate was further indicated by experiments in *E. coli*[202] in which pyruvate competed favorably with glucose for conversion to valine.

More direct evidence for the biosynthesis of valine via pyruvate was obtained by Strassman *et al.*,[216] who obtained valine from yeast after growth on glucose plus trace quantities of labeled lactate or acetate. Since lactate was incorporated heavily and acetate only slightly, it appeared that valine was derived directly from pyruvate rather than from a member of the tri-

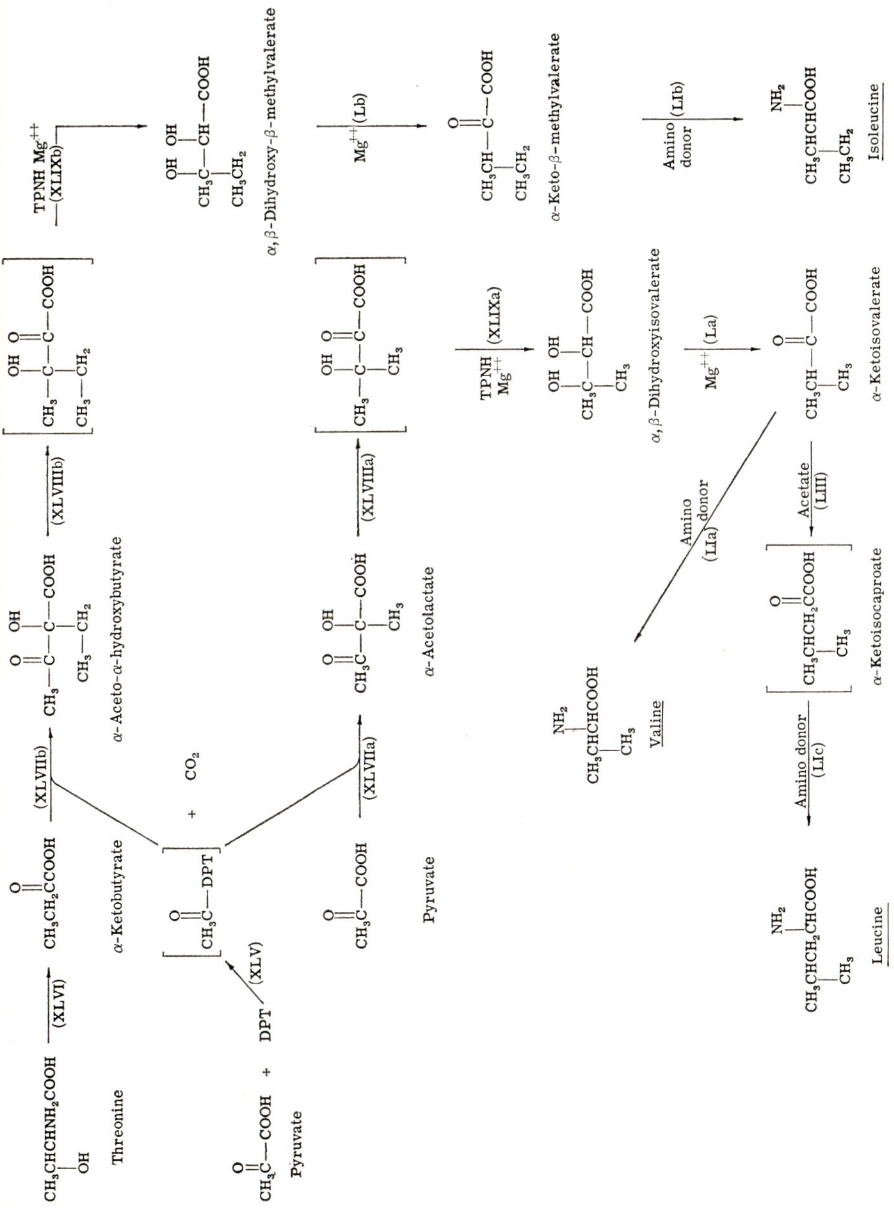

FIG. 8. The biosynthesis of isoleucine, valine, and leucine.

carboxylic cycle. However, the results of degradation, summarized in Table IV, indicated that the 3-carbon chain of pyruvate could not have remained intact. These workers therefore proposed a condensation of pyruvate and acetaldehyde (derived from pyruvate by decarboxylation) to yield acetolactate, followed by an alkyl shift (Table IV). Degradation studies by Adelberg[217] with *Neurospora* provided similar evidence for the same intramolecular rearrangement in this organism.

TABLE IV

DISTRIBUTION OF LACTATE CARBONS IN VALINE[216]

Valine carbon atom	Lactate carbon atom		
	$^{3}CH_3$	$^{2}CHOH$	$^{1}COOH$
1	1	3	99
2	4	49	0
3	4	47	1
4,4′	91	1	

Valine numbering

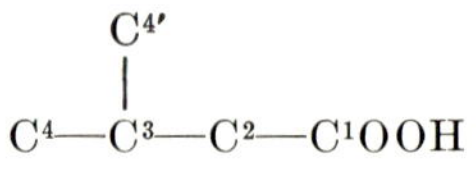

Inferred distribution of lactate carbons 1, 2, and 3

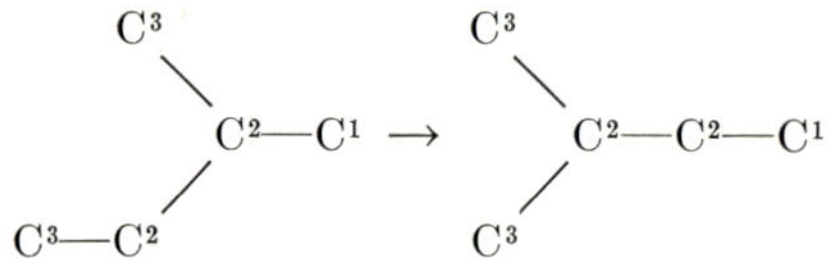

Although the isotopic data from yeast and *Neurospora* supported an intramolecular rearrangement, acetolactate was not the only intermediate that could undergo this shift. To identify the intermediate, enzymic studies were essential.

2. ACETOLACTATE FORMATION

Acetolactate was supported as an intermediate by its identification in the culture fluids of a valine auxotroph of *E. coli*.[218] Its enzymic synthesis was then studied in extracts of this mutant as well as of several wild-type strains of *E. coli*.[24] The over-all reaction (XLIV) required diphosphothiamine and a divalent cation (Mg^{++}) for maximal activity.

$$2 \text{ Pyruvate} \xrightarrow[\text{Mg}^{++}]{\text{diphosphothiamine}} \text{acetolactate} + CO_2$$

(XLIV)

Although no mutants blocked in this enzyme have been found, its relation to valine biosynthesis was strengthened by the finding that its formation is repressed by valine. In addition, valine inhibits the action of the enzyme, thus providing the pathway to valine with a regulatory feedback mechanism.[24]

In *A. aerogenes* (in contrast to *E. coli*) two enzyme systems forming acetolactate from pyruvate have been demonstrated. One was shown earlier by Juni[219] to account, along with acetolactate decarboxylase, for the acetoin formation that is characteristic of this organism. It is formed and can function only at pH 6.0 or lower, and its primary value appears to lie in providing a neutral end product of glucose dissimilation.[220] Since it is not inhibited by valine it can carry out this function in a medium containing a complete mixture of amino acids. The other enzyme system appears to be identical to that found in *E. coli*;[221] it functions best at pH 8.0 and both its activity and its formation are inhibited by valine. This enzyme thus appears to have an exclusively biosynthetic function. It is not formed at low pH, and so the pH 6 enzyme presumably functions in valine biosynthesis under these conditions.*

Although the enzyme system has not been fractionated, it seems likely that two reactions are involved. The first would consist of the generation of an "active acetaldehyde," presumably as an acetal-diphosphothiamine (DPT) complex (Reaction XLV, Fig. 8). The second reaction (XLVIIa) would be the transfer of the acetal group to a second molecule of pyruvate.

The structure of the DPT complex has not been established, but an acetal-diphosphothiamine compound has recently been synthesized chemically[222] and isolated from *E. coli*.[223] This compound is active in several enzymic reactions involving pyruvate and diphosphothiamine, including the formation of acetolactate by the pH 6 enzyme system in *A. aerogenes*.

* *Note added in proof:* Radhakrishnan and Snell[221a] have recently reported that *E. coli* also forms acetolactate at pH 6.0. It seems likely that these workers have observed the formation of racemic acetolactate which is formed by most pyruvic oxidase systems in the presence of high concentrations of pyruvate. The acetolacetate requires a much lower pyruvate concentration if the enzyme is assayed in the absence of free valine, the inhibitor of the reaction. (Such an assay is most readily accomplished by preparing extracts from cells of a valine-isoleucine auxotroph grown on limiting valine.) The acetolactate formed by *E. coli* at pH 8.0 is the same optically active isomer as that formed by *A. aerogenes* at pH 8.0 as well as pH 6.0.

It contains at C-2 of the thiazole ring an α-hydroxyethyl group which is presumably the activated acetal.

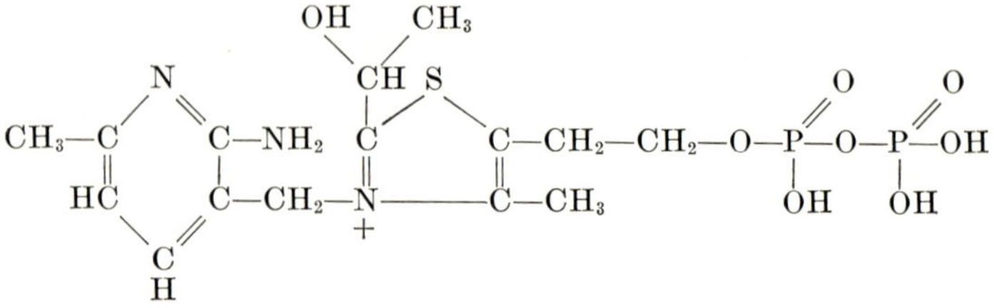

B. The Origin of the Carbon Chain of Isoleucine

1. Isotope Competition Studies

Early studies revealed a class of mutants that responded to isoleucine or α-amino- or α-ketobutyrate,[89] suggesting the origin of 4 of the 6 carbons of

TABLE V

Distribution of Acetate and Lactate Carbons in Isoleucine[224]

Isoleucine carbon atom	Acetate		Lactate
	Methyl	Carboxyl	Methyl
1	17	47	5
2	39	1	15
3	4	3	5
4	3	3	41
5	19	0	27
6	18	46	7

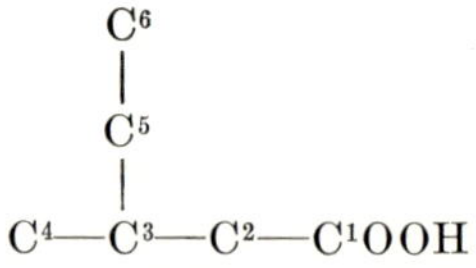

isoleucine. Quite compatible with this view was the finding[202] that exogenous aspartate, homoserine, threonine, α-aminobutyrate, and α-ketobutyrate could each be converted to isoleucine in competition with labeled glucose. However, the early results of Ehrensvärd and co-workers,[215] working with yeast grown on acetate as the sole carbon source, had revealed that isoleucine contained no 4-carbon sequence with the distribution of acetate methyl and carboxyl carbons observed in aspartate and threonine.

This apparent discrepancy was clarified by Strassman *et al.* working with yeast[224] and by Adelberg working with *Neurospora*.[225] The results obtained in the yeast experiments are given in Table V. It will be noted that the labeling of carbons 3 and 4 in isoleucine resembles the labeling in those carbons of valine (Table IV). Furthermore, just as valine lacks the intact carbon chain of pyruvate, isoleucine lacks that of α-ketobutyrate. Strassman *et al.* therefore proposed that isoleucine biosynthesis proceeded via a pathway analogous to that proposed for valine: condensation of acetaldehyde and α-ketobutyrate to yield α-aceto-α-hydroxybutyrate, followed by alkyl rearrangement to yield the carbon chain of isoleucine.

2. The Formation of α-Ketobutyrate

Isotopic and nutritional studies suggested that α-ketobutyrate arose via the deamination of threonine, and this idea was verified by the finding that the enzyme for this reaction was present in wild-type *E. coli* but absent from mutants responding to either α-ketobutyrate or isoleucine.[226] It was also observed that this reaction, the first step leading specifically to isoleucine (Reaction XLVI, Fig. 8), was inhibited by that amino acid. This inhibition by the end product provides a sensitive control mechanism which prevents oversynthesis of isoleucine (and, incidentally, excessive breakdown of threonine[23]).

The enzyme catalyzing the reaction, which has been studied only in crude extracts, requires pyridoxal phosphate. The same enzyme can also catalyze the deamination of serine to yield pyruvate.

This biosynthetic enzyme is distinct from an adaptive, and presumably catabolic, L-threonine-L-serine deaminase that is found in *E. coli* grown anaerobically in glucose-free nutrient broth.[226, 227] The adaptive enzyme can still be formed after a genetic block in formation of the biosynthetic L-threonine deaminase. Under the special circumstance of anaerobic incubation in a mixture of amino acids this mutant can grow without isoleucine, presumably because of the formation and action of this ordinarily catabolic enzyme.

The action of the catabolic enzyme is not inhibited by isoleucine. Hence it can deaminate serine and threonine in nutrient broth, whereas the biosynthetic L-threonine deaminase would not function in such an isoleucine-containing mixture. The formation of a biosynthetic and a degradative serine-threonine deaminase in the same organism parallels the formation of two distinct acetolactate-forming systems (see Section VII, A, 2).

3. The Formation of α-Acetohydroxybutyrate

The enzymic synthesis of acetohydroxybutyrate was first observed with *E. coli*.[228] In parallel with the mechanism proposed for acetolactate formation, acetohydroxybutyrate formation would consist of generation of the acetal-diphosphothiamine compound in Reaction (XLV) followed by transfer of the acetal group to α-ketobutyrate (XLVIIb). Evidence presently available indicates that this synthesis is catalyzed by the same enzyme system that catalyzes the corresponding step in valine biosynthesis (acetolactate formation). While there is no mutant available to test this point critically, it is significant that valine inhibits not only acetolactate formation (see above) but also acetohydroxybutyrate formation in extracts. In fact, in a strain (K-12) of *E. coli* in which this feedback control is particularly sensitive, growth is inhibited by valine unless isoleucine is also present.

C. Common Steps in Valine and Isoleucine Biosynthesis

Following the condensation reactions that yield acetolactate and acetohydroxybutyrate, probably by the same enzymes, the subsequent reactions in both valine and isoleucine biosynthesis are catalyzed by a single sequence of enzymes, the substrates differing from each other only by a $—CH_2—$ group. Mutants blocked at any one of the steps require *both* isoleucine and valine (or the appropriate precursors).

1. Reduction and Rearrangement

The enzymic conversion of acetolactate and acetohydroxybutyrate has been studied in *E. coli* and *A. aerogenes*[229, 230] as well as in yeast[231-234] and *Neurospora*.[235, 236] The only products identified are the α,β-dihydroxy analogs of valine and isoleucine. These compounds had been isolated earlier by Adelberg and co-workers[9, 10, 237] from culture fluids of a mutant of *Neurospora* that required isoleucine and valine. The over-all reaction (XLVIII a and b, and XLIX a and b) requires TPNH and Mg^{++}.

Although it has not been possible to separate the rearrangement (XLVIII a and b) and the reduction (XLIX a and b) steps by enzyme purification, it seems likely that rearrangement precedes reduction. This view is supported by the fact that the same enzyme preparations can convert either acetolactate or the rearranged, unreduced compound (α-keto-β-hydroxyisovalerate) to the same product, α,β-dihydroxyisovalerate.[230, 235] However, this α-keto compound has not been identified as a product of an enzyme reaction, and hence it may occur as an enzyme-bound intermediate.

An *A. aerogenes* mutant blocked in this reaction is doubly auxotrophic for isoleucine and valine, thus indicating the obligatory role of this step in the biosynthesis of these amino acids.[230]

2. Keto Acid Formation

Another class of mutants auxotrophic for both isoleucine and valine is blocked in the conversion of the dihydroxy acids to the keto acids (Reactions L a and b).[238] The enzyme requires magnesium ions.

3. The Amination Reaction

This reaction was discussed in Section II, B, 1. It should be emphasized here that the pattern of double auxotrophy is continued in mutants blocked in this reaction (catalyzed by transaminase B). However, the requirement for valine is only a relative one, because of the presence of a transaminase that transfers amino groups between alanine, α-aminobutyrate, and valine (Reactions LI a and b). As was emphasized in Section V, A, the ultimate source of the amino group for this transaminase is unknown.

D. The Formation of Leucine

1. The Formation of α-Ketoisocaproate

Isotopic evidence[202, 239-242] has shown that the isobutyl group of leucine is derived from the isobutyl group of α-ketoisovalerate. The alpha and carboxyl carbons of leucine are derived from the methyl and carboxyl carbons of acetate (or the methyl and carbonyl of pyruvate), respectively. The studies of Webb[241] indicate that the carboxyl group of α-ketoisovalerate becomes an active 1-carbon unit (similar to formate), since it was found to be distributed into the methyl group of thymine and into carbons 2 and 8 of purines. The over-all reaction is therefore as follows:

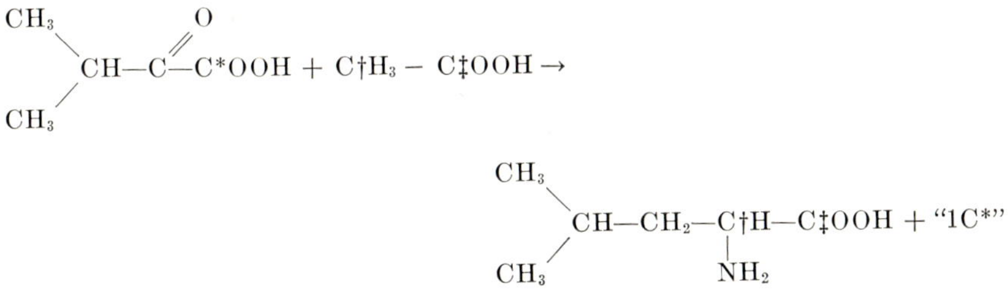

The transformation has not yet been observed in extracts.

The leucine auxotrophs isolated from *S. typhimurium* can be grouped genetically into four distinct complementation groups.[243] Since all of these, as well as the available leucine auxotrophs of *E. coli* and *A. aerogenes*, respond to α-ketoisocaproate, it seems probable that there are at least four enzymes specifically concerned with the sequence of reactions (LIII, Fig. 8) between α-ketoisovalerate and α-ketoisocaproate. However, the intermediates in these reactions do not appear to pass freely through the permeability barrier of the cell, since there has been no cross-feeding among mutants blocked in different genetic loci in the pathway. This fact is reflected in the lack of present knowledge of the intermediates; as stressed in Section I, A, 2, the identification of accumulated intermediates has often provided the key and the stimulus to a detailed analysis of a pathway. These steps are clearly in need of intensive enzymic study.*

2. The Amination Reaction

As described in Section II, C, 2, two transaminases in *E. coli* have been shown to transfer amino groups between glutamate and leucine (Reaction LI c). As shown in Table I, a mutant lacking transaminase B has only about one-tenth of the wild-type glutamate-leucine transaminase activity; yet this organism could still synthesize leucine.† It would thus appear that this reduced level of transaminase is sufficient for growth or that some alternative mechanism accounts for α-ketoisocaproate amination.

IX. Histidine

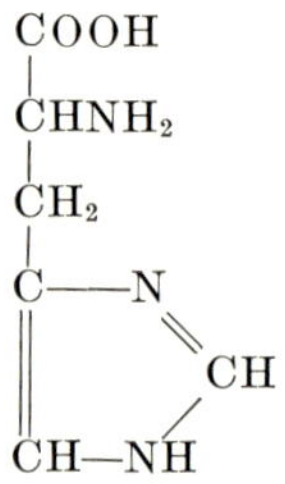

Histidine

The pathway of biosynthesis of histidine has been fairly well elucidated. The later steps became known first, largely as a result of the identification of compounds accumulated by various mutants. Study of the enzymic interconversions of these compounds followed. The earlier steps, leading to the formation of the imidazole ring, were revealed only after extensive isotopic and enzymic experiments.

* *Note added in proof:* Recently, a re-examination of the *S. typhimurium* mutants revealed that three of the genetic groups accumulate a compound utilized by the fourth group.[243a] The accumulated compound was identified as β-carboxy-β-hydroxy-isocaproate, which had been proposed by Strassman *et al.*[239] as an intermediate in leucine biosynthesis. Its synthesis from α-ketoisovalerate and acetyl coenzyme A, by extracts of strains which accumulated it, and the absence of the property in an extract of the strain which could utilize it as a leucine precursor, demonstrate the obligatory role of the compound. The compound has also been demonstrated in certain leucine auxotrophs of *N. crassa*. Its conversion to α-ketoisocaproate by extracts of *S. typhimurium*[243b] and of yeast[243c] has been demonstrated but the mechanism of the over-all reaction has not been studied.

† Likewise leucine auxotrophs, capable of responding to α-ketoisocaproate, continue to respond to this compound after loss of transaminase B by a second mutation.

A. Compounds Accumulated by Histidine Auxotrophs

The pathway to histidine is unique among amino acids in that the carboxyl group is formed last. It arises by oxidation of the primary alcohol group of L-histidinol. Analysis of this pathway was initiated by the finding that this compound is accumulated by certain histidine auxotrophs of *E. coli* and utilized as a growth factor by others.[244]

Various histidine auxotrophs of *Neurospora* were later found to accumulate in their culture filtrates various imidazole derivatives: Histidinol, 4-(trihydroxypropyl)-imidazole (imidazole glycerol), and 4-(2′-keto-3′-hydroxypropyl)-imidazole (imidazole acetol).[14, 245, 246] Furthermore, these mutants accumulated within their mycelia the corresponding phosphorylated compounds, with the phosphate on the terminal position of the side chain.

All these compounds were nutritionally inert for the *Neurospora* auxotrophs, and hence could not be arranged in a sequence by the usual criteria. However, the following biosynthetic sequence (LIV) was established by obtaining various doubly blocked mutants by genetic recombination[247] and observing their accumulations. In such strains the accumulated compound would be determined by the earlier of its two blocks.

$$\text{Imidazole glycerol (phosphate)} \rightarrow \text{imidazole acetol (phosphate)} \rightarrow \text{histidinol (phosphate)} \rightarrow \text{histidine} \quad \text{(LIV)}$$

This sequence was confirmed by the results of subsequent enzymic studies, which also showed that the phosphorylated derivatives were the true intermediates (see below).

B. Enzymic Conversions of Imidazole Derivatives

1. Phosphorylated Precursors

Imidazole glycerol phosphate is converted to imidazole acetol phosphate by a dehydrase that requires Mn^{++} and a sulfhydryl compound (Reaction LV, Fig. 9).[248, 249]

Imidazole acetol phosphate, in turn, is converted to histidinol phosphate by a transaminase (Reaction LVI).[250] Glutamate is the best of the amino donors tested, but the enzyme also reacts with α-aminoadipate, arginine, and histidine.

In the conversion of histidinol phosphate to histidine the phosphate is first split off by a specific phosphatase (LVII).[251]

Although the enzymes catalyzing Reactions (LV), (LVI), and (LVII) have been purified only from *Neurospora*, the reactions have also been demonstrated in extracts of *Salmonella typhimurium*.[252]

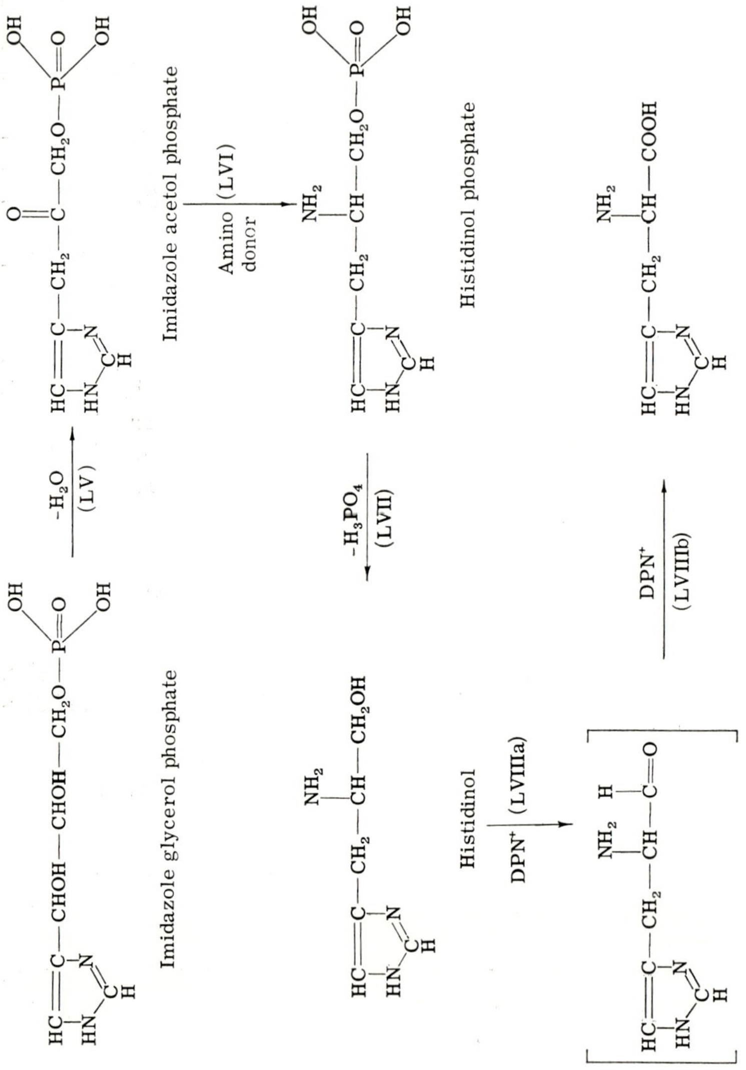

FIG. 9. Histidine biosynthesis: reactions of the imidazole derivatives.

2. Histidinol to Histidine

Histidinol is converted to histidine by a DPN-linked dehydrogenase which has been found in *E. coli* and other bacteria.[253] In the over-all reaction (LVIII a and b) two molecules of DPNH are formed. Although the expected aldehyde intermediate, L-histidinal, has not been detected in incubation mixtures, added histidinal is converted by the enzyme in the presence of D N$^+$ to histidine (Reaction LVIII b), or in the presence of DPNH to histidinol (the reverse of Reaction LVIII a).[254] Since there was no separation of these activities during purification from a bacterial extract (*Arthrobacter histidinolvorans*), and since both activities were missing in extracts of an *E. coli* auxotroph that accumulated histidinol, it appears that a single enzyme catalyzes the two successive dehydrogenations, and that histidinal is an enzyme-bound intermediate.

C. Origin of the Imidazole Ring

Structural considerations, including the configuration of imidazole glycerol phosphate (D-erythro), prompted the suggestion that the 5-carbon chain of histidine might be derived from D-ribose or D-ribulose (Fig. 10).[14] However, the proof of this origin had to await later enzymic work, since mutants blocked early in histidine biosynthesis failed to accumulate any detectable precursors of imidazole glycerol phosphate. Meanwhile, another possible biosynthetic relationship was indicated early by the observation that the purine requirement of *L. casei* was spared by histidine.[255]

1. Isotope Distribution Studies

It at first seemed likely on structural grounds that the purine-histidine relationship was based on the common imidazole ring. This view was supported by the finding that labeled formate was readily incorporated in yeast into carbon 2 of the imidazole ring of histidine,[256, 257] as well as into carbons 8 ("imidazole") and 2 of the purine ring. However, Mitoma and Snell[258] later showed that this carbon of histidine could be derived in *L. casei* from carbon 2 but not carbon 8 of guanine. Furthermore, Magasanik[259] found with *E. coli* that nitrogen 1 of histidine was also derived from the purine ring, and Neidle and Waelsch[77, 260, 261] demonstrated that the purine nitrogen involved was number 1. It was thus clear that histidine derived an adjacent pair of atoms, N-1 and C-2, from the nonimidazole portion of the purine ring.

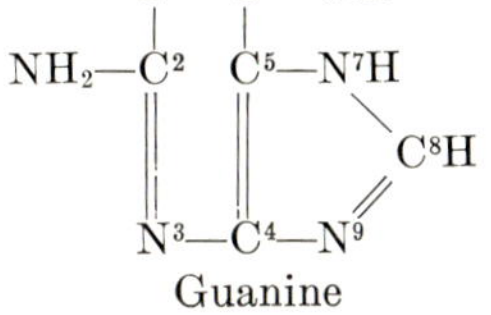

Guanine

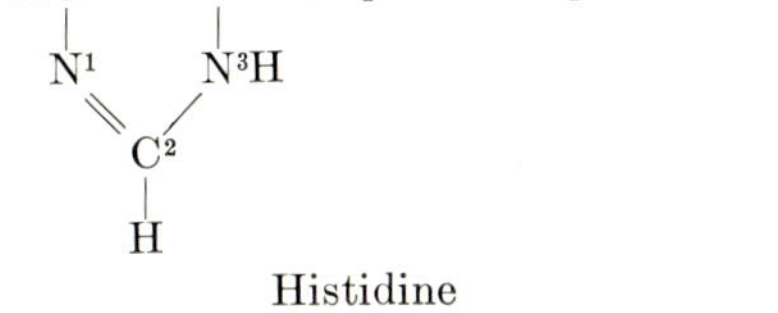

Histidine

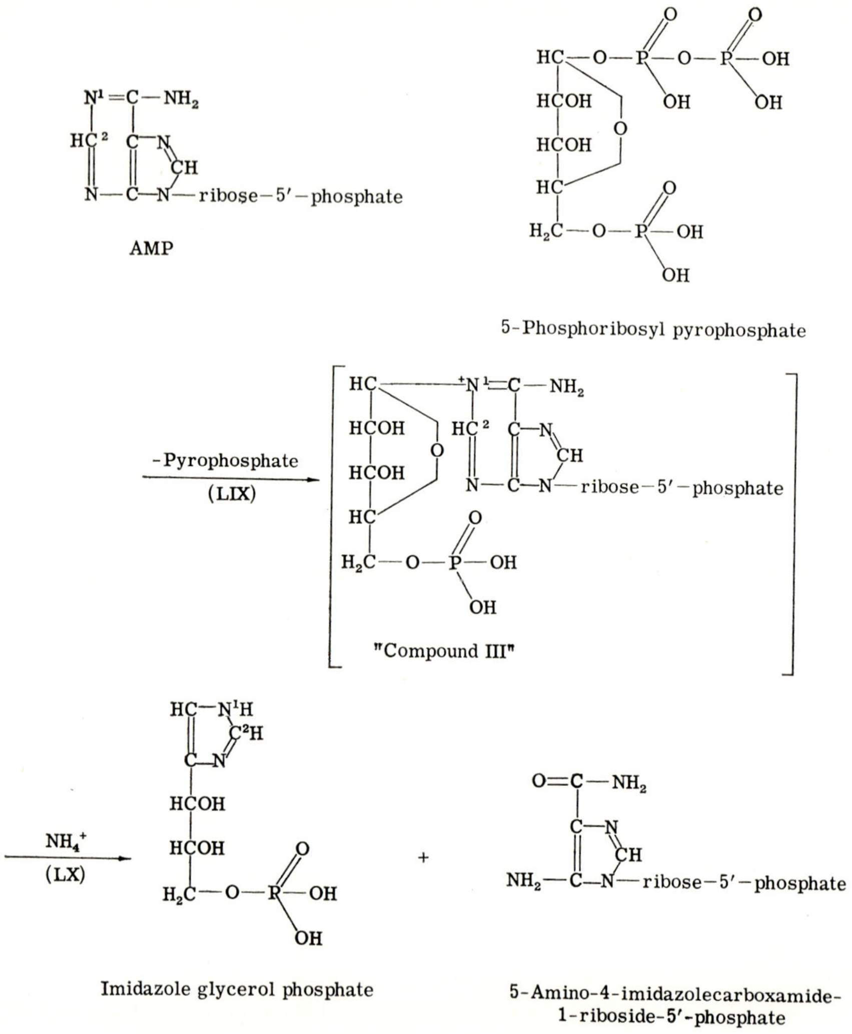

FIG. 10. Histidine biosynthesis: the origin of the imidazole ring.

2. The Cyclic Role of the Purine Nucleus in Histidine Biosynthesis

The enzymatic basis for the conversion of carbon-2–nitrogen 1 of a purine to carbon-2–nitrogen 1 of histidine was demonstrated by Moyed and Magasanik in extracts of *E. coli*, *A. aerogenes*, and *S. typhimurium*.[78, 79]

Incubation of these extracts with ATP and ribose-5-phosphate resulted in the formation of an unstable substance, "Compound III," provisionally identified as 1-(5′-phosphoribosyl)-adenosine-5′-phosphate. When the incubation mixture also included glutamine (later shown to be replaceable by NH_4^+) it yielded both imidazole glycerol phosphate and 4-amino-5-imidazole carboxamide (AICAR). The AICAR presumably is the product of disruption of the adenine ring with loss of N-1 and C-2; it could also be produced by treatment of Compound III with dilute acid. The imidazole glycerol phosphate would be the product of cyclization of the remaining moiety of Compound III, with uptake of nitrogen.

Experiments with labeled substrates and partially purified enzyme preparations support the reactions shown in Fig. 10. In particular, the ribose-5-phosphate and ATP can be replaced by 5-phosphoribosyl pyrophosphate and AMP.*

Fractionation indicated that at least two steps are involved in Reaction (LX), one of which is the ammoniolysis of Compound III. The inactivity of ammonium sulfate observed earlier was found to be due to an inhibitory effect of sulfate on Reaction (XXXI).[79] Ammonium acetate was as active as glutamine; and the latter was shown to require the presence of a glutaminase. In contrast, Neidle and Waelsch[77] observed that glutamine was a better source than ammonia for nitrogen 3 of histidine in growing bacteria; but the discrepancy may reflect a difference in rate of penetration into the intact cell.

Since AICAR is an intermediate in the biosynthesis of the purine ring, its formation from Compound III indicates a cyclic role for the purine ring in histidine biosynthesis. This cycle leads from AICAR to AMP and, via Compound III, back to AICAR again. The operation of the cycle is controlled by histidine by a negative feedback mechanism in which histidine inhibits the first reaction (LIX) in the sequence leading specifically to histidine.[262] This cycle and its control by histidine can thus readily explain the sparing effect of histidine on the purine requirement of *L. casei*, as well as on that of certain purine auxotrophs that have proved to be blocked beyond AICAR formation.

* *Note added in proof:* Recent work of Ames *et al.*[79a] has shown that ATP rather than AMP reacts with 5-phosphoribosyl pyrophosphate in *S. typhimurium* and that the product of this reaction, *N*-1-(5′-phosphoribosyl)-ATP, is, in fact, the first intermediate in histidine biosynthesis. It is now known that the compound referred to here as compound III is a later intermediate in which a ring has opened and pyrophosphate has been split off.[79]

X. The Aromatic Amino Acids: Tyrosine, Phenylalanine, and Tryptophan

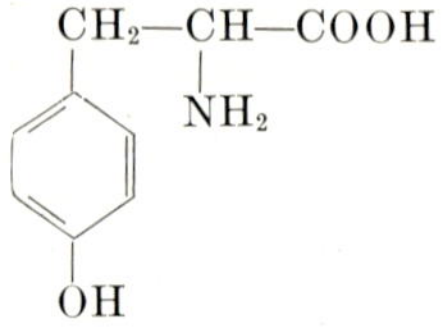

Tyrosine

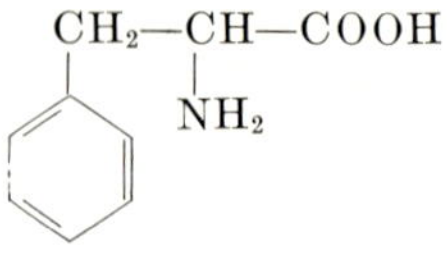

Phenylalanine

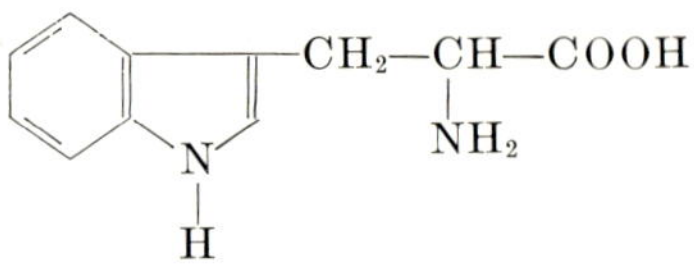

Tryptophan

The synthesis of benzenoid compounds from aliphatic precursors presents a certain challenge to the organic chemist; hence its mechanism in biological systems has long been a problem of interest to biochemists. In the past decade one such pathway, leading to the aromatic amino acids,* has been nearly completely elucidated in bacteria.[263] The same sequence leads also to the formation of a variety of more complex aromatic products (e.g., the B ring of flavonoids; lignin) in higher plants.

In *E. coli*, in contrast to mammals, the aromatic amino acids do not appear to serve as precursors of further essential metabolites. In *Neurospora* tryptophan is the source of nicotinamide,[264] but *E. coli* employs quite a different pathway, derived from aspartate.[265]

A. The Common Pathway

1. Shikimic Acid and Its Immediate Precursors

The key to the aromatic pathway was provided by the isolation of a series of mutants that required all three aromatic amino acids as well as trace quantities of *p*-aminobenzoate and *p*-hydroxybenzoate. This multiple aromatic requirement suggested a block in the biosynthesis of the benzene ring. Empirical screening showed that some of these mutants of *E. coli*[266] and *Neurospora*[267] could respond also to shikimic acid (Fig. 11), a hydroaromatic compound that had been isolated from a plant 75 years earlier; and other mutants of *E. coli* were found to accumulate this compound.[266] The inferred function of shikimate as an intermediate was later verified by enzymic studies.

Two groups of bacterial mutants blocked before shikimate were each found to accumulate a precursor recognized as a growth factor for mutants with earlier blocks. The immediate precursor of shikimate was identified as 5-dehydroshikimic acid,[268] and its precursor in turn as 5-dehydroquinic

* While for purposes of convenience we shall refer to this as the aromatic pathway, it should be noted that the benzene rings in some metabolites (e.g., riboflavin, the A ring of flavonoids) arise by other routes.

acid[269] (Fig. 11).* After the identification of these compounds, the enzymes responsible for their conversion to shikimate were demonstrated in extracts of wild-type *E. coli*. Furthermore, each enzyme was shown to be absent from a mutant whose nutritional behavior and accumulation indicated a block in that reaction.†

Another possible member of this sequence was quinic acid, which was just as active as 5-dehydroquinate for mutants of *Aerobacter*.[6] However, quinate has been excluded as an intermediate on several grounds: (*a*) quinic dehydrogenase, which interconverts quinate and 5-dehydroquinate, could not be detected in many organisms, including *E. coli*, that have the aromatic pathway;[270] (*b*) mutants of *E. coli* blocked before 5-dehydroquinate neither accumulate nor respond to quinate; and (*c*) no mutants of *Aerobacter* could be obtained that were blocked between quinate and 5-dehydroquinate. The conversion of quinate to dehydroquinate in *Aerobacter* is therefore considered a side reaction, presumably associated with the fact that this organism (in contrast to *E. coli*) can utilize quinate as a carbon source.

Dehydroquinase, which catalyzes the dehydration of 5-dehydroquinate to 5-dehydroshikimate (Reaction LXIII, Fig. 11), has no demonstrated cofactor requirement.[273] Its equilibrium constant favors the reaction in the direction of biosynthesis. 5-Dehydroshikimic reductase, which catalyzes the formation of shikimate from 5-dehydroshikimate (Reaction LXIV), is TPN-linked.[271] Both these enzymes have been demonstrated in extracts of a variety of microbes and higher plants, and each is absent from a mutant that accumulates its substrate.‡

2. The Reactions Preceding 5-Dehydroquinate

Cross-feeding tests failed to extend the pathway earlier than dehydroquinate, although enzymic treatment of culture filtrates did reveal an earlier intermediate. Before it was identified the early steps in aromatic biosynthesis were clarified by collaborative isotopic and enzymic studies of Sprinson and Davis.

* Actually 5-dehydroquinic acid is prevented by a permeability barrier from supporting growth of mutants with earlier blocks; but secondary mutants that have become permeable to the compound are readily selected.[6]

† It is of interest that bacteria with incomplete blocks in various reactions of the aromatic pathway exhibit preferential synthesis among products of a common precursor:[272] with increasing capacity for aromatic biosynthesis the requirement is lost sequentially for *p*-hydroxybenzoate, *p*-aminobenzoate, tryptophan, phenylalanine, and finally tyrosine.

‡ It is of interest that the interconversion of quinate and 5-dehydroquinate, presumably functioning in the oxidative direction for degradative purposes, employs DPN as cofactor,[270] whereas the closely parallel interconversion of 5-dehydroshikimate and shikimate, functioning in the reductive direction for biosynthetic purposes, employs TPN.[271]

In the isotopic work accumulated shikimate was selected for degradation because all its atoms could be individually determined, in contrast to the symmetrical ring of tyrosine and phenylalanine, which had been studied earlier by Gilvarg and Bloch[274, 275] and Ehrensvärd.[276, 277] The carbon atoms of variously labeled glucose were found[15] to be distributed in shikimate in the manner summarized in Fig. 12. The carboxyl-C1-C2 portion of shikimate evidently arose from a 3-carbon intermediate of glycolysis, since the shikimate carboxyl was derived about equally from carbons 1 and 6 of glucose, shikimate C-1 from 2 and 5 of glucose, and shikimate C-2 from 3 and/or 4 of glucose. The remaining 4-carbon portion of shikimate had a

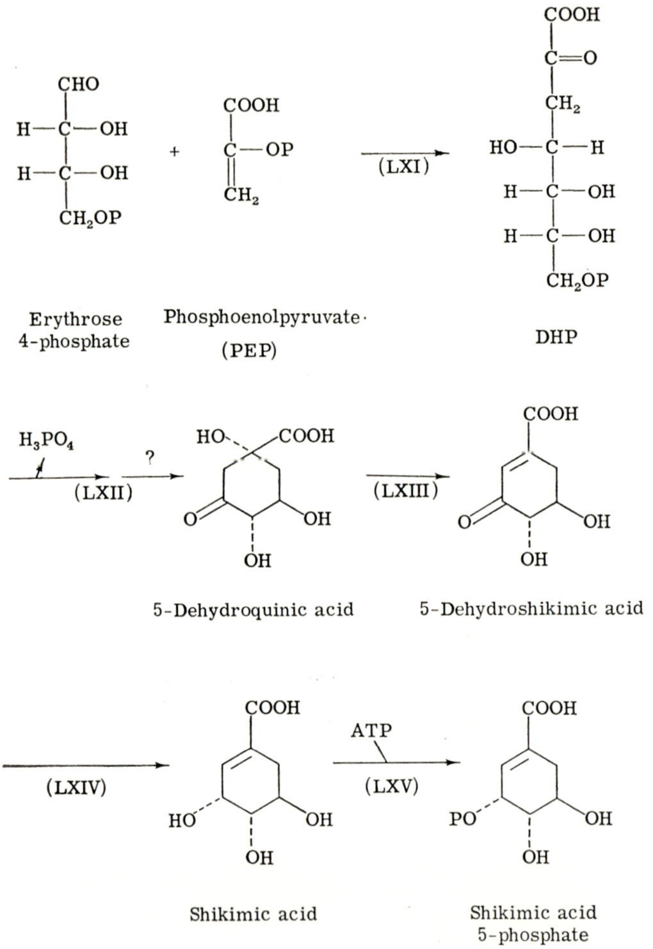

FIG. 11. The common pathway

more complex pattern of labeling which strongly suggested, on the basis of a detailed analysis of the mechanisms of formation of erythrose-4-phosphate,[15] that carbon atoms 3 to 6 of shikimate arose from 1 to 4, respectively, of this tetrose.

When enzymic explorations of this area were undertaken, the specific and sensitive bioassay for 5-dehydroshikimate made it possible to show that extracts of a mutant blocked after this compound formed it from various phosphorylated carbohydrates (e.g., glucose-6-phosphate), despite the numerous competing reactions;[278] the yield was about 5%. The yield from sedoheptulose-1,7-diphosphate, in contrast, approached 100%[279]; but the isotopic results described above indicated that its 7-carbon chain could not be cyclized intact.* Sedoheptulose diphosphate turned out to be not an

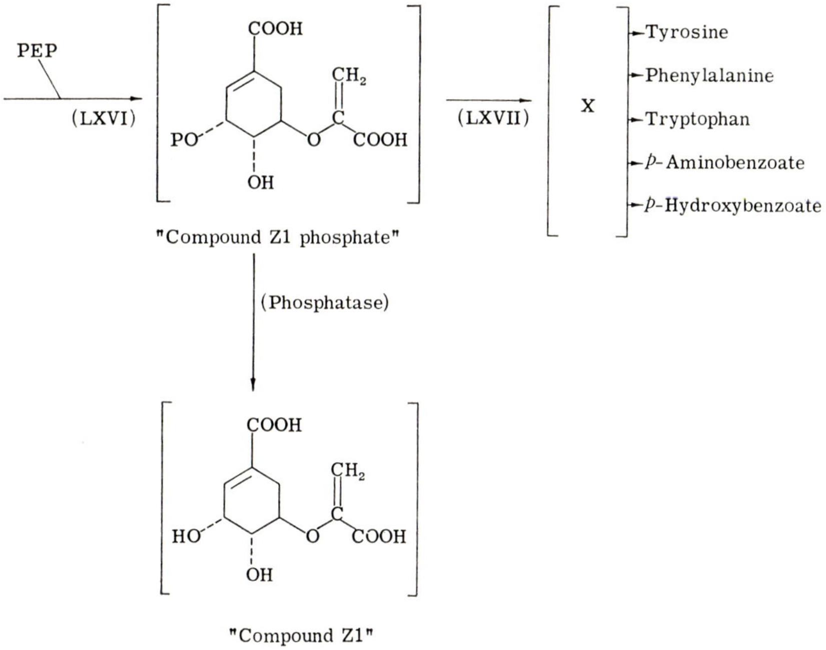

in aromatic biosynthesis.

intermediate but simply an efficient source (by the action of aldolase) of triose phosphate and erythrose-4-phosphate. The latter compound, in the

* The glycolytic moiety would be attached in sedoheptulose diphosphate by the carbon derived from 3,4 of glucose, but in shikimate its end was attached by the carbon derived from 1,6 of glucose (Fig. 12). In analyzing these early reactions of aromatic biosynthesis the isotopic and the enzymic studies contributed much to each other's interpretation.

presence of phosphoenolpyruvate (readily derived from triose phosphate), was almost quantitatively converted to 5-dehydroshikimate.[280] Furthermore, the bacterial extracts first formed, at a much higher rate, a compound identified[281] as a noncyclic condensation product, 3-deoxy-D-arabinoheptulosonic acid 7-phosphate (DHP) (formerly called 2-keto-3-deoxyheptonic acid 7-phosphate, KDHP). This compound was subsequently also identified in culture filtrates of mutants blocked before 5-dehydroquinate.[282]

Purified DHP synthetase (Reaction LXI, Fig. 11) has shown no cofactor requirement.[281] Because of the presence of phosphatase, crude bacterial extracts also form the dephosphorylated product of DHP.[283]

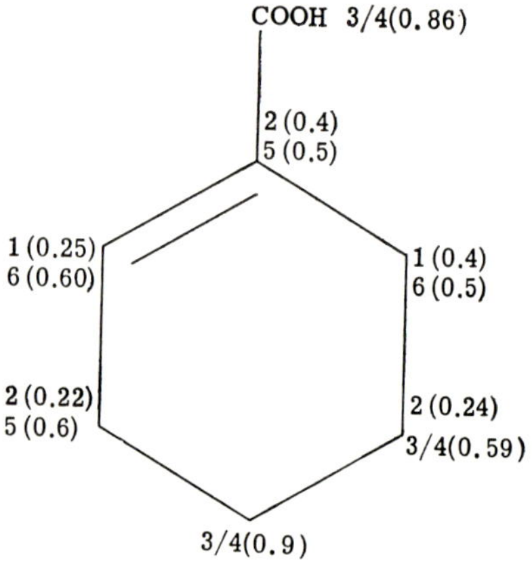

FIG. 12. Major contributions of glucose carbon atoms to shikimate biosynthesis.[15] Beside each carbon atom of shikimate each number in parentheses represents the fraction of that atom derived from a given carbon atom of glucose, whose number is denoted beside the parentheses.

The enzyme(s) responsible for the dephosphorylation and cyclization of DHP (Reaction LXII, Fig. 11) requires Co^{++} and DPN.[281] Since the reaction involves no net electron transfer, the DPN is presumably concerned with the conversion of the 6-hydroxyl and the phosphorylated 7-hydroxyl of DHP into the keto group of 5-dehydroquinate.

3. The Reactions Following Shikimic Acid

Beyond shikimic acid nutritional tests failed to reveal any accumulation; but two derivatives were recognized because they yielded shikimic acid on acid hydrolysis.[11] The earlier of these has been identified as shikimic acid 5-phosphate;[12] the other has been denoted as "Compound Z1."[11] The nutritional inactivity of shikimic phosphate presumably reflects a permeability barrier, since in enzymic experiments this compound has proved to be an intermediate between shikimate and "Compound Z1,"[284] anthranilic acid,[285] and *p*-aminobenzoate.[286] Shikimic acid 5-phosphate could be replaced by shikimate plus ATP, but the kinase (Reaction LXV) has not been characterized.

Compound Z1 is an enol-pyruvic ether, attached to shikimate in the 3 or 5 position.[287] Extracts of *E. coli* mutants can form it from shikimic acid 5-phosphate and phosphoenolpyruvate.[288] Recently such extracts have been shown by Levin and Sprinson[289] also to produce a compound whose hydrolysis yields phosphate in addition to shikimate and pyruvate. These findings suggest the structures and reactions (LXV–LXVII) depicted in Fig. 11, in which Compound Z1, like a number of other accumulated compounds described in this chapter, is considered an artifact of dephosphorylation. The common pathway presumably extends beyond the phosphorylated parent of Z1, since mutants that accumulate Z1 have a quintuple aromatic requirement.

B. The Formation of Tyrosine and Phenylalanine

1. The Formation of Prephenic Acid

The origin of the side chain of tyrosine from a 3-carbon glycolytic fragment was indicated by the finding that its β-carbon is derived almost equally from C-1 and C-6 of glucose.[275] The branch point must occur above pyruvate, since in yeast growing in the presence of glucose labeled acetate does not appear in tyrosine, even though it is incorporated (via pyruvate) into alanine.[274] This conclusion was verified by the finding that phosphoenolpyruvate, but not pyruvate, is a substrate for the enzymic formation of Compound Z1 (see above).

The mechanism of aromatization was revealed in large measure through the finding that phenylalanine auxotrophs of *E. coli* accumulate an unstable precursor which breaks down spontaneously in the slightly acid culture medium to yield phenylpyruvic acid, a growth factor for the same mutants.[13, 290] This precursor has been isolated and identified and given the name prephenic acid.[291] It is a cyclohexadiene with a carboxyl and a pyruvic acid side chain on carbon 1 and a hydroxyl on carbon 4 (Fig. 13). This structure is consistent with the isotopic evidence that carbon atom 1 of shikimate corresponds to carbon atom 1 of tyrosine. The structure of prephenate also permits a reasonable explanation (Fig. 14) of its ready aromatization by protonic attack (or by an enzyme in Reaction LXIX) to yield phenylpyruvate.

A plausible mechanism (Fig. 15) can also be suggested for the formation of prephenate (Reaction LXVIII), if the tentative structure of the phosphorylated parent of Compound Z1 is correct (see preceding section). The conversion would involve a shift of the pyruvate chain from an enol-ether at position 3 to a C—C bond at position 1. Although this formulation is attractive, it should be emphasized that the structure of the precursor and

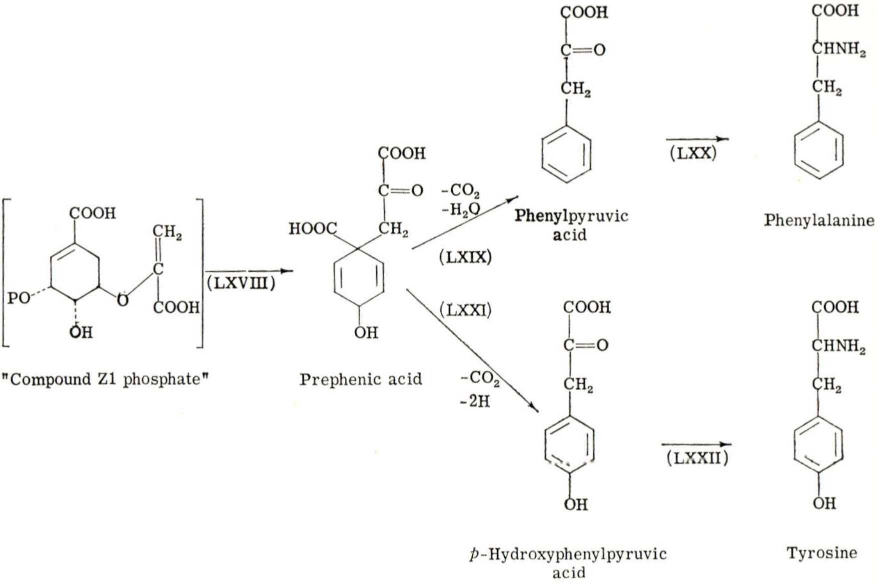

FIG. 13. The biosynthesis of tyrosine and phenylalanine.

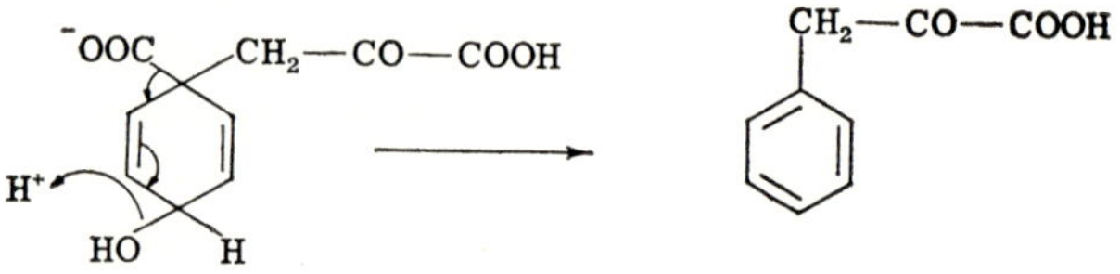

FIG. 14. The aromatization of prephenic acid by acid.

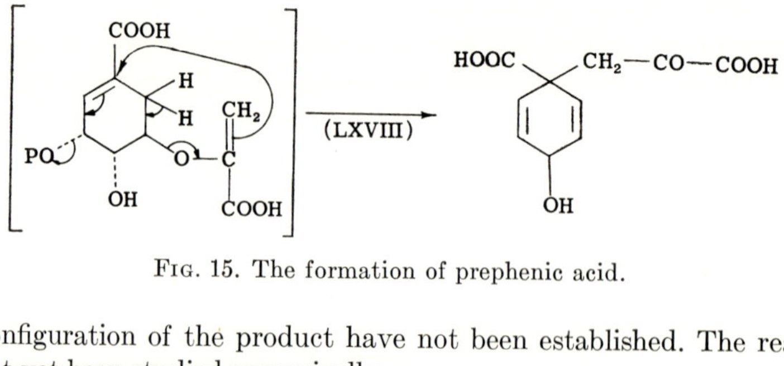

FIG. 15. The formation of prephenic acid.

the configuration of the product have not been established. The reaction has not yet been studied enzymically.

Prephenic acid has no nutritional activity for mutants with early blocks in aromatic biosynthesis, presumably because of a permeability barrier. Hence the following enzymic studies have been especially important in establishing the function of this compound.

2. The Formation of Phenylalanine

The conversion of prephenate to phenylpyruvate (Reaction LXIX, Fig. 13) is undoubtedly irreversible since it involves aromatization. It is catalyzed by extracts of wild-type *E. coli*, but not by those of a phenylalanine auxotroph that accumulates prephenate.[291] Although this transformation appears quite complex, including both dehydration and decarboxylation, the ease of its catalysis by mild acid (Fig. 14) suggests that one enzyme (prephenic aromatase) could trigger the whole sequence.

The conversion of phenylpyruvate to phenylalanine (Reaction LXX, Fig. 13) is presumably catalyzed by transaminases A and B in *E. coli* (Table II). While this transamination could conceivably be nonspecific, the obligatory participation of prephenic aromatase in biosynthesis establishes the keto acid produced by it as a true intermediate.

3. The Formation of Tyrosine

The presence of a hydroxyl group on position 4 of prephenate, and the finding that this compound is accumulated also by a tyrosine auxotroph, suggested that it might be a precursor of tyrosine as well as phenylalanine. In investigating this possibility it was helpful to use a mutant of *E. coli* blocked in the competing conversion of prephenate to phenylpyruvate. Extracts of this mutant were shown[292] to contain an enzyme, prephenic dehydrogenase, that converts prephenate to *p*-hydroxyphenylpyruvate (Reaction LXXI, Fig. 13). In this reaction DPN is the electron acceptor. The dehydrogenation is accompanied by a decarboxylation; and the resulting aromatization must make the reaction irreversible. Prephenic dehydrogenase could not be detected in extracts of a tyrosine auxotroph.

The conversion of *p*-hydroxyphenylpyruvate to tyrosine (Reaction LXXII), like the corresponding reaction of phenylpyruvate, is an obligatory biosynthetic reaction presumably catalyzed by transaminases A and B in *E. coli*.

C. The Formation of Tryptophan

A key to the biosynthesis of tryptophan was provided as early as 1940, before the development of the mutant approach, by observations on tryptophan-requiring bacteria encountered in nature. Fildes[293] found that a variety of species could use indole in place of tryptophan, and Snell[294] observed similar activity with anthranilic acid. Similar responses were noted later with mutants of *Neurospora*[295] and *E. coli*; and some mutants were found to accumulate these precursors. This evidence, followed by the enzymic conversion of anthranilate to indole to tryptophan, seemed to establish these precursors firmly as intermediates. It is therefore particularly interesting that recent work has excluded free indole as an intermediate.

The reactions following anthranilate (Fig. 14) have been largely worked out by Yanofsky. Those leading from the common aromatic pathway to anthranilate are only beginning to be explored, and will be discussed last.

1. The Conversion of Anthranilic Acid to Indole-3-Glycerol Phosphate

The main features of the pathway from anthranilate to indole were suggested by studies with labeled compounds and were established by enzymic

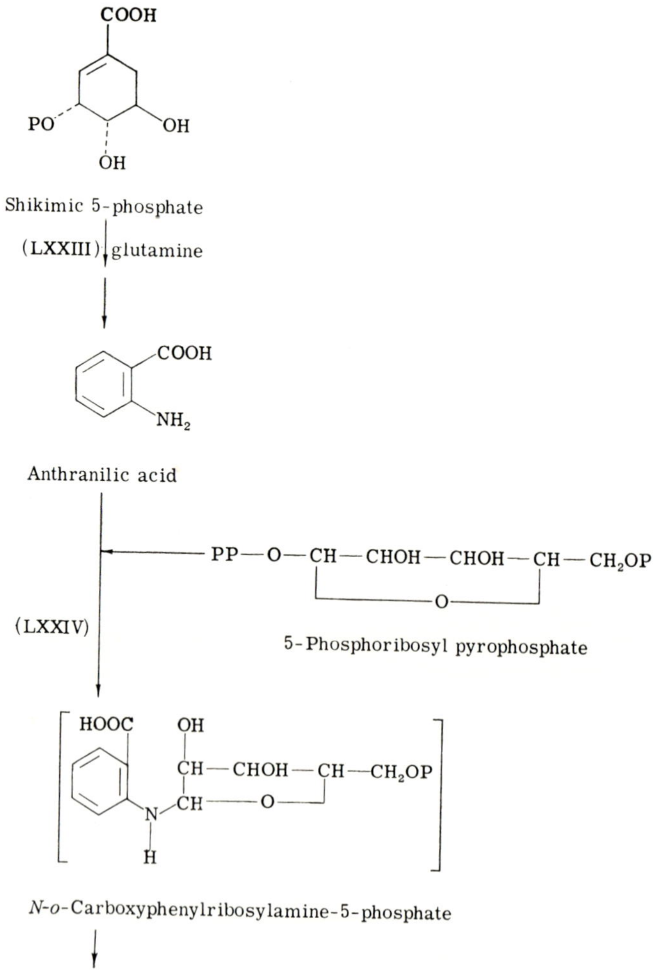

FIG. 16. The biosynthesis

observations; confirmation by identification of accumulated indole-3-glycerol came later, even though several laboratories had already recognized but not identified this accumulation (see Fig. 16).

The first development was the demonstration by means of isotopes that in the conversion of anthranilate to tryptophan the carboxyl was eliminated[296] but the nitrogen was retained.[297] Furthermore, Yanofsky[298] showed that *E. coli* converted 4-methylanthranilate to 6-methylindole, indicating

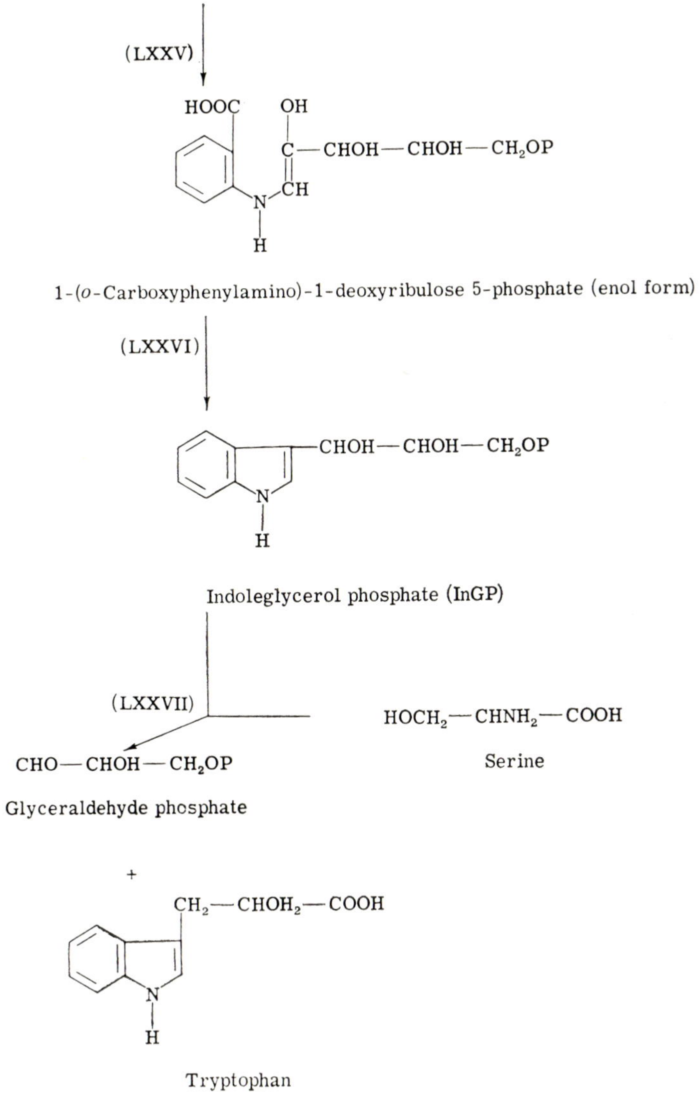

of tryptophan.

that in forming the pyrrole ring a 2-carbon fragment replaced the carboxyl on C-1 of anthranilate.*

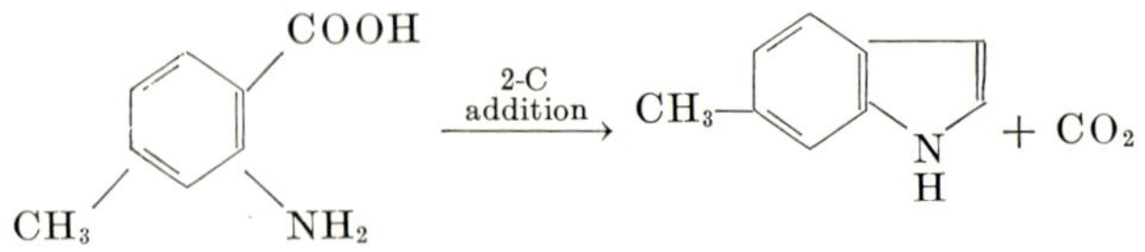

4-Methylanthranilic acid 6-Methylindole

Results obtained with *E. coli* grown on labeled glucose pointed to C-1 and C-2 of ribose as precursors of the 2-carbon fragment.[299]

Study of the conversion of anthranilate to indole with extracts of *E. coli* showed 5-phosphoribosyl-1-pyrophosphate (PRPP) to be a particularly effective substrate.[300] Furthermore, fractionation of these extracts made it possible to obtain a product that was identified as indole-3-glycerol phosphate (InGP). Extracts of mutants blocked between anthranilate and indole fell into two classes: one group could not form InGP, while the others could not convert it to indole and triose phosphate.[301, 302] Finally, indole-3-glycerol was identified as an accumulation product in culture filtrates of certain bacterial mutants blocked before indole.[301, 303, 304]

These findings established InGP with reasonable certainty as an intermediate. As with the imidazoleglycerol phosphate accumulated by histidine auxotrophs (Section IX, A), neither indoleglycerol phosphate nor its dephosphorylated derivative could serve as a growth factor, the former presumably because of impermeability and the latter because it cannot be rephosphorylated.

To account for the formation of InGP from anthranilate and PRPP, Yanofsky[301] postulated the two intermediates depicted in Reactions (LXXIV–LXXVI) of Fig. 16. This pathway was strongly supported by the finding of Doy and Gibson[305] that certain mutants of *Aerobacter* and *E. coli* blocked between anthranilate and indole accumulate the dephosphorylation product of the second of these intermediates, i.e., anthranilate with ribulose phosphate substituted on the nitrogen [1-(*o*-carboxyphenylamino)-1-deoxyribulose-5-phosphate]. This compound would be derived by Amadori rearrangement (Reaction LXXV, Fig. 16) from the first of the postulated intermediates, the corresponding ribose phosphate derivative (*N*-*o*-carboxyphenylribosylamine-5-phosphate; anthranilic ribonucleotide). A search for accumulation of this ribosylamine derivative (with or without phosphate) has been unsuccessful;[306] but this failure is not surprising since the ribosylamine, in contrast to the ribulose derivative, would be unstable,

* This is an interesting example of the use of a nonisotopic label, made possible by the imperfect specificity of the enzymes concerned.

hydrolyzing spontaneously in neutral aqueous solution to yield anthranilate and ribose.[305, 306] At present, then, the ribose phosphate derivative must be considered a plausible but not an experimentally proved intermediate.*

A striking feature of the enzymes that convert anthranilate to indole is their lack of specificity: they will also act on anthranilate substituted with a methyl group or a halogen in the 4 or the 5 position.[301]

2. The Conversion of Indolegylcerol Phosphate to Tryptophan[309]

As noted above, nutritional responses have long pointed to indole as an intermediate in tryptophan biosynthesis; and this conclusion seemed to be firmly established by the observation that certain tryptophan auxotrophs had lost "InGP hydrolase" [Reaction (1) below][301, 310] while others had lost "tryptophan synthetase"† [Reaction (2)].[301, 310, 311]

Reaction (1): InGP + $H_2O \rightleftharpoons$ indole + triose phosphate ("InGP hydrolase")

Reaction (2): Indole + L-serine → L-tryptophan + H_2O ("tryptophan synthetase")

Reaction (3): InGP + L-serine → L-tryptophan + triose phosphate

However, an unusual connection between these two enzyme activities was indicated by several observations. First, in both *Neurospora* and *E. coli* not all tryptophan auxotrophs blocked in the utilization of indole accumulated this compound; some accumulated indoleglycerol, and their extracts were found to have lost, presumably in a single step, the capacity to carry out not just Reaction (2) and (3) but also Reaction (1).[310]

Immunochemical evidence along the same lines was made possible by the

* The incompleteness of our knowledge of this portion of the pathway is emphasized by reports[307, 308] that *Saccharomyces* mutants accumulate compounds in which anthranilate is condensed with a hexose rather than a pentose. Such compounds, if intermediates, would not require any major modification of the pathway, since in the reaction subsequent to formation of the indole nucleus (see below) the carbons outside the ring are discarded.

Note added in proof: The postulated derivatives of anthranilate have been strongly supported as intermediates by the results of recent enzymic studies on anthranilate-accumulating mutants of *E. coli*,[308a] which were found to fall into two groups. Group I proved to be blocked immediately after anthranilate: its extracts could not catalyze a reaction of this compound with PRPP. The other class was blocked in the next reaction: its extracts converted anthranilate and PRPP to an exceedingly labile compound, with properties consistent with anthranilic ribonucleotide. This compound broke down within a few minutes in the reaction mixture to regenerate anthranilate; but a mixture of the two extracts converted anthranilate plus PRPP to later, stable members of the sequence. Incidentally, these findings also support anthranilate as a true intermediate.

† Also denoted by earlier authors as tryptophan desmolase and as indole-serine carboligase.

serological uniqueness of the enzymes responsible for these reactions, which is demonstrated by the fact that extracts of certain mutants blocked in all three reactions failed to absorb antibody to "tryptophan synthetase." (The antibody was detected by inhibition of the enzymic reaction; see Volume I of this treatise, p. 432.) In contrast, with some mutants of *E. coli* that had lost Reaction (2) but retained Reaction (1), and with all such mutants of *Neurospora*, the extracts contained a serologically cross-reacting material (CRM) which absorbed antibody to the missing enzyme.[312, 313]

These facts suggested that the two "enzymes" might be two activities of a single enzyme. This would transfer indole from InGP to serine in one step (Reaction 3) when both substrates were available, but could also form or utilize free indole, in Reaction (1) or (2), when the system was incomplete. The mutants that had retained only Reaction (1) or (2) would be interpreted as having an altered enzyme, with loss of reactivity toward one but not toward another of its substrates.

This interpretation was verified for *Neurospora* by several kinds of evidence, including (*a*) the low rate of Reaction (1) compared with (3), (*b*) failure to trap indole during Reaction (3), and (*c*) inability of Reaction (3) to be restored by mixing a *Neurospora* preparation capable only of (1) with an *E. coli* preparation capable only of (2).[314] Furthermore, a frequent class of mutants in *Neurospora* has lost all three enzymic activities; and most of these are undoubtedly point mutations rather than deletions since they are revertible and are associated with the formation of an altered protein (CRM). These several lines of evidence lead to the conclusion that in the normal biosynthetic Reaction (3) indole remains enzyme-bound; and the term "tryptophan synthetase" is now used to refer to the enzyme that carries out this reaction.

In *E. coli* tryptophan synthetase also acts as a single enzyme, but the problem has been complicated by the remarkable further finding that it can be chromatographically separated into two protein fractions, A and B, which separately show little enzyme activity.[309, 315] Furthermore, *E. coli* point mutants blocked in tryptophan synthetase fall into four classes, forming either no A, altered A, no B, or altered B.

Component A contributes the combining site for InGP, since the purified component carries out Reaction (1) slowly (i.e., about 1% as rapidly as the complete enzyme). Component B contributes the combining site for serine and pyridoxal phosphate, since it is stabilized by the latter compound and carries out Reaction (2) slowly. On mixing the two components full activity is restored, presumably as a result of spontaneous association. Furthermore, the enzymically inactive protein A (CRM-A) formed by certain mutants can restore the full activity of normal protein B in Reaction (2); and CRM-B can restore the full activity of protein (A) in Reaction (1). Hence tryptophan auxotrophs that make CRM-A and normal B grow faster on indole than those that make no A and normal B.

Despite this separation of tryptophan synthetase into two genetically and chemically distinct proteins in *E. coli*, the fact that they must combine with each other for full activity implies that in this organism also, as in *Neurospora*, free indole is not an intermediate.

Tryptophan synthetase is now under intensive genetic investigation, particularly in relation to the formation of altered proteins and the action of suppressor mutations.[309]

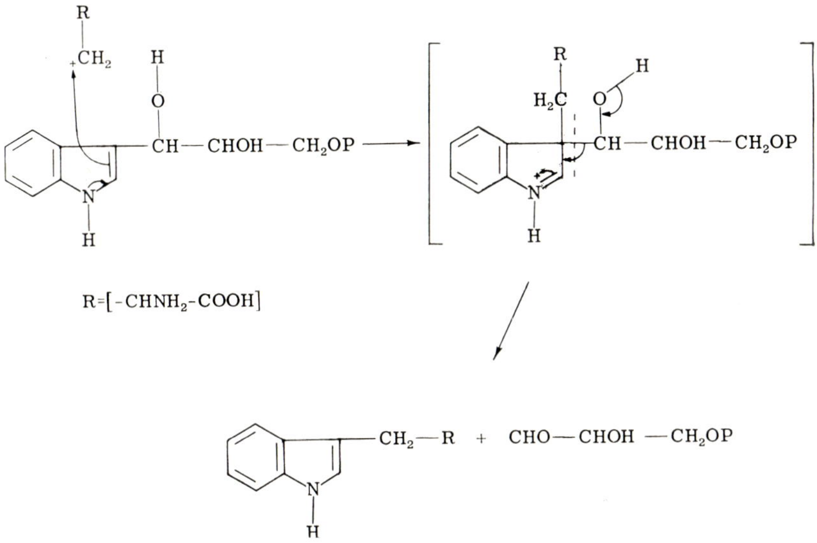

FIG. 17. Proposed mechanism of attack of serine on InGP. Pyridoxal phosphate (not shown) presumably promotes the action of serine as a cationoid reagent.

The one-step conversion of indoleglycerol phosphate to tryptophan, by exchange of the side chain with serine, is a remarkably economical reaction, compared with the sequence leading to histidine (Fig. 9). Theoretical considerations[316] lead to the prediction that the presence of the NH group in the ring of indole favors this reaction by electrophilic attack (Fig. 17).

It should be noted that tryptophan synthetase is distinct from the degradative enzyme tryptophanase, which converts tryptophan to pyruvate and ammonia in addition to indole.[317]

3. The Formation of Anthranilic Acid

Srinivasan[285] has shown that extracts of an *E. coli* mutant blocked after anthranilate can convert shikimic acid 5-phosphate (see Reaction LXXIII,

Fig. 16) almost quantitatively to this compound in the presence of glutamine. The reaction requires either DPN or TPN, and also other undefined cofactors. The intermediates are unknown but presumably include later common intermediates in aromatic biosynthesis; for mutants blocked after shikimic acid 5-phosphate or still later, after Compound Z1, exhibit a quintuple aromatic requirement and hence would be expected to be blocked in a reaction of the common pathway.* Compound Z1 is inactive in the enzymatic formation of anthranilate;[285] but this fact is consistent with the evidence of Sprinson (Section X, 4, 3) that this compound is an artifact produced by dephosphorylation.

The reactions indicated as (LXXIII–LXXV) in Fig. 16 still offer much challenge. In this connection it would be particularly important to establish unequivocally whether or not the benzene ring of tryptophan is formed from the intact ring of shikimate, for the results reported for certain isotopic studies on tryptophan[318-321] are incompatible with this process.

XI. The Formation of D-Amino Acids

While the amino acids isolated from proteins throughout the biological kingdom have been uniformly of the L-configuration, a role of D-amino acids in bacterial metabolism was suggested by an early observation of Snell on bacterial nutrition.[322] He found that in the absence of vitamin B_6 *Lactobacillus arabinosus* required D-alanine for growth. (This vitamin was later shown to participate in amino acid racemization.) D-Amino acids were found in the capsule of members of the genus *Bacillus* and in various excreted polypeptide antibiotics; but these products of the cell hardly seemed essential for its growth. The essential role of D-amino acids in bacteria finally became clear with the discovery of D-alanine and D-glutamate in the mucopeptides of the cell walls of a wide variety of bacteria.[323, 324]

These D-amino acids apparently arise from L-amino acids by racemization. Alanine racemase has been found in a large number of bacterial species,[188] and a glutamic racemase, which apparently does not involve free α-ketoglutarate, has been described for *L. arabinosus*[325] and *Mycobacterium avium*.[326]

In addition, a novel transamination reaction involving D-amino acids in *B. subtilis* and *B. anthracis* was observed by Thorne and co-workers.[327, 328, 329] Certain strains of these organisms are remarkable for their ability to form capsules consisting of a polyglutamic acid, predominantly in the D-configuration. Partial purification of extracts yielded a preparation that converted α-ketoglutarate to D-glutamate in the presence of D-alanine or

* An alternative explanation would be that shikimic acid 5-phosphate is a branch point, and a genetic block in one of the branches (to prephenate or to anthranilate) results in inhibition, by an accumulated compound, of the other branch(es).

D-aspartate. This enzyme, coupled with the racemase for alanine, would appear to account for the origin of the D-glutamate in these organisms. The existence of transaminases specific for a group of D-amino acids suggests the possibility of a general mechanism for the formation of D-amino acids analogous to that established for L-amino acids. This view is supported by the recent isolation of a mutant of *B. subtilis* that requires either D-glutamate, D-aspartate, or D-alanine.[330]

The conversion of L- to *meso*-diaminopimelate, an essential reaction in the biosynthesis of lysine, has been shown to be catalyzed by a specific diaminopimelic racemase (p. 202).

XII. The Formation of Amino Acids as Major Excretion Products

As noted in the introduction to this chapter, the pathways of amino acid biosynthesis are to a great extent under precise control by feedback mechanisms, which regulate the rate of synthesis of each end product to correspond to its rate of utilization by the growing cell. These mechanisms are described in another part of this treatise. It seems relevant, however, to consider briefly certain instances of escape from such controls, which have resulted in the overproduction and substantial excretion of certain amino acids.

One method of overcoming the controls is the cultivation of auxotrophic mutants on a growth-limiting amount of the required end product. As stressed throughout this chapter, this procedure frequently results in the heavy accumulation of metabolic intermediates. Such an accumulation of diaminopimelate by a lysine auxotroph[30] has made possible a commercial process[167] in which a second organism is used to convert the diaminopimelate to lysine.*

For commercial production a one-step "fermentation" would clearly be the more desirable; and one approach is suggested by the observation of a few auxotrophic mutants that not only accumulate the precursor of the blocked reaction but also convert it to the end product of a related pathway. Thus tyrosine can be excreted by some phenylalanine auxotrophs and phenylalanine by some tyrosine auxotrophs.[89]

Another approach, not based on auxotrophic mutations, has been under intensive investigation in recent years in Japan. Random screening of soil samples, as in the search for antibiotics, has yielded a number of organisms that excrete amino acids, including lysine.[331] One industrial firm reported[332]

* It is of interest to note that considerably heavier accumulations are possible than those customarily seen in the laboratory. In the original work on diaminopimelate accumulation[30] a yield of 0.5 g./liter was considered quite respectable; but by varying the conditions of cultivation and the medium, and by selection of secondary mutants, it has proved commercially possible to obtain yields 10 times as high.[167]

production of 100 metric tons of glutamic acid per month by use of an organism, called *Micrococcus glutamicus*, that excretes L-glutamic acid in an amount equal to about one-fifth of the glucose consumed.[333] Interestingly, an arginine auxotroph derived from this organism accumulates glutamate in the usual manner as long as there is an excess of arginine present; but when arginine is limiting the glutamate is replaced by ornithine.[334]

Such overproduction of an amino acid could conceivably be due to either development of an accessory pathway or loss of the control mechanism in the normal pathway. In the few cases in which there is pertinent evidence the latter appears to be the mechanism. For example, an organism that excretes valine was found to give a positive color test for acetolactate,[335] but to lack the decarboxylase that converts this compound to acetoin (see Section VIII, A, 2).[336] Apparently the "biosynthetic" acetolactate-forming system in this organism, like the "catabolic" one in *A. aerogenes*, is not inhibited by valine.

Organisms that excrete amino acids can also be isolated by a more rationally directed method: selection of mutants resistant to growth inhibition by amino acid analogs. In one systematic search resistant mutants of *E. coli* were obtained for a variety of amino acid analogs, and each was shown to excrete the corresponding amino acid.[337] It was initially thought that the overproduction of the amino acid, due to a loss of feedback control, was itself responsible for reversing the inhibition. More recent work, however, has shown that at least in some cases the excretion is merely a byproduct of the change responsible for resistance.[25, 26] In these cases the analog mimics the inhibitory effect of the end product, in the wild-type organism, on the initial enzyme of the sequence. The mutant has an altered enzyme, selected for resistance to the analog; and since this enzyme is also resistant to feedback by the normal end product, this product is excreted.

The development of inexpensive microbial processes for the production of such amino acids as lysine, tryptophan, and methionine would obviously have great significance for the problem of human nutrition in areas of the world with an essentially vegetarian diet.

XIII. General Considerations

A. Comparative Biochemistry

While investigations of amino acid biosynthesis have concentrated heavily on *E. coli* (and the closely related *Aerobacter*) and on *Neurospora*, limited comparative studies have provided evidence for a good deal of unity throughout the biological kingdom in this as in other areas of intermediary metabolism. Only a few exceptions have been established:

1. Lysine arises in bacteria from the 7-carbon compound diaminopime-

late, which is also a component of the cell wall mucopeptide of certain bacteria; whereas lysine arises in molds by a 6-carbon pathway that includes α-aminoadipate.

2. Ornithine, and hence arginine, arises in *E. coli* via a series of *N*-acetyl derivatives of glutamate, whereas *Neurospora* and certain Gram-positive bacteria appear to employ a different pathway which has not yet been established.

3. In *E. coli* and *Neurospora* tyrosine and phenylalanine arise only from a common precursor, but tyrosine can be formed from phenylalanine in some bacteria,[338, 339] as in mammals. Whether this conversion has biosynthetic significance is not clear.

Such exceptions to biochemical unity may prove very valuable in deciding vexing problems of microbial taxonomy. Thus Stanier[340] has suggested that the presence of diaminopimelate might be a useful guide for distinguishing bacteria from other protista. Unfortunately negative evidence would have little value, since diaminopimelate, while restricted to bacteria, is not found in all species. However, a more universally applicable modification of this approach is suggested by the evidence, from studies both on enzyme distribution[166] and on isotope incorporation,[170] that the diaminopimelate pathway to lysine is the one present in all bacteria that can synthesize lysine.

B. General Features of Biosynthesis

In the biosynthesis of the amino acids certain general patterns have emerged. Thus, as in other aspects of intermediary metabolism, the reversibility of many reactions apparently has little physiological significance: nature develops biosynthetic and degradative sequences whose energetics force a flow in the appropriate direction. Examples include the different mechanisms of splitting cystathionine in the formation of methionine from cysteine and in degrading methionine; the different reactions catalyzed by tryptophan synthetase and by tryptophanase; and the entirely unrelated biosynthetic and degradative pathways observed for a number of amino acids, including histidine, tyrosine, and threonine.

In promoting reactions in a biosynthetic direction ATP naturally contributes frequently, often through an as yet unknown mechanism. In some cases the expenditure of ATP energy is seen in the participation of such compounds as phosphoenolpyruvate, 5-phosphoribosyl-1-pyrophosphosphate, or carbamyl phosphate. Another means of ensuring the forward flow of biosynthesis involves the oxidative and the reductive steps. The sequences leading to amino acids conform to the generalization that among pyridine nucleotide-linked reactions the biosynthetic reductions involve TPN, which is predominantly in the reduced form in normally metabolizing cells, while the biosynthetic oxidations involve DPN, which is found to be more extensively oxidized in the cell.[46] A striking example is the TPN-linked formation of shikimate by reduction, for biosynthetic purposes, compared with the structurally very similar DPN-linked oxidation of quinate for degradative purposes.

With the recognition of a large number of intermediates in the biosynthesis of the amino acids and other essential cell constituents, another generalization has emerged:[341] namely, that in these pathways all the low-molecular water-soluble intermediates have one or more groups that are largely ionized at neutrality. Usually it is a carboxyl group, but in a few compounds (e.g., histidinol) it is an amino group. Examples of this principle can be seen in the transformations of imidazole glycerol phosphate, in which the phosphate group is retained until the amino group appears, and the transformations of shikimate, in which the carboxyl group is retained until an acidic side chain is added. The one apparent exception in amino acid biosynthesis, indole, has now been shown to remain enzyme-bound and thus not to be a true intermediate. The nature of the advantage of ionizable intermediates is not evident, but one possibility is more efficient retention in the cell.

C. Special Features of Amino Acid Biosynthesis

Now that the majority of the steps in amino acid biosynthesis are known, it can reasonably be estimated that about 100 enzymes account for the formation of all the amino acids from their branch points in the less specialized pathways. Table VI summarizes these estimates and lists the branch points for the various pathways.

In surveying the approximately 75 reactions in these pathways that are known, one is led to admire the elegance and economy, from the point of view of the organic chemist, of the sequences that have been selected in the course of evolution. Among the interesting and novel chemical features that have appeared are the following: (*1*) The alkyl shift in the biosynthesis of isoleucine and valine. (*2*) The presence of a sequence of enzymes shared in these two pathways (in which all the pairs of intermediates differ by one methylene group), in contrast to the high degree of specificity of the dehydrogenases, phosphatases, decarboxylases, etc., found in the other pathways. (*3*) The spontaneous cyclization of glutamic semialdehyde in the formation of proline, and the prevention of such cyclization by the use of an *N*-acetyl blocking group in the ornithine pathway. An *N*-succinyl group has a similar function in the diaminopimelate pathway. (*4*) The use of an acyl phosphate plus TPNH to bring about the difficult reduction of the carboxyl group of glutamate and aspartate; and the elimination of a carboxyl group in the formation of lysine and the aromatic amino acids. (*5*)

TABLE VI

BIOSYNTHETIC ORIGINS OF THE AMINO ACIDS

Amino acid	Atom	Origin	Probable number of enzymes
Glutamate	All C	α-Ketoglutarate	1
Aspartate	All C	Oxalacetate	1
Alanine	All C	Pyruvate	1
Proline	All C	Glutamate	3
Arginine	C 1–5	Glutamate to ornithine	7+
	Guanidino N—C	CO_2, NH_3, via carbamyl phosphate	
	Guanidino N	Aspartate	
Threonine	All C	Aspartate (via semialdehyde and homoserine)	6+
Methionine	C 1–4	Aspartate (via semialdehyde and homoserine)	8+ (4 shared with threonine)
	S	Cysteine	
	Methyl	"1-carbon"	
m-Diaminopimelate	C 1–4	Aspartate (via semialdehyde)	7+
	C 5–7	Pyruvate	
Lysine (bacteria only)	All C	*m*-Diaminopimelate	1
Isoleucine	C 1,2,5,6	Threonine	6+
	C 3,4	Pyruvate-2,3	
Valine	C 1,2,4	Pyruvate	6+ (5 shared with isoleucine)
	C 3,4′	Pyruvate-2,3	
Leucine	C 3,4,5,5′	Pyruvate	10+ (5+ shared with valine)
	C 1,2	Acetate	
Serine		Probably *P*-glycerate	?
Glycine		?	?
Cysteine	All C	Serine	?4+
	S	Sulfide	
Tyrosine	Side chain	*P*-Enolpyruvate	10+
	Ring	Erythrose-4-phosphate + *P*-enolpyruvate, via shikimate	
Phenylalanine	Same as tyrosine		10+ (9+ shared with tyrosine)
Tryptophan	Side chain	Serine	13+ (7+ shared with ty, ph)
	Benzene ring	Like tyrosine	
	C2,3 of indole	C 1,2 of ribose-5-phosphate	
	N of indole	Amide of glutamine	
Histidine	Imidazole N 1, C 2	N 1, C 2 of adenylic acid	6+
	Imidazole N 3	NH_3 or glutamine	
	Remaining C	Ribose-5-phosphate	

The use of concerted electron shifts in attaching the pyruvate side chain to the necessary (although hindered) position in prephenate, and in bringing about the aromatization of this compound. (*6*) Formation of tryptophan from indoleglycerol phosphate by a one-step exchange of the side chain with serine, contrasted with formation of histidine from imidazoleglycerol phosphate by a four-step series of transformations of the same side chain. Theoretical considerations[316] lead to the prediction that while the —NH— of both rings would strongly favor exchange by electrophilic attack (Fig. 17), the additional C═N of the imidazole ring would oppose this process. Hence it appears that nature has not neglected the possibility of a shorter pathway in histidine biosynthesis; it has rather taken advantage of a possible short cut in tryptophan biosynthesis. Similar short cuts are seen in the one-step synthesis of lysine from a compound required for other purposes; in the use of carbamyl phosphate to provide an N—C fragment for both pyrimidine and citrulline biosynthesis; and in the use of adenylic acid to provide a more reduced N—C fragment for histidine biosynthesis.

Economy in biosynthesis is illustrated not only by the nature of the sequences evolved but also by the presence of feedback mechanisms that prevent or decrease the formation of both unnecessary enzymes and unnecessary building blocks. In addition, the estimates summarized in Table VI reveal an interesting economy in mammalian biochemistry. The mammal requires in its diet about half the amino acids. Since these are, without exception, the ones with long sequences (Table VI), the mammal has dispensed with much more than half the enzymes of amino acid biosynthesis

D. Biosynthetic versus Degradative Enzymes; Control Mechanisms

The topic of feedback control, including both end product repression of the formation of the several enzymes of a pathway and end product inhibition of the action of the first enzyme (i.e., the branch point) of that pathway, is considered elsewhere in this treatise. However, certain aspects deserve mention here in connection with the problem of determining the true participants in a biosynthetic pathway. Throughout this chapter we have emphasized the importance of determining whether a substance that could serve as a precursor also did serve as a normal intermediate; and several misleading precursors have been recognized (indole, quinic acid, histidine as a source of glutamate, several α-keto acids). The same problem extends to characterizing the enzymes of biosynthesis, and makes it important to pay heed to effects of the conditions of growth on the enzymic composition of cells.

In a particularly striking case (see p. 211) *Aerobacter* makes two different enzymes that convert pyruvate to acetolactate.[221] The reactions catalyzed are identical; but one enzyme, on the pathway to valine, is subject to both repression and feedback inhibition by valine, while the other is not subject to these controls but is subject to induction by growth at low pH on glucose. Similarly, in *E. coli* two different enzymes can each convert threonine to α-ketobutyrate, one functioning in the biosynthesis of isoleucine and the other in the degradation of threonine.[226] It is clear that in attempts to increase the yield of enzymes concerned with the metabolism of a given substance, induction with that substance, or the selection of mutants that degrade it readily, may yield results that are not relevant to a biosynthetic pathway.

On the other hand, altering the amount of the end product of a sequence will often influence in a predictable way the levels of the biosynthetic enzymes of that sequence. Thus growth in the presence of an excess of the end product will usually yield a batch of cells with decreased amounts (or even none) of the enzymes of that pathway. Conversely, the intracellular level of an amino acid can be lowered to less than the normal steady-state value by growing an auxotroph on a limiting supply of the required compound, or by supplying the wild type with an extensive enrichment that lacks the end product under investigation. Under these circumstances the cell becomes "derepressed," and has been observed to synthesize as much as 50 times the normal amount of the enzymes of the derepressed pathway, relative to the rest of the protein of the cell.[17-19] Mutations leading to escape from repression have a similar effect. Such devices are of real value to enzymologists in providing exceptionally rich sources of desired biosynthetic enzymes.

References

[1] H. A. Krebs and H. L. Kornberg, *Ergeb. Physiol.* **49**, 212 (1957).

[1a] B. D. Davis, H. L. Kornberg, A. Nagler, P. Miller, and E. Mingioli, *Federation Proc.* **18**, 211 (1959).

[2] B. D. Davis, *Advances in Enzymol.* **16**, 247 (1955).

[3] C. Gilvarg and B. D. Davis, *J. Biol. Chem.* **222**, 307 (1956).

[4] F. C. Neidhardt and B. Magasanik, *J. Bacteriol.* **73**, 253 (1957).

[5] E. A. Adelberg, *Bacteriol. Revs.* **17**, 253 (1953).

[6] B. D. Davis and U. Weiss, *Arch. Exptl. Pathol. Pharmakol.* **220**, 1 (1953).

[7] G. W. Beadle and E. L. Tatum, *Proc. Natl. Acad. Sci. U. S.* **27**, 499 (1941).

[8] H. E. Umbarger and B. Magasanik, *J. Biol. Chem.* **189**, 287 (1951).

[9] E. A. Adelberg, D. M. Bonner, and E. L. Tatum, *J. Biol. Chem.* **190**, 837 (1951).

[10] E. A. Adelberg and E. L. Tatum, *Arch. Biochem.* **29**, 235 (1950).

[11] B. D. Davis and E. S. Mingioli, *J. Bacteriol.* **66**, 129 (1953).

[12] U. Weiss and E. S. Mingioli, *J. Am. Chem. Soc.* **78**, 2894 (1956).

[13] B. D. Davis, *Science* **118**, 251 (1953).

[14] B. N. Ames, *in* "Amino Acid Metabolism" (W. D. McElroy and B. Glass, eds.), Johns Hopkins Press, Baltimore, Maryland, 1955.

[15] P. R. Srinivasan, H. T. Shigura, M. Sprecher, D. B. Sprinson, and B. D. Davis, *J. Biol. Chem.* **220**, 477 (1956).

[16] R. B. Roberts, P. H. Abelson, D. B. Cowie, E. T. Bolton, and R. J. Britten, Studies of Biosynthesis in *Escherichia coli*. *Carnegie Inst. Wash. Publ. No.* **607,** 1955.
[17] H. J. Vogel, *in* "The Chemical Basis of Heredity" (W. D. McElroy and B. Glass, eds.), p. 276. John Hopkins Press, Baltimore, Maryland, 1956.
[18] R. A. Yates and A. B. Pardee, *J. Biol. Chem.* **227,** 677 (1957).
[19] L. Gorini and W. K. Maas, *Biochim. et Biophys. Acta* **25,** 208 (1957).
[20] H. E. Umbarger, *Science* **123,** 848 (1956).
[21] R. A. Yates and A. B. Pardee, *J. Biol. Chem.* **221,** 757 (1956).
[22] H. J. Strecker, *J. Biol. Chem.* **225,** 825 (1957).
[23] H. E. Umbarger and B. Brown, *J. Biol. Chem.* **233,** 415 (1958).
[24] H. E. Umbarger and B. Brown, *J. Biol. Chem.* **233,** 1156 (1958).
[25] H. S. Moyed and M. Friedman, *Science* **129,** 968 (1959).
[26] H. S. Moyed, *J. Biol. Chem.* **235,** 1098 (1960).
[27] A. J. Kluyver and H. J. L. Donker, *Chem. Zelle u. Gewebe* **13,** 134 (1926).
[28] H. J. Vogel, *in* "Amino Acid Metabolism" (W. D. McElroy and B. Glass, eds.), p. 335. Johns Hopkins Press, Baltimore, Maryland, 1955.
[29] H. K. Mitchell and M. B. Houlahan, *J. Biol. Chem.* **174,** 883 (1948).
[30] B. D. Davis, *Nature* **169,** 534 (1952).
[31] B. D. Davis, unpublished observations.
[32] H. Green and B. D. Davis, cited by B. D. Davis, *in* "Enzymes: Units of Biological Structure and Function" (O. H. Gaebler, ed.), p. 509. Academic Press, New York, 1956.
[33] S. Ochoa, *Advances in Enzymol.* **15,** 183 (1952).
[34] J. M. Wiame, *Advances in Enzymol.* **18,** 241 (1957).
[35] J. M. Wiame, personal communication.
[36] J. A. Boezi and R. D. DeMoss, *Bacteriol. Proc.* (*Soc. Am. Bacteriologists*) **59,** p. 124 (1959).
[37] R. P. Cook and B. Woolf, *Biochem. J.* **22,** 474 (1928).
[38] J. H. Quastel and B. Woolf, *Biochem. J.* **20,** 545 (1926).
[39] E. F. Gale, *Biochem. J.* **32,** 1583 (1938).
[40] E. F. Gale, *Bacteriol. Revs.* **4,** 135 (1940).
[41] E. F. Gale and H. M. R. Epps, *Biochem. J.* **36,** 600 (1942).
[42] A. C. I. Warner, *Biochem. J.* **64,** 1 (1956).
[43] J. R. S. Fincham, *J. Gen. Microbiol.* **5,** 793 (1951).
[44] J. R. S. Fincham, *J. Biol. Chem.* **182,** 61 (1950).
[45] Y. S. Halpern and H. E. Umbarger, *J. Bacteriol.* **80,** 285 (1960).
[46] G. E. Glock and P. McLean, *Biochem. J.* **61,** 388 (1955).
[47] J. M. Wiame and A. Piérard, *Nature* **176,** 1073 (1956).
[48] A. Piérard and J. M. Wiame, *Biochim. et Biophys. Acta* **37,** 490 (1960).
[49] S. C. Shen, M. M. Hong, and A. E. Braunstein, *Biochim. et Biophys. Acta* **36,** 290 (1959).
[50] M. M. Hong, S. C. Shen, and A. E. Braunstein, *Biokhimiya* **24,** 929 (1959).
[51] S. C. Shen, M. M. Hong, and A. E. Braunstein, *Biokhimiya* **24,** 957 (1959).
[52] M. M. Hong, S. C. Shen, and A. E. Braunstein, *Biochim. et Biophys. Acta* **36,** 288 (1959).
[53] J. M. Wiame, J. Collette, and S. Bourgeois, *Arch. intern. physiol. et biochem.* **63,** 271 (1955).
[54] A. E. Braunstein and M. G. Kritzmann, *Enzymologia* **2,** 129 (1937).
[55] L. I. Feldman and I. C. Gunsalus, *J. Biol. Chem.* **187,** 821 (1950).

[56] D. Rudman and A. Meister, *J. Biol. Chem.* **200,** 591 (1953).
[57] A. Meister, *Advances in Enzymol.* **16,** 185 (1955).
[58] R. A. Altenbern and R. D. Housewright, *J. Biol. Chem.* **204,** 159 (1953).
[59] A. Meister, "Biochemistry of the Amino Acids." Academic Press, New York, 1957.
[60] D. E. Metzler, J. Olivard, and E. E. Snell, *J. Am. Chem. Soc.* **76,** 644 (1954).
[61] D. E. Metzler, M. Ikawa, and E. E. Snell, *J. Am. Chem. Soc.* **76,** 648 (1954).
[62] A. Meister, H. A. Sober, and E. A. Peterson, *J. Biol. Chem.* **206,** 89 (1954).
[63] M. R. Stetten and R. Schoenheimer, *J. Biol. Chem.* **153,** 113 (1944).
[64] L. Levintow and A. Meister, *J. Biol. Chem.* **209,** 265 (1954).
[65] L. Levintow and A. Meister, *Federation Proc.* **15,** 299 (1956).
[66] H. Waelsch, *Advances in Enzymol.* **13,** 237 (1952).
[67] B. D. Davis, *in* "Symposium Microbiological Metabolism," *Intern. Congr. Microbiol., 6th Congr., Rome,* p. 23 (1953).
[68] Y. S. Halpern, personal communication.
[69] H. McIlwain, P. Fildes, G. P. Gladstone, and B. C. J. G. Knight, *Biochem. J.* **33,** 223 (1959).
[70] P. Ayengar, E. Roberts, and G. B. Ramasarma, *J. Biol. Chem.* **193,** 781 (1951).
[71] J. H. Mueller and P. A. Miller, *J. Biol. Chem.* **181,** 39 (1949).
[72] E. Sandheimer and D. C. Wilson, *Arch. Biochem. Biophys.* **61,** 313 (1956).
[73] M. N. Camien and M. S. Dunn, *J. Biol. Chem.* **217,** 125 (1955).
[74] D. A. Goldthwait, G. R. Greenberg, and R. A. Peabody, *Biochim. et Biophys. Acta* **18,** 148 (1955).
[75] B. Levenberg and J. M. Buchanan, *J. Am. Chem. Soc.* **78,** 504 (1956).
[76] B. M. Pogell and R. M. Gryder, *J. Biol. Chem.* **228,** 701 (1957).
[77] A. Neidle and H. Waelsch, *J. Biol. Chem.* **234,** 586 (1959).
[78] H. S. Moyed and B. Magasanik, *J. Am. Chem. Soc.* **79,** 4812 (1957).
[79] B. Magasanik, personal communication.
[79a] B. N. Ames, R. G. Martin, and B. J. Garry, *J. Biol. Chem.* **236,** 2018 (1961).
[80] W. J. Williams and C. B. Thorne, *J. Biol. Chem.* **210,** 203 (1954).
[81] W. J. Williams and C. B. Thorne, *J. Biol. Chem.* **211,** 631 (1954).
[82] W. J. Williams, J. Litwin, and C. B. Thorne, *J. Biol. Chem.* **212,** 427 (1955).
[83] P. J. Samuels, *Biochem. J.* **55,** 441 (1953).
[84] J. E. Snoke and K. Bloch, *J. Biol. Chem.* **199,** 407 (1952).
[85] S. Mandeles and K. Bloch, *J. Biol. Chem.* **214,** 639 (1955).
[86] A. Meister, *J. Biol. Chem.* **200,** 571 (1953).
[87] A. Meister, H. A. Sober, S. V. Tice, and P. E. Frazer, *J. Biol. Chem.* **197,** 319 (1952).
[88] A. Meister and P. E. Frazer, *J. Biol. Chem.* **210,** 37 (1954).
[89] B. D. Davis, *Experientia* **6,** 41 (1950).
[90] H. J. Vogel and B. D. Davis, *J. Am. Chem. Soc.* **74,** 109 (1952).
[91] D. Shemin and D. Rittenberg, *J. Biol. Chem.* **158,** 71 (1945).
[92] A. Meister, A. N. Radhakrishnan, and S. Buckley, *J. Biol. Chem.* **229,** 789 (1957).
[93] T. Yura, personal communication.
[94] T. Yura and H. J. Vogel, *J. Biol. Chem.* **234,** 335 (1959).
[95] H. A. Krebs and K. Henseleit, *Z. Physiol. Chem.* **210,** 33 (1932).
[96] S. Ratner, *Advances in Enzymol.* **15,** 319 (1954).
[97] H. J. Vogel and B. D. Davis, unpublished observations.
[98] B. D. Davis, *Advances in Enzymol.* **16,** 261 (1955).
[99] H. J. Vogel, *Proc. Natl. Acad. Sci. U. S.* **39,** 578 (1953).
[100] H. J. Vogel, P. H. Abelson, and E. T. Bolton, *Biochim. et Biophys. Acta* **11,** 584 (1953).

[101] L. Gorini, personal communication.
[102] W. K. Maas, G. D. Novelli, and F. Lipmann, *Proc. Natl. Acad. Sci. U. S.* **39,** 1004 (1953).
[103] S. Udaka and S. Kinoshita, *J. Gen. Appl. Microbiol.* **4,** 272 (1958).
[104] H. J. Vogel and D. M. Bonner, *J. Biol. Chem.* **218,** 97 (1956).
[105] J. R. S. Fincham, *Biochem. J.* **53,** 313 (1953).
[106] A. Meister, *J. Biol. Chem.* **206,** 587 (1954).
[107] A. Meister, personal communication.
[108] W. I. Scher and H. J. Vogel, *Proc. Natl. Acad. Sci. U. S.* **43,** 796 (1957).
[109] H. J. Vogel and D. M. Bonner, *Proc. Natl. Acad. Sci. U. S.* **40,** 688 (1954).
[110] J. P. Aubert, J. Millet, E. Pineau, and G. Milhaud, *Compt. rend. acad. sci.* **249,** 1956 (1959).
[111] R. L. Metzenberg, M. Marshall, and P. P. Cohen, *J. Biol. Chem.* **233,** 1560 (1958).
[112] M. E. Jones, L. Spector, and F. Lipmann, *J. Am. Chem. Soc.* **77,** 819 (1955).
[113] M. E. Jones, L. Spector, and F. Lipmann, *Proc. 3rd. Intern. Congr. Biochem. Brussels* p. 278 (1955).
[114] J. S. Loutit, *Australian J. Exptl. Biol. Med. Sci.* **30,** 287 (1952).
[115] D. Kanazir, H. D. Barner, J. G. Flaks, and S. S. Cohen, *Biochim. et Biophys. Acta* **34,** 341 (1959).
[116] J. Fresco, W. K. Maas, and B. D. Davis, unpublished observations.
[117] S. Ratner, *in* "Essays in Biochemistry" (S. Graff, ed.), p. 216. Wiley, New York, 1956.
[118] J. B. Walker and J. Myers, *J. Biol. Chem.* **203,** 143 (1953).
[119] L. Gorini, *Bull. soc. chim. biol.* **40,** 1949 (1958).
[120] L. Gorini, *Proc. Natl. Acad. Sci. U. S.* **46,** 682 (1958).
[121] H. L. Ennis and L. Gorini, *Federation Proc.* **18,** 222 (1959).
[122] S. Udaka and S. Kinoshita, *J. Gen. Appl. Microbiol.* **4,** 283 (1958).
[123] H. Tabor, S. M. Rosenthal, and C. W. Tabor, *J. Biol. Chem.* **233,** 907 (1958).
[124] C. Gilvarg, *Federation Proc.* **19,** 948 (1960).
[125] See reference 67.
[126] S. Black and N. G. Wright, *J. Biol. Chem.* **213,** 27 (1955).
[127] G. N. Cohen, M. L. Hirsch, S. B. Wiesendanger, and B. Nisman, *Compt. rend. acad. sci.* **238,** 1746 (1954).
[128] S. Black and N. G. Wright, *J. Biol. Chem.* **213,** 39 (1955).
[128a] E. R. Stadtman, G. N. Cohen, G. Le Bras, and H. de Robichon-Szulmajster, *J.Biol. Chem.* **236:** 2033 (1961).
[129] H. J. Teas, N. H. Horowitz, and M. Fling, *J. Biol. Chem.* **172,** 651 (1948).
[130] M. Fling and N. H. Horowitz, *J. Biol. Chem.* **190,** 277 (1951).
[131] H. J. Teas, *J. Bacteriol.* **59,** 93 (1950).
[132] P. H. Abelson, E. T. Bolton, and E. Aldous, *J. Biol. Chem.* **198,** 173 (1952).
[133] P. H. Abelson, E. Bolton, R. Britten, D. B. Cowie, and R. B. Roberts, *Proc. Natl. Acad. Sci. U. S.* **39,** 1020 (1953).
[134] S. Black and N. G. Wright, *J. Biol. Chem.* **213,** 51 (1955).
[135] B. Nisman, G. N. Cohen, S. B. Wiesendanger, and M. L. Hirsch, *Compt. rend. acad. sci.* **238,** 1342 (1954).
[136] G. N. Cohen and M. L. Hirsch, *J. Bacteriol.* **67,** 182 (1954).
[137] E. H. Wormser and A. B. Pardee, *Arch. Biochem. Biophys.* **78,** 416 (1958).
[138] Y. Watanabe and K. Shimura, *J. Biochem. Tokyo* **43,** 283 (1956).
[139] Y. Wanatabe, S. Konishi, and K. Shimura, *J. Biochem. Tokyo* **42,** 837 (1955).
[140] M. Flavin and C. Slaughter, *Biochim. et Biophys. Acta* **36,** 554 (1959).

[141] F. Binkley, W. P. Anslow, Jr., and V. du Vigneaud, *J. Biol. Chem.* **143**, 559 (1942).
[142] N. A. Horowitz, *J. Biol. Chem.* **171**, 255 (1947).
[143] G. A. Fischer, *Biochim. et Biophys. Acta* **25**, 50 (1957).
[144] S. Wijesundera and D. D. Woods, *J. Gen. Microbiol.* **9**, iii (1953).
[145] Y. Matsuo and D. M. Greenberg, *J. Biol. Chem.* **230**, 561 (1958).
[146] J. S. Harris and H. I. Kohn, *J. Pharmacol.* **73**, 383 (1941).
[147] W. Shive and E. C. Roberts, *J. Biol. Chem.* **162**, 463 (1946).
[148] K. C. Winkler and P. G. de Haan, *Arch. Biochem.* **18**, 97 (1948).
[149] B. D. Davis and E. S. Mingioli, *J. Bacteriol.* **60**, 17 (1950).
[150] F. T. Hatch, S. Takeyama, and J. M. Buchanan, *Federation Proc.* **18**, 243 (1959).
[151] R. L. Kisliuk and D. D. Woods, *J. Gen. Microbiol.* **18**, xv (1958).
[152] C. W. Helleiner, R. L. Kisliuk, and D. D. Woods, *J. Gen. Microbiol.* **18**, xv (1958).
[152a] J. R. Guest and K. M. Jones, *Biochem. J.* **75**, 12P (1960).
[153] J. Szulmajster and D. D. Woods, *Biochem. J.* **75**, 3 (1960).
[154] H. A. Barker, H. Weissbach, and R. D. Smyth, *Proc. Natl. Acad. Sci. U. S.* **44**, 1093 (1958).
[155] F. T. Hatch, S. Takeyama, R. E. Cathou, A. R. Larrabee, and J. M. Buchanan, *J. Am. Chem. Soc.* **81**, 6525 (1959).
[156] E. Work, *Nature* **165**, 74 (1950).
[157] E. Work, *Biochem. J.* **49**, 17 (1951).
[158] M. R. J. Salton, *Biochim. et Biophys. Acta* **10**, 512 (1953).
[159] C. S. Cummins and H. Harris, *J. Gen. Microbiol.* **14**, 583 (1956).
[160] R. E. Strange, *Bacteriol. Revs.* **23**, 1 (1959).
[161] N. Bauman and B. D. Davis, *Science* **126**, 170 (1957).
[162] P. Meadow, D. S. Hoare, and E. Work, *Biochem. J.* **66**, 270 (1957).
[163] L. E. Rhuland, *J. Bacteriol.* **73**, 778 (1957).
[164] K. McQuillen, *Biochim. et Biophys. Acta* **27**, 410 (1958).
[165] D. L. Dewey and E. Work, *Nature* **169**, 533 (1952).
[166] P. Meadow and E. Work, *Biochem. J.* **72**, 400 (1959).
[167] L. E. Casida, Jr., U. S. Patent No. 2,771,396 (1956).
[168] L. D. Wright and E. L. Cresson, *Proc. Soc. Exptl. Biol. Med.* **82**, 354 (1953).
[169] E. Windsor, *J. Biol. Chem.* **192**, 607 (1951).
[170] H. J. Vogel, *Biochim. et Biophys. Acta* **34**, 282 (1959).
[171] J. J. Perry and J. W. Foster, *J. Bacteriol.* **69**, 337 (1955).
[172] H. H. Martin and J. W. Foster, *J. Bacteriol.* **76**, 167 (1958).
[173] J. F. Powell and R. E. Strange, *Biochem. J.* **65**, 700 (1957).
[174] E. Work, *in* "Amino Acid Metabolism" (W. D. McElroy and B. Glass, eds.), p. 462. Johns Hopkins Press, Baltimore, Maryland, 1955.
[175] D. L. Dewey, D. S. Hoare, and E. Work, *Biochem. J.* **58**, 523 (1954).
[176] R. F. Denman, D. S. Hoare, and E. Work, *Biochim. et Biophys. Acta* **16**, 442 (1955).
[177] P. Meadow and E. Work, *Biochim. et Biophys. Acta* **29**, 180 (1958).
[178] C. Gilvarg, *J. Biol. Chem.* **233**, 1501 (1958).
[179] L. E. Rhuland and J. A. Soda, *J. Bacteriol.* **78**, 400 (1959).
[180] L. E. Rhuland, *Nature* **185**, 224 (1960).
[181] C. Gilvarg, *J. Biol. Chem.* **234**, 2955 (1959).
[182] B. Peterkofsky and C. Gilvarg, *Federation Proc.* **18**, 301 (1959).
[183] C. Gilvarg, *Biochim. et Biophys. Acta* **24**, 216 (1957).
[184] M. Antia, D. S. Hoare, and E. Work, *Biochem. J.* **65**, 448 (1957).
[185] D. E. Green, L. F. Leloir, and V. Nocito, *J. Biol. Chem.* **161**, 559–82 (1945).
[186] A. Meister and S. V. Tice, *J. Biol. Chem.* **187**, 173 (1950).

[187] A. Meister, H. A. Sober, and S. V. Tice, *J. Biol. Chem.* **189,** 577 (1951).
[188] W. A. Wood and I. C. Gunsalus, *J. Biol. Chem.* **190,** 403 (1951).
[189] R. R. Roepke, R. L. Libby, and M. H. Small, *J. Bacteriol.* **48,** 401 (1944).
[190] B. L. Strehler, *J. Bacteriol.* **59,** 105 (1950).
[191] B. E. Wright and T. C. Stadtman, *J. Biol. Chem.* **219,** 863 (1956).
[192] B. E. Wright, *J. Biol. Chem.* **219,** 873 (1956).
[193] R. L. Kisliuk and W. Sakami, *J. Biol. Chem.* **214,** 47 (1955).
[194] G. R. Greenberg, L. Jaenicke, and M. Silverman, *Biochim. et Biophys. Acta* **17,** 588 (1955).
[195] J. C. Rabinowitz and W. E. Pricer, Jr., *J. Am. Chem. Soc.* **78,** 5702 (1956).
[196] H. I. Nakada, B. Friedmann, and S. Weinhouse, *J. Biol. Chem.* **216,** 583 (1955).
[197] B. D. Davis, *J. Bacteriol.* **62,** 221 (1951).
[198] B. Magasanik, *J. Biol. Chem.* **213,** 557 (1955).
[199] S. Simmonds and D. A. Miller, *J. Bacteriol.* **74,** 775 (1957).
[200] B. E. Wright, *Arch. Biochem.* **31,** 332 (1951).
[201] H. L. Kornberg and H. A. Krebs, *Nature* **179,** 988 (1957).
[202] P. H. Abelson, *J. Biol. Chem.* **206,** 335 (1954).
[203] J. F. Nyc and I. Zabin, *J. Biol. Chem.* **215,** 35 (1955).
[204] H. R. V. Arnstein and D. Keglevic, *Biochem. J.* **62,** 199 (1956).
[205] H. J. Sallach, *in* "Amino Acid Metabolism" (W. D. McElroy and B. Glass, eds.), p. 782. Johns Hopkins Press, Baltimore, Maryland, 1955.
[206] R. A. Smith, C. W. Schuster, S. Zimmerman, and I. C. Gunsalus, *Bacteriol. Proc.* (*Soc. Am. Bacteriologists*) **56,** 107 (1956).
[207] J. Hanford and D. D. Davies, *Nature* **182,** 532 (1958).
[208] L. F. Borkenhagen and E. P. Kennedy, *Biochim. et Biophys. Acta* **28,** 222 (1958).
[209] F. C. Neuhaus and W. L. Byrne, *Biochim. et Biophys. Acta* **28,** 223 (1958).
[210] L. Gorini and B. D. Davis, unpublished observations.
[211] J. O. Lampen, R. R. Roepke, and M. J. Jones, *Arch. Biochem.* **13,** 55 (1951).
[212] K. Schlossmann and F. Lynen, *Biochem. Z.* **328,** 591 (1957).
[213] D. B. Cowie, E. T. Bolton, and M. K. Sands, *J. Bacteriol.* **62,** 63 (1951).
[213a] R. S. Bandurski, L. G. Wilson, and C. Squires, *J. Am. Chem. Soc.* **78,** 6408 (1956).
[213b] P. W. Robbins and F. Lipmann, *J. Am. Chem. Soc.* **78,** 6409 (1956).
[213c] J. B. Ragland, *Arch. Biochem. Biophys.* **84,** 541 (1959).
[214] W. K. Maas and H. J. Vogel, *J. Bacteriol.* **65,** 388 (1953).
[215] G. Ehrensvärd, L. Reio, E. Saluste, and R. Stjernholm, *J. Biol. Chem.* **189,** 93 (1951).
[216] M. Strassman, A. J. Thomas, and S. Weinhouse, *J. Am. Chem. Soc.* **75,** 5135 (1953).
[217] E. A. Adelberg, *in* "Amino Acid Metabolism" (W. D. McElroy and B. Glass, eds.), p. 419. Johns Hopkins Press, Baltimore, Maryland, 1956.
[218] H. E. Umbarger, B. Brown, and E. J. Eyring, *J. Am. Chem. Soc.* **79,** 2980 (1957).
[219] E. Juni, *J. Biol. Chem.* **195,** 715 (1952).
[220] E. F. Gale, "The Chemical Activities of Bacteria." Academic Press, New York, 1948.
[221] Y. S. Halpern and H. E. Umbarger, *J. Biol. Chem.* **234,** 3067 (1959).
[221a] A. N. Radhakrishnan and E. E. Snell, *J. Biol. Chem.* **235:** 2316 (1960).
[222] L. O. Krampitz, G. Greull, and I. Suzuki, *Federation Proc.* **18,** 266 (1959).
[223] G. L. Carlson and G. M. Brown, *J. Biol. Chem.* **235,** PC3 (1960).
[224] M. Strassman, A. J. Thomas, L. A. Locke, and S. Weinhouse, *J. Am. Chem. Soc.* **76,** 4241 (1954).
[225] E. A. Adelberg, *J. Am. Chem. Soc.* **76,** 4241 (1954).
[226] H. E. Umbarger and B. Brown, *J. Bacteriol.* **73,** 105 (1957).

[227] W. A. Wood and I. C. Gunsalus, *J. Biol. Chem.* **181,** 171 (1949).
[228] R. I. Leavitt and H. E. Umbarger, *J. Bacteriol.* **80,** 18 (1960).
[229] H. E. Umbarger, *Federation Proc.* **17,** 326 (1958).
[230] H. E. Umbarger, B. Brown, and E. J. Eyring, *J. Biol. Chem.* **235,** 1425 (1960).
[231] M. Strassman, K. F. Lewis, M. E. Corsey, J. B. Shatton, and S. Weinhouse, *Federation Proc.* **17,** 317 (1958).
[232] M. Strassman, J. B. Shatton, M. E. Corsey, and S. Weinhouse, *J. Am. Chem. Soc.* **80,** 1771 (1958).
[233] M. Strassman, J. B. Shatton, and S. Weinhouse, *J. Biol. Chem.* **235,** 700 (1960).
[234] Y. Watanabe, K. Hayashi, and K. Shimura, *Biochim. et Biophys. Acta* **31,** 583 (1959).
[235] R. P. Wagner, A. N. Radhakrishnan, and E. E. Snell, *Proc. Natl. Acad. Sci. U. S.* **44,** 1047 (1958).
[236] A. N. Radhakrishnan and E. E. Snell, *Federation Proc.* **18,** 306 (1959).
[237] J. R. Sjolander, K. Folkers, E. A. Adelberg, and E. L. Tatum, *J. Am. Chem. Soc.* **76,** 1085 (1954).
[238] J. W. Myers and E. A. Adelberg, *Proc. Natl. Acad. Sci. U. S.* **40,** 493 (1954).
[239] M. Strassman, L. A. Locke, A. J. Thomas, and S. Weinhouse, *J. Am. Chem. Soc.* **78,** 1599 (1956).
[240] M. Webb, *J. Gen. Microbiol.* **18,** xiv (1958).
[241] M. Webb, *Biochem. J.* **70,** 472 (1958).
[242] M. E. Rafelson, Jr., *Arch. Biochem. Biophys.* **72,** 376 (1957).
[243] P. Margolin, *Genetics* **44,** 525 (1959).
[243a] C. Jungwirth, P. Margolin, H. E. Umbarger, and S. R. Gross. *Biochem. Biophys. Research Comm.* **5,** 435, (1961).
[243b] S. R. Gross, C. Jungwirth, P. Margolin, and H. E. Umbarger, unpublished observations.
[243c] M. Strassman, personal communication.
[244] H. J. Vogel, B. D. Davis, and E. S. Mingioli, *J. Am. Chem. Soc.* **73,** 1897 (1951).
[245] B. N. Ames and H. K. Mitchell, *J. Am. Chem. Soc.* **74,** 252 (1952).
[246] B. N. Ames and H. K. Mitchell, *J. Biol. Chem.* **212,** 687 (1955).
[247] F. Maas, M. B. Mitchell, B. N. Ames, and H. K. Mitchell, *Genetics* **37,** 217 (1952).
[248] B. N. Ames, *Federation Proc.* **15,** 210 (1956).
[249] B. N. Ames, *J. Biol. Chem.* **228,** 131 (1957).
[250] B. N. Ames and B. L. Horecker, *J. Biol. Chem.* **220,** 113 (1956).
[251] B. N. Ames, *J. Biol. Chem.* **226,** 583 (1957).
[252] B. N. Ames, B. Garry, and L. A. Herzenberg, *J. Gen. Microbiol.* **22,** 369 (1960).
[253] E. Adams, *J. Biol. Chem.* **209,** 829 (1954).
[254] E. Adams, *J. Biol. Chem.* **217,** 325 (1955).
[255] H. P. Broquist and E. E. Snell, *J. Biol. Chem.* **180,** 59 (1949).
[256] L. Levy and M. J. Coon, *J. Biol. Chem.* **192,** 807 (1951).
[257] H. Tabor, A. H. Mehler, O. Hayaishi, and J. White, *J. Biol. Chem.* **196,** 121 (1952).
[258] C. Mitoma and E. E. Snell, *Proc. Natl. Acad. Sci. U. S.* **41,** 891 (1955).
[259] B. Magasanik, *J. Am. Chem. Soc.* **78,** 5449 (1956).
[260] A. Neidle and H. Waelsch, *J. Am. Chem. Soc.* **78,** 1767 (1956).
[261] A. Neidle and H. Waelsch, *Federation Proc.* **16,** 225 (1957).
[262] B. Magasanik, *Med. J. Osaka Univ.* **8,** 71 (1958).
[263] B. D. Davis, *Harvey Lectures* **50,** 230 (1954–55).
[264] C. Yanofsky *in* "Amino Acid Metabolism" (W. D. McElroy and B. Glass, eds.), p. 931. Johns Hopkins Press, Baltimore, Maryland, 1955.

[265] M. V. Ortega and G. M. Brown, *J. Am. Chem. Soc.* **81,** 4437 (1959).
[266] B. D. Davis, *J. Biol. Chem.* **191,** 315 (1951).
[267] E. L. Tatum, S. R. Gross, G. Ehrensvärd, and L. Garnjobst, *Proc. Natl. Acad. Sci. U. S.* **40,** 271 (1954).
[268] I. I. Salamon and B. D. Davis, *J. Am. Chem. Soc.* **75,** 5567 (1953).
[269] U. Weiss, B. D. Davis, and E. S. Mingioli, *J. Am. Chem. Soc.* **75,** 5572 (1953).
[270] S. Mitsuhashi and B. D. Davis, *Biochim. et Biophys. Acta* **15,** 268 (1954).
[271] H. Yaniv and C. Gilvarg, *J. Biol. Chem.* **213,** 787 (1955).
[272] B. D. Davis, *J. Bacteriol.* **64,** 729 (1952).
[273] S. Mitsuhashi and B. D. Davis, *Biochim. et Biophys. Acta* **15,** 54 (1954).
[274] C. Gilvarg and K. Bloch, *J. Biol. Chem.* **193,** 339 (1951).
[275] C. Gilvarg and K. Bloch, *J. Biol. Chem.* **199,** 689 (1952).
[276] G. Ehrensvärd and L. Reio, *Arkiv. Kemi* **5,** 229 (1953).
[277] G. Ehrensvärd and L. Reio, *Arkiv Kemi* **5,** 327 (1953).
[278] E. B. Kalan, B. D. Davis, P. R. Srinivasan, and D. B. Sprinson, *J. Biol. Chem.* **223,** 907 (1956).
[279] P. R. Srinivasan, D. B. Sprinson, E. B. Kalan, and B. D. Davis, *J. Biol. Chem.* **223,** 913 (1956).
[280] P. R. Srinivasan, M. Katagiri, and D. B. Sprinson, *J. Biol. Chem.* **234,** 713 (1959).
[281] P. R. Srinivasan and D. B. Sprinson, *J. Biol. Chem.* **234,** 716 (1959).
[282] E. B. Kalan, F. Leitner, and B. D. Davis, unpublished observations.
[283] J. Hurwitz and A. Weissbach, *J. Biol. Chem.* **234,** 710 (1959).
[284] E. B. Kalan and B. D. Davis, unpublished observations.
[285] P. R. Srinivasan, *J. Am. Chem. Soc.* **81,** 1772 (1959).
[286] B. Weiss and P. R. Srinivasan, *Proc. Natl. Acad. Sci. U. S.* **45,** 1491 (1959).
[287] C. Gilvarg, unpublished observations.
[288] E. B. Kalan and B. D. Davis, unpublished observations (cited in Footnote 9a of reference 12).
[289] J. G. Levin and D. B. Sprinson, unpublished observations.
[290] M. Katagiri and R. Sato, *Science* **118,** 250 (1953).
[291] U. Weiss, C. Gilvarg, E. S. Mingioli, and B. D. Davis, *Science* **119,** 774 (1954).
[292] I. Schwinck and E. Adams, *Biochim. et Biophys. Acta* **36,** 102 (1959).
[293] P. Fildes, *Brit. J. Exptl. Pathol.* **21,** 315 (1940).
[294] E. E. Snell, *Arch. Biochem.* **2,** 389 (1943).
[295] E. L. Tatum, D. Bonner, and G. W. Beadle, *Arch. Biochem.* **3,** 477 (1944).
[296] J. F. Nyc, H. K. Mitchell, E. Leifer, and W. H. Langham, *J. Biol. Chem.* **179,** 783 (1949).
[297] C. W. H. Partridge, D. M. Bonner, and C. Yanofsky, *J. Biol. Chem.* **194,** 269 (1952).
[298] C. Yanofsky, *Science* **121,** 138 (1955).
[299] C. Yanofsky, *J. Biol. Chem.* **217,** 345 (1955).
[300] C. Yanofsky, *Biochim. et Biophys. Acta* **16,** 594 (1955).
[301] C. Yanofsky, *J. Biol. Chem.* **223,** 171 (1956).
[302] C. Yanofsky, *Biochim. et Biophys. Acta* **20,** 438 (1956).
[303] F. Lingens, H. J. Burkhardt, H. Hellmann, and F. Kaudewitz, *Z. Naturforsch.* **12b,** 493 (1957).
[304] J. S. Gots and S. H. Ross, *Biochim. et Biophys. Acta* **24,** 429 (1957).
[305] C. H. Doy and F. Gibson, *Biochem. J.* **72,** 586 (1959).
[306] C. H. Doy, unpublished observations.
[307] L. W. Parks and H. C. Douglas, *Biochim. et Biophys. Acta* **23,** 207 (1957).
[308] F. Lingens, M. Hildinger, and H. Hellmann, *Biochim. et Biophys. Acta* **30,** 668 (1958).

[308a] C. H. Doy, A. Rivera, Jr., and P. R. Srinivasan, *Biochem. Biophys. Research Comm.* **4,** 83 (1961).
[309] C. Yanofsky, *Bacteriol. Revs.* **24,** 221 (1960).
[310] C. Yanofsky, *J. Biol. Chem.* **224,** 783 (1957).
[311] H. K. Mitchell and J. Lein, *J. Biol. Chem.* **175,** 481 (1948).
[312] S. R. Suskind, C. Yanofsky, and D. M. Bonner, *Proc. Natl. Acad. Sci. U. S.* **41,** 577 (1955).
[313] P. Lerner and C. Yanofsky, *J. Bacteriol.* **74,** 494 (1957).
[314] C. Yanofsky and M. Rachmeler, *Biochim. et Biophys. Acta* **28,** 640 (1958).
[315] F. Gibson, M. J. Jones, and H. Teltscher, *Biochem. J.* **64,** 132 (1956).
[316] R. B. Woodward, personal communication.
[317] W. A. Wood and I. C. Gunsalus, *J. Biol. Chem.* **190,** 403 (1951).
[318] M. E. Rafelson, G. Ehrensvärd, M. Bashford, E. Saluste, and C. G. Heden, *J. Biol. Chem.* **211,** 725 (1954).
[319] M. E. Rafelson, *J. Biol. Chem.* **212,** 953 (1955).
[320] M. E. Rafelson, G. Ehrensvärd, and L. Reio, *Exptl. Cell Research* (*Suppl.*) **3,** 281 (1955).
[321] M. E. Rafelson, *J. Biol. Chem.* **213,** 479 (1955).
[322] E. E. Snell, *J. Biol. Chem.* **158,** 497 (1945).
[323] M. Ikawa and E. E. Snell, *Biochim. et Biophys. Acta* **19,** 576 (1956).
[324] M. R. J. Salton, *Nature* **180,** 388 (1957).
[325] P. Ayengar and E. Roberts, *J. Biol. Chem.* **197,** 453 (1952).
[326] K. Itoh, *Nagoya J. Med. Sci.* **21,** 181 (1958).
[327] C. B. Thorne, C. G. Gomez, and R. D. Housewright, *J. Bacteriol.* **69,** 357 (1955).
[328] C. B. Thorne and D. M. Molnar, *J. Bacteriol.* **70,** 420 (1955).
[329] C. B. Thorne, *in* "Amino Acid Metabolism" (W. D. McElroy and B. Glass, eds.), p. 41. Johns Hopkins Press, Baltimore, Maryland, 1955.
[330] H. Momose and Y. Ikeda, *Nature* **186,** 567 (1960).
[331] S. Kinoshita, K. Nakagama, and S. Kitada, *J. Gen. Appl. Microbiol.* **4,** 128 (1958).
[332] S. Kinoshita, K. Tanaka, S. Udaka, and S. Akita, *in* "Proceedings of the International Symposium on Enzyme Chemistry, Tokyo and Kyoto, 1957" (K. Ichihara, ed.), p. 464. Academic Press, New York, 1958.
[333] S. Kinoshita, S. Udaka, and M. Shimono, *J. Gen. Appl. Microbiol.* **3,** 193 (1957).
[334] S. Udaka and S. Kinoshita, *J. Gen. Appl. Microbiol.* **4,** 283 (1958).
[335] Z. Sugisaki, *J. Gen. Appl. Microbiol.* **5,** 138 (1959).
[336] S. Udaka and S. Kinoshita, *J. Gen. Appl. Microbiol.* **5,** 159 (1960).
[337] E. A. Adelberg, *J. Bacteriol.* **76,** 326 (1958).
[338] S. Dagley, M. E. Fewster, and F. C. Happold, *J. Gen. Microbiol.* **8,** 1 (1953).
[339] C. Mitoma and L. C. Leeper, *Federation Proc.* **13,** 266 (1954).
[340] R. Y. Stanier, *in* "Cellular Metabolism and Infections" (E. Racker, ed.), p. 3. Academic Press, New York, 1954.
[341] B. D. Davis, *Arch. Biochem. Biophys.* **78,** 497 (1958).

CHAPTER 5

The Synthesis of Vitamins and Coenzymes

J. G. MORRIS

I. General Introduction 253
II. Experimental Methods of Approach 254
A. With Organisms Able to Synthesize the Vitamin 254
B. With Vitamin-Dependent Organisms 257
III. Individual Vitamins and Coenzymes—Synthetic Pathways 259
A. Thiamine (Thiamine Pyrophosphate, ThPP) 259
B. Riboflavin (Riboflavin Monophosphate and FAD) 262
C. Nicotinic Acid (Diphosphopyridine Nucleotide, DPN, and Triphosphopyridine Nucleotide, TPN) 265
D. The Vitamin B_6 Group (Pyridoxal-5′-Phosphate) 269
E. Pantothenic Acid (Coenzyme A) 270
F. Biotin 274
G. *p*-Aminobenzoic Acid and the Folic Acid Group 276
H. Vitamin B_{12} 280
I. Miscellaneous 283
References 287

I. General Introduction

The failure of an organism to synthesize an essential metabolite is reflected by a requirement either for that substance, or for products of its metabolism and function; thus the growth factor requirements of an organism are a measure of the failure of its biosynthetic capacity. Among growth factors there may easily be distinguished a number of organic compounds which, since they function catalytically, are required in only minute amounts. The distinction of being classified as vitamins is conferred upon such of these substances as are required by man.

Vitamins such as A and D, which are functional only in the specialized organs of higher animals, are not formed by bacteria, although certain of their precursors may incidentally by synthesized by microorganisms. However, all those B vitamins which function coenzymically are as essential to unicellular organisms as to multicellular forms, and their biosynthesis may often best be studied in bacteria. Such studies assume that the formation of any given vitamin or coenzyme follows the same pathway in very different organisms. Indeed, divergences from a common route of biosynthesis are seemingly infrequent, although not unknown; the best established example is the synthesis of nicotinic acid which is formed in certain plants, molds, and bacteria by a route different from that followed in other microorganisms and mammals.[1-3]

Many bacteria synthesize vitamins in excess of their requirements and excrete the surplus into the environment. Indeed, the excretion of B vitamins by intestinal microorganisms, coupled with autolytic liberation of vitamins from dead cells, may furnish such a large fraction of an animal's vitamin supply that it is difficult to render the animal vitamin-deficient by mere dietary vitamin deprivation. Provision of vitamins by the gut flora is particularly important in the ruminant. The contribution from this source in other animals, always considerable, may be further amplified by refection.[4-6]

Bacteria vary widely in their ability to form vitamins and coenzymes.[7] Many intermediates along the routes of synthesis utilized by prototrophic organisms are growth factors for closely related auxotrophic strains.[8] Symbiotic cooperation between two (or more) bacterial species may thus allow each to synthesize a vitamin which neither is individually capable of forming. Certain microorganisms excrete such large amounts of a vitamin that media in which they have been grown represent commercially valuable sources of that substance. The accumulation of riboflavin (easily detected because of its intense color, fluorescence, and relative insolubility) has been found to accompany growth of many microorganisms; *Eremothecium ashbyii* and *Ashbya gossypii* produce up to 1 mg. riboflavin/ml. of medium under certain conditions of culture.[9] The production of comparatively large amounts of riboflavin is also characteristic of such bacteria as *Bacillus megatherium*, *Mycobacterium smegmatis*, and *Clostridium acetobutylicum*.[10] Certain bacteria produce high concentrations of vitamin B_{12} derivatives, e.g., *Propionibacterium shermanii*.[11]

II. Experimental Methods of Approach

A. With Organisms Able to Synthesize the Vitamin

1. Growing Cultures

Once a chemically defined growth medium has been developed, some insight into the path of vitamin synthesis can be obtained by studying the effects of different substrates, of suspected precursors, and of inhibitors on vitamin production. It is most profitable to study organisms which synthesize abnormally large amounts of the vitamin in question, since in such cases its formation represents a major commitment of substrate and is more likely to respond markedly to the provision of precursors. Isotope-labeling studies are also best carried out with such bacteria, since the synthesis of an abnormally large amount of vitamin is often a prerequisite for its isolation. Direct incorporation of labeled precursor into the vitamin molecule may then be demonstrable, or conversely, dilution of isotope derived from an unspecific substrate on the addition of unlabeled precursor.

When no information is available as to the possible nature of precursors, a specifically labeled main carbon source may be provided, and the location of isotope in the synthesized vitamin determined. It may then become possible to infer the nature of intermediates produced by recognized metabolic routes. Such methods were successfully employed to determine the route of synthesis of riboflavin in *Eremothecium ashbyii*, *Ashbya gossypii*, and *Candida flareri*. The observations that purines differentially enhance flavogenesis in these organisms[12-14a] and that $HC^{14}OOH$ and $C^{14}O_2$ are incorporated into riboflavin according to a pattern also characteristic of incorporation of these one-carbon units into purines,[15] culminated in the use of uniformly C^{14}-labeled adenine and 8-C^{14}-adenine to prove that the pyrimidine moiety of the isoalloxazine structure arises from purine by the loss of the C-8 atom.[13, 16, 17]

Several potent inhibitors of bacterial growth whose action can be overcome noncompetitively by vitamins or coenzymes have proved to be specific antimetabolites of vitamin precursors. Studies of the nature of substances which counteract the effect of these inhibitors have led in several instances to the discovery of the site of action of the antimetabolite, and have resulted in the identification of many intermediates along the routes of vitamin synthesis.[18] Thus salicylic acid in low concentrations was observed to inhibit growth only of those bacteria which synthesize pantothenic acid.[19, 20] Its inhibitory action on *Escherichia coli* was overcome noncompetitively by pantothenic acid, pantoic acid, and α-oxo-β,β-dimethyl-γ-hydroxybutyric acid (ketopantoic acid) and competitively by α-oxoisovaleric acid (ketovaline) or valine. Similar studies with *E. coli* employing cysteic and hydroxyaspartic acids as antimetabolites suggested that the remaining β-alanine fragment of the pantothenic acid molecule might arise from aspartic acid,[21, 22] while the growth inhibitors α- and γ-hydroxy-β,β-dimethylbutyric acid and β,γ-dihydroxy-β-methylbutyric acid were found to prevent the ultimate condensation of β-alanine with pantoic acid[23] (see Fig. 1). Thus, even in the absence of evidence from other sources, the route of biogenesis of pantothenic acid in *E. coli* might have been revealed by antimetabolite studies alone.

Although a number of antimetabolites function at stages in the conversion of endogenous vitamin to coenzyme, many more seem to inhibit uptake of exogenous vitamin. However, this is not necessarily the sole reason why so many antimetabolites are inhibitory only to organisms that are dependent upon an exogenous supply of the growth factor. The antimetabolite may only enter the cells of organisms that possess a specific mechanism for the uptake of the growth factor; or the growth factor may not itself be an intermediate in the *de novo* biosynthesis of vitamin or coenzyme and may enter into the normal route of biosynthesis only as the consequence of an unusual series of reactions subject to inhibition by the antimetabolite.[24] Lack of specificity of action of an antimetabolite can lead to erroneous conclusions; a case in point is provided by the interpretations offered of the path of nicotinic acid synthesis in *E. coli* based on the use of indole-3-acrylic acid as an inhibitor[25] and later shown to be incorrect. Certain antimetabolites may also cause secondary injurious effects, particularly at high concentrations.[24]

Another use of antimetabolites which could be of value in studies of bacterial vitamin synthesis is in the selection of mutants that have developed resistance to a competitive inhibitor by increased synthesis of the substrate of the blocked reaction. When the inhibitor acts upon an enzyme for which endogenously produced vitamin is substrate (for example, an

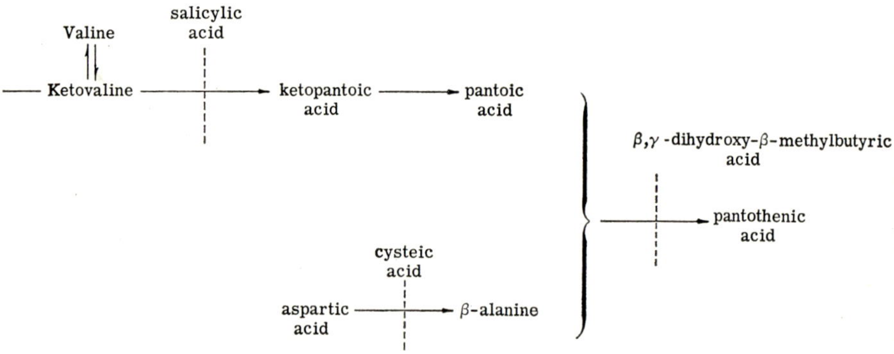

FIG. 1. The route of biogenesis of pantothenic acid in *Escherichia coli* as indicated by antimetabolite studies.

enzyme concerned with the elaboration of coenzyme), the vitamin could be accumulated and excreted into the culture medium. Although this is by no means the only mechanism by which drug resistance might be developed, it is possible that some strains may be resistant to the action of chosen antimetabolites as a result of increased vitamin synthesis. Thus, sulfonamide-resistant staphylococci have been described whose increased drug resistance is due to enhanced production of *p*-aminobenzoic acid.[26]

2. Washed Suspensions

Incubation of washed suspensions of bacteria in a chemically defined medium designed to evoke maximal synthesis of vitamin is frequently a useful technique. The density of organisms can be greater than that obtainable during growth, and vitamin production is sometimes commensurately increased. Usually little or no multiplication occurs, and the conditions are much better defined than at any time during growth. At some stage or other in their study of bacterial vitamin synthesis most investigators have had recourse to washed suspensions, usually to examine incorporation into the vitamin of known or suspected precursors. Pantothenic acid synthesis by suspensions of *E. coli* provided with β-alanine and glucose was so studied.[27]

Little synthesis of folic acid by washed suspensions of *E. coli* and *Staphylococcus aureus* could be shown unless the organisms initially had a low content of this factor.[28] The need for preliminary vitamin depletion of normal organisms, if subsequent synthesis is to be evoked, may be a general phenomenon.

3. Cell-Free Preparations

A limit is set to the utility of intact bacteria by their impermeability to certain substrates. In such cases, and when individual enzymes concerned in the biosynthetic pathway are to be examined, disrupted preparations are employed. It is not to be expected that extracts will of necessity accomplish syntheses requiring the proper functioning of a number of sequential reactions, but it is at least reassuring to be able to demonstrate in such preparations all the constituent enzymes of the supposed pathway of biosynthesis in the intact organism.

B. With Vitamin-Dependent Organisms

1. Growing Cultures

a. Natural Auxotrophs. Some natural bacterial auxotrophs respond specifically to a particular vitamin; others respond also to one or more precursors of the vitamin. From such data, it is occasionally possible to arrange requirements in a sequence (usually of increasing complexity) which indicates the route of vitamin biosynthesis. Thus β-alanine supports growth of certain strains of *Corynebacterium diphtheriae*, whereas most strains of this species require pantothenic acid;[29] pantoic acid replaces pantothenic acid for *Clostridium septicum*.[30] Both species synthesize pantothenic acid during growth on these precursors, and their culture filtrates are in consequence able to supply the vitamin to *Lactobacillus helveticus*, which does not respond to β-alanine and/or pantoic acid.[30]

Care must be exercised in the interpretation of such nutritional data, for products of vitamin function can often replace the vitamin itself in supporting growth of an organism. Thus oleic acid and aspartic acid will substitute for biotin as growth factors for *Lactobacillus casei*, other lactobacilli, and yeasts, but are not biosynthetic precursors of biotin.[31, 32]

b. Induced Auxotrophs. As first became evident from the study of mutant strains of *Neurospora*, a series of induced auxotrophs, all blocked in the synthesis of the same growth factor, can be of great service in studies of vitamin biosynthesis.[33] The number of different mutants responsive to a given growth factor indicates the minimum number of steps in its synthesis, and the occasional accumulation of substrates of blocked reactions facilitates the identification of intermediates.

Such studies are now particularly feasible in bacteria, since the introduction of the penicillin-penicillinase selection technique[34-36] has made the isolation of many different vitamin auxotrophs a simple matter. The value of this technique may be illustrated by results obtained in studies of pantothenic acid biosynthesis.[27] A relation between valine and pantothenic acid was revealed by the isolation of a mutant strain of *Aerobacter aerogenes* (A 4-9) which responds to any one of the following compounds: valine, ketovaline, ketopantoate, pantoate, and pantothenate. The situation was clarified by the finding that an *Escherichia coli* mutant, 99-4, which responds to ketopantoate, pantoate, or pantothenate but not to ketovaline, in fact excretes valine (see Fig. 2). It appears therefore, that the *Aerobacter*

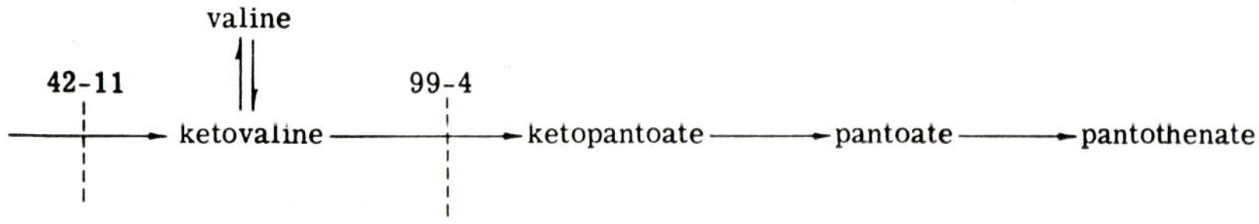

FIG. 2. The biosynthesis of pantothenic acid in *Escherichia coli.*

mutant, A 4-9, may carry a partial block or biochemical lesion in the synthesis of the ketovaline that serves as a common precursor of valine and pantothenic acid in all these organisms.

As in studies with natural auxotrophs, care must be exercised in interpretation, lest products of the function of a vitamin be misconstrued as vitamin precursors, merely because they support growth of the mutant organism. The relative concentrations of a suspected precursor or product needed to support, say, half-maximal growth of the organism may be diagnostic. If the precursor is utilized solely for the one biosynthesis, then (questions of permeability and degradation aside) its required concentration should be of the same order of magnitude as the required vitamin concentration. Thus all known vitamin B_{12} auxotrophs of *E. coli* grow in the absence of the vitamin if supplied with methionine; but since the amount of amino acid required is some 10^6 times the vitamin requirement, it is reasonable to suppose that methionine is not a precursor of vitamin B_{12} in this organism, but rather a product of its function.[37]

Other pitfalls of biochemical work with induced mutants are discussed in an excellent review.[38]

2. Washed Suspensions of Auxotrophs

When suspensions of prototrophic bacteria are used to investigate part or all of a route of vitamin synthesis, the influence of a supplied intermediate is often diminished by major alternative utilization, degradation, or dilution by endogenous precursors. These difficulties may often be circumvented by the judicious use of auxotrophic strains. Thus clear evidence of stimulation of pantothenic acid synthesis by ketovaline was difficult to obtain with wild-type *Escherichia coli*, which could synthesize the vitamin from glucose and β-alanine. In contrast, suspensions of the auxotrophic strain 42-11, with a block in ketovaline synthesis, produced no pantothenate at all when incubated with glucose and β-alanine, synthesis of the vitamin being strictly dependent on and proportional to the supply of ketovaline.[27]

Bacteria deficient in various factors are readily obtained by growing suitable auxotrophs on suboptimal concentrations of the factor, or on factor-replacement media. Suspensions of these deficient organisms may then be used to investigate the part which this factor plays in the biosynthesis under study. This technique was first used to study the effects of diphosphopyridine nucleotide deficiency in *Haemophilus parainfluenzae*.[39]

3. Cell-Free Preparations of Auxotrophs

Confirmation of the participation of an enzyme in vitamin biosynthesis is obtained when the enzyme in question can be demonstrated to be present in cell-free extracts of the prototroph but absent from identical preparations of the vitamin-dependent auxotroph.

III. Individual Vitamins and Coenzymes—Synthetic Pathways

It will not be possible to discuss critically all the relevant observations which have contributed to the present knowledge of the routes of biosynthesis of the vitamins and coenzymes; but an attempt will be made to summarize the various pathways, and to indicate the general means by which this information has been obtained.

A. Thiamine (Thiamine Pyrophosphate, ThPP)

Comparatively few bacteria, among which *Lactobacillus fermenti*, *Clostridium botulinum*, and *Clostridium tetani* are perhaps the best known examples, have been reported to possess a growth requirement for thiamine.[40] Many bacteria synthesize this vitamin in excess of their needs and excrete it: e.g., *Escherichia coli*, *Proteus vulgaris*, *Aerobacter aerogenes*, *Bacillus mesentericus*, and *Alcaligenes faecalis*.[41, 42] Thus thiamine production by an intestinal flora may render the host almost independent of any dietary supply of the vitamin[43, 44] (see Fig. 3).

Very little is known of the route of synthesis of thiamine in microorganisms although, from results obtained with mutant strains of *Neurospora*,[45, 46] it appears certain that the pyrimidine and thiazole moieties are separately formed and subsequently united. Thus in several microorganisms the requirement for thiamine can be satisfied by provision of the pyrimidine and/or thiazole fragments.[47-48a] Indeed we seem to be better informed as to the mechanism of linkage of the two fragments than as to their origin. Thus, thiamine is formed from these compounds by cell-free extracts of baker's yeast only when adenosine triphosphate (ATP) and Mg^{++} are concurrently supplied;[48, 49] it has been suggested that 2-methyl-4-amino-5-hydroxymethylpyrimidine is phosphorylated prior to its coupling with the 4-methyl-5-β-hydroxyethylthiazole.[48] However, the requirement for ATP is not changed by provision of the monophosphate ester of the pyrimidine.[48b] This has been clarified by the finding that the actual substrates for the condensation reaction are the pyrophosphate ester of the

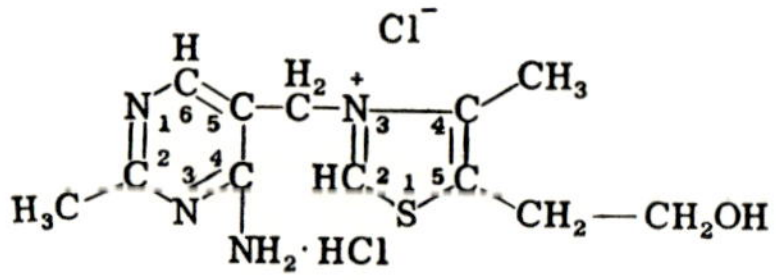

FIG. 3. Structure of thiamine chloride hydrochloride [3-(2′-methyl-4′-amino-5′-methylpyrimidyl)-4-methyl-5,β-hydroxyethylthiazolium chloride. HCl].

pyrimidine and the monophosphate ester of the thiazole, the synthesis yielding not free thiamine but its monophosphate ester.[48c, d, e, f]

$$\text{pyrimidine.PP} + \text{thiazole.P} \xrightarrow{Mg^{++}} \text{thiamine.P} + \text{PP}$$

Pyrimidine monophosphate has been isolated from extracts of yeast, and its further phosphorylation by ATP demonstrated in extracts,[48c] but it is not known whether it is an obligatory intermediate in the normal synthesis of the pyrophosphate.

Nothing specific is known of the mechanism of synthesis of the pyrimidine moiety. If it is formed according to the normal route of biosynthesis of this class of compound, then an *N*-ribose substituent must be removed and the methyl, amino, and hydroxymethyl side chains added after ring closure. Few thiazole-containing compounds of biological importance are known; the only one whose synthesis has been studied is penicillamine, in the formation of which cysteine and valine may be concerned.[50] It is thus of some interest that either cystine or thiazolidine carboxylic acid can support some growth of several thiazole auxotrophs of *Escherichia coli* and *Neurospora crassa*.[50a] The hypothesis has been advanced [47, 51, 52] that the thiazole moiety of thiamine could arise in yeast from 2-amino-3-(4′-methylthiazole-5′-)propionic acid, by a reaction analogous to those concerned with fusel oil production during fermentation (see Fig. 4). Some possibly relevant observations have emerged from a study of the nutritional requirements of *Bacillus paraalvei*. This organism grows well in a defined medium containing 15–18 amino acids and thiamine, which is replaceable by thiazole. Growth is possible in the absence of thiamine or thiazole, but the four amino acids, phenylalanine, valine, isoleucine, and cystine, then become essential for good growth. It was suggested that these amino acids are in some way concerned with the formation of the thiazole ring.[53, 54] However, synthesis of thiamine during growth in its absence was not assayed and the amino acids may merely reduce the requirement for the vitamin to an amount which the organism can itself produce.

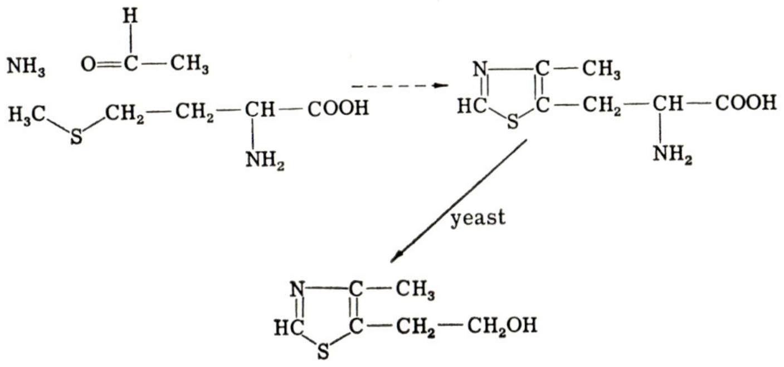

FIG. 4. Possible origin of thiazole moiety of thiamine from 2-amino-3-(4′ methylthiazole-5′-)propionic acid.

Pyrophosphorylation of the thiazole hydroxyethyl side chain converts thiamine into its coenzymic form (cocarboxylase, ThPP). The reaction is accomplished in a single stage by the enzyme thiaminokinase.[55-58]

$$\text{thiamine} + \text{ATP} \rightarrow \text{ThPP} + \text{AMP}$$

Thiamine monophosphate is definitely not an intermediate in this reaction, not being converted to ThPP by partially purified preparations of the enzyme.[48c, 58a] Therefore, if thiamine monophosphate is indeed the primary product of the biosynthetic sequence it must undergo dephosphorylation prior to its conversion to cocarboxylase.[47c] Recently it has been found that several nucleoside triphosphates can replace ATP as the pyrophosphate donor in the thiaminokinase reaction; under certain conditions they are even more effective than ATP.[58b]

Although its formation has been studied mainly in yeast, ThPP has been obtained from thiamine by bacterial synthesis using *Propionibacterium pentosaceum*.[59] ThPP has been shown to be some 30% more active than thiamine itself as a growth factor for *Lactobacillus fermenti*.[60] ThPP is also active for a strain of *Neisseria gonorrhoeae* which is completely unable to utilize thiamine; rather surprisingly, in this bacterium thiamine monophosphate has 80% of the activity of the pyrophosphate.[61] The α-hydroxyethyl derivative of ThPP has also recently been isolated from *Escherichia coli*.[61a]

There exists a complex interrelationship between thiamine and vitamin B_6 in many microorganisms (chiefly yeasts and molds). Studies with a strain of *Saccharomyces cerevisiae* appear to suggest that pyridoxine and the pyrimidine moiety of thiamine may be interconvertible via a common pyrimidine.[62] However, it is known that 2-methyl-4-amino-5-hydroxymethylpyrimidine monophosphate (the so-called toxopyrimidine phosphate) is an antimetabolite of pyridoxal phosphate,[63-65] while members of the B_6 group and other similar compounds may competitively interfere with thiamine biosynthesis.[66] In the absence of more convincing evidence there seems to be little necessity to invoke yet another mechanism to explain the interaction of these two vitamins.

B. Riboflavin (Riboflavin Monophosphate and FAD)

Although a number of bacteria show a growth requirement for riboflavin (see Fig. 5), including species of lactobacilli, streptococci, propionibacteria, and clostridia,[67] there have been few reports of possible precursors substituting for the vitamin. It was early suggested that 2-amino-4,5-dimethyl-1′-ribitylaminobenzene might be such a precursor, for when provided together with alloxan, it stimulated growth of riboflavin-requiring *Lactobacillus casei*[68] and was utilized for the synthesis of riboflavin by *Mycobacterium tuberculosis*.[69] (see Fig. 6). Antimetabolite studies indicated that

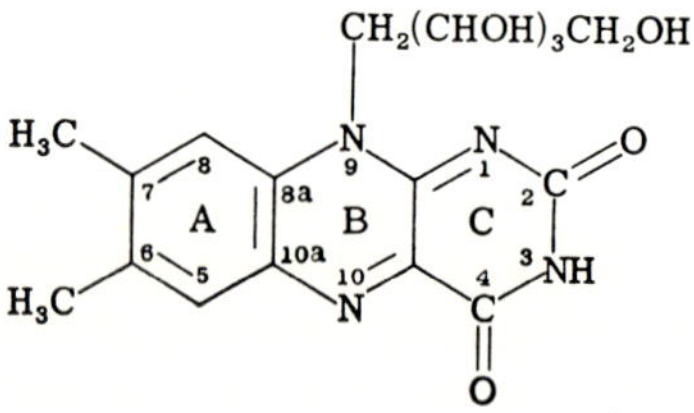

Fig. 5. Structure of riboflavin [6,7-dimethyl-9-(1′-D-ribityl)isoalloxazine].

1,2-dimethyl-4,5-diaminobenzene might be a common precursor of both riboflavin and vitamin B_{12}.[18, 70] But isotopic studies with growing cultures of *Eremothecium ashbyii* and *Ashbya gossypii* described earlier (see p. 255) confirmed that the pyrimidine portion of the riboflavin molecule arose from purines, previously observed to enhance differentially riboflavin production by these organisms.[12, 17] The preferential inhibition of flavinogenesis by azaxanthine would suggest that purines are converted into riboflavin via xanthine (which is the purine most closely related to rings B and C of the riboflavin molecule[71]).

If both N atoms of ring B are derived from purine, there remains the problem of the origin of the *o*-xylene ring A. Isotope from $C^{14}H_3$ COOH is incorporated into positions 6, 7, 8a, and 10a, while that from glucose-1-C^{14} and glucose-6-C^{14} appears in the two methyl groups and positions 5 and 8. Label from $C^{14}H_3$.COOH is, however, incorporated into all positions in

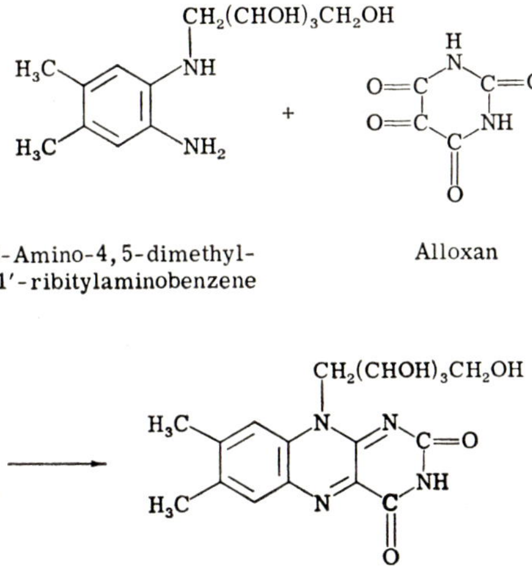

2-Amino-4,5-dimethyl-1'-ribitylaminobenzene

Alloxan

Riboflavin

FIG. 6. Possible precursors of riboflavin.

ring A, suggesting that glucose is not incorporated into the *o*-xylene ring via C_2 units that are in equilibrium with exogenous acetate.[72, 73] Uniformly C^{14}-labeled shikimic acid is not incorporated.[74] However, isotope from 2-C^{14}-labeled acetylmethylcarbinol (AMC) added to growing cultures of *Eremothecium ashbyii* is introduced into ring A of riboflavin. As half of the radioactivity proved to be localized in the methyl groups and none was detectable in positions 6 and 7, it was supposed that the remaining isotope was confined to positions 5 and 8.[74] Since both AMC and diaminouracil are known metabolites of *Eremothecium ashbyii*[74, 75] it is possible that one molecule of AMC can condense with 4-ribitylamino-5-aminouracil to form 6,7-dimethyl-8-(D-ribityl)lumazine (compound G). This compound has been isolated from cultures of *E. ashbyii*, *Ashbya gossypii*, and *Clostridium acetobutylicum*[76-78a] and is probably a riboflavin precursor.[17, 78, 79]

Compound G might then condense with a second molecule of AMC to yield riboflavin.[74, 80] Enzymic conversion of compound G to riboflavin (see Fig. 7) has in fact been observed in cell-free extracts of *Escherichia coli*, *Lactobacillus plantarum*, *Clostridium acetobutylicum*, *Eremothecium ashbyii*, *Neurospora crassa*, and beef liver, with pyruvate, acetaldehyde, or acetate acting as the source of the necessary C-4 unit.[81, 81a] Katagiri *et al.*[81]

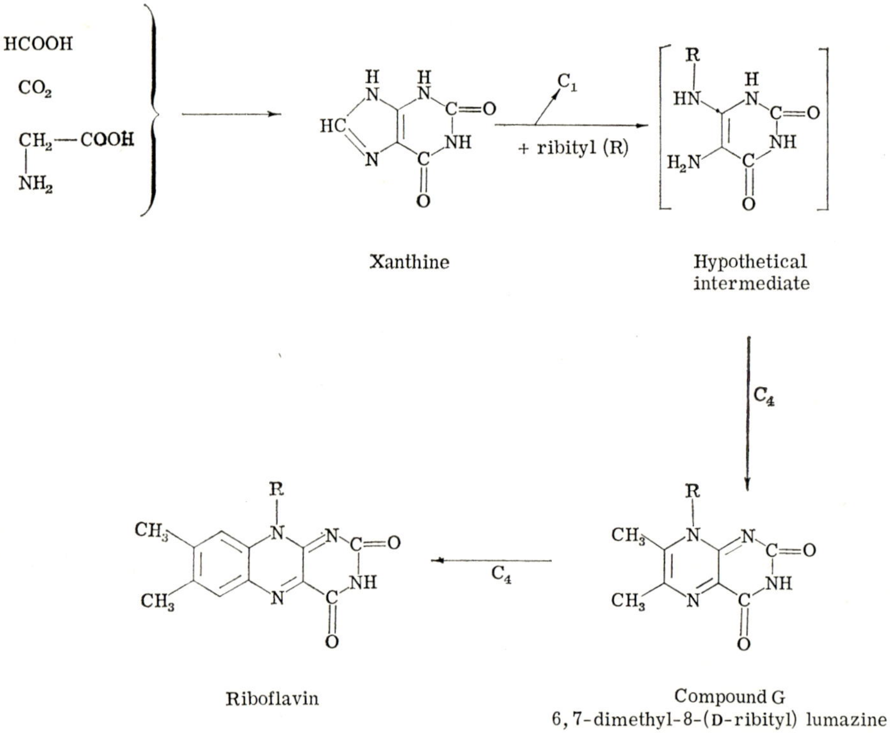

FIG. 7. Possible steps in enzymic formation of compound G and riboflavin.

found ATP and reduced diphosphopyridine nucleotide (DPNH) to be required for the formation of riboflavin from compound G and acetate by partly purified extracts of *Clostridium acetobutylicum* and *Escherichia coli*. The synthesis proceeded better under these conditions than when pyruvate was provided together with DPN and ThPP, and it was considered probable that two active acetate molecules were involved rather than one C_4 compound. Recently, with extracts of *Escherichia coli* and *Ashbya gossypii* utilizing 6,7-dimethyl-8-ribityllumazine for the biosynthesis of riboflavin, Plaut[81c] could show no stimulatory effect of acetoin, acetate, pyruvate, acetoacetate, or acetyl CoA. On the contrary, his observations suggested that all the carbon atoms of the *o*-xylene moiety of riboflavin are derived from this lumazine, two or more molecules of this therefore being required for the formation of one molecule of the vitamin.

Although 6,7-dimethyllumazine could be produced by extracts of *Eremothecium ashbyii* and *Escherichia coli* incubated with 4,5-diaminouracil and acetate,[81, 81d] it could not be further converted to 6,7-dimethylalloxazine by these preparations. The ribityl side chain probably arises early in the biosynthetic sequence from pentose phosphate, which in turn has its origin in 6-phosphogluconate.[81b] Perhaps synthesis actually takes place from xanthosine which would explain why no stimulation of riboflavin formation by unsubstituted diaminopyrimidines can be obtained.[17, 71, 79] Of several blue and violet fluorescent substances produced by organisms concurrently synthesizing riboflavin,[78, 82, 83] one, the compound V of *Eremothecium ashbyii* and *Ashbya gossypii* cultures, has been characterized as 6-methyl-7-hydroxy-8-*N*-ribityllumazine.[84-85a] These compounds seem to be side products of the reaction sequence rather than intermediates in the biosynthesis of the vitamin.[78, 86, 86a]

Riboflavin is metabolically functional in two coenzymic forms. The first of these, riboflavin-5′-phosphate, also somewhat erroneously called flavin mononucleotide (FMN), can be formed in yeast [87] and bacteria[87a] by the action of flavokinase.

$$\text{riboflavin} + \text{ATP} \rightarrow \text{FMN} + \text{ADP}$$

A specific phosphotransferase in *Escherichia coli* var. *neapolitanus* also effects the phosphorylation of riboflavin, at the expense of glucose-1-phosphate.[87b]

Flavin adenine dinucleotide (FAD), the second coenzymic form, is produced by the union of FMN and AMP via their respective phosphate groups, a reaction achieved by donation of AMP from ATP, and catalyzed by the enzyme FAD pyrophosphorylase present in yeast [88a] and bacteria.[88]

$$\text{FMN} + \text{ATP} \rightarrow \text{FAD} + \text{PP}$$

C. Nicotinic Acid (Diphosphopyridine Nucleotide, DPN, and Triphosphopyridine Nucleotide, TPN)

The route of biosynthesis of nicotinic acid from tryptophan via formylkynurenine, kynurenine, 3-hydroxykynurenine, and 3-hydroxyanthranilic acid in *Neurospora* was revealed by studies of mutant strains; this work has been well summarized elsewhere.[89] The same pathway has been reported to operate in animal tissues,[90] other fungi, e.g. *Trichophyton equinum*[91] and *Phycomyces blakesleeanus*,[92] and in the bacterium *Xanthomonas pruni*.[93] It appears, however, that this is not the sole route of formation of nicotinic acid (see Fig. 8) in all organisms. Studies of mutant strains of *Aspergillus nidulans* suggest that two pathways of nicotinic acid formation may be utilized by this organism. Both probably proceed via 3-hydroxyanthranilic acid, the duplicity apparently occurring in the synthesis of this compound.[2] Marnay concluded from growth inhibition studies with indole-3-acrylic acid that the route of nicotinic acid formation in *Escherichia coli* is identical with that in *Neurospora*. However, Yanofsky found that when uniformly C^{14}-labeled indole and tryptophan were supplied to suitable tryptophan auxotrophs of *Escherichia coli* and *Bacillus subtilis*, the nicotinic acid formed was unlabeled,[3] thus confirming earlier results with *Escherichia coli*.[94] A strain of *Escherichia coli* (presumably wild-type) was found to hydroxylate anthranilic acid directly in the 3-position by an enzymic reaction requiring a *p*-aminobenzoic acid (PABA) derivative as cofactor and proceeding at a rate compatible with its being a step in the major pathway of nicotinic acid synthesis by the organism.[95] The occurrence of this reaction could perhaps explain the above results with *Aspergillus nidulans* and *Escherichia coli*. If the direct anthranilic to 3-hydroxyanthranilic acid shunt is the

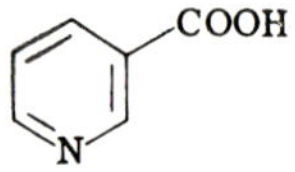

FIG. 8. Structure of nicotinic acid (pyridine-3-carboxylic acid).

only way of forming nicotinic acid in these organisms, mutants requiring either anthranilic acid or tryptophan and nicotinic acid should occur; but such mutants have not yet been described. If there is an alternative route available, this cannot proceed via tryptophan; otherwise some C^{14} would have been incorporated into the nicotinic acid synthesized during Yanofsky's experiments. Some evidence has been presented that nicotinic acid may be derived, at least in *Escherichia coli*, from C_3 and C_4 units. Thus, optimal nicotinic acid synthesis by washed suspensions of this organism requires the provision of ribose, adenine, succinate, glycerol, and ammonium chloride. Furthermore, label from succinate-2,3-C^{14} and glycerol 1,3-C^{14} is incorporated into nicotinic acid, whereas pyruvate-2-C^{14} is not utilized. The requirement for ribose and adenine suggests that the end product of synthesis is not the free vitamin but rather a nucleoside or nucleotide of nicotinic acid or nicotinamide.[95a, 59b] There is also some evidence that in plants an alternative route to nicotinic acid not involving tryptophan may exist.[1, 96, 97]

In those cases where nicotinic acid is derived from 3-hydroxyanthranilic acid, it is only one of several pyridine derivatives which may be formed from this aromatic compound. Several steps intervene between 3-hydroxyanthranilic acid and nicotinic acid and the biosynthesis has as yet not been unequivocally accomplished *in vitro*. The probable sequence of reactions is shown in Fig. 9.[90, 98-101] 3-Hydroxyanthranilic acid is converted to a mixture of "aminoacroleinfumaric" and "aminoacroleinmaleic" acids in equilibrium via the imino acid. The maleic acid derivative may possibly be enzymically decarboxylated to a product which undergoes spontaneous ring closure to give nicotinic acid.[90] The conversion of 3-hydroxyanthranilic

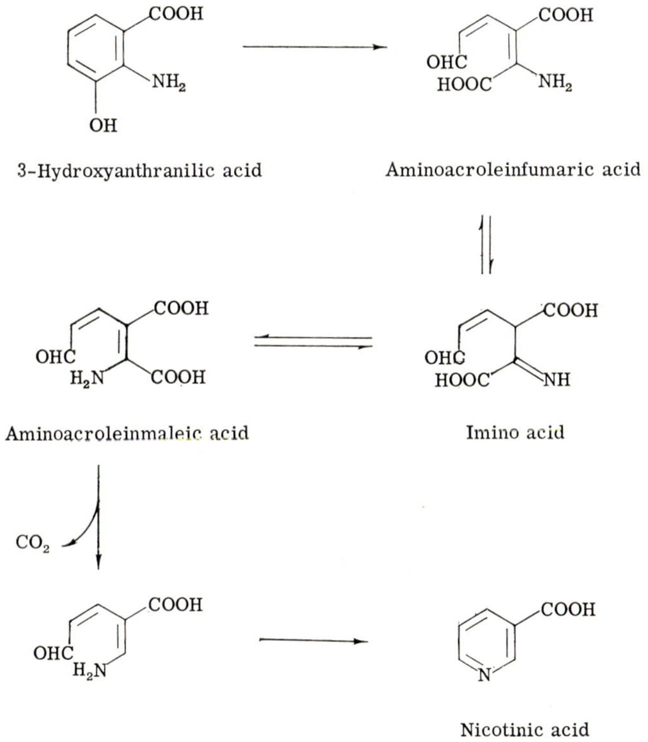

FIG. 9. Possible steps in biosynthesis of nicotinic acid via 3-hydroxyanthranilic acid.

acid to aminoacroleinfumaric acid may occur in more than one stage. Quinolinic and picolinic acids possibly arise as side products of the above sequence, the former as a consequence of spontaneous cyclization of aminoacroleinmaleic acid, the latter through the agency of picolinic carboxylase.[98, 99] Some evidence has been obtained that α-aminomethyl-*cis,trans*-muconic acid may be used for nicotinic acid biosynthesis by *Xanthomonas pruni*,[102] although it is uncertain whether this constitutes evidence of C-2:C-3 cleavage of 3-hydroxyanthranilic acid in nicotinic acid formation by this organism.

Although some bacteria (e.g., *Pasteurella spp.*) which respond to nicotinamide are unable to utilize the free acid,[7] this is possibly a question of the relative permeability of these compounds, since for other organisms (e.g., strains of *Corynebacterium diphtheriae*) the acid is more effective than the amide. In all organisms so far examined, nicotinic acid and nicotinamide appear to be the starting materials for the synthesis of the pyridine nucleotides (DPN and TPN)[103, 104, 105, 106] which are the coenzymic forms of the vitamin (see Fig. 10). Work, chiefly carried out with mam-

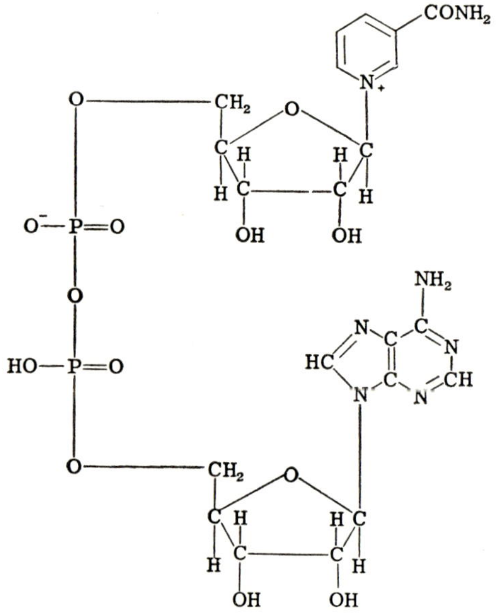

FIG. 10. Structure of diphosphopyridine nucleotide (DPN).

malian tissues, has revealed that DPN synthesis from nicotinic acid occurs in three steps,[103]

$$\text{nicotinic acid} + \text{phosphoribosylpyrophosphate} \rightleftharpoons \text{desamido-NMN} + \text{PP}$$
$$\text{desamido-NMN} + \text{ATP} \leftrightarrows \text{desamido-DPN} + \text{PP}$$
$$\text{desamido-DPN} + \text{ATP} + \text{glutamine} \rightarrow \text{DPN} + \text{AMP} + \text{glutamate} + \text{PP}$$

Although these reactions have been most fully investigated in animal tissues, there is evidence that they may also occur in microorganisms. All three enzymes have been partially purified from yeast autolysates[103, 105a] Nicotinic acid mononucleotide has been isolated from *Saccharomyces cerevisiae*[105b] and *Fusarium*.[105c] Desamido-DPN has been found in *Penicillium chrysogenum* and *Fusarium*[107] and in the bacterium *Haemophilus parainfluenzae*[103] (although this organism shows a growth requirement for the pyridine nucleotides which may be replaced by NMN or nicotinamide riboside but not by nicotinic acid or nicotinamide plus ribose.)[7] The final amination reaction is essentially irreversible and the whole sequence of reactions may be driven by hydrolysis of the inorganic pyrophosphate formed at each stage. Evidence has been obtained[107a] that the biosynthesis of DPN in *Escherichia coli* proceeds from nicotinic acid via nicotinic acid nucleotide intermediates though possibly with ammonium ions being utilized in place of glutamine.

Triphosphopyridine nucleotide (the extra phosphate group being situated on position 2′- of the adenosyl component) is synthesized by a kinase,[108] also purified from yeast, which brings about the reaction

$$\text{DPN} + \text{ATP} \rightarrow \text{TPN} + \text{ADP}$$

D. The Vitamin B_6 Group (Pyridoxal-5′-Phosphate)

Three compounds, pyridoxine (pyridoxol), pyridoxal, and pyridoxamine, together comprise the B_6 group of vitamins[109] (see Fig. 11). These occur intracellularly mainly as the 5′-phosphate esters of pyridoxamine and pyridoxal, the latter compound being the chief coenzymic form of the vitamin.

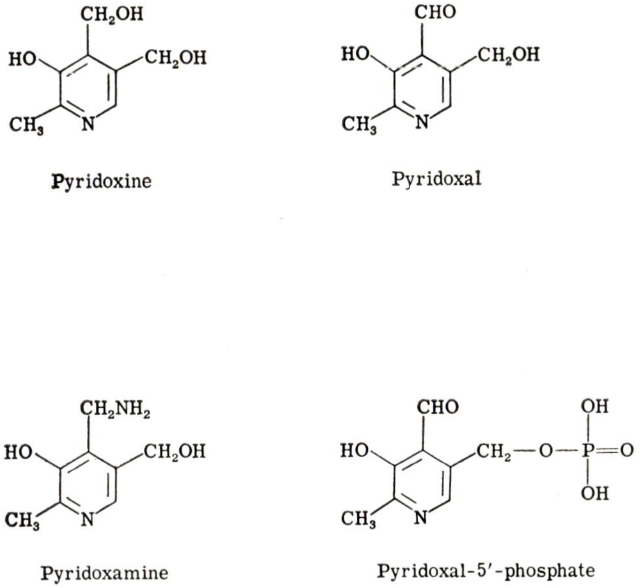

Fig. 11. Structure of the vitamin B_6 group.

The investigation of vitamin B_6 biosynthesis has suffered from the unavailability of microorganisms able to synthesize large amounts of the vitamin. The biosynthesis is a complete mystery; it is not even known which member of the group is the primary product.

Dalgliesh[90] supposed that a compound arising from decarboxylation of a compound such as α,β-dihydroxy-β-methylglutaric acid might condense with pyruvate and ammonia to give the necessary skeleton. The three-fragment synthesis envisaged originally by Snell and Guirard[110] and used chemically with great success[111] has remained an attractive possibility, if only because the postulated four-carbon fragment (giving rise to C-4 and C-5 of the pyridine ring with their C_1 substituents) could resemble a direct product of carbohydrate metabolism.

The hypothesis that alanine might be incorporated into the B_6 molecule arose from the observation that DL-alanine substituted for the vitamin in supporting growth of *Streptococcus faecalis* R in a rich basal medium,[110] and was abandoned when it was subsequently discovered that D-alanine is an essential metabolite of this organism and a product of vitamin B_6 function[112-114]

Certain mutant strains of *Escherichia coli* have been found to respond either to vitamin B_6, or to one of the amino acids serine or glycine together with glycolaldehyde.[115] Normal amounts of vitamin B_6 are formed during growth of these organisms in the absence of added vitamin, and glycolaldehyde is an obligatory requirement for synthesis of vitamin by washed suspensions of endogenously depleted organisms.[116] The nature of the metabolic lesion in these strains has not been determined, and it cannot be concluded that any of these alternative growth requirements is concerned with the synthesis of vitamin B_6 by wild-type *E. coli*.[116]

Conversion of all other forms of the vitamin into pyridoxal-5′-phosphate during growth of bacteria on media supplemented with these compounds has been demonstrated.[109, 117] Reversible interconversion of pyridoxal-5′-phosphate and pyridoxamine-5′-phosphate by extracts of *E. coli* has been reported,[118] while transamination between pyridoxamine phosphate and pyruvate has been demonstrated in cell-free preparations of *Clostridium welchii*.[119] Direct conversion of free pyridoxamine to pyridoxal probably occurs in *E. coli*[129] and *Acetobacter rancens*.[121] A pyridoxal kinase, which can phosphorylate the free compounds, has been partly purified from yeast.[122] Little is known about the reactions concerned in the utilization of pyridoxine. Some bacteria (e.g., several lactobacilli and clostridia) show a specific growth requirement for pyridoxal or pyridoxamine; this may reflect either inability to take up pyridoxine, or inability to convert it to pyridoxal-5′-phosphate. A few lactobacilli require the phosphate esters themselves, pyridoxamine-5′-phosphate being some twenty-five times more active than pyridoxal-5′-phosphate for *Lactobacillus lactis* (Dorner).[123]

E. Pantothenic Acid (Coenzyme A)

The route of pantothenic acid biosynthesis in microorganisms has been established mainly by nutritional studies,[29, 30] antimetabolite studies with wild-type *Escherichia coli*,[19-22, 124] and work with pantothenate auxotrophs.[27] Much of the experimental evidence has been presented earlier in this chapter and the whole has been the subject of a comprehensive review.[125]

The pantoic acid moiety arises from α-oxoisovaleric acid ("ketovaline") by way of its hydroxymethylation to give ketopantoic acid (see Fig. 12). A strain of *Bacterium linens* was found to show a growth requirement for

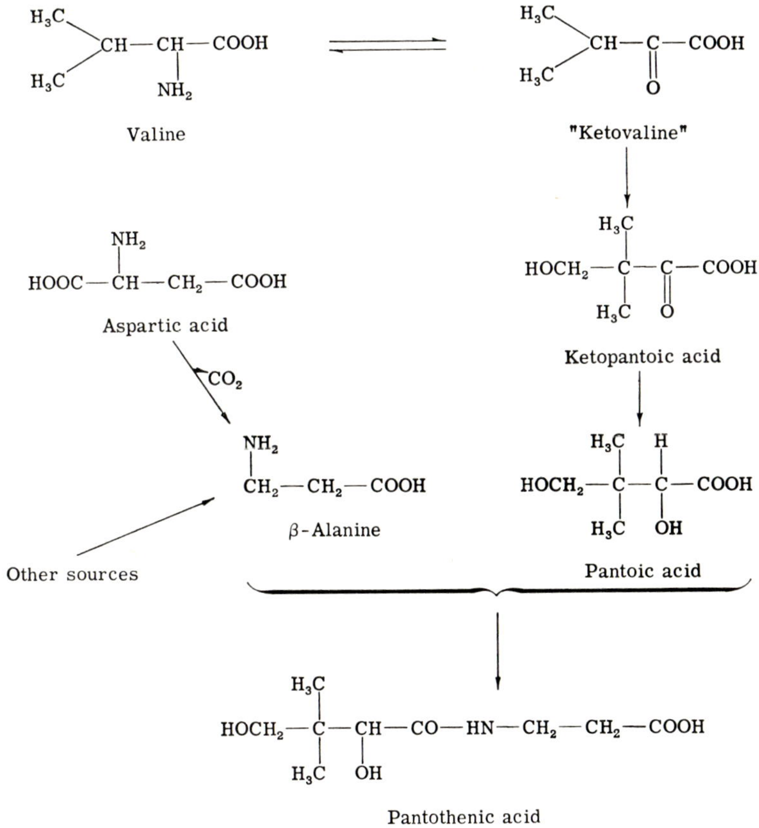

FIG. 12. Route of pantothenic acid biosynthesis.

PABA, which could be replaced by pantothenic, pantoic, or ketopantoic acid; this suggested that PABA might play some role in this hydroxymethylation.[126, 127] However, the problem is complicated by the apparent function of pantothenate in PABA biosynthesis in this organism. No folic acid requirement could be demonstrated for a partially purified enzyme from *E. coli* capable of utilizing formaldehyde for the hydroxymethylation of ketovaline;[128] but a coenzyme form of PABA could be active in making the necessary C_1 fragment available from normal donors such as serine.[125, 128, 129]

Although there seems to be one general route of pantoic acid formatino, β-alanine can be formed in several different ways:[125] by decarboxylation of aspartic acid, as in *Rhizobium trifolii*[130] and other organisms;[21, 22, 131-131c] by a glutamic acid-dependent mechanism which might be transamination of formylacetic acid arising from propionic acid;[125, 132, 133] from spermine or spermidine as in *Pseudomonas aeruginosa*;[134] from uracil via β-ureidopropionic acid as in *Clostridium uracilicum*;[135] or from propionyl-CoA by way of acrylyl-CoA and β-alanyl-CoA as in *Clostridium propionicum*.[136] β-Alanine is formed from aspartic acid by washed suspensions of *E. coli* only in the pH range from 5 to 6; at other pH values, alternative routes may be of major significance.[125]

Once formed, pantoic acid and β-alanine are coupled by an enzyme, pantothenate synthetase, which has been partially purified from *E. coli* extracts[137] and shown to participate in the following reactions.[138, 139]

$$\text{enzyme (E)} + \text{ATP} + \text{pantoate} \xrightleftharpoons{K^+;\ Mg^{++}} \text{E–pantoyladenylate} + \text{PP}$$

$$\text{E–pantoyladenylate} + \beta\text{-alanine} \rightarrow \text{pantothenate} + \text{AMP} + \text{E}$$

This was in fact the first demonstration of what was later to be recognized as a common mechanism of peptide bond formation.

Pantothenic acid forms a comparatively minor fraction of the ultimate functional molecule, coenzyme A;[140] consequently, there are many successive steps in the synthesis of the coenzyme from the vitamin. A large number of bacterial species have been shown to respond to different "higher forms" of pantothenic acid, and the elucidation of the biochemical pathway that leads from vitamin to coenzyme (see Fig. 13) owes much to the identification of these growth factors. The path of CoA biosynthesis summarized below was deduced from results obtained chiefly with animal tissue preparations[141] and for some time was considered to be the only route of formation of the coenzyme.

(1) $$\text{pantothenic acid} + \text{cysteine} \xrightarrow{\text{ATP}} \text{pantothenylcysteine}$$

(2) $$\text{pantothenylcysteine} \xrightarrow{\nearrow CO_2} \text{pantetheine}$$

(3) $$\text{pantetheine} + \text{ATP} \rightarrow 4'\text{-phosphopantetheine} + \text{ADP}$$

(4) $$4'\text{-phosphopantetheine} + \text{ATP} \rightarrow \text{dephospho-CoA} + \text{PP}$$

(5) $$\text{dephospho-CoA} + \text{ATP} \rightarrow \text{CoA} + \text{ADP}$$

Lactobacillus arabinosus converts exogenous pantothenate to endogenous CoA during growth;[142] washed suspensions of this organism and of *Proteus morganii* required cysteine to accomplish the reaction.[143] Pantothenylcysteine is almost as active as pantetheine, and much more active than free pantothenate for the growth of *Acetobacter suboxydans* dried cells of which can decarboxylate pantothenylcysteine to pantetheine.[143] This decarboxylase activity has also been demonstrated in *Lactobacillus helveticus* and *Lactobacillus bulgaricus*.[144]

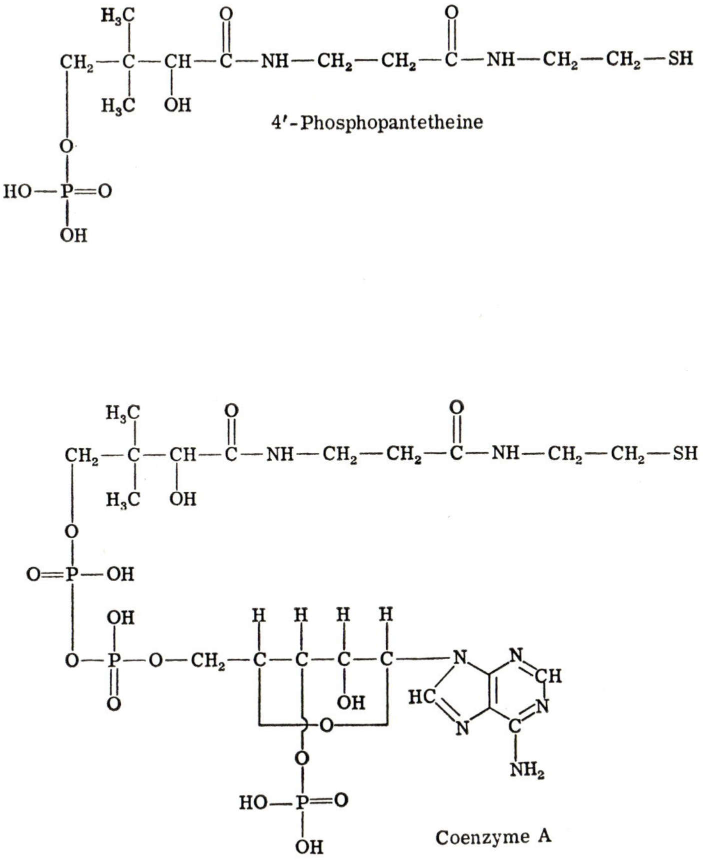

FIG. 13. Structure of 4′-phosphopantetheine and of coenzyme A.

There are, however, indications of another pathway, in which phosphorylation of pantothenate precedes condensation with cysteine,[145, 146] i.e.,

$$\text{pantothenate} + \text{ATP} \rightarrow \text{4}'\text{-phosphopantothenate} + \text{ADP}$$

$$\text{4}'\text{-phosphopantothenate} + \text{cysteine} \xrightarrow{+\,\text{CTP}} \text{4}'\text{-phosphopantothenylcysteine}$$

$$\text{4}'\text{-phosphopantothenylcysteine} \xrightarrow{\nearrow CO_2} \text{4}'\text{-phosphopantetheine}$$

4′-Phosphopantothenate has in fact been isolated from the growth media of several microorganisms including *Neurospora crassa*, *Streptobacterium plantarum*, and *Saccharomyces carlsbergensis*,[147, 148] and is formed by cell-free extracts of *Proteus morganii*, *Acetobacter suboxydans*, *Escherichia coli*, and *Lactobacillus arabinosus*.[145, 146, 149] Brown[150] concluded that this second route of CoA biosynthesis proceeding via 4′-phosphopantothenate and 4′-phosphopantothenylcysteine is used by *Proteus morganii*, *Escherichia coli*, and mammalian tissues, and may also operate in yeast, *Lactobacillus arabinosus*, and *Neurospora crassa*. The first route may be utilized by *Acetobacter suboxydans*, *Lactobacillus helveticus*, and *Lactobacillus bulgaricus* although enzymic synthesis of pantothenylcysteine has yet to be demonstrated.

Coenzyme A is the major bound form of pantothenic acid present in microorganisms, but some phosphopantetheine is also found, together with lesser amounts of phosphopantothenic acid and pantetheine.[151] However, coenzyme A itself has little activity for most microorganisms known to require pantothenic acid for growth,[140] presumably because of its inability to enter the cell.

F. Biotin

Biotin (see Fig. 14) is the simplest compound able to counteract the nutritional deficiency induced in animals by the ingestion of raw egg white, which contains a biotin-binding protein, avidin. The normal requirement of animals for the vitamin is apparently satisfied by its production by the intestinal flora.[6] Many bacteria are able to synthesize and excrete biotin[41] though some require it for growth.[152] What little is known of the biosynthesis of this vitamin in fact derives from nutritional studies with microorganisms.

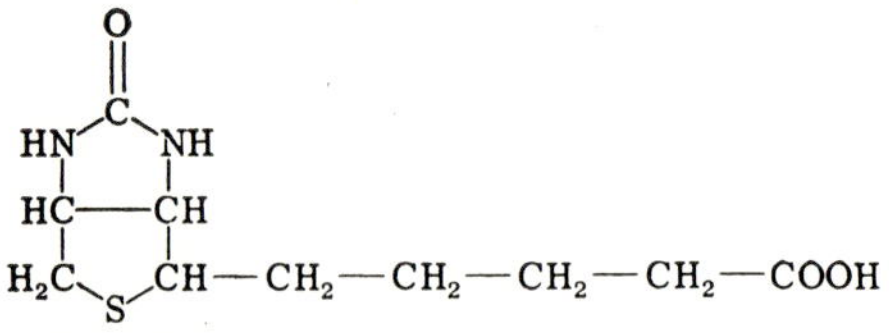

FIG. 14. Structure of biotin [*cis*-hexahydro-2-oxo-1*H*-thieno(3,4)imidazole-4-valeric acid].

The suggestion that pimelic acid (see Fig. 15) might be a biosynthetic precursor of biotin was made as a result of the observation that biotin can substitute for pimelic acid in supporting growth of the Allen strain of *Corynebacterium diphtheriae*.[153] Pimelic acid was subsequently found to replace the biotin requirement of several thermophilic bacteria,[154] to increase biotin and biotin sulfoxide production by *Aspergillus niger*,[155-157] and to be converted into desthiobiotin and other compounds with biotin activity by *Corynebacterium xerose*.[158] Bacterial growth inhibition by ε-(2,4-dichlorosulfanilido) caproic acid (see Fig. 15) can only be observed with bacteria which show a growth requirement for biotin; its inhibitory action is overcome competitively by pimelic acid, and noncompetitively by biotin.[159] Reasonably strong evidence also suggests that desthiobiotin is a product of pimelic acid utilization and a precursor of biotin, although it cannot yet be definitely asserted that this compound is a normal intermediate on the direct biosynthetic pathway. Desthiobiotin replaces biotin with varying efficiency for a number of organisms. It is even more active than the vitamin for *Propionibacterium pentosaceum*,[160] and is converted to biotin by *Saccharomyces cerevisiae.* [161-163] It is synthesized from pimelic acid by *Corynebacterium xerose*[158] and is accumulated by a biotin-requiring mutant of *Penicillium chrysogenum.*[164] Growth of *Escherichia coli* and of a biotin-synthesizing strain of *Mycobacterium tuberculosis* is inhibited by 4-(imidazolidone-2-)caproic acid; this inhibition is overcome competitively by desthiobiotin (see Fig. 16) and noncompetitively by bio-

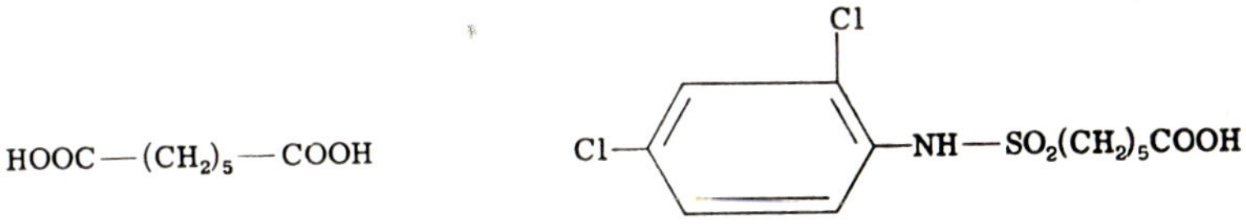

Pimelic acid

ε-(2,4-Dichlorosulfanilido)-caproic acid

FIG. 15. Structure of pimelic acid and of ε-(2,4-dichlorosulfanilido)caproic acid.

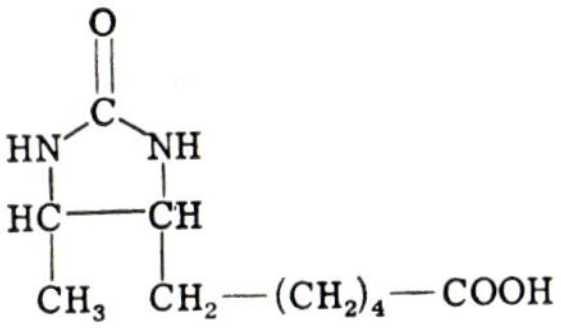

FIG. 16. Structure of desthiobiotin.

tin.[165-166] In *Lactobacillus arabinosus* and *Lactobacillus casei* desthiobiotin actively inhibits growth. It is possible that these organisms have a metabolic block in the conversion of desthiobiotin to biotin, and that excess desthiobiotin interferes with normal biotin utilization.[161, 167]

Oxybiotin (which is the furane analog of the naturally occurring thiophene compound) may be active per se for yeast, *Rhizobium trifolii*, and *Lactobacillus pentosus.*[168-170] Its utilization by propionibacteria is delayed, which suggests that in these organisms it is utilized only after conversion to another substance.[160]

There are many reports of protein-bound forms of biotin.[171-173] In common with other B vitamins that are required in extremely small amounts, such as lipoic acid and B_{12}, the active form of biotin may be bound to protein. A peptidase (biotinidase), which releases the vitamin from proteins, has been described.[174] The simplest peptide that has been reported to have high biological activity is ε-*N*-biotinyl-L-lysine, or biocytin.[175-177] This replaces biotin as a growth factor for *Streptococcus faecalis* R, *Lactobacillus acidophilus*, *Lactobacillus delbrueckii*, and some yeasts and molds.[177] It is about 65% more active than the vitamin for *Micrococcus sodonensis*.[178] Since biocytin has not been demonstrated to be more active than biotin in any cell-free system, its higher activity for *M. sodonensis* may result from its more ready entry into the cell. In contrast, biocytin is without activity for *Lactobacillus arabinosus*, *Lactobacillus pentosus*, and *Leuconostoc mesenteroides* P60.[177]

Biotin-L-sulfoxide, formed by *Aspergillus niger*,[157, 179] is as effective as biotin in supporting the growth of *Neurospora crassa*.[157] It is more active than the D-isomer for *Lactobacillus casei*, but less active than this form for *Saccharomyces cerevisiae* and *Lactobacillus arabinosus*.[180] Since both isomers are inactive for other microorganisms, they are probably converted into biotin by those organisms able to use them.

The possible existence of a biologically active phosphate ester of biotin has been suggested, but synthetic biotinyl phosphate is rapidly hydrolyzed at pH 7.5, and is no more effective than biotin in activating the aspartic deaminase of *Bacterium cadaveris*.[181]

G. *p*-Aminobenzoic Acid and the Folic Acid Group

p-Aminobenzoic acid (PABA) was shown to be a bacterial growth factor as a consequence of studies on the mode of action of the sulfonamide drugs,[182] and it was suggested that the reaction inhibited by these drugs is the conversion of PABA into a more complex metabolically active substance. It now appears that the main functional form, coenzyme F, is one of a number of compounds closely resembling tetrahydropteroylglutamic acid in structure; these compounds are named the folic acid group of factors, after the parent vitamin folic acid, i.e., pteroylglutamic acid, PtG (see Fig. 17). The structure of coenzyme F itself is still undetermined.

So far as is known, PABA is synthesized from the common precursor of aromatic compounds, shikimic acid.[183] It is formed by extracts of baker's yeast supplied with shikimic acid-5-phosphate and L-glutamine.[183a] Besides PABA, the molecule of PtG is composed of glutamic acid and a substituted pteridine. Although the three fragments may be separately synthesized and only then linked together, it is possible that PABA or PABA-glutamate combines with a pteridine precursor or a reduced pteridine. The detailed evidence at present available has been summarized in a recent review.[184]

Free PABA can be utilized as a folic acid precursor by many bacteria, as indicated by early studies with washed suspensions of *Escherichia coli*, *Staphylococcus aureus*, and *Streptobacterium plantarum*.[28, 185, 186] This conversion has now been confirmed by the demonstration that the radioactivity of C^{14}-carboxyl-labeled PABA is incorporated into folic acid derivatives during growth of *Enterococcus stei*.[187, 188] Although pteroic acid (see Fig. 18) can serve as a precursor of folic acid in some microorganisms, such

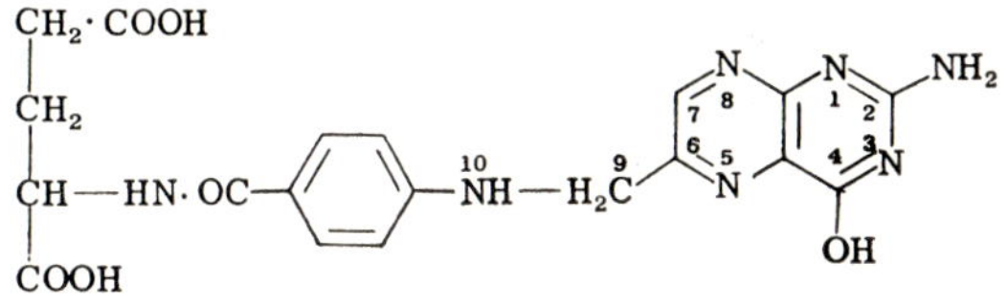

FIG. 17. Structure of pteroylglutamic acid (PtG).

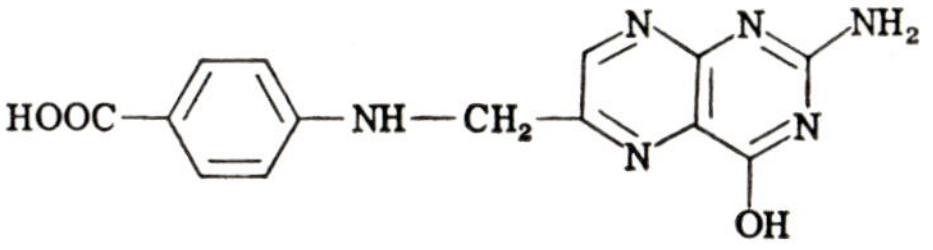

FIG. 18. Structure of pteroic acid.

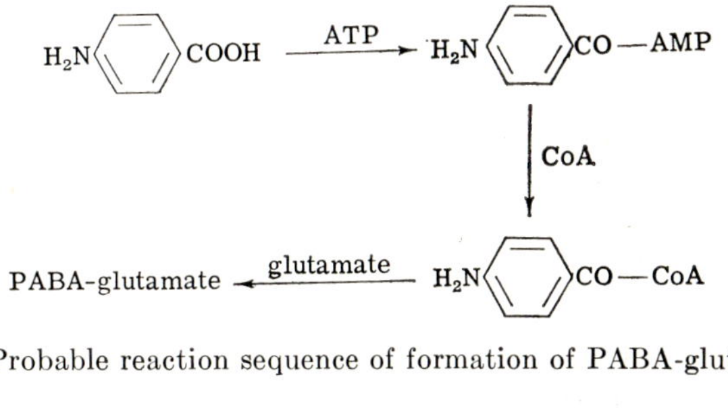

FIG. 19. Probable reaction sequence of formation of PABA-glutamate.

as *Streptococcus faecalis* R for which its N^{10}-formyl derivative (rhizopterin) is a growth factor,[189] it is probably not a normal metabolite in organisms able to synthesize folic acid from simpler precursors, such as *Mycobacterium avium*.[190] The results of nutritional studies with PABA-glutamate are difficult to interpret. In only one organism, *Lactobacillus plantarum* 10S, is this peptide more active than PABA itself in overcoming sulfonamide inhibition, and then its action is competitive.[184] Yet PABA-glutamate is formed by cell-free extracts of *Mycobacterium avium* according to the probable reaction sequence shown in Fig. 19.[184, 190, 191]

The preliminary activation of PABA as well as the production of PABA-glutamate is inhibited by sulfonamides (although it is not known whether

PABA-glutamate overcomes sulfonamide inhibition of growth of *M. avium*). The results obtained with growing cultures do not suggest that free PABA-glutamate is a normal intermediate in folic acid biosynthesis. It is conceivable, however, that an acyl-PABA-glutamate or other derivative capable of direct condensation with a substituted pteridine may yet prove to be a precursor of the vitamin.

Next, there is the problem of how the pteridine fragment is introduced into the folic acid molecule. Most studies with substituted pteridines have been carried out with microorganisms that require PtG or pteroic acid for growth, so that it is difficult to assess the relevance of the results obtained. The only study with growing cultures of organisms able to synthesize folic acid from simple precursors has been made with two yeasts, *Pichia membranaefaciens* and *Candida albicans*.[192] This study showed that radioactivity from 6-C^{14}-labeled 2-amino-4,6-dihydroxypteridine (xanthopterin), and 2-amino-4,7-dihydroxypteridine-6-carboxylic acid appears in a fraction corresponding to PtG or its polyglutamyl conjugates. Folic acid synthesis from PABA, glucose, and glutamate by washed suspensions of *Staphylococcus aureus* strains is increased twofold when 2-amino-4-hydroxypteridine-6-aldehyde is provided, although this compound inhibits folic acid synthesis by a PABA auxotroph of *Escherichia coli*; the 6-methyl and 6-carboxyl derivatives are inactive in both cases.[28] The 6-carboxyl derivative is, however, used for folic acid biosynthesis by a sulfathiazole-resistant PABA mutant of *E. coli*.[193]

Folic acid (mainly PtG) is synthesized from 2-amino-4-hydroxypteridine-6-carboxylic acid and PABA-glutamate by cell-free preparations of *Mycobacterium avium* supplied also with ATP, but xanthopterin is used only in the presence of additional cofactors.[184, 190, 191] In other bacteria, xanthopterin does not appear to be a normal precursor of folic acid.[193a] Pteroic acid and PtG are synthesized from PABA and PABA-glutamate, respectively, by dialyzed cell-free extracts of *Lactobacillus arabinosus*. In this system, chemically or enzymically degraded PtG can be used as source of pteridine; 2-amino-4-hydroxypteridine-6-aldehyde and the 6-hydroxymethyl compound can also be used, but their effectiveness is greatly increased by preliminary chemical reduction.[194] It has been suggested, in fact, that tetrahydrofolic acid (PtH_4G) is the primary product when these dialyzed extracts are incubated with ATP, Mg^{++}, PABA-glutamate, and 2-amino-4-hydroxy-6-hydroxymethyltetrahydropteridine.[194] Similar observations have been made with cell-free extracts of *Escherichia coli*;[194a] in this system, PABA is about 40 times as effective as PABA-glutamate for the synthesis of tetrahydropteroic acid.[194b] The ATP may serve for the preliminary phosphorylation of the 2-amino-4-6-hydroxymethylpteridine.[194c] In contrast to the results obtained with extracts of *Mycobacterium avium*,[184] in the *Es-*

cherichia coli extracts sulfonamides appeared to inhibit not so much the synthesis of PABA-glutamate as the formation of tetrahydropteroic acid from the reduced pteridine and PABA.[194b]

The route of biosynthesis of the pteridine moiety itself is still obscure. It has been suggested that pteridines may arise from diaminopyrimidines[195, 196] or from purines.[197, 198] Perhaps the two possibilities are reconciled by the finding, discussed previously, that pteridine side products accumulate during the utilization of purines for riboflavin biosynthesis by growing cultures of *Eremothecium ashbyii.* Thus the differential stimulation of flavogenesis in this organism by aminopterin might be the result of diversion of pteridine precursors from folic acid biosynthesis into the route of riboflavin formation.[71] Results obtained with washed suspensions of *Leuconostoc mesenteroides* P60[199] suggest that CO_2 may be concerned in the synthesis of the pteridine ring of CoF in this organism.

Several forms of CoF appear to exist; they may be polyglutamyl derivatives of reduced PtG, and some are perhaps phosphorylated.[200-202] It is possible that PtG itself is not a normal intermediate in the *de novo* synthesis of CoF, but can be converted to such an intermediate by some organisms, e.g., *Lactobacillus plantarum, Streptococcus faecalis* R, and *Enterococcus stei,* although not by others, e.g., *Lactobacillus arabinosus* and *Escherichia coli.*[184, 187, 203] Many bacteria in fact contain enzymes able to convert PtG to reduced biologically active derivatives including citrovorum factor (N^5-PtH_4G-CHO).[203a] An enzyme has been isolated from *Clostridium sticklandii* which catalyzes the reduction of folic acid derivatives to their dihydro analogs[204, 205] the reduction being linked with pyruvate oxidation, thus,

$$\text{CoASH} + \text{pyruvate} + \text{PtG} \rightarrow \text{acetyl-SCoA} + CO_2 + PtH_2G$$

In mammalian tissues it seems that the reduction of folic acid to tetrahydrofolic derivatives is accomplished by a two-stage pyridine nucleotide-linked reaction sequence.[201, 206-208] The first two hydrogen atoms are probably attached to positions 7 and 8 of the pteridine ring, the subsequent pair occupying positions 5 and 6.[208] Aminopterin and amethopterin may exert their inhibitory action at this stage in the formation of CoF in animal tissues,[206, 209, 210] although the formation of PtH_2G by *Clostridium sticklandii* is not affected by these antimetabolites.[201]

Once formed, PtH_4G can be formylated by the enzyme tetrahydrofolate formylase which has been obtained from *Micrococcus aerogenes*[202] and *Clostridium cylindrosporum,*[211] and which accomplishes the reaction in two stages.[202, 212]

$$PtH_4G + \text{ATP} \rightleftharpoons PtH_4G\text{-P} + \text{ADP}$$
$$PtH_4G\text{-P} + \text{HCOOH} \rightleftharpoons N^{10}\text{-}PtH_4G\text{-CHO} + \text{P}$$
$$\text{Sum: } PtH_4G + \text{ATP} + \text{HCOOH} \rightleftharpoons N^{10}\text{-}PtH_4G\text{-CHO} + \text{ADP} + \text{P}$$

Enzymes are known which convert N^{10}-PtH_4G-CHO to N^5-PtH_4G-CH_2OH, others which form this hydroxymethyl derivative from PtH_4G and formaldehyde, and still others which convert N^5-PtH_4G-CHO to N^{10}-PtH_4G-CHO or N^5-PtH_4G-CH_2OH.[201]

H. Vitamin B_{12}

There exists a vast array of naturally occurring compounds which to a greater or lesser degree show the biological activity ascribed to vitamin

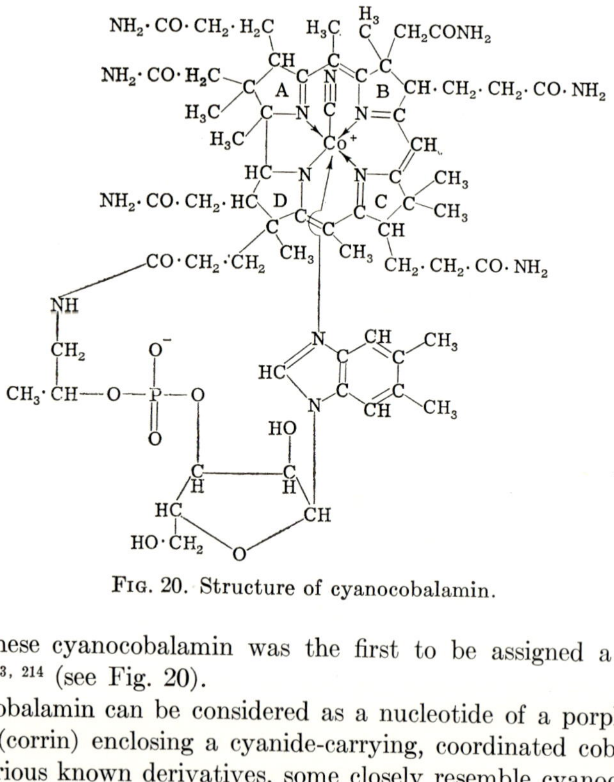

Fig. 20. Structure of cyanocobalamin.

B_{12}. Of these cyanocobalamin was the first to be assigned a chemical structure[213, 214] (see Fig. 20).

Cyanocobalamin can be considered as a nucleotide of a porphyrin-like structure (corrin) enclosing a cyanide-carrying, coordinated cobalt atom. Of the various known derivatives, some closely resemble cyanocobalamin in biological activity, whereas others are inactive or inhibitory. They may be classified chemically according to whether (*a*) the cyanide group has been replaced, (*b*) the nucleotide fragment is wholly absent or is substituted, (*c*) the corrin carboxamide groups are substituted (generally the propionamide groups).

It now appears possible that cyanocobalamin itself is no more than a derivative produced during extraction of naturally occurring forms of vitamin B_{12}. Using more careful extraction methods, Barker and his colleagues have recently isolated and characterized a number of compounds which have coenzymic function, whereas cyanocobalamin itself does not. The first of these compounds to be described was adenylcobamide coenzyme (AC-coenzyme) isolated from *Clostridium tetanomorphum*[214a] and later *Clostridium sticklandii*.[213a] This compound has a distinctive absorption spectrum and is sensitive to light and to mild hydrolysis. In the presence of cyanide it gives rise to an adenine-containing compound and pseudo-

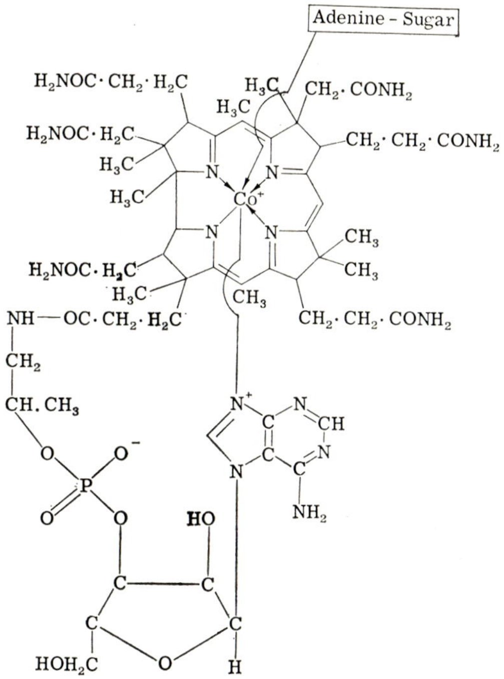

FIG. 21. Tentative structure proposed for adenylcobamide coenzyme (AC-coenzyme).[214c]

vitamin B_{12} (i.e. a cyanocobalamin derivative in which the nucleotide base is adenine). From these and other properties the structure shown in Fig. 21 has been tentatively proposed for the AC-coenzyme.[214b, c, d] Benzimidazolecobamide coenzyme (BC-coenzyme) and 5,6-dimethylbenzimidazolecobamide coenzyme (DBC-coenzyme) were obtained by growing *Clostridium tetanomorphum* in the presence of the corresponding bases. The adenine nucleoside is retained in these compounds. The DBC-coenzyme was also isolated from *Propionibacterium shermanii*, where it accounted for at least 60% of the vitamin B_{12} content of the cells.[215a, b]

The cobamides are synthesized exclusively by microorganisms,[215c] and culture filtrates of streptomycetes, of propionibacteria,[216a] and of *Bacillus megaterium* are commercially exploited as sources of the vitamin. Studies of the route of biosynthesis, although not as yet far advanced, have been much aided by the existence of bacteria and fungi with blocks at different points in the complex sequence of biosynthetic reactions. Many of these will accept derivatives of the normal intermediates as precursors of what are consequently analogs of cyanocobalamin.[213] Thus some bacteria (e.g., *Escherichia coli* auxotroph 113-3) are unable to synthesize corrin but are able to complete the synthesis of cobamide if provided with factor B (the cobamide molecule without the nucleotide fragment).[213, 215-217] Others, such as *Corynebacterium diphtheriae*, produce various abnormal or "incomplete" derivatives of the porphyrinlike structure (differing from factor B) which they may transform into B_{12} analogs in the presence of added base.[218] Failure to synthesize cobamide may also result from inability to form the correct heterocyclic base, with consequent accumulation of factor B and B_{12} analogs. This situation occurs in *Propionibacterium shermanii*.[11, 219] When *P. shermanii* is furnished with 5,6-dimethylbenzimidazole, the formation of factor B and of B_{12} analogs is greatly depressed, in favor of the production of normal vitamin B_{12}.

Several pieces of evidence suggest that the porphyrinlike structure of vitamin B_{12} is synthesized from the same precursors as are the normal porphyrins. The accumulation of coproporphyrin by cultures of *Propionibacterium shermanii* can be decreased by provision of cobalt ions, suggesting diversion of coproporphyrin precursors into synthesis of the B_{12} analogs formed by this organism.[220] Furthermore, C^{14}-labeled acetate and glycine are incorporated into B_{12}-corrin by cultures of *P. shermanii* growing at the expense of glucose.[220a] Even more conclusive evidence is provided by the incorporation of C^{14}-labeled δ-aminolevulinic acid[221] and porphobilinogen[222] into the vitamin B_{12} synthesized by an unnamed bacterium. The methyl groups that are a distinctive feature of corrin, and which can not be derived from δ-aminolevulinic acid, may be introduced by transmethylation of the methyl groups of methionine.[223] The D_g-1-amino-2-propanol fragment, which links the nucleotide to a propionic side chain on the pyrrole ring D, may arise from the decarboxylation of threonine; N^{15} from labeled L-threonine added to the culture medium of *Streptomyces griseus* is very effectively incorporated into this fragment.[224] Cell-free extracts of *Streptomyces olivaceus* can form vitamin B_{12} when provided with ATP, porphyrin, 5,6-dimethylbenzimidazole, D-ribose, cobaltous ions, and DL-threonine.[224a] As for the 5,6-dimethylbenzimidazole base, it has been suggested by Woolley on the basis of antimetabolite studies that it could arise from 1,2-dimethyl-4,5-diaminobenzene,[70] although it is possible that this is not the usual precursor.

The conversion of factor B to vitamin B_{12} by *Escherichia coli* proceeds better at the expense of free 5,6-dimethylbenzimidazole than of its α-riboside or α-ribotide.[225] Consequently, it has been suggested that the first step in the conversion is phosphorylation of factor B:[225]

$$\text{Factor B} \rightarrow \text{factor B-phosphate} \rightarrow \text{factor B-phosphate-ribose} \downarrow \text{cobalamin}$$

Factor B-phosphate is in fact utilized by *Escherichia coli* to form vitamin B_{12}, although rather less well than factor B itself.[226] A monophosphate ester and a guanosine diphosphate ester of factor B have been isolated from a strain of *Nocardia rugosa*,[227] and it has been suggested that they are precursors of the vitamin.[227a, b] An auxotrophic strain of *Bacillus megaterium* synthesized mainly a monocarboxylic acid derivative of factor B, which could be converted by the parent wild-type strain into vitamin B_{12}, and which was therefore suggested to be a normal intermediate in the synthesis of cobamides by the organism.[228] Studies with *Escherichia coli* and *Corynebacterium diphtheriae* indicate, however, that monocarboxylic acids are formed by the degradation of vitamin B_{12} and its analogs.[218, 229] Benzimidazolylcobamide coenzyme is synthesized by a partially purified extract of *P. shermanii* when this is supplied with benzimidazolylaquocobamide, ATP, a flavin, a reduced sulfhydryl compound, DPNH, and manganous ions. The finding that benzimidazolylcyanocobamide is only two thirds as effective as the aquo form, suggests that the cyano group must be displaced from the vitamin before conversion to the coenzyme can occur. The ATP gives rise to the adenyl group of the coenzyme and in extracts of *Clostridium tetanomorphum* is apparently also the source of the nucleoside sugar (at present, of unknown structure).[229a, b, c]

Vitamin B_{12} may require further activation, perhaps by conversion into a protein-bound form, before it exerts its biological activity. This could explain the results obtained by Kisliuk and Woods[230, 231] during their investigation of the role of vitamin B_{12} in the synthesis of methionine from homocysteine by cell-free extracts of a strain of *Escherichia coli*. When serine or formaldehyde is used as C_1-donor in the presence of PtH_4G, a requirement is shown for a product of cyanocobalamin metabolism (not replaceable by DBC-coenzyme) whose synthesis is competitively inhibited by the analog, B_{12}-anilide.[232]

I. Miscellaneous

1. Lipoic Acid

Higher animals do not appear to require preformed lipoid acid[233] although several microorganisms require it in very small amounts for growth.

These organisms include strains of *Corynebacterium bovis*,[234] *Streptococcus cremoris*,[235] *Butyribacterium rettgeri*, when growing in a lactate medium,[236] and *Lactobacillus casei*, when growing in an acetate-free medium.[237] Natural materials contain many compounds which exhibit biological activity, and which are utilized with varying degrees of efficiency by such dependent organisms. All these substances appear to be derivatives of (+)α-lipoic acid[233, 238] (see Fig. 22). Thus β-lipoic acid is a monosulfoxide derivative,[238] probably formed by oxidation of α-lipoic acid during chemical isolation. It is reconverted to α-lipoic acid by growing cultures of *Lactobacillus casei*.[239] Other active derivatives include dihydrolipoic acid (6,8-dithioloctanoic acid) linked through one or both S atoms to an unidentified compound,[240] a biologically active analog synthesized by the mold *Allomyces macrogynus*,[241] and possibly esters, amides, and mixed disulfides formed with other cell constituents.[233] Nothing is known of the route of biogenesis of any of these compounds.

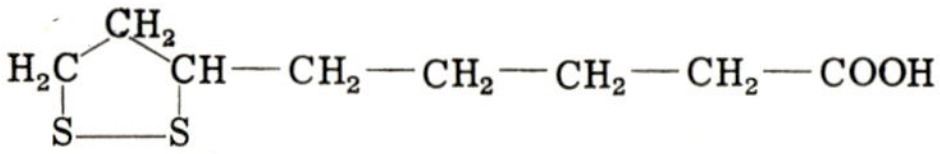

FIG. 22. Structure of α-lipoic acid (6,8-dithio-*n*-octanoic acid).

Transformations of lipoic acid and lipoic acid derivatives have been obtained with various bacterial[242-244] and mammalian preparations,[245-247] but do not necessarily reflect formation of the actual coenzyme. The finding that dihydrolipoamide is more active than dihydrolipoic acid itself as a substrate for dihydrolipoic dehydrogenase and dihydrolipoic transacetylase suggests that in its coenzymic form, lipoic acid is carboxyl-linked with a basic group,[233, 248] although it does not now appear that this group is thiamine pyrophosphate, as was once supposed.[249, 250] Present evidence suggests that lipoic acid in fact functions in an activated, protein-bound form, the protein perhaps being one of the enzymes concerned with the lipoic acid-dependent reaction sequence. The establishment of the lipoic acid-protein bond seems to require initial carboxyl activation of the acid with intermediary formation of lipoyl-adenylate[251, 252] prior to the condensation of the carboxyl group with the ε-NH_2 group of lysine.[252a] An enzyme which has been partially purified from extracts of *Streptococcus faecalis*[252] and yeast[253] cleaves the bond, with liberation of lipoic acid.

2. Vitamin K

Several derivatives of menadione (2-methyl-1,4-naphthoquinone) with vitamin K activity are formed in nature. Those of bacterial origin are of the vitamin K_2 series and differ from the 3-phytylmenadione (vitamin K_1) produced by plants[254] in possessing a polyisoprenoid side chain[255] (see Fig. 23). The quantitatively most important member of the vitamin K_2 group isolated from putrified fish meal[255] and *Bacillus brevis*[259] bears a terpenoid side chain of 7 isoprenoid units, while other homologs found in bacteria carry 6 or 10 unit side chains.[259a] Phthiocol (3-hydroxymenadione) probably does not exist as such in bacteria, although it may arise as a degradation product during isolation of vitamin K_2.[257, 258] Synthesis of vitamin K_2 has been demonstrated in many species of bacteria, including aerobic bacilli such as *Bacillus brevis*,[259] and in *Staphylcoccus aureus*, *Sarcina lutea*, *Escherichia coli*, etc.[260, 261] *Mycobacterium tuberculosis* and *Mycobacterium phlei* contain surprisingly large amounts of these compounds.[256, 262] Vitamin K synthesis by the intestinal flora may contribute a considerable fraction of the host's requirement.[263]

The route of biogenesis of vitamin K_2 is unknown, although it is possible

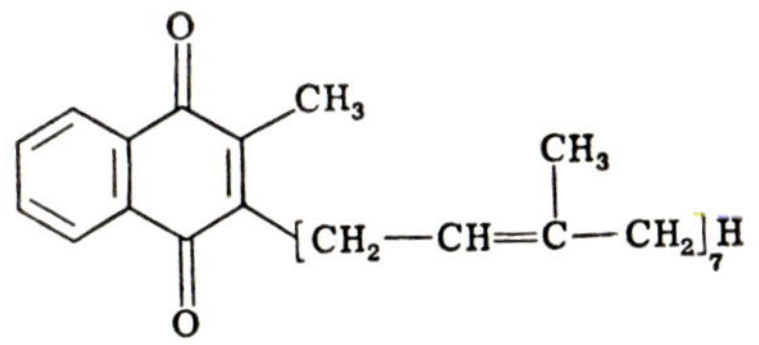

FIG. 23. Structure of vitamin $K_{2(35)}$ (2-methyl-3-all *trans*-farnesyl-geranylgeranyl-1,4-naphthoquinone).

that mevalonic acid may be a precursor of the side chain as of other isoprenoid compounds.[264] Growth of at least two bacteria, *Mycobacterium paratuberculosis* (Johne's bacillus).[265, 266] and the rumen organism, *Fusiformis nigrescens*,[267, 268] is stimulated by menadione, suggesting that these organisms may be unable to form the naphthoquinone portion of the vitamin K_2 molecule.

3. Coenzyme Q or Ubiquinone

Several compounds, isolated both from microorganisms and from tissues of higher plants and animals, possess the chemical structure shown in Fig. 24. They are derivatives of 2,3-dimethoxy-5-methylbenzoquinone carrying a polyisoprenoid side chain on position 6. Their close structural resemblance suggests a common metabolic function, thought to be concerned with oxidative and photosynthetic phosphorylation.[269] These compounds, the various forms of coenzyme Q or ubiquinone, are distinguished from each other by the length of the polyisoprenoid side chain.[270] The problem of nomenclature is not yet settled. Workers who use the generic name coenzyme Q indicate the number of isoprene units as a suffix; those who use the generic name ubiquinone record the number of carbon atoms in the side chain. Thus the compound most commonly found in the tissues of higher plants and animals is known either as coenzyme Q_{10} or as ubiquinone (50).

Only one instance of partial growth requirement for a member of this group of substances has been reported, in the case of mycobacterium for which vitamin K_1 was also growth stimulatory.[271] However, a variety of these compounds has been isolated from bacteria, e.g., CoQ_7 from *Chromatium*, CoQ_8 from *Azotobacter vinelandii*, *Escherichia coli*, and *Hydrogenomonas*, CoQ_9 from *Pseudomonas fluorescens* and *Rhodospirillum rubrum*.[272] The facultative, nonphotosynthetic aerobes so far examined, e.g., *Escherichia coli* and *Saccharomyces cerevisiae*, only form coenzyme Q when grown aerobically, and the obligate anaerobe, *Clostridium perfringens*, contains neither vitamin K nor coenzyme Q.[272]

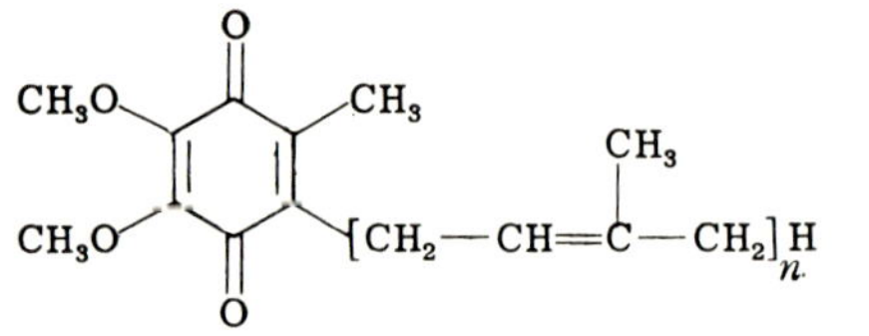

FIG. 24. Structure of coenzyme Q_n, the *n*-polyisoprenoid derivative of 2,3-dimethoxy-5-methylbenzoquinone.

One member of the coenzyme Q group formerly characterized by its absorption at 254 mμ (and somewhat confusingly called coenzyme Q_{254}), has been found only in association with the photosynthetic apparatus of algae and higher plants, and in recognition of this specificity has been given the trivial name of "plastoquinone." It is of interest therefore that, although they may contain other forms of coenzyme Q, plastoquinone has never been detected in photosynthetic bacteria.[272] Observations on the synthesis of ubiquinone (50) in animal tissues have implicated mevalonic acid as a precursor of the side chain.[273]

4. INOSITOL

Although myoinositol is a structural component of the lipids of mycobacteria[274] and is synthesized by many bacteria, e.g., *Aerobacter aerogenes*, *Proteus vulgaris*, and *Pseudomonas fluorescens*,[41] others appear neither to produce it, nor to be dependent upon its provision.[275] In fact, although it is a common growth factor for yeasts and molds, no bacterium has yet been found to demonstrate a requirement for inositol, and no coenzymic function has with certainty been attributed to it or any derivative.

Acknowledgment

The author is greatly indebted to Professor D. D. Woods, F.R.S., for guidance in the preparation of this chapter.

References

[1] J. Grimshaw and L. Marion, *Nature* **181,** 112 (1958).
[2] G. Pontecorvo, *Advances in Genetics* **5,** 141 (1953).
[3] C. Yanofsky, *J. Bacteriol.* **68,** 577 (1954).
[4] V. A. Najjar and R. Barrett, *Vitamins and Hormones* **3,** 23 (1945).
[5] S. K. Kon and J. W. G. Porter, *Vitamins and Hormones* **12,** 53 (1954).
[6] O. Mickelsen, *Vitamins and Hormones* **14,** 1 (1956).
[7] B. C. J. G. Knight, *Vitamins and Hormones* **3,** 105 (1945).
[8] D. D. Woods, *J. Gen. Microbiol.* **9,** 151 (1953).
[9] F. W. Tanner, Jr. and J. M. Van Lanen, *J. Bacteriol.* **54,** 38 (1947).
[10] J. M. Van Lanen and F. W. Tanner, Jr., *Vitamins and Hormones* **6,** 163 (1948).
[11] J. Pawelkiewicz, *Acta Biochim. Polon.* **1,** 313 (1954).
[12] J. A. MacLaren, *J. Bacteriol.* **63,** 233 (1952).
[13] W. S. McNutt, *J. Biol. Chem.* **210,** 511 (1954).
[14] T. W. Goodwin and S. Pendlington, *Biochem. J.* **57,** 631 (1954).
[14a] T. W. Goodwin and D. McEvoy, *Biochem. J.* **71,** 742 (1959).
[15] G. W. E. Plaut, *J. Biol. Chem.* **208,** 513 (1954).
[16] W. S. McNutt, Jr., *J. Biol. Chem.* **219,** 365 (1956).
[17] E. R. Stadtman, The biosynthesis and degradation of riboflavin. *In* "Symposium XI on Vitamin Metabolism." *Intern. Congr. Biochem. 4 Congr., Vienna, 1958,* p. 19.
[18] D. W. Woolley, "A Study of Antimetabolites." Wiley, New York, 1952.
[19] G. Ivánovics, *Z. physiol. Chem., Hoppe-Seyler's* **276,** 33 (1942).
[20] W. K. Maas, *J. Bacteriol.* **63,** 227 (1952).
[21] J. M. Ravel and W. Shive, *J. Biol. Chem.* **166,** 407 (1946).
[22] W. Shive and J. Macow, *J. Biol. Chem.* **162,** 451 (1946).
[23] V. H. Cheldelin and C. A. Schink, *J. Am. Chem. Soc.* **69,** 2625 (1947).
[24] D. D. Woods, *in* "Symposium on Nutrition and Growth Factors." *Intern. Congr. Microbiol., 6th Congr., Rome, 1953,* p. 1.
[25] C. Marnay, *Bull. soc. chim. biol.* **33,** 174 (1951).
[26] M. Landy, N. W. Larkum, E. J. Oswald, and F. Streightoff, *Science* **97,** 265 (1943).
[27] W. K. Maas and H. J. Vogel, *J. Bacteriol.* **65,** 388 (1953).
[28] J. Lascelles and D. D. Woods, *Brit. J. Exptl. Pathol.* **33,** 288 (1952).
[29] J. H. Mueller and A. W. Klotz, *J. Am. Chem. Soc.* **60,** 3086 (1938).
[30] F. J. Ryan, L. K. Schneider, and R. Ballentine, *J. Bacteriol.* **53,** 417 (1947).
[31] V. R. Williams and E. A. Fieger, *J. Biol. Chem.* **166,** 335 (1946).
[32] A. E. Axelrod, M. Mitz, and K. Hofmann, *J. Biol. Chem.* **175,** 265 (1948).
[33] H. K. Mitchell, *Vitamins and Hormones* **8,** 127 (1950).
[34] B. D. Davis, *J. Am. Chem. Soc.* **70,** 4267 (1948).
[35] J. Lederberg and N. Zinder, *J. Am. Chem. Soc.* **70,** 4267 (1948).
[36] E. A. Adelberg and J. W. Myers, *J. Bacteriol.* **65,** 348 (1953).
[37] B. D. Davis and E. S. Mingioli, *J. Bacteriol.* **60,** 17 (1950).
[38] H. K. Mitchell, *in* Symposium on Nutrition and Growth Factors, *Intern. Congr. Microbiol., Rome, 1953,* p. 75.
[39] A. Lwoff and M. Lwoff, *Proc. Roy. Soc.* **B122,** 352 (1937).

[40] F. A. Robinson, "The Vitamin B Complex," p. 110. Chapman and Hall, London, 1951.
[41] R. C. Thompson, *Univ. Texas Publ. No.* **4237,** 87 (1942).
[42] P. R. Burkholder and I. McVeigh, *Proc. Natl. Acad. Sci.* **28,** 285 (1952).
[43] V. A. Najjar and L. E. Holt, *Science* **98,** 456 (1943).
[44] F. A. Robinson, "The Vitamin B Complex," p. 74. Chapman and Hall, London, 1951.
[45] E. L. Tatum and T. T. Bell, *Am. J. Botany* **33,** 15 (1946).
[46] D. L. Harris, *Arch. Biochem. Biophys.* **57,** 240 (1955).
[47] W. H. Schopfer, "Plants and Vitamins," pp. 110, 114. Chronica Botanica, Waltham, Massachusetts, 1949.
[48] R. J. Williams, R. E. Eakin, E. Beerstecher, and W. Shive, "The Biochemistry of Vitamins," p. 686. Reinhold, New York, 1950.
[48a] D. E. Wright, *Nature* **183,** 262 (1959).
[48b] I. G. Leder, *Federation Proc.* **18,** 270 (1959).
[48c] G. W. Camiener and G. M. Brown, *J. Am. Chem. Soc.* **81,** 3800 (1959).
[48d] I. G. Leder, *Biochem. Biophys. Research Commun.* **1,** 63 (1959).
[48e] Y. Nose, K. Ueda, and T. Kawasaki, *Biochim. et Biophys. Acta* **34,** 277 (1959).
[48f] G. W. Camiener and G. M. Brown, *J. Biol. Chem.* **235,** 2404, 2411 (1960).
[49] D. L. Harris and J. Yavit, *Federation Proc.* **16,** p. 192 (1957).
[50] H. R. V. Arnstein and H. Margreiter, *Biochem. J.* **68,** 339 (1958).
[50a] H. Nakayama, *Vitamins* (*Kyoto*) **11,** 169 (1956).
[51] J. Bonner, *Science* **85,** 163 (1937).
[52] C. R. Harington and R. C. G. Moggridge, *Biochem. J.* **34,** 685 (1940).
[53] H. Katznelson, *J. Biol. Chem.* **167,** 615 (1947).
[54] H. Katznelson and A. G. Lochhead, *J. Bacteriol.* **53,** 83 (1947).
[55] H. Weil-Malherbe, *Biochem. J.* **33,** 1937 (1939).
[56] E. P. Steyn-Parvé, *Biochim. et Biophys. Acta* **8,** 310 (1952).
[57] O. Forsander, *Soc. Sci. Fennica, Commentationes Phy.-Math.* **19** (22), (1956).
[58] H. G. K. Westenbrink, Biochemical features of thiamine metabolism. *In* "Symposium XI on Vitamin Metabolism." *Intern. Congr. Biochem. 4th Congr. 1958,* p. 73.
[58a] N. Shimazono, Y. Mano, R. Tanaka, and Y. Kajiro, *J. Biochem.,* (*Tokyo*) **46,** 959 (1959).
[58b] Y. Kajiro and N. Shimazono, *J. Biochem.,* (*Tokyo*) **46,** 963 (1959).
[59] M. Silverman and C. H. Werkman, *Proc. Soc. Exptl. Biol. Med.* **40,** 369 (1939).
[60] H. P. Sarett and V. H. Cheldelin, *J. Biol. Chem.* **155,** 153 (1944).
[61] C. E. Lankford and P. K. Skaggs, *Arch. Biochem.* **9,** 265 (1946).
[61a] G. L. Carlson and G. M. Brown, *J. Biol. Chem.* **235,** PC3 (1960).
[62] W. Moses and M. A. Joslyn, *J. Bacteriol.* **66,** 204 (1953).
[63] R. Abderhalden, *Z. Vitamin-, Hormon- u. Fermentforsch.* **6,** 295 (1954).
[64] K. Makino, T. Kinoshita, Y. Aramaki, and S. Shintani, *Nature* **174,** 275 (1954).
[65] K. Makino and M. Koike, *Nature* **174,** 1056 (1954).
[66] D. L. Harris, *Arch. Biochem. Biophys.* **41,** 294 (1952); ibid **60,** 35 (1956).
[67] F. A. Robinson, "The Vitamin B Complex," p. 203. Chapman and Hall, London, 1951.
[68] H. P. Sarett, *J. Biol. Chem.* **162,** 87 (1946).
[69] M. I. Smith and E. W. Emmart, *J. Immunol.* **61,** 259 (1949).
[70] D. W. Woolley, *Proc. Soc. Exptl. Biol. Med.* **75,** 745 (1950).
[71] E. G. Brown, T. W. Goodwin, and S. Pendlington, *Biochem. J.* **61,** 37 (1955).
[72] G. W. E. Plaut, *J. Biol. Chem.* **211,** 111 (1954).

[73] G. W. E. Plaut, *in* "Symposium on Vitamin Metabolism," p. 20. National Vitamin Foundation, New York, 1956.
[74] T. W. Goodwin and D. H. Treble, *Biochem. J.* **70,** 14P (1958).
[75] T. W. Goodwin and D. H. Treble, *Biochem. J.* **67,** 10P (1957).
[76] T. Masuda, *Pharm. Bull., (Tokyo)* **3,** 434 (1956); ibid. **5,** 28 (1957).
[77] G. F. Maley and G. W. E. Plaut, *Federation Proc.* **17,** p. 268 (1958).
[78] H. Katagiri, I. Takeda, and K. Imai, *J. Vitaminol., (Osaka)* **4,** 207, 211 (1958).
[78a] G. F. Maley and G. W. E. Plaut, *J. Biol. Chem.* **234,** 641 (1959).
[79] S. Kuwada, T. Masuda, T. Kishi, and M. Asai, *J. Vitaminol., (Osaka)* **4,** 217 (1958).
[80] T. Masuda, *Pharm. Bull., (Tokyo)* **5,** 136 (1957).
[81] H. Katagiri, I. Takeda, and K. Imai, *J. Vitaminol., (Osaka)* **4,** 278, 285 (1958).
[81a] G. F. Maley and G. W. E. Plaut, *J. Am. Chem. Soc.* **81,** 2025 (1959).
[81b] G. W. E. Plaut and P. L. Broberg, *J. Biol. Chem.* **219,** 131 (1956).
[81c] G. W. E. Plaut, *J. Biol. Chem.* **235,** PC41 (1960).
[81d] H. Katagiri, I. Takeda, and K. Imai, *J. Vitaminol., (Osaka)* **5,** 81, 287 (1959).
[82] T. Masuda, *Pharm. Bull., (Tokyo)* **5,** 598 (1957).
[83] H. S. Forrest and W. S. McNutt, *J. Am. Chem. Soc.* **80,** 739 (1958).
[84] T. Masuda, T. Kishi, and M. Asai, *Pharm. Bull., (Tokyo)* **6,** 291 (1958).
[85] G. W. E. Plaut and G. F. Maley, *Arch. Biochem. Biophys.* **80,** 219 (1959).
[85a] G. W. E. Plaut and G. F. Maley, *J. Biol. Chem.* **234,** 3010 (1959).
[86] W. S. McNutt and H. S. Forrest, *J. Am. Chem. Soc.* **80,** 951 (1958).
[86a] W. S. McNutt, *J. Am. Chem. Soc.* **82,** 217 (1960).
[87] E. B. Kearney and S. Englard, *J. Biol. Chem.* **193,** 821 (1951).
[87a] H. Katagiri, H. Yamada, and K. Imai, *J. Vitaminol., (Osaka)* **5,** 129 (1959).
[87b] H. Katagiri, H. Yamada, and K. Imai, *J. Biochem.,* **46,** 1119 (1959).
[88] H. Katagiri, H. Yamada, and K. Imai, *J. Vitaminol., (Osaka)* **5,** 307 (1959).
[88a] A. W. Schrecker and A. Kornberg, *J. Biol. Chem.* **182,** 795 (1950).
[89] D. M. Bonner and C. Yanofsky, *J. Nutrition* **44,** 603 (1951).
[90] C. E. Dalgliesh, Biosynthesis of niacin and pyridoxine. *In* "Symposium XI on Vitamin Metabolism." *Intern. Congr. Biochem. 6th Congr., Vienna, 1958,* p. 32.
[91] L. K. Georg, *Proc. Soc. Exptl. Biol. Med.* **72,** 653 (1949).
[92] W. H. Schopfer and M. L. Boss, *Helv. Physiol. et Pharmacol. Acta* **7,** C20-2 (1949); *Chem. Abstr.* **43,** 9186d (1949).
[93] D. Davis, L. M. Henderson, and D. Powell, *J. Biol. Chem.* **189,** 543 (1951).
[94] P. Ellinger and M. M. Abdel Kader, *Biochem. J.* **45,** 276 (1949).
[95] W. G. McCullogh, *Abstr. 131st Meeting Am. Chem. Soc.* p. 26C (1957).
[95a] M. V. Ortega and G. M. Brown, *J. Am. Chem. Soc.* **81,** 4437 (1959).
[95b] M. V. Ortega and G. M. Brown, *J. Biol. Chem.* **235,** 2939 (1960).
[96] T. Terroine, *Compt. rend.* **226,** 511 (1948).
[97] F. L. Crane, *Plant Physiol.* **29,** 395 (1954).
[98] A. H. Mehler, *J. Biol. Chem.* **218,** 241 (1956).
[99] A. H. Mehler and E. L. May, *J. Biol. Chem.* **223,** 449 (1956).
[100] A. Miyake, A. H. Bokman, and B. S. Schweigert, *J. Biol. Chem.* **211,** 391 (1954).
[101] O. Wiss, H. Simmer, and H. Peters, *Z. physiol. Chem., Hoppe-Seyler's* **304,** 221 (1956).
[102] J. O. Harris and F. Binns, *Nature* **179,** 475 (1957).
[103] P. Handler, Metabolism of nicotinic acid and the pyridine nucleotides. *In* "Symposium XI on Vitamin Metabolism." *Intern. Congr. Biochem., 4th Congr. Vienna, 1958,* p. 39.
[104] D. E. Hughes and D. H. Williamson, *Biochem. J.* **51,** 330 (1952).

[105] D. E. Hughes and D. H. Williamson, *Biochem. J.* **55,** 851 (1953).
[105a] J. Preiss and P. Handler, *J. Biol. Chem.* **233,** 493 (1959).
[105b] R. W. Wheat, *Arch. Biochem. Biophys.* **82,** 83 (1959).
[105c] A. Ballio and S. Russi, *Arch. Biochem. Biophys.* **85,** 567 (1959).
[106] T. P. Singer and E. B. Kearney, *Advances in Enzymol.* **15,** 79 (1954).
[107] G. Serlupi-Crescenzi and A. Ballio, *Nature* **180,** 1203 (1957).
[107a] J. Imsande, *J. Biol. Chem.* **236,** 1494 (1961).
[108] A. Kornberg, *J. Biol. Chem.* **182,** 580 (1950).
[109] E. E. Snell, *Vitamins and Hormones* **16,** 77 (1958).
[110] E. E. Snell and B. M. Guirard, *Proc. Natl. Acad. Sci. U.S.* **29,** 66 (1943).
[111] A. Cohen, J. W. Haworth, and E. G. Hughes, *J. Chem. Soc.* p. 4374 (1952).
[112] E. E. Snell, *J. Biol. Chem.* **158,** 497 (1945).
[113] W. A. Wood and I. C. Gunsalus, *J. Biol. Chem.* **190,** 403 (1951).
[114] E. E. Snell, N. S. Radin, and M. Ikawa, *J. Biol. Chem.* **217,** 803 (1955).
[115] J. G. Morris and D. D. Woods, *J. Gen. Microbiol.* **20,** 576 (1959).
[116] J. G. Morris, *J. Gen. Microbiol.* **20,** 597 (1959).
[117] W. D. Bellamy, W. W. Umbreit, and I. C. Gunsalus, *J. Biol. Chem.* **160,** 461 (1945).
[118] R. B. Beechey and F. C. Happold, *Biochem. J.* **66,** 520 (1957).
[119] A. Meister, H. A. Sober, and S. V. Trice, *J. Biol. Chem.* **189,** 577 (1951).
[120] C. F. Gunsalus and J. Tonzetich, *Nature* **170,** 162 (1952).
[121] J. Hurwitz, *Natl. Vitamin Foundation, Nutrition Symposium Ser. No.* **13,** 49 (1956).
[122] J. Hurwitz, *J. Biol. Chem.* **205,** 935 (1953).
[123] D. Hendlin, M. C. Caswell, V. J. Peters, and T. R. Wood, *J. Biol. Chem.* **186,** 647 (1950).
[124] W. K. Maas and B. D. Davis, *J. Bacteriol.* **60,** 733 (1950).
[125] W. K. Maas, The biosynthesis of pantothenic acid. *In* "Symposium XI on Vitamin Metabolism." *Intern. Congr. Biochem. 4th Congr., Vienna, 1958,* p. 161.
[126] M. Purko, W. O. Nelson, and W. A. Wood, *J. Bacteriol.* **66,** 561 (1953).
[127] M. Purko, W. O. Nelson, and W. A. Wood, *J. Biol. Chem.* **207,** 51 (1954).
[128] E. N. McIntosh, M. Purko, and W. A. Wood, *J. Biol. Chem.* **228,** 499 (1957).
[129] A. Matsuyama, *Nippon Nôgei-kagaku Kaishi* **29,** 973 (1955); *Chem. Abstr.* **51,** 11459b (1957).
[130] A. I. Virtanen and T. Laine, *Enzymologia* **3,** 266 (1937).
[131] D. Billen and H. C. Lichstein, *J. Bacteriol.* **58,** 215 (1949).
[131a] S. R. Mardeshev and R. N. Etingof, *Biokhimiya* **13,** 402 (1948).
[131b] R. A. Altenbern and H. S. Ginoza, *J. Bacteriol.* **68,** 570 (1954).
[131c] P. R. Vagelos, J. M. Earl, and E. R. Stadtman, *J. Biol. Chem.* **234,** 490 (1959).
[132] E. Roberts and H. M. Bregoff, *J. Biol. Chem.* **201,** 303 (1953).
[133] F. P. Kupiecki and M. J. Coon, *J. Biol. Chem.* **229,** 743 (1957).
[134] S. Razin, U. Bachrach, and I. Gery, *Nature* **181,** 700 (1958).
[135] L. L. Campbell, Jr., *J. Bacteriol.* **73,** 225 (1957).
[136] E. R. Stadtman, *J. Am. Chem. Soc.* **77,** 5765 (1955).
[137] W. K. Maas, *J. Biol. Chem.* **198,** 23 (1952).
[138] W. K. Maas and G. D. Novelli, *Arch. Biochem. Biophys.* **43,** 236 (1953).
[139] W. K. Maas, *Federation Proc.* **15,** p. 305 (1956).
[140] F. Lipmann, *in* "The Vitamins" (W. H. Sebrell, Jr. and R. S. Harris, eds.), Vol. 2, Chapter 11, p. 598, Academic Press, New York, 1954.
[141] G. D. Novelli, The turnover of pantothenic acid. *In* "Symposium XI on Vitamin Metabolism" *Intern. Congr. Biochem. 4th Congr., Vienna, 1958,* p. 169.
[142] W. S. Pierpoint and D. E. Hughes, *Biochem. J.* **56,** 130 (1954).
[143] G. M. Brown and E. E. Snell, *J. Am. Chem. Soc.* **75,** 2782 (1953).

[144] G. M. Brown, *J. Biol. Chem.* **226,** 651 (1957).
[145] G. B. Ward, G. M. Brown, and E. E. Snell, *J. Biol. Chem.* **213,** 869 (1955).
[146] W. S. Pierpoint, D. E. Hughes, J. Baddiley, and A. P. Mathias, *Biochem. J.* **61,** 368 (1955).
[147] G. M. Brown, M. Ikawa, and E. E. Snell, *J. Biol. Chem.* **213,** 864 (1955).
[148] T. Wieland, W. Maul, and E. F. Möller, *Biochem. Z.* **327,** 85 (1955).
[149] G. M. Brown, *Federation Proc.* **17,** p. 197 (1958).
[150] G. M. Brown, *J. Biol. Chem.* **234,** 370 (1959).
[151] G. M. Brown, *J. Biol. Chem.* **234,** 379 (1959).
[152] P. György, *in* "The Vitamins" (W. H. Sebrell, Jr. and R. S. Harris, eds.), Vol. 1, Chapter 4, p. 616. Academic Press, New York, 1954.
[153] V. du Vigneaud, K. Dittmer, E. Hague, and B. Long, *Science* **96,** 186 (1942).
[154] L. L. Campbell, Jr. and O. B. Williams, *J. Bacteriol.* **65,** 146 (1953).
[155] R. E. Eakin and E. A. Eakin, *Science* **96,** 187 (1942).
[156] L. D. Wright, E. L. Cresson, and C. A. Driscoll, *Proc. Soc. Exptl. Biol. Med.* **89,** 234 (1955).
[157] L. D. Wright and C. A. Driscoll, *J. Am. Chem. Soc.* **76,** 4999 (1954).
[158] D. S. Genghof, *Arch. Biochem. Biophys.* **62,** 63 (1956).
[159] D. W. Woolley, *J. Biol. Chem.* **183,** 495 (1950).
[160] H. C. Lichstein, *Arch. Biochem. Biophys.* **58,** 423 (1955).
[161] K. Dittmer, D. B. Melville, and V. du Vigneaud, *Science* **99,,** 203 (1944).
[162] J. L. Stokes and M. Gunness, *J. Biol. Chem.* **157,** 121 (1945).
[163] L. H. Leonian and V. G. Lilly, *J. Bacteriol.* **49,** 291 (1945).
[164] E. L. Tatum, *J. Biol. Chem.* **160,** 455 (1945).
[165] L. L. Rogers and W. Shive, *J. Biol. Chem.* **169,** 57 (1947).
[166] H. Pope, *J. Bacteriol.* **63,** 39 (1952).
[167] V. G. Lilly and L. H. Leonian, *Science* **99,** 205 (1944).
[168] K. Hofmann and T. Winnick, *J. Biol. Chem.* **160,** 449 (1945).
[169] A. E. Axelrod, B. C. Flinn, and K. Hofmann, *J. Biol. Chem.* **169,** 195 (1947).
[170] K. K. Krueger and W. H. Peterson, *J. Bacteriol.* **55,** 693 (1948).
[171] J. P. Bowden and W. H. Peterson, *J. Biol. Chem.* **178,** 533 (1949).
[172] K. Hofmann, D. F. Dickel and A. E. Axelrod, *J. Biol. Chem.* **183,** 481 (1950).
[173] H. C. Lichstein, *J. Biol. Chem.* **212,** 217 (1955).
[174] R. W. Thoma and W. H. Peterson, *J. Biol. Chem.* **210,** 569 (1954).
[175] L. D. Wright, E. L. Cresson, H. R. Skeggs, T. R. Wood, R. L. Peck, D. E. Wolf, and K. Folkers, *J. Am. Chem. Soc.* **74,** 1996 (1952).
[176] R. L. Peck, D. E. Wolf, and K. Folkers, *J. Am. Chem. Soc.* **74,** 1999 (1952).
[177] P. György *in* "The Vitamins" (W. H. Sebrell, Jr. and R. S. Harris, eds.), Vol. 1, Chapter 4. p. 553. Academic Press. New York, 1954.
[178] S. Aaronson, *J. Bacteriol.* **69,** 67 (1955).
[179] L. D. Wright, E. L. Cresson, J. Valiant, D. E. Wolf, and K. Folkers, *J. Am. Chem. Soc.* **76,** 4163 (1954).
[180] D. B. Melville, D. S. Genghof, and J. M. Lee, *J. Biol. Chem.* **208,** 503 (1954).
[181] V. R. Williams and S. E. Cauthen, *Federation Proc.* **16,** p. 271 (1957).
[182] D. D. Woods, *Brit. J. Exptl. Pathol.* **21,** 74 (1940).
[183] B. D. Davis, *Harvey Lectures, Ser.* **50,** 230 (1956).
[183a] B. Weiss and P. R. Srinivasan, *Proc. Natl. Acad. Sci.* **45,** 1491 (1959).
[184] D. D. Woods, The biosynthesis and breakdown of folic acid. *In* "Symposium XI on Vitamin Metabolism." *Intern. Congr. Biochem., 4th Congr., Vienna, 1958,* p. 87.
[185] R. H. Nimmo-Smith, J. Lascelles, and D. D. Woods, *Brit. J. Exptl. Pathol.* **29,** 264 (1948).

[186] D. D. Woods, *Ann. N.Y. Acad. Sci.* **52,** 1199 (1950).
[187] F. Weygand, A. Wacker, A. Trebst, and O. P. Swoboda, *Z. Naturforsch.* **11b,** 689 (1956).
[188] A. Wacker, M. Ebert, and K. Holm, *Z. Naturforsch.* **13b,** 141 (1958).
[189] J. L. Stokes and A. Larsen, *J. Bacteriol.* **50,** 219 (1945).
[190] N. Katunuma and T. Shoda, *Kôso Kagaku Shimpojiumu* **12,** 124 (1957); *Chem. Abstr.* **52,** 6488d (1958).
[191] N. Katunuma, T. Shoda, and H. Noda, *J. Vitaminol.,* (*Osaka*) **3,** 77 (1957).
[192] F. Korte, H. Weitkamp, and H. G. Schicke, *Ber. deut. chem. Ges.* **90,** 1100 (1957).
[193] K. Ishii and M. G. Sevag, *Bacteriol. Proc.* (*Soc. Am. Bacteriologists*) p. 71 (1957).
[193a] F. Korte, H. Barkemyer, and G. Synnatschke, *Z. physiol. Chem., Hoppe-Seyler's* **314,** 106 (1959).
[194] T. Shiota, *Arch. Biochem. Biophys.* **80,** 155 (1959).
[194a] G. M. Brown, *Federation Proc.* **18,** 19 (1959).
[194b] G. M. Brown, *Physiol. Revs.* **40,** 331 (1960).
[194c] T. Shiota and M. N. Disraely, *Bacteriol. Proc.* (*Soc. Am. Bacteriologists*) p. 174 (1960).
[195] R. B. Angier, E. L. R. Stokstad, J. H. Mowat, B. L. Hutchings, J. H. Boothe, C. W. Waller, J. Semb, Y. SubbaRow, D. B. Cosulich, M. J. Fahrenbach, M. E. Hultquist, E. Kuh, E. H. Northey, D. R. Seeger, J. P. Sickels, and J. M. Smith, *J. Am. Chem. Soc.* **70,** 25 (1948).
[196] H. S. Forrest and J. Walker, *Nature* **161,** 721 (1948).
[197] A. Albert, *Biochem. J.* **65,** 124 (1957).
[198] I. Ziegler-Günder, H. Simon, and A. Wacker, *Z. Naturforsch.* **11b,** 82 (1956).
[199] M. J. Cross, *Abstr. Proc. Intern. Congr. Microb., 6th Congr., Rome, 1953* p. 121.
[200] B. E. Wright, *J. Am. Chem. Soc.* **77,** 3930 (1955).
[201] B. E. Wright, Folic acid coenzyme forms and function. *In* "Symposium XI on Vitamin Metabolism." *Intern. Congr. Biochem., 4th Congr., Vienna, 1958,* p. 266.
[202] H. R. Whiteley, M. J. Osborn, and F. M. Huennekens, *J. Am. Chem. Soc.* **80,** 759 (1958).
[203] D. D. Woods, *in* "Chemistry and Biology of Pteridines" (G. E. W. Wolstenholme and M. P. Cameron, eds.), pp. 220–236, 300. Churchill, London, 1954.
[203a] F. M. Huennekens and M. J. Osborn, *Advances in Enzymol.* **21,** 369 (1959).
[204] B. E. Wright and M. L. Anderson, *Biochim. et Biophys. Acta* **28,** 370 (1958).
[205] B. E. Wright, M. L. Anderson, and E. C. Herman, *J. Biol. Chem.* **230,** 271 (1958).
[206] S. Futterman, *J. Biol. Chem.* **228,** 1031 (1957).
[207] J. M. Peters and D. M. Greenberg, *Nature* **181,** 1669 (1958).
[208] J. M. Peters and D. M. Greenberg, *Biochim. et Biophys. Acta* **32,** 273 (1959).
[209] M. J. Osborn and F. M. Huennekens, *J. Biol. Chem.* **233,** 969 (1958).
[210] M. J. Osborn, M. Freeman, and F. M. Huennekens, *Proc. Soc. Exptl. Biol. Med.* **97,** 429 (1958).
[211] J. C. Rabinowitz and W. E. Pricer, Jr., *Federation Proc.* **17,** p. 293 (1958).
[212] G. R. Greenberg and L. Jaenike, *in* "The Chemistry and Biology of Purines" (G. E. W. Wolstenholme and C. M. O'Connor, eds.), p. 204. Churchill, London, 1957.
[213] J. E. Ford and S. H. Hutner, *Vitamins and Hormones* **13,** 101 (1955).
[213a] T. C. Stadtman, *J. Bacteriol.* **79,** 904 (1960).
[214] S. K. Kon and J. Pawelkiewicz, Biosynthesis of vitamin B_{12} analogues. *In* "Symposium XI on Vitamin Metabolism." *Intern. Congr. Biochem. 4th Congr., Vienna, 1958,* p. 115.
[214a] H. A. Barker, H. Weissbach, and R. D. Smyth, *Proc. Natl. Acad. Sci. U.S.* **44,** 1093 (1958).

[214b] H. A. Barker, R. D. Smyth, H. Weissbach, A. Munch-Petersen, J. I. Toohey, J. N. Ladd, B. E. Volcani, and R. M. Wilson, *J. Biol. Chem.* **235,** 181 (1960).
[214c] H. Weissbach, J. N. Ladd, B. E. Volcani, R. D. Smyth, and H. A. Barker, *J. Biol. Chem.* **235,** 1462 (1960).
[214d] J. N. Ladd, H. P. C. Hogenkamp, and H. A. Barker, *Biochem. Biophys. Research Commun.* **2,** 143 (1960).
[215] J. E. Ford and J. W. G. Porter, *Brit. J. Nutrition* **7,** 326 (1953).
[215a] H. Weissbach, J. Toohey, and H. A. Barker, *Proc. Natl. Acad. Sci.* **45,** 521 (1959).
[215b] H. A. Barker, R. D. Smyth, H. Weissbach, J. I. Toohey, J. N. Ladd, and B. E. Volcani, *J. Biol. Chem.* **235,** 480 (1960).
[215c] D. Perlman, *Advances in Appl. Microb.* **1,** 87 (1959).
[216] J. E. Ford and E. S. Holdsworth, *Biochem. J.* **56,** xxxv (1954).
[216a] D. Perlman and J. M. Barrett, *J. Bacteriol.* **78,** 171 (1959).
[217] B. H. Peterson, B. Hall, and O. D. Bird, *J. Bacteriol.* **71,** 91 (1956).
[218] J. Pawelkiewicz and K. Zodrow, *Acta Biochim. Polon.* **4,** 203 (1957).
[219] J. Pawelkiewicz and K. Nowakowska, *Acta Biochim. Polon.* **2,** 259 (1955).
[220] J. Pawelkiewicz and K. Zodrow, *Acta Biochim. Polon.* **3,** 225 (1956).
[220a] G. V. Pronyakova, *Doklady Akad. Nauk. S.S.S.R.* **123,** 331 (1958).
[221] J. W. Corcoran and D. Shemin, *Biochim. et Biophys. Acta* **25,** 661 (1957).
[222] S. Schwartz, K. Ikeda, I. M. Miller, and C. J. Walton, *Science* **129,** 40 (1959).
[223] R. Bray and D. Shemin, *Biochim. et Biophys. Acta* **30,** 647 (1958).
[224] A. I. Krasna, C. Rosenblum, and D. B. Sprinson, *J. Biol. Chem.* **225,** 745 (1957).
[224a] T. Muto, K. Kawasaki, M. Koyama, H. Yamagami, and Y. Sahashi, *Bull. Agr. Chem. Soc. Japan* **22,** 437 (1958).
[225] H. Dellweg, E. Becher, and K. Bernhauer, *Biochem. Z.* **327,** 422 (1956).
[226] H. Dellweg and K. Bernhauer, *Arch. Biochem. Biophys.* **69,** 74 (1957).
[227] A. Di Marco, G. Boretti, A. Migliacci, P. Julita, and A. Minghetti, *Boll soc. ital. biol. sper.* **33,** 1513 (1957).
[227a] R. Barchelli, G. Boretti, A. Di Marco, P. Julita, A. Migliacci, A. Minghetti, and C. Spalla, *Biochem. J.* **74,** 382 (1960).
[227b] G. Baretti, A. Di Marco, L. Fuoco, M. P. Marnati, A. Migliacci, and C. Spalla, *Biochim. et Biophys. Acta* **37,** 379 (1960).
[228] M. Juillard, *Zentr. Bakteriol. Parasitenk. Abt. II,* **110,** 701 (1957).
[229] H. Dellweg, E. Becher, and K. Bernhauer, *Biochem. Z.* **328,** 81 (1956).
[229a] R. O. Brady and H. A. Barker, *Biochem. Biophys. Research Commun.* **4,** 464 (1961).
[229b] H. Weissbach, B. Redfield, and A. Peterkofsky, *J. Biol. Chem.* **236,** PC40 (1961).
[229c] A. Peterkofsky, B. Redfield, and H. Weissbach, *Biochem. Biophys. Research Commun.* **5,** 213 (1961).
[230] R. L. Kisliuk and D. D. Woods, *J. Gen. Microbiol.* **18,** xxv (1958).
[231] D. D. Woods, Vitamin B_{12} in the synthesis of methionine by *Escherichia coli*. *In* "Symposium XI on Vitamin Metabolism." *Intern. Congr. Biochem. 4th Congr. Vienna, 1958* p. 302.
[232] J. R. Guest, *Biochem. J.* **72,** 5P (1959).
[233] L. J. Reed, *Advances in Enzymol.* **18,** 319 (1957).
[234] E. L. R. Stokstad, C. E. Hoffmann, and M. Belt, *Proc. Soc. Exptl. Biol. Med.* **74,** 571 (1950).
[235] V. L. Lytle and D. J. O'Kane, *J. Bacteriol.* **61,** 240 (1951).
[236] L. Kline, L. Pine, I. C. Gunsalus, and H. A. Barker, *J. Bacteriol.* **64,** 467 (1952).
[237] E. E. Snell and H. P. Broquist, *Arch. Biochem.* **23,** 326 (1949).
[238] J. A. Brockman, Jr., E. L. R. Stokstad, E. L. Patterson, J. V. Pierce, and M. E. Macchi, *J. Am. Chem. Soc.* **76,** 1827 (1954).

[239] G. R. Seaman, *J. Biol. Chem.* **200,** 813 (1953).

[240] E. L. Patterson, H. P. Broquist, M. H. von Saltza, A. Albrecht, E. L. R. Stokstad, and T. H. Jukes, *Am. J. Clin. Nutrition* **4,** 269 (1956).

[241] L. Machlis, *Arch. Biochem. Biophys.* **70,** 413 (1957).

[242] I. C. Gunsalus, *in* "The Mechanism of Enzyme Action" (W. D. McElroy and H. B. Glass, eds.), p. 455. Johns Hopkins Press, Baltimore, Maryland, 1954.

[243] I. C. Gunsalus, L. S. Barton, and W. Gruber, *J. Am. Chem. Soc.* **78,** 1763 (1956).

[244] R. A. Smith, I. F. Frank, and I. C. Gunsalus, *Federation Proc.* **16,** p. 251 (1957).

[245] D. R. Sanadi, M. Langley, and F. White, *Biochim. et Biophys. Acta* **29,** 218 (1958).

[246] D. R. Sanadi, M. Langley, and R. L. Searls, *J. Biol. Chem.* **234,** 178 (1959).

[247] D. R. Sanadi, M. Langley, and F. White, *J. Biol. Chem.* **234,** 183 (1959).

[248] D. R. Sanadi and R. L. Searls, *Biochim. et Biophys. Acta* **24,** 220 (1957).

[249] L. J. Reed and B. G. DeBusk, *J. Biol. Chem.* **199,** 873, 881 (1952).

[250] L. J. Reed and B. G. DeBusk, *J. Am. Chem. Soc.* **75,** 1261 (1953).

[251] L. J. Reed, F. R. Leach, and M. Koike, *J. Biol. Chem.* **232,** 123 (1958).

[252] L. J. Reed, M. Koike, M. E. Levitch, and F. R. Leach, *J. Biol. Chem.* **232,** 143 (1958).

[252a] H. Nawa, W. T. Brady, M. Koike, and L. J. Reed, *J. Am. Chem. Soc.* **81,** 2908 (1959).

[253] G. R. Seaman, *J. Biol. Chem.* **234,** 161 (1959).

[254] D. W. MacCorquodale, L. C. Cheney, S. B. Binkley, W. F. Holcomb, R. W. McKee, S. A. Thayer, and E. A. Doisy, *J. Biol. Chem.* **131,** 357 (1939).

[255] O. Isler, R. Rüegg, L. H. Chopard dit Jean, A. Winterstein, and O. Wiss, *Helv. Chim. Acta* **41,** 786 (1958).

[256] H. Noll, *J. Biol. Chem.* **232,** 919 (1958).

[257] L. F. Fieser, W. P. Campbell, and E. M. Fry, *J. Am. Chem. Soc.* **61,** 2213 (1939).

[258] J. Francis, J. Madinaveitia, H. M. Macturk, and G. A. Snow, *Nature* **163,** 365 (1949).

[259] M. Tishler and W. L. Sampson, *Proc. Soc. Exptl. Biol. Med.* **68,** 136 (1949).

[259a] O. Isler and O. Wiss, *Vitamins and Hormones* **17,** 53 (1959).

[260] H. J. Almquist, L. F. Pentler, and F. Mecchi, *Proc. Soc. Exptl. Biol. Med.* **38,** 336 (1938).

[261] H. Dam, J. Glavind, S. Orla-Jensen, and M. D. Orla-Jensen, *Naturwissenschaften* **29,** 287 (1941).

[262] G. A. Snow, *Congr. intern. biochim., 2e Congr., Paris, 1952,* p. 95.

[263] H. J. Almquist, *Physiol. Revs.* **21,** 194 (1941).

[264] Various Authors, *in* "The Biosynthesis of Terpenes and Sterols" (G. E. W. Wolstenholme and M. O'Connor, eds.). Churchill, London, 1959.

[265] F. W. Twort and G. L. Y. Ingram, *Proc. Roy. Soc.* **84B,** 517 (1911).

[266] D. W. Woolley and J. R. McCarter, *Proc. Soc. Exptl. Biol. Med.* **45,** 357 (1940).

[267] M. Lev, *Nature* **181,** 203 (1958).

[268] M. Lev, *J. Gen. Microbiol.* **20,** 697 (1959).

[269] Y. Hatefi, *Biochim. et Biophys. Acta* **34,** 183 (1959).

[270] R. A. Morton, *Nature* **182,** 1674 (1958).

[271] J. O. Norman and R. P. Williams, *Biochem. Biophys. Research Commun.* **2,** 372 (1960).

[272] R. L. Lester and F. L. Crane, *J. Biol. Chem.* **234,** 2169 (1959).

[273] U. Gloor and O. Wiss, *Arch. Biochem. Biophys.* **83,** 216 (1959).

[274] R. J. Anderson, *J. Am. Chem. Soc.* **52,** 1607 (1930).

[275] D. W. Woolley, *J. Exptl. Med.* **75,** 277 (1942).

CHAPTER 6

Biosynthesis of Purine and Pyrimidine Nucleotides

BORIS MAGASANIK

I. Introduction 295
II. Formation of Ribose Phosphates 296
III. Biosynthesis of Purine Nucleotides 298
A. Precursors 298
B. Enzymic Reactions 299
C. Synthesis in Enterobacteriaceae 305
D. Interconversions in Enterobacteriaceae 307
E. Patterns of Utilization in Other Bacteria 312
F. Regulation 313
G. Formation of Deoxyribonucleotides 317
H. Relation to Vitamins and Amino Acids 318
IV. Biosynthesis of Pyrimidine Nucleotides 320
A. Pathway to Uridine Phosphates 320
B. Cytidine Phosphates and Deoxycytidine Phosphates 322
C. Thymidine Phosphates and 5-Hydroxymethyldeoxycytidine Phosphates 323
D. Regulation 325
E. Interconversions 327
F. Problem of an Alternative Pathway 328
G. Relation to Vitamins and Amino Acids 329
V. Conclusion 330
References 331

I. Introduction*

The purine and pyrimidine nucleotides are major constituents of bacteria. The nucleic acids, which are composed entirely of these nucleotides, account for one fifth to one quarter of the dry weight of a bacterial cell[1]; in addition, many of the coenzymes consist entirely or in part of purine or pyrimidine nucleotides.

* Abbreviations used in this chapter: RNA—ribonucleic acid; DNA—deoxyribonucleic acid; DPN, DPNH—oxidized and reduced diphosphopyridine nucleotide; TPN, TPNH—oxidized and reduced triphosphopyridine nucleotide; AMP, dAMP, ADP, ATP—adenosine-5′-phosphate, deoxyadenosine-5′-phosphate, adenosine diphosphate, adenosine triphosphate; GMP, dGMP, GDP, GTP—guanosine-5′-phosphate, deoxyguanosine-5′-phosphate, guanosine diphosphate, guanosine triphosphate; IMP, IDP, ITP—inosine-5′-phosphate, inosine diphosphate, inosine triphosphate; UMP, dUMP, UDP, UTP—uridine-5′-phosphate, deoxyuridine-5′-phosphate, uridine diphosphate, uridine triphosphate; CMP, dCMP, CDP, CTP—cytidine-5′-phosphate, deoxycytidine-5′-phosphate, cytidine diphosphate, cytidine triphosphate; dTMP, dTTP—thymidine-5′-phosphate, thymidine triphosphate; FAD—flavin adenine dinucleotide; P_i, PP_i—inorganic orthophosphate, inorganic pyrophosphate.

The major components of DNA are the deoxyribonucleotides of adenine, guanine, cytosine, and thymine. The T-even bacteriophage of *Escherichia coli* contain the deoxyribonucleotide of 5-hydroxymethylcytosine in place of that of cytosine.[2]

The major components of RNA are the ribonucleotides of adenine, guanine, cytosine, and uracil. Recently discovered minor constituents include ribonucleotides of 6-methylaminopurine, 6-hydroxy-2-methylaminopurine, 6,6-dimethylaminopurine, 2-methyladenine, 1-methylguanine, 5-methylcytosine, thymine, and finally, a ribonucleotide of uracil in which the ribosyl moiety is attached at position 5 rather than 1 of the pyrimidine ring.[3-8] These minor components appear to be constituents of the soluble RNA which accounts for about 15% of the total RNA of the cell.[9]

The intracellular pools contain the soluble mono-, di-, and triphosphates of adenosine, guanosine, cytidine, uridine, and inosine in readily measurable quantities. These compounds have been found to play the role of coenzymes in a variety of enzymic reactions. In addition, the ribonucleotides of adenine and of uracil are present as part of complex coenzymes.[10]

Many bacteria are able to produce their purine and pyrimidine nucleotides either by synthesis *de novo* or from exogenously supplied bases or nucleosides; other bacteria are incapable of synthesis *de novo* and require an exogenous supply of these compounds. The biosynthetic pathways leading from small presursors to the ribonucleotides of adenine, guanine, cytosine, and uracil, and to the deoxyribonucleotides of thymine and hydroxymethylcytosine have been almost completely elucidated in the last ten years; the pathways of synthesis of the minor constituents of the nucleic acids have not yet been explored.

II. Formation of Ribose Phosphates

The normal precursor of the sugar component of the purine and pyrimidine nucleotides is glucose. Both the ribosyl and the deoxyribosyl moieties of the nucleotides are produced by the two pathways which lead from glucose to ribose-5-P. The reactions of these pathways have been discussed in Vol. II of this treatise because of the role they play in the energy-yielding catabolism of glucose in different bacteria.[11, 12] It is, however, important to realize that one of the most essential functions of these reactions is the biosynthesis of ribose phosphates.

The reactions shown in Fig. 1 produce ribose-5-P by a direct oxidative route; the reactions shown in Fig. 2 produce ribose-5-P by a more circuitous route which does not involve any oxidative reactions. This second series of reactions converts 3 molecules of glucose to 3 molecules of ribose-5-P and one of dihydroxyacetone-3-P, which can in turn condense with another molecule of triose phosphate formed in a similar manner, to give a molecule of hexose phosphate. Consequently, the over-all conversion of glucose to ribose-5-P occurs here without the concomitant production of any other carbon compound: 5 molecules of glucose produce 6 molecules of ribose-5-P.

Tracer studies with glucose and acetate have shown that both pathways are used in *E. coli*.[13, 14] The operation of the oxidative pathway requires TPN^+ as coenzyme in the two dehydrogenation steps (Fig. 1); the operation of the nonoxidative pathway requires thiamine pyrophosphate as a coenzyme in the reactions between fructose-6-P and glyceraldehyde-3-P, and between sedoheptulose-7-P and glyceraldehyde-3-P, which are cata-

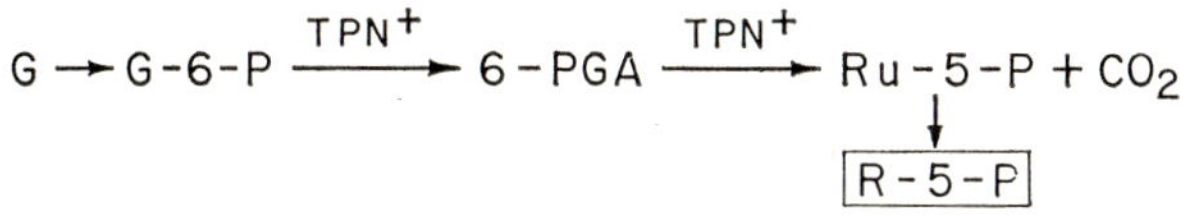

FIG. 1. Biosynthesis of ribose-5-phosphate: oxidative pathway. G, glucose; PGA, phosphogluconic acid; Ru, ribulose; R, ribose.

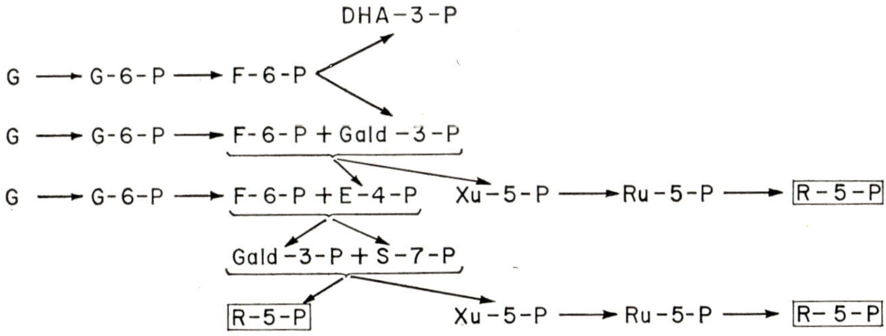

FIG. 2. Biosynthesis of ribose-5-phosphate: nonoxidative pathway. G, glucose; F, fructose; DHA, dihydroxyacetone; Gald, glyceraldehyde; E, erythrose; Xu, xylulose; Ru, ribulose; R, ribose; S, sedoheptulose.

lyzed by transketolase. It is possible that this requirement for thiamine pyrophosphate has some bearing on the rather obscure relationship of thiamine to purine biosynthesis which will be discussed later.

The ribose-5-P formed by the reactions shown in Figs. 1 and 2 can be converted by an enzyme which is found widespread in microbial and animal tissues to 5-phosphoribosyl pyrophosphate.[15]

$$\text{ribose-5-P} + \text{ATP} \rightleftharpoons \text{PP-ribose-P} + \text{AMP}$$

This phosphate ester of ribose is almost invariably used by the cell whenever a ribosylphosphate is to be attached to a nitrogen atom; it is an essential participant in the formation of the precursors of the purine and pyrimidine nucleotides and in the conversion of purine and pyrimidine bases to the corresponding nucleotides.

Another phosphate ester of ribose which may play a role in the biosynthesis of the nucleotides is ribose-1-P; this compound can react enzymically with purine bases and with uracil to form the corresponding nucleosides.[16, 17] Ribose-1-P may be formed from ribose-5-P by phosphoribomutase, an enzyme that has been found in yeast and animal tissues.[18]

The deoxyribosyl moiety of the deoxyribonucleotides is presumably formed by the reduction of a ribose derivative, and not by an independent pathway. When *E. coli* is grown on glucose labeled in carbon 1, the label is found distributed in identical patterns along the carbon chains of the ribose and deoxyribose of the nucleic acids.[13] Similar results have been obtained with *E. coli* grown on carboxyl-labeled acetate which gives rise to glucose labeled in carbons 3 and 4.[14]

It seems likely that the deoxynucleotides are formed directly by reduction from ribonucleotides; although a more detailed discussion of these conversions will be postponed until later in this chapter, the interesting involvement of vitamin B_{12} in this process may be mentioned here. It has long been known that deoxyribonucleosides can replace vitamin B_{12} in the nutrition of *Lactobacillus leichmanii*.[19] More recently it has been shown that when this organism is grown on thymidine in the absence of vitamin B_{12}, the deoxyribosyl portion of both the purine and pyrimidine deoxyribonucleotides is provided by the sugar moiety of the thymidine; in the presence of vitamin B_{12}, on the other hand, thymidine does not act as the donor of the deoxyribosyl moiety.[20] Finally it was discovered that ribose is not incorporated into the DNA of this organism when its growth is supported by deoxycytidine in the absence of vitamin B_{12}; but addition of vitamin B_{12} to the medium brings about the incorporation of ribose into DNA at a rate which indicates that this sugar is now the sole source of the deoxyribose of the DNA.[21] Vitamin B_{12} thus appears to be essential for the reduction of ribonucleotides to deoxyribonucleotides in this *Lactobacillus*. There is, however, no evidence that it plays a similar role in *E. coli*: a mutant of *E. coli* which requires vitamin B_{12} can grow without either vitamin B_{12} or deoxyribonucleosides in a medium supplemented with methionine.[22]

III. Biosynthesis of Purine Nucleotides

A. Precursors

The fundamental work on the origin of the carbon and nitrogen atoms of the purine ring was carried out in intact pigeon and in pigeon liver slices in the years 1946 to 1956 (see a recent review by Buchanan[23]). The choice of this material for the study of the problem of purine biosynthesis was astute: in birds the major excretory product of nitrogen is the purine, uric acid. Therefore, the reactions leading to the formation of the purine ring occur in bird liver at a very rapid rate, facilitating the identification of the precursors and intermediates of purine biosynthesis. It could be shown subsequently that the processes which lead to the formation of uric acid in birds are identical with those responsible for the formation of the purine ring in other tissues, where the purine nucleotides are primarily precursors of the nucleic acids. Using isotopically labeled compounds, Buchanan and his collaborators demonstrated that carbon 6 of the purine ring is derived from carbon dioxide, carbons 2 and 8 from formate, and carbons 4 and 5, and nitrogen 7, respectively, from the carboxyl carbon, the α-carbon, and the nitrogen atom of glycine; they also showed that the nitrogen of aspartic acid is the precursor of nitrogen 1, and the amide-nitrogen of glutamine of nitrogens 3 and 9 (Fig. 3). Not only formate, but also the α-carbon of glycine, the β-carbon of serine, and the imidazole carbon 2 of histidine can serve as the precursors of purine carbons 2 and 8 in animals. It appears

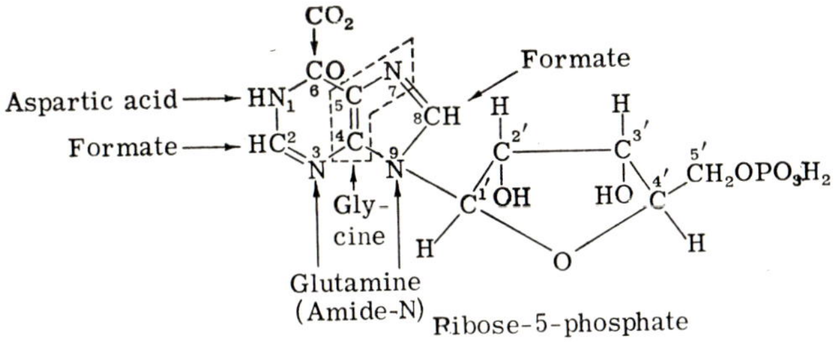

FIG. 3. Origin of purine nucleotide atoms (from Buchanan[23]).

that these carbon atoms are drawn from a pool of single carbon fragments that may receive contributions from formate, glycine, serine, and histidine.

Although capable of the synthesis *de novo* of purine nucleotides, animals as well as many species of bacteria are also capable of using preformed purine bases and nucleosides. Different organisms vary greatly in their ability to use and to interconvert these exogenously supplied purine derivatives.

B. Enzymic Reactions

The key finding that permitted the eventual identification of the intermediates in purine synthesis was the observation that *E. coli* whose growth was partly inhibited by sulfanilamide excreted 4-amino-5-imidazolecarboxamide into the culture medium.[24, 25] This compound may be considered as hypoxanthine with carbon 2 missing (Fig. 4). Its formation during sulfanilamide inhibition suggested this compound to be the immediate precursor of the purine, hypoxanthine. The last step in the formation of the purine ring seemed therefore to be the addition of a single carbon unit to aminoimidazolecarboxamide.

The role of the carboxamide as a purine precursor was supported by the observation that this compound could replace a purine base as a growth factor in mutants of *Ophiostoma multiannulatum* and of *E. coli*.[26-28] However, pigeon liver homogenates capable of incorporating glycine into hypoxanthine failed to incorporate glycine into the carboxamide.[29, 30] An explanation for this failure was found in the observation of Greenberg that IMP, rather than hypoxanthine, is the first purine compound formed in the pigeon liver system.[29] Furthermore, Greenberg demonstrated that the compound actually excreted by sulfanilamide-inhibited *E. coli* is the riboside of aminoimidazolecarboxamide.[31] These findings indicated that 5-amino-1-ribosyl-4-imidazolecarboxamide-5′-P rather than the free

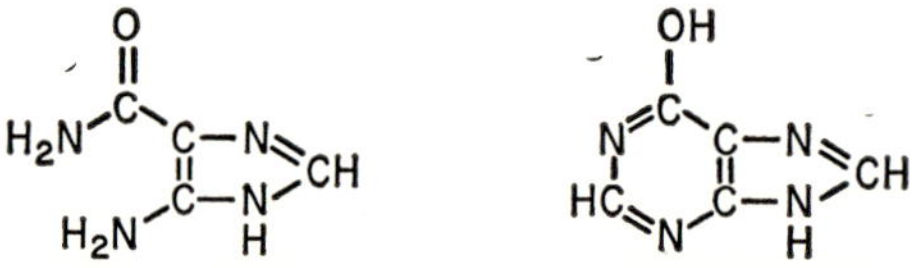

FIG. 4. 4-Amino-5-imidazolecarboxamide and hypoxanthine.

base is the actual intermediate in purine biosynthesis, and that the ribosyl phosphate moiety is introduced prior to the completion of the purine ring.

In the years following this discovery, Greenberg and his colleagues, and Buchanan and his colleagues, succeeded in elucidating all of the enzymic reactions which together can account for the biosynthesis of IMP.[23] Further enzymic reactions capable of converting IMP to AMP and to GMP were discovered by other workers.[23] The enzymes involved in the synthesis of IMP were all isolated from pigeon or chicken liver; however, some of them have been demonstrated in bacteria, and, as will be shown later, there is no doubt that except for minor details the purine nucleotides are formed by the same pathway in animals and bacteria.

This pathway leading to IMP is shown in Fig. 5, taken from the excellent review published by Buchanan in 1960.[23] This review, as well as others written by Buchanan and Hartman, should be consulted for a detailed description of the enzymic reactions involved.[32, 33] Here, it is only necessary to describe an additional recently discovered reaction which may be of importance in the synthesis of IMP, and to point out which of the enzymic steps have also been demonstrated in bacteria.

The first specific intermediate in purine biosynthesis is 5-phosphoribosylamine, which can be formed by the reaction of glutamine with 5-phosphoribosyl pyrophosphate; the enzyme which catalyzes this reaction has been found not only in animal tissues but also in extracts of *Salmonella typhimurium*.[34] However, this reaction is not the only one capable of producing this compound: 5-phosphoribosylamine can also be formed more directly from ribose-5-P and NH_3 when ATP is present.[35] The enzyme responsible for this reaction has been found in extracts of Enterobacteriaceae and of chicken liver, and can be separated from the enzyme which catalyzes the formation of 5-phosphoribosylamine from PP-ribose-P and glutamine. The activities of the two enzyme systems seem to be roughly equal in both animal and bacterial tissues. At present, the physiological significance of the existence of these two reactions leading to the same product is not clear; it is possible, as a later discussion will attempt to show, that they are essential for the precise regulation of the synthesis of AMP and GMP.

The enzymes responsible for the conversion of formylglycinamide ribose phosphate to formylglycinamidine ribose phosphate, for the cleavage of 5-amino-1-ribosyl-4-imidazolesuccinocarboxamide-5′-P, and for the conversion of 5-amino-1-ribosyl-4-imidazolecarboxamide-5′-P to IMP, have been demonstrated in extracts of Enterobacteriaceae.[36-38] They closely resemble the corresponding enzymes of bird liver. Thus, the first of these enzymes is very susceptible to inhibition by azaserine and 6-diazo-5-oxo-L-norleucine; this susceptibility accounts for the fact that azaserine is a growth inhibitor for bacteria, and that bacteria poisoned with azaserine excrete a formylglycinamide ribose derivative into the medium.[39] The enzyme which cleaves the succinocarboxamide, just as the corresponding avian enzyme, is also able to cleave adenylosuccinate.[37] Finally, in both the avian and the bacterial tissues, a single enzyme entity seems to be responsible for the formylation of the aminoimidazolecarboxamide ribose phosphate and the cyclization of the resulting formamido compound to IMP.[38]

The enzymic reactions by which AMP and GMP can be formed from IMP are shown in Fig. 6. The enzymes catalyzing these reactions have been isolated from animal tissues and from bacteria.[23] It is in the case of the xanthosine-5′-P aminase that a minor but clear difference between the animal enzyme and the bacterial enzyme was discovered: for the animal enzyme glutamine is the preferred amino donor, while for the bacterial enzyme NH_3 is the exclusive amino donor.[40-42]

The AMP formed by this series of reactions can be converted to ADP by reaction with ATP; ATP is in turn generated from ADP by the energy-yielding reactions of glycolysis and oxidative phosphorylation. The GMP is converted to GDP, and this in turn to GTP by reactions in which ATP is the phosphate donor; IMP can be converted to IDP and ITP in a similar manner. All the higher phosphates of the purine nucleotides participate

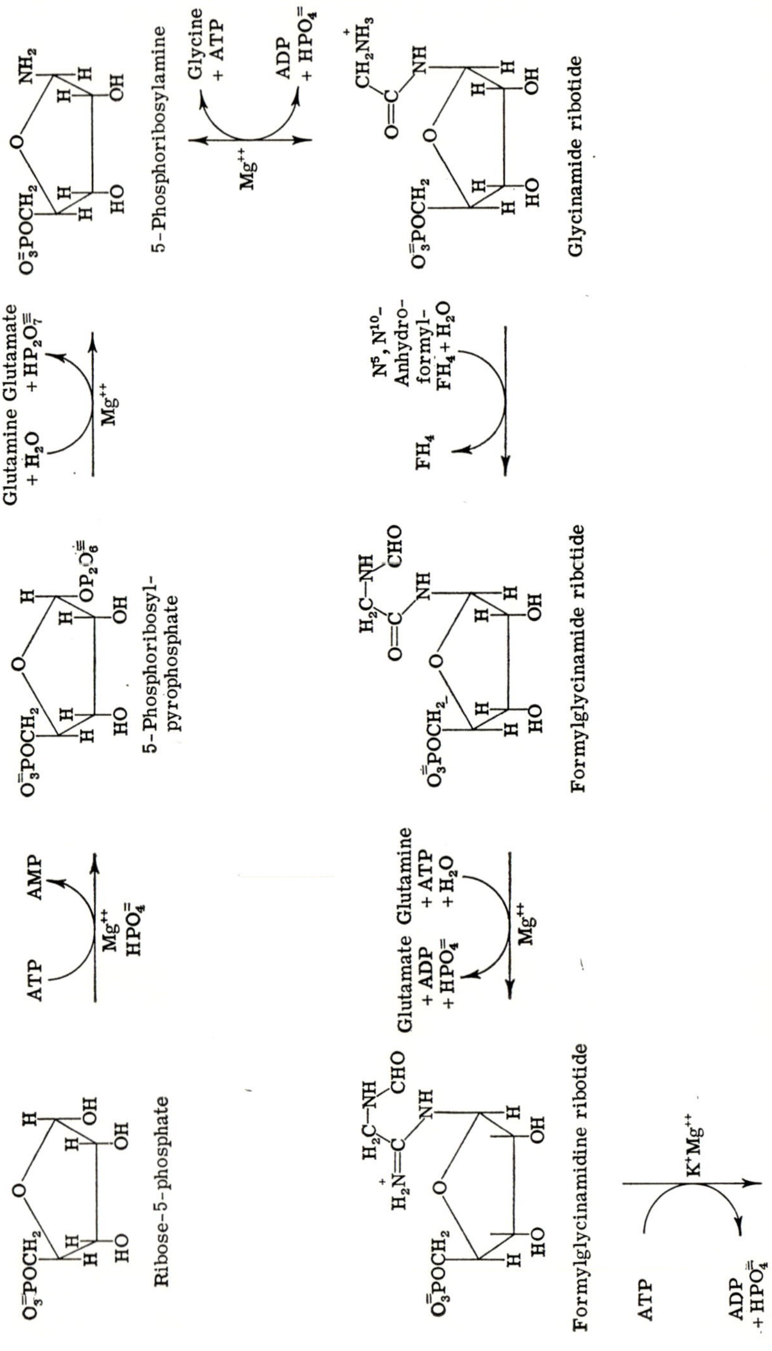
Ribose-5-phosphate
ATP
AMP
Mg++ HPO4=
5-Phosphoribosyl-pyrophosphate
Glutamine + H2O
Glutamate + HP2O7≡
Mg++
5-Phosphoribosylamine
Glycine + ATP
ADP + HPO4=
Mg++
Glycinamide ribotide
N5, N10-Anhydro-formyl-FH4 + H2O
FH4
Formylglycinamide ribotide
Glutamate + ADP + HPO4=
Glutamine + ATP + H2O
Mg++
Formylglycinamidine ribotide
ATP
ADP + HPO4=
K+Mg++

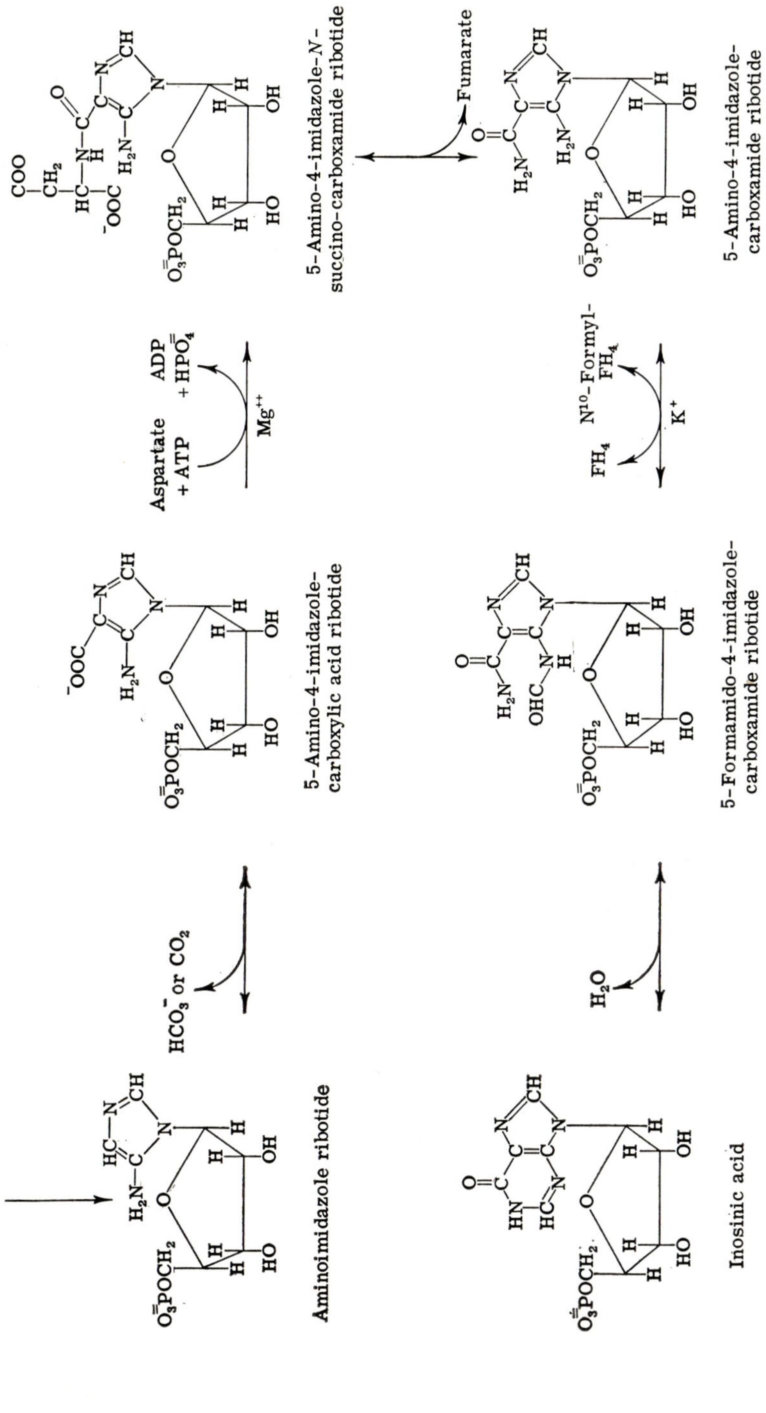

FIG. 5. Biosynthesis of inosine-5′-phosphate (from Buchanan[23]).

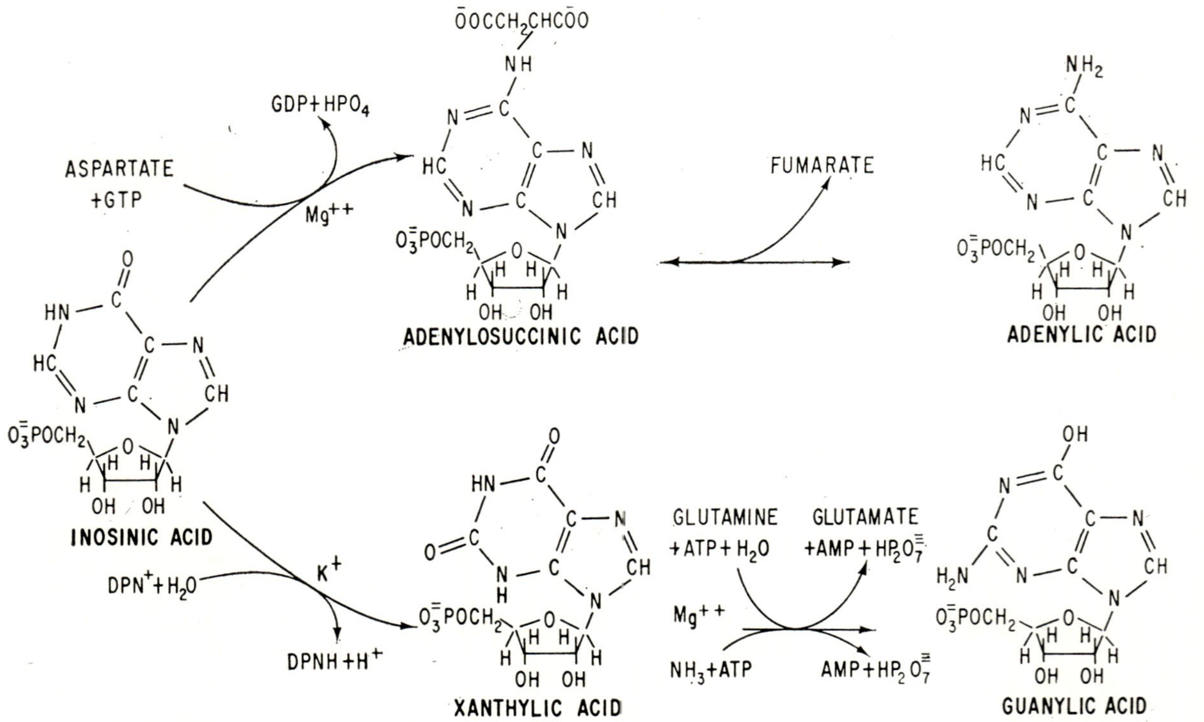

FIG. 6. Conversion of inosine-5′-phosphate to adenosine-5′-phosphate and guanosine-5′-phosphate (from Buchanan[23]).

as coenzymes in many reactions, and those of adenylic acid and guanylic acid are reactants in the synthesis of different types of RNA.[10]

The series of reactions shown in Figs. 5 and 6 can account for the synthesis *de novo* of AMP and GMP. The formation of these nucleotides by this pathway is in accord with the results of the experiments demonstrating the utilization of glycine, CO_2, formate, aspartate, and glutamine for purine biosynthesis. The conclusive evidence that this pathway is essential for purine biosynthesis rests on experiments with purine-requiring mutants of bacteria; these experiments are discussed in the next section.

C. Synthesis in Enterobacteriaceae

Enterobacteriaceae are capable of forming both AMP and GMP by synthesis *de novo*, as shown by the fact that these organisms can be cultivated in media containing glucose or related carbon compounds as only sources of carbon, and ammonia as the only source of nitrogen. It is possible to obtain auxotrophic mutants requiring purine bases for growth by the usual methods of selection in the presence of penicillin. These mutants can be divided into classes according to their response to the exogenously supplied purines.[43]

Mutants of class 1 will grow when supplied with hypoxanthine, adenine, xanthine, or guanine. Mutants of class 2 can grow on either xanthine or guanine; mutants of class 3 have an absolute requirement for guanine, and those of class 4 an absolute requirement for adenine. The conversion of the bases to the corresponding nucleotides will be discussed in the next section.

The existence of a class of mutants capable of responding to any one of the four common purines proves that Enterobacteriaceae are capable of interconverting the purine derivatives. These mutants, class 1, are unable to synthesize IMP *de novo*. They may be subdivided into several groups according to the compounds which they excrete into their culture fluids. These are 5-amino-1-ribosyl-4-imidazolecarboxamide, 5-amino-1-ribosylimidazole (both diazotizable amines which are readily differentiated by the colors produced in the Bratton-Marshall test), and compounds giving the orcinol reaction characteristic of pentoses but not containing a diazotizable amino group.[27, 38, 44-46]

The mutants which excrete the carboxamide lack the ability to convert the corresponding nucleotide to IMP.[38] It could be shown that extracts prepared from such mutants can neither formylate the aminoimidazolecarboxamide ribonucleotide nor convert the formylated compound to IMP; this finding supports the claim that a single protein may be responsible for both activities.

The mutants which excrete ribosylaminoimidazole are presumably blocked in the conversion of the corresponding phosphate ester to 5-amino-1-ribosyl-4-imidazolecarboxylate-5′-P.[45] The enzyme responsible for this reaction has not yet been demonstrated in bacterial extracts.

The mutants excreting a ribose derivative which does not contain a diazotizable amino group are presumably blocked before the aminoimidazole. One mutant of this group has been found to lack the enzyme required for the conversion of glycinamide ribose-5′-P to the corresponding amidine.[46] It is of interest that some, or perhaps all, of the mutants of this group have in addition to their purine requirement a requirement for thiamine. This seems to imply that these early steps of purine biosynthesis are also essential for the biosynthesis of thiamine.[43]

The mutants of the last two groups can use 4-amino-5-imidazolecarboxamide in place of a purine base for growth. The carboxamide must be supplied in considerably higher concentration than a purine base to allow rapid growth.[27, 28]

The excretion of these compounds, which are usually not the true nucleotide intermediates but rather the nucleosides produced from them by the action of nonspecific phosphatases, provides strong evidence that the enzymic reactions shown in Fig. 5 are indeed essential steps in the biosynthesis of IMP. This evidence is not weakened by the fact that in general the excreted compounds cannot support the growth of mutants blocked at earlier steps; the organisms do not appear to possess the enzymes required to convert the nucleosides of the IMP precursors into the corresponding nucleotides, and the nucleotides themselves may not penetrate the intact cell readily enough to serve as growth factors. The demonstration that the loss of enzymes catalyzing one of the reactions leading from 5-phosphoribosylamine to IMP is associated with a nutritional requirement for a purine base confirms the essential nature of these reactions for the biosynthesis of purine nucleotides.

Mutants of class 2 require xanthine or guanine for growth and have been shown to lack the enzyme responsible for the conversion of IMP to xanthosine-5′-P.[47] Mutants of class 3 require guanine specifically, excrete xanthosine, and lack the enzyme responsible for the conversion of xanthosine-5′-P to GMP.[40, 48] Mutants of these two classes can form AMP by synthesis *de novo*, but GMP from exogenous guanine only (class 3) or from exogenous xanthine or guanine (class 2).[49] The two enzymes, IMP dehydrogenase and xanthosine-5′-P aminase, are therefore essential for the biosynthesis of GMP as well as for the conversion of adenine or hypoxanthine to GMP; otherwise the mutants should be able to grow on exogenously supplied adenine or hypoxanthine (see Fig. 6).

Among the mutants of class 4 that have a specific requirement for adenine, there is one group which excretes 5-amino-1-ribosyl-4-imidazole-*N*-

succinocarboxamide-5′-P.[37] It has been shown that these mutants lack adenylosuccinase, the enzyme responsible for the cleavage of the succinyl derivatives of both 5-amino-1-ribosyl-4-imidazolecarboxamide-5′-P and AMP.[37] These mutants are therefore blocked in the synthesis *de novo* of IMP and in the conversion of IMP to AMP (see Figs. 5 and 6). Other mutants of class 4 fail to excrete a diazotizable amine and have been shown to lack the enzyme required for the conversion of IMP to adenylosuccinate[50]; mutants of this group can form guanine but not adenine by synthesis *de novo*. The fact that the loss of either of these enzymes results in a specific requirement for adenine proves the essential nature of these steps

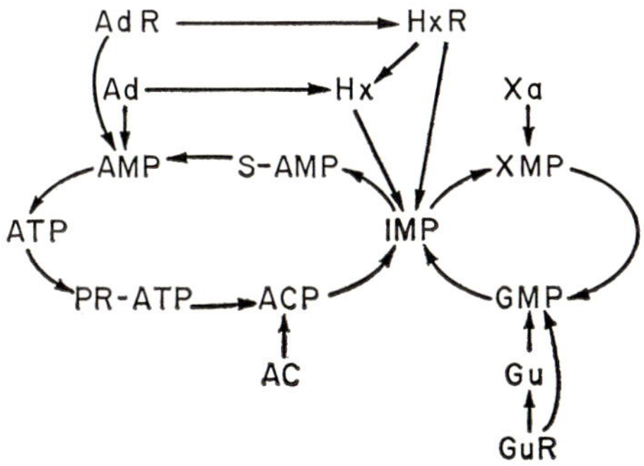

FIG. 7. Interconversion of purine bases, nucleosides, and nucleotides. Ad adenine; Gu, guanine; Hx, hypoxanthine; Xa, xanthine; R, riboside; S-AMP, adenylosuccinate; PR-ATP, 1-(5′-phosphoribosyl) adenosine triphosphate; AC, 5-amino-4-imidazolecarboxamide; ACP, 5-amino-1-ribosyl-4-imidazolecarboxamide-5′-phosphate; XMP, xanthosine-5′-phosphate.

for the production of AMP by either endogenously formed or exogenously supplied derivatives of hypoxanthine, xanthine, or guanine (see Fig. 6).

D. INTERCONVERSIONS IN ENTEROBACTERIACEAE

It has been pointed out in the preceding section that the ability of auxotrophs blocked in the pathway of IMP synthesis to grow in media supplemented with any one of the four common purine bases provides clear evidence that the Enterobacteriaceae are able to interconvert these compounds. Although many enzymes capable of catalyzing the interconversions of the purine bases, nucleosides, and nucleotides have been found in bacteria, it cannot be assumed without further evidence that they actually play such a role in the growing cell. Such evidence can be obtained through experiments with auxotrophic mutants and tagged purines; the reactions which were thus revealed as being essential for these interconversions are shown schematically in Fig. 7.

The utilization of the purine bases is a clear indication of the ability of the organisms to convert the bases to nucleotides; indeed, enzymes responsible for the conversion of the four purine bases to the corresponding ribonucleotides have been demonstrated in extracts of these organisms.[51, 52] A single enzyme appears to be responsible for the reaction between PP-ribose-P and hypoxanthine or guanine that leads to IMP or GMP, respectively. Another enzyme mediates the reaction of PP-ribose-P with adenine or 4-amino-5-imidazolecarboxamide that produces AMP or 5-amino-1-ribosyl-4-imidazolecarboxamide-5′-P, respectively; while a third enzyme mediates the reaction of PP-ribose-P with xanthine that produces xanthosine-5′-P.[32]

$$\text{purine base} + \text{PP-ribose-P} \rightleftharpoons \text{purine mononucleotide} + \text{PP}_i$$

The essential nature of these enzymes, the nucleotide pyrophosphorylases, for the utilization of the purine bases is demonstrated by the following observations. These nucleotide pyrophosphorylases convert toxic analogs of the purine bases to the corresponding nucleotides that are the actual growth inhibitors. Bacteria can mutate to resistance against such an analog by losing the appropriate nucleotide pyrophosphorylase; the fact that this loss leads simultaneously to the inability of incorporating the corresponding natural purine base demonstrates clearly that this enzymic reaction is the only mechanism by which the base can be converted to its nucleotide. Thus, mutation to resistance against 2,6-diaminopurine involves the loss of AMP pyrophosphorylase and of the ability to utilize adenine, and mutation to resistance against 6-mercaptopurine involves the loss of the IMP, GMP pyrophosphorylase and of the ability to utilize hypoxanthine and guanine.[51-55]

The conversion of xanthine and of guanine to AMP must proceed through the steps leading from IMP to AMP (Fig. 6); for, otherwise, the adenine-requiring mutants which are blocked in these steps should be able to grow on guanine or xanthine. Therefore, guanine and xanthine must first be converted to IMP. This conversion cannot take place by the reversal of the reactions leading from IMP to GMP, for these reactions are irreversible, and indeed, mutants blocked between IMP and GMP can nevertheless convert guanine to AMP.[40, 47, 49] Furthermore, although mutants blocked between IMP and xanthosine-5′-P can convert xanthine to AMP, mutants blocked between xanthosine-5′-P and GMP cannot carry out this conversion.[47, 56] It is therefore clear that GMP must be an intermediate in the conversion of xanthine to IMP, and that xanthine cannot be an intermediate in the conversion of guanine to IMP. Finally, the observation that mutants blocked between 5-amino-1-ribosyl-4-imidazolecarboxamide-5′-P and IMP are not hindered in their conversion of guanine to IMP makes it unlikely that the conversion involves the cleavage of the purine ring.[56] All these facts point to the direct formation of IMP from a derivative of guanine; in fact, an enzyme has been discovered which catalyzes the irreversible reductive deamination of GMP to IMP.[57]

$$\text{GMP} + \text{TPNH} + \text{H}^+ \rightarrow \text{IMP} + \text{TPN}^+ + \text{NH}_3$$

The loss of this enzyme by mutation makes it impossible for the cell to convert guanine to AMP;[56] thus, no other pathway for this conversion appears to exist (Fig. 7).

The conversion of adenine to GMP presumably does not involve the reversal of the steps leading from IMP to AMP (Fig. 6), for the equilibrium of these reactions strongly favors AMP synthesis, and mutants blocked in one of these reactions are nevertheless capable of converting adenine to GMP.[37, 56] On the other hand, as mentioned earlier, the steps leading from IMP to GMP (Fig. 6) are essential for the conversion of adenine to GMP, for otherwise the guanine-requiring mutants blocked in these reactions should be able to grow on adenine. It is evident that adenine is converted to GMP by way of IMP. The finding that in this conversion carbon 2 of adenine is partially lost and the amino nitrogen of adenine becomes one of the ring nitrogens of GMP provides strong evidence that a portion of the adenine is converted to GMP by way of a derivative of 5-amino-1-ribosyl-4-imidazolecarboxamide.[56] The pathway of the conversion of AMP to the carboxamide was discovered by Moyed and Magasanik and shown to be essential for the biosynthesis of histidine;[58, 59] the steps involved in this pathway have been elucidated by Ames and his co-workers and are shown in Fig. 8.[60] The products of this series of reactions are imidazoleglycerol phosphate which is an essential precursor of histidine, and 5-amino-1-ribosyl-4-imidazolecarboxamide-5′-P which can react with N^{10}-formyltetrahydrofolic acid to produce IMP (Fig. 5); IMP in turn can be converted to GMP (Figs. 6 and 7). It is apparent that in this conversion carbon 2 of adenine is transferred to histidine and thus does not appear in GMP, while the amino nitrogen of adenine becomes the amide nitrogen of the carboxamide and thus appears finally as nitrogen 1 of the guanine ring.

This pathway, however, does not alone account for the conversion of adenine to GMP, for about one half of the GMP formed has retained carbon 2 of adenine.[56] The most likely mechanism for the conversion in which the integrity of the purine ring is preserved is the direct deamination of a derivative of adenine to one of hypoxanthine. It has recently been possible to show that the enzymes which together are actually responsible for this conversion of adenine to IMP in *Salmonella typhimurium* are adenine deaminase and IMP pyrophosphorylase.[61] As mentioned earlier, bacteria can become resistant to 6-mercaptopurine through the loss of the pyrophosphorylase that catalyzes the reaction of hypoxanthine with PP-ribose-P to produce IMP[52, 55]; a mercaptopurine-resistant mutant of this kind, which is unable to convert hypoxanthine to cellular AMP and GMP, was

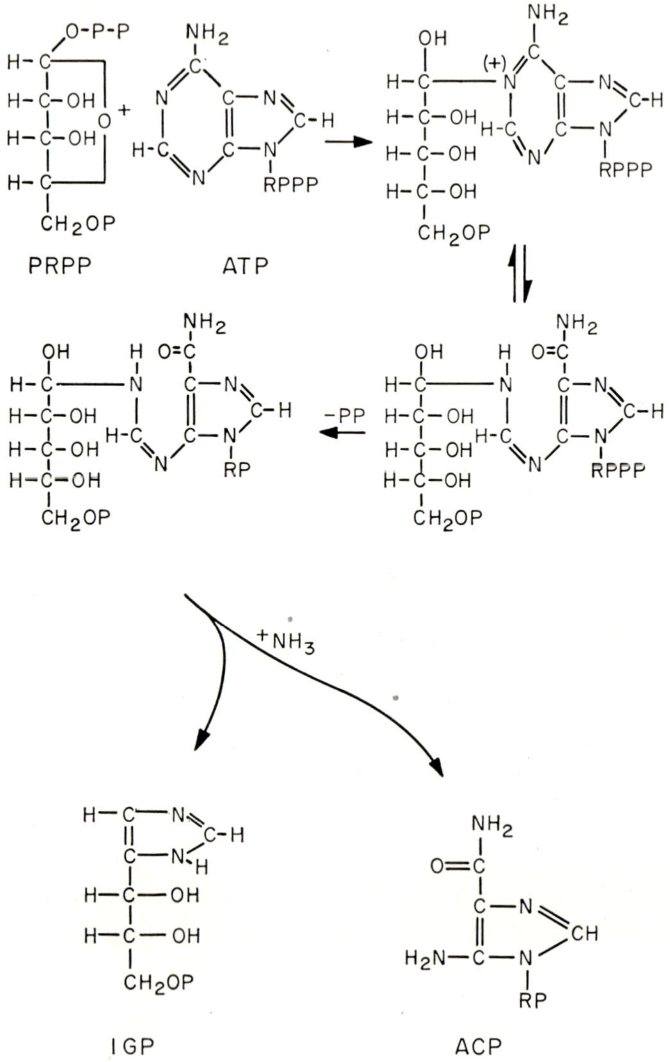

FIG. 8. Conversion of adenosine triphosphate to imidazoleglycerol phosphate (IGP) and 5-amino-1-ribosyl-4-imidazolecarboxamide-5′-phosphate (ACP).

tested for its ability to convert exogenous adenine to cellular AMP and GMP in the presence of histidine, which strongly inhibits the conversion of adenine to IMP by way of the carboxamide (see Section III, E). It was found that the mutant, in contrast to the parent strain, failed to convert exogenous adenine to GMP although it incorporated it readily into AMP. It must therefore be concluded that the IMP pyrophosphorylase, which the mutant lacks, is essential for the conversion of adenine to GMP when the pathway through the carboxamide is blocked by histidine. Under this condition adenine is converted to IMP, which in turn is converted to GMP, by the following reactions:

$$\text{adenine} \xrightarrow{H_2O} \text{hypoxanthine} \xrightarrow{\text{PP-ribose-P}} \text{IMP} + \text{PP}_i + \text{NH}_3$$

The fact that the mercaptopurine-resistant mutant can form AMP from exogenous adenine as readily as the parent organism, but nevertheless fails to convert it to GMP makes it unlikely that *S. typhimurium*, and perhaps the other Enterobacteriaceae, possess a functional AMP deaminase. It is also of interest that both the mercaptopurine-resistant mutant and the parent strain convert inosine and adenosine very readily to GMP[61]; inosine apparently can be converted to IMP without passing through hypoxanthine, and adenosine can be readily deaminated to inosine.[62] The observations that a mutant of *S. typhimurium* blocked between IMP and AMP grows well on adenine but only very poorly on adenosine, and that most of the adenosine supplied to this mutant is soon found as hypoxanthine in the medium suggest that adenosine is converted to inosine much more readily than to AMP or adenine; however, the small amount of cellular AMP formed from adenosine under these conditions contained both the purine and the ribosyl moieties of the nucleoside. Adenosine can therefore be converted directly to AMP, although at a slower rate than to inosine.[61] It has been mentioned previously that a single enzyme is responsible for the reactions of hypoxanthine and of guanine with PP-ribose-P that yield the corresponding nucleotides. The mercaptopurine-resistant mutant which lacks this enzyme can convert guanosine, but not guanine, to cellular GMP. This finding indicates that guanosine can be converted to GMP without the cleavage of the ribosidic bond.[61] It has been reported that in *E. coli* only the base moieties of the nucleosides of adenine and guanine are incorporated into nucleotides[63]; in these experiments the nucleosides were supplied at high levels and it seems likely that adenosine and guanosine were cleaved to yield hypoxanthine and guanine, respectively, which competed successfully with the nucleosides for incorporation into nucleotides. Xanthosine cannot be used at all; it is apparently neither phosphorylated to xanthosine-5′-P nor cleaved to xanthine.[64]

Other purine bases which can be converted to cellular nucleotides of adenine and guanine are isoguanine (6-amino-2-hydroxypurine) and 2,6-diaminopurine.[49, 64] Isoguanine can be deaminated to xanthine[65]; that it is incorporated into cellular AMP and GMP by way of xanthine is shown by its ability to support the growth of mutants blocked before xanthosine-5′-P but not of those blocked between xanthosine-5′-P and GMP.[64] 2,6-Diaminopurine is not always a growth inhibitor, but under appropriate conditions can be a source of guanine nucleotides. The pattern of incorporation of 2,6-diaminopurine into the cellular nucleotides closely resembles that of guanine.[49] It is apparently converted to GMP, but too slowly to support the growth of mutants in which it has to serve as the source of both of the cellular purine nucleotides; it can, however, spare the purine requirement of such mutants, and it can singly support the growth of mutants blocked only in the synthesis of GMP.[64]

E. Patterns of Utilization in Other Bacteria

Different species of microorganisms show great variation in their ability to form purine nucleotides by synthesis *de novo* and in their ability to use exogenously supplied purine bases. *Lactobacillus casei* does not require a purine base when grown in the presence of folic acid; in the absence of folic acid, its purine requirement can be met by any one of the four usual purine bases, and exogenously supplied adenine or exogenously supplied guanine is readily converted to both AMP and GMP.[66, 67] On the other hand, the yeast, *Torulopsis utilis*, which does not require a purine base for growth, incorporates exogenously supplied adenine into both AMP and GMP, but exogenously supplied guanine only into GMP.[68, 69] This observation seems to indicate that the yeast lacks the GMP-reductase that is required for the conversion of guanine to adenine nucleotides (Fig. 7).

The converse pattern is found in *Staphylococcus aureus* which also can grow in the absence of exogenously supplied purine bases; this organism incorporates exogenous guanine into both AMP and GMP but exogenous adenine only into AMP. A mutant of this organism requires adenine together with hypoxanthine, xanthine, or guanine, and is not able to convert exogenously supplied guanine to AMP.[70] The parent seems to lack the ability to convert AMP to IMP by way of the carboxamide (which is not surprising in view of the fact that these reactions are part of the mechanism of histidine biosynthesis and that staphylococci are unable to form histidine), as well as the ability to convert adenine directly to hypoxanthine. The mutant has presumably lost in addition adenylosuccinase, the enzyme whose double function is the formation of IMP and the conversion of IMP to AMP. It is evident that in this case neither adenine alone nor guanine alone would be able to meet the purine requirement.

Other microorganisms are apparently unable to synthesize purine nucleotides *de novo* and have to be supplied with a purine base. Of particular interest are *Tetrahymena geleii* and *L. leichmanii*, both organisms with a specific requirement for guanine.[71-74] It can be shown that in *T. geleii* guanine serves as the sole source of both purine nucleotides; adenine can be incorporated only into cellular AMP but not into cellular GMP.[71-73] It would appear that this organism lacks at least one, or perhaps most, of the enzymes required for the synthesis of IMP *de novo*, as well as one or both of the enzymes required for the conversion of IMP to GMP. However, it presumably possesses GMP reductase as well as the two enzymes which convert IMP to AMP.

A different pattern is found in *Corynebacterium diphtheriae* which requires either hypoxanthine alone or a mixture of adenine and guanine.[75] Apparently, this organism possesses the enzymes necessary for the conversion of hypoxanthine via IMP to AMP and to GMP, but lacks the enzymes required for the formation of IMP by synthesis *de novo*, from adenine, and from guanine.

The different patterns of purine biosynthesis and utilization which are encountered among the many species of microorganisms are all variations on a single theme. There exists but one scheme for the synthesis and interconversion of the purines; the variations result from the lack of one or more of the enzymes of the great array responsible for these reactions.

F. Regulation

The synthesis of purine nucleotides from small precursors is controlled, as are most other biosynthetic sequences, by feedback inhibition. It can be readily shown that the addition of a purine base to the culture medium of an organism capable of forming the purine nucleotides by synthesis *de novo* results in the preferential utilization of the preformed purine base.[63] Enterobacteriaceae use adenine somewhat better than guanine. Furthermore, adenine serves more readily for the synthesis of cellular AMP than of GMP, and conversely, guanine serves better for the synthesis of GMP than of AMP.[63] When both purine bases are supplied, each is used preferentially for the synthesis of the corresponding nucleotide, although some interconversion occurs.[49]

Another way to study the feedback inhibition exerted by adenine and guanine is to measure the effect of these compounds on the excretion of purine precursors or of compounds derived from them by mutants lacking one of the reactions essential for purine synthesis. Such studies have been carried out by Gots, who was able to show that the excretion of 5-amino-1-ribosyl-4-imidazolecarboxamide by a mutant blocked between the corresponding nucleotide and IMP was strongly inhibited by purine bases (see Fig. 7).[76] This inhibition also occurred in the case of a mutant with an additional amino acid requirement where the added purine, because of the lack of the amino acid, failed to restore growth. It is impossible to tell from these studies which purine is responsible for the inhibition, since the mutant is capable of interconverting the various purines.

Studies with mutants in which a second block prevented the conversion of guanine to AMP have given divergent results. Using a mutant lacking adenylosuccinase, the enzyme required both for the synthesis of the aminoimidazolecarboxamide and for the conversion of guanine derivatives via IMP to AMP (Figs. 5–7), Gots and Goldstein noted that guanine failed to halt the excretion of 5-amino-1-ribosyl-4-succinocarboxamide-5′-P, and concluded that an adenine derivative was responsible for the feedback inhibition.[77] On the other hand, Magasanik and Karibian used a mutant lacking both the inosinicase required for the conversion of aminoimidazolecarboxamide-ribose-P to IMP, and the GMP-reductase required for the conversion of GMP via IMP to AMP (Figs. 5–7), and found that guanine did inhibit the excretion of the carboxamide; they concluded that a guanine derivative is capable of exerting feedback inhibition on the synthesis *de novo* of purine nucleotides.[56] It is possible that actually both adenine and guanine derivatives play a part in this regulation. It has been reported in Section III, B that the first reaction of purine biosynthesis, the formation of 5-phosphoribosylamine, which by analogy with other systems should be the target for feedback inhibition, can be catalyzed by two separate enzymes.[35] It is possible that one of these enzymes is inhibited by an adenine derivative and the other by a guanine derivative; in this manner, purine biosynthesis would only come to a halt when both nucleotides are present in excess. However, for technical reasons it has so far not been possible to demonstrate such inhibitions of the bacterial enzymes, although the inhibitory effect of ATP on the avian enzyme responsible for the formation of 5-phosphoribosylamine from glutamine and PP-ribose-P has been reported.[78] It should be pointed out that an analogous situation, that is, the existence of two separate enzymes catalyzing the first step of a pathway that branches further on, has been discovered by Stadtman *et al.* who were able to show that one of two aspartate kinases is inhibited by the ultimate product of one branch, threonine, and the other by that of the other branch, lysine.[79]

The pivotal position in the complex array of reactions responsible for the interconversions of the purine nucleotides and for the biosynthesis of histidine is occupied by IMP (Fig. 7). The flow of nucleotides through irreversible pathways to and from IMP is controlled by histidine, GMP, and ATP, as shown in Fig. 9.[56]

The function of the purine nucleotide cycle comprising reactions 2–7 (Fig. 9) is the biosynthesis of histidine; it is a catalytic function, since each turn of the cycle produces one molecule of the histidine precursor imidazoleglycerol phosphate from PP-ribose-P, aspartate, N^{10}-formyltetrahydrofolic acid, and ammonia. No purine nucleotide is lost, and the fumarate and tetrahydrofolic acid produced can be converted again to aspartate and N^{10}-formyltetrahydrofolic acid and these can participate in another

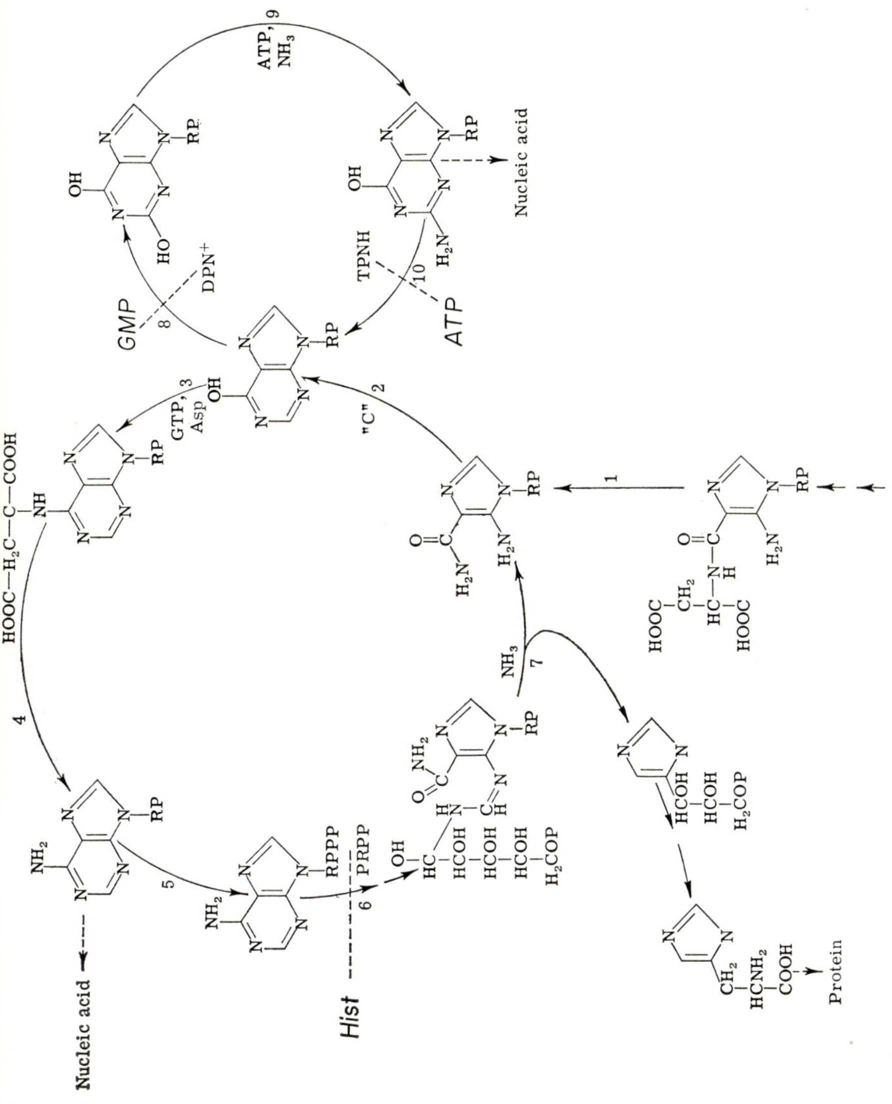

FIG. 9. Control of purine nucleotide interconversions and of histidine biosynthesis.

round of histidine biosynthesis. It is only in the case of mutants blocked between the aminoimidazolecarboxamide ribonucleotide and IMP (Fig. 9, reaction 2) that the AMP used for the formation of histidine cannot be reconverted to a purine nucleotide, and is lost as the carboxamide into the medium. It can indeed be shown that such mutants require more purine for growth than other mutants in which the cycle is not interrupted. The fact that histidine reduces the purine requirement of such mutants is explained by the inhibition which histidine exerts on the enzymic reaction of PP-ribose-P with ATP (Figs. 8 and 9). This reaction, which is the first step in the pathway specifically leading to histidine, is the target of the feedback inhibition exerted by its ultimate product.

The operation of this purine nucleotide cycle (Fig. 9, reactions 2–7) is thus controlled by histidine. When histidine is present in excess the cycle ceases to function, and consequently AMP can no longer be converted to GMP through reactions 5, 6, 7, 2, 8, and 9, as shown in Fig. 9. It has been shown in Section III, D that, under these conditions, exogenous adenine is converted to GMP exclusively via hypoxanthine and IMP while exogenous adenosine is converted to GMP via inosine and IMP; endogenous AMP apparently cannot be converted directly to IMP (Fig. 7).

The feedback effect exerted by histidine may be important not only for the control of histidine biosynthesis but also for the maintenance of the intracellular pools of adenine nucleotides. It has been found that a mutant in which the enzyme which catalyzes the formation of 5-amino-1-ribosyl-4-imidazolecarboxamide-5′-P and imidazoleglycerol phosphate (Fig. 9, reaction 7) is defective, and which consequently cannot maintain a high enough level of histidine to exert feedback inhibition, excretes the substrate of this enzyme into the medium, and thus loses enough of the endogenously produced AMP to be greatly deficient in its ability to grow. A supplement of either adenine, which restores the loss, or histidine, which prevents the loss, enables this mutant to grow.[80]

In contrast to the "big purine nucleotide cycle" (Fig. 9, reactions 2–7), the "small cycle" (Fig. 9, reactions 8–10) does not fulfill a catalytic function in biosynthesis, but appears to serve solely as a mechanism for the conversion of IMP to GMP and of GMP to IMP.[56] Theoretically, the three irreversible enzymes of this cycle could catalyze the irreversible cyclic conversion of IMP via xanthosine-5′-P and GMP to IMP, and obtain the energy for the operation of this useless merry-go-round by the splitting of ATP. Actually, the operation is controlled in such a way that AMP and GMP are produced by these reactions as needed by the cell. The conversion of IMP to xanthosine-5′-P is inhibited by GMP; consequently, IMP will not proceed to GMP when the intracellular level of this nucleotide is high, but will be free to enter the path leading to AMP. Similarly, the reduction of GMP to IMP is inhibited by ATP; consequently, GMP will replenish the cell's supply of IMP which can be converted to AMP only when the intracellular level of adenine nucleotides is low.[57]

The inhibitions exerted by histidine, ATP, and GMP on reactions essential for the interconversion of purine nucleotides have all been demonstrated directly in experiments using cell extracts.[57-60] In addition, experiments with intact cells, which have been partly outlined in this section and whose detailed description has been published, prove quite conclusively that these feedback inhibitions actually control the flow of nucleotides in the growing cell.[56]

It remains to mention other mechanisms which may also be of importance for the control of purine nucleotide metabolism. One of these may be the reciprocal role as energy donors, of GTP in the conversion of adenylosuccinate to AMP, and of ATP in the conversion of xanthosine-5′-P to GMP. This relationship may help coordinate the synthesis of the two nucleotides. Derivatives of the purine bases may also control through repression the synthesis of enzymes in their pathway. So far this has only been demonstrated for inosinicase and for IMP dehydrogenase, both of which are controlled by derivatives of guanine.[38]

G. Formation of Deoxyribonucleotides

It seems very likely, although conclusive evidence has not yet been obtained, that the purine ribonucleotides are the precursors of the purine deoxyribonucleotides. Support for this idea comes from observations showing that the pattern of incorporation of precursors into purine ribonucleotides and purine deoxyribonucleotides is the same.[13, 14] Recent experiments by Volkin with a mutant of *E. coli*, which specifically requires adenine because it lacks adenylosuccinase (see Fig. 6), suggest that AMP is a precursor of dAMP.[81] He found that this organism is unable to grow when adenine is replaced by deoxyadenosine. This is in good accord with the observation that deoxyadenosine is converted rapidly via hypoxanthine deoxyriboside to hypoxanthine[82]; the mutant blocked between IMP and AMP is thus unable to use the deoxyriboside as a source of AMP. However, when the mutant is infected with bacteriophage T2 and is incubated for a short period in a medium containing adenine (to permit the synthesis of the small amount of specific RNA which is required for the formation of the phage), it is subsequently able to synthesize the phage DNA when the adenine of the medium is replaced by deoxyadenosine. This finding indicates that AMP is normally a precursor of dAMP, and that deoxyadenosine can be converted to dAMP directly but to AMP only by way of hypoxanthine.

The mechanism responsible for the conversion of the purine ribonucleotides to deoxyribonucleotides is not known with certainty, although recent work suggests a direct reduction. A purine-requiring mutant of *Neurospora crassa* was found to incorporate uniformly labeled adenosine into both RNA and DNA; the distribution of label was essentially the same in either polymer.[83] The ribosyl and deoxyribosyl moieties of the isolated AMP, GMP, dAMP, and dGMP contained about one half the concentration of C^{14} as did the bases[83]; it seems likely that this loss of the glycosyl moieties is not an essential part of the reductive step but represents merely a side reaction of adenosine, perhaps its conversion via inosine to hypoxanthine which always occurs very readily.[62] The only direct evidence for the enzymic reduction of a purine ribonucleotide to a deoxyribonucleotide comes from experiments with extracts of 5-day-old chick embryos; these extracts were able to convert with maintenance of the glycosyl linkage GMP to dGMP in the presence of ATP.[84]

Certain lactobacilli and *E. coli* also contain trans-*N*-deoxyribosylase, an enzyme (or perhaps several enzymes with different specificities) able to catalyze the transfer of the deoxyribosyl group of pyrimidine and purine deoxyribonucleosides to guanine, adenine, hypoxanthine, thymine, uracil, and cytosine.[85, 86] There is no evidence that this enzyme plays a role in the synthesis *de novo* of purine or pyrimidine deoxyribonucleotides. Its role may be to catalyze the transfer of the deoxyribosyl moiety of exogenously supplied deoxyribonucleosides to purine or pyrimidine bases; this transfer is essential for the growth of bacteria that are unable to synthesize deoxyribonucleotides.[85]

The deoxynucleoside monophosphates of adenine and guanine can be phosphorylated to the corresponding triphosphates with ATP. The kinases which catalyze these reactions have been found in extracts of *E. coli*.[87, 88] A great increase in deoxyguanylate kinase occurs upon infection with certain phages of the T-group.[88]

H. Relation to Vitamins and Amino Acids

The importance for the biosynthesis of purine nucleotides of tetrahydrofolic acid which in many microorganisms is produced from *p*-aminobenzoic acid is well known. Two of the reactions on the pathway leading to IMP require the transfer of a formyl group from tetrahydrofolic acid to a purine precursor (Fig. 5). The two formylation reactions involve different tetrahydrofolic acid derivatives: the glycinamide is formylated by N^5,N^{10}-anhydroformylfolate-H_4, while the carboxamide is formylated by N^{10}-formylfolate-H_4.[89]

As mentioned earlier, a variety of compounds may supply the formyl group; however, it is not certain how this single carbon fragment is formed from glucose by organisms which grow in an unsupplemented medium. In bacteria, the two purine carbons derived by the transfer of a single carbon fragment, carbons 2 and 8, can be produced from carbon 2 of glycine.[90, 91] Formate and carbon 3 of serine appear to give rise more readily to carbon 8 than to carbon 2 of the purine ring.[91, 92] Carbon 2 of histidine, which is an excellent source of carbons 2 and 8 in animals, cannot be used by Enterobacteriaceae because either, like *E. coli*, they lack the ability to degrade histidine entirely, or, like *Aerobacter aerogenes*, they degrade histidine with the conversion of its carbon 2 to formamide. Pseudomonads, which degrade histidine with the conversion of its carbon 2 to formate, can incorporate carbon 2 of histidine into carbon 8 of purine but not into carbon 2.[91]

Other vitamins which play a role in purine metabolism are nicotinamide, biotin, and thiamine. Nicotinamide is a constituent of DPN^+ and TPNH which are essential for the interconversion of IMP and GMP.[47, 57]

A role for biotin is suggested by the finding that biotin-deficient yeasts excrete 5-amino-1-ribosylimidazole (see Fig. 5)[93, 94]; the excretion is prevented by aspartic acid.[95] The biotin-starved cell is presumably hampered in its ability to synthesize aspartic acid[96]; lack of aspartic acid will prevent the conversion of 5-amino-1-ribosyl-4-imidazolecarboxylate-5′-P to its product, the succinocarboxamide-5′-P (see Fig. 5), and the labile imidazole carboxylic acid will decompose and will be excreted as the aminoimidazole riboside.[32, 95]

The role of thiamine is obscure. It has already been mentioned that mutants blocked at early stages of purine synthesis generally have an absolute requirement for thiamine; in addition, thiamine can overcome the inhibitory effect exerted by adenine on the growth of many organisms.[43]

The roles of aspartic acid and of glutamine have already been mentioned. The amino nitrogen of the former is the source of nitrogen 1 and of the amino nitrogen of adenine; the amide nitrogen of the latter is the source of nitrogens 3 and 9 of the purine ring. Nitrogen 9 is perhaps also derived in part from ammonia which in bacteria is the sole source of the amino group of GMP (Fig. 6).

Glycine provides carbons 4 and 5 and nitrogen 7 of the purine ring (Fig. 3); its α-carbon may also be a source of carbons 2 and 8.

The intimate relation of purine nucleotides and histidine has been discussed in considerable detail in Sections III, D and F as well as earlier in this section. It may be added that the first evidence for a role of purine nucleotides as precursors of a portion of the histidine ring came from experiments of Broquist and Snell, who showed that histidine spared the purine requirement of *L. casei*.[97] Later, Mitoma and Snell demonstrated the incorporation of carbon 2 of guanine into histidine in this organism, and showed that folic acid was not required for this transfer but was required for the incorporation of formic acid into histidine.[98] A similar incorporation of carbon 2 of guanine into histidine was later demonstrated in *E. coli*, where it could be shown that the transfer involved not only this carbon, but also one of the attached nitrogen atoms which was subsequently identified as nitrogen 1 of the purine ring.[99-101] It is quite evident that these observations are all in perfect agreement with the results of the enzyme studies outlined in Fig. 8.

IV. Biosynthesis of Pyrimidine Nucleotides

A. Pathway to Uridine Phosphates

The enzymic reactions which are essential for the formation of the pyrimidine ribonucleotides in higher organisms and in microorganisms are all known and have recently been discussed in two reviews[102, 103]; the detailed consideration of the reactions leading to UMP, which are shown in Fig. 10, may be found there.

It is of interest that, as in the case of purine biosynthesis, a microorganism provided at an early stage in the investigation of pyrimidine biosynthesis the key to the solution of the problem: it was found that orotic acid could replace pyrimidine ribonucleosides in the nutrition of certain lactobacilli, streptococci, and mutants of *Neurospora*.[104-106] Subsequent investigations, in which mammalian tissues were used, showed clearly that these tissues were able to synthesize orotic acid from aspartic acid, and to convert orotic acid to pyrimidine nucleotides.[102] The further observations that in *Lactobacillus bulgaricus* the orotic acid required for growth could be replaced by carbamylaspartate and that both of these compounds could be converted to pyrimidine nucleotides suggested the sequence of reactions leading from asparate to orotic acid which is shown in Fig. 10.[107-109]

An enzyme capable of synthesizing carbamylphosphate from CO_2 and ATP was discovered in *Streptococcus faecalis*, as well as an enzyme capable of catalyzing the reaction of this compound with aspartate to give carbamylaspartate.[110] Corresponding enzymes have since been found in other microorganisms including *E. coli*, and in mammalian tissues where, however, acetylglutamate is required for the production of carbamylphosphate.[102, 103, 111] The aspartate transcarbamylase of *E. coli* has recently been highly purified.[112]

Enzymes linking carbamylaspartate and orotic acid were discovered by Lieberman and Kornberg in *Zymobacterium oroticum* and some corynebacteria which had been cultivated in media containing orotic acid as the major source of energy; extracts of these organisms contained enzymes that catalyzed the reversible conversions of carbamylaspartate to dihydroorotic acid and of dihydroorotic acid to orotic acid (see Fig. 10).[113-115] These reactions serve these aerobic organisms for the degradation of orotic acid to aspartate.[116] Indeed, like many other catabolic reactions, they are catalyzed by inducible enzymes which are formed in large amounts only when orotic acid is present in the growth medium.[117] Additional evidence was therefore required to show that these reactions which are able to produce orotic acid from carbamylaspartate actually serve this purpose, and

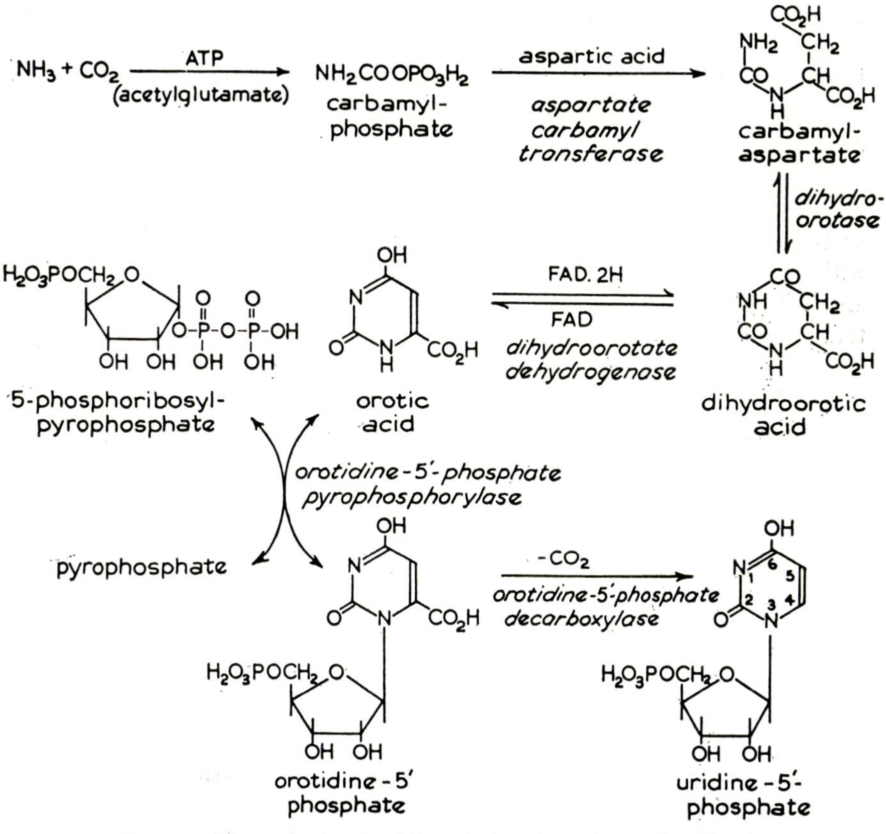

FIG. 10. Biosynthesis of uridine-5′-phosphate (from Crosbie[103]).

that they are essential for pyrimidine biosynthesis. This evidence was provided by Yates and Pardee, who showed that dihydroorotase and dihydroorotate dehydrogenase are present in extracts of *E. coli* grown in the absence of orotic acid, and that the loss by mutation of dihydroorotate dehydrogenase results in a growth requirement for uracil, cytosine, or orotic acid[117]; they also found that another pyrimidine-requiring mutant which was unable to grow on orotic acid excreted orotic acid, dihydroorotic acid, and carbamylaspartic acid into its culture fluid.[117]

The entire sequence of reactions leading from aspartate, CO_2, and ATP to orotic acid is thus seen to be essential for pyrimidine biosynthesis. However, the first of these reactions, the formation of carbamylphosphate, is also essential for the biosynthesis of arginine whose precursor, citrulline, is produced by the reaction of carbamylphosphate with ornithine. Mutants which require both a pyrimidine and arginine or citrulline for growth presumably are unable to form carbamylphosphate[118] (see Chapter 4 by Umbarger and Davis). The first reaction of the pathway leading specifically to pyrimidines, the condensation of aspartate with carbamylphosphate, is catalyzed by an enzyme quite distinct from that which catalyzes the condensation of ornithine with carbamylphosphate.[112]

The enzymes responsible for the conversion of orotic acid to UMP were discovered by Lieberman, Kornberg, and Simms in yeast. A highly specific orotidine-5′-P pyrophosphorylase catalyzes the condensation of orotic acid with PP-ribose-P to orotidine-5′-P, which is decarboxylated irreversibly by another enzyme to UMP (Fig. 10).[119] It was in connection with these studies that Kornberg and his co-workers discovered the existence of PP-ribose-P and its reaction with purine and pyrimidines to form the corresponding nucleotides.[15] The enzymes which catalyze the conversion of orotic acid to UMP have also been demonstrated in extracts of *E. coli*.[120] The loss of the decarboxylase is presumably responsible for the excretion of orotidine by a pyrimidine-requiring mutant of *Neurospora*.[121]

UMP is converted to UDP and then to UTP by reactions with ATP, catalyzed by specific kinases.[122, 123] The higher phosphates are used for the synthesis of different types of RNA and of coenzymes, and UTP is also the precursor of the cytidine phosphates.[10]

B. Cytidine Phosphates and Deoxycytidine Phosphates

An enzyme capable of converting a uridine phosphate to a cytidine phosphate was discovered by Lieberman, who showed that partially purified extracts of *E. coli* could catalyze the following reaction[124]:

$$UTP + NH_3 + ATP \rightarrow CTP + ADP + P_i$$

UTP could be replaced by UDP, although the reaction then proceeded at a lower rate; this may be explained by the presence of a nucleotide diphosphate kinase in these extracts, which catalyzed the conversion of UDP to UTP.[123] The proof that the UTP aminase is essential for the biosynthesis of cytidine phosphates would require the demonstration that its loss by mutation produces a requirement for a cytosine derivative. However, no organism with a specific requirement for cytosine or cytidine has been discovered. This may reflect the lack of enzymes capable of converting cytosine or cytidine directly to CMP; these compounds could thus provide

CMP only after prior conversion to UMP and they would therefore not be able to support the growth of mutants which had lost the UTP aminase. It is possible that CMP would support the growth of such mutants, for it has been shown that exogenous CMP gives rise to the cytidylic acid of the bacterial RNA much more readily than to the uridylic acid.[125]

A UTP aminase has also been found in animal tissues. In contrast to the *E. coli* enzyme, this UTP aminase uses glutamine rather than ammonia as the aminating agent, and, in addition, requires guanine nucleotides.[126] A recent brief report describes a similar glutamine-requiring UTP aminase in extracts of *E. coli*.[127] It is at present not certain whether *E. coli* possesses two UTP aminases or a single enzyme whose affinity for glutamine was lost in the course of the purification carried out by Lieberman.[124, 127]

CMP can be phosphorylated to CDP, and CDP to CTP by reactions with ATP catalyzed by kinases found in extracts of animal and bacterial tissues.[128-130] CMP can also be converted to deoxycytidine phosphates by enzymes recently discovered in extracts of *E. coli* which appear to catalyze the following reactions[131]:

$$\text{CMP} \xrightarrow{\text{ATP, Mg}^{++}} \text{CDP} \xrightarrow{\text{TPNH}} \text{dCDP}$$

Deoxy-CDP as well as dCMP can be converted to dCTP by the action of kinases which have been found in extracts of *E. coli*.[87, 88] The deoxycytidylate kinase activities increase greatly upon infection of the organism with phage T5, which contains deoxycytidylic acid in its DNA.[88] The enzyme activities do not increase in cells infected with T-even phages, which contain 5-hydroxymethyldeoxycytidylic acid in place of deoxycytidylic acid[88]; rather, an enzyme which hydrolyzes dCTP to dCMP and pyrophosphate is produced in organisms infected with these phages. It is presumably due to the action of this enzyme that the dCTP of the infected cell is not incorporated into phage DNA, but is instead converted to dCMP, which is the precursor of the two deoxypyrimidine nucleotides of the phage DNA.[132, 133]

All of the enzymic steps required for the synthesis *de novo* of the cytidine and deoxycytidine phosphates have thus been demonstrated in bacterial tissues. However, there is no definite evidence that the reactions leading from UMP to the cytidine phosphates are essential for the synthesis of these compounds.

C. Thymidine Phosphates and 5-Hydroxymethyldeoxycytidine Phosphates

It is reasonably certain that a derivative of uracil is the normal precursor of thymidylic acid: the thymine of the DNA of uracil-requiring mutants of *E. coli* is derived from the uracil supplied in the growth medium.[134, 135]

In addition, the enzymic conversion of dUMP to dTMP has been demonstrated in extracts of *E. coli*.[136]

The question therefore arises how ribose derivatives of uracil are converted to dUMP. A possible route to this compound is the deamination of dCMP. An enzyme capable of carrying out this deamination has been found in extracts of animal tissues and in extracts of *E. coli* infected with phage T2, but not in extracts of uninfected *E. coli*.[137-140] The lack of this enzyme in uninfected *E. coli* cells seems to militate against assigning to it a role in the normal synthesis of thymidylic acid by this organism.[139] However, it cannot be excluded that the enzyme is present in normal *E. coli* but failed to be discovered by the methods used for the demonstration of the enzyme formed after phage infection, either because it possesses properties which differ from those of the phage-induced enzyme, or simply because it is present in a much smaller amount. This is not unlikely in view of the finding that another one of the enzymes required for thymidylate synthesis, thymidylate synthetase, increases sevenfold upon infection of *E. coli* with phage T2.[140]

Another reaction which may be responsible for the formation of dUMP is the direct reduction of UMP, which seems to occur in chick embryo extracts.[141] At present, it is impossible to decide whether dUMP is formed directly from UMP or by way of CMP. The finding that in *E. coli* exogenously supplied CMP competes effectively with exogenously supplied uracil for incorporation into thymidylate, but not for incorporation into uridylate, favors the idea that CMP is an intermediate in thymidylate synthesis.[125] The entire sequence of reactions involved in the conversion of UMP to dUMP consists presumably of seven steps, although a single step is perhaps sufficient:

$$\begin{array}{ccccccc} \text{UMP} & \longrightarrow & \text{UDP} & \longrightarrow & \text{UTP} & \longrightarrow & \text{CTP} \\ \downarrow & & & & & & \downarrow \\ \text{dUMP} & \leftarrow & \text{dCMP} & \leftarrow & \text{dCDP} & \leftarrow & \text{CDP} \end{array}$$

No kinases capable of phosphorylating dUMP to dUTP have been discovered; this lack accounts for the failure of deoxyuridylic acid to appear in DNA in place of thymidylic acid.[87]

The conversion of dUMP to dTMP by extracts of *E. coli* has recently been elucidated: an enzyme, thymidylate synthetase, has been isolated which catalyzes the reaction of dUMP with 5,10-methylene tetrahydrofolic acid to give dTMP and dihydrofolic acid (Fig. 11).[136, 140, 142] The methylene group is transferred to carbon 5 of dUMP to give a hydroxymethyl derivative which is presumably reduced by tetrahydrofolic acid to dTMP. Experiments in which an extract of *S. faecalis* were used show that the dihydrofolic acid formed is reduced to tetrahydrofolic acid by TPNH.[143]

The essential role of thymidylate synthetase in dTMP synthesis has been demonstrated: mutants of *E. coli* which have lost the ability to produce this enzyme require thymine for growth.[144]

It has been mentioned earlier that infection with phage T2 results in a great increase in the level of thymidylate synthetase. The enzyme formed by the phage-infected cell may not be identical with the one produced by uninfected bacteria; its structure may be determined by a phage gene rather than by a bacterial gene. This concept is supported by the discovery that mutants which lack thymidylate synthetase acquire upon infection with phages T2 and T5 the ability to form this enzyme.[144]

A kinase capable of converting dTMP to dTTP has been demonstrated

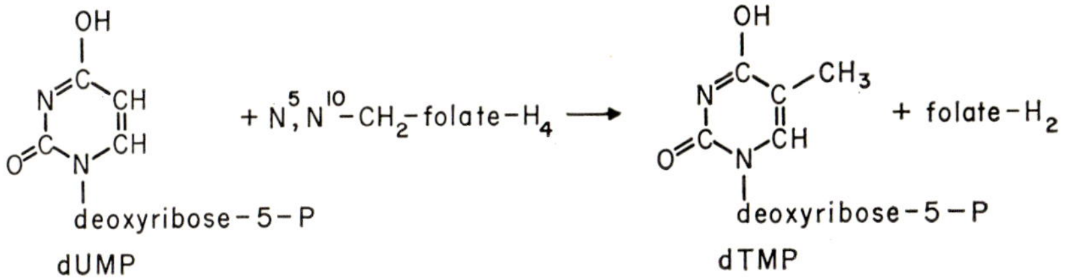

$$\text{Folate-H}_2 + \text{TPNH} + \text{H}^+ \longrightarrow \text{folate-H}_4 + \text{TPN}^+$$

Fig. 11. Conversion of deoxyuridine-5′-phosphate to thymidine-5′-phosphate.

in extracts of *E. coli*.[87, 88] The level of this enzyme is greatly increased upon infection with certain T-phages.[88]

Still another enzyme is produced by *E. coli* at a rapid rate following infection with phage T2; this is an enzyme which cannot be detected at all in uninfected cells and which is specifically required for the synthesis of 5-hydroxymethyldeoxycytidylic acid, a compound found in phage T2 DNA but not in bacterial DNA. This enzyme, deoxycytidylate hydroxymethylase, catalyzes the conversion of dCMP to 5-hydroxymethyldeoxycytidine-5′-P, as shown in Fig. 12.[145, 146] The phage-infected cell also produces a specific kinase which converts the hydroxymethyldeoxynucleotide to the corresponding triphosphate.[132, 147]

D. Regulation

The synthesis *de novo* of pyrimidine nucleotides is controlled by feedback inhibition. The excretion of orotic acid by a pyrimidine-requiring mutant of *A. aerogenes* begins immediately upon the removal of uracil from the medium; the excretion can be inhibited by the addition of uracil, even in cases where, because of another deficiency, the addition of uracil fails to restore growth. Uracil or a compound derived from it thus appears to inhibit one of the steps required for the synthesis of orotic acid.[148]

The compound responsible for this inhibition and the enzyme affected were discovered by Yates and Pardee: they found that in *E. coli* cytidine-5′-P inhibits aspartate transcarbamylase.[149] As in all cases of feedback inhibition, of which this was one of the first to be discovered, the ultimate product of the biosynthetic sequence inhibits the first enzyme of the sequence (see Fig. 10). The physiological advantage of this device is well illustrated in this case: inhibition at an earlier point, for example, inhibition of the synthesis of carbamylphosphate, would have led to an interference with arginine synthesis; inhibition at a later point, for example, inhibition of dihydroorotase, would have lead to the useless accumulation of carbamylaspartate.

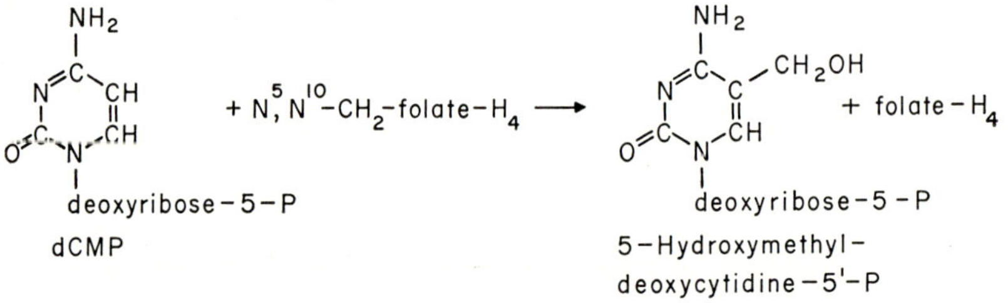

FIG. 12. Conversion of deoxycytidine-5′-phosphate to 5-hydroxymethyldeoxycytidine-5′-phosphate.

It has been pointed out in an earlier section (III, F) that in the pathway leading to the purine nucleotides IMP occupies the pivotal position, and that the many branches of the pathway leading to and from IMP are controlled by feedback inhibition. In the pathway leading to the pyrimidine nucleotides there is no pivotal position; rather, the reactions are arranged in a straight line leading from aspartic acid to UMP, from there to CMP, then to dCMP, and finally to dTMP. A linear system of this kind may perhaps be controlled quite simply by the feedback loop between a cytidine phosphate and aspartate, which is already known, and by a second feedback loop between a thymidine phosphate and a cytidine phosphate, which yet remains to be discovered.

Another regulatory mechanism in pyrimidine biosynthesis, which has also been studied by Yates and Pardee, is the repression exerted by derivatives of uracil on the formation of aspartate transcarbamylase, dihydroorotase, and dihydroorotate dehydrogenase; a uracil-requiring mutant of *E. coli*, which lacks the ability to convert orotic acid to UMP, produces these three enzymes at a rapid rate when incubated in a medium free from uracil.[150]

E. Interconversions

The ability of uracil and of cytosine, and of the corresponding nucleosides, to support the growth of mutants of Enterobacteriaceae which have lost the ability to synthesize UMP shows that these compounds can be converted to UMP. The enzyme responsible for the conversion of uracil to UMP is UMP-pyrophosphorylase, which catalyzes the condensation of uracil with PP-ribose-P.[151]

$$\text{uracil} + \text{PP-ribose-P} \rightleftharpoons \text{UMP} + \text{PP}_i$$

It has been shown that this enzyme is essential for the synthesis of UMP from uracil in *E. coli*; a mutant of this organism which has become resistant to 5-fluorouracil by the loss of the UMP-pyrophosphorylase fails to incorporate uracil into nucleic acids.[152]

In animal tissues, uracil is converted to UMP in two steps:[153-155]

$$\text{uracil} + \text{ribose-1-P} \rightleftharpoons \text{uridine} + \text{P}_i$$
$$\text{uridine} + \text{ATP} \rightarrow \text{UMP} + \text{ADP}$$

Although *E. coli* possesses the first of these enzymes and also incorporates exogenous uridine into cellular pyrimidine nucleotides without cleavage of the glycosidic bond, it is apparently unable to use these reactions for the synthesis of UMP from uracil, for otherwise the loss of UMP-pyrophosphorylase should not prevent the uptake of uracil.[63, 156, 157] The bacteria may lack the ability to produce the necessary ribose-1-P. The importance of the UMP-pyrophosphorylase for the utilization of uracil is also demonstrated by the finding that only those lactobacilli which possess this enzyme can satisfy their pyrimidine requirement with uracil; lactobacilli which require orotic acid and cannot use uracil possess the pyrophosphorylase which converts orotic acid to its nucleotide, but lack UMP-pyrophosphorylase.[152]

No enzyme capable of converting cytosine directly to CMP has been described. It is likely that the ability of many microorganisms to incorporate cytosine into pyrimidine nucleotides can be attributed to their possession of a cytosine deaminase which converts cytosine to uracil.[158] Similarly, cytidine is very readily converted to uridine by a cytidine deaminase found in extracts of *E. coli*; the same enzyme also converts deoxycytidine to deoxyuridine.[159] The only enzyme which has been shown to be capable of converting cytidine to CMP is a nucleoside phosphotransferase, which can transfer the phosphate group of 5′-nucleotides to carbon 5 of the ribosyl moiety of nucleosides.[160] It is not known whether this enzyme plays a role in the bacterial metabolism of nucleosides.

Thymine is not very well utilized by either bacteria or higher organisms. However, mutants of *E. coli* which have lost the ability to convert UMP to dTMP can grow in media supplied with thymine, and incorporate the exogenously supplied thymine into DNA; the ability to use thymine is not lost by reverse mutations restoring the ability of the mutants for synthesis of dTMP *de novo*.[161] The mechanism of the conversion of thymine to dTMP is not known; it may occur by way of thymidine, which could be produced from thymine either by reaction with deoxyribose-1-P (although it is not clear how the cell would obtain this sugar phosphate) or by the transfer of the deoxyribosyl group of another deoxyriboside to thymine by means of a trans-*N*-deoxyribosylase.[85, 86, 162]

Thymidine is readily utilized by some lactobacilli which require its deoxyribosyl moiety in media not supplemented with vitamin B_{12}.[19] In *E. coli*, thymidine is incorporated without cleavage of the glycosidic bond into the thymidylic acid of the DNA, but does not supply the deoxyribosyl moiety of the deoxycytidylic acid.[125] A kinase present in extracts of *E. coli* converts thymidine to dTMP.[163]

A recent study by Lichtenstein, Barner, and Cohen has revealed some interesting facts about the utilization of deoxycytidine and of deoxycytidylic acid by *E. coli*.[125] The pyrimidine moiety of deoxycytidine was incorporated without the deoxyribosyl moiety equally into all of the pyrimidine nucleotides of RNA and DNA. Similarly, the pyrimidine moiety of dCMP was incorporated alone into uridylic acid and, to a lesser extent, into cytidylic acid and deoxycytidylic acid; on the other hand, the pyrimidine and deoxyribosyl moieties of dCMP were incorporated together into thymidylic acid. The phosphate of exogenous dCMP was not at all incorportated into nucleic acids. The authors suggest that deoxycytidine is readily deaminated by cytidine deaminase to deoxyuridine and that deoxyuridine is cleaved by pyrimidine deoxyribosephosphorylase to uracil, which is then used for the synthesis of all the pyrimidine nucleotides. DeoxyCMP penetrates the cell slowly after dephosphorylation at the cell surface and produces intracellularly deoxycytidine which is deaminated to deoxyuridine; however, the latter, perhaps because it is present in low concentration, is converted via dUMP to dTMP faster than to uracil.

The interconversions observed in *E. coli* are shown schematically in Fig. 13.

F. Problem of an Alternative Pathway

The possibility that pyrimidine nucleotides might also be formed by a pathway not involving orotic acid was suggested by experiments with

Neurospora. It has been reported that a pyrimidine-requiring mutant of *Neurospora* could grow in media supplemented with propionic acid, L-aminobutyric acid, or dihydrouracil[164]; propionate is incorporated by this mutant into pyrimidine nucleotides, and both aminobutyrate and propionate reduce the incorporation of uracil into pyrimidine nucleotides.[165] None of these compounds is able to support the growth of pyrimidine-requiring mutants of *E. coli*.

It seems very likely, as discussed in detail in recent reviews, that the mold may be able to use dihydrouracil and propionate because it is capable of degrading uracil via dihydrouracil to β-alanine by a series of not too irreversible steps.[102, 103] Uracil catabolism by such a pathway has been dis-

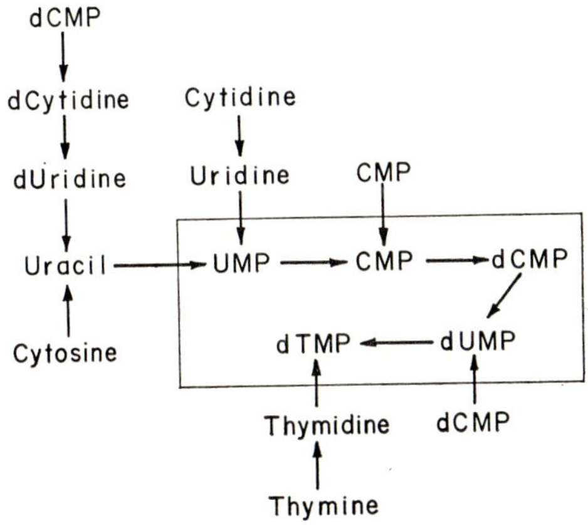

FIG. 13. Interconversion of pyrimidine bases, nucleosides, and nucleotides. The compounds enclosed in the rectangle are endogenously produced.

covered in animals and in a *Clostridium*.[102, 166] *Neurospora* may form β-alanine from propionic acid, and propionic acid from succinic acid; as succinate is a normal product of mold metabolism, this series of reactions would permit a mutant whose regular pathway of pyrimidine synthesis is blocked to form uracil nevertheless by synthesis *de novo*.[165] The fact that the mutant cannot grow unless an intermediate of this potential alternative pathway is provided in high concentration shows clearly that the enzymes of this route are not able to assume the burden of providing the mold with pyrimidine nucleotides in a quantity sufficient for growth.

G. Relation to Vitamins and Amino Acids

The importance of aspartic acid as a precursor of a major portion of the pyrimidine ring has been discussed in Section IV, A. The other amino acids which play a role in the biosynthesis of pyrimidine nucleotides are serine and glycine; they can provide the single carbon fragment necessary for the formation of thymidylic acid, but they apparently are not intermediates in the production of this carbon from glucose.[103] Arginine is indirectly related to pyrimidines by the fact that carbamylphosphate is required for its synthesis; mutants requiring both uracil and arginine for growth presumably are unable to synthesize this intermediate.

Nicotinamide and riboflavin are required for the oxidation of dihydroorotic acid to orotic acid; the enzyme contains tightly bound FAD which is reduced in the course of the reaction and is reoxidized by DPN.[103]

The formation of thymidylic acid requires the participation of a folic acid derivative[136]; consequently, organisms which lack folic acid, either because of a nutritional deficiency or because of poisoning by sulfonamides or folic acid antagonists, are unable to synthesize thymidylic acid.[102]

Vitamin B_{12} may also play a role in the formation of thymidylic acid from dUMP. It has been shown that *L. leichmanii* grown in a medium supplemented with deoxycytidine is capable of incorporating formic acid into the thymidylic acid of its DNA only when the medium also contains vitamin B_{12}[167]; serine, glycine, methionine, or formaldehyde are able to provide the methyl group of thymidylic acid in the absence of vitamin B_{12}. It seems therefore that the vitamin is essential for the reduction of a formyltetrahydrofolic acid to the N^5,N^{10}-methylene tetrahydrofolic acid which provides the methyl group of thymidylic acid.[167, 168]

V. Conclusion

During the last fifteen years all of the steps of the pathways leading to the purine and pyrimidine ribonucleotides have been discovered. Although today some uncertainty still exists regarding the exact steps by which the deoxyribonucleotides are produced, there is little doubt that these reactions will soon become completely known.

The enzymic reactions responsible for the polymerization of the nucleotides to the nucleic acids have also been discovered (see Chapter 11 by Gale). The entire metabolic chain of reactions leading from single carbon compounds and nitrogen compounds to these major macromolecular constituents of the cell is thus known.

The studies with bacteria have contributed greatly to our understanding of these fundamental life processes. In particular, the technique peculiar to work with microorganisms, the investigation of the nutrition and the metabolism of nutritionally exacting organisms and of auxotrophic mutants, has been of great value; it was through such studies that many of the metabolic intermediates were discovered and that definitive proof of the essential nature of many of the enzymic reactions was obtained. In addition, studies with bacteria have led to the recognition of the mechanisms responsible for the regulation of the vast array of reactions which are required for the biosynthesis and the interconversion of the purine nucleotides and of the pyrimidine nucleotides.

References

1 S. E. Luria, *in* "The Bacteria" (I. C. Gunsalus and R. Y. Stanier, eds.), Vol. I, p. 1. Academic Press, New York, 1960.
2 G. R. Wyatt and S. S. Cohen, *Biochem. J.* **55,** 774 (1953).
3 M. Adler, B. Weissmann, and A. B. Gutman, *J. Biol. Chem.* **230,** 717 (1958).
4 D. B. Dunn and J. D. Smith, *Biochem. J.* **68,** 627 (1958).
5 J. W. Littlefield and D. B. Dunn, *Nature* **181,** 254 (1958).
6 H. Amos and M. Korn, *Biochim. et Biophys. Acta* **29,** 444 (1958).
7 F. F. Davis and F. W. Allen, *J. Biol. Chem.* **227,** 907 (1957).
8 W. E. Cohn, *Biochim. et Biophys. Acta* **32,** 569 (1959).
9 D. B. Dunn, *Biochim. et Biophys. Acta* **34,** 286 (1959).
10 V. R. Potter, "Nucleic Acid Outlines," Vol. I, p. 235. Burgess, Minneapolis, Minnesota, 1960.
11 W. A. Wood, *in* "The Bacteria" (I. C. Gunsalus and R. Y. Stanier, eds.), Vol. II, p. 59. Academic Press, New York, 1961.
12 L. O. Krampitz, *in* "The Bacteria" (I. C. Gunsalus and R. Y. Stanier, eds.), Vol. II, p. 209. Academic Press, New York, 1961.
13 M. R. Loeb and S. S. Cohen, *J. Biol. Chem.* **234,** 360 (1959).
14 F. K. Bagatell, E. M. Wright, and H. Z. Sable, *J. Biol. Chem.* **234,** 1369 (1959).
15 A. Kornberg, I. Lieberman, and E. S. Simms, *J. Biol. Chem.* **215,** 389 (1955).
16 H. M. Kalckar, *J. Biol. Chem.* **167,** 477 (1947).
17 L. M. Paege and F. Schlenk, *Arch. Biochem. Biophys.* **40,** 42 (1952).
18 A. J. Guarino and H. Z. Sable, *J. Biol. Chem.* **215,** 515 (1955).
19 E. E. Snell, E. Kitay, and W. S. McNutt, Jr., *J. Biol. Chem.* **175,** 473 (1948).
20 M. Downing and B. S. Schweigert, *J. Biol. Chem.* **220,** 521 (1956).
21 W. H. Spell, Jr. and J. S. Dinning, *J. Am. Chem. Soc.* **81,** 3804 (1959).
22 B. D. Davis and E. S. Mingioli, *J. Bacteriol.* **60,** 17 (1950).
23 J. M. Buchanan, *in* "The Nucleic Acids" (E. Chargaff and J. N. Davidson, eds.), Vol. III, p. 303. Academic Press, New York, 1960.
24 M. R. Stetten and C. L. Fox, Jr., *J. Biol. Chem.* **161,** 333 (1945).
25 W. Shive, W. W. Ackermann, M. Gordon, M. E. Getzendaner, and R. E. Eakin, *J. Am. Chem. Soc.* **69,** 725 (1947).
26 N. Fries, S. Bergstroem, and M. Rottenberg, *Physiol. Plantarum* **2,** 210 (1949).
27 J. S. Gots, *Arch. Biochem.* **29,** 222 (1950).
28 E. D. Bergmann, R. Ben-Ishai, and B. E. Volcani, *J. Biol. Chem.* **194,** 521 (1952).
29 G. R. Greenberg, *J. Biol. Chem.* **190,** 611 (1951).
30 M. P. Schulman and J. M. Buchanan, *J. Biol. Chem.* **196,** 513 (1952).
31 G. R. Greenberg and E. L. Spilman, *J. Biol. Chem.* **219,** 411 (1956).
32 J. M. Buchanan and S. C. Hartman, *Advances in Enzymol.* **21,** 199 (1959).
33 S. C. Hartman and J. M. Buchanan, *Ann. Rev. Biochem.* **28,** 365 (1959).
34 S. C. Hartman, personal communication.
35 D. P. Nierlich and B. Magasanik, *J. Biol. Chem.* **236,** PC32 (1961).
36 T. C. French, quoted in[33].
37 J. S. Gots and E. A. Gollub, *Proc. Natl. Acad. Sci. U. S.* **43,** 826 (1957).
38 A. P. Levin and B. Magasanik, *J. Biol. Chem.* **236,** 184 (1961).

[39] A. J. Tomisek, H. J. Kelley, and H. E. Skipper, *Arch. Biochem. Biophys.* **64,** 437 (1956).
[40] H. S. Moyed and B. Magasanik, *J. Biol. Chem.* **226,** 351 (1957).
[41] U. Lagerkvist, *J. Biol. Chem.* **233,** 143 (1958).
[42] R. Abrams and M. Bentley, *Arch. Biochem. Biophys.* **79,** 91 (1959).
[43] B. Magasanik, *Ann. Rev. Microbiol.* **11,** 221 (1957).
[44] J. S. Gots, *J. Biol. Chem.* **228,** 57 (1957).
[45] S. H. Love and J. S. Gots, *J. Biol. Chem.* **212,** 647 (1955).
[46] T. C. French, personal communication.
[47] B. Magasanik, H. S. Moyed, and L. B. Gehring, *J. Biol. Chem.* **226,** 339 (1957).
[48] B. Magasanik and M. S. Brooke, *J. Biol. Chem.* **206,** 83 (1954).
[49] M. E. Balis, M. S. Brooke, G. B. Brown, and B. Magasanik, *J. Biol. Chem.* **219,** 917 (1956).
[50] J. S. Gots, personal communication.
[51] C. E. Carter, *Biochem. Pharmacol.* **2,** 105 (1959).
[52] G. P. Kalle, J. S. Gots, and C. Abramson, *Federation Proc.* **19,** 310 (1960).
[53] C. N. Remy and M. S. Smith, *J. Biol. Chem.* **228,** 325 (1957).
[54] G. P. Kalle and J. S. Gots, *Proc. Am. Assoc. Cancer Research* **3,** 31 (1959).
[55] R. W. Brockman, C. Sparks, M. S. Simpson, and H. E. Skipper, *Biochem. Pharmacol.* **2,** 77 (1959).
[56] B. Magasanik and D. Karibian, *J. Biol. Chem.* **235,** 2672 (1960).
[57] J. Mager and B. Magasanik, *J. Biol. Chem.* **235,** 1474 (1960).
[58] H. S. Moyed and B. Magasanik, *J. Am. Chem. Soc.* **79,** 4812 (1957).
[59] H. S. Moyed and B. Magasanik, *J. Biol. Chem.* **235,** 149 (1960).
[60] B. N. Ames, R. G. Martin, and B. J. Garry, *J. Biol. Chem.* **236,** 2019 (1961).
[61] E. F. Zimmerman and B. Magasanik, unpublished observations.
[62] A. L. Koch and G. Vallee, *J. Biol. Chem.* **234,** 1213 (1959).
[63] R. B. Roberts, P. H. Abelson, D. B. Cowie, E. T. Bolton, and R. J. Britten, "Studies of Biosynthesis in *Escherichia coli.*" *Carnegie Inst. Wash. Publ.* **No. 607** (1955).
[64] M. S. Brooke and B. Magasanik, *J. Bacteriol.* **68,** 727 (1954).
[65] S. Friedman and J. S. Gots, *Arch. Biochem. Biophys.* **32,** 227 (1951).
[66] M. E. Balis, G. B. Brown, G. B. Elion, G. H. Hitchings, and H. Vanderwerff, *J. Biol. Chem.* **188,** 217 (1951).
[67] M. E. Balis, D. H. Levin, G. B. Brown, G. B. Elion, H. Vanderwerff, and G. H. Hitchings, *J. Biol. Chem.* **196,** 729 (1952).
[68] S. E. Kerr, K. Seraidarian, and G. B. Brown, *J. Biol. Chm.* **188,** 207 (1951).
[69] S. E. Kerr and F. Chernigoy, *J. Biol. Chem.* **200,** 887 (1953).
[70] R. C. Wood and E. Steers, *J. Bacteriol.* **77,** 760 (1959).
[71] M. Flavin and S. Graff, *J. Biol. Chem.* **191,** 55 (1951).
[72] M. Flavin and S. Graff, *J. Biol. Chem.* **192,** 485 (1951).
[73] M. R. Heinrich, V. C. Dewey, and G. W. Kidder, *J. Am. Chem. Soc.* **75,** 1741 (1953).
[74] F. Weygand, A. Wacker, and H. Dellweg, *Z. Naturforsch.* **7b,** 156 (1952).
[75] A. Dalby and E. Holdsworth, *J. Gen. Microbiol.* **15,** 335 (1956).
[76] J. S. Gots, *J. Biol. Chem.* **228,** 57 (1957).
[77] J. S. Gots and J. Goldstein, *Science* **130,** 622 (1959).
[78] J. B. Wyngaarden and D. M. Ashton, *J. Biol. Chem.* **234,** 1492 (1959).
[79] E. R. Stadtman, G. N. Cohen, G. Le Bras, and H. de Robichon-Szulmajster, *J. Biol. Chem.* **236,** 2033 (1961).
[80] A. Shedlovsky and B. Magasanik, *Federation Proc.* **19,** 51 (1960) and unpublished observations.

[81] E. Volkin, *Proc. Natl. Acad. Sci. U. S.* **46,** 1336 (1960).
[82] R. J. Mans and A. L. Koch, *J. Biol. Chem.* **235,** 450 (1960).
[83] W. S. McNutt, Jr., *J. Biol. Chem.* **233,** 193 (1958).
[84] P. Reichard, *Biochim. et Biophys. Acta* **41,** 368 (1960).
[85] W. S. McNutt, Jr., *Biochem. J.* **50,** 384 (1952).
[86] A. H. Roush and R. F. Betz, *J. Biol. Chem.* **233,** 261 (1958).
[87] I. R. Lehman, M. J. Bessman, E. S. Simms, and A. Kornberg, *J. Biol. Chem.* **233,** 163 (1958).
[88] M. J. Bessman, *J. Biol. Chem.* **234,** 2735 (1959).
[89] S. C. Hartman and J. M. Buchanan, *J. Biol. Chem.* **234,** 1812 (1959).
[90] A. L. Koch, *J. Biol. Chem.* **217,** 931 (1955).
[91] H. R. B. Revel and B. Magasanik, *J. Biol. Chem.* **233,** 439 (1958).
[92] A. L. Koch and H. R. Levy, *J. Biol. Chem.* **217,** 947 (1955).
[93] N. Chamberlain, N. S. Cutts, and C. Rainbow, *J. Gen. Microbiol.* **7,** 54 (1952).
[94] A. G. Moat, C. N. Wilkins, Jr., and H. Friedman, *J. Biol. Chem.* **223,** 985 (1956).
[95] H. Friedman and A. G. Moat, *Arch. Biochem. Biophys.* **78,** 146 (1958).
[96] H. C. Lichstein, *Vitamins and Hormones* **9,** 27 (1951).
[97] H. P. Broquist and E. E. Snell, *J. Biol. Chem.* **180,** 59 (1949).
[98] C. Mitoma and E. E. Snell, *Proc. Natl. Acad. Sci. U. S.* **41,** 891 (1955).
[99] B. Magasanik, H. S. Moyed, and D. Karibian, *J. Am. Chem. Soc.* **78,** 1510 (1956).
[100] B. Magasanik, *J. Am. Chem. Soc.* **78,** 5449 (1956).
[101] A. Neidle and H. Waelsch, *J. Biol. Chem.* **234,** 586 (1959).
[102] P. Reichard, *Advances in Enzymol.* **21,** 263 (1959).
[103] G. W. Crosbie, *in* "The Nucleic Acids" (E. Chargaff and J. N. Davidson, eds.), Vol. III, p. 323. Academic Press, New York, 1960.
[104] F. W. Chattaway, *Nature* **153,** 250 (1944).
[105] H. J. Rogers, *Nature* **153,** 251 (1944).
[106] H. S. Loring and J. G. Pierce, *J. Biol. Chem.* **153,** 61 (1944).
[107] L. D. Wright, K. A. Valentik, D. S. Spicer, J. W. Huff, and H. R. Skeggs, *Proc. Soc. Exptl. Biol. Med.* **75,** 293 (1950).
[108] L. D. Wright, C. S. Miller, H. R. Skeggs, J. W. Huff, L. L. Weed, and D. W. Wilson, *J. Am. Chem. Soc.* **73,** 1898 (1951).
[109] L. L. Weed and D. W. Wilson, *J. Biol. Chem.* **207,** 439 (1954).
[110] M. E. Jones, L. Spector, and F. Lipmann, *J. Am. Chem. Soc.* **77,** 819 (1955).
[111] P. Reichard and G. Hanshoff, *Acta Chem. Scand.* **10,** 548 (1956).
[112] M. Shepherdson and A. B. Pardee, *J. Biol. Chem.* **235,** 3233 (1960).
[113] I. Lieberman and A. Kornberg, *J. Biol. Chem.* **207,** 911 (1954).
[114] E. S. Reynolds, I. Lieberman, and A. Kornberg, *J. Bacteriol.* **69,** 250 (1955).
[115] I. Lieberman and A. Kornberg, *Biochim. et Biophys. Acta* **12,** 223 (1953).
[116] I. Lieberman and A. Kornberg, *J. Biol. Chem.* **212,** 909 (1955).
[117] R. A. Yates and A. B. Pardee, *J. Biol. Chem.* **221,** 743 (1956).
[118] J. S. Loutit, *Australian J. Exptl. Biol. Med. Sci.* **30,** 287 (1952).
[119] I. Lieberman, A. Kornberg, and E. S. Simms, *J. Biol. Chem.* **215,** 403 (1955).
[120] R. B. Hurlbert and P. Reichard, *Acta Chem. Scand.* **9,** 251 (1955).
[121] A. M. Michelson, W. Drell, and H. K. Mitchell, *Proc. Natl. Acad. Sci. U. S.* **37,** 396 (1951).
[122] I. Lieberman, A. Kornberg, and E. S. Simms, *J. Biol. Chem.* **215,** 429 (1955).
[123] P. Berg and W. K. Joklik, *J. Biol. Chem.* **210,** 657 (1954).
[124] I. Lieberman, *J. Biol. Chem.* **222,** 765 (1956).
[125] J. Lichtenstein, H. D. Barner, and S. S. Cohen, *J. Biol. Chem.* **235,** 457 (1960).
[126] R. B. Hurlbert and H. O. Kammen, *J. Biol. Chem.* **235,** 443 (1960).

[127] R. B. Hurlbert and K. P. Chakraborty, *Federation Proc.* **20,** 361 (1961).
[128] J. L. Strominger, L. A. Heppel, and E. S. Maxwell, *Arch. Biochem. Biophys.* **52,** 488 (1954).
[129] E. Herbert and V. R. Potter, *J. Biol. Chem.* **222,** 453 (1956).
[130] F. Maley, *Federation Proc.* **17,** 267 (1958).
[131] P. Reichard and L. Rutberg, *Biochim. et Biophys. Acta* **37,** 554 (1960).
[132] A. Kornberg, S. B. Zimmerman, S. R. Kornberg, and J. Josse, *Proc. Natl. Acad. Sci. U. S.* **45,** 772 (1959).
[133] J. F. Koerner and M. S. Smith, *Federation Proc.* **18,** 264 (1959).
[134] A. M. Moore and J. B. Boylen, *Arch. Biochem. Biophys.* **54,** 312 (1955).
[135] M. Green and S. S. Cohen, *J. Biol. Chem.* **225,** 387 (1957).
[136] M. Friedkin and A. Kornberg, *in* "The Chemical Basis of Heredity" (W. D. McElroy and B. Glass, eds.), p. 609. Johns Hopkins Press, Baltimore, Maryland, 1957.
[137] E. Scarano, *Biochim. et Biophys. Acta* **29,** 459 (1958).
[138] G. F. Maley and F. Maley, *J. Biol. Chem.* **234,** 2975 (1959).
[139] K. Kleck, H. R. Mahler, and D. Fraser, *Arch. Biochem. Biophys.* **86,** 85 (1960).
[140] J. G. Flaks and S. S. Cohen, *J. Biol. Chem.* **234,** 2981 (1959).
[141] P. Reichard, *Biochim. et Biophys. Acta* **27,** 434 (1958).
[142] A. J. Wahba and M. Friedkin, *J. Biol. Chem.* **236,** PC11 (1961).
[143] B. M. McDougall and R. L. Blakley, *Biochim. et Biophys. Acta* **39,** 176 (1960).
[144] H. D. Barner and S. S. Cohen, *J. Biol. Chem.* **234,** 2987 (1959).
[145] J. G. Flaks and S. S. Cohen, *J. Biol. Chem.* **234,** 1501 (1959).
[146] J. G. Flaks, J. Lichtenstein, and S. S. Cohen, *J. Biol. Chem.* **234,** 1507 (1959).
[147] R. Somerville, K. Ebisuzaki, and G. R. Greenberg, *Proc. Natl. Acad. Sci. U. S.* **45,** 1240 (1959).
[148] M. S. Brooke, D. Ushiba, and B. Magasanik, *J. Bacteriol.* **68,** 534 (1954).
[149] R. A. Yates and A. B. Pardee, *J. Biol. Chem.* **221,** 757 (1956).
[150] R. A. Yates and A. B. Pardee, *J. Biol. Chem.* **227,** 677 (1957).
[151] I. Crawford, A. Kornberg, and E. S. Simms, *J. Biol. Chem.* **226,** 1093 (1957).
[152] R. W. Brockman, J. M. Davis, and P. Stutts, *Biochim. et Biophys. Acta* **40,** 22 (1960).
[153] E. S. Canellakis, *J. Biol. Chem.* **227,** 329 (1957).
[154] P. Reichard and O. Sköld, *Biochim. et Biophys. Acta* **28,** 376 (1958).
[155] P. Reichard and O. Sköld, *Nature* **183,** 939 (1959).
[156] L. M. Paege and F. Schlenk, *Arch. Biochem. Biophys.* **40,** 42 (1952).
[157] H. Amos and B. Magasanik, *J. Biol. Chem.* **229,** 653 (1957).
[158] J. Kream and E. Chargaff, *J. Am. Chem. Soc.* **74,** 4274 (1952).
[159] T. P. Wang, H. Z. Sable, and J. O. Lampen, *J. Biol. Chem.* **184,** 17 (1950).
[160] G. Brawerman and E. Chargaff, *Biochim. et Biophys. Acta* **15,** 549 (1954).
[161] L. V. Crawford, *Biochim. et Biophys. Acta* **30,** 428 (1958).
[162] W. E. Razzell and H. G. Khorana, *Biochim. et Biophys. Acta* **28,** 562 (1958).
[163] A. Kornberg, I. R. Lehman, and E. S. Simms, *Federation Proc.* **15,** 291 (1956).
[164] J. L. Fairley, R. L. Herrmann, and J. M Boyd, *J. Biol. Chem.* **234,** 3229 (1959).
[165] J. M. Boyd and J. L. Fairley, *J. Biol. Chem.* **234,** 3232 (1959).
[166] L. L. Campbell, Jr., *J. Bacteriol.* **73,** 225 (1957).
[167] J. S. Dinning, B. K. Allen, R. S. Young, and P. L. Day, *J. Biol. Chem.* **233,** 674 (1958).
[168] J. S. Dinning and R. S. Young, *J. Biol. Chem.* **234,** 3241 (1959).

CHAPTER 7

Tetrapyrrole Synthesis in Microorganisms

JUNE LASCELLES

I. Introduction ... 335
A. General Distribution of Tetrapyrroles ... 335
B. Chemistry of Porphyrins ... 336
II. Tetrapyrroles as Growth Factors ... 339
A. Requirement for Hemin ... 339
B. Streptomycin Resistance and Hemin Requirement ... 341
C. Iron-Binding Growth Factors ... 342
III. Excretion of Porphyrins by Cultures ... 342
A. Photosynthetic Bacteria ... 342
B. Other Microorganisms ... 345
IV. The Path of Tetrapyrrole Synthesis ... 347
A. General Background ... 348
B. Tetrapyrrole Synthesis by Microorganisms ... 349
V. Synthesis of Heme and Hemoproteins ... 358
A. Insertion of Iron into the Tetrapyrrole Nucleus ... 358
B. Synthesis of Hemoproteins ... 358
VI. Synthesis of the Chlorophylls ... 362
A. Algae ... 362
B. Photosynthetic Bacteria ... 362
C. Effect of Oxygen and Light Intensity on Bacteriochlorophyll Synthesis ... 364
VII. Prodigiosin ... 365
VIII. The Regulation of Synthesis of Tetrapyrroles ... 366
References ... 368

I. Introduction

A. GENERAL DISTRIBUTION OF TETRAPYRROLES

There is no need to elaborate on the ubiquity and importance in nature of substances containing the tetrapyrrole ring structure. The ability to synthesize such compounds from simple precursors is widespread among microorganisms, since most of them contain tetrapyrroles and only a very few require a preformed source for growth.

In its biologically active state, the tetrapyrrole nucleus (or the modified versions of it found in the chlorophylls) is coordinated with a metal and combined with specific proteins. In this chapter, however, considerable prominence is given to those organisms which accumulate free porphyrins such as coproporphyrin III (Fig. 1). These compounds occur relatively rarely in nature and, as far as we know, have no function.[1,2] In vertebrates, for instance, they are normally found only in traces in the tissues, blood, and excreta, but certain pathological states (e.g., the porphyria diseases) are characterized by the excretion of considerable quantities of free porphyrins; this phenomenon is also induced by drugs such as Sedormid.[3-6]

The accumulation of porphyrins is presumably the result of some metabolic block in the normal biosynthetic pathway leading to the biologically active tetrapyrrole derivatives such as the cytochromes and chlorophylls. Organisms which excrete porphyrins are thus useful tools for study of the formation of the tetrapyrrole nucleus. Also information concerning the conditions which lead to their accumulation might ultimately contribute to an understanding of the control mechanisms operating in tetrapyrrole synthesis and their possible relationship to the porphyria diseases.

B. Chemistry of Porphyrins

The chemistry and properties of tetrapyrroles in general are treated in full in the monographs of Fischer and Orth[7] and of Lemberg and Legge[8]; there is also much useful information in the reviews of Granick.[1,2] The chlorophylls and their derivatives are discussed in great detail by Rabinowitch.[9] A few salient features of the porphyrins are described here, mainly to enable microbiologists to recognize these pigments when formed by cultures of microorganisms.

The structure of some of the porphyrins and derivatives mentioned in this chapter are shown in Figs. 1 and 2. There are a number of theoretically possible arrangements of the various side chains about the tetrapyrrole nucleus. For example, there are four isomeric forms of uroporphyrin, depending upon the positions of the acetic and propionic acid side chains, and fifteen isomeric forms of protoporphyrin with its methyl, vinyl, and propionic acid groups. Nature, however, seems to have chosen only one isomeric series, since all physiologically active tetrapyrroles have the same configuration as uroporphyrin III or protoporphyrin IX (hereafter termed protoporphyrin). Free porphyrins of series I have, however, been found and enzyme systems which form predominantly uroporphyrin I have been described. The porphyrins occurring most commonly in nature are coproporphyrin III, uroporphyrin I and III, and protoporphyrin; chromatography has revealed traces of other porphyrins but these have yet to be identified.

Accumulation of porphyrins by bacteria can be recognized by the red-brown pigmentation, which usually appears in the extracellular environment. The presence of these compounds can be confirmed by their red fluorescence in ultraviolet light, which provides a sensitive and characteristic test. The porphyrins have characteristic spectra with maximum absorption in the region of the Soret band at 395–410 mμ, depending on the compound in question. In the visible region of the spectrum, the neutral porphyrins have four bands between 500 and 700 mμ, the position of the maxima varying with the different porphyrins and depending on the solvent. The visible spectrum gives important clues to structure; for example, substances with the dihydro- or tetrahydroporphyrin ring of the chloro-

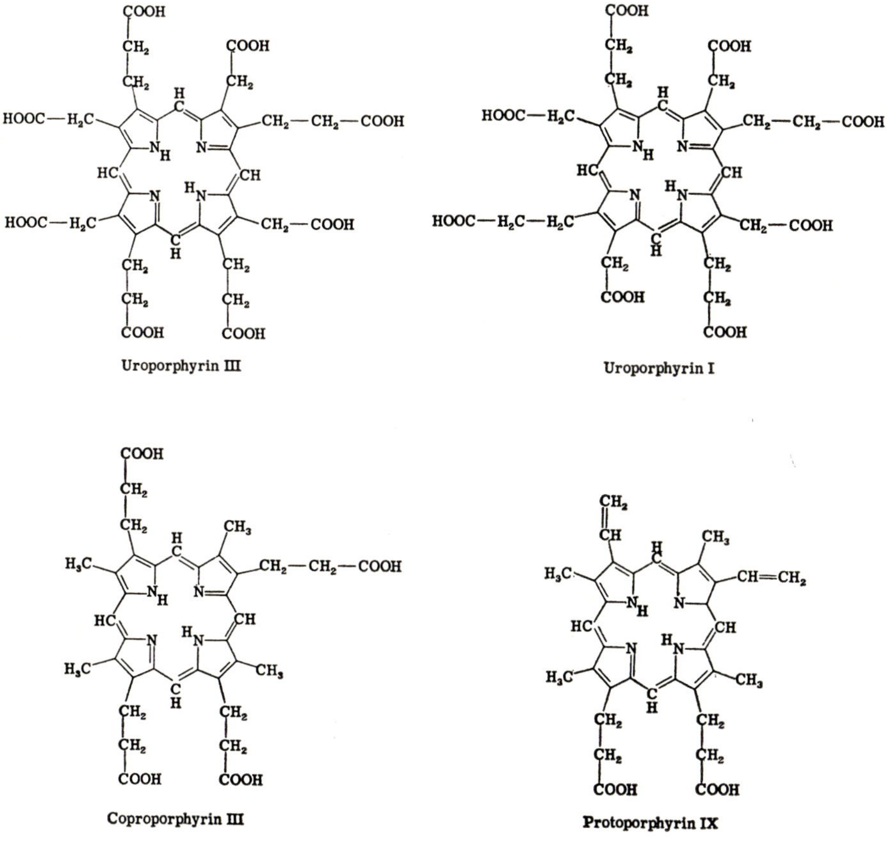

FIG. 1. Some naturally occurring porphyrins.

phylls can be readily distinguished spectrally from porphyrins of the etio type such as coproporphyrin.[1]

The most common method for quantitative separation of porphyrins involves extraction from organic solvents with increasing concentrations of HCl; the more carboxyl groups in the side chains about the tetrapyrrole nucleus, the lower the concentration of HCl required to remove the pigment from the solvent phase. A number of methods have been developed for separation of porphyrins and their methyl esters by paper chromatography; some of these are able to resolve mixtures of type I and type III isomers.[10]

The following nomenclature is used in this chapter. Compounds whose

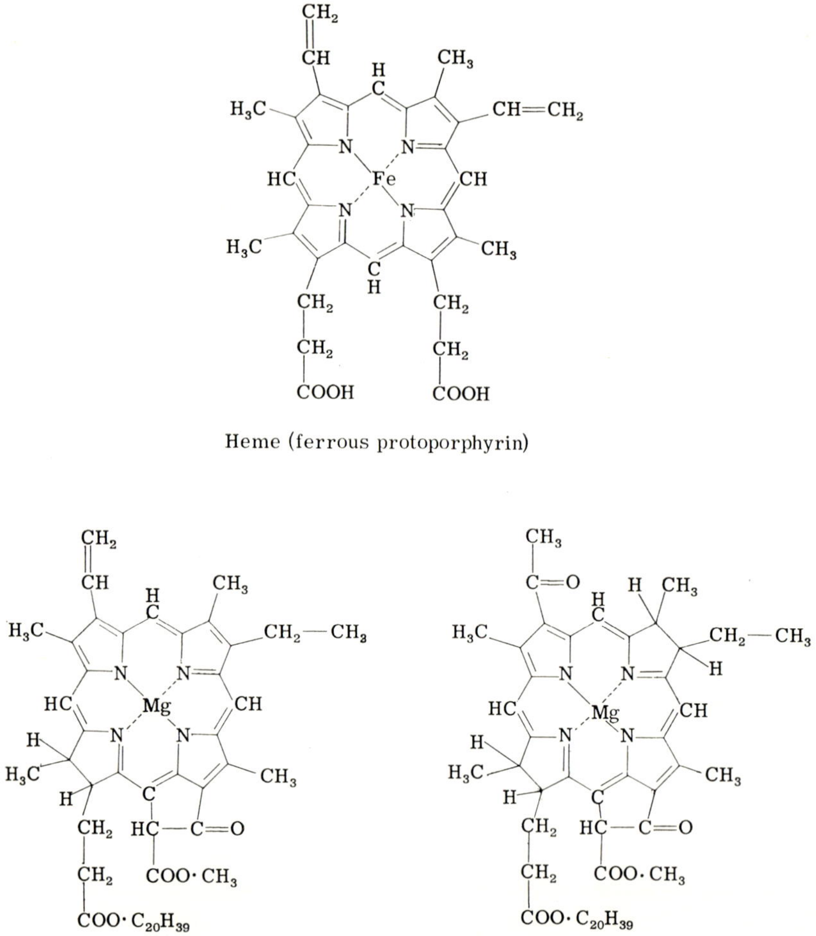

Heme (ferrous protoporphyrin)

Chlorophyll *a*

Bacteriochlorophyll

FIG. 2. Heme, chlorophyll *a*, and bacteriochlorophyll.

structure is based upon the tetrapyrrole nucleus are collectively referred to as tetrapyrroles. Substances containing an iron porphyrin prosthetic group such as catalase and the cytochromes are referred to collectively as heme compounds or hemoproteins. Hematin and hemin are used only in reference to the specific compounds ferric protoporphyrin and ferric protoporphyrin chloride, respectively. Hematin is found in some organisms and may arise by oxidation of the unstable ferrous protoporphyrin or heme while hemin is most commonly used as a source of iron protoporphyrin for growth of exacting organisms. The term chlorophylls includes those found in plants as well as in bacteria; specific compounds are designated by name, e.g., bacteriochlorophyll.

II. Tetrapyrroles as Growth Factors

Studies of the growth factor requirements of different microorganisms, particularly those of induced mutants, has contributed immeasurably to the elucidation of the path of biosynthesis of many essential metabolites. Furthermore, mutants blocked at various stages in a particular pathway

TABLE I

BACTERIA REQUIRING HEMIN FOR GROWTH

Organism	Hemin required for maximum growth (μg./ml.)	Other active tetrapyrroles	Reference
Haemophilus influenzae	0.05–0.3	Protoporphyrin; iron hemato-, deutero-, and mesoporphyrins	16 17
Micrococcus pyogenes var. 511	0.03–0.06	—	34, 37
Bordetella pertussis	2.5	—	25
Pasteurella pestis	1–5	—	24
Mycobacterium tuberculosis (strains resistant to isonicotinic acid hydrazide)	2–5	—	26, 27
Fusiformis (Bacteroides) melaninogenicus	1	—	31a

may accumulate precursors of the metabolite whose formation is impeded. Such methods of approach have, however, contributed little to the understanding of the mechanism of tetrapyrrole synthesis; so far, no induced mutant has been reported to have a requirement for preformed tetrapyrrole. Most microorganisms are able to make these substances from simple carbon and nitrogen sources, and, like the higher animals and plants, do not need preformed tetrapyrroles for growth. The few microbes which require hemin or derivatives for growth are mostly pathogenic and have, in general, limited biosynthetic ability (see Table I).

A. REQUIREMENT FOR HEMIN

1. HAEMOPHILUS INFLUENZAE

Although a requirement for tetrapyrroles is rare among bacteria, the observation that *Haemophilus influenzae* needed a substance associated with blood pigment (factor "X" of Thjötta and Avery[11,12]) was one of the earliest discoveries in the field of bacterial nutrition.[13] Crystalline hemoglobin provided factor X[11,12] and it was soon found that hemin (ferric protoporphyrin chloride) had the full activity of the blood pigment.[14,15]

Some years later Granick and Gilder[16,17] made an extensive study of the specificity of the requirement for hemin. They found protoporphyrin to be as active as hemin for all strains examined, but, with the exception of one strain, no other free porphyrin was effective; instead, other porphyrins acted as competitive inhibitors of growth in the presence of protoporphyrin or hemin. The rough strain "Turner" differed from all other strains in utilizing mesoporphyrin for growth.

Although protoporphyrin is the only free porphyrin utilized by most strains of *Haemophilus*, the requirement for iron porphyrins is less specific. The iron complexes of hematoporphyrin, mesoporphyrin, and deuteroporphyrin promote growth; but *Haemophilus* does not convert these compounds to iron protoporphyrin, since organisms grown in their presence do not reduce nitrate, a property found only in cultures grown on protoporphyrin or hemin.[17] The activity of protoporphyrin alone of the free porphyrins suggests that the vinyl groups are needed for the insertion of iron; *Haemophilus* has the mechanism necessary for this reaction but is presumably blocked at an earlier stage in the formation of the porphyrin nucleus.

Experiments with organisms grown with limiting hemin show that it plays a role in the respiratory metabolism of *Haemophilus*, presumably after incorporation into cytochromes.[18]

2. The Trypanosomidae

Many, although not all, of this group of parasitic flagellates fail to multiply unless hemin is provided in the medium.[19,20] Protoporphyrin is utilized by these organisms in place of hemin but no other porphyrin or iron porphyrin of a large number tested is active.[21,22] It seems that synthesis in these organisms, as in *Haemophilus influenzae*, fails at a stage prior to the formation of protoporphyrin.

Hemin is presumably used by the flagellates for the formation of cytochromes, since their respiration is greatly decreased after growth with limiting concentrations of the factor, and is restored by addition of hemin or protoporphyrin.[23]

3. Other Organisms

A requirement for hemin has been observed in the pathogens *Pasteurella pestis*[24] and *Bordetella pertussis*[25], and in strains of *Mycobacterium tuberculosis* resistant to isonicotinic acid hydrazide.[26,27] These bacteria require 10 to 100 times as much hemin as does *H. influenzae* (Table I). It is probable that exogenous hemin fulfils a different function in these organisms, not serving in the formation of hemoproteins. It may serve as a catalyst for the destruction of hydrogen peroxide. Thus, hemin is required for growth of *P. pestis* from small inocula under aerobic conditions, and Herbert[24] has suggested that it is needed to destroy peroxides produced in the early stages of growth when synthesis of catalase may be limited. Hemin may exert a similar action in the isonicotinic acid hydrazide-resistant strains of *M. tuberculosis*, since these strains are deficient in catalase[28,29] although not in cytochromes.[30]

Hemin has also been reported to be required for growth of the strict anaerobe, *Fusiformis* (*Bacteroides*) *melaninogenicus* (*nigrescens*) but its function is unknown.[31,31a] This interesting organism also requires vitamin K,[32,31b] a factor which may participate in the electron transfer system in animal tissues and possibly also in microorganisms. When grown on blood agar, the colonies are black due to conversion of the hemoglobin to ferric protoporphyrin (hematin) which is adsorbed onto the cells; these observations led to the suggestion that the specific name should be altered from *melaninogenicus* to *nigrescens*.[33]

B. Streptomycin Resistance and Hemin Requirement

There appears to be a relation between resistance to inhibition by streptomycin and the ability to form heme compounds in some bacteria. Jensen and Thofern[34-36] have isolated a streptomycin-resistant strain of *Micrococcus pyogenes* var. *aureus* which needs hemin for growth on a casein hydrolyzate-glucose medium, whereas the parent sensitive strain does not. The incorporation of iron into the tetrapyrrole nucleus is apparently blocked at some stage in this variant, since it cannot utilize protoporphyrin or other porphyrins[34] but is able to convert δ-aminolevulinic acid to free porphyrins.[37] The requirement for hemin is not absolute, since growth occurs in its absence in meat broth; but heme compounds are formed only in organisms grown with hemin.[34-36,38] The meat broth probably provides acetate and nucleic acid derivatives, which have been shown to replace hemin for growth on a medium containing amino acids and glucose.[37]

Several streptomycin-resistant strains of *Escherichia coli* characterized by a low respiratory activity have been isolated.[39,40] They grow only poorly aerobically with or without hemin, but under anaerobic conditions hemin increases growth. Unlike the resistant *Micrococcus*, the strains of *E. coli* respond also to protoporphyrin, but to no other porphyrin of a number tested, and may therefore be blocked at an earlier stage in the synthesis of porphyrins. Furthermore, free porphyrins could not be detected in cultures of these drug-resistant mutants, whereas they were found in the normal strains, although in low concentration.

These observations with entirely different organisms suggest that investigation of the metabolism of porphyrins and their precursors in streptomycin-resistant mutants may throw light on the mechanism of action of the drug and on the development of resistance to it.

C. Iron-Binding Growth Factors

The recent isolation of several iron-containing substances of unknown structure which are active as growth factors for certain microorganisms has led to the suggestion that they may function as carriers of iron in microbial metabolism.[41] Such substances include ferrichrome, isolated from the smut fungus *Ustilago sphaerogena*,[42,43] coprogen, isolated from dung,[44,45] and terregens factor, prepared from bacterial culture filtrates.[46a] All preparations are active as growth factors for the fungus *Pilobolus* as well as for *Arthrobacter terregens*. A strain of *Microbacterium lacticum* with an absolute requirement has been recommended for the assay of these substances.[46b,46c] These organisms also respond to hemin although it is much less active than the factors. It is likely that hemin is acting as a source of organically bound iron and not as a source of the pyrrole ring, but the iron complexes may play a role in the insertion of the metal into the tetrapyrrole nucleus in the synthesis of heme compounds.

III. Excretion of Porphyrins by Cultures

In his classical work on the distribution and chemistry of porphyrins and related compounds, Hans Fischer found porphyrins in cultures of various yeasts and tubercle bacilli. Since that time other bacteria have been found to accumulate porphyrins during growth; these compounds are usually found in the culture fluids. The accumulation of these pigments does not seem to be a common phenomenon. This is not surprising since microbes growing under optimal conditions tend to synthesize and to use their metabolites economically and do not accumulate them unless there is some type of block in their utilization.

A. Photosynthetic Bacteria

Many photosynthetic bacteria, particularly members of the Athiorhodaceae, excrete considerable quantities of free porphyrin into the medium during growth. The appearance of a pink extracellular pigment in cultures of several *Rhodopseudomonas* species was first observed by van Niel.[47] This pigment was later shown to be free porphyrin.[48, 48a] It is formed in substantial amounts by many strains of the four *Rhodopseudomonas* species, but only trace amounts have been found in cultures of *Rhodospirillum rubrum* (see Table II). Porphyrins have also been found in cultures of red sulfur bacteria (Thiorhodaceae), particularly when grown with malate, and also in cultures of green sulfur bacteria.[48]

1. Relation of Porphyrin Excretion to the Synthesis of Bacteriochlorophyll

The exceptional activity of the photosynthetic bacteria in forming free porphyrins during growth is probably a reflection of their ability to form more complex derivatives of the tetrapyrrole nucleus. The cells of these organisms contain heme compounds at levels similar to that found in aerobic microorganisms.[49] However, by far the major tetrapyrrole derivative is bacteriochlorophyll in purple bacteria, or chlorobium chlorophyll in the green sulfur bacteria. The concentration of chlorophyll in cells grown anaerobically in the light is of the order of 10 to 50 mμmoles/mg. dry weight of cells. In other microorganisms the only tetrapyrrole compounds are the hemes, mostly present as prosthetic groups of cytochromes and catalase; the concentration of total heme varies among different organisms but is of the general order of 0.1 to 1 mμmole/mg. dry weight of cells, far less than the amount of bacteriochlorophyll in the photosynthetic bacteria.

TABLE II
Excretion of Porphyrins by Cultures of Athiorhodaceae[a]

Organism	Concentration of iron (mμmoles/ml.)	Coproporphyrin III (mμmoles/ml.)	Bacteriochlorophyll (mμmoles/ml.)
Rhodopseudomonas spheroides	0	82	13
N. C. I. B. 8253	10	3	36
Rhodopseudomonas capsulatus	0	19	11
Strain 2.3.11	10	6	23
Rhodopseudomonas palustris	0	12	8
Strain 2.1.7	10	2	24
Rhodopseudomonas gelatinosa	0	105	10
Strain 2.2.13	10	66	29
Rhodospirillum rubrum	0	0.3	10
Strain-S1	10	0.2	22

[a] All strains came originally from the collection of Professor C. B. van Niel. They were grown for 3 days anaerobically in the light on an iron-deficient medium containing malate and glutamate as carbon source,[107] with addition of *p*-aminobenzoic acid for *Rhodopseudomonas palustris*. Coproporphyrin III accounted for at least 95% of the total porphyrin in all cases.

Excretion of porphyrins by these bacteria is closely associated with the formation of bacteriochlorophyll. In *Rhodopseudomonas spheroides*, porphyrins are found only in cultures grown anaerobically in the light, the conditions under which bacteriochlorophyll is formed. Cultures grown aerobically in the dark form very little photopigment and accumulate no porphyrins.[48] The connection between the synthesis of free porphyrins and of bacteriochlorophyll is shown also by kinetic studies of their synthesis by growing cultures of various rhodopseudomonads. Porphyrins begin to accumulate only in the later stages of growth when synthesis of bacteriochlorophyll has ceased.[48,50] This is particularly striking in cultures grown with limiting amounts of iron; only traces of porphyrin are detectable during the period of exponential growth and synthesis of bacteriochlorophyll, but soon as this pigment has reached its maximum level there is a rapid increase in extracellular prophyrin.

2. Effect of Iron

Excretion of porphyrins by photosynthetic bacteria is strikingly influenced by the concentration of iron in the medium. With all organisms examined so far, maximum accumulation occurs in media containing suboptimal concentrations of iron.[48] Increasing the concentration suppresses the excretion of porphyrin but increases the synthesis of bacteriochlorophyll (see Table II). The effect of iron on the formation of chlorophyll pigments by photosynthetic bacteria was first observed by Larsen[51] with the green bacterium *Chlorobium thiosulfatophilum*, which contains chlorobium chlorophyll; his findings have been amply confirmed with purple bacteria containing bacteriochlorophyll. Iron is also essential for synthesis of chlorophyll by green plants.[9]

In the cultures of Athiorhodaceae examined so far the quantity of porphyrins excreted does not bear any obvious relation to the amount of bacteriochlorophyll formed; usually the amount of porphyrin formed under conditions of iron deficiency is considerably in excess of the bacteriochlorophyll synthesized in cultures containing optimal concentrations of iron.

The inverse relationship between excretion of porphyrins and formation of bacteriochlorophyll, influenced by the concentration of iron, suggests further that the porphyrins are associated with the synthesis of the photopigment.

3. Types of Porphyrin

The porphyrin most commonly found in cultures of photosynthetic bacteria is coproporphyrin III. In iron-deficient cultures of *R. spheroides* it comprises about 97% of the total, the remainder consisting of uroporphyrins I and III with traces of other porphyrins, containing 3, 5, and 7 carboxylic acid groups, detectable by chromatography. A similar mixture of porphyrins accumulates in iron-deficient cultures of *R. capsulatus*, but small amounts of magnesium protoporphyrin are detectable in cultures grown with optimal iron.[52]

Dihydroporphyrins have been found in the culture fluids of a blue-green mutant strain of *R. spheroides* unable to form colored carotenoids.[53a] The main components of this mixture of dihydroporphyrins show a close similarity in chemical and physical properties to pheophorbide *a* (derived from plant chlorophyll *a*, see Fig. 8), derivatives of bacteriochlorophyll being only minor constituents. Magnesium protoporphyrin has also been found in culture fluids from this organism, in this laboratory. These compounds are only found in cultures grown with optimal iron and it is likely that they are either intermediates or stabilized by-products of intermediates in the synthesis of bacteriochlorophyll. Recently a protochlorophyll-like pigment has been isolated from a mutant strain of *R. spheroides* which does not form bacteriochlorophyll.[53b] Its spectrum and other properties suggest that it is closely related in structure to the photochlorophyll of some higher plants and that it is another possible intermediate in the biosynthesis of the bacterial pigment.

B. Other Microorganisms

When compared with the synthesis of porphyrins by photosynthetic bacteria the amount of porphyrin formed by other organisms looks rather insignificant. There is always the possibility that modification of the growth medium may increase the accumulation of these substances—so far, this aspect has not been examined in detail. The examples given in Table III[54-67] have been collated from various sources; the organisms were grown under a variety of conditions, not necessarily those which lead to maximum porphyrin formation.

As with the photosynthetic bacteria, coproporphyrin III is the most common porphyrin appearing in cultures of other microorganisms (Table III), but *Saccharomyces anamensis* is exceptional in that it forms the type I isomer. Protoporphyrin IX has been found in the cells of *Tetrahymena vorax*[67] and *T. geleii*.[68]

1. Effect of Iron

An influence of iron has been consistently observed in studies of porphyrin formation by cultures of various microorganisms. Considerable attention has been given to *Corynebacterium diphtheriae* since the discovery of porphyrins in the culture fluids,[69] followed by the observation that formation of these pigments accompanied toxin formation.[70] Pappenheimer[54] found that the amount of porphyrin and toxin formed was dependent on the concentration of iron in the medium, the yield of both decreasing upon addition of iron beyond a certain critical level.

Since then many more instances of suppression of porphyrin formation by iron have been reported. Organisms showing this effect include *Propionibacterium shermanii*,[64] *Bacillus subtilis*,[62] *Micrococcus lysodeikticus*,[63] and *Corynebacterium erythogenes*.[56] Nor is the action of iron limited to bacteria; formation of porphyrins by the yeast, *Saccharomyces anamensis*,[65] and by the ciliate, *Tetrahymena vorax*,[67] is suppressed by addition of iron.

The simplest interpretation of the action of iron is that it is required as a substrate for the synthesis of intracellular heme compounds from porphyrins or precursors and that under conditions of iron deficiency the porphyrins accumulate. This view is supported by the observation that in

TABLE III

PORPHYRIN FORMATION BY CULTURES[a]

Organism	Type of porphyrin	Amount (mg./liter)	Reference
Rhodopseudomonas spheroides	Copro III; traces of uro I	56	48
Corynebacterium erythogenes	Copro III; uro I	49	56
Arthrobacter globiformis	Copro	33	57
Corynebacterium diphtheriae	Copro III; traces of uro I	7	54, 55
Propionibacterium shermanii	Copro III	9	64
Micrococcus lysodeikticus	Copro III	6	63
Bacillus subtilis	Copro III	5	62
Bacillus cereus (anaerobic)	Copro	0.14	61
Mycobacterium karlinski and other	Copro III	3	58
Mycobacterium spp.			59, 60, 60a
Saccharomyces anamensis	Copro I	298	65
Saccharomyces cerevisiae (anaerobic)	Copro	0.09	66
Tetrahymena vorax	Uro III; copro III; proto	6	67

NOTE: Copro = coproporphyrin; uro = uroporphyrin; proto = protoporphyrin

[a] The amounts of porphyrin have been calculated from the published values and are approximate only since the extraction procedures used were not always quantitative. In cases where the porphyrins remain within the cells, the amounts are given as milligram per kilogram wet weight of cells; otherwise the values are for extracellular porphyrin/liter of culture fluid. Unless specified, the isomeric type has not been determined.

cultures of *C. diphtheriae* (strain "Toronto" P.W. 8) there is a mole-for-mole relation between the amount of iron added (and completely taken up by the cells) and the amount of porphyrin which fails to appear.[54] Also, cells grown with excess iron contain more cytochrome b and catalase although total hemes were not determined quantitatively.[71] More recent work with another strain of this organism (G 12/6) has not given such clear-cut results.[72] The amount of iron recovered from the cells of this strain accounts for only 50 to 60% of that added to the medium, and the molar ratio of iron taken up by the cells and the amount of porphyrin which fails to appear is 1:2. Furthermore, cytochrome b, the main heme derivative in the cells, accounts for only 16% of the total intracellular iron which leads to the conclusion that the heme compounds of the cells grown with excess iron account for only about 10% of the porphyrin which fails to appear.

There is little information on the heme content of other organisms which excrete porphyrin; it seems likely that the amount of excreted pigment (about 10 μmoles/liter) is generally in excess of the hemes, which are probably present at concentrations of the order 0.1 to 1 μmole/g. cells. The accumulation of excess quantities of precursors when a biosynthetic pathway is blocked by a variety of means is a well-recognized phenomenon and these discrepancies in no way eliminate the porphyrins either as intermediates, or, more likely, as stabilized by-products of intermediates in the synthesis of intracellular hemes.

There is evidence, to be discussed later, that iron may function catalytically in the formation of the vinyl side chains of protoporphyrin, a necessary step in the synthesis of heme compounds. Interference with this reaction by iron deficiency could also lead to the accumulation of precursors of protoporphyrin.

2. Other Factors Influencing Porphyrin Formation

The relation between synthesis of cytochrome and accumulation of porphyrin may be influenced by other factors besides iron. In *Bacillus cereus* porphyrins are found in cultures grown anaerobically, but are not detectable in aerobic cultures. Cells grown under the latter conditions contain cytochromes a, b, and c, whereas only cytochrome b is found in cells grown anaerobically and the total hematin content is reduced to one-fifth of the aerobic level.[73,74] Normal and "petite colonie" strains of *Saccharomyces cerevisiae* also accumulate porphyrins in anaerobic culture but not aerobically.[66] As with *B. cereus* the appearance of porphyrins is correlated with the absence of cytochrome c, which is found only in aerobically grown cells. Traces of porphyrin have also been observed in anaerobic but not aerobic cultures of *Staphylococcus aureus*.[74]

Saccharomyces anamensis accumulates porphyrin intracellularly when grown on a glucose-urea medium under essentially anaerobic conditions but heme compounds are not detectable.[65] Addition of riboflavin to the medium suppresses porphyrin formation but results in the synthesis of heme compounds, which is increased by the further addition of iron.

IV. The Path of Tetrapyrrole Synthesis

The biosynthesis of the tetrapyrrole ring is being explored in many laboratories with enzyme systems from animals, plants, and microbes, although it is only within the last eighteen months or so that microorganisms have been favored with the attentions of enzymologists in this field. The main purpose of this section is to discuss the path of tetrapyrrole synthesis in microorganisms; but this question cannot be considered in isolation, since the major advances in the field as a whole have come mainly from work with animal tissues. It is now known that the biosynthetic pathway in microbes is similar to that occurring in animal and plant tissues and before considering the work with microorganisms a review of the general state of knowledge is provided (see Fig. 3). More detailed information of the field as a whole can be found in the reviews of Granick,[2] Shemin,[75] Rimington,[6,76] and of various authors in the Ciba Foundation Symposium.[77]

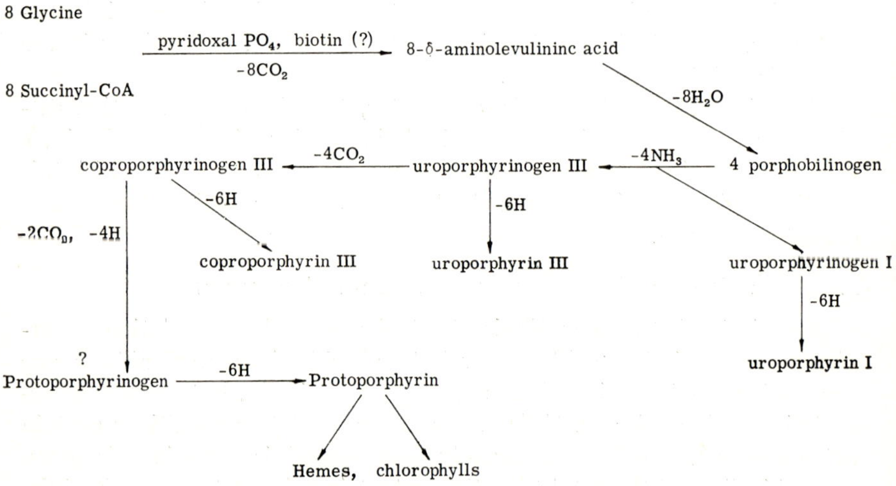

FIG. 3. The path of tetrapyrrole synthesis.

A. General Background

1. Experiments with Isotopes

The first clues to the origin of the pyrrole ring came from studies of the incorporation into heme of isotopically labeled compounds administered to whole animals; these showed that acetate and glycine were precursors.[78,79] A major technical advance was provided by the discovery that synthesis of heme occurred *in vitro* with the nucleated red blood cells from ducks[80]; since that time whole cells or hemolyzed preparations of avian red blood cells have been widely used.

The ability to synthesize heme is not, however, confined to avian erythrocytes. Immature mammalian red blood cells, which are nucleated, are also active, whereas the mature unnucleated cells are inactive; the production of immature cells is increased by bleeding or by administration of phenylhydrazine.[81]

The isotope incorporation studies showed that glycine provides not only the pyrrole nitrogen but also the methene bridge carbons and 1 carbon atom in each pyrrole ring, in the α-position under the vinyl and propionic acid side chains. Of the carbon skeleton of glycine, only the α-carbon atom contributes to the tetrapyrrole ring and, over-all, 8 molecules of glycine are used for the formation of 1 molecule of heme. Similar studies tracing the incorporation of labeled acetate and intermediates of the tricarboxylic acid cycle led to the conclusion that the initial reaction in pyrrole synthesis is a condensation of glycine and an unsymmetrical C_4-compound derived from the tricarboxylic acid cycle.[2,75]

2. δ-Aminolevulinic Acid, Porphobilinogen, and Porphyrinogens

Within the last few years important advances have been made in clarifying the nature of the steps between the initial condensation reaction and formation of the complete tetrapyrrole ring. The recognition of the monopyrrole, porphobilinogen (PBG), and of δ-aminolevulinic acid (ALA) as intermediates were major "breakthroughs" in this field. Their arrival on the scene also marked a broadening of experimental approach. Isotope incorporation techniques had been used exclusively to follow the incorporation of glycine and intermediates of the tricarboxylic acid cycle into heme, using whole blood or hemolyzates as the enzyme system. The enzymes concerned with the conversion of ALA and PBG into the porphyrin nucleus have yielded more readily to classical methods of enzymology and the utilization of these substances has been investigated with enzyme preparations from a variety of sources. Furthermore, actual *de novo* synthesis of free porphyrins or intermediates has been measured, without relying entirely on isotope incorporation techniques.

Evidence for ALA as an intermediate in the synthesis of heme was first obtained by isotopic techniques[82-84] More recently enzyme preparations from chick erythrocytes have been described which catalyze a net synthesis of ALA from glycine and succinyl-CoA[85-88a,88b] (see Fig. 4). The role of ALA as an intermediate has been amply confirmed by the demonstration of its conversion to PBG or to free porphyrins by enzyme preparations from animals, plants, and bacteria. The purification and properties of ALA dehydrase, which converts ALA to PBG, has been described;[89] this is the first enzyme to be isolated which catalyzes a single step in the biosynthesis of porphyrins (see Fig. 5).

Porphobilinogen was first shown to be converted to porphyrins by Falk, *et al.*[90] using chick hemolysates. This substance had been known for many years to be excreted in the urine in cases of acute porphyria, although its isolation and identification has been achieved only recently.[91,92] It is now well established as an intermediate in porphyrin and heme formation.[93,94]

Information concerning the mechanism of conversion of PBG to porphyrins is now accumulating. Crude enzyme systems such as hemolysates and frozen-thawed *Chlorella* preparations convert it to a mixture of uro-, copro-, and protoporphyrin with traces of porphyrins containing from 3 to 7 carboxyl groups detectable by chromatography; the ratio of the products varies with the type of enzyme preparation and also with the concentration

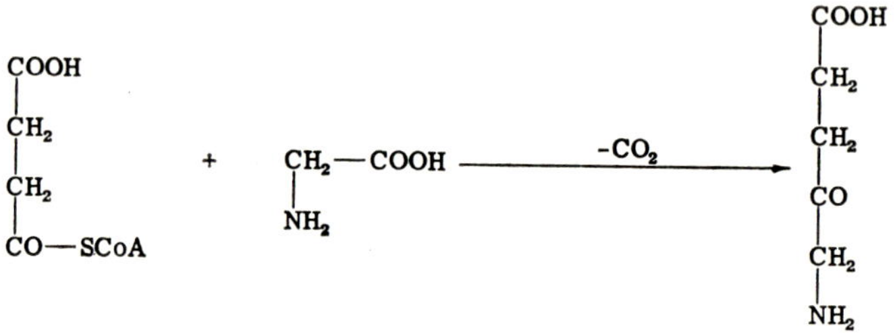

Succinyl-CoA + glycine δ-Aminolevulinic acid

FIG. 4. Initial steps in tetrapyrrole synthesis.

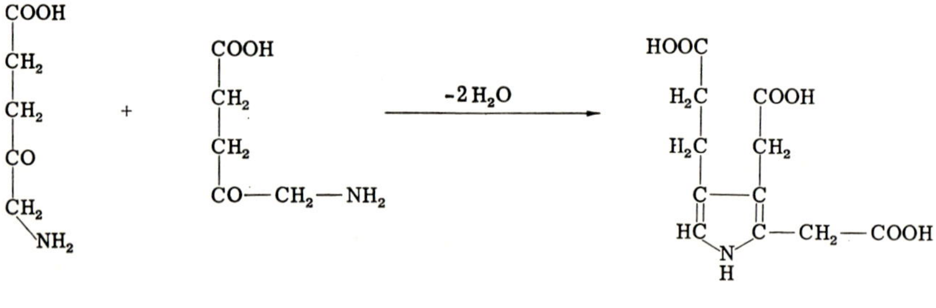

2 δ-Aminolevulinic acid Porphobilinogen

FIG. 5. Reaction catalyzed by δ-aminolevulinic acid dehydrase.

of substrates.[94-96] Work with more purified preparations has shown that the porphyrins arise from the corresponding porphyrinogens, which are colorless compounds containing 6 extra atoms of hydrogen; these substances are readily autoxidized to porphyrins (see Fig. 6) but their conversion may also be enzymically catalyzed. There is now strong evidence that the porphyrins are not intermediates in the synthesis of heme. Isotopically labeled uro-, copro-, and protoporphyrin are not incorporated into heme by hemolysates which incorporate labeled glycine ALA and PBG.[97] Uroporphyrinogen III is, however, converted to heme[98] and it is probable that the porphyrinogens rather than the porphyrins are the actual intermediates in heme synthesis.

Enzyme fractions have been obtained from avian and mammalian red cells which catalyze the conversion of PBG to uroporphyrinogen III.[99-101] When these preparations are heated to 50 to 60°C. uroporphyrinogen I is formed in place of the type III isomer. Similar enzymes have been isolated from plant tissues. Bogorad[102] has obtained a fraction from acetone powders of spinach leaf, PBG deaminase, which catalyzes the reaction

$$4PBG \rightarrow \text{uroporphyrinogen I} + 4NH_3$$

A preparation from wheat germ converts PBG to uroporphyrinogen III and Bogorad[103] has suggested that the type III isomer is formed by the action of two enzymes, PBG deaminase and PBG isomerase; the latter may

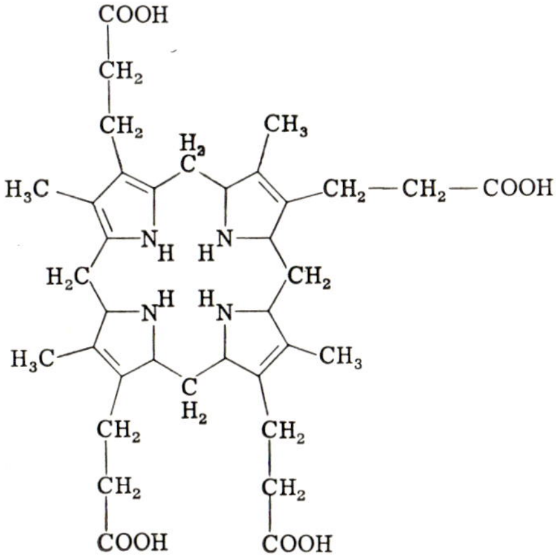

FIG. 6. Coproporphyrinogen III.

react with PBG plus some product of the action of the deaminase on PBG. The mechanism of formation of the two isomers is, however, far from clear.

3. Interconversion of Porphyrinogens

Fractions from red cells and plant tissues catalyze the decarboxylation of uroporphyrinogen III to coproporphyrinogen III, but the purified preparations do not utilize uroporphyrin.[96,104,105] The intermediate stages are not known, although porphyrins with 7, 6, and 5 carboxyl groups (probably derived from the corresponding porphyrinogens) have been detected by chromatography of the reaction mixtures.[105]

The synthesis of protoporphyrin has been studied only in crude systems that form a mixture of porphyrins. It is likely that coproporphyrinogen III rather than the corresponding porphyrin is a precursor.[101] The propionic acid side chains of coproporphyrinogen III may be activated before oxidative decarboxylation to the vinyl level, since protoporphyrin formation is stimulated by addition of mitochondria and ATP to *Chlorella* preparations[95] and to hemolysates.[104,104a] Whether protoporphyrinogen is first formed is not known, nor is it clear at what stage the metal is introduced into the porphyrin ring in the formation of heme and the chlorophylls.

B. Tetrapyrrole Synthesis by Microorganisms

The high rates of biosynthetic activity of microorganisms offer good opportunities for elucidating biosynthetic pathways. In addition, most microorganisms can be grown in a chemically defined medium and the effect of factors such as B-group vitamins can be readily determined. It is therefore surprising that these advantages have been realized only very recently by those working in this field of investigation.

1. The Biosynthetic Pathway in Various Microorganisms

(*a*) *Corynebacterium diphtheriae.* Evidence that bacteria form the tetrapyrrole nucleus by a pathway similar to that operating in animal tissues was first obtained with *Corynebacterium diptheriae.*[106] When grown in the presence of N^{15}-glycine, this organism incorporates considerable amounts of N^{15} into the intracellular hemes as well as into the excreted coproporphyrin III and uroporphyrin I.

(*b*) *Photosynthetic bacteria.* The ability of *Rhodopseudomonas spheroides* to form large amounts of porphyrin during growth suggested its use for study of the precursors of the pigment.[107] Cell suspensions, harvested after anaerobic growth in the light on a defined medium containing malate and glutamate but no added iron, convert glycine and α-ketoglutarate to porphyrins. The requirement for these substrates is specific; no other amino acid replaces glycine, and of the intermediates of the tricarboxylic acid cycle, only succinate replaces α-ketoglutarate although it is less active. Besides these substrates an energy source is necessary; synthesis occurs only anaerobically in the light and the addition of fumarate or other oxidizable compounds is necessary for maximum activity. The yield of porphyrin in these systems is high, accounting for up to 20% of the glycine and α-ketoglutarate added. In addition, the pigment is found almost exclusively in the extracellular fluid, which facilitates analysis. Coproporphyrin III accounts for about 95% of the total porphyrin, uroporphyrin III for about 3%, and traces of other porphyrins with 7 and 5 carboxyl groups are detectable by chromatography. The suspensions also convert ALA to porphyrin when incubated anaerobically in the light with phosphate and Mg^{++} only, and during the early stages of incubation PBG is also detectable. The mixture of porphyrins formed from ALA by iron-deficient cells has the same composition as that formed from glycine and α-ketoglutarate.

Oxygen suppresses synthesis of porphyrins from glycine and α-ketoglutarate whether the suspensions are incubated in the light or in the dark. Its inhibitory effect is reminiscent of its action in preventing synthesis of bacteriochlorophyll (see Section VI,C). On the other hand, porphyrin formation from ALA occurs aerobically, although the amount formed is only slight unless an oxidizable substrate such as fumarate or succinate is added; but even then, the total yield is only about one-third of that found under anaerobic conditions in the light.

Suspensions of a strain of *R. capsulatus* accumulate a colorless porphyrin precursor when incubated anaerobically in the light with glycine and succinate.[52] On exposure to air the precursor is rapidly converted to coproporphyrin III. It was shown to be coproporphyrinogen III by comparison with an authentic sample of this compound, prepared by catalytic reduction of the corresponding porphyrin. The observation that *R. spheroides* accumulates porphyrins whereas *R. capsulatus* accumulates the reduced porphyrinogens suggests that the latter organism is deficient in systems which catalyze the oxidation of porphyrinogens to porphyrins under anaerobic conditions in the light. The nature of these systems is quite unknown.

The experiments with whole cells suggested that the path of porphyrin formation is similar to that in animal tissues. Further work with cell-free preparations of *R. spheroides* have confirmed this. Synthesis of ALA from glycine and succinyl-CoA has been demonstrated by several groups of workers.[108-112] The succinyl-CoA may be added at substrate levels of concentration or can be generated either from succinate, ATP, and CoA by the succinic thiokinase present in the extracts or from α-ketoglutarate, CoA, and diphosphopyridine nucleotide by the action of α-ketoglutarate dehydrogenase. The enzyme responsible for the condensation of glycine and succinyl-CoA has now been purified 80-fold from extracts of *R. spheroides*[88b] It requires pyridoxal phosphate as cofactor.[88b,109-111] Biotin may also participate in this reaction, since it is inhibited by avidin.[112a] In addition, growth of *R. spheroides* in a biotin-deficient medium diminishes the activity of the enzyme.

The activity with respect to synthesis of ALA from these substrates is four to five times higher in extracts from organisms grown anaerobically in the light than in those from organisms cultured aerobically in the dark; this may be connected with the ability of the photosynthetically grown organisms to form porphyrins and bacteriochlorophyll.[112b] In both types of extract the enzyme system is in the supernatant fraction remaining after centrifugation of sonic extracts for 90 minutes at 105,000 *g*. The high activity of the system for synthesizing ALA in *R. spheroides* compared to those so far studied in animal tissues makes it a useful one for work on purification.

δ-Aminolevulic acid dehydrase, which catalyzes conversion of ALA to

PBG, has also been observed in extracts of *R. spheroides*. This activity is again higher in light- than in dark-grown organisms.[112b] The conversion of PBG to uro- and coproporphyrin with intermediate formation of porphyrinogens has been shown with acetone-dried preparations of the same organism.[113] By various treatments preparations can be obtained which convert PBG to (*a*) uroporphyrin I only, (*b*) uroporphyrin III (with some type I isomer), and (*c*) a mixture of uro- and coproporphyrin III. Two enzymes (PBG deaminase and uroporphyrinogen decarboxylase) have been separated and partly purified from such extracts; they catalyze, respectively, the deamination of PBG to uroporphyrinogen I and the decarboxylation of all four isomers of uroporphyrinogen to the corresponding coproporphyrinogen.[114]

(*c*) *Micrococcus lysodeikticus.* Townsley and Neilands[63] have shown conversion of ALA to PBG, uro-, and coproporphyrin III by lysed preparations of *M. lysodeikticus*, grown in an iron-deficient medium; the conversion is favored by anaerobic conditions.

(*d*) *Yeast.* There is some evidence that the tetrapyrrole nucleus is formed by the usual pathway in yeast. A mutant strain of *Saccharomyces cerevisiae* has been isolated which contains no detectable heme compounds when grown without glycine. Upon addition of this amino acid to the medium there is considerable synthesis of hemes including cytochrome c and catalase; protoporphyrin acts similarly to glycine.[115] Incorporation of glycine-2-C^{14} into the heme moiety of cytochrome c has also been observed in *S. cerevisiae*.[116] The fact that ALA dehydrase is not detectable in preparations of brewers' yeast suggests that a different pathway may also operate (see Gibson[77]).

(*e*) *Tetrahymena vorax.* Suspensions of this ciliate, grown in an iron-deficient medium, form a mixture of uro- and coproporphyrin III and protoporphyrin when incubated with glycine and succinate or with ALA.[67] The utilization of these substrates suggests again that the path of biosynthesis is identical with that found in other organisms.

2. Effect of Iron

The suppression by iron of porphyrin accumulation in cultures of various microorganisms has already been discussed (Section III). A similar phenomenon has been observed with cell suspensions; under these conditions, it is possible to obtain more exact information about the mode of action of iron.

With suspensions of *R. spheroides* synthesis of porphyrins from glycine and α-ketoglutarate under anaerobic conditions in the light is almost completely abolished by addition of iron salts, but at the same time the concentration of intracellular bacteriochlorophyll shows an 8-fold increase.[107]

The amount of iron required to produce this effect is small; porphyrin formation is reduced to about one-third of the value without added iron by 2 mμmoles/ml., and completely abolished by 5 mμmoles/ml. This suggests that iron acts catalytically: the amount of porphyrin that fails to appear in the presence of these low concentrations of iron is 50 to 100 times greater than the amount of iron added. The increased amounts of bacteriochlorophyll and of intracellular heme compounds that are formed in the presence of added iron are roughly equivalent to 10% and 1%, respectively, of the extracellular porphyrins that would have been produced in the absence of iron.

With ALA as substrate a rather different picture emerges. The addition of iron does not suppress porphyrin synthesis from ALA by suspensions of *R. spheroides*, but it does influence the composition of the mixture of porphyrins formed. Coproporphyrin III is still the predominant component, but considerable amounts of protoporphyrin are formed, as well as free extracellular iron protoporphyrin. Cobalt ions inhibit formation of protoporphyrin by the cell suspensions without affecting synthesis of coproporphyrin, suggesting that the locus of inhibition is at a stage between coproporphyrin (or the porphyrinogen) and protoporphyrin.[37]

No bacteriochlorophyll is synthesized with ALA as substrate, possibly because precursors of the phytol side chain are lacking.

Similar effects of iron have been observed with *Tetrahymena vorax*. Although addition of iron to suspensions of deficient cells does not influence porphyrin formation from either glycine or from ALA, profound effects are found with organisms that have been grown with added iron.[67] The iron-rich cells do not form free porphyrins from glycine and succinate, but they do convert ALA to a mixture of porphyrins containing predominantly protoporphyrin, whereas uroporphyrin III predominates with iron-deficient organisms.

Is it possible to reconcile the apparently distinct actions of iron (*a*) in promoting synthesis of protoporphyrin from ALA, and (*b*) in suppressing the accumulation of porphyrin from glycine and intermediates of the tricarboxylic acid? Since these effects have been observed in organisms as dissimilar as *R. spheroides* and *T. vorax*, it is reasonable to assume that they are general ones. The influence of iron on the composition of the mixture of porphyrins formed from ALA suggests that it may participate in the conversion of the propionic acid side chains of coproporphyrin to the vinyl side chains of protoporphyrin. This reaction is an oxidative decarboxylation, the mechanism of which is quite unknown; it may involve activation of the carboxyl groups, a possibility supported by the fact that, in addition to iron, an energy source is required for conversion of ALA to protoporphyrin by suspensions of *R. spheroides*. Whether iron participates in the activation or in the oxidation can be determined only when the enzyme systems have been purified. It is difficult to assess the observation that addition of iron to lysates of *M. lysodeikticus*[63] results in disappearance of added coproporphyrinogen III, since the reaction occurred in heat-treated preparations and may therefore have been largely nonenzymic.

The suppression of porphyrin accumulation with glycine but not with ALA as substrate might seem to suggest that iron inhibits the initial condensation reaction. This assumption cannot, however, be seriously entertained since synthesis of functionally active tetrapyrroles is actually favored by iron. The simplest way of accounting for the effects of iron is to assume that it acts primarily in the formation of protoporphyrin, an intermediate in the synthesis of both heme compounds and bacteriochlorophyll. In its presence, precursors of porphyrins would then be diverted toward synthesis of these substances, provided that sources of the nonpyrrole moieties were available. Since in *R. spheroides*, at any rate, the amount of coproporphyrin accumulating in the absence of iron is at least ten times greater than the amount of hemes and bacteriochlorophyll formed with iron, it is necessary, in addition, to invoke some kind of regulatory mechanism. An intracellular heme compound, possibly a hemoprotein, may regulate the activity of one or more of the enzymes concerned in the initial stages of porphyrin synthesis.

3. B-Group Vitamins

Tetrahymena vorax is a useful organism for examining the effect of B-group vitamins since it both accumulates excess porphyrins and requires a number of vitamins for growth. Synthesis of porphyrins can therefore be studied with suspensions of organisms deficient in the various factors.[67] The amount of porphyrin formed from glycine by suspensions of cells grown with suboptimal concentrations of pantothenate and pyridoxal is only about 20% of that synthesized by cells grown with an adequate vitamin supply. Subsequent work with enzyme preparations from various sources has shown clearly that the coenzyme forms of these vitamins participate in the initial condensation resulting in the formation of ALA (see above). Deficiency of pyridoxal or of pantothenate does not reduce the ability of the cells to convert ALA to porphyrins. However, the composition of the mixture formed by pantothenate-deficient cells differs from that formed by normal organisms in that the proportion of protoporphyrin is considerably reduced. This suggests that pantothenate, like iron, participates in the conversion of coproporphyrin III (or coproporphyrinogen III) to protoporphyrin. Possibly, the propionic acid side chains are oxidized to the vinyl level as their CoA esters in a fashion analogous to the oxidation of fatty acids. Of the other B-group vitamins examined with *T. vorax*, only riboflavin exerts a striking action; deficient organisms formed no porphyrin from glycine or from ALA. Hence it is possible that flavoproteins may participate in the over-all oxidation necessary to obtain porphyrins from ALA.

R. spheroides needs only nicotinic acid, thiamine, and biotin for growth; suspensions of organisms deficient in any one of these vitamins form very little porphyrin from glycine and α-ketoglutarate, whereas synthesis from ALA is unaffected. Nicotinic acid and thiamine are presumably required because of their participation (in coenzyme form) in the formation of succinyl-CoA from α-ketoglutarate. Biotin may be directly concerned in the enzymic condensation of glycine and succinyl-CoA.[112a] Obviously, the experiments with whole cells can provide only initial indications about the role of vitamins; detailed information will require studies with more refined systems.

4. Importance of the Tricarboxylic Acid Cycle

The tricarboxylic acid appears to play an important role in supplying intermediates for the early stages of pyrrole synthesis, not only providing succinyl-CoA but probably also contributing to the synthesis of glycine. The existence of the cycle has not been proved in all of the many organisms known to form cytochromes and other pyrrole derivatives,[117,118] but it seems reasonable to assume that it operates in full or in part in most of these microbes. It is perhaps significant that the cycle has yet to be demonstrated in the clostridia and some of the lactic acid bacteria, which lack cytochromes and catalase. On the other hand, the strongly fermentative propionibacteria contain cytochromes[119] and have also enzymes of the tricarboxylic acid cycle.[120] The strictly anaerobic *Desulfovibrio desulfuricans* is rich in cytochromes[121] but the existence of the full cycle has yet to be shown. The observation that suspensions can form CO_2 and acetate from α-ketoglutarate, succinate, fumarate, malate, or pyruvate[122,123] suggests that some, at least, of the necessary enzymes are present.

Acetobacter suboxydans and *A. melanogenum* provide a clear exception to the general rule that the ability to form the tetrapyrrole nucleus is associated with the presence of a tricarboxylic acid cycle. These organisms are rich in cytochromes[118] but lack many key enzymes of the cycle, including α-ketoglutarate dehydrogenase.[124] It is not known whether they form pyrroles by the established pathway or by some other route. The problem of pyrrole synthesis is not the only biochemical mystery in these organisms. Strains of *A. suboxydans* and *A. melanogenum* can use NH_3 as the sole source of nitrogen to form amino acids[125, 126] and must therefore have mechanisms which bypass the normal tricarboxylic acid cycle for synthesizing compounds such as glutamic and aspartic acids.

V. Synthesis of Heme and Hemoproteins

A. Insertion of Iron into the Tetrapyrrole Nucleus

The problem of how iron is inserted into the tetrapyrrole nucleus by microorganisms is still unsolved. The fact that all organisms so far shown to need preformed tetrapyrrole for growth respond only to protoporphyrin or hemin and not to uro- or coproporphyrin suggests that incorporation of iron occurs at the level of protoporphyrin. Suspensions of *R. spheroides* form free hematin as well as copro- and protoporphyrin when incubated with ALA, ferric citrate, and an energy source; but other iron porphyrins are not detectable.[37] Synthesis of coproheme has been demonstrated with lysed preparations of *Micrococcus lysodeikticus* incubated aerobically with coproporphyrinogen III and ferrous sulfate[63] but there is no evidence that this reaction is part of the biosynthetic sequence leading to heme or hemoproteins.

Mauzerall and Granick[105] have pointed out that the porphyrinogens cannot bind metals and that incorporation of iron must hence occur during or after the oxidation of these compounds to porphyrins. Ferrous iron coordinates spontaneously with protoporphyrin under physiological conditions.[101] It is therefore surprising that some organisms require hemin specifically for growth, being unable to utilize protoporphyrin and iron. Recently a soluble enzyme fraction prepared from rat liver mitochondria has been shown to catalyze the formation of heme from protoporphyrin and ferrous ions.[127,128] Organisms that respond only to hemin may lack this enzyme system, or they may lack some factor essential for the mobilization or stabilization of ferrous iron. Possibly the iron-binding substances (e.g., ferrichrome) required by some microorganisms may participate in the formation of heme.

B. Synthesis of Hemoproteins

1. Effect of Nutritional Factors

There is little information about the effect of nutritional factors on synthesis of hemoproteins in bacterial cultures apart from many observations that their level is influenced by the concentration of iron.[71, 129–132] In *Aerobacter indologenes* grown with limiting iron, some observations suggest that cytochromes may be formed preferentially to catalase and other enzymes in whose synthesis iron is involved; cyanide-sensitive respiration was the criterion employed to detect the presence of cytochromes.[129] Respiration of *Aerobacter aerogenes* in the presence of glucose is unaffected by growth in an iron-deficient medium, even though cytochrome a_2 becomes barely detectable spectroscopically and cytochrome a_1 and b_2 are decreased.[133]

The cytochrome spectrum of *Bacillus subtilis* varies with the physiological age of the organism[134] and also with the growth rate and the carbon source.[135b] The cytochromes of pseudomonads show similar variations.[135a] There is evidence that glucose, in relatively high concentration, represses hemoprotein formation in yeast.[136,140] Until more is known of the composition and interrelationships of the various cytochromes the bearing of these observations on the synthesis of cytochromes cannot be evaluated.

2. Effect of Oxygen

The level of many enzymes in microorganisms is influenced by the degree of aeration during growth.[137] Hemoproteins are particularly affected; indeed, in many cases oxygen acts as a regulator of their synthesis.

(*a*) *Formation of Heme Compounds in Response to Oxygen.* The dependence of hemoprotein synthesis upon oxygen was first observed in yeast.[66,138-141a,141b] Anaerobically grown *Saccharomyces cerevisiae* contains only cytochromes a_1 and b_1, but these are replaced by cytochromes a, a_3, b, and c when the organisms are aerated in buffered glucose. The changes in cytochrome content are accompanied by an increase in the rate of respiration; hence the phenomenon has been termed "adaptation respiratoire." The mutant strain, "petite colonie," forms only cytochrome c on aeration and there is no rise in the respiratory rate. In normal yeast there is a considerable increase upon aeration in enzymes concerned in the tricarboxylic acid cycle.[66,142a] This may be due in part to the formation of a complete electron transport system to oxygen, since the succinic dehydrogenase activity is highly active in both anaerobic and aerobic cells.[142b] Other hemoproteins increased by aeration are cytochrome peroxidase and catalase.[143]

Protein synthesis seems to be linked to these changes induced by aeration since the phenomenon is inhibited by *p*-fluorophenylalanine and other amino acid analogues.[140,144,145a] Formation of nucleic acid also accompanies the adaptation since incorporation of radioactive adenine and uracil is increased during the adaptation to form hemoproteins.[145b]

Slonimski[66] found that the total heme content of anaerobically grown yeast does not differ significantly from that found after aeration, suggesting that the adaptation involves synthesis of the protein moiety of the enzymes. Experiments with more sensitive isotopic techniques have shown, however, that net synthesis of heme does occur during adaptation of both normal and "petite colonie" yeast but incorporation of C^{14}-glycine into the protein of cytochrome c may be considerably greater than that fixed into the heme moiety.[116] It is not known whether transfer of heme from the cytochromes a_1 and b_1, originally present in the organisms, occurs during adaptation in oxygen.

A similar phenomenon has been found in *Pasteurella pestis*.[146-148] Cytochromes are not detectable in anaerobically grown cells of this organism and there are only traces of free hematin; also, the rate of respiration of glucose and other substrates is low, and the oxidation is incomplete. Aeration of suspensions of such organisms results in synthesis of both cytochrome b and free hematin, accompanied by an increase in respiratory activity. A nitrogen source is necessary for these changes to occur, and the process is inhibited by irradiation with ultraviolet light. As in yeast, adaptation in response to oxygen involves not only synthesis of hematin and cytochrome b, but also of many enzymes of the tricarboxylic acid cycle that are only slightly active in anaerobically grown cells.

Only cytochrome b is found in anaerobically grown *Bacillus cereus*, whereas aerobically grown cells contain cytochromes a and c in addition to b.[73] The total amount of heme in aerobically grown cells is about five times as great as in anaerobically grown cells. Observations on the respiratory activity of *B. subtilis* grown under various conditions suggest that synthesis of the cytochromes of this organism is controlled by oxygen, much as in *B. cereus*.[149]

Development of cytochromes in response to aeration has also been observed in *E. coli*[150] and related bacteria.[151a]

In *Rhodopseudomonas spheroides* catalase is synthesized rapidly upon aeration, being present only in traces in cells grown anaerobically in the light.[151b,151c]

(*b*) *Suppression of Hemoprotein Synthesis by Oxygen.* In contrast to the observations described so far, there are many examples of hemoprotein synthesis being suppressed by oxygen.

Formation of cytochromes by pseudomonads seems to be particularly sensitive to oxygen. The cytochrome content of *P. fluorescens* represents 4.1% of the total bacterial protein in cells grown in an atmosphere containing 1% oxygen; this figure falls to 0.3% for cells grown in an atmosphere containing 20% oxygen. The content of cytochrome peroxidase and of catalase is likewise lowered.[132] Similar observations have been made with a pseudomonad growing in continuous culture; more cytochrome is formed when the growth-limiting factor is oxygen than when it is succinate.[152]

Striking changes in the level of cytochromes have been found in denitrifying bacteria. The cells of such organisms are exceptionally rich in cytochromes when grown anaerobically with nitrate as acceptor[153] and there is good evidence that "nitrate respiration" occurs via hemoproteins.[37,154,155,156a] The concentration of cytochromes in the denitrifiers *P. aeruginosa*, *P. stutzeri*, and *Micrococcus denitrificans* is four to five times higher in cells grown anaerobically on nitrate than in cells grown with strong aeration, while growth with a lower degree of aeration results in an intermediate cytochrome content.[154,156b] It is well known that oxygen suppresses the development of nitratase activity[157] and this effect may be related to its apparent inhibitory effect on cytochrome formation. In the Athiorhodaceae, growth under a high oxygen tension also decreases the cytochrome content.[151b]

Oxygen appears to suppress cytochrome synthesis in some organisms and to stimulate it in others. A thorough study has yet to be made of the relationship between oxygen tension and the production of hemoproteins in organisms where formation of these pigments is induced by exposure to oxygen. There is some evidence that the level of cytochromes in such organisms may also be influenced by the concentration of available oxygen; in *A. aerogenes* growing in continuous culture the concentration of cytochrome a_2 reaches a maximum at a critical concentration of oxygen, above or below which the pigment content decreases.[158] Hence it is possible that a single basic mechanism of regulation by oxygen is operative in all cases.

3. Formation of Hemoproteins by Organisms Requiring Tetrapyrroles for Growth

Organisms requiring preformed tetrapyrroles for growth offer good material for the study of the synthesis of hemoproteins. There are many observations showing that the contents of cytochrome and catalase, and the ability to reduce nitrate are functions of the concentration of hemin (or protoporphyrin) in the growth medium.[16,18,34,40,159]

Synthesis of hemoproteins has been studied with suspensions of the streptomycin-resistant strain of *Micrococcus pyogenes*, which needs hemin for growth.[34] This organism can be grown in the absence of hemin in complex medium or in a semidefined medium containing acetate or pyruvate and nucleic acid derivatives.[34,37] Cells which have grown under these conditions contain no detectable cytochromes, and are devoid of respiratory activity, catalase, and nitratase. They probably manufacture the apoenzymes, however, since incubation of deficient cells with hemin and glucose results in restoration of these activities within a matter of minutes.[35,37] Further experiments with hemin-deficient organisms[160] suggest that synthesis of catalase from added hemin requires an energy source and is favored by reducing conditions; coenzyme A stimulates enzyme formation, possibly by virtue of its reducing activity. It would be interesting to know whether an intracellular pool of free amino acids is required for synthesis of the hemoproteins.

In a streptomycin-resistant strain of *E. coli* which needs hemin for growth at maximal rate, suspensions of hemin-deficient cells have been shown to synthesize catalase. Restoration of catalase activity to the level found in normal organisms can be obtained by the addition of hemin to cell-free extracts of organisms grown without this substance. The reactivation is inhibited by prior incubation of the extracts with free porphyrins containing two or less free carboxyl groups, but not by prior incubation with uro- or coproporphyrin.

VI. Synthesis of the Chlorophylls

A. ALGAE

The work of Granick and colleagues (reviewed by Granick[2,77]) has shown clearly that the early steps of heme and chlorophyll synthesis share a common pathway, which probably branches at the stage of protoporphyrin. Induced mutants of *Chorella* that are unable to synthesize chlorophyll accumulate a variety of porphyrins. Some of the accumulated substances have been isolated and identified; these include hematoporphyrin IX, protoporphyrin IX, magnesium protoporphyrin, and magnesium vinylpheoporphyrin a_5. Other dicarboxylic porphyrins lacking vinyl groups or containing only one vinyl group, as well as porphyrins with three to eight carboxyl groups, have been detected chromatographically.

There are large gaps in our knowledge of the path of chlorophyll synthesis (see Fig. 7). In particular, the steps leading to the formation of the isocyclic ring and the stage of attachment of the phytol group are obscure; there is some evidence that esterification with phytol may occur at the final stage, after reduction of ring D of the pyrrole nucleus.[161] Preparations obtained from higher plants carry out the conversion of protein-bound protochlorophyll to chlorophyll *a*,[162a] but the path of chlorophyll synthesis has yet to be studied at the enzymic level.

B. PHOTOSYNTHETIC BACTERIA

Bacteriochlorophyll, found in photosynthetic bacteria belonging to the Thiorhodaceae and Athiorhodaceae, has a tetrahydropyrrole ring structure, whereas chlorophyll *a* is a dihydropyrrole (Fig. 2).[7,162b] This difference has a profound effect on the absorption spectrum, and is of considerable ecological significance.[163]

Two distinct molecular species of chlorophyll have been recognized in different strains of the green sulfur bacterium, *Chlorobium thiosulfatophilum*.[164a] One is spectroscopically similar to plant chlorophyll *a* but shows different chromatographic properties.[164b,165a] The other type of chlorobium chlorophyll has a markedly different absorption spectrum.[164a] There is evidence that both pigments lack the isocyclic ring found in all chlorophylls of known structure.[164a,165b]

Evidence that the early stages of synthesis of bacteriochlorophyll and porphyrins involve a common path has already been discussed. Knowledge of the specific steps in chlorophyll synthesis beyond protoporphyrin is virtually nonexistent. Some of them may be identical with the steps that lead to formation of chlorophylls. This is suggested by the accumulation of

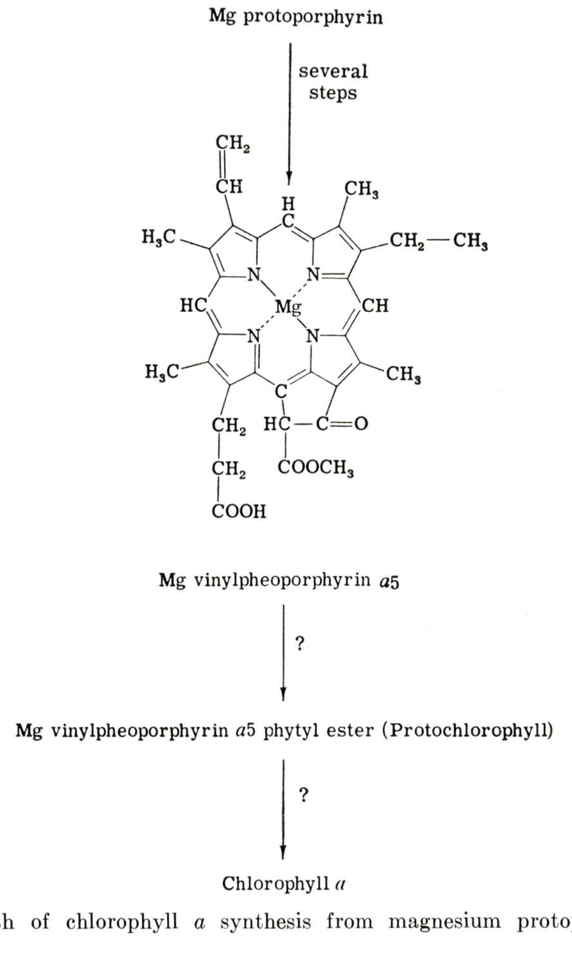

FIG. 7. Path of chlorophyll *a* synthesis from magnesium protoporphyrin.

magnesium protoporphyrin in cultures of some of the Athiorhodaceae.[52] The close similarity of the biosynthetic pathways is also suggested by the excretion of compounds with the same dihydropyrrole ring system as plant chlorophyll by cultures of the blue-green mutant strain of *R. spheroides*, which does not form colored carotenoids.[53a] Some of these compounds have been isolated and their absorption spectrum, solubility, and positive reaction in the phase test (denoting the presence of the isocyclic ring) shows that they are closely related to pheophorbide *a* (see Fig. 8); of the two major components of the mixture of pigments, one is probably pheophorbide *a*, while the other may differ in some substituent group only. Only one compound with the tetrahydropyrrole ring system of bacteriochlorophyll was detected—this being probably bacteriopheophorbide. The demonstration of a protochlorophyll-like pigment in another mutant of *R. spheroides* unable to make bacteriochlorophyll again points to a close similarity in the biosynthetic pathways leading to plant and bacterial chlorophylls.[166a] It is interesting that the excreted substances all appear to lack the phytol group, suggesting that esterification occurs at the final stages after reduction of the pyrrole rings.

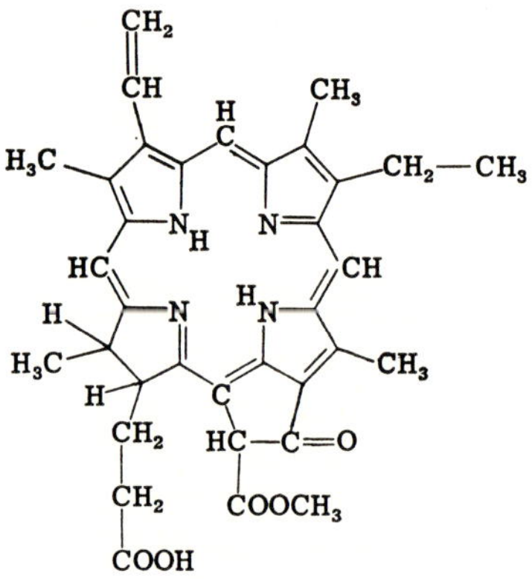

FIG. 8. Pheophorbide *a*.

Synthesis of bacteriochlorophyll by suspensions of *R. spheroides* is inhibited by chloramphenicol, *p*-fluorophenylalanine, and 8-azaguanine, suggesting that formation of the pigment is closely connected with synthesis of protein and nucleic acid.[112b] The suspensions excrete small amounts of a pheophorbidelike pigment in the presence of these analogs (unpublished personal observations).

C. Effect of Oxygen and Light Intensity on Bacteriochlorophyll Synthesis

It has been known for many years that the formation of bacteriochlorophyll and carotenoids by the Athiorhodaceae is favored by anaerobic growth in the light; cultures of these organisms capable of aerobic growth in the dark are only slightly pigmented.[47] Kinetic studies with cultures of *R. spheroides* and *Rhodospirillium rubrum* have shown that synthesis of bacteriochlorophyll and carotenoids is affected both by oxygen and by the light intensity.[50] Synthesis of photopigments by cultures growing under continuous illumination ceases abruptly when oxygen is introduced into the gas phase. The inhibition by oxygen is reversible and pigment synthesis is resumed when cultures are returned to anaerobic conditions. In addition, the light intensity influences the amount of pigment formed under anaerobic conditions. The rate of pigment synthesis is maximal at low light intensities, and declines as the light intensity is increased.

Cultures of *R. spheroides* and other members of the Athiorhodaceae become pigmented on prolonged incubation in the dark under aerobic conditions, a fact which suggests that they can form bacteriochlorophyll and carotenoids in the dark provided that the tension of oxygen is low. This has been confirmed by experiments with cell suspensions of *R. spheroides* harvested after aerobic growth in the dark.[112b] When such cells are incubated with appropriate substrates in the dark, they form bacteriochlorophyll provided that the oxygen tension is low; pigment synthesis is inhibited under high tensions of oxygen.

VII. Prodigiosin

Prodigiosin is the red pigment long considered to be characteristic of *Serratia marcescens* and related species. It now appears to have a wider biological distribution, since an apparently identical pigment has been isolated from actinomycetes.[166b] As a result of their chemical studies, Wrede and Rothhaas[167] suggested that prodigiosin is a tripyrrylmethene, an assumption which appeared to be supported by its spectral similarities to synthetic tripyrrylmethenes.[168b] Such a structure has now been shown to be untenable, and a dipyrrylmethene structure (see Fig. 9) has been proposed in its place.[168a] This structure is based principally on the evidence derived from biosynthetic studies: a dipyrrole is accumulated by a mutant of *S. marcescens* unable to synthesize prodigiosin (*vide infra*), and this dipyrrole can be converted to prodigiosin by a second mutant strain. The dipyrrole likewise reacts chemically with 2-methyl, 3-amyl pyrrole (originally isolated by Wrede and Rothhaas[167] after degradation of prodigiosin) to form prodigiosin, under conditions which are characteristic for the synthesis of dipyrrylmethenes.

Prodigiosin may be a mixture of components closely related in structure but perhaps differing in their side chains or the degree of aggregation of the molecule.[169,170-173a] Chromatography of extracts of *S. marcescens* has revealed at least four compounds with similar absorption spectra but differing R_f values. Young cultures contain two chromatographically distinct red substances; in older cultures a blue and an orange-red pigment are also found.

Study of the incorporation of isotopically labeled compounds in prodigiosin originally suggested that it is synthesized in a similar fashion to the porphyrins.[168b] Both glycine and acetate are precursors; the glycine nitrogen supplies all three nitrogens of the pigment molecule and the α-carbon atom is also extensively incorporated. However, a common pathway for the synthesis of prodigiosin and tetrapyrroles seems now to be excluded, since C^{14}-labeled ALA is not incorporated, and proline is a more effective precursor than glycine.[173b]

Several mutant strains of *S. marcescens* which excrete substances capable of promoting prodigiosin synthesis by white mutants have been described.[174-178] One of these substances has been isolated and identified

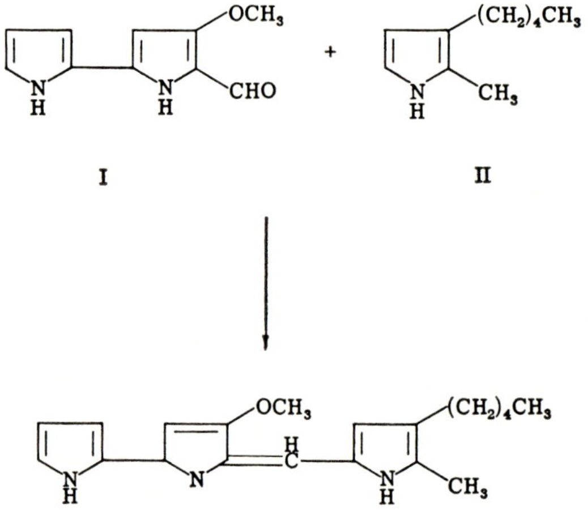

FIG. 9. The pathway of synthesis of prodigiosin from the dipyrrole precursor (I) and 2-methyl-3-amyl-pyrrole (II).

(Compound I, Fig. 9).[178] It contains C^{14} when formed by cultures grown with glycine-2-C^{14}; the isolated substance so labeled with C^{14} is converted to prodigiosin of the same specific activity by a white mutant. This white mutant has now been shown to excrete methyl amyl pyrrole, which condenses with compound I to form prodigiosin (Fig. 9).[178a]

VIII. The Regulation of Synthesis of Tetrapyrroles

Oxygen exerts an important general influence on tetrapyrrole synthesis: formation of cytochromes and other hemoproteins is regulated by the tension of oxygen, and in the photosynthetic bacteria oxygen suppresses synthesis of both bacteriochlorophyll and free porphyrins. Nor is the effect of oxygen confined to microorganisms. The hemoglobin content of many invertebrates is increased by growth under conditions of low oxygenation,[179a] while in man it is well known that the concentration of this pigment becomes raised at high altitudes, where the partial pressure of oxygen is diminished.

Glycine and succinyl-CoA are utilized in a variety of biosynthetic pathways. Hence it seems likely that the level of activity of ALA synthetase could play an important part in regulating the flow of these metabolites towards tetrapyrrole formation. A control at this level is suggested by the fact that the differential rate of synthesis of ALA synthetase in growing cultures of *R. spheroides* is affected by environmental factors in the same way as is the differential rate of synthesis of bacteriochlorophyll.[179b] In cultures growing under highly aerobic conditions, when little or no bacteriochlorophyll is formed, the rate of synthesis of the enzyme is about one-third of that in cultures growing anaerobically in the light. The synthesis both of the enzyme and of bacteriochlorophyll is repressed by high light intensities under anaerobic conditions. The formation of ALA dehydrase, the next enzyme in the sequence leading to tetrapyrroles, is similarly affected.

The mechanism by which oxygen or a high light intensity exerts its effect on the synthesis of these enzymes is unknown. The proposal of Cohen-Bazire *et al.*[50] that formation of photopigments in the Athiorhodaceae is regulated by the state of oxidation of a carrier in the electron transport system could apply also to the regulation of enzyme synthesis.

The available concentration of the terminal electron acceptor could also influence the intracellular concentration of key intermediates required for tetrapyrrole synthesis. Under the reducing conditions engendered by a low tension of oxygen (or, in photosynthetic bacteria, by a low light intensity) the rate of oxidation of succinate, for instance, may be diminished, thus making it more readily available for tetrapyrrole formation. With organisms possessing a tricarboxylic acid cycle there is evidence that oxidation of succinate becomes a rate-limiting step under anerobic conditions.[180] The fact that oxygen is needed for synthesis of hematin and hemoproteins in some organisms does not necessarily invalidate this suggestion, since in these organisms oxygen is also needed for full development of enzymes of the tricarboxylic acid cycle and the cycle must operate, in part at least, for tetrapyrrole synthesis to occur. The main point of the present argument is that there may be more opportunity to divert key intermediates toward biosynthesis when the cycle is not operating at its full rate.

Another way in which oxygen may influence tetrapyrrole synthesis is by inhibiting the *action* rather than the *formation* of an enzyme in the biosynthetic pathway. Such an effect has been observed with hemolysates of chicken blood, in which conversion of glycine to porphyrins and heme is inhibited by high tensions of oxygen.[181b]

Regulation of tetrapyrrole synthesis may also be compared with other biosynthetic pathways that are controlled by the concentration of the end product, acting either by enzyme repression or by negative feedback mechanisms.[181a,182] Both ALA and heme repress the formation of ALA synthetase by growing cultures of *R. spheroides*, and this might have some physiological significance.[179b] The excretion of coproporphyrin III by iron-deficient cultures of *R. spheroides*[48] and *C. diphtheriae*[183] occurs only in the final stages of growth and the amount of porphyrin formed is considerably in excess of the amount of hemoprotein and bacteriochlorophyll formed in the presence of iron, which suggests that some type of negative feedback control operates. Future research in this direction may be profitable.

REFERENCES

1 S. Granick and H. Gilder, *Advances in Enzymol.* **7,** 305 (1947).
2 S. Granick, *in* "Chemical Pathways of Metabolism" (D. M. Greenberg, ed.), Vol. II, Chapter 16, p. 287. Academic Press, New York, 1954.
3 C. J. Watson and E. A. Larson, *Physiol. Revs.* **27,** 478 (1947).
4 J. E. Falk, *Biochem. Soc. Symposia (Cambridge, Engl.) No.* **12,** 17 (1954).
5 R. Lemberg, *Australasian Ann. Med.* **4,** 5 (1955).
6 C. Rimington, *Ann. Rev. Biochem.* **26,** 561 (1957).
7 H. Fischer and H. Orth, "Die Chemie des Pyrrols." Edwards Brothers, Ann Arbor, Michigan, 1943.
8 R. Lemberg and J. W. Legge, "Hematin Compounds and Bile Pigments." Interscience, New York, 1949.
9 E. I. Rabinowitch, "Photosynthesis and Related Processes," Vols. I, II, III. Interscience, New York, 1945, 1951, 1956.
10 J. E. Falk, *Brit. Med. Bull.* **10,** 211 (1954).
11 T. Thjötta and O. T. Avery, *J. Exptl. Med.* **34,** 97 (1921).
12 T. Thjötta and O. T. Avery, *J. Exptl. Med.* **34,** 455 (1921).
13 R. Pfeiffer, *Z. Hyg. Infektionskrankh.* **13,** 357 (1893).
14 P. Fildes, *Brit. J. Exptl. Pathol.* **2,** 16 (1921).
15 O. Olsen, *Zentr. Bakteriol. Parasitenk. Abt. I. Orig.* **85,** 12 (1921).
16 S. Granick and H. Gilder, *J. Gen. Physiol.* **30,** 1 (1946).
17 H. Gilder and S. Granick, *J. Gen. Physiol.* **31,** 103 (1947).
18 A. Lwoff and M. Lwoff, *Ann. inst. Pasteur* **59,** 129 (1937).
19 M. Lwoff, *Ann. inst. Pasteur* **51,** 55 (1933).
20 M. Lwoff, *Ann. inst. Pasteur* **51,** 707 (1933).
21 A. Lwoff, *Bull. soc. chim. biol.* **30,** 817 (1948).
22 M. Lwoff, *in* "Biochemistry and Physiology of Protozoa" (A. Lwoff, ed.) Vol. I, p. 129. Academic Press, New York, 1951.
23 A. Lwoff, *Zentr. Bakteriol. Parasitenk. Abt. I. Orig.* **130,** 498 (1934).
24 D. Herbert, *Brit. J. Exptl. Pathol.* **30,** 509 (1949).
25 E. Rowatt, *J. Gen. Microbiol.* **17,** 279 (1957).
26 M. W. Fisher, *Am. Rev. Tuberc.* **69,** 797 (1954).
27 R. Knox, *J. Gen. Microbiol.* **12,** 191 (1955).
28 G. Middlebrook, *Am. Rev. Tuberc.* **69,** 471 (1954).
29 R. Bönicke, *Naturwissenschaften* **41,** 430 (1954).
30 A. Andrejew, C. Gernez-Rieux, and A. Tacquet, *Ann. inst. Pasteur* **93,** 281 (1957).

[31a] R. J. Gibbons and J. B. MacDonald, *J. Bacteriol.* **80,** 164 (1960).
[31b] R. J. Evans, *J. Gen. Microbiol.* **5,** xix (1951).
[32] M. Lev, *Nature* **181,** 203 (1958); *J. Gen. Microbiol.* **20,** 697 (1959).
[33] H. Schwabacher, D. R. Lucas, and C. Rimington, *J. Gen. Microbiol.* **1,** 109 (1947).
[34] J. Jensen and E. Thofern, *Z. Naturforsch.* **8b,** 599 (1953).
[35] J. Jensen and E. Thofern, *Z. Naturforsch.* **8b,** 604 (1953).
[36] J. Jensen and E. Thofern, *Z. Naturforsch.* **9b,** 596 (1954).
[37] J. Lascelles, *J. Gen. Microbiol.* **15,** 404 (1956).
[38] M. Kiese, H. Kurz, and E. Thofern, *Biochem. Z.* **330,** 541 (1958).
[39] M. Beljanski, *Compt. rend.* **240,** 374 (1955).
[40] M. Beljanski and M. Beljanski, *Ann. inst. Pasteur* **92,** 396 (1957).
[41] J. B. Neilands, *Bacteriol. Revs.* **21,** 101 (1957).
[42] J. Neilands, *J. Am. Chem. Soc.* **74,** 4846 (1952).
[43] J. A. Garibaldi and J. B. Neilands, *J. Am. Chem. Soc.* **77,** 2429 (1955).
[44] C. W. Hesseltine, C. Pidacks, A. R. Whitehill, M. Bohonos, B. L. Hutchings, and J. H. Williams, *J. Am. Chem. Soc.* **74,** 1362 (1952).
[45] C. W. Hesseltine, A. R. Whitehill, C. Pidacks, M. Ten Hagen, N. Bohonos, N. L. Hutchings, and J. H. Williams, *Mycologia* **45,** 7 (1953).
[46a] M. O. Burton, F. J. Sowden, and A. G. Lochhead, *Can. J. Biochem. Physiol.* **32,** 400 (1954).
[46b] A. L. Demain and D. Hendlin, *J. Gen. Microbiol.* **21,** 72 (1959).
[46c] D. Hendlin and A. L. Demain, *Nature* **184,** 1894 (1959).
[47] C. B. van Niel, *Bacteriol. Revs.* **8,** 1 (1944).
[48] J. Lascelles, *Ciba Foundation Symposium on Porphyrin Biosynthesis and Metabolism*, p. 265 (1955).
[48a] K. K. Voinovskaya and A. A. Krasnovskii, *Biokhimiya* **20,** 123 (1955).
[49] M. D. Kamen, *in* "Enzymes: Units of Biological Structure and Function" (O. H. Gaebler, ed.), p. 483. Academic Press, New York, 1956.
[50] G. Cohen-Bazire, W. R. Sistrom, and R. Y. Stanier, *J. Cellular Comp. Physiol.* **49,** 25 (1957).
[51] H. Larsen, *J. Bacteriol.* **64,** 187 (1952).
[52] R. Cooper, *Biochem. J.* **63,** 25P (1956).
[53a] W. R. Sistrom, M. Griffiths, and R. Y. Stanier, *J. Cellular Comp. Physiol.* **48,** 459 (1956).
[53b] R. Y. Stanier and J. H. C. Smith, *Biochim. et Biophys. Acta* **41,** 478 (1960).
[54] A. M. Pappenheimer, Jr., *J. Biol. Chem.* **167,** 251 (1947).
[55] C. H. Gray and L. B. Holt, *Biochem. J.* **43,** 191 (1948).
[56] W. Hodgkiss, J. Liston, T. W. Goodwin, and M. Jamikorn, *J. Gen. Microbiol.* **11,** 438 (1954).
[57] J. G. Morris, *J. Gen. Microbiol.* **22,** 564 (1960).
[58] H. Fischer and H. Fink, *Z. physiol. Chem. Hoppe-Syeler's* **150,** 243 (1925).
[59] C. M. Todd, *Biochem. J.* **45,** 386 (1949).
[60] M. Bariéty, A. Gajdos, M. Gajdos-Torok, and J. Sifferlen, *Ann. inst. Pasteur* **93,** 553 (1957).
[60a] D. S. P. Patterson, *Biochem. J.* **76,** 189 (1960).
[61] P. Schaeffer, *Compt. rend.* **231,** 381 (1950).
[62] J. A. Garibaldi and J. B. Neilands, *Nature* **177,** 526 (1956).
[63] P. M. Townsley and J. B. Neilands, *J. Biol. Chem.* **224,** 695 (1957).
[64] J. Pawelkiewicz and K. Zodrow, *Acta Biochim. Polon.* **3,** 225 (1956).
[65] W. Stich and H. Eisgruber, *Z. physiol. Chem. Hoppe-Seyler's* **287,** 19 (1951).

[66] P. P. Slonimski, "La Formation des Enzymes Respiratoires chez la Levure." Masson, Paris, 1953.
[67] J. Lascelles, *Biochem. J.* **66,** 65 (1957).
[68] M. A. Rudzinska and S. Granick, *Proc. Soc. Exptl. Biol. Med.* **83,** 525 (1953).
[69] S. Campbell Smith, *Lancet* **1,** 529 (1930).
[70] C. B. Coulter and F. M. Stone, *J. Gen. Physiol.* **14,** 583 (1931).
[71] A. M. Pappenheimer, Jr. and E. D. Hendee, *J. Biol. Chem.* **171,** 701 (1947).
[72] G. D. Clarke, *J. Gen. Microbiol.* **18,** 698 (1958).
[73] P. Schaeffer, *Biochim. et Biophys. Acta* **9,** 261 (1952).
[74] P. Schaeffer, *Biochim. et Biophys. Acta* **9,** 362 (1952).
[75] D. Shemin, *in* "Currents in Biochemical Research" (D. E. Green, ed.), p. 518. Interscience, New York, 1956.
[76] C. Rimington, *Brit. Med. Bull.* **15,** 19 (1959).
[77] Various authors in *Ciba Foundation Symposium on Porphyrin Biosynthesis and Metabolism* (1955).
[78] K. Bloch and D. Rittenberg, *J. Biol. Chem.* **159,** 45 (1945).
[79] D. Shemin and D. Rittenberg, *J. Biol. Chem.* **166,** 621 (1946).
[80] D. Shemin, I. M. London, and D. Rittenberg, *J. Biol. Chem.* **183,** 757 (1950).
[81] I. M. London, D. Shemin, and D. Rittenberg, *J. Biol. Chem.* **183,** 749 (1950).
[82] D. Shemin and C. S. Russell, *J. Am. Chem. Soc.* **75,** 4873 (1953).
[83] A. Neuberger and J. J. Scott, *Nature* **172,** 1093 (1953).
[84] D. Shemin, C. S. Russell, and T. Abramsky, *J. Biol. Chem.* **215,** 613 (1955).
[85] W. G. Laver, A. Neuberger, and S. Udenfriend, *Biochem. J.* **70,** 4 (1958).
[86] K. D. Gibson, W. G. Laver, and A. Neuberger, *Biochem. J.* **70,** 71 (1958).
[87] S. Granick, *J. Biol. Chem.* **232,** 1101 (1958).
[88a] E. G. Brown, *Biochem. J.* **70,** 313 (1958).
[88b] G. Kikuchi, A. Kumar, and D. Shemin, *Federation Proc.* **18,** 259 (1959).
[89] K. D. Gibson, A. Neuberger, and J. J. Scott, *Biochem. J.* **61,** 618 (1955).
[90] J. E. Falk, E. I. B. Dressel, and C. Rimington, *Nature* **172,** 292 (1953).
[91] R. G. Westall, *Nature* **170,** 614 (1952).
[92] G. H. Cookson and C. Rimington, *Biochem. J.* **57,** 476 (1954).
[93] E. I. B. Dresel and J. E. Falk, *Biochem. J.* **63,** 80 (1956).
[94] J. E. Falk, E. I. B. Dresel, A. Benson, and B. C. Knight, *Biochem. J.* **63,** 78 (1956).
[95] L. Bogorad and S. Granick, *Proc. Natl. Acad. Sci. U. S.* **39,** 1176 (1953).
[96] L. Bogorad, *J. Biol. Chem.* **233,** 516 (1958).
[97] E. I. B. Dresel and J. E. Falk, *Biochem. J.* **63,** 388 (1956).
[98] R. A. Neve, R. F. Labbe, and R. A. Aldrich, *J. Am. Chem. Soc.* **78,** 691 (1956).
[99] H. L. Booij and C. Rimington, *Biochem. J.* **65,** 4P (1957).
[100] W. H. Lockwood and A. Benson, *Biochem. J.* **75,** 372 (1960).
[101] S. Granick and D. Mauzerall, *J. Biol. Chem.* **232,** 1119 (1958).
[102] L. Bogorad, *J. Biol. Chem.* **233,** 501 (1958).
[103] L. Bogorad, *J. Biol. Chem.* **233,** 510 (1958).
[104] C. Rimington and H. L. Booij, *Biochem. J.* **65,** 3P (1957).
[104a] S. Sano, S. Inoue, Y. Tanabe, C. Sumiya, and S. Koike, *Science* **129,** 275 (1959).
[105] D. Mauzerall and S. Granick, *J. Biol. Chem.* **232,** 1141 (1958).
[106] J. H. Hale, W. A. Rawlinson, C. H. Gray, L. B. Holt, C. Rimington, and W. Smith, *Brit. J. Exptl. Pathol.* **31,** 96 (1950).
[107] J. Lascelles, *Biochem. J.* **62,** 78 (1956).
[108] G. Kikuchi, D. Shemin, and B. J. Bachmann, *Biochim. et Biophys. Acta* **28,** 219 (1958).
[109] G. Kikuchi, A. Kumar, P. Talmage, and D. Shemin, *J. Biol. Chem.* **233,** 1214 (1958).

[110] K. D. Gibson, *Biochim. et Biophys. Acta* **28,** 451 (1958).
[111] E. Sawyer and R. A. Smith, *Soc. Am. Bacteriologists Proc.* **P49,** 111 (1958).
[112a] A. Neuberger, *Biochem. J.* **78,** 1 (1961).
[112b] Lascelles, *Biochem. J.* **72,** 508 (1959).
[113] H. Heath and D. S. Hoare, *Biochem. J.* **72,** 14 (1959).
[114] D. S. Hoare and H. Heath, *Biochem. J.* **73,** 679 (1959).
[115] M. Yčas and T. J. Starr, *J. Bacteriol.* **65,** 83 (1953).
[116] M. Yčas and D. L. Drabkin, *J. Biol. Chem.* **224,** 921 (1957).
[117] W. Frei, L. Riedmüller, and F. Almasy, *Biochem. Z.* **274,** 253 (1934).
[118] L. Smith, *Bacteriol. Revs.* **18,** 106 (1954).
[119] P. Chaix and C. Fromageot, *Trav. membres soc. chim. biol.* **24,** 1125 (1952).
[120] E. A. Delwiche and S. F. Carson, *J. Bacteriol.* **65,** 318 (1953).
[121] J. R. Postgate, *J. Gen. Microbiol.* **14,** 545 (1956).
[122] J. P. Grossman and J. R. Postgate, *J. Gen. Microbiol.* **12,** 429 (1955).
[123] J. C. Senez, *Bull. soc. chim. biol.* **36,** 541 (194).
[124] M. R. R. Rao, *Ann. Rev. Microbiol.* **11,** 317 (1957).
[125] M. R. R. Rao and J. L. Stokes, *J. Bacteriol.* **65,** 405 (1953).
[126] I. O. Foda and R. H. Vaughn, *J. Bacteriol.* **65,** 79 (1953).
[127] G. Nishida and R. F. Labbe, *Biochim. et Biophys. Acta* **31,** 519 (1959).
[128] R. F. Labbe and N. Hubbard, *Biochim. et Biophys. Acta* **41,** 185 (1960).
[129] W. S. Waring and C. H. Werkman, *Arch. Biochem.* **4,** 75 (1944).
[130] J. R. Postgate, *J. Gen. Microbiol.* **15,** 186 (1956).
[131] W. E. van Heyningen, *Brit. J. Exptl. Pathol.* **36,** 381 (1955).
[132] H. M. Lenhoff, D. J. D. Nicholas, and N. O. Kaplan, *J. Biol. Chem.* **220,** 983 (1956).
[133] A. Tissières, *Biochem. J.* **50,** 279 (1951).
[134] P. Chaix and G. Roncoli, *Biochim. et Biophys. Acta* **6,** 268 (1950).
[135a] K. Titani, S. Minakani, and H. Mitsui, *J. Biochem.* (*Tokyo*) **47,** 290 (1960).
[135b] P. Chaix and J. F. Petit, *Biochim. et Biophys. Acta* **25,** 481 (1957).
[136] C. F. Strittmatter, *J. Gen. Microbiol.* **16,** 169 (1957).
[137] E. F. Gale, *Bacteriol. Revs.* **7,** 139 (1943).
[138] B. Ephrussi and P. P. Slonimski, *Biochim. et Biophys. Acta* **6,** 256 (1950).
[139] P. P. Slonimski, *in* "Adaptation in Microorganisms" (E. F. Gale and R. Davies, eds.), p. 76. Cambridge Univ. Press, London and New York, 1953.
[140] P. P. Slonimski, *3rd Intern. Congr. Biochem. Brussells,* p. 242, 1955.
[141a] C. H. Chin, *Nature* **165,** 926 (1950).
[141b] T. Heyman-Blanchet and P. Chaix, *Biochim. et Biophys. Acta* **35,** 85 (1959).
[142a] H. M. Hirsch, *Biochim. et Biophys. Acta* **9,** 674 (1952).
[142b] C. R. Hebb, J. Slebodnik, T. P. Singer, and P. Bernath, *Arch. Biochem. Biophys.* **83,** 10 (1959).
[143] H. Chantrenne, *Biochim. et Biophys. Acta* **14,** 157 (1954).
[144] H. Chantrenne and C. Courtois, *Biochim. et Biophys. Acta* **14,** 397 (1954).
[145a] R. Kattermann and P. P. Slonimski, *Compt. rend. acad. sci.* **250,** 220 (1960).
[145b] H. Chantrenne, *Arch. Biochem. Biophys.* **65,** 414 (1956).
[146] E. Englesberg, A. Gibor, and J. B. Levy, *J. Bacteriol.* **68,** 146 (1954).
[147] E. Englesberg, J. B. Levy, and A. Gibor, *J. Bacteriol.* **68,** 178 (1954).
[148] E. Englesberg and J. B. Levy, *J. Bacteriol.* **69,** 418 (1955).
[149] N. D. Gary and R. C. Bard, *J. Bacteriol.* **64,** 501 (1952).
[150] F. Moss, *Australian J. Exptl. Biol. Med. Sci.* **30,** 531 (1952).
[151a] S. Hollmann and E. Thofern, *Naturwissenschaften* **42,** 586 (1955).
[151b] R. K. Clayton, *Arch. Mikrobiol.* **33,** 260 (1959).
[151c] R. K. Clayton, *J. Biol. Chem.* **235,** 405 (1960).

[152] R. F. Rosenberger and M. Kogut, *J. Gen. Microbiol.* **19**, 228 (1958).
[153] W. Verhoeven, *in* "Inorganic Nitrogen Metabolism" (W. D. McElroy and B. Glass, eds.), p. 61. Johns Hopkins Press, Baltimore, Maryland, 1956.
[154] W. Verhoeven and Y. Takeda, *ibid* p. 159.
[155] R. Sato, *ibid* p. 163.
[156a] A. Nason and H. Takahashi, *Ann. Rev. Microbiol.* **12**, 203 (1958).
[156b] T. Higashi, *J. Biochem.* (*Tokyo*) **47**, 326 (1960).
[157] A. J. Kluyver and W. Verhoeven, *Antonie van Leeuwenhoek Ned. Tydschr. J. Microbiol. Serol.* **20**, 337 (1954).
[158] F. O. Moss, *Australian J. Exptl. Biol. Med. Sci.* **34**, 395 (1956).
[159] W. Smith, J. H. Hale, and C. H. O'Callaghan, *J. Pathol. Bacteriol.* **65**, 229 (1953).
[160] J. Jensen, *J. Bacteriol.* **73**, 324 (1957).
[161] J. B. Wolff and L. Price, *Arch. Biochem. Biophys.* **72**, 293 (1957).
[162a] J. H. C. Smith and D. W. Kupke, *Nature* **178**, 751 (1956).
[162b] J. H. Golden, R. P. Linstead, and G. M. Whitham, *J. Chem. Soc.* **2**, 1725 (1958).
[163] R. Y. Stanier and G. Cohen-Bazire, *in* "Microbial Ecology" (R. E. O. Williams and C. C. Spicer eds.), p. 56. Cambridge University Press, London and New York, 1957.
[164a] R. Y. Stanier and J. H. C. Smith, *Biochim. et Biophys. Acta* **41**, 478 (1960).
[164b] H. Larsen, *Kgl. Norske Videnskab. Selskabs Skrifter* **26**, 1 (1953).
[165a] T. W. Goodwin, *Biochim. et Biophys. Acta* **18**, 309 (1955).
[165b] A. S. Holt and H. V. Morley, *J, Amer. Chem. Soc.* **82**, 500 (1960).
[166a] R. Y. Stanier and J. H. C. Smith, *Carnegie Inst. Wash. Year Book* **58**, 336 (1959).
[166b] E. Deitzel, *Hoppe-Seyler's Z. physiol. Chem.* **284**, 262 (1949).
[167] F. Wrede and A. Rothhaas, *Hoppe-Seyler's Z. physiol. Chem.* **226**, 95 (1934).
[168a] H. H. Wasserman, J. E. McKeon, L. Smith, and P. Forgione, *J. Am. Chem. Soc.* **82**, 506 (1960).
[168b] R. Hubbard and C. Rimington, *Biochem. J.* **46**, 220 (1950).
[169] C. M. Weiss, *J. Cellular Comp. Physiol.* **34**, 467 (1949).
[170] R. P. Williams, J. A. Green, and D. A. Rappoport, *J. Bacteriol.* **71**, 115 (1956).
[171] J. A. Green, D. A. Rappoport, and R. P. Williams, *J. Bacteriol.* **72**, 483 (1956).
[172] G. W. Monk, *J. Bacteriol.* **74**, 71 (1957).
[173a] R. P. Williams, *Bacteriol. Revs.* **20**, 282 (1956).
[173b] G. S. Marks and L. Bogorad, *Proc. Natl. Acad. Sci. U.S.* **46**, 25 (1960).
[174] M. T. M. Rizki, *Proc. Natl. Acad. Sci. U.S.* **40**, 1135 (1954).
[175] M. T. M. Rizki, *J. Bacteriol.* **76**, 607 (1958).
[176] R. P. Williams and J. A. Green, *J. Bacteriol.* **72**, 537 (1956).
[177] J. A. Green and R. P. Williams, *J. Bacteriol.* **74**, 633 (1957).
[178] U. V. Santer and H. J. Vogel, *Biochim. et Biophys. Acta* **19**, 578 (1956).
[178a] H. H. Wasserman, J. E. McKeon and U. V. Santer, *Biochem. Biophys. Research Communs.* **3**, 146 (1960).
[179a] H. Munro Fox, *Proc. Roy. Soc.* **B143**, 203 (1955).
[179b] J. Lascelles, *J. Gen. Microbiol.* **23**, 487 (1960).
[180] J. M. Wiame, *Advances in Enzymol.* **18**, 241 (1957).
[181a] H. J. Vogel, *in* "The Chemical Basis of Heredity" (W. D. McElroy and B. Glass, eds.), p. 276. Johns Hopkins Press, Baltimore, Maryland, 1957.
[181b] J. E. Falk, R. J. Porra, A. Brown, F. Moss, and H. E. Larminie, *Nature* **184**, 1217 (1959).
[182] Various authors, *Ciba Foundation Symposium on Regulation of Cell Metabolism*, 1959.
[183] M. Yoneda and A. M. Pappenheimer, Jr., *J. Bacteriol.* **74**, 256 (1957).

CHAPTER 8

Synthesis of Polymeric Homosaccharides

SHLOMO HESTRIN*

I. Historical Retrospect.... 373
II. Dextrans.... 374
A. Structure and Occurrence.... 374
B. Mechanism of Synthesis of Dextran from Sucrose.... 375
C. Origin of Branching.... 377
D. Synthesis of Dextran from Dextrins.... 378
III. Amyloses and Glycogen.... 378
IV. Cellulose.... 380
A. Structure and Occurrence.... 380
B. Mechanism of Synthesis.... 380
V. Levan.... 382
A. Structure and Occurrence.... 382
B. Transfructosylation Reactions Catalyzed by Levansucrase.... 383
References.... 386

I. Historical Retrospect

Important clues to the mode of the synthesis of microbial polysaccharides became available early in this century. Cremer[1] reported in 1899 that fermentation of glucose in a cell-free extract of yeast is accompanied at low concentration of orthophosphate by a transient synthesis of a compound with properties of glycogen. Beijerinck[2] was able to show by 1910 that a system, termed "viscosaccharase," enters the extracellular phase of cultures of bacilli growing on a sucrose-gelatin medium, and there catalyzes a polymerative degradation of sucrose with resulting synthesis of a fructose polymer (levan) that appears in the form of a co-acervate (*emulsions erscheinung*). It is difficult at our present distance in time to understand why the afore-mentioned fundamental observations were either overlooked during a period of many years, or held in tacit disbelief in scientific circles.

Modern studies of polysaccharide synthesis in soluble systems stem directly from the much later discovery that starch and glycogen can be converted reversibly into α-glucose-1-phosphate by a phosphorolytic process.[3] A decisive additional advance was later initiated by the realization early in the nineteen forties that phosphate is but one of a number of radicals whose attachment to C-1 in a glycose unit facilitates the enzymic polymerization of glycose in an aqueous milieu.[4, 5]

The properties of all the known enzymic polymerizations which lead to

* Deceased.

formation of polysaccharide are consistent with the concept that the growth of glycose chains is effected by successive addition of glycosyl units one at a time to the nonreducing terminal of a chain at an acceptor site that is regenerated in each added glycosyl unit. These reactions fit into a common scheme of the general form below:

$$(n)\ \text{HG-OR} + \text{HA} \rightarrow \text{HG}_{(n)}\text{A} + (n)\ \text{ROH}$$

where HG-OR designates a donor system, HG is the transferred glycosyl group, (n) is a molar quantity, HA is the acceptor system and may in suitable instances be simply a molecule of HG-OR itself, $\text{HG}_{(n)}\text{A}$ is a formed polysaccharide whose molecule contains n residues of G, and ROH represents a nonpolymeric product (orthophosphate, uridine diphosphate, or glycose).

The range of polysaccharide types produced by microorganisms is so enormous that it has not seemed feasible to cover this large area of knowledge within the space of the present chapter. Our discussion will therefore be restricted to selected examples of synthesis of typical macromolecular homosaccharides by bacteria.

II. Dextrans

A. Structure and Occurrence

Polyglucans of the type of dextran contain numerous interglycosidic linkages of the type α-1,6. The members of this class form a heterogenous family in which structures range widely with respect to the size of the molecule, degree of the branching, and type of interglycosidic linkage at the branch points. Much work has been devoted to these polymers in view of their use as source materials for production of a plasma-volume expander (for a comprehensive review see Neely[6]).

Early workers considered that all or most branch linkages in dextran are of the type α-1,4. A dextran produced from amylodextrins by certain species of *Acetobacter* conforms to this description.[7] In other forms, however, all or most of the branch linkages in dextran are of the type α-1,3. In the dextran formed by *Leuconostoc mesenteroides* strain B512, for example, branch linkages, all of which are of type α-1,3, represent 5% of the total number of interglycosidic linkages.[8, 9] In *Betacoccus arabinosaceus*, the proportion of the α-1,3 branch linkages in dextran appears to vary with the magnesium content of the growth medium.[10] A linear form of dextran is produced by a strain of *Streptococcus bovis*.[11]

Many naturally occurring dextrans have molecular weights in the order of 10^6 to 10^7. Dextran, which has a relatively small molecular weight and bears sucrose at one of the chain terminals, is formed by a species of *Streptococcus* (DS strain 50).[12]

Purified dextrans are poorly antigenic in the rabbit but readily evoke antigens in man.[13] Immunochemical procedures and enzyme degradation studies combined with the periodate oxidation technique and classical methylation procedure may be expected in the future to throw new light on the fine structures of the various dextrans.

B. Mechanism of Synthesis of Dextran from Sucrose

Our knowledge concerning the enzymic mechanism of the synthesis of dextrans stems largely from the pioneering work of Hehre,[14] who first separated a dextran-synthesizing system "dextransucrase" in a cell-free form from a culture of *Leuconostoc mesenteroides* in 1941. The major reaction catalyzed by "dextransucrase" was shown to conform to an equation of the form:

$$(n)\ \text{sucrose} \rightarrow (\text{anhydroglucose})_n + (n)\ \text{fructose}$$

where molar value n equals the number of the anhydroglucose units in a molecule of the formed polymer. The chemical and serological properties of the polymeric product resembled those of the dextran that was produced from sucrose in a growing culture of the organism. This reaction proceeds apparently irreversibly in the direction of polymer production until the entire available sucrose supply is exhausted.[15]

The production of dextransucrase in *Leuconostoc* depends on sucrose supply. The system is largely extracellular, and can readily be recovered from the culture in a cell-free form simply by centrifuging or filtering. Synthesis of enzyme proceeds most rapidly in cultures at pH 6.7. However, this pH also favors a process of an as yet unknown nature which is responsible for large losses of enzyme activity. A standard dextransucrase unit (DSU) has been defined as that amount which utilizes 1 mg. of sucrose (releasing 0.52 mg. fructose) per hour in a system containing 10% sucrose in acetate buffer at pH 5.0 at 30°C. The pH activity function of the enzyme exhibits a broad optimum (pH range, 5.0–6.7).[16]

Factors governing molecular weight of dextran are of particular interest both from an academic and from practical points of view. The polymer synthesized by this enzyme acting in standard conditions of reaction has a molecular weight in the range 10^6 to 10^7. If the synthesis is conducted in the presence of an added appropriate oligosaccharide, e.g., maltose or fructose, the molecular weight distribution of the formed polymer assumes a bimodal form, in which the "high mode" has a molecular weight of 10^6 to 10^7, and the "low mode" has a molecular weight which is broadly distributed in a range $<10^5$. Compounds whose addition to the system led to increased formation of low-mode dextran served directly as acceptors of glucosyl. They gave rise to oligosaccharides in which one or several added α-1,6-linked glucose residues formed a chain which extended from an appropriate site in the added compound. A bimodal molecular weight distribution of dextran also occurred in reaction mixtures in which sucrose was maintained at a very high concentration in absence of an added alternate acceptor. Elevation of the enzyme concentration and decrease of reaction temperature below 30°C. increased the proportion of low-mode fraction in the product.[17,18]

Dextran fractions prepared from partial hydrolyzates of native dextran were efficient acceptors of the glucosyl unit transferred from sucrose by dextransucrase. When a synthesis of dextran from sucrose is conducted in the presence of a degraded dextran, the amount of product recovered as high-mode dextran is sparse whereas the yield of product in the low-mode is markedly increased.[17,19]

It has been suggested[6,20] in the light of the above findings that dextransucrase may be able to catalyze two kinetically distinct kinds of polymerization processes, leading one to the product in the high mode and the other selectively to products of a much lower polymerization degree. In the former case, it is supposed that the initiator is sucrose and that the latter gives rise on interaction with one or several units of transferred glycosyl to a chain product from which the enzyme cannot readily escape. Further elongation of the chain could then occur with conservation of activation energy and the growth of the chain could be rapid and extensive. In the reaction that leads to products of low mode, on the other hand, the initial acceptor is thought to be one of relatively high reactivity and to give rise to a chain from which the enzyme is readily dissociable. In this case, the catalytic activity of the enzyme molecule is directed to a large number of simultaneously growing chains. At limiting donor supply in a system of this nature, polymerization degree of the product recovered can be expected, as observed, to be a relatively low value.

The bond in sucrose split by dextransucrase is that between C-1 of glucose and the interglycosidic oxygen.[21] Thus dextransucrase is indeed a glucosylase rather than a glucosidase.

The geometry of the donor site in dextransucrase must be of a kind which favors the reaction with sucrose and precludes reaction with any one of a variety of closely related oligosaccharides. Thus, galsucrose (α-galactopyranosyl-β-fructofuranoside), which differs from sucrose only by an epimerization at C-4 in the aldose moiety, is completely resistant to attack by dextransucrase. Furthermore, neither xylsucrose (α-D-xylopyranosyl-β-fructofuranoside), raffinose (6^G-D-galactosylsucrose), nor lactsucrose (4^G-D-galactosylsucrose) shows activity.[22] Even the attachment of a glycosyl group to the *fructose* moiety of sucrose at either C-1, 3, or 6 results in a total suppression of reaction. Finally, it should also be noted that α-glucose-1-phosphate is not utilized by this enzyme as a donor system.[4]

At temperatures above 30°C., a preparation of dextransucrase mediated a slow reaction by which maltose was converted into trisaccharide and glucose. It has been suggested that the dextransucrase protein is in a partly unfolded state at this temperature and that the donor specificity of the system is altered thereby.[6,23] Shifts in physical parameters of dextran have been observed when dextransucrase was allowed to act on dextran in absence of any sucrose.[24] Whether such changes were indeed brought about by dextransucrase itself or by an enzymic contaminant remains to be determined.

Acceptor activity exhibited by polyglucoses in the presence of dextransucrase depends among other factors on chain length. Low molecular weight dextrans form a particularly efficient class of acceptors, which is by far superior in this respect to free glucose. Transfer of glucosyl occurs to the primary carbinol at C-6 of the nonreducing terminal glucose residue in both maltose and isomaltose. Epimerization in glucose at C-2 to yield mannose or at C-4 to yield galactose markedly decreases or abolishes the acceptor activity of C-6. Introduction of a methyl group at C-1 in α-glucose appears to enhance the acceptor activity of the C-6 site in glucose. When efficient alternate acceptors, such as α-methyl glucoside, maltose, or isomaltose, are added to reaction system containing sucrose as donor, the over-all rate at which the donor undergoes decomposition is markedly enhanced.[16,18,25,26]

Secondary as well as primary carbinol groups appear to be capable of serving in the dextransucrase system as acceptor sites. It has been found, for instance, that dextransucrase attaches a glucosyl residue to secondary carbinol at C-5 in fructopyranose.[27] Also, it has been observed that transfer of glucosyl occurs in the presence of dextransucrase to the secondary carbinol at C-2 of the glucose moiety in raffinose, and to secondary carbinol at C-2 of the reducing moiety both in cellobiose and lactose.[28]

C. Origin of Branching

The occurrence of branching in dextran poses the question whether the different linkage types are produced by several enzyme species, or by a single enzymic entity. The first alternative has been favored in view of the finding that a *Betacoccus arabinosaceus* dextransucrase preparation, which ordinarily formed a highly branched dextran, was converted on aging into a system that produced an essentially linear product.[29] It has also been found that the *Streptococcus bovis* dextransucrase has general properties resembling those of the *Leuconostoc* enzyme but forms an essentially linear dextran.[30]

Studies on the acceptor specificity of dextransucrase have been made difficult by the presence of dextran in the available dextransucrase preparations. This glucose polymer enters into a tight complex with the dextransucrase protein, and as yet no method of freeing the latter completely from this contaminant has been devised in the case of the enzyme of *Leuconostoc*. In cultures of *Streptococcus bovis*, however, dextransucrase has been recovered from a glucose growth medium which contained no added sucrose. Enzyme prepared from this source had a relatively low polysaccharide content (4% w./w.; 0.04 mg. of carbohydrate/unit of enzyme) yet proved to be active on sucrose, even in absence of any added dextran or any added alternate acceptor system other than sucrose itself.[30]

The structure of dextransucrase, as that of other polysaccharide-synthesizing enzymes with the notable exception of phosphorylase, is still a largely unknown territory. A recent kinetic study has suggested that the catalytic site of dextransucrase involves at least two distinct groups whose dissociation constants are those of carboxyl and imidazole, respectively.[31,32]

D. Synthesis of Dextran from Dextrins

A remarkable metabolic bypath for synthesis of dextran has been discovered in strains of acetic acid bacteria—*Acetobacter viscosum* and *Acetobacter capsulatum*—which cause ropiness when they are grown in amylaceous media. The phenomenon of ropiness was found to be due to the action of a unique enzyme ("dextrandextrinase"), which utilizes dextrin (rather than sucrose) as a donor system of dextran synthesis.[33] Dextran was formed by this enzyme from short amylose chains (produced from native amylose by action of acid or α-amylase) but not from native amylose, amylopectin, glycogen, cyclized amyloses (Schardinger dextrins), maltose, sucrose, and α-glucose-1-phosphate. The product of reaction consisted of chains of α-1,6-linked glucose units which were branched via α-1,4. Accordingly, the essential feature of the action of this enzyme would appear to consist in a chain rearrangement whereby α-1,4 linkages in a dextrin are replaced in part by α-1,6 linkages. The α-1,3 linkage was not formed by dextrandextrinase in these reaction conditions.[7] The rearranging process leading from dextrin to dextran appears to be irreversible. This makes it probable that the conversion of dextrin into dextran involves a considerable loss of free energy.[4]

III. Amyloses and Glycogen

A variety of pathways whereby bacteria are able to synthesize linear amyloses as well as branched polyglucans of the type of glycogen (main linkage, α-1,4; branch linkage, α-1,6) are known.

In *Corynebacterium*, certain streptococci, and in *Neisseria perflava*, α-glucose-1-phosphate was found to serve as donor system and to give rise to a mixture consisting of a linear amylose fraction and an amylopectin-like fraction.[34, 35]

Neisseria strains are also able to form a starch-like polysaccharide specifically from sucrose via a pathway which does not involve the intermediary formation of α-glucose-1-phosphate.

$$(n)\ \text{sucrose} \rightarrow \text{anhydroglucose}_{(n)} + (\text{n})\ \text{fructose}$$

The enzyme catalyzing this sucrose conversion has been named *amylosucrase*. The reaction is difficultly reversible. It resembles formally the transformation catalyzed by dextransucrase, except of course with regard to the types of the interglycosidic linkages that are generated in the formed polymer.[36]

Still another bacterial pathway which can be directed to synthesis of long chains of amylose involves maltose as the donor system, and a catalyst which has been named *amylomaltase*. This enzyme mediates a reversible reaction of the form:

$$g_{(2)} + g_{(n)} \rightleftharpoons g_{(1)} + g_{(n+1)}$$

where $n \geqq 2$, $g_{(2)}$ represents maltose, $g_{(1)}$ represents glucose, and $g_{(n)}$ and $g_{(n+1)}$ are α-1,4-linked chains containing, respectively, n and $n + 1$ glucose units. If glucose is continuously removed in this system by a chemical transformation such as oxidation by notatin, the reaction in the direction to the left is prevented, while reaction in the direction to the right continues and leads eventually to product of a high polymerization degree. If, on the other hand, glucose is allowed to accumulate in the system, the mixture at equilibrium consists of a polymer homologous dextrin series in which products of relatively low molecular weight are predominant.[37-40]

A quite remarkable amylase system secreted by *Bacillus macerans* catalyzes a reversible homologizing reaction of the general form:

$$g_{(n)} + g_{(m)} \rightleftharpoons g_{(n+x)} + g_{(m-x)}$$

where x, n, and m may have a variety of values. In the special case in which $m = 2$ and $x = 1$, the above reaction becomes equivalent to that mediated by amylomaltase. The special feature of the *B. macerans* system relates to its activity with respect to cyclodextrins (Schardinger dextrins). In the presence of the *B. macerans* amylase, a cyclodextrin can serve as a donor system in a reversible reaction, which might be pictured schematically as in Scheme I:

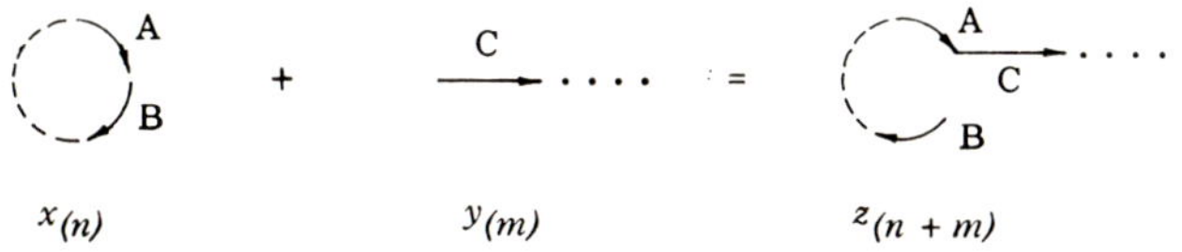

Scheme I

where arrows (A, B, and C) represent anhydroglucose units, $x_{(n)}$ designates a donor system (cyclic dextrin) of polymerization degree (n), y is an appropriate acceptor system of polymerization degree (m), and z is a product of polymerization degree ($n + m$) formed by transfer of a polyglycosyl radical from a C-4 site in the donor to C-4 in the acceptor (see French[41] for comprehensive review).

IV. Celluose

A. Structure and Occurrence

Whereas cellulose is ubiquitous in higher plants, it is rare in the animal kingdom, and among true bacteria occurs only among the *Acetobacter*. Typical strains of *Acetobacter xylinum* growing on glucose convert the latter into cellulose which accumulates in the extracellular phase of the culture medium in the form of a tough pellicle. The latter consists mostly of fluid containing cells held in a mesh of intertwined cellulose microfibrils.[42] Cellulose chains lie in parallel array within the long axis of the microfibrils. It has been estimated that the latter are only about two or four molecules deep.[43] Bubbles of metabolically formed carbon dioxide which tend to cling to the cellulose mesh eventually raise this structure to the surface of a liquid medium, thus ensuring ready supply of atmospheric oxygen to the organism.[44]

Cellulose formed by *A. xylinum* consists of anhydroglucose units which are linked β-1,4, as in typical green plant cellulose. The polymerization degree is $\gtrsim$500. Infrared absorption spectrum, X-ray diffraction pattern, and other characteristics of bacterial cellulose are indistinguishable from those of typical cotton α-cellulose.[45,46]

Strains of *Acetobacter xylinum* with high cellulose-forming ability may mutate to forms which show feeble or no cellulose-synthesizing activity.[44] Mutants of this class are observed frequently in an inoculum taken from an agitated liquid culture or from standard stock which has been maintained for long periods in a type culture collection. Shimwell[47] was able to demonstrate mutation in the direction to cellulose production on the part of celluloseless mutants. It is probable that species of *Acetobacter*, to which an independent taxonomic status has been assigned, are in fact mutants of a common cellulose-forming parent type, *A. xylinum*.[48]

B. Mechanism of Synthesis

Washed, cellulose-free suspensions of *A. xylinum* cells readily form cellulose from glucose under aerobic conditions. Freeze-dried cells retain this synthesizing capacity.[46] When the growth medium contained fructose, the harvested cells were found to be able to form cellulose aerobically also from fructose.[49] Gluconic acid, which is rapidly formed from glucose by *A. xylinum* cells, is an excellent precursor of cellulose. At suitable conditions of pH, furthermore, both 2- and 5-ketogluconic acid are converted into cellulose by this organism.[50] A variant which converts C^{14}-pyruvate into cellulose has recently been isolated from a strain of *A. xylinum* grown on succinate.[49] That glycerol is an efficient precursor of cellulose has long been known.[51]

Evidence has been obtained which indicates that *A. xylinum*, in the manner of several related *Acetobacter* spp., can oxidize glucose and gluconate via a pentose cycle.[52] If intracellular hexose phosphate that is undergoing oxidation via the pentose cycle is an intracellular precursor of the extracellularly deposited cellulose, it might be expected that the specifically labeled C^{14}-glucose and C^{14}-fructose would give rise to cellulose with C^{14} distributions which differ in a characteristic manner from those of C^{14} in the added hexose. C^{14} rearrangements anticipated on the basis of the hypothesis that hexose phosphate giving rise to cellulose had been generated in part in the cell by means of a pentose cycle were indeed found to occur.[53,54] These results are consistent with the idea that a phosphorylated hexose that is formed endogenously plays an intermediary role in the conversion of exogenously supplied substrates into cellulose. In a cell that is unable to generate adenosine triphosphate by glycolysis, the dependence of the cellulose synthesis on oxygen can readily be rationalized on this basis in terms of the necessary role of an oxidative process in the synthesis of the needed phosphorylated intermediate.

Since the enzyme system synthesizing cellulose is known to be anchored to the cell structure, the implication of an intracellular metabolite as an intermediate in the process of cellulose synthesis requires either (*a*) that the immediate cellulose precursor is polymerized within the cell and that the formed macromolecule is then excreted, or alternatively (*b*) that the immediate precursor is able to pass from the cell interior through the cellular membrane and undergoes polymerization on the outer surface of the latter. A critical choice between these alternatives is not as yet possible. The finding that exogenously supplied hexose phosphates (including α-glucosyl-1-phosphate, glucose-6-phosphate, and uridine diphosphoglucose) are neither oxidized nor converted into cellulose by *A. xylinum* cells still does not rule out the possibility that these compounds are converted into cellulose when they are formed endogenously.[43,46,50-55]

A system of subcellular particles prepared from sonic extracts of *A. xylinum* incorporated C^{14}-glucose from uridine diphosphoglucose into cellulose. This reaction was accelerated by cellulodextrins.[56] When the alkali-insoluble product of this reaction was partially hydrolyzed by acid, cellobiose, and higher homologs recovered from the partial hydrolyzate proved to be radioactive. These data showed that a β-glucosyl group was attached by this system to a C-4 position in a nonreducing terminal glucose residue, as would be required for a synthesis of cellulose. However, it remains undetermined whether the reaction observed consisted only of transfer of a single glucosyl group exclusively to preexisting acceptor sites or had comprised a polymerative series of such transfers.

Klungsöyer detected an enzyme system in extracts of *A. xylinum* cells that catalyzes a disproportionation reaction by which soluble cellodextrins are converted in part into an insoluble cellulose fraction.[57,58] The possibility that this reaction plays an intermediary role in the process of cellulose synthesis from simple sugars is indeed an attractive one.

Another clue which remains to be explored has been provided by the observation that the frequency of microfibrils in a heat-labile extract of *A. xylinum* increases when this system is maintained in the presence of glucose.[59,60] The question whether *chemical* reaction of polymerization occurs under these conditions has not been definitely answered. An alternative explanation might be that the observed increase in fibril frequency represented a crystallization of preformed molecules of polymer and that the extracts contain a heat labile factor that promotes such crystallization. Claim[61] to the effect that massive net synthesis of cellulose is mediated from glucose in *A. xylinum* homogenates that are fortified by adenosine triphosphate has not been substantiated.[43]

When a fibril-free suspension of carefully dispersed and washed *A. xylinum* cells is brought into contact with glucose, the cellulose which first appears in the medium presents an amorphous structure, as might be expected if the material released from the cellular surface consisted of randomly coiled diffusible molecules. This implies that the cellulose micro fibrils are probably not pushed out fully formed from the bacterial surface, but rather arise secondarily in solution by crystallization.[43]

V. Levan

A. Structure and Occurrence

Polyfructans of the type of levan are characterized by the presence of rows of linkages of the type β-2,6 in the molecule. Levans occur extracellularly in cultures of many kinds of bacteria in medium containing sucrose or a related aldosylfructoside (e.g., raffinose) as carbon source. Typically, their presence is revealed by a striking increase of the viscosity and by a marked decrease in the transparency of the medium to light. Production of the high polymer is attended by oligosaccharide production and by substrate hydrolysis. Molecular weights of the formed bacterial levans are often of the order 10^6 to 10^8. The enzyme system, levansucrase, that mediates this synthesis may either be secreted into the medium or remain anchored to the cell. Since the molecule of levan appears outside the cell, and is of very large size, a cell-bound levansucrase must be presumed to be located on or near the outer surface of the cell. Mostly, the formed levan diffuses freely into the solution. However, under special conditions, whose determinants are as yet poorly defined, such movement of polymer may be impeded by an intervening phase separation, as a result of which the cells come to be retained within a levan-rich zoogleal mass.[2,43,62-64]

Methylation analyses of the bacterial levans formed by a range of microorganisms (*Bacillus*, *Pseudomonas*, *Aerobacter*, *Corynebacterium*) have indicated in every case that the structure is branched. Rows of fructofuranose residues are bonded by main linkages of type β-2,6 and branch linkages of the type β-2,1.[65-67] The basal chain length of several levans has been shown on the basis of methylation analysis to be about 10. However, a levan formed by a *Corynebacterium* sp. probably has an average chain length nearer 15.[68] Immunochemical analysis has revealed quantitative differences in the reaction of an antiserum with levans from different sources.[69]

Additional insight into the branching patterns of this polymer family can be anticipated from applications of enzymic tools to the elucidation of levan structure.[63,70-72] An extracellular hydrolase system "levanpolyase," which is induced by levan in a range of microorganisms, is of particular interest in this connection.[73] This enzymic agent converts levan into two polymer homologous series of oligosaccharides. The β-2,1-branch linkages, whose presence in levan could be deduced on the basis of methylation analysis, are retained in one of these series. Lengths of chains in the limit hydrolyzate ranged continuously from 2 to about 16 fructose residues. The observed variation in the length of the recovered "limit levulans" is taken to be a reflection of the feature that the internode distance in levan is correspondingly variable.

Levansucrase production requires the presence of sucrose in the growth medium in some bacterial strains but appears to be a constitutive process in others.[63,74,75]

B. Transfructosylation Reactions Catalyzed by Levansucrase

Particular attention has hitherto been devoted to two representatives of the levansucrase class of enzymes: an endocellular constitutive levansucrase formed by *Aerobacter levanicum*[66,76,77,80] and an exocellular adaptive levansucrase formed by *Bacillus subtilis*.[75,78,79] Enzyme preparations available from these sources have now been purified by a factor of about 2×10^2 to a basis of protein in the crude cell extract. Only minor differences in behavior between these levansucrases have been observed. The pH activity function of the enzyme from both above-mentioned sources is nearly identical, revealing a maximum in the neighborhood of pH 5.8.

A salient characteristic of a levansucrase preparation is the ability to synthesize giant macromolecules of levan from micromolecules. The donor in the substrate system is typically an appropriate β-D-fructofuranosyl-α-D-aldoside, e.g., sucrose. Levansucrase cleaves sucrose at the bond between C-2 of the fructose moiety and the interglycoside oxygen. Hence levansucrase is a fructosylase, not a fructosidase.[21] In addition to levan synthesis, the levansucrase preparations also effect transfer of fructosyl from donor to alternate hydroxyl sites in sugars and to water. As a result oligosaccharidic products, notably 1^F-fructosylsucrose and free fructose, are accumulated in the reaction mixture contemporaneously with the formation of levan.[76,81] If short levan chains are introduced in high concentration into sucrose solution in the presence of levansucrase, the production of new levan macromolecules and liberation of fructose are largely suppressed thereby, their place being taken by a process which consists in the addition of a few or only one fructose residue to terminals in the added chain population.[79,82]

The levansucrase-mediated transfer of fructosyl from a donor to the hydroxyl site in an acceptor is a difficultly reversible process. Transfer of fructosyl from sucrose to the growing levan chains likewise progresses in the direction of hydroxyl substitution until practically all sucrose has been exhausted. This process, too, is difficultly reversible.[66,75,83] If, however, an appropriate free aldose is present in a reaction mixture containing fructosyl donor (fructosylaldoside) in the presence of levansucrase, a rapid and readily reversible transfer of fructosyl from donor to the anomeric oxygen in the added aldose occurs.[77,84] Since the presence of free aldose in the system markedly retards the rate of levan production, a steady-state equilibrium which approaches a true equilibrium, as defined by the equation below, may be maintained in such a system over a large part of the observed over-all reaction.

$$[A'F]\,[A]/[AF]\,[A'] = K$$

where A′F and AF are β-fructofuranosylaldosides and A and A′ are free aldoses. The value of K has been estimated on this basis to be close to unity in systems involving a range of aldose pairs. This has suggested that the members of a family of β-fructofuranosylaldosides are all at much the same energy level.[82] The $-\Delta F$ of hydrolysis of sucrose—the type member of this family—has been estimated, on the basis of enzyme equilibrium data, to be 6.6 kcal.[85] (cf., however, ref. 86).

Reactivity of neutral aldopyranose in the acceptance of a levansucrase-transferred fructosyl unit at C-1 has been shown to be conditional upon the presence of an unsubstituted *trans* pair of hydroxyl groups at C-2 and C-3 in the aldose. Large changes at other carbon sites in the aldose structure, e.g., substitution of a mono- or polyglucose radical at C-4 or O-6 or epimerization at C-4, exert only a relatively small influence on the rate of observed fructosyl transfer to C-1.[77,84]

Sucrose derivatives which bear a substituent on carbons 1 or 6 in the fructose moiety do not serve with this enzyme as donors. Substitution of a glycosyl group at C-2 of the glucose moiety in sucrose likewise suppresses the donor activity. However, analogs of sucrose, which differ from sucrose in the aldose moiety at carbons other than C-2 and 3, e.g., levansucrase-synthesized analogs such as xylsucrose (β-fructofuranosyl-α-D-xylopyranoside), galsucrose (β-fructofuranosyl-α-D-galactopyranoside), and raffinose (β-fructofuranosyl-α-melibioside), serve readily as donor systems and give rise to levan in the presence of levansucrase.[77]

Types of single transfer of a fructose unit from a high-energy donor system (fr ~ aldoside) to an appropriate acceptor (water, alcoholic hydroxyl, or carbonyl in an aldose acceptor) as catalyzed by levansucrase may be summarized schematically as shown in Scheme II:

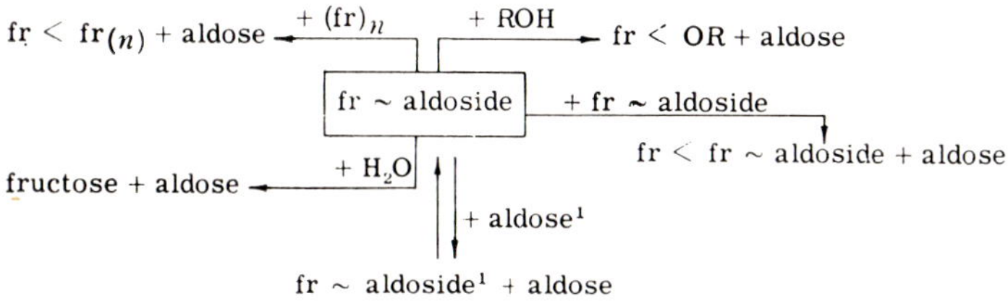

Scheme II

where ~ designates a high-energy glycosidic bond such as occurs in sucrose, and < is a glycosidic bond of lower energy such as occurs in oligolevan ($fr_{(n)}$).

The findings are in good accord with the conclusion that synthesis of levan is a stepwise process in which a single fructosyl residue is transferred to the end of a growing chain at each successive step in chain growth. According to this view, it is conceivable that levansucrase shares with some other transfructosylases an identical mechanism of donor activation, even though the production of high polymer is unique only to the former enzyme type. Presumably, the initiation step ordinarily involves two sucrose molecules, as shown in Scheme III.

fr ~ aldoside + fr ~ aldoside + enzyme

↓

enzyme.fr < fr ~ aldoside + aldose

Scheme III

Most of the molecules of the primary product (Scheme III) of the above initiation reaction presumably undergo dissociation, releasing adsorbed enzyme with attendant dissipation of activation energy. Further growth of this dissociated product would be slow. However, the primarily formed enzyme-trisaccharide complex, being in an activated state, could probably grow in chain length more rapidly than the free trisaccharide. The growing structure provides a network from which neither enzyme nor activated fructosyl can readily escape.[79] In this situation, chain propagation would be rapid and extensive, and would result as observed in the predominant emergence of a product of very high molecular weight.

References

1 M. Cremer, *Ber.* **32,** 2002 (1899).
2 M. W. Beijerinck, *Folia Microbiol.* **1,** 377 (1912).
3 C. F. Cori, G. Schmidt, and G. T. Cori, *Science* **89,** 464 (1939).
4 E. J. Hehre, *Advances in Enzymol.* **2,** 297 (1951).
5 S. Hestrin, *J. Cellular Comp. Physiol.* **54,** *Suppl. 1,* 127 (1959).
6 W. B. Neely, *Advances in Carbohydrate Chem.* **15,** 341 (1960).
7 S. A. Barker, E. J. Bourne, G. T. Bruce, and M. Stacey, *J. Chem. Soc.* p. 4414 (1958).
8 J. C. Rankin and A. Jeanes, *J. Am. Chem. Soc.* **76,** 4435 (1954).
9 J. W. Sloan, B. H. Alexander, R. L. Lohmar, Jr., I. A. Wolff, and C. E. Rist, *J. Am. Chem. Soc.* **76,** 4429 (1954).
10 S. A. Barker, E. J. Bourne, A. E. James, W. B. Neely, and M. Stacey, *J. Chem. Soc.* p. 2096 (1955).
11 R. W. Bailey, *Biochem. J.* **71,** 23 (1959).
12 E. J. Hehre, *J. Biol. Chem.* **222,** 739 (1956).
13 E. A. Kabat, *J. Immunol.* **84,** 82 (1960).
14 E. J. Hehre, *Science* **93,** 237 (1941).
15 E. J. Hehre, *J. Biol. Chem.* **163,** 221 (1946).
16 E. J. Hehre, *in* "Methods in Enzymology" (S. P. Colowick and N. O. Kaplan, eds.), Chapter 21, p. 178. Academic Press, New York, 1955.
17 H. M. Tsuchiya, N. N. Hellman, H. J. Koepsell, J. Corman, C. S. Stringer, S. P. Rogovin, M. O. Bogard, G. Bryant, V. H. Feger, C. A. Hoffman, F. R. Senti, and R. W. Jackson, *J. Am. Chem. Soc.* **77,** 2412 (1955).
18 H. J. Koepsell, H. M. Tsuchiya, N. N. Hellman, A. Kazenko, C. A. Hoffman, E. S. Sharpe, and R. W. Jackson, *J. Biol. Chem.* **200,** 793 (1953).
19 H. Nadel, C. I. Randles, and G. L. Stahly, *Appl. Microbiol.* **1,** 217 (1953).
20 F. A. Bovey, *J. Polymer Sci.* **35,** 191 (1959).
21 F. Eisenberg, Jr., and S. Hestrin, *Federation Proc.* **19,** 123 (1960); and unpublished experiments.

[22] D. S. Feingold, G. Avigad, and S. Hestrin, *J. Biol. Chem.* **224,** 295 (1957).
[23] W. B. Neely, *J. Am. Chem. Soc.* **81,** 4416 (1959).
[24] F. A. Bovey, *J. Polymer Sci.* **35,** 167 (1959).
[25] H. M. Tsuchiya, *in* Colloque International sur la "Biochimie des Glucides: Structure, Specifité," Paris, 1960. *Bull. soc. chim. biol.* **42,** 1777 (1960).
[26] R. W. Bailey, S. A. Barker, E. J. Bourne, and M. Stacey, *J. Chem. Soc.* p. 3536 (1957).
[27] F. H. Stodola, E. S. Sharpe, and H. J. Koepsell, *J. Chem. Soc.* p. 3084 (1953).
[28] S. A. Barker, E. J. Bourne, P. M. Grant, and M. Stacey, *Nature* **178,** 1221 (1956).
[29] R. W. Bailey, S. A. Barker, E. J. Bourne, and M. Stacey, *J. Chem. Soc.* p. 3530 (1957).
[30] R. W. Bailey, *Biochem. J.* **72,** 42 (1959).
[31] W. B. Neely, *J. Am. Chem. Soc.* **80,** 2010 (1958).
[32] W. B. Neely, *Arch. Biochem. Biophys.* **79,** 297 (1959).
[33] E. J. Hehre and D. M. Hamilton, *J. Biol. Chem.* **192,** 161 (1951).
[34] A. S. Carlson and E. J. Hehre, *J. Biol. Chem.* **177,** 281 (1949).
[35] E. J. Hehre and D. M. Hamilton, *J. Bacteriol.* **55,** 197 (1948).
[36] E. J. Hehre, *J. Biol. Chem.* **177,** 267 (1949).
[37] J. Monod and A. M. Torriani, *Compt. rend. acad. sci.* **227,** 240 (1948).
[38] M. Doudoroff, W. Z. Hassid, E. W. Putnam, A. L. Potter, and J. Lederberg, *J. Biol. Chem.* **179,** 921 (1949).
[39] J. Monod and A. M. Torriani, *Ann. inst. Pasteur* **78,** 65 (1950).
[40] S. A. Barker and E. J. Bourne, *J. Chem. Soc.* p. 209 (1952).
[41] D. French, *Advances in Carbohydrate Chem.* **12,** 190 (1957).
[42] K. Mühlethaler, *Biochim. et Biophys. Acta* **3,** 527 (1949).
[43] S. Hestrin, *in* "Biological Structure and Function" (First Intern. IUB/IUBS Symposium, Stockholm, 1960), (T. W. Goodwin and O. Lindberg, eds.), Vol. I, pp. 315–325. Academic Press, New York, 1961.
[44] M. Schramm and S. Hestrin, *J. Gen. Microbiol.* **11,** 123 (1954).
[45] B. G. Ranby, *Arkiv. Kemi* **4,** 249 (1952).
[46] S. Hestrin and M. Schramm, *Biochem. J.* **58,** 345 (1954).
[47] J. L. Shimwell, *J. Inst. Brewing* **62,** 339 (1956).
[48] J. L. Shimwell, *Antonie van Leeuwenhoek, J. Microbiol. Serol.* **25,** 49 (1959).
[49] Z. Gromet-Elhanan, Ph.D. Thesis, University of Jerusalem, 1960.
[50] M. Schramm, Z. Gromet, and S. Hestrin, *Biochem. J.* **67,** 669 (1957).
[51] J. Barsha and H. Hibbert, *Can. J. Research* **10,** 170 (1934).
[52] Z. Gromet, M. Schramm, and S. Hestrin, *Biochem. J.* **67,** 679 (1957).
[53] M. Schramm, Z. Gromet, and S. Hestrin, *Nature* **179,** 28 (1957).
[54] F. W. Minor, G. A. Greathouse, and H. G. Shirk, *J. Am. Chem. Soc.* **77,** 1244 (1955).
[55] J. Weigl, *Arch. Microbiol.* **38,** 350 (1961).
[56] L. Glaser, *J. Biol. Chem.* **232,** 627 (1958).
[57] S. Klungsöyr, *Nature* **185,** 104 (1960).
[58] C. P. Jackson and K. Ramamutri, *Nature* **187,** 942 (1960).
[59] J. R. Colvin, *Nature* **183,** 1135 (1959).
[60] A. M. Brown and J. A. Gascoigne, *Nature* **187,** 1010 (1960).
[61] G. A. Greathouse, *J. Am. Chem. Soc.*, **79,** 4503 (1957).
[62] S. Hestrin, *in* "Symposium Metabolismo Microbico," p. 63. Istituto Superiore di Sanita, 1953.
[63] A. Fuchs, Ph.D. Thesis, Delft, Holland, 1959.
[64] G. Lindeberg, *Proc. 2nd Intern. Symposium Food Microbiol. Cambridge, Engl.* p. 157 (1957).

[65] D. J. Bell and R. Dedonder, *J. Chem. Soc.* p. 2866 (1954).
[66] S. Hestrin, D. S. Feingold, and G. Avigad, *Biochem. J.* **64,** 340 (1956).
[67] D. S. Feingold and M. Gehatia, *J. Polymer Sci.* **23,** 783 (1957).
[68] G. Avigad and D. S. Feingold, *Arch. Biochem. Biophys.* **70,** 178 (1957).
[69] E. J. Hehre, *Trans. N. Y. Acad. Sci.* [*2*] **10,** 188 (1948).
[70] S. Hestrin and J. Goldblum, *Nature* **172,** 1046 (1953).
[71] R. Zelikson (Ben-Tovim), M.Sc. Thesis, University of Jerusalem, 1958.
[72] J. R. Loewenberg and R. T. Reese, *Can. J. Microbiol.* **3,** 643 (1957).
[73] R. Zelikson and S. Hestrin, *Biochem. J.* **79,** 71 (1961).
[74] Y. Henis, *J. Gen. Microbiol.* **15,** 462 (1956).
[75] R. Dedonder, *in* Colloque International sur la "Biochimie des Glucides: Structure, Specifité." Paris 1960. *Bull. soc. chim. biol.* **42,** 1745 (1960).
[76] D. S. Feingold, G. Avigad, and S. Hestrin, *Biochem. J.* **64,** 351 (1956).
[77] S. Hestrin and G. Avigad, *Biochem. J.* **69,** 388 (1958).
[78] R. Dedonder and C. Peaud-Lenoël, *Bull. soc. chim. biol.* **39,** 483 (1957).
[79] C. Peaud-Lenoël, *Bull. soc. chim. biol.* **39,** 757 (1957).
[80] S. Avineri-Shapiro and S. Hestrin, *Biochem. J.* **39,** 167 (1945).
[81] R. Dedonder and M. Noblesse, *Ann. inst. Pasteur* **85,** 356 (1953).
[82] S. Hestrin, *in* "Carbohydrate Chemistry of Substances of Biological Interest." *Proc. Intern. Congr. Biochem. 4th Congr.* p. 181 (1958).
[83] C. Peaud-Lenoël and R. Dedonder, *Compt. rend. acad. sci.* **241,** 1518 (1955).
[84] C. Peaud-Lenoël, *Bull. soc. chim. biol.* **39,** 747 (1957).
[85] H. M. Kalckar, *in* "The Mechanism of Enzyme Action" (W. D. McElroy and B. Glass, eds.), p. 675. Johns Hopkins Press, Baltimore, Maryland, 1955.
[86] K. Burton and H. A. Krebs, *Biochem. J.* **54,** 94 (1953).

CHAPTER 9

The Biosynthesis of Homopolymeric Peptides

RILEY D. HOUSEWRIGHT

I. Introduction . . . 389
A. Definition and Scope . . . 389
B. Taxonomic Considerations . . . 390
II. Morphological Evidence of Biosynthesis . . . 391
III. Nutritional Requirements for Polyglutamic Acid Biosynthesis . . . 394
A. *Bacillus subtilis* . . . 394
B. *Bacillus anthracis* . . . 396
IV. Enzyme Systems for Polyglutamate Biosynthesis . . . 397
A. Transaminases . . . 397
B. Racemases . . . 398
C. Glutamine Synthesizing System . . . 398
D. Transamidation . . . 399
E. Transpeptidation . . . 399
F. Polyglutamic Acid Synthetases . . . 400
V. Hydrolysis of Homopolymers . . . 400
A. Microorganisms . . . 400
B. Tissue Extract Enzymes . . . 401
C. Proteolytic Enzymes . . . 402
VI. Composition and Structural Properties of the Peptides . . . 402
A. Are there Separate Homopolymeric or Copolymeric Forms? . . . 402
B. Alpha or Gamma Linkage . . . 404
C. Other Physicochemical Properties . . . 405
VII. Biological Activity . . . 407
A. Influence on Infection . . . 407
B. Antigenicity . . . 408
C. Plasma Volume Expander . . . 408
VIII. Biosynthetic Homopolymers and the Genetic Code . . . 409
References . . . 409

I. Introduction*

A. DEFINITION AND SCOPE

The distinctive attribute of the polypeptides to be considered in this chapter is that they are composed of a single amino acid. Polyglutamic acid

* The following symbols are used throughout: PGA: polyglutamic acid; A-PGA: *B. anthracis* polyglutamic acid; S-PGA: *B. subtilis* polyglutamic acid; M-PGA: *B. megaterium* polyglutamic acid; Na-PGA: sodium salt of polyglutamic acid; H-PGA: free polyglutamic acid; D-PGA: polyglutamic acid composed entirely of D-glutamic acid; L-PGA: polyglutamic acid composed entirely of L-glutamic acid; DL-PGA:

(PGA) is the only peptide known to be composed of a single amino acid that has been studied in relation to its biosynthesis, chemical and physical properties, and biological activity. This polymer will be considered in some detail. Folic acid, synthetic peptides of glutamic acid, and other interesting substances having unusual sequences of amino acids[1,2] will be considered only for comparative purposes.

The first naturally occurring high molecular weight PGA was detected, isolated, and characterized in the period from 1921–1937.[3-6] The information on PGA available in 1953 was summarized by Bricas and Fromageot.[7] In *Bacillus anthracis* and *Bacillus megaterium*, PGA is a major component of the capsule, whereas the peptides from *Bacillus subtilis* and certain related species are found free of cells in the culture medium. Ample work has demonstrated that these peptides contain only glutamic acid and that the one derived from *B. anthracis* contains only the D-isomer.[5,8,9] Both D- and L-isomers are obtained upon acid hydrolysis of the peptides from *B. subtilis* and *B. megaterium*.[10,11]

Research during the past decade has defined media and conditions required for high yields of PGA from *B. anthracis* and *B. subtilis*. The extracellular product from *B. subtilis* has proved to be two peptides, each composed of a single isomer, and not a single PGA composed of both isomers. Methods have been developed for separation of the isomeric homopolymers and for obtaining different proportions of either of them while maintaining a high total yield. Many of the enzymic steps for peptide biosynthesis are now well known, including those forming the unnatural amino acid constituent. The PGA components have been shown consistently to be gamma linked. The precise role of PGA in anthrax infections remains obscure even though a vast amount of information has been gathered.

B. Taxonomic Considerations

Considering that anthrax was among the first diseases shown to be caused by bacteria one would expect *B. anthracis* to have clear taxonomic status. Such has not been the case, but at least it now has regained a separate species status.[12]

There still is a high degree of uncertainty in the classification of organisms reported as *B. subtilis*, *B. licheniformis*, *B. mesentericus*, and *B. pumilus*.[13] The sixth edition of "Bergey's Manual" (p. 709) suggested that "since the

---

polyglutamic acid copolymer containing both D- and L-glutamic acid; ATP: adenosine triphosphate; DNFB: dinitrofluorobenzene; RNA: ribonucleic acid.

European *B. mesentericus* is only a stage of growth of *B. subtilis*, the former name should be dropped." American strains so named were said to be identical with *B. pumilus*. *B. licheniformis* appears as a separate species in the seventh edition of "Bergey's Manual," whereas it had been combined with *B. subtilis* in the sixth edition and was mentioned separately only in the appendix. Among the better known strains called *B. subtilis*, the Marburg strain retains the name but the Ford strain now is called *B. licheniformis*.[14]

In view of these complications, and in an effort to avoid further confusion, each reference to a given species, or strain, in this text will be the same as used in the original reference. All work on *B. subtilis* in our laboratories at Fort Detrick has been with the 9945 or 9945A strain. The American Type Culture Collection now lists this organism as *B. licheniformis* 9945 or 9945A.

II. Morphological Evidence of Biosynthesis

Capsules are readily demonstrated in *B. anthracis* and certain related organisms by microscopic examination of cells. Colonial appearance correlates with the presence of capsular substance, as shown in Figs. 1–4. This well-defined capsule contains PGA. Immunological and genetic studies demonstrate considerable variation in both the appearance and the chemical composition of the capsule.[15-18]

One unanswered question is why the peptide remains as a structural component in one organism and diffuses freely into the growth medium of another. Microchemical reactions with intact capsular PGA from *B. anthracis* and from *B. megaterium* show no free carboxyl groups.[19] The water solubility of the peptide is diminished when carboxyl groups are bound. Ivanovics and Horvath[19] conclude that the carboxyl groups are bound as secondary acidamides in the native peptide and that even mild hydrolysis effects decapsulation and converts the peptide into a substance containing free carboxyl groups (Fig. 5, p. 394). Although the proposed structure appears reasonable, there is meager evidence to support it.

Tomcsik and co-workers[20-23] have concluded, primarily from immunological reactions, that the cross-septa of *B. megaterium* and certain other species are composed of polysaccharides and that PGA fills the free space between the septa. In contrast, Ivanovics and Horvath[24] suggest, from microscopic and biochemical studies of *B. megaterium*, a capsular framework composed of bundles of D-PGA micelles firmly bound to the cell surface and interseptal spaces filled with a granular polysaccharide.

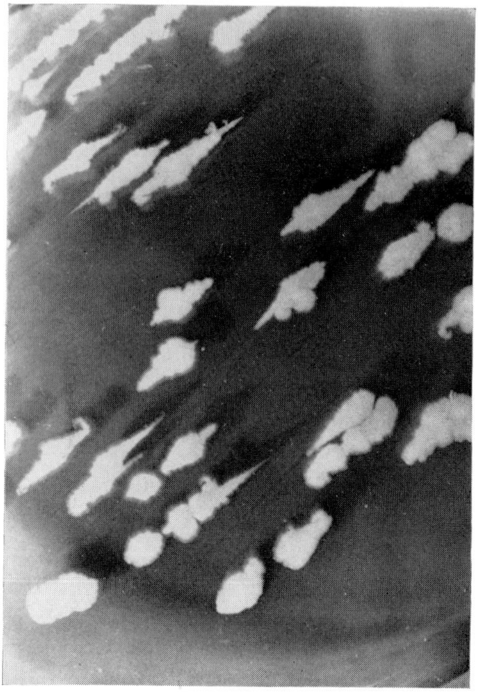

FIG. 1. *Bacillus anthracis* M-36. Rough, noncapsulated colonies on nutrient agar grown in air (similar to rough variant under CO_2 or air); not producing PGA.

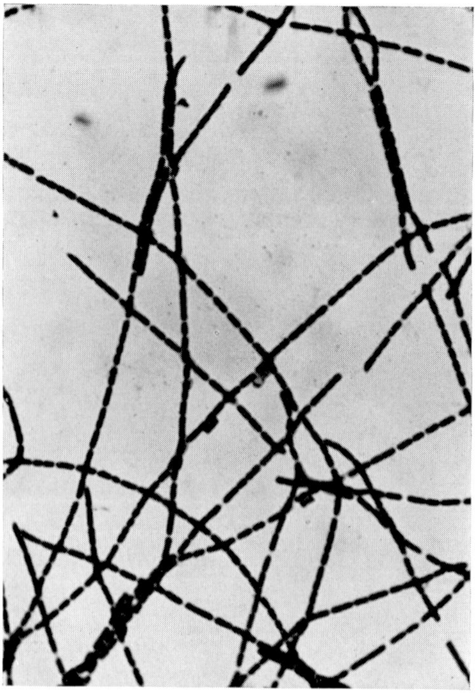

FIG. 2. *Bacillus anthracis* M-36 cells from rough colonies; not producing PGA.

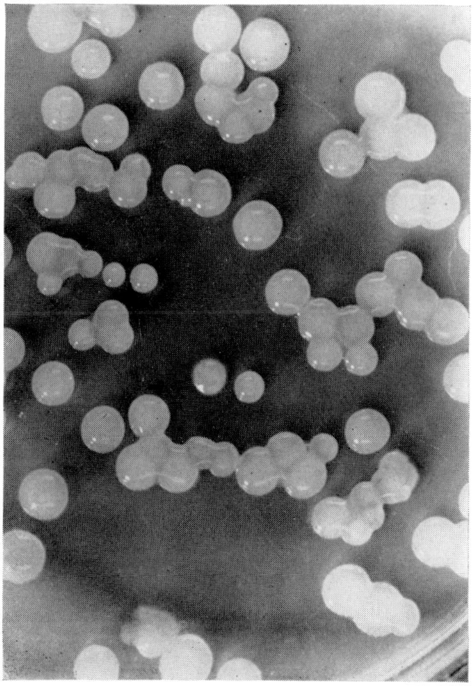

FIG. 3. *Bacillus anthracis* M-36. Mucoid, encapsulated colonies on nutrient agar grown under 50% CO_2 and air; producing PGA.

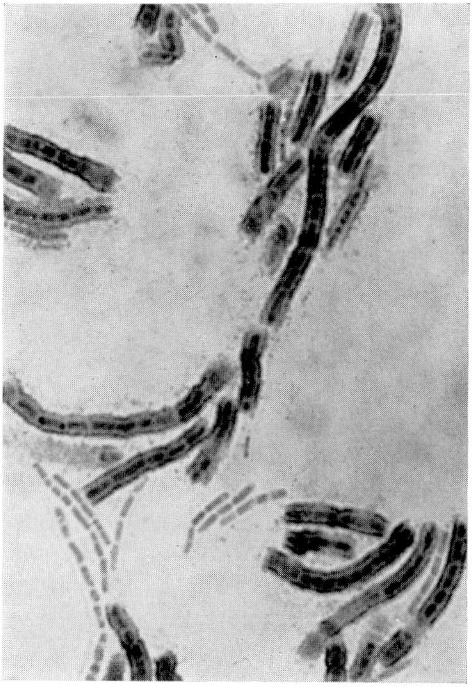

FIG. 4. *Bacillus anthracis* M-36. PGA-producing cells from mucoid colonies.

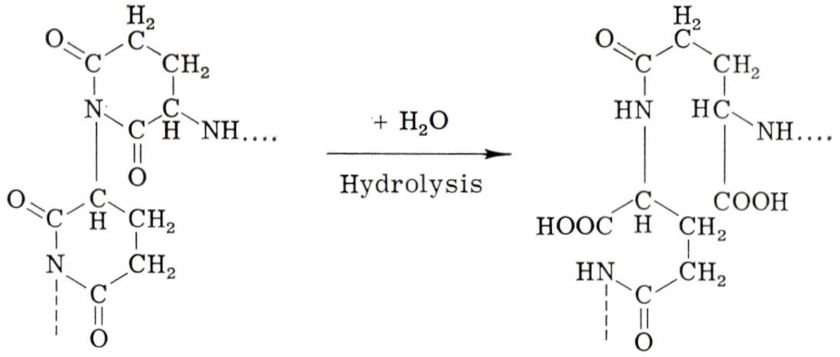

Intact PGA Isolated PGA

FIG. 5. Structure suggested by Ivanovics and Horvath[19] for the intact capsular and for isolated PGA.

III. Nutritional Requirements for Polyglutamic Acid Biosynthesis

A. *Bacillus subtilis*

1. Carbon and Nitrogen Requirements

The minimal nutritional requirements for multiplication of *B. subtilis* are filled by simple carbon, nitrogen, and metallic ion sources. The conditions for maximal PGA formation, however, are more exacting. Sauton's medium, used for many years and considered adequate,[5, 25] yields 1–2 g. PGA/liter; the improved medium[26, 27] regularly supports the production of 20 g./liter.

Free L-glutamic acid is required for high yields even though more than half the glutamic acid of PGA arises from other medium constituents.[28-30] The pH pattern during biosynthesis is important; it can be controlled by the amount of citric acid or glycerol in the medium. A high utilization rate of these substrates, particularly glycerol, 80 g./liter, provides the acid pH (5.5–6.0) necessary for optimum S-PGA synthesis.

2. Inorganic Salt Requirements

The effects of inorganic salts are among the most significant findings in all the studies on PGA formation. The composition and concentration of inorganic salts in the growth medium influence not only the total S-PGA synthesis but also which isomer is formed (Fig. 6).

The requirement of K^+, Mn^{++}, Mg^{++}, Fe^{+++}, and PO_4^{---} for growth and multiplication of *B. subtilis* has been known for several years. Some strains require Zn^{++}.[31] High levels of nitrogen, as ammonium ion, are not required for multiplication but are necessary for high yields of peptide.[26] NH_4Cl is superior to equimolar NH_4OH in promoting peptide formation, because of better pH control resulting from the HCl formed upon utilization of the NH_3.

The amount of Mn^{++} required for maximum cell yield is insufficient for maximum peptide synthesis. By increasing the Mn^{++} from 1.5×10^{-7} to 6.15×10^{-4} M, the yield of peptide is increased from 8 to 20 g./liter without significant change in cell production. Of particular interest is the effect of Mn^{++} on the proportions of the two glutamic acid isomers in the PGA formed. With increasing concentrations of Mn^{++}, the proportion of D-PGA increases until a concentration is reached which inhibits growth. Manganese concentrations that support high peptide production do not increase cell yield but do prolong cell viability[27] thus providing a longer period of

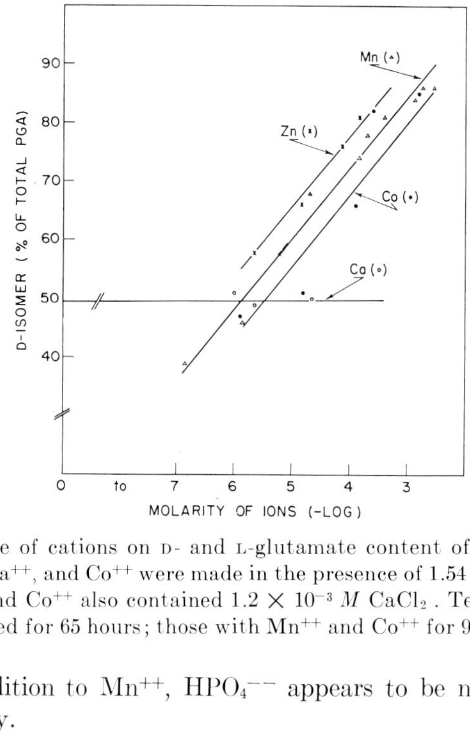

FIG. 6. Influence of cations on D- and L-glutamate content of capsular peptide. Tests with Zn^{++}, Ca^{++}, and Co^{++} were made in the presence of 1.54×10^{-6} M $MnSO_4$; those with Zn^{++} and Co^{++} also contained 1.2×10^{-3} M $CaCl_2$. Tests with Ca^{++} and Zn^{++} were incubated for 65 hours; those with Mn^{++} and Co^{++} for 90 hours.

synthesis. In addition to Mn^{++}, HPO_4^{--} appears to be necessary to prolong cell viability.

Calcium stimulates peptide synthesis in the presence of an amount of Mn^{++} sufficient for high cell yield but suboptimal for peptide synthesis without affecting the proportions of the D- and L-isomers of glutamic acid in the PGA. In the presence of an optimum Ca^{++} level, increasing Mn^{++} concentrations do not change PGA yield significantly but do increase the proportion of D-PGA, much as in the absence of calcium.

Co^{++}, but not Zn^{++}, will replace Mn^{++} for growth. With the concentration of Mn^{++} optimum for growth and Ca^{++} optimum for S-PGA production, Co^{++} and Zn^{++} each increases the proportion of D-glutamic acid in the peptide in the same way as does Mn^{++}. Upon increasing the concentration of either ion, the proportion of D-PGA increases from about 50 to 85% of the total. The effects of Zn^{++}, Co^{++}, and Mn^{++} are not additive.

Several other inorganic ions, found essential either for growth or peptide production, have been found not to influence the proportion of the two glutamic isomers in the peptide. Among these ions are K^+, NH_4^+, Fe^{+++}, Mg^{++} and PO_4^{---}.

B. *Bacillus anthracis*

1. Amino Acid and Energy Requirements

Gladstone,[32] and later Brewer *et al.*,[33] described synthetic media which support excellent cell yields of *B. anthracis*. Brewer's medium contains 17 amino acids, inorganic salts, glucose, D-ribose, guanine, and thiamine. Only the amino acids leucine, valine, and methionine were found essential for growth of the virulent strains 994 and M-36 and the avirulent M strain.[9] Omission of any one of these 3 amino acids resulted in poor cell yield. Of the remaining 14 amino acids, omission of any one did not affect growth, whereas omission of 2 or more decreased total growth significantly. Three amino acids, not essential for growth, exerted an appreciable effect on peptide synthesis. These are DL-isoleucine, DL-phenylalanine, and L-glutamic acid.

Glucose is required for maximum A-PGA production by the 994, M-36, and M strains. The two virulent strains (994 and M-36) produce higher yields of PGA when glucose is autoclaved in the medium than when it is sterilized separately, either by filtration or autoclaving. The converse is true for the avirulent M strain.

2. Carbon Dioxide Requirements

Carbon dioxide enhances capsule formation in some but not all strains of *B. anthracis*.[15, 35] Some strains grow well without CO_2 but produce little PGA; other strains require high levels of L-glutamic acid but no CO_2 for a maximum PGA formation.[9] The relationship between virulence and CO_2 requirement for PGA synthesis is considered in a later section of this chapter.

Large amounts of $C^{14}O_2$ are fixed into the α-carboxyl of glutamic acid in the peptide. If this were the only function of CO_2, one might expect a total replacement by adding glutamic acid. An increased glutamic acid content in the medium does reduce, but not completely replace, $C^{14}O_2$ incorporation into the peptide. Carbon dioxide does not serve as the principal carbon source. Thus, the function of CO_2 in PGA synthesis is still incompletely understood.

IV. Enzyme Systems for Polyglutamate Biosynthesis

Knowledge of the nutritional requirements for PGA synthesis is important for its contribution to an understanding of the enzyme systems involved and the conditions required for such synthesis. Biosynthesis of any peptide involves the free amino acids and other nitrogen components in the medium, the origin and interconversion of amino acids, and finally the enzymic mechanisms for joining the components into a peptide chain. The medium optimum for production of PGA contains a high concentration of L-glutamic acid. Origin of D-glutamic acid from the L-isomer or from other sources is a matter of considerable interest.

A. Transaminases

About 10 years ago our standard references denied the existence of all but two or three L-amino acid transaminases. This conclusion resulted from work on partially purified enzyme systems, usually from mammalian sources. Such purification may have resulted from separation of the coenzyme and led to loss of several active enzyme systems that were described more accurately in the early publications on transaminases.[36] The use of fewer and less specific methods of purification has resulted in the description of a wide variety of L-amino acid transaminases in microorganisms[37, 37a] and of an alanine racemase in many microbial cells.[37b] A completely new series of enzymes, the D-amino acid transaminases, also were found in *B. subtilis*[38] and *B. anthracis*.[39, 40]

Sonic extracts of these bacilli contain both L- and D-transaminases. Fractions which are specific for D-amino acids have been obtained by ammonium sulfate precipitation. Such partially purified preparations from *B. subtilis*[38] catalyze the synthesis of D-glutamic acid by transamination between α-ketoglutarate and D-aspartic acid, D-alanine, D-methionine, or D-serine, whereas the L-isomers are inactive as amino donors. Alanine racemase was present in the purified transaminase preparations; consequently, both D- and L-alanine served as amino donors to α-ketoglutarate. The resulting glutamic acid was, however, the D-isomer; thus indicating that only the D-isomer of alanine was actually active in the transamination step. D-Aspartic and D-glutamic acids transaminated with pyruvate to form alanine but the corresponding L-isomers were inactive. The balance obtained between the reactants and products provide strong evidence that the reaction between D-glutamic acid and pyruvate is a transamination.

The D-transaminase activity of extracts from *B. anthracis* differs from those in *B. subtilis*.[39] Of the several amino acids tested, with the exception of glutamic acid and alanine, only the L-isomers transaminate with α-ketoglutarate and only the D-isomers with pyruvate. Studies on ammonium sulfate fractionation and stability of the enzymes from *B. anthracis* have demonstrated that the D-transaminases are distinct from those responsible for transamination with L-amino acids. One of the most active reactions was that between D-phenylalanine and pyruvate, forming D-alanine and phenylpyruvate. A balance was obtained showing that for each mole of alanine and phenylpyruvate formed one mole of phenylalanine and one mole of pyruvate were used. This reaction is of particular interest because nutritional studies also have indicated that phenylalanine is connected in some way with PGA synthesis by *B. anthracis*.[9, 40]

B. Racemases

A single racemase and a variety of L- and D-transaminases could account for the synthesis of either isomer of a large number of amino acids. Glutamic acid racemases have received attention in the past; the evidence appears to support their occurrence in certain organisms.[41-43] Both L- and D-glutamic acids-C^{14} are incorporated into PGA as D-glutamic acid by *B. subtilis*. Thus far, glutamic acid racemase has not been demonstrated in *B. subtilis* or in *B. anthracis*.[40] Kögl *et al.*[29, 30] postulated but were unable to demonstrate occurrence of a glutamic acid racemase to account for incorporation of L-glutamate carbon in D-glutamate PGA.

Alanine racemase and both L- and D-amino acid transaminases have been found in numerous bacteria including *B. subtilis*. The presence of alanine racemase has been used to explain the synthesis of D-glutamic acid from the L-isomer in the absence of a glutamic acid racemase. The following reactions, which have been demonstrated in extracts of *B. subtilis* and *B. anthracis*,[38, 39] could result in the indirect conversion of L-glutamic acid to the D-isomer:

(1) L-glutamic acid + pyruvic acid $\rightleftharpoons$ α-ketoglutaric acid + L-alanine

(2) L-alanine $\rightarrow$ DL-alanine

(3) D-alanine + α-ketoglutaric acid $\rightleftharpoons$ pyruvic acid + D-glutamic acid

Thus alanine would be a key link in D-amino acid synthesis from the L-isomer.

C. Glutamine Synthesizing System

Cell-free extracts of *B. subtilis*, strain 9945A, synthesize glutamine from glutamic acid and NH_4Cl when ATP and Mn^{++} or Mg^{++} are added.[44] Both Mn^{++} and Mg^{++} are active in the synthesis of L-glutamine, whereas, only Mn^{++} is effective in the synthesis of D-glutamine. The glutamine synthesized has the same configuration as the glutamic acid used in the reaction. These enzyme preparations do not racemize glutamic acid or glutamine. Glutamine synthetase has been isolated from several sources (see review by Meister[45]).

D. Transamidation

The cell-free enzyme preparation that has received most attention in the biosynthesis of γ-linked polyglutamic acid is obtained by $(NH_4)_2SO_4$ or isoelectric precipitation of filtrates from cultures of *B. subtilis*.[46] This preparation is identical with one responsible for hydrolysis of the peptides of both *B. subtilis* and *B. anthracis*.

Two γ-glutamyl transfer enzymes have been known for several years: (*1*) a glutamotransferase[47, 48] which transfers the γ-glutamyl radical from glutamine to various amines, and (*2*) a γ-glutamyl transpeptidase[49, 50] which transfers the γ-glutamyl radical of glutathione to amino acids or peptides. Direct evidence for the transfer of the γ-glutamyl radical from glutamine to amino acids was not reported in these two systems. Williams and Thorne[46, 51-53] have studied the biosynthesis of low molecular weight γ-PGA by transfer reactions. Some of their findings are summarized in the following paragraphs.

When L-glutamine and D-glutamic acid are incubated with the enzyme from *B. subtilis*, a new compound is formed which migrates as a dipeptide of glutamic acid on paper chromatograms. Behavior in a variety of solvents, ion exchange resins, and hydrolysis experiments indicates the compound to be γ-glutamyl-glutamic acid. Evidence for the transfer of the γ-glutamyl radical from L-glutamine to D-glutamic acid by transamidation was obtained from degradation of the dipeptide with DNFB. With both the biosynthetically and chemically prepared dipeptide, removal of the carboxyl end group released DNFB L-glutamic acid equal to one-half of the total glutamic acid present, i.e., L-glutamine donated the free glutamyl amino end group. In other experiments with glutamic acid-1-C^{14} and DNFB, D-glutamic acid was found to be incorporated as the free γ-carboxyl end group, but not to appear significantly as the free amino end group of the biosynthetic γ-glutamyl-glutamic acid. The biosynthetic dipeptides are γ-linked. Metallic ions or other cofactor requirements have not been demonstrated for this enzyme.

More dipeptide was formed from L-glutamine and D-glutamic acid than from other combinations of the optical isomers of these two compounds. The significance of this observation is difficult to understand in view of the strong evidence for the existence of a D-PGA and L-PGA and not a single polymer composed of both isomers. There also is substantial evidence for biosynthesis of γ-linked tri, tetra-, pentapeptides, and possibly longer ones resulting from the same enzyme and substrates.

E. Transpeptidation

Incubation of synthetic γ-D-glutamyl-D-glutamic acid with the *B. subtilis* enzyme leads to the formation of equimolar amounts of glutamic acid and a tripeptide of glutamic acid.[46] The use of D-glutamic acid-1-C^{14} and end-group analysis with DNFB have indicated the reaction to be a transpeptidation in which the free amino end group of one mole of dipeptide is transferred to a second mole of dipeptide to form a mole each of tripeptide and free glutamic acid. Evidence for longer peptides also has been obtained.

The enzyme preparation used in these studies catalyzes the following reactions:

1. Hydrolysis of glutamyl polypeptides from *B. subtilis* and *B. anthracis*.
2. Hydrolysis of D- and L-glutamine.
3. Transamidation between glutamine and glutamic acid.
4. Transpeptidation with γ-D-glutamyl-D-glutamic acid.

Reactions 3 and 4 form peptides containing two, three, or more molecules of glutamic acid, but polymers approaching the molecular weight of those produced by living cultures are not formed by these enzymes.

F. Polyglutamic Acid Synthetases

Cell-free enzymes catalyzing synthesis of high molecular weight PGA have been isolated from cellular extracts of *B. subtilis*, strain 9945A (Leonard and Housewright, unpublished data). The glutamic acid-incorporating enzymes, polyglutamic acid synthetases, require for activity adenosine triphosphate, Mn^{++}, glutamic acid, and β-mercaptoethanol. PGA synthesis was quantitated by measurement of DL-glutamic acid-1-C^{14} incorporation. The PGA synthesized, like the natural product, is undialyzable, precipitates with copper or egg white lysozyme, but not with trichloroacetic acid. It has the same electrophoretic and chromatographic pattern as S-PGA and yields only glutamic acid upon acid or enzymic hydrolysis. Figure 7 shows a possible scheme for the origin of PGA.

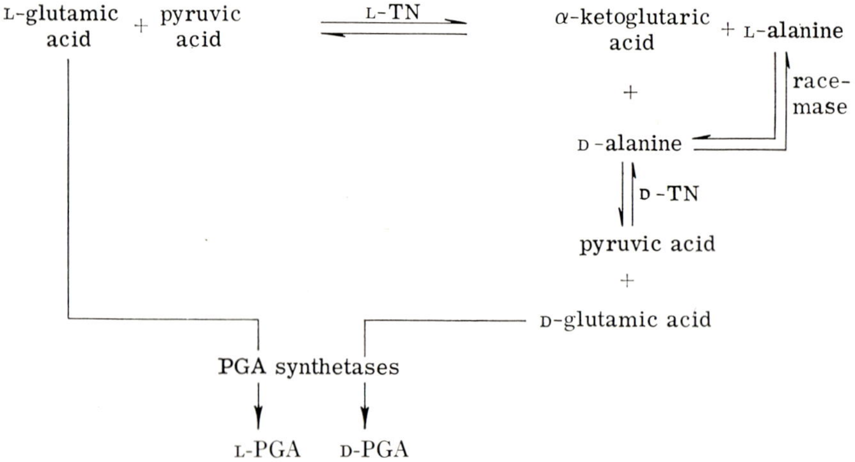

Fig. 7. Possible scheme for origin of PGA by enzymes found in *B. subtilis*. L-TN: L-amino acid transaminase; racemase: alanine racemase; D-TN: D-amino acid transaminase; PGA synthetases: polyglutamic acid synthetases.

V. Hydrolysis of Homopolymers

A. Microorganisms

1. *Bacillus subtilis*

Studies of PGA production by *B. subtilis*[26] show that after attaining maximum yield a decrease in peptide occurs with a concomitant increase in free glutamic acid. With *B. anthracis* no decrease in peptide is noted. Cell-free filtrates of *B. subtilis* contain a peptidase which hydrolyzes both S-PGA and A-PGA. S-PGA is hydrolyzed more rapidly than the A-PGA and yields both intermediate peptides and free glutamic acid, while A-PGA yields only the free glutamic acid. The hydrolysis products of S-PGA include both D- and L-glutamic acid.[26]

2. *Flavobacterium polyglutamicum*

Volcani and Margalith[54] have made a detailed study of the hydrolysis of several kinds of PGA. A new species, *Flavobacterium polyglutamicum*, hydrolyses the γ-L-glutamyl bond in polypeptides. Sonic extracts of this organism contain an enzyme similar to one found in human tissues[55] which attacks only the polymer of L-configuration. The peptidase from *B. subtilis*[26] hydrolyzes both the γ-linked D- and L-PGA as well as the α-linked L-PGA, but not the D-isomer, as has been cited incorrectly.[54] The *F. polyglutamicum* enzyme resembles chicken pancreas conjugase. Both split the γ-peptides of pteroylglutamic acid and require a minimum of three terminal glutamic acid residues[54, 56] for action. Neither enzyme hydrolyzes glutathione. These enzymes differ in reaction rates, requirement for activation by calcium ions, and effect of some inhibitors.

The *F. polyglutamicum* enzyme has the properties of a γ-L-glutamyl carboxypeptidase, as indicated by a requirement for three terminal L-glutamic residues, absence of intermediate peptides, and the successive removal of single amino acids from the carboxyl end of the peptide chain. The enzyme is, however, active as a dipeptidase.

B. Tissue Extract Enzymes

Bovarnick and co-workers[57-60] have studied the potential of PGA from *B. subtilis* as a plasma volume extender. They considered that both the physicochemical properties as a polyelectrolyte and nondiffusibility through a semipermeable membrane should provide a high osmotic efficiency. Initial experiments demonstrated satisfactory pharmacological and immunological properties, but revealed serious deficiencies of rapid excretion and enzymic degradation by several human tissue extracts including liver, kidney, spleen, brain, and erythrocytes.[55] Diffusion and sedimentation measurements on their preparation indicated a peptide length approximating the serum albumin molecule but a diameter only about one-third that of serum albumin. This finding led to the suggestion that the rate of hydrolysis and of excretion by the kidney might be decreased by increasing the diameter of the molecule through cross-linking. This was effected and the conjugate provided higher blood levels in man for longer time intervals because of reduced renal excretion as well as reduced enzymic degradation.

Torii[61] and Utsumi *et al.*[34] found dog liver extract to decompose PGA derived from *B. megaterium* and *B. anthracis*. They concluded that these polyglutamylpeptides possess differences in structure. The significance of these data and other evidences of linkage and isomer content are considered in a later section of this chapter.

C. Proteolytic Enzymes

The attack of proteolytic enzymes on native γ-PGA has not been studied in detail. Green and Stahmann[62] have synthesized α-L-PGA (molecular weight about 10,300) and measured hydrolysis by several proteolytic enzymes. Pancreas extract was active at pH 5 and 6 (68 and 24% hydrolysis at 50 hours, respectively) but not at pH 4. Papain was active (34% in 48 hours) as was carboxypeptidase (40–50%) at pH 5.0 but not at pH 7.5. Crystalline pepsin was inactive (pH 5, 4, and 2), as were trypsin and chymotrypsin (pH 5 and 7.5). On the other hand, trypsin does rapidly hydrolyze the positively charged polylysine.

VI. Composition and Structural Properties of the Peptides

A. Are There Separate Homopolymeric or Copolymeric Forms?

The peptide formed by *B. anthracis* is generally considered to be composed mainly, if not entirely, of D-glutamic acid. In contrast, preparations from *B. subtilis* have been reported to contain both isomers of glutamate in various proportions.[10, 63] The recent data on the effect of metallic ions[27] now furnish a satisfactory explanation for the reports of widely different proportions of D- and L-glutamic acid upon hydrolysis of *B. subtilis* peptides. Knowing the conditions required for a high yield of either isomer should now permit study of why and how certain metals determine the proportion of each isomer in the peptide.

Whether the two glutamate isomers obtained upon hydrolysis of *B. subtilis*[64] PGA are derived from a single copolymeric peptide or from two separate homopolymeric peptides of different configuration is known from recent isolation and purification of peptides from culture filtrates. Peptides with a high percentage of L-glutamic acid are obtained from media containing low concentrations of $MnSO_4$ (1.54×10^{-7} M). Precipitation of the peptides either by $CuSO_4$ or by ethanol and subsequent precipitation from HCl to obtain the peptide in acid form yields a product containing close to 50% of each isomer. Addition of ethanol to the supernatant from the acid precipitation yields an additional peptide containing at least 90% L-glutamate. A similar treatment of filtrates from cultures containing high levels of $MnSO_4$ (6.15×10^{-4} M) produces a peptide containing 94% D-glutamate. The yields of DL-, L-, and D-peptides suggested that preparations composed of 50% of each isomer result from coprecipitation of the D- and L-peptides from acid solutions, with the isomer present in excess remaining in the supernatant solution. This explanation is compatible with the high D-peptide content of cultures grown in high Mn^{++} concentrations and the high L-peptide content of cultures with low Mn^{++} levels. Detailed study of the solubilities of the various isolated peptides showed all to be soluble in neutral solution. In acid solution the D- and L-peptides are each soluble, but mixtures of them are insoluble. The two isolated peptides, mixed in equal proportions, precipitate from acid solution as a product containing about 50% of each isomer. Mixtures of unequal proportions yield the peptide present in lowest concentration quantitatively with an equal amount of the other, leaving the remainder of the excess peptide in the supernate. These results constitute strong evidence that the *B. subtilis* peptides isolated previously from acid solution, and found to contain about 50% of each isomer, were in fact mixtures of two peptides, one composed mostly, if not entirely, of the L-isomer and the other containing principally or only D-glutamate. Further support for these conclusions came from the determinations of the specific activities of precipitates obtained from mixtures of several peptides labeled and unlabeled in the α-COOH carbon with C^{14}.[64] Volcani and Margalith[54] supplied additional evidence for two separate peptides by use of an enzyme which quantitatively releases L-glutamic acid from S-PGA preparations containing both isomers of glutamic acid. All these studies provide strong evidence for the existence of two separate peptides, each composed of at least 90–95% of a single glutamate isomer. Specific *B. subtilis* peptides may contain a very small per cent ($<5\%$) of a second isomer, but this organism does not produce a mixed polyglutamic acid with large proportions of both D- and L-glutamic acid in the same peptide chain.

Torii[65] recently confirmed the observation that *B. subtilis* and *B. megaterium* produce peptides which yield both D- and L-glutamic acid, whereas that produced by *B. anthracis* yields only D-glutamic acid. Torii also offered further evidence for the γ-linkage in peptides from all three species and, with Utsumi *et al.*,[11] substantiated the conclusions of Thorne and Leonard[64] on the chemical structure of the *B. subtilis* peptide. Physicochemical and serological evidence led Utsumi *et al.*[11] to the conclusion that the *B. megaterium* peptide is a γ-copolymer of L- and D-glutamic acid, whereas the *B. anthracis* peptide is a γ-linked homopolymer of D-glutamic acid.

B. Alpha or Gamma Linkage

1. Chemical and Physical Studies

Bovarnick[25] suggested a γ-linkage for the S-PGA because it did not give the biuret test and was not racemized by alkali. Bruckner, Kovacs, and

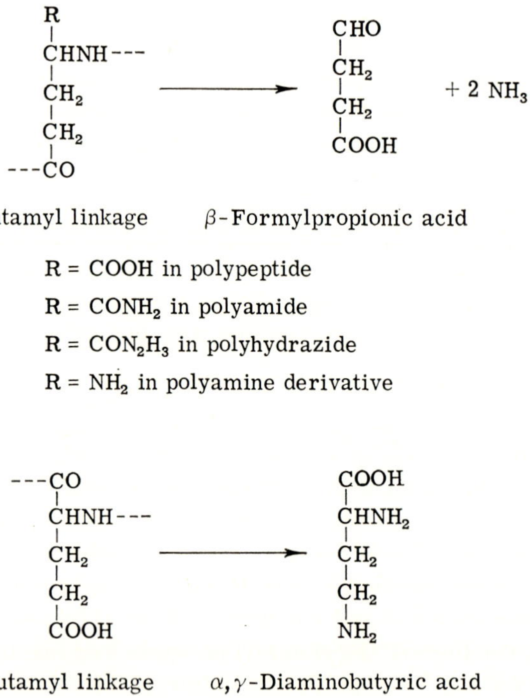

Fig. 8. Hydrolytic products of polyamine derivatives of γ- and α-linked peptides.

their co-workers[66, 67] prepared the polyamide and polyhydrazide derivatives and subjected them to Hofmann (or Curtius) degradation to produce the polyamine derivatives. Hydrolysis yielded only β-formylpropionic acid, whereas synthetic α-polyglutamic acid yielded α-γ-diaminobutyric acid on acid hydrolysis (Fig. 8). These results are consistent with the serological and chemical data.[5]

According to Waley[68] synthetic γ- and α-linked peptides can be differentiated by their infrared spectra. S-PGA and A-PGA resembled the γ-linked peptide. Partial hydrolysis, solubility, and ionization constant determinations also established similarities between the synthetic γ-linked and natural products, and dissimilarities from the synthetic α-linked peptide.

Thus the conclusion of Hanby and Rydon,[8] principally from electrometric titrations, that the capsular substance of *B. anthracis* is composed of α-linked peptide chains containing 50–100 D-glutamic acid residues joined together by γ-peptide chains, has not been substantiated by the subsequent work.

2. Enzymic Decompositon and Synthesis

A peptidase from *B. subtilis*[46] hydrolyzes S-PGA, A-PGA, both the synthetic γ-L-glutamyl-glutamic and α-L-glutamyl-glutamic acids, as well as the γ-D-glutamyl-glutamic but not the α-D-glutamyl-glutamic acid.

A sonic extract from *F. polyglutamicum* possessing a carboxypeptidase activity to three terminal γ-linked L-glutamates (see Section V, A) has been tested by Volcani and Margalith[54] for activity on a wide variety of both synthetic and natural peptides, some with γ- and others with α-linkages. The extract was highly active on γ-linked peptides containing L-glutamic acid and hydrolyzed S-PGA to about 50% but was essentially inactive on the A-PGA. Glutathione and some γ-linked L-dipeptides were unsuitable substrates. A number of α-linked peptides were not suitable substrates.

3. Immunological Reactivity and Specificity

Utsumi *et al.*[11, 69] used quantitative precipitin reactions to provide evidence for the γ-linkage of peptides from *B. subtilis*, *B. anthracis*, and *B. megaterium*. They were able to distinguish also the D-glutamyl peptide of *B. anthracis*, to separate the D- and L-peptide chains of *B. subtilis*, and to establish the copolymeric nature of the *B. megaterium* peptide. Except for the DL-glutamate copolymer from *B. megaterium*, these observations confirmed earlier chemical and enzymic evidence. The poor antigenicity and nonspecific reactions of the glutamyl polymers lead one to suggest that the conclusion on the heteropolymeric nature of the *B. megaterium* peptide, reached solely on an immunological basis, should be considered as tentative.

C. Other Physicochemical Properties

Synthetic, *in vivo*, and *in vitro*-produced PGA have been examined for viscosity, partial specific volume, refractive index increment, sedimentation in the ultracentrifuge, light-scattering properties, optical rotation, infrared spectroscopy, end-group analysis, electrophoretic mobility, behavior as surface films, and diffusion patterns.[5, 8, 68, 71, 72]

1. Molecular Weight

The molecular weight of PGA from natural sources has been reported to vary from above 4000 to about 250,000.[6, 8, 25, 68, 72] Different PGA preparations show a considerable spread in the molecular weights as determined by end-group analysis.[8] On a single preparation light-scattering data (Kent *et al.*[72]) showed a spread between the number average molecular weight (88,000) and the weight average molecular weight (238,000). Considering the active peptidases present in many *B. subtilis* cultures,[26] and the variety of methods of isolation, the spread in molecular weights is not surprising. The data of Kent *et al.*,[72] indicating the polydisperse nature of a single preparation, emphasized the need to interpret physicochemical studies in the light of the heterogeneity in chain length. The isolated PGA undoubtedly differs physically from *in situ* polymers found in bacterial capsules. The high solubility of the isolated product, as contrasted with the intact peptide, has already been mentioned. Moderate amounts of heat result in a marked drop in intrinsic viscosity and molecular weight. Much higher temperatures ordinarily are used for removing the capsular peptide during isolation. These differences point out the need for more detailed *in situ* studies, particularly by X-ray diffraction which has been neglected thus far.

2. Branching

The clear evidence for γ-linked residues and the failure in most studies to detect α-linked residues precludes extensive branching of the polymer chain. This concept is substantiated by agreement between molecular weights as determined by end-group analysis and by certain other physicochemical methods.

3. Random Coil Behavior

The nonionized acidic form, H-PGA, is a random coil, spherically symmetrical, occupying a volume about 100 times that occupied by a globular protein of the same molecular weight. There is a small amount of hypercoiling and cross-linking through hydrogen bonds between strands of the coil.

As PGA becomes ionized electrostatic repulsion causes the molecule to expand. The expansion is slow until the hydrogen bonds are broken, then expansion is rapid until the volume occupied by the fully ionized form, Na-PGA, is about 1000 times that of a globular protein of the same molecular weight. Expansion of the H-PGA also occurs upon rupture of hydrogen bonds by urea and guanadine.[72a]

4. Metallic Ion Effects

The Na-PGA is a very bulky molecule whose size and behavior are highly sensitive to concentration, charge, and nature of the medium, including the other ions present. Heavy metal ions, which are tightly bound to the carboxyl groups, decrease the volume of the molecule in proportion to the firmness of their binding. Differences in the densities at which zero sedimentation is reached in the presence of NaCl and KCl in high density solvent (chloral hydrate) suggest that water of hydration is displaced by chloral hydrate in the presence of potassium but not in the presence of sodium.[72] No evidence of helix formation has been reported.

VII. Biological Activity

A. Influence on Infection

Interest in the mechanism of establishment of anthrax infections led to the characterization of A-PGA and M-PGA.[4, 5, 73, 74] Later studies have considered the role of PGA in this infection.[75-77] The capsule long has been associated with antiphagocytic activity i.e., only encapsulated cells are virulent, although some capsule-forming strains are avirulent. Watson *et al.*[10] have observed a neutralization of the anthracidal factor of leukocytes and other tissues by A-PGA. This peptide also blocks the staphylococcidal activity of luekin, a basic protein obtained from rabbit polymorphonuclear leukocytes.[78] Bail's[79] aggressin hypothesis has found support by some[80-83] but not by other investigators. Smith and Gallop[80] prepared an A-PGA *in vivo* which possessed antiphagocytic activity at 0.8% concentration but was not active as an immunizing antigen.

The CO_2 requirement for capsule (and PGA) formation as a criterion for virulence has been examined in some detail.[15, 17, 35, 84, 85]

Most strains fall into one of three types:

1. The normal virulent type produces A-PGA when grown in $NaHCO_3$ or CO_2 but not when grown in air.
2. An avirulent type produces A-PGA when grown in air, and the presence of $NaHCO_3$ or CO_2 has little or no effect on the amount produced.
3. Another avirulent type does not produce A-PGA under any conditions of growth thus far tested.

These observations form the basis for an *in vitro* test of virulence. Exceptions to these three general types have been found,[15, 17] thus the reliability of the method must await a determination of the frequency of exceptions. If, as now appears, the frequency is low, cultures producing mucoid colonies in air are avirulent; those producing rough colonies in air and CO_2 are avirulent; and those that produce rough colonies in air and mucoid colonies in CO_2 are virulent.

Burnet[86] considered the S-R variation in bacteria and its relationship to bacteriophage in 1929. McCloy[87] observed more recently that nonencapsulated cells of *B. anthracis* are lysed by phage while encapsulated (PGA-containing) cells are resistant. Thorne[88] and others used this observation as a basis for detecting small numbers (1–10) of mucoid cells in the presence of large numbers (10^8) of rough cells. Phage treatment of mixed cell types followed by culture on $NaHCO_3$ agar in CO_2 results in drastic reduction in numbers of R-type cells and the contrasting appearance of mucoid colonies provides a simple method of separation and enumeration of the mucoid type. Reversion from the avirulent R-type to the virulent S-type was demonstrated by this method. The changes from S to R and R to S occur *in vivo* as well as *in vitro*.[88, 89] The proportion of R-type cells isolated from tissues of animals dying of anthrax varies with the strain and increases in direct proportion to the survival time of the animal. R-type cells produce toxin which may well contribute to the outcome of the disease. The inability of small numbers of the R-type to establish an infection in healthy animals appears related to the inability to produce PGA and other components of the capsule. R-type cells, however, apparently are able to survive and produce toxin during late stages of the disease.

Sterne[15] in 1937 suggested virulence to be determined by the capsule (which contains large amounts of PGA) and another factor called "A" factor. Today the capsule and toxin are known to be important factors in virulence; loss of either is accompanied by loss of virulence. The biochemical properties of virulent and avirulent strains of *B. anthracis* have been reviewed recently by Thorne.[88]

B. Antigenicity

Immunological studies, already mentioned in connection with the structure of PGA, have provided much information on the antigenic behavior of the peptide. A detailed review of this subject, except for characterization of the biosynthetic product, is beyond the scope of this chapter.

Isolated PGA is a poor antigen[11, 90]; it behaves as a hapten. Antibodies produced in response to repeated injection of killed encapsulated cells have been used to identify species and the different forms of PGA. The high molecular weight and acidic properties of PGA cause precipitation nonspecifically with basic histones, basic serum components,[10] and other basic substances such as lysozyme. These facts should be considered in all immunological work on PGA, as emphasized by Leonard and Thorne.[70]

C. Plasma Volume Expander

Cross-linked preparations of PGA have been prepared and tested as plasma volume expanders.[57-60] Polymers made from backbones of molecular weight 40,000 or over and side chains of 3,000 or over have a plasma volume retention half-life of 15 hours or longer and show an oncotic efficiency *in vivo* similar to polymer-serum albumin solutions *in vitro*.[60]

The antigenicity of branched PGA polymers has been studied in human volunteers and evidence of poor antigenicity provided by testing for skin reactions, precipitins, and complement fixation. Many attempts to immunize rabbits and guinea pigs with both linear and cross-linked polymers of PGA were completely negative.

VIII. Biosynthetic Homopolymers and the Genetic Code

During the writing of this chapter Nirenberg and Matthaei[91-93] have produced polyphenylalanine by adding polyuridylic acid to a protein-synthesizing system from *Escherichia coli* supernatant and ribosomes. The use of other homopolyribonucleotides or copolymers of mixed nucleotide species is likely to result in the biosynthesis of other homopolypeptides. Homopolynucleotides and copolymers containing two or more nucleotides in known ratios have been synthesized by polynucleotide phosphorylases from *Azotobacter vinelandii*[94, 95] and *Micrococcus lysodeikticus*.[96] These biosynthetic compounds are active as messenger RNA and have been shown to determine the incorporation of amino acids into proteins. Triplet code letters have been assigned to most of the common amino acids although the nucleotide sequence has yet to be determined except for phenylalanine.[91-96] The proposed code is in agreement with amino acid replacement data for several nitrous acid mutants of tobacco mosaic virus.

Crick *et al.*[97] have suggested the following features of the genetic code: (*a*) A group of three bases codes one amino acid; (*b*) the code is not of the overlapping type; (*c*) the sequence of bases read from a fixed starting point; (*d*) the code probably is "degenerate," i.e., one particular amino acid can be coded by one of several triplets of bases.

These recent studies provide direct evidence of genetic and biochemical nature to support the view that the sequence of amino acids in an α-linked polypeptide chain is determined by the bases along the nucleic acid chain.

The synthesis of γ-linked PGA by polyglutamic acid synthetase isolated from *B. subtilis* is not sensitive to ribonuclease or chloramphenicol. Therefore, messenger RNA probably is not involved directly in the synthesis of this unusual polymer (Leonard and Housewright, unpublished data).

ACKNOWLEDGMENTS

The author is particularly grateful to his associates, Dr. C. B. Thorne and Mrs. Carmen Gomez Leonard, for their suggestions on the manuscript, and to Dr. Myles Maxfield for discussions which provided a better understanding of the physical properties of PGA. These colleagues are not responsible for any errors.

REFERENCES

[1] S. Ratner, M. L. Blanchard, and D. E. Green, *J. Biol. Chem.* **164**, 691 (1946).
[2] P. Haas and T. G. Hill, *Biochem. J.* **25**, 1472 (1931).
[3] E. Kramar, *Zentr. Bakteriol. Parasitenk. Abt. I Orig.* **87**, 401 (1921–1922).

[4] J. Tomcsik and H. Szongott, *Z. Immunitätsforsch.* **77,** 86 (1933).
[5] G. Ivanovics and V. Bruckner, *Z. Immunitätsforsch.* **90,** 304 (1937).
[6] G. Ivanovics and L. Z. Erdos, *Z. Immunitätsforsch.* **90,** 5 (1937).
[7] E. Bricas and C. Fromageot, *Advances in Protein Chem.* **8,** 64 (1953).
[8] W. E. Hanby and H. N. Rydon, *Biochem. J.* **40,** 297 (1946).
[9] C. B. Thorne, C. G. Gomez, G. R. Blind, and R. D. Housewright, *J. Bacteriol.* **65,** 472 (1953).
[10] D. W. Watson, W. J. Cromartie, W. L. Bloom, R. J. Heckley, W. J. McGhee, and N. Weissman, *J. Infectious Diseases* **80,** 121 (1947).
[11] S. Utsumi, M. Torii, O. Kurimura, H. Yamamuro, and T. Amano, *Biken's J.* **2,** 165 (1959).
[12] R. S. Breed, E. G. D. Murray, and N. R. Smith, eds., "Bergey's Manual of Determinative Bacteriology," 7th ed., Williams & Wilkins, Baltimore, Maryland, 1957.
[13] G. Ivanovics and J. Foldes, *Acta Microbiol.* **5,** 89 (1958).
[14] N. R. Smith, R. Gordon, and F. E. Clark, *Agr. Monograph* **16,** 72 (1952).
[15] M. Sterne, *Onderstepoort J. Vet. Sci. Animal Ind.* **8,** 271 (1937).
[16] J. Tomcsik and S. Guex-Holzer, *Schweiz. Z. Pathol. Bakteriol.* **14,** 515 (1951).
[17] H. P. Chu, *J. Hyg.* **50,** 433 (1952).
[18] J. Tomcsik and J. B. Baumann-Grace, *Proc. Soc. Exptl. Biol. Med.* **101,** 570 (1959).
[19] G. Ivanovics and St. Horvath, *Acta Physiol.* **4,** 400 (1952).
[20] J. Tomcsik, *Experientia* **7,** 459 (1951).
[21] J. Tomcsik and S. Guex-Holzer, *Schweiz. Z. Pathol. Bakteriol.* **16,** 882 (1953).
[22] J. Tomcsik and S. Guex-Holzer, *Schweiz. Z. Pathol. Bakteriol.* **17,** 221 (1954).
[23] J. B. Baumann-Grace and J. Tomcsik, *Schweiz. Z. allgem. Pathol. u. Bakteriol.* **21,** 5 (1958).
[24] G. Ivanovics and St. Horvath, *Acta Physiol.* **4,** 175 (1953).
[25] M. Bovarnick, *J. Biol. Chem.* **145,** 415 (1942).
[26] C. B. Thorne, C. G. Gomez, H. E. Noyes, and R. D. Housewright, *J. Bacteriol.* **68,** 307 (1954).
[27] C. G. Leonard, R. D. Housewright, and C. B. Thorne, *J. Bacteriol.* **76,** 499 (1958).
[28] C. B. Thorne, D. M. Molnar, and C. G. Gomez, *Bacteriol. Proc.* (*Soc. Am. Bacteriologists*) **107,** 1956.
[29] F. Kögl, P. Emmelot, and D. H. W. den Boer, *Ann. Chem. Liebigs* **589,** 1 (1954).
[30] F. Kögl, P. Emmelot, and D. H. W. den Boer, *Ann. Chem. Liebigs* **589,** 15 (1954).
[31] R. E. Feeney and J. A. Garibaldi, *Arch. Biochem.* **15,** 13 (1947).
[32] G. P. Gladstone, *Brit. J. Exptl. Pathol.* **20,** 189 (1939).
[33] C. R. Brewer, W. G. McCullough, R. C. Mills, W. G. Roessler, E. T. Herbst, and A. F. Howe, *Arch. Biochem.* **10,** 65 (1946).
[34] S. Utsumi, M. Torii, H. Yamamuro, O. Kurimura and T. Amano, *J. Bikens* **4,** 151 (1961).
[35] G. Ivanovics, G. *Zentr. Bakteriol. Parasitenk. Abt. I, Orig.* **138,** 449 (1937).
[36] A. E. Braunstein, *Advances in Protein Chem.* **3,** 1 (1947).
[37] R. D. Housewright and C. B. Thorne, *J. Bacteriol.* **60,** 89 (1950).
[37a] L. I. Feldman and I. C. Gunsalus, *J. Biol. Chem.* **187,** 821 (1950).
[37b] W. A. Wood and I. C. Gunsalus, *J. Biol. Chem.* **190,** 403 (1951).
[38] C. B. Thorne, C. G. Gomez, and R. D. Housewright, *J. Bacteriol.* **69,** 357 (1955).
[39] C. B. Thorne and D. M. Molnar, *J. Bacteriol.* **70,** 420 (1955).
[40] C. B. Thorne, "Amino Acid Metabolism," p. 41. Johns Hopkins Press, Baltimore Maryland, 1955.
[41] S. A. Narrod and W. A. Wood, *Arch. Biochem. Biophys.* **35,** 462 (1952).

[42] P. Ayengar and E. Roberts, *J. Biol. Chem.* **197,** 453 (1952).
[43] L. Glaser, *J. Biol. Chem.* **235,** 2095 (1960).
[44] C. G. Leonard, R. D. Housewright, and C. B. Thorne, *Biochim. et Biophys. Acta,* in press.
[45] A. Meister, *Physiol. Revs.* **36,** 103 (1956).
[46] W. J. Williams and C. B. Thorne, "Amino Acid Metabolism," p. 107. Johns Hopkins Press, Baltimore, Maryland, 1955.
[47] N. Grossowicz, E. Wainfan, E. Borek, and H. Waelsch, *J. Biol. Chem.* **187,** 111 (1950).
[48] P. K. Stumpf and W. E. Loomis, *Arch. Biochem.* **25,** 451 (1950).
[49] C. S. Hanes, F. J. R. Hird, and F. A. Isherwood, *Nature* **166,** 288 (1950).
[50] C. S. Hanes, F. J. R. Hird, and F. A. Isherwood, *Biochem. J.* **51,** 25 (1952).
[51] W. J. Williams and C. B. Thorne, *J. Biol. Chem.* **210,** 203 (1954).
[52] W. J. Williams and C. B. Thorne, *J. Biol. Chem.* **211,** 631 (1954).
[53] W. J. Williams, J. Litwin, and C. B. Thorne, *J. Biol. Chem.* **212,** 427 (1955).
[54] B. E. Volcani and P. Margalith, *J. Bacteriol.* **74,** 646 (1957).
[55] J. Kream, B. A. Borek, C. J. DiGrado, and M. Bovarnick, *Arch. Biochem. Biophys.* **53,** 333 (1954).
[56] W. Dabrowska, A. Kazenko, and M. Laskowski, *Science* **110,** 95 (1949).
[57] M. Bovarnick, S. Fieber, M. R. Bovarnick, and J. Kazlowski, *Proc. Soc. Exptl. Biol. Med.* **83,** 253 (1953).
[58] M. Bovarnick, F. Eisenberg, Jr., D. J. O'Connell, J. Victor, and P. Owades, *J. Biol. Chem.* **207,** 593 (1954).
[59] B. J. Kessler, C. J. DiGrado, C. Benante, M. Bovarnick, R. H. Silber, and A. J. Zambito, *Proc. Soc. Exptl. Biol. Med.* **88,** 651 (1955).
[60] W. S. Rosenthal, D. J. O'Connell, D. R. Axelrod, and M. Bovarnick, *J. Exptl. Med.* **103,** 667 (1956).
[61] M. Torii, *Med. J. Osaka Univ.* **6,** 725 (1955).
[62] M. Green and M. A. Stahmann, *J. Biol. Chem.* **197,** 771 (1952).
[63] M. Torii, *Med. J. Osaka Univ.* **6,** 1043 (1956).
[64] C. B. Thorne and C. G. Leonard, *J. Biol. Chem.* **233,** 1109 (1958).
[65] M. Torii, *J. Biochem.* **46,** 189 (1959).
[66] V. Bruckner, J. Kovacs, and G. Denes, *Nature* **172,** 508 (1953).
[67] V. Bruckner, M. Kajtar, J. Kovacs, H. Hagy, and J. Wein, *Tetrahedron* **2,** 211 (1958).
[68] S. G. Waley, *J. Chem. Soc.* p. 517 (1955).
[69] S. Utsumi, M. Torii, O. Kurimura, H. Yamamuro, and T. Amano, *Biken's J.* **1,** 201 (1958).
[70] C. G. Leonard and C. B. Thorne, *J. Immunol.* **87,** 175 (1961).
[71] H. Edelhock and J. B. Bateman, *J. Am. Chem. Soc.* **79,** 6093 (1957).
[72] L. H. Kent, B. R. Record, and R. G. Wallis, *Phil. Trans. Roy. Soc. London Ser. A* **250,** 1 (1957).
[72a] H. Edelhoch and R. E. Lippoldt, *Biochim. et Biophys. Acta* **45,** 205 (1960).
[73] H. Preiss, *Zentr. Bakteriol. Parasitenk.* **44,** 209 (1907).
[74] H. Preiss, *Zentr. Bakteriol. Parasitenk.* **47,** 585 (1908).
[75] W. J. Cromartie, W. L. Bloom, and D. W. Watson, *J. Infectious Diseases* **80,** 1 (1947).
[76] W. J. Cromartie, D. W. Watson, W. L. Bloom, and R. J. Heckley, *J. Infectious Diseases* **80,** 14 (1947).
[77] W. L. Bloom, D. W. Watson, W. J. Cromartie, and M. Freed, *J. Infectious Diseases* **80,** 41 (1947).

[78] R. C. Skarnes and D. W. Watson, *J. Exptl. Med.* **104,** 845 (1956).
[79] O. Bail and E. Weil, *Arch. Hyg.* **73,** 218 (1911).
[80] H. Smith and R. C. Gallop, *Brit. J. Exptl. Pathol.* **37,** 144 (1956).
[81] H. Smith, *Ann. Rev. Microbiol.* **12,** 77 (1958).
[82] H. Smith and J. Keppie, *Symposium Soc. Gen. Microbiol.* **5,** 126 (1955).
[83] H. Smith, J. Keppie, and J. L. Stanley, *Brit. J. Exptl. Pathol.* **36,** 460 (1955).
[84] C. B. Thorne, C. G. Gomez, and R. D. Housewright, *J. Bacteriol.* **63,** 363 (1952).
[85] K. L. Burdon, E. Comstock, and R. D. Wende, *Bacteriol. Proc.* (*Soc. Am. Bacteriologists*) 106 (1955).
[86] F. M. Burnet, *J. Pathol. Bacteriol.* **32,** 15 (1929).
[87] E. W. McCloy, *J. Hyg.* **49,** 114 (1951).
[88] C. B. Thorne, *Ann. N.Y. Acad. Sci.* **88,** 1024 (1960).
[89] E. S. Beck, M. K. Gorelick, and H. Shimanuki, *Bacteriol. Proc.* (*Soc. Am. Bacteriologists*) 43 (1960).
[90] P. H. Mauer, *Proc. Soc. Exptl. Biol. Med.* **96,** 394 (1957).
[91] M. W. Nirenberg and J. H. Matthaei, *Intern. Congr. Biochem. 5th Congr.* 1961.
[92] M. W. Nirenberg and J. H. Matthaei, *Proc. Natl. Acad. Sci. U.S.* **47,** 1588 (1961).
[93] M. W. Nirenberg, J. H. Matthaei, and O. W. Jones, *Proc. Natl. Acad. Sci. U.S.* In press.
[94] P. Lengyel, J. F. Speyer, and S. Ochoa, *Proc. Natl. Acad. Sci. U.S.* **47,** 1936 (1961).
[95] J. F. Speyer, P. Lengyel, C. Basilio, and S. Ochoa, *Proc. Natl. Acad. Sci. U.S.* **48,** 63 (1962).
[96] R. G. Martin, J. H. Matthaci, O. W. Jones, and M. W. Nirenberg, *Biochem. Biophys. Res. Commun.* **6,** 410 (1962).
[97] F. H. C. Crick, L. Barnett, S. Brenner, and R. J. Watts-Tobin, *Nature* **192,** 1227 (1961).

CHAPTER 10

Biosynthesis of Bacterial Cell Walls

JACK L. STROMINGER*

I. Introduction . . . 413
II. Structure of Bacterial Cell Walls . . . 415
A. The Basal Structure of Bacterial Cell Walls: the Glycopeptide . . . 415
B. The Special Structure of Bacterial Cell Walls . . . 418
III. Isolation of Intermediates in Bacterial Cell Wall Synthesis . . . 426
A. Structure of a Uridine Nucleotide That Accumulates in Penicillin-Inhibited *Staphylococcus aureus* . . . 426
B. Methods of Measurement of Nucleotide Accumulation and of Isolation of Nucleotides . . . 429
C. Accumulation of Nucleotides in *Staphylococcus aureus* . . . 431
D. Isolation of Uridine and Cytidine Nucleotide Cell Wall Precursors from Other Microorganisms . . . 439
IV. Enzymic Synthesis of the Nucleotide Cell Wall Precursors . . . 444
A. Enzymic Synthesis of the Uridine Nucleotide Intermediates . . . 444
B. Mechanism of Action of Oxamycin . . . 453
C. Enzymic Synthesis of Cytidine Nucleotide Intermediates . . . 457
V. Biosynthesis of the Cell Wall . . . 457
A. Participation of Nucleotide Intermediates . . . 457
B. Autolysins and Cell Wall Synthesis . . . 458
C. Cell Wall Synthesis and a Mechanism of Antibiotic Action . . . 459
D. Mutants Blocked in Cell Wall Synthesis . . . 461
VI. Some Additional Problems: The Role of Thymidine and Guanosine Nucleotides in Cell Wall Synthesis . . . 461
References . . . 465

I. Introduction†

Although detailed information is now available about the mechanisms by which several homopolymers (compounds polymerized from a single monomer) are synthesized, relatively less is known about the mechanisms

* Work from the author's laboratory has been generously supported by research grants from the National Institutes of Health, USPHS (NIAMD A-1158 and NIAID E-1902) and from the National Science Foundation (G-7619 and G-18742).

† Abbreviations used in this chapter: DPN, DPNH—oxidized and reduced diphosphopyridine nucleotide; RNA—ribonucleic acid; AMP, ADP, ATP—adenosine-5′-phosphate, adenosine diphosphate, adenosine triphosphate; CMP, CDP, CTP—cytidine-5′-phosphate, cytidine diphosphate, cytidine triphosphate; UMP, UDP, UTP—uridine-5′-phosphate, uridine diphosphate, uridine triphosphate; DNP—dinitrophenyl; DAP—diaminopimelic acid; TDP—thymidine diphosphate; GNAc—*N*-acetylglucosamine; Ala—alanyl; Glu—glutamyl; Lys—lysyl; P, PP—orthophosphate, pyrophosphate; Pi, PPi—inorganic orthophosphate, inorganic pyrophosphate.

of synthesis of heteropolymers. Heteropolymers are relatively high molecular weight compounds built up from two or more different monomers. Obviously, this definition applies to a variety of compounds of different types. In the simplest examples, the molecules all belong to the same group of organic compounds. This would include polysaccharides containing two or more sugars, proteins containing up to twenty different amino acids, and both ribo- and deoxyribonucleic acids, containing four or more nucleotides. In other examples, however, the polymer may contain molecules belonging to several different types of organic substances. This would include polymers of sugars and amino acids (glycoproteins), lipid and amino acids (lipoproteins), and lipid and sugars (glycolipids).

All of these types of compounds occur in microorganisms. Bacterial cell walls are particularly complex heteropolymers. They contain polysaccharide, both small peptides and proteins and unusual lipids. Although the chemistry of these substances is incompletely known, it is evident that each of these three types of molecules may be contained within a polymer (a "glycoproteolipid").

Nucleoside monophosphate acid anhydrides are important carriers of small molecules for incorporation into polymers. Thus, amino acid derivatives of adenylic acid are intermediates in protein synthesis (equations 1a–c); sugar derivatives of uridine, guanosine, or thymidine diphosphate are intermediates in the biosynthesis of polysaccharides (equations 2a and b); and alcohol derivatives of cytidine diphosphate are intermediates in the synthesis of phospholipids (equations 3a and b).

$$\text{ATP} + \text{amino acid} \rightleftharpoons \text{amino acyl-AMP} + \text{PP} \tag{1a}$$

$$\text{Amino acyl-AMP} + \text{RNA} \rightleftharpoons \text{amino acyl-RNA} + \text{AMP} \tag{1b}$$

$$\text{Amino acyl-RNA} \rightarrow \rightarrow \text{protein} \tag{1c}$$

$$\text{UTP} + \text{sugar-1-phosphate} \rightleftharpoons \text{UDP-sugar} + \text{PP} \tag{2a}$$

$$\text{UDP-sugar} + \text{acceptor} \rightleftharpoons \text{acceptor-sugar} + \text{UDP} \tag{2b}$$

$$\text{CTP} + \text{alcohol phosphate} \rightleftharpoons \text{CDP-alcohol} + \text{PP} \tag{3a}$$

$$\text{CDP-alcohol} + \text{acceptor} \rightleftharpoons \text{acceptor-P-alcohol} + \text{CMP} \tag{3b}$$

As will become apparent, these mechanisms are all involved in the biosynthesis of bacterial cell walls.*

In addition to these relatively simple substances, more complex types of molecules can also be activated as nucleotide derivatives. The complex molecule may be formed after a simple molecule is first added to the nucleotide (as in the case of formation of UDP-glucose and its subsequent

* For a full discussion of nucleotides as metabolic intermediates, see refs. 1 and 2.

transformation to UDP-glucuronic acid), or it may be formed before addition to the nucleotide (as in the case of formation of CDP-diglyceride from diglyceride phosphate).

At the present time it is believed that nucleotide derivatives of these types are precursors of the bacterial cell wall. A great deal is known about the structure and mechanism of synthesis of the nucleotide intermediates, and these data will form the central theme of this chapter. Almost nothing is now known about the mechanisms by which these intermediates are used in cell wall synthesis. Bacterial cell wall synthesis is, however, a subject in ferment since it has become apparent that some antibiotics owe their selective toxicity for microorganisms to interference at some point in these synthetic reactions. The mechanism of action of one of these antibiotics (oxamycin or D-cycloserine), which interferes with the synthesis of one of the nucleotide intermediates, is relatively completely understood. Other inhibitors of cell wall synthesis appear to interfere with the reactions which lead directly to polymer synthesis; their precise mechanisms of action are unknown. Knowledge of intermediates in cell wall synthesis was initially almost entirely due to a study of the effects of antibiotics and other inhibitors on microorganisms. These substances, therefore, have been and will be in the future important tools with which to explore this interesting aspect of microbial metabolism. Current interest in this field has also been catalyzed by the realization that other aspects of the interaction of man and microorganisms, such as specific toxic and antigenic responses of higher animals to bacterial infections and the defense mechanisms of these animals to such infections, involve components of bacterial cell walls.

II. Structure of Bacterial Cell Walls

This subject has been discussed in detail by Salton[3] in Vol. I of this treatise. The purpose of the brief recapitulation here is to emphasize some recently discovered aspects of cell wall structure which relate to cell wall synthesis, especially in *Staphylococcus aureus*.

A. The Basal Structure of Bacterial Cell Walls: the Glycopeptide

Qualitative analyses of the cell walls of all the Gram-positive bacteria so far examined have revealed a strikingly similar and relatively simple composition. The few compounds that are regularly found in these walls are believed to be components of a structure which is termed the basal structure.[4-6] These components are acetylglucosamine, a lactic acid ether of acetylglucosamine (acetylmuramic acid), alanine, glutamic acid, and either lysine or diaminopimelic acid. As will be seen, additional components may be linked to the basal structure, but that portion of it which contains the amino sugars and amino acids may be called the glycopeptide or mucopeptide (without prejudice to the important question as to whether these components are linked in the same or in a different manner in different microorganisms).

Some of the current concept of the structure of the glycopeptide is derived by inference from the structure of the nucleotide intermediates, which will be discussed below. Thus, in *Staphylococcus aureus* a part of the structure is represented as acetylglucosamine-lactyl·L-Ala·D-Glu·

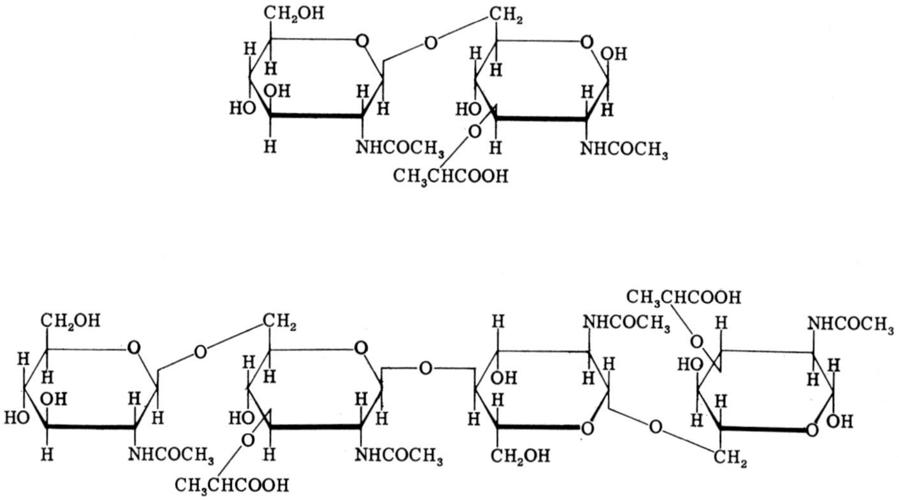

FIG. 1. The proposed structure of a disaccharide (β-1,6-acetylglucosaminyl-acetylmuramic acid) and tetrasaccharide from cell walls of *Micrococcus lysodeikticus* (Salton and Ghuysen[17]).

L-Lys·D-Ala·D-Ala,[7] and in *E. coli* a similar structure exists in which *meso*-diaminopimelic acid replaces L-lysine in the peptide sequence.[8] Other evidence regarding structure is derived from a study of the products of hydrolysis by egg white lysozyme[9-16] and by lytic enzymes from other sources (e.g., F_1 and F_2 enzymes from *Streptomyces*,[17] bacteriophage lysozymes[18, 19]). The use of a variety of specific hydrolytic enzymes in analysis of cell wall structure is certain to become increasingly important.

The disaccharide, β-1,6-acetylglucosaminyl-acetylmuramic acid, and a corresponding tetrasaccharide (as well as products of greater complexity) are formed in small amounts during lysozyme digestion of cell walls from *Micrococcus lysodeikticus*[9-15] (Fig. 1). In this organism, some of the acetylmuramic acid residues are apparently not substituted by a peptide. Other acetylmuramic acid residues may be substituted by peptides of different sizes and complexity and in other organisms by peptides containing different amino acids. For example, the basal layer of the cell wall of *Escherichia coli* has recently been isolated in relatively pure form.[20] The principal products of hydrolysis of this preparation by egg white lysozyme or by a lysozyme from bacteriophage T2 lysates of *E. coli* have been identified[19] as the disaccharides (in which the composition but not the sequence of the peptides is known):

```
GNAc—GNAc            GNAc—GNAc
|                    |
lactyl               lactyl
|                    |
Ala                  Ala
|         and        |
Glu                  Glu
|                    |
DAP                  DAP
                     |
                     Ala
```

The occurrence of GNAc-lactyl [Ala-Glu-DAP-Ala] in the structure is paralleled by a report of the occurrence of a uridine nucleotide containing this fragment in some *E. coli* strains,[21] but presents interesting biosynthetic problems (cf. below).*

Similar attempts have been made recently to obtain such disaccharide peptides from cell walls of *S. aureus*.[22] After removal of the ribitol phosphate polymer (see below) from the cell wall, the residue of the cell wall of *S. aureus* was solubilized by lysozyme digestion; the cell wall itself was insensitive to this enzyme. All of the product, however, remained in a high molecular weight compound and no small fragments could be detected in these hydrolyzates. Further hydrolysis with a β-acetylglucosaminidase released virtually all of the acetylglucosamine as the monomer. The other products were separated into three components on columns of Sephadex G-25 and G-50. Unlike some of the products obtained from cell walls of *M. lysodeikticus* and *E. coli*, all three products from *S. aureus* cell walls were high molecular weight compounds; no monosaccharide or disaccharide peptides were detected.

In addition to the glycopeptide, the residue of the wall, after removal of the ribitol phosphate polymer, also contains a polyglycine component.[22] Reaction of this residue and the various soluble products derived from it

* See footnote on page 442, however.

(as well as cell wall itself) with dinitrofluorobenzene showed that neither the α- nor the ϵ-amino group of lysine is available to the reagent, and only traces of DNP-glycine could be found. These data, therefore, led to the hypothesis that the glycopeptide of *S. aureus* is polymerized by polyglycine cross-links between peptide chains (see Fig. 3) as well as by the primary polysaccharide backbone. Other investigators have also recognized the paucity of free amino groups in cell walls,[13, 23, 24] despite the presence of diamino acids, and have similarly postulated the presence of linked peptides. Such structures must be considered in terms of possible biosynthetic pathways.

With regard to the polysaccharide polymer, at the present writing most workers in the field believe that all bacterial cell walls contain a "backbone" consisting of alternating acetylglucosamine and acetylmuramic acid residues, most of which are substituted by a peptide. It is, however, by no means certain that the linkage of the sugars is the same in all cases, e.g., β-1,6-, α-1,6-, β-1,4-, α-1,4-, and other linkages could all occur in the structure. Indeed, it has not been proved that strict alternation of the sugars always occurs. Most, but not necessarily all, of the acetylmuramic acid residues in the walls examined so far are substituted by peptide chains. The peptide chains on different acetylmuramic residues are not necessarily identical (as indicated above for *E. coli*; see below for the case of *S. aureus*).

B. The Special Structure of Bacterial Cell Walls

Analyses of cell wall preparations from various species have revealed that *none* of these preparations contains *only* the components of the basal structure. The additional component(s) may be referred to as special structure, again without prejudice to the question of the manner of linkage to the basal structure in different species or to the question of anatomical location with respect to the basal structure.

In some microorganisms the special structure may be relatively simple.[4-6, 25, 26] For example, in *S. aureus* it contains only glycine (about 5 residues per glutamic acid residue) and a ribitol phosphate polymer. In *M. lysodeikticus* the additional components are glycine and glucose. In other organisms these additional components may be more complex. For example, in corynebacteria complex polysaccharides are present and in streptococci both the group-specific polysaccharides and the type-specific proteins are cell wall components. These surface components are responsible for the characteristic serological behavior of the microorganism.

In the Gram-negative bacteria, the wall has an even greater complexity, containing from 8–30% lipid (or waxes) as well as protein and polysaccharide. In Gram-negative cells these "additional" components include the lipopolysaccharides (or bacterial endotoxins)[27] which contain chemical groupings responsible for characteristic toxic reactions as well as specific serological reactions, and in acid-fast bacteria they include the complex glycolipids (waxes).[28]

Quantitatively, in the simplest cases (e.g., in some strains of *S. aureus*) the basal structure comprises as much as 80% of the weight of the cell wall while the special structure comprises only 20%. In the most complex examples, however, the quantitative situation may be reversed, and the special structure may comprise more than 80% of the cell wall weight (e.g., in *E. coli*). These quantitative considerations are important in the interpretation of isotopic data to be discussed below.

In several instances more detailed information has become available regarding structure of several of these components. Only the polyol phosphate polymers, however, and the lipopolysaccharide (O-antigen) of Gram-negative bacteria will be considered here.

1. The Polyol Phosphate Component of Gram-Positive Bacteria

a. Isolation and Structure. Cell walls from many of the Gram-positive microorganisms contain considerable amounts of organic phosphate (e.g., *S. aureus*) although some contain little, if any, phosphate (e.g., *M. lysodeikticus*). The nature of the phosphate component was first investigated by Mitchell and Moyle,[29] who identified glycerol phosphate and an unknown polyol phosphate in hydrolyzates of cell walls of *S. aureus*, strain Duncan. Subsequently, Baddiley and co-workers identified the polyol phosphate, ribitol phosphate, as a constituent of many bacterial cell walls,[30] recognized the nature of the polymers containing ribitol phosphate or glycerol phosphate (called teichoic acids), and carried out detailed investigations of their distribution and composition.[31] These polymers were extracted from cell walls by cold 10% trichloroacetic acid. In some cases the polymer contained only ribitol phosphate units and in others apparently only glycerol phosphate. In a few organisms both polyols were found in teichoic acid preparations, but it is not known whether they occur in the same or different molecules. A sugar, glucose or acetylglucosamine, and possibly other sugars, may be attached to the ribitol. D-Alanine is linked as an ester. The glycerophosphate polymers which have been isolated contained no sugar component but they did contain esterified D-alanine.

The structure of the repeating unit of the polymer was first investigated in the case of *Bacillus subtilis*.[32] Evidence has been obtained that this unit is 4-*O*-*β*-glucosyl-D-ribitol-5-phosphate (Fig. 2a); the D-alanine, originally believed to be esterified to one of the hydroxyl groups of the glucose residue, is now believed to be located on the ribitol.[33] The polymer itself contains 1,5-phosphodiester linkages between the ribitol units.

More recently, the structure of the polymer obtained from *S. aureus*, strain Copenhagen, has been investigated in this laboratory.[34] As previously observed,[31] the ribitol phosphate polymer in this species contains acetylglucosamine residues. Glucosaminyl ribitol phosphates were isolated following alkaline hydrolysis of the polymer, and the acetyl group was replaced by selective *N*-acetylation. This product has been characterized as a mixture *predominantly* of 4-*O*-β-acetylglucosaminyl-D-ribitol-1- and 2-phosphates (Fig. 2b). The β-linkage of the acetylglucosamine was deduced from the fact that both this monomer and the polymer itself were hydrolyzed extensively by a β-acetylglucosaminidase from pig epididymis (cf. also below). Evidence that the phosphodiester linkage is 1,5 in this

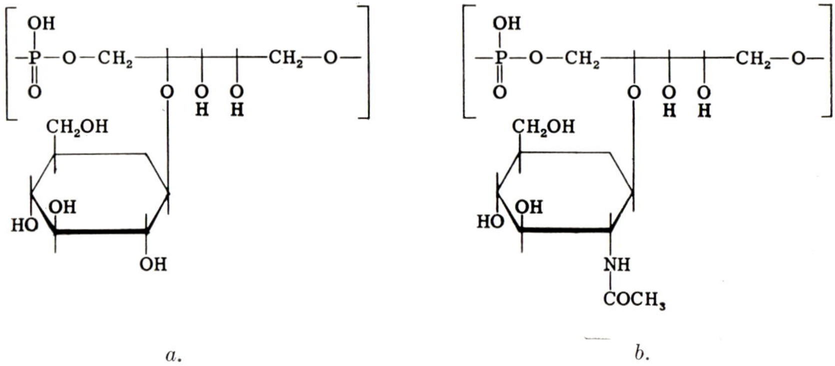

FIG. 2. Structure of the repeating units of the ribitol phosphate polymers from (*a*) *Bacillus subtilis* (4-*O*-β-glucosyl-D-ribitol-5-phosphate[32]) and (*b*) *Staphylococcus aureus* (4-*O*-β-acetylglucosaminyl-D-ribitol-5-phosphate[34]). A small percentage of the acetylglucosamine residues are α-linked in *S. aureus*, strain Copenhagen; these are the immunologically active groups.[34, 36] The D-alanine ester is not shown.

organism, as in *B. subtilis*, was also obtained.* The polymer itself contained near 0.5 esterified D-alanine residues per repeating unit. Very little D-alanine was liberated in the free form during acid extraction of this polymer from cell walls, and this amount is the maximum of D-alanine which can be extracted with the polymer. The polymer appears to contain about 16 repeating units (mol. wt. about 7000).

Of particular interest is the fact that the teichoic acid is the antigenic determinant in this strain of *S. aureus*.[36] Studies of the chemical basis of the serological reaction have led also to more detailed knowledge of the fine structure of the polymer.[34] Rabbits were immunized against formalized or heat-killed cells of *S. aureus*, strain Copenhagen. The resulting antisera, but not those from normal rabbits, agglutinated cell walls of this strain.

* Similar conclusions have been reached regarding the structure of the ribitol phosphate polymer from *S. aureus*, strain H.[35]

It was found by the specific inhibition technique that the major antibody present had a specificity for the acetylglucosamine linkage in the teichoic acid. Thus, acetylglucosamine, teichoic acid, and all hydrolysis products of teichoic acid which contain the acetylglucosamine residue (e.g., acetylglucosaminyl ribitol) specifically inhibited agglutination. D-Ribitol-5-phosphate and a large number of compounds derived from or related to the glycopeptide of the cell wall did not inhibit agglutination. In addition, α-phenylacetylglucosaminide was an excellent inhibitor, while β-phenylacetylglucosaminide was inactive in this respect. Immunochemical data, therefore, suggested that the acetylglucosamine linkages in the teichoic acid antigen were α, while enzymic data had suggested they were β. This paradox was resolved by the finding that the polymer contains one α-acetylglucosaminyl ribitol for each seven β-acetylglucosaminyl ribitols. Careful measurement of the hydrolysis of this teichoic acid by β-acetylglucosaminidase revealed that the reaction stopped when 85% of the acetylglucosamine residues had been removed. The product was isolated and found to contain a ratio of ribitol phosphate: acetylglucosamine of 1:0.12. The residual acetylglucosamine could not be removed by β-acetylglucosaminidase, but was slowly hydrolyzed by a preparation from rat epididymis, which contains an α-acetylglucosaminidase. This limit product of β-acetylglucosaminidase hydrolysis was the best inhibitor of cell wall agglutination obtained (Table I).* An antigen from *S. aureus* strains had in fact been isolated 25 years ago by Julianelle and Wieghard,[39] and was called "type A polysaccharide." From their description of the method of isolation and chemical properties, it is clear that they had isolated a teichoic acid. Moreover, the wheal and erythema observed by Julianelle and Hartmann[40] in immune human subjects on intradermal injection of the "type A polysaccharide" has been reproduced with teichoic acid.[41]†

b. Relationship of the Polyol Phosphate Polymer to the Glycopeptide. The manner in which the various components of the special structure in

* Cell walls of *S. aureus*, strain Copenhagen, were also agglutinated by antisera prepared against Group A streptococcal carbohydrate (in which the antigenic determinant is a β-acetylglucosamine residue)[37] or against an artificial antigen (azoprotein) containing the β-*p*-aminophenylacetylglucosamine residue[38] (both kindly given by Dr. M. McCarty). These cross-reactions were both inhibited by teichoic acid and by β-phenylacetylglucosaminide but not by α-phenylacetylglucosaminide or by the limit product of β-acetylglucosaminidase hydrolysis of teichoic acid. The β-acetylglucosaminyl ribitol units of teichoic acid are apparently able to react with appropriate antisera, although they do not induce antibody formation in rabbits.

† The finding that teichoic acid is an antigen in *S. aureus* has been confirmed with another strain, but in this case it has been suggested that the determinant is not an α-acetylglucosaminyl-ribitol grouping.[38a] Moreover, an isolated surface antigen from still another strain does not appear to be a teichoic acid; e.g., among other properties, its phosphate content is very low.[38b]

different bacteria are linked to the glycopeptide has never been clearly defined. This problem can be exemplified by consideration of the question of the linkage of the ribitol phosphate polymer to the glycopeptide in *S. aureus*. The extraction of this polymer from the wall in 10% trichloroacetic acid at 0° C. is a slow process. The half-time is of the order of 40 hours.

TABLE I

IDENTIFICATION OF THE SEROLOGICALLY ACTIVE GROUP IN THE TEICHOIC ACID FROM THE CELL WALL OF *Staphylococcus aureus*, STRAIN COPENHAGEN[34, 36]

Additions	Final conc.	Degree of agglutination			
		5 min.	1 hr.	2 hr.	24 hr.
None	—	4+	4+	4+	4+
Acetylglucosamine	0.01 *M*	0	0	1+	3+
α-Phenylacetylglucosaminide	0.0025 *M*	0	0	1+	3+
β-Phenylacetylglucosaminide	0.0025 *M*	4+	4+	4+	4+
Teichoic acid	0.0013 *M*	0	0	1+	3+
Alanine-free teichoic acid	0.0008 *M*	0	0	1+	3+
Teichoic acid (limit β-acetylglucosaminidase)	0.0003 *M*	0	0	0	0
Acetylglucosaminyl ribitol phosphates	0.003 *M*	0	0	2+	4+
Acetylglucosaminyl ribitol	0.003 *M*	0	0	2+	4+
D-Ribitol-5-phosphate	0.01 *M*	4+	4+	4+	4+
Normal rabbit serum substituted for antiserum	—	0	0	0	0

NOTE: Rabbit antisera prepared against *S. aureus*, strain Copenhagen, agglutinated purified cell walls of this strain. To investigate the chemical basis of the immunological reaction, specific inhibition of agglutination by haptenes derived from or related to the cell wall was examined. All compounds active as inhibitors and several inactive ones are listed. A large number of other compounds not listed did not inhibit agglutination. The data shown indicate that an α-acetylglucosaminyl ribitol linkage in the teichoic acid is the serologically active group.

After 3 weeks 5% of the original phosphate remains in the glycopeptide,[22] although it is probable that this residual phosphate is bound in a different form from teichoic acid. At 38°, 60°, or 100° separation of the polymers is accelerated. One can visualize either that some tight ionic bonding is involved (although, as already mentioned, the glycopeptide contains extremely few free amino groups which could interact with phosphate) or that some covalent linkage actually exists. Some labile bond might be cleaved in trichloroacetic acid, faster at higher temperatures.

The problem is defined further by the fact that D-alanine is a component of both the teichoic acids and of the glycopeptide. The possibility that esterification of the terminal D-alanine of glycopeptides provides a means of linking glycopeptide to teichoic acids and to other types of special structure can, therefore, be considered. Various procedures are employed for removal of special structure from glycopeptide in various bacterial cell walls. For example, hot formamide has been employed to remove the group-specific carbohydrate from streptococcal cell walls,[42] and hot phenol or acetic acid has been employed to separate the lipopolysaccharide in Gram-negative bacteria.[27, 43] In some of these cases the preparations obtained contain small amounts of amino acids, the linkage and significance of which has not been established.

In the case of *S. aureus*, strain Copenhagen, two D-alanine residues occur in the cell wall for each L-alanine, D-glutamic acid, or L-lysine residue.[44] Hence, it was suggested that most, if not all, of the peptide in this glycopeptide had the sequence L-Ala·D-Glu·L-Lys·D-Ala·D-Ala, the sequence found in a uridine nucleotide cell wall precursor.[7] The occurrence of D-alanine in teichoic acid prompted an investigation of the amounts of D-alanine in this polymer and in the glycopeptide.[22] Exactly half of the D-alanine was in each polymer. The glycopeptide, therefore, had only one D-alanine residue per each residue of L-alanine, D-glutamic acid, and L-lysine. A uridine nucleotide cell wall precursor containing only a single D-alanine residue does not occur in this organism. There are two prominent interpretations of these data (Fig. 3). The glycopeptide might contain a mixture of peptides of sequence L-Ala·D-Glu·L-Lys·D-Ala·D-Ala and L-Ala·D-Glu·L-Lys in approximately equivalent amounts, or in some other sequences and proportions, the average of which yields the ratios found (Fig. 3, A). Alternatively, all of the peptide in the glycopeptide might originally have had the sequence L-Ala·D-Glu·L-Lys·D-Ala·D-Ala, with the terminal D-alanine esterified to the acetylglucosamine of the ribitol phosphate polymer (Fig. 3, B). Thus, the two polymers would be linked by a peptide bridge. In this view the separation by trichloroacetic acid would be the consequence of peptide bond cleavage and separation of the terminal D-alanine residue of the peptide with the ribitol phosphate polymer. Further investigation will be required to decide clearly between the various possibilities.*†

* Extraction of teichoic acid from cell walls of *L. arabinosus* leaves a wall structure intact, as visualized by electron microscopy. In addition, many of the D-alanine residues in wall preparations from several bacteria were found to have free amino groups.[44a] If, as believed by these authors, the teichoic acid is ionically bound to the wall, these substances should be regarded as microcapsular. These data would favor the representation of the wall of *S. aureus*, strain Copenhagen, shown in Fig. 3A, but the experiments reported do not exclude attachment of teichoic acid at the surface by some covalent linkage, cleaved by acid. In *S. aureus*, at least, the difficulty of extraction with acid and its marked temperature coefficient require some explanation; these facts are not consistent with a simple dissociation of ionic bonds in acid.

† A teichoic acid, apparently containing a novel polyol, has been isolated recently from *B. megaterium*, strain KM, and evidence has been obtained to indicate that it is covalently linked to a portion of the glycopeptide.[44b]

This discussion of the complexities of structure of the cell wall of one microorganism and of the relationship of the various polymers which it contains is intended to emphasize possible complexities in considering biosynthetic pathways. The biosynthesis of the cell walls of various bacteria

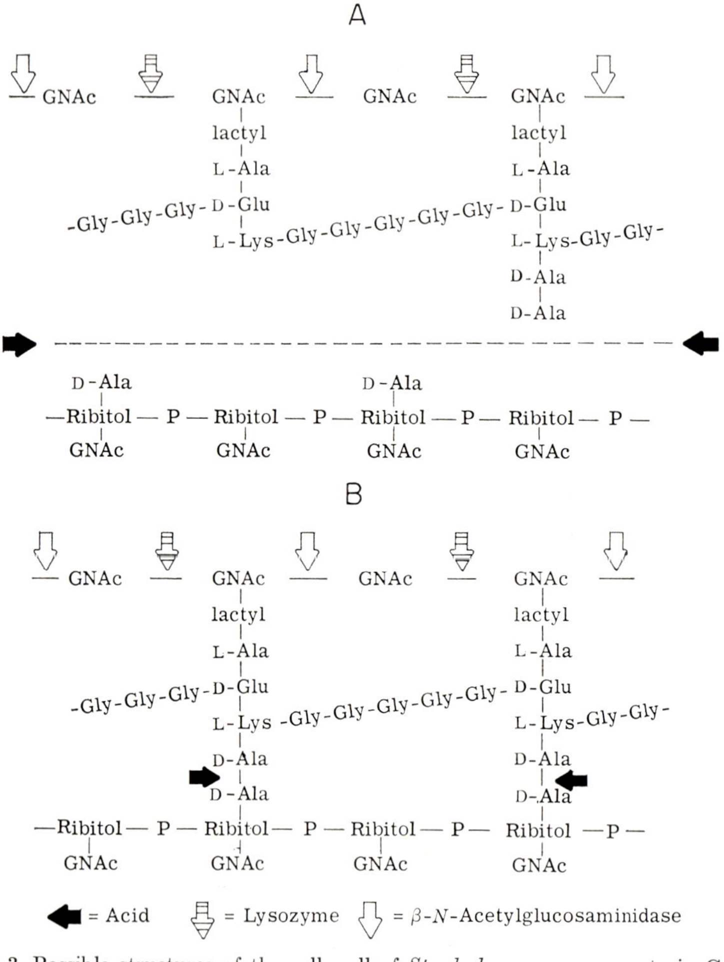

FIG. 3. Possible structures of the cell wall of *Staphylococcus aureus*, strain Copenhagen.[22] In both structures the glycopeptide (polymer of *N*-acetylglucosamine and *N*-acetylglucosamine-lactyl-peptide) is shown cross-linked by a polyglycine component, linking the lysine of one peptide chain and the glutamic acid of another. For convenience, the polyglycine cross-link is shown between two peptide units in the same polysaccharide chain; it could equally well connect parallel chains, thus giving another dimension to the wall. In structure A, two different peptides occur in the glycopeptide, and the means of attachment of the ribitol phosphate polymer to the rest of the wall is not specified. In structure B, the ribitol phosphate polymer is shown attached to the glycopeptide through the peptide. Points of cleavage during structural studies are indicated by arrows.

is one matter if the various polymers which they contain are synthesized independently, but it becomes a totally different problem if these syntheses are interrelated.

2. The Lipopolysaccharide (Endotoxin) of Gram-Negative Bacteria

These substances are usually extracted from cell walls (or from whole cells) with 45% phenol at 90° C.[27] The extracted substances are toxic in animals as well as antigenic, and can be separated by acid hydrolysis into a lipid (which carries the groupings responsible for toxicity) and a polysaccharide (which has the antigenically active groups). In addition, small amounts of a few amino acids are contained in the extracted lipopolysaccharide.[43] * Both the lipids and the polysaccharides of different bacteria are very complex substances. The lipid appears to be similar in the few cases of closely related bacteria from which it has been obtained. It is a polymer of glucosamine phosphate in which all of the available amino and hydroxyl groups are esterified by long-chain fatty acids.[45, 46]

Considerable work has been done on the composition of the polysaccharides in different species.[47-54] Hydrolyzates of different polysaccharides contain up to seven different sugars. It is possible that one sugar is linked in several different ways,[50, 51] and there is no evidence that there is any small repeating unit or regularity of structure in these substances.† The gross sugar composition, however, has led to a chemical classification of different species of *Salmonella* which coincides with their serological grouping according to the Kauffmann-White scheme.[47] On this basis, strains of *Escherichia coli* also fall within a number of different chemotypes[48] and, similarly, serological classification of strains of *Pasteurella pseudotuberculosis* fits with chemical analysis of their polysaccharides.[52, 54]

A new class of sugars, the 3,6-dideoxyhexoses,[55, 56] of which five of eight possible isomers are now known, was discovered as components of these polysaccharides. 6-Deoxysugars, such as rhamnose, fucose, and 6-deoxytalose,[57] also occur frequently in them. Immunochemical data have indicated that the dideoxysugars are frequently found as end-groups in polysaccharide chains and, furthermore, that where they occur, they appear to be the chemical components responsible for the specific immunological reactions.[50, 51] These polysaccharides also contain the unusual sugars, aldoheptoses, of which several isomers are now known.[53]

Much current interest is centered on the role of nucleotides in the synthesis of these novel sugars and of the polysaccharides themselves (see below).

* A great deal of the amino acid-containing material can be removed by brief treatment with ammonia or dilute NaOH without other gross alterations of lipopolysaccharide structure (J. L. Strominger, O. Lüderitz, and O. Westphal, unpublished). The nature of the linkages that are broken in alkali has not been established.

† See, however, a recent report by Robbins and Uchida.[51a]

III. Isolation of Intermediates in Bacterial Cell Wall Synthesis

Although very little is known about the mechanism of synthesis of any of the components of *special structure* in different bacteria, considerable information is now available concerning the mechanism of synthesis of the *basal structure*. Knowledge of the nucleotide intermediates involved first came from study of the effects of penicillin on *Staphylococcus aureus*. Uridine nucleotides with an unusual structure accumulated under these circumstances.[58, 59] Later evidence was obtained that these nucleotides are precursors of the bacterial cell wall.[44, 60] Recent studies of the effects of other inhibitors and enzymic studies of the mechanism of nucleotide synthesis have added to knowledge in this area.

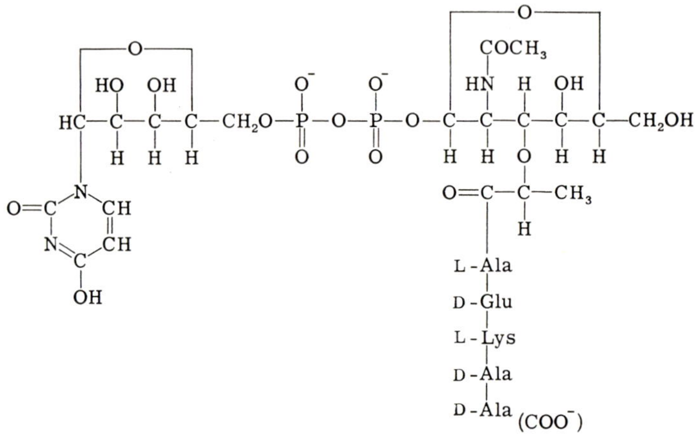

FIG. 4. Structure of UDP-GNAc-lactyl-peptide, a uridine nucleotide from *Staphylococcus aureus*.

A. STRUCTURE OF A URIDINE NUCLEOTIDE THAT ACCUMULATES IN PENICILLIN-INHIBITED *Staphylococcus aureus*

Before discussing the general phenomenon of nucleotide accumulation, it will be useful to describe the structure of the most complex of the known intermediates (Fig. 4). This nucleotide (UDP-acetylmuramyl-peptide) is one of the three uridine nucleotides that accumulate in penicillin-inhibited *S. aureus*, and contains a substituted sugar fragment linked to uridine diphosphate through its aldehyde group.[59] The structure is analogous to uridine diphosphoglucose, the "coenzyme" of the glucose-galactose transformation. The general structure of these compounds was first elucidated by Leloir and his collaborators[61] and was soon followed by a description of the nucleotides from *S. aureus*. Since then, a large number of UDP-sugar compounds have been isolated, including UDP-galactose, UDP-glucuronic acid, UDP-galacturonic acid, UDP-xylose, UDP-arabinose, UDP-acetylglucosamine, UDP-acetylgalactosamine, UDP-acetylglucosamine-6-phospho-1-galactose, and UDP-acetylgalactosamine-4-sulfate. These nucleotides have a number of important metabolic functions, one of which is the activation of sugar fragments for synthetic reactions (e.g., glycogen, sucrose, cellulose and, as we shall see, bacterial cell walls) (see refs. 1 and 2).

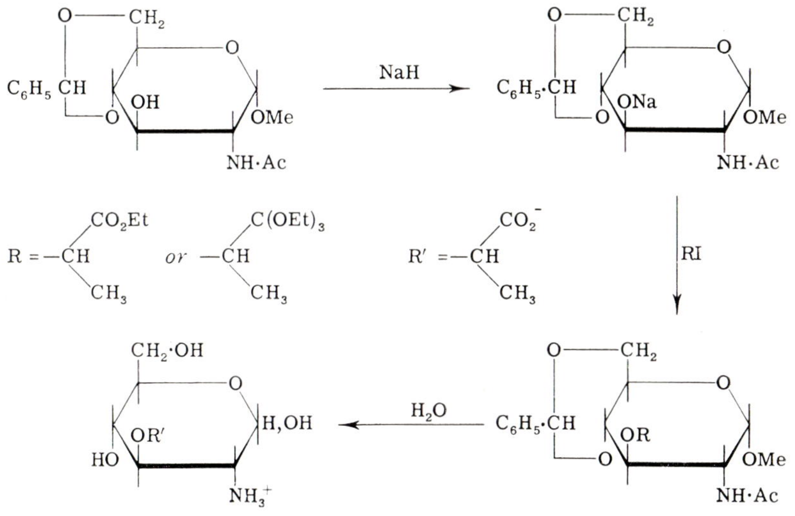

FIG. 5. Chemical synthesis of muramic acid (Strange and Kent[64]).

1. STRUCTURE OF THE SUGAR

The sugar, acetylmuramic acid, is a 3-*O*-lactic acid-ether of *N*-acetylglucosamine. It was first isolated as an unknown acetylamino sugar in the uridine nucleotide.[59] An unknown amino sugar was also detected in the hydrolyzate of a peptide liberated during sporulation in *Bacillus cereus*.[62] After crystallization from a hydrolyzate of spore peptide,[63] chemical studies led to a proposal of its structure.[64-66] This basic structure has been confirmed by synthesis[64] (Fig. 5). Since a racemic intermediate (ethyl α-DL-iodopropionate) was employed in the synthesis, two isomers, which differed in optical rotation and could be separated by paper chromatography, were obtained. The isomer with the higher rotation ($\alpha^{20}{}_D = +109$) is believed to contain a D-lactic acid residue and that with the lower rotation ($\alpha^{20}{}_D = +52$) a L-lactic acid residue. More recently, a stereospecific synthesis has been employed, confirming these assignments, and α- and β-phenyl glycosides of each isomer have been prepared.[67] The muramic acid from spore peptide is identical to the isomer containing the lactic acid which has been assigned the D-configuration. Its mode of synthesis is also different from that of the L-lactic acid formed during during glycolysis.

Muramic acid has been isolated from a large number of bacterial cell walls, and identified as the compound isolated from spore peptide. Similarly, the unknown amino sugar in the uridine nucleotide is identical to muramic acid isolated from spore peptide or from the cell wall of *S. aureus*.[44]

2. Structure of the Peptide

On hydrolysis in 0.1 *N* HCl or 0.01 *N* HCl for a few minutes, the acetylmuramyl-peptide fragment is liberated from the nucleotide. The peptide is presumed to be linked to the lactic acid of the amino sugar through its *N*-terminal end, since UDP-GNAc-lactyl·L-Ala, one of the precursors of UDP-GNAc-lactyl-peptide, contains no free amino group and only a single carboxyl group.[59]

Hydrolysis of the nucleotide in 6 *N* HCl yields three ninhydrin positive compounds in addition to muramic acid, viz: alanine, glutamic acid, and lysine in the molar ratio 3:1:1. The lysine residue has the L-configuration, but the glutamic acid residue has the D-configuration.[59] One of the three alanine residues in the peptide has the L-configuration and the other two residues have the D-configuration.[68] The occurrence of D-amino acids in the peptide is an unusual structural feature.

In 1 *N* HCl at 38° C., the acetylmuramyl-peptide fragment is partially degraded to a number of fragments, including one which contains all of the amino acids but no amino sugar.[7] Reaction of this fragment with phenylisothiocyanate led to recovery of alanine as the phenylthiohydantoin. Alanine, is, therefore, the *N*-terminal amino acid of the peptide. Alanyl·alanine was also formed during the limited acid hydrolysis of the acetylmuramyl-peptide fragment. These data, together with the isolation of the nucleotide, UDP-GNAc-lactyl·L-Ala·D-Glu (see below), led to formulation of the peptide sequence, L-Ala·D-Glu·L-Lys·D-Ala·D-Ala.[7, 68] Dinitrophenylation of the acetylmuramyl-peptide fragment followed by acid hydrolysis yielded ε-DNP-lysine[22]; hence, the ε-amino group of lysine is free in the peptide. Whether the γ- or the α-carboxyl group of glutamic acid is involved in the peptide structure has not yet been determined.*

One may ask whether any special significance can be attached to the occurrence of the D-amino acids in the nucleotide and in the basal structure of the cell wall. The peptide containing these amino acids is resistant to digestion by the ordinary proteolytic enzymes. Furthermore, in a peptide

* Dr. Eiji Ito (personal communication) has now shown that a γ-glutamyl linkage occurs in the peptide.

containing all L- or all D-amino acids, the side chains of adjacent amino acids projects from the peptide backbone (arranged in linear fashion) in opposite directions. The side chain of a D-amino acid, however, projects in the same direction as the side chain of an adjacent L-amino acid. Interaction between the methyl groups of poly-DL-alanine has been shown to stabilize that polypeptide.[69] A similar interaction could occur between the side chains of the first four amino acids in the nucleotide, which form a peptide with alternating L- and D-amino acids, L-Ala·D-Glu·L-Lys·D-Ala. If any of the functional groups in dicarboxylic or diamino acids in such peptides were free, strong salt linkages could provide additional stabilizing forces. The occurrence of the D-amino acids could, therefore, result in an unusually stable peptide, stable both biologically and chemically. From an evolutionary standpoint, such a structure might provide a selective advantage to those organisms which contained it.

B. Methods of Measurement of Nucleotide Accumulation and of Isolation of Nucleotides

1. Measurement

The first method employed for measurement of nucleotide accumulation was isolation of the nucleotides by fractional precipitation of barium salts with ethanol and measurement by their absorption of ultraviolet light.[58] A more precise and rapid method is measurement of accumulation of acetylamino sugar esters by a colorimetric method (Table II).[70] This method can be applied to unpurified bacterial extracts, which ordinarily contain very little material that reacts in the colorimetric procedure. A blank is obtained by omitting the acid hydrolysis step, required to break the linkage between UDP and the sugar. Essentially all the acetylamino sugar esters that accumulate in penicillin-inhibited *S. aureus* are present as nucleotide derivatives. In other situations, however, the possibility that some or all of the accumulated compound may be another easily hydrolyzable derivative of an acetylamino sugar, e.g., acetylglucosamine-1-phosphate, should be considered (although this complication has not yet been encountered).

2. Isolation

Ion exchange chromatography on Dowex 1 has been employed to prepare the nucleotides (Fig. 6).[70] They are recovered from the column eluate by charcoal adsorption and elution. Compounds so obtained can be further purified by paper chromatography in neutral ethanol-ammonium acetate. Uridine nucleotides prepared from extracts of inhibited cells by ion exchange chromatography usually do not require further purification for ordinary purposes.

TABLE II

Colorimetric Estimation of Accumulation of Uridine Nucleotide in *Staphylococcus aureus*[70]

Tube no.[a]	Optical density (550 mμ)	*N*-Acetylamino sugar: Total μmoles	*N*-Acetylamino sugar: μmoles/liter of culture at OD 1.0 (650 mμ)
	Normal culture		
1	0.033	—	—
2	0.052	0.024	0.5
3	0.097	0.057	1.1
4	1.004	—	—
	Pencillin-treated culture		
1	0.034	—	—
2	0.190	0.219	4.4
3	1.449	1.77	35.7
4	1.146	—	—

Note: An 80-ml. culture in logarithmic phase was centrifuged. An extract was prepared from one half of the cells. The other half was resuspended in fresh growth medium containing penicillin (1 μg./ml.). After vigorous shaking for 2 hours at 37° C., the cells were centrifuged. An extract was prepared and analyzed for *N*-acetylamino sugar.

[a] Tube 1 is a reagent blank. Tube 2 is obtained by omitting an acid-heating step. As *N*-acetylamino sugar-1 esters do not react in the test without acid hydrolysis, Tube 2 measures only the nonesterified *N*-acetylamino sugars. Tube 3 is the measure of *N*-acetylamino sugar esters (complete procedure). Tube 4 is an internal standard (containing 0.43 μmole of *N*-acetylglucosamine), analyzed as Tube 2. Tube 2 is used as the blank for Tubes 3 and 4.

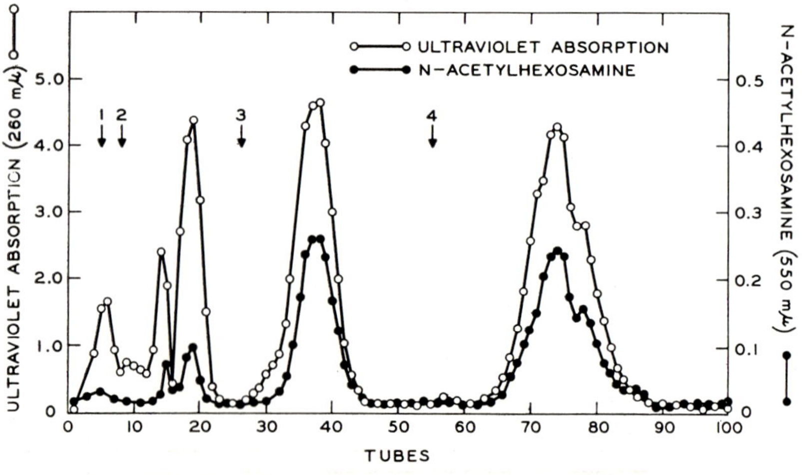

Fig. 6. Separation of nucleotides from extract of penicillin-inhibited *S. aureus* by chromatography on Dowex-1 chloride. The cell wall precursors occur in the peaks at tubes 38, 73, and 77.

The nucleotides from inhibited cells can also be separated on filter paper. Originally, paper electrophoresis followed by paper chromatography was employed in a two-dimensional separation.[71] Considerable "streaking" occurred, however, during the electrophoresis due to the large number of compounds in the extract. More recently, two-dimensional paper chromatography has been employed with isobutyric acid—0.5 N NH_4OH in the first dimension and neutral ethanol ammonium acetate in the second dimension.[71] Isobutyric acid-ammonia has a very high capacity for salts and extract of cells recovered from 100–500 ml. of culture can be handled easily on a single sheet on Whatman No. 3MM filter paper. The resulting chromatogram is photographed in ultraviolet light by placing it between the light source and a sheet of reflex document copying paper (e.g., Remington G-91). By comparison with an appropriate control, a qualitative estimate of accumulation of all ultraviolet absorbing materials can be rapidly obtained (see Fig. 7). This procedure has also been employed for preparative purposes. Whereas originally measurement and isolation of the nucleotides from a single experimental flask required 2 to 3 weeks, measurement of 6 to 8 flasks now requires only part of a day and isolation can be accomplished in 2 to 3 days.

C. Accumulation of Nucleotides in *Staphylococcus aureus*

1. Penicillins

Three uridine nucleotides accumulate in some strains of *S. aureus* treated with penicillin. They are UDP-GNAc-lactic, UDP-GNAc-lactyl·L-Ala, and UDP-GNAc-lactyl·L-Ala·D-Glu·L-Lys·D-Ala·D-Ala (Fig. 7). Concentrations of penicillin near the growth inhibitory level (0.03 μg./ml. for the strain used) permit observable accumulation but the maximum is obtained at higher concentrations (100 μg./ml.).[70] The time for half-maximum accumulation is about 15 min. after the addition of the drug, and maximum concentrations of nucleotide are reached at about 90 min. (Fig. 8).[70] Thus, the immediacy of appearance of the lesion after addition of the drug and its close relation to the growth inhibitory concentration suggest that nucleotide accumulation is closely related to the point of attack of penicillin within the cell.

Several strains of *S. aureus* have, however, shown variability in the degree of nucleotide accumulation obtained with penicillin. With the strain employed by the author, maximum accumulation was formerly about 40 μmoles per liter of culture at half-maximum growth (about 2.5 g. of wet, packed cells). At the present writing it is about 20 μmoles while a much greater degree of accumulation can be induced by oxamycin (see below). A similar variability in nucleotide accumulation induced by penicillin, for

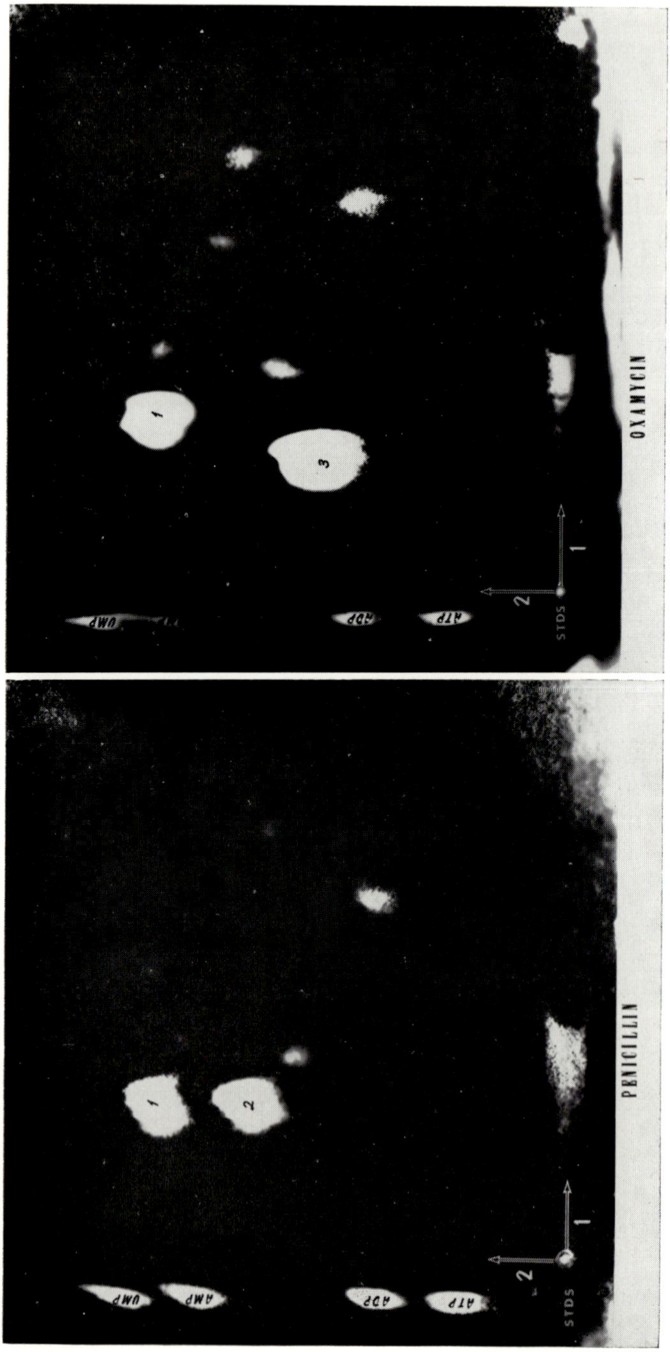

FIG. Separation of nucleotides from penicillin- and oxamycin-inhibited *Staphylococcus aureus* by two-dimensional paper chromatography. Compound 1 is a mixture of UDP-GNAc-lactic and UDP-GNa-lactyl·L-Ala. Compound 2 is UDP-GNAc-lactyl·L-Ala·D-Glu·L-Lys·D-Ala·D-Ala. Compound 3 is UDP-GNAc-lactyl·L-Ala·D-Glu·L-Lys.

which no explanation is available, has also been observed by Abraham and Newton.[72]

As will become evident, the three nucleotides that accumulate with penicillin are part of a metabolic sequence. If one postulates that the point of the block is beyond the most complex intermediate that accumulates (UDP-GNAc-lactyl·L-Ala·D-Glu·L-Lys·D-Ala·D-Ala), accumulation of the early intermediates (UDP-GNAc-lactic and UDP-GNAc-lactyl·L-Ala) is not fully explained. Blocks produced between UDP-GNAc-lactyl·L-Ala and UDP-GNAc-lactyl-peptide do not necessarily lead to accumulation of the early intermediates. It may, therefore, be necessary to

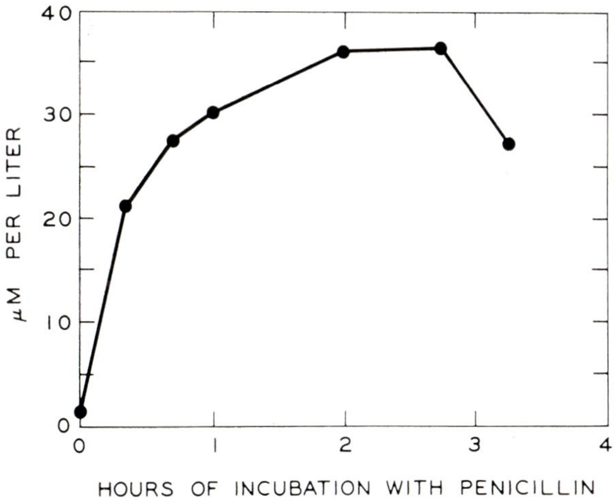

FIG. 8. Time course of accumulation of uridine nucleotides in penicillin-treated *Staphylococcus aureus*, strain Copenhagen.[70]

postulate that penicillin blocks the metabolic pathway in some manner at two different points. Such a hypothesis is attractive since, if one of the blocks was induced at lower penicillin concentrations than the other, it might provide a basis for explanation of the "zone phenomenon," i.e., the killing rate for some microorganisms is higher at low penicillin concentrations than at high penicillin concentrations.[73] In the strain of *S. aureus* employed, however, the nucleotides which accumulated at low penicillin concentrations were the same as those which accumulated at high penicillin concentrations. The possible nature of the lesion induced by penicillin will be discussed further below.

2. Bacitracin

Strains of microorganisms that develop resistance to penicillin also develop resistance to bacitracin. Recent investigations of the chemical structure of bacitracin have indicated the presence of a sulfur-containing ring very similar to the thiazolidine ring in penicillin. These similarities led to the finding that bacitracin also induces uridine nucleotide accumulation at concentrations closely related to its growth inhibitory concentration.[72] The nucleotides which accumulate are identical to those which accumulate with penicillin. It is noteworthy, however, that the degree of accumulation induced by bacitracin remained unchanged during a period when the amount of accumulation which could be obtained with penicillin was declining to one third of its original value.

Penicillin is irreversibly fixed by bacterial cells, and cannot be recovered as penicillin.[74] It is possible that some chemical reaction with a component of the bacterial cell occurs which involves alteration of the penicillin molecule. The occurrence of sulfur-containing rings in both penicillin and bacitracin suggests the possibility that this reaction might involve the sulfur atom.

3. Novobiocin

Novobiocin also induces accumulation of the same uridine nucleotides which accumulate in the presence of bacitracin and penicillin.[68] The compounds were isolated and their structures determined. However, the biology of the phenomenon has not yet been extensively explored.

4. Oxamycin

Oxamycin (D-cycloserine, D-4-amino-5-isoxazolidone) also induces uridine nucleotide accumulation in *S. aureus*.[75] The principal nucleotide which accumulates is a compound not detected with any other inhibitor.[76] It was detected by its distinct position on two-dimensional paper chromatography (Fig. 7b). Isolation and analysis of the compound indicated that it contained one residue each of D-glutamic acid, L-lysine, and L-alanine per residue of uridine, but lacked D-alanine. Its structure may be represented as UDP-GNAc-lactyl·L-Ala·D-Glu·L-Lys. The mechanism of its accumulation will be discussed further below.*

5. Lysine Deprivation

Staphylococci are heterotrophic organisms with a multitude of growth requirements. Anaerobically, for example, they require uracil. Some strains include a requirement for peptides among the complex amino acid requirements. Despite the severe growth requirements of the organism, uridine nucleotide accumulation could be induced in a simple medium containing only phosphate buffer, glucose, and $MgCl_2$ together with DL-alanine, L-

* Vancomycin[76a, 76b] and ristocetin[76c] are the most recent additions to the list of antibiotics which are inhibitors of cell wall synthesis and induce uridine nucleotide accumulation.

glutamic acid, and L-lysine. That is, these three amino acids could satisfy the nitrogen requirement in the synthesis of uracil and of the amino sugar as well as of the amino acids in the peptide. Moreover, the degree of accumulation induced by penicillin was not diminished when lysine was omitted from the medium. Later, it was found that in lysine-deficient medium penicillin was unnecessary for nucleotide accumulation, i.e., lysine deprivation itself was an inducer of nucleotide accumulation (Table III).[77] Furthermore, during the course of this investigation a requirement for D-glutamic acid in the medium appeared which had not been apparent in early experiments. Isolation and analysis of the principal nucleotide accumulated in lysine deprivation showed the presence of one residue of L-alanine and one of D-glutamic acid per residue of uridine. Its structure may be represented as UDP-GNAc-lactyl·L-Ala·D-Glu. The maximum amount of this compound that can now be accumulated by lysine deprivation is small, about 8 to 10 μmoles per liter of culture, somewhat less than the amount that can now be accumulated by inhibition with penicillin. As in the case of penicillin, in earlier experiments the accumulation of up to 40 μmoles was observed.

TABLE III

ACCUMULATION OF URIDINE NUCLEOTIDES BY *Staphylococcus aureus* AS THE CONSEQUENCE OF LYSINE DEPRIVATION[77]

Medium	Without penicillin	With penicillin (100 μg./ml.)
Growth	1	26.4
"Deficient"[a] plus DL-Ala, D-Glu, L-Lys	6.2	15.5
"Deficient" plus DL-Ala, D-Glu	16.4	18.0
"Deficient" plus DL-Ala, L-Glu	4.8	—

NOTE: Data are expressed as μmoles of uridine nucleotide per liter culture at half-maximal growth.

[a] Phosphate buffer, $MgCl_2$, glucose.

Some lysine is avialable to the organism under conditions of lysine deprivation, as evidenced by the accumulation along with UDP-GNAc-lactyl·L-Ala·D-Glu of a small amount of UDP-GNAc-lactyl·L-Ala·D-Glu L-Lys·D-Ala·D-Ala. When cells which had accumulated UDP-GNAc-lactyl·L-Ala·D-Glu were incubated in the presence of a few milligrams of lysine for 10 min., this compound was replaced by UDP-GNAc-lactyl·L-Ala·D-Glu·L-Lys·D-Ala·D-Ala (Fig. 9).[77] Clearly, in whole cells UDP-GNAc-lactyl·L-Ala·D-Glu is a precursor of the peptide-containing nucleotide.

These data fit well with observations of the effects of amino acid deprivation on the cellular integrity of *Streptococcus faecalis*. Among several

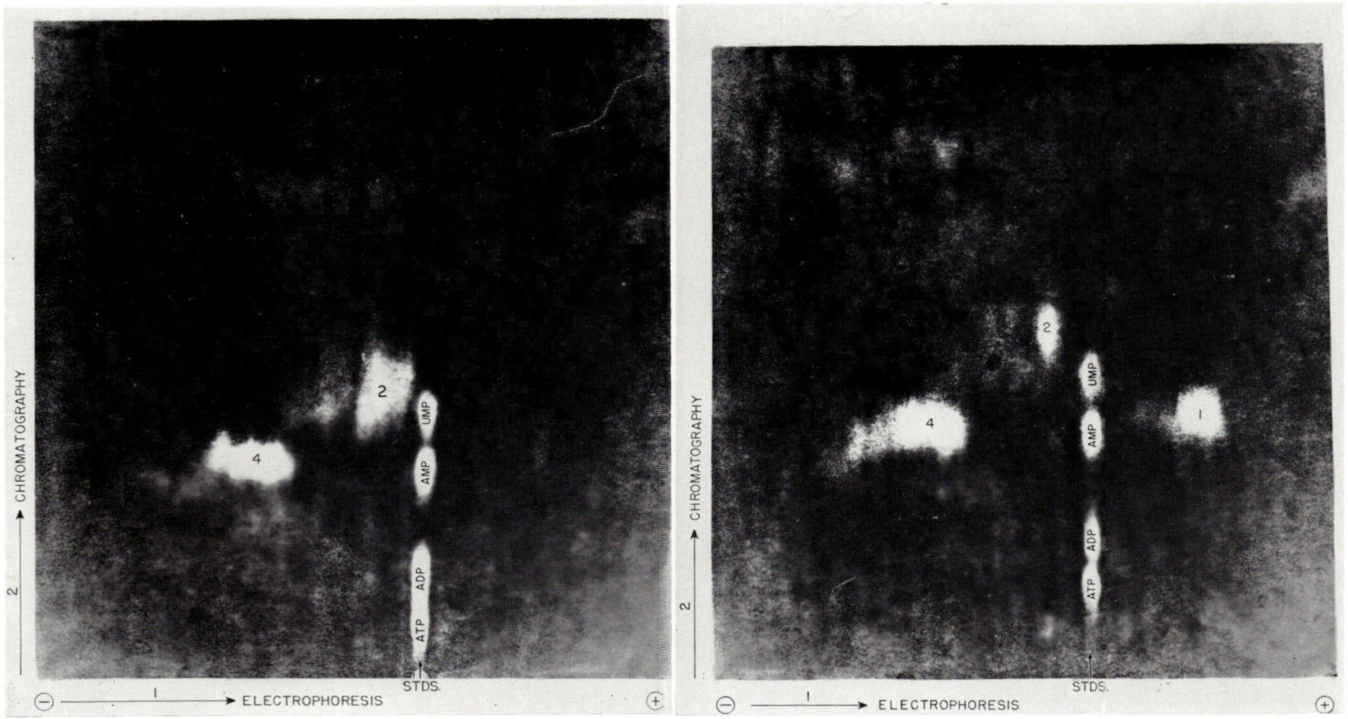

FIG. 9. Separation of nucleotides from lysine-deprived *Staphylococcus aureus* by a combination of paper electrophoresis and chromatography.[77] The separation on the left is from a culture deprived of lysine for 90 min. That on the right is from a similar culture to which a few mg. of lysine were added 10 min. before harvesting. Compound 1 is UDP-GNAc-lactyl·L-Ala·D-Glu. Compound 2 is UDP-GNAc-lactyl·L-Ala·D-Glu·L-Lys·D-Ala·D-Ala. Compound 4 is predominantly AMP, the only ultraviolet-absorbing compound seen in a control culture.

amino acids, depletion of lysine only was followed by autolysis.[78, 79] This phenomenon is the consequence of the structural importance of lysine as a component of the cell wall in *S. faecalis*, and it might be presumed that UDP-GNAc-lactyl·L-Ala·D-Glu, or some related intermediate, accumulates in this organism prior to lysis.

6. Gentian Violet

Gentian violet, the dye that is irreversibly fixed by some bacteria as the basis of the Gram stain, is a mixture of closely related dyes, the major component of which is crystal violet. In *S. aureus*, strain H, crystal violet induced uridine nucleotide accumulation, but the compounds isolated were early intermediates in cell wall synthesis, UDP-GNAc and UDP-GNAc-lactic.[80] These data were very important in that they indicated that UDP-GNAc must be a biological precursor of the peptide-containing intermediates. Several other dyes did not have this effect. Later, with a gentian violet preparation manufactured more than 25 years ago, accumulation of these uridine nucleotides was also observed in *S. aureus*, strain Copenhagen.[71] In addition, UDP-GNAc-pyruvate was detected as a minor component among the uridine nucleotides, and several cytidine nucleotides accumulated (Fig. 10). These cytidine derivatives were identified as cytidine monophosphate and CDP-ribitol (Fig. 11)* and will be further discussed below. The accumulation of uridine nucleotides, but *not* of cytidine nucleotides, was observed in a parallel experiment with *S. aureus*, strain H, using the same gentian violet preparation.

More recently, many samples of gentian violet and crystal violet of recent manufacture were found not to induce uridine or cytidine nucleotide accumulation in *S. aureus*, strain Copenhagen.[81] Accumulation is believed to be induced by an impurity which has been detected in the old gentian violet preparation, but, for the present, one cannot assume that uridine and cytidine nucleotide accumulation are induced by the same compound.

* A large degree of accumulation (i.e., 20–30 μmoles per liter of culture) of CDP-ribitol along with uridine nucleotides is induced by this gentian violet preparation (but *not* by penicillin) only in *S. aureus*, strain Copenhagen. Gentian violet does not induce massive CDP-ribitol accumulation in *S. aureus*, strain H. Park (personal communication) has, however, observed the presence of CDP-ribitol in penicillin or chloramphenicol-treated cultures of *S. aureus*, strain H, in the amount of about 1 μmole per liter of culture. Although this amount is 5–10 times the level found in his normal culture, it is not comparable to the much larger accumulation of uridine nucleotides induced simultaneously by penicillin in this strain. Very recently, one strain of *S. aureus* (strain 209P) has been reported to accumulate, under the influence of penicillin, massive amounts of CDP-ribitol along with the uridine nucleotides.[80a] These differences between various inhibitors on the one hand and various strains on the other are curious since the cytidine nucleotide and the uridine nucleotides are each cell wall precursors, although of different parts of the cell wall.

CDP-ribitol[82] is a presumed precursor of the polyribitol phosphate component of the cell wall. The accumulation of this compound in gentian violet-treated *S. aureus* is particularly useful for preparative purposes since about 20–30 μmoles per liter of culture at half-maximal growth can be ob-

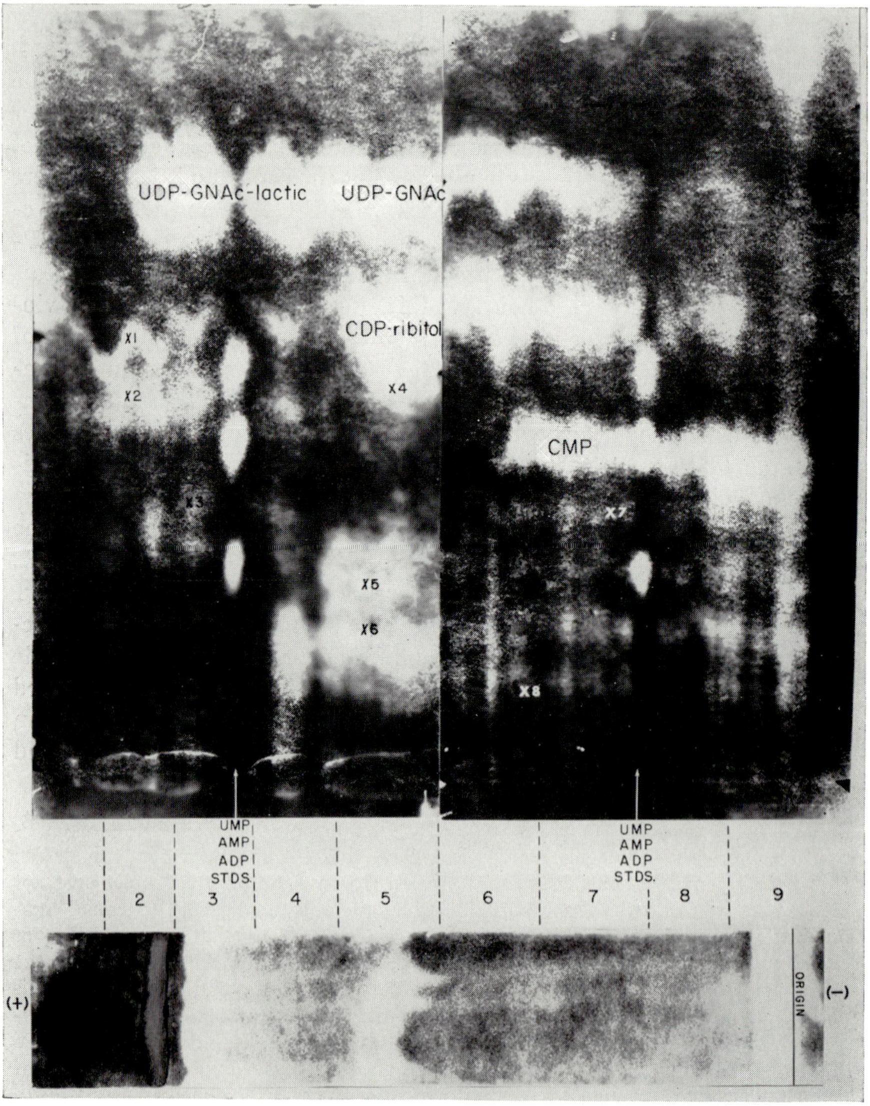

FIG. 10. Separation of nucleotides from gentian violet-inhibited *Staphylococcus aureus*, strain Copenhagen, by a combination of paper electrophoresis and chromatography.[71] The electrophoretic strip is shown below. X1 is UDP-GNAc-enolpyruvate-ether.

tained in this manner,[71] compared to about 0.1–0.5 μmole in untreated cells. The amount of cytidine nucleotides obtained is approximately equivalent to the amount of uridine nucleotides.

In addition to the uridine and cytidine nucleotides that have been described, a number of other ultraviolet-absorbing materials accumulate in gentian violet-treated cells in smaller amounts. Such compounds have also been observed with novobiocin and occasionally with penicillin, but they have not yet been identified.

7. Uracil Analogs

Flurouracil[83-85] and azauracil[86, 87] induce nucleotide accumulation and osmotically unstable forms in *S. aureus* and *E. coli*. Partial structural studies of the accumulated nucleotides have been carried out.

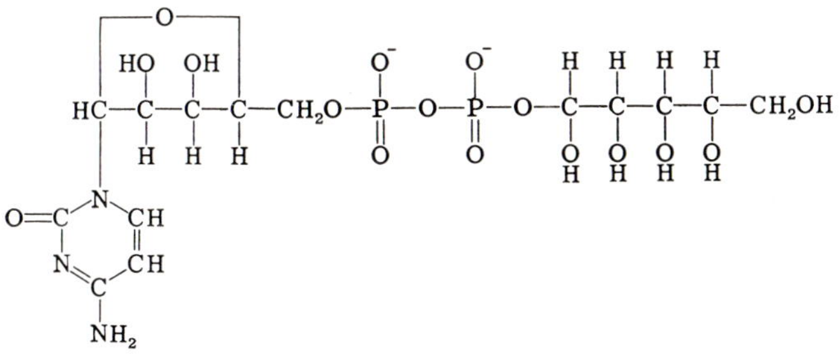

Fig. 11. Structure of CDP-ribitol.[82]

8. Amino Acids

Large amounts of glycine induce nucleotide accumulation in *S. aureus*.[167] This substance as well as several other amino acids had previously been reported to induce spheroplast formation in *E. coli*.[87a] The induction of spheroplasts in *Alcaligenes faecalis* by several D-amino acids has been extensively studied,[162] and, in the case of D-methionine, formation of the globular forms is accompanied by incorporation of the D-amino acid into the cell wall peptide, presumably in place of some normal constituent.[163] Nucleotide accumulation in this circumstance has not been investigated, however.

D. Isolation of Uridine and Cytidine Nucleotide Cell Wall Precursors from Other Microorganisms

1. Uridine Nucleotide Intermediates

a. Staphylococcus aureus, Strain 209P. This organism apparently has a more complex cell wall than some other strains of *S. aureus*.[88] This fact appears to be reflected in the occurrence of derivatives of uridine diphospho-acetylmuramic acid containing amino acids not found in the nucleotides from *S. aureus*, strains H or Copenhagen, e.g., aspartic acid and glycine.[89, 90] However, a definitive report of the purity and structure of the compounds from *S. aureus* 209P has not yet appeared.

b. Escherichia coli, Strain K_{12}. Several strains in this line have been investigated, viz, *E. coli*, strain Y-10 (an amino acid auxotroph of *E. coli*, strain K_{12}), and *E. coli*, strain W-3656 (a DAP-requiring mutant of *E. coli* Y-10). In the steady state each of these strains contains a relatively high concentration of a uridine nucleotide which, on analysis, contained two residues of phosphate, one each of acetylmuramic acid, diaminopimelic acid, and glutamic acid, and three of alanine.[8] Paper chromatography indicated that the DAP was the *meso* isomer. Enzymic analysis indicated that two of the three alanine residues were D-alanine residues and the other was an L-alanine residue, while the glutamic acid was entirely in the D-configuration.

In the absence of DAP, the DAP-requiring mutant continued to grow for 30 min. and then an abrupt lysis ensued, a phenomenon which had been previously described for another DAP-less *E. coli* mutant.[91-94] In hypertonic sucrose medium, lysis was prevented and the organisms were converted to spheroplasts. This phenomenon is analogous to the lysis of *S. faecalis* in the absence of lysine,[78, 79] and is presumably due to the loss of integrity of the cell wall.

A small accumulation of acetylamino sugar esters could be detected in *E. coli* W-3656 30 min. after DAP-deprivation (the point of maximum growth just prior to lysis) (Fig. 12). Uridine nucleotides were isolated from the cell extract of a large culture harvested at this time. The DAP-containing nucleotide had disappeared and was replaced by a uridine nucleotide containing one residue each of L-alanine and D-glutamic acid. This nucleotide substituted for UDP-GNAc-lactyl·L-Ala·D-Glu in an enzymic reaction in *S. aureus* in which lysine is added to the uridine nucleotide (see below) and was also identical to the nucleotide from *S. aureus* by other methods.

Since UDP-GNAc-lactyl·L-Ala·D-Glu accumulates following DAP-deprivation, it may be inferred that DAP is the next amino acid in the peptide sequence. The structure of the DAP-containing nucleotide can, therefore, be represented as UDP-GNAc-lactyl·L-Ala·D-Glu·*meso*-DAP·D-Ala·D-Ala (Fig. 13). This structure is analogous to that of the nucleotide from *S. aureus*, differing only in that L-lysine is replaced by *meso*-DAP.

The amino sugar in these nucleotides has been identified by paper chromatography of the sugar and of its ninhydrin degradation product and by characteristic colorimetric reactions. These properties do not exclude the possibility that this acetylamino sugar is a stereoisomer of the muramic acid isolated from spore peptide. They do, however, exclude the occurrence of a related compound, acetylneuraminic acid.

c. Escherichia coli, Strain K_{235}. In addition to several of the compounds

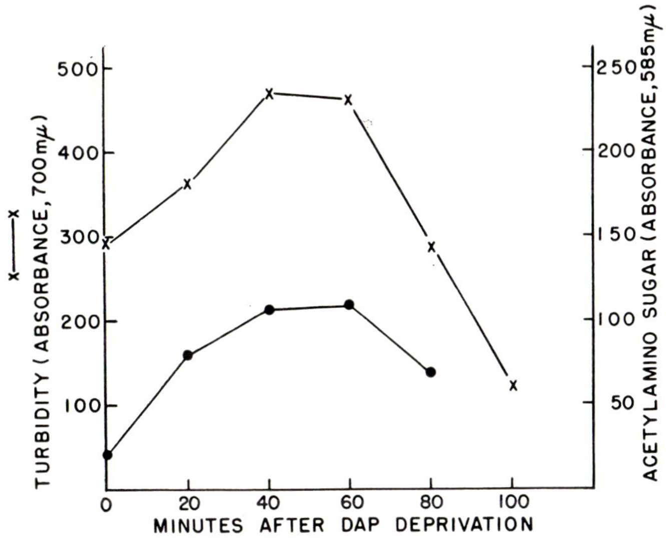

FIG. 12. Changes in turbidity and content of acetylamino sugar (as uridine nucleotides) following diaminopimelic acid deprivation of a DAP-requiring mutant of *Escherichia coli*.

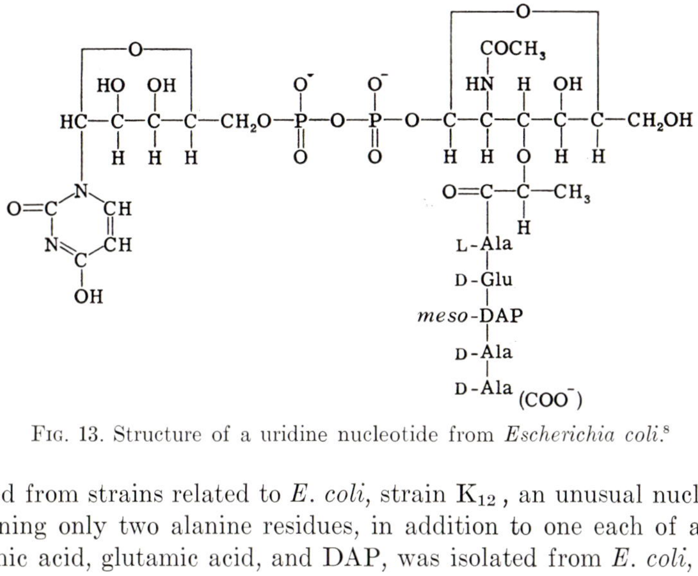

FIG. 13. Structure of a uridine nucleotide from *Escherichia coli*.[8]

isolated from strains related to *E. coli*, strain K_{12}, an unusual nucleotide containing only two alanine residues, in addition to one each of acetylmuramic acid, glutamic acid, and DAP, was isolated from *E. coli*, strain

K_{235}, as well as a similar nucleotide containing only a single alanine residue.[21] No data were obtained on the structure of the peptides or on the optical forms of the amino acids.*

E. coli, strain K_{235}, contains a polymer of acetylneuraminic acid (coliminic acid).[95] Acetylneuraminic acid is related structurally to acetylmuramic acid. The 3-carbon fragment, however, is at the oxidation level of pyruvic acid, the methylene carbon of which is attached to C-1 of the amino sugar. The amino sugar has the configuration of acetylmannosamine and can also be linked through C-3 to the 2-position of pyruvic acid.[96] CMP-acetylneuraminic acid has been isolated from this strain and is presumably a precursor of coliminic acid.[97]

Culture filtrates of this organism have been reported in another laboratory to contain peptides linked to acetylneuraminic acid, both free as the sugar-peptide and as a UDP-acetylneuraminyl-peptide.[98] In addition, a peptide linked to an unidentified "acetylhexosamine" was reported. These compounds were reported to have an unusual composition, including the following hydrolysis products: (*a*) Lysine, alanine, glutamic acid, and aspartic acid, together with acetylhexosamine; (*b*) DAP, alanine, glutamic acid, aspartic acid, and neuraminic acid; (*c*) DAP, alanine, glutamic acid, neuraminic acid, and UDP. Other unusual properties were reported. Detailed studies of the structure and purity of these materials have not yet appeared. The definitive isolation of UDP-acetylneuraminic acid and of peptide derivatives thereof would be of unusual interest. Preparations containing acetylneuraminic acid and peptides have also been reported to occur in ox brain.[99]

d. Streptococcus faecalis. Two peptide derivatives of UDP-acetylmuramic acid have been isolated from penicillin-treated cultures of one strain of this organism.[100] One of these compounds yielded the same analyses as the nucleotide, UDP-GNAc-lactyl·L-Ala·D-Glu·L-Lys·D-Ala·D-Ala from *S. aureus*, and was also chromatographically identical to this compound. The second compound differed on analysis only in containing two L-alanine residues. After reaction with dinitrofluorobenzene and hydrolysis, compound 1 yielded ε-DNP-lysine, as did the corresponding nucleotide from *S. aureus*. Compound 2 yielded only DNP-alanine after similar treatment. This compound, therefore, must contain a branched peptide (Fig. 14), although proof of the proposed structure of the peptide by degradation was not carried out. Some strains of *S. faecalis* have been reported to contain

* *Note added in proof:* A sample of this compound, reported to contain two alanine residues, kindly given us by Dr. D. Comb, has been degraded in our laboratory recently, however, and contained 3.0 alanine residues, as does the nucleotide isolated from other *E. coli* strains. It also had chromatographic mobilities identical to the previously isolated nucleotide.

equimolar amounts of D- and L-alanine in the cell wall,[101, 102] as does this novel uridine nucleotide.

e. Streptococcus hemolyticus. UDP-acetylmuramic acid has been isolated from a type A *Streptococcus.*[103] A UDP-acetylmuramyl-peptide which contained four alanine residues was also isolated from a similar strain.[104] This work was not completed because of difficulties encountered subsequently in isolating the compound again.

2. Accumulation of Uridine Nucleotides Induced by Penicillin in Microorganisms Other Than *Staphylococcus aureus*

Attempts have been made to demonstrate uridine nucleotide accumulation in several other bacteria. As measured by accumulation of acetylamino

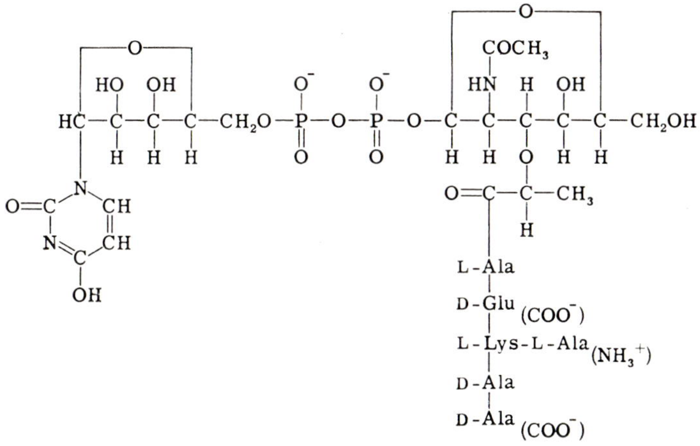

FIG. 14. Structure of a uridine nucleotide from penicillin-treated *Streptococcus faecalis.*[100]

sugar esters, accumulation was demonstrated in a penicillin-sensitive *Lactobacillus helveticus.*[70] More recently, this phenomenon has also been observed with strains of *Streptococcus faecalis, S. salivarius,* and *S. hemolyticus.*[100]

With a number of microorganisms, it has not been possible to demonstrate a penicillin-induced nucleotide accumulation by the colorimetric method. These include several strains of *E. coli, B. subtilis, B. megaterium,* and another strain of *S. faecalis.*[41] In this context it may also be mentioned that uridine nucleotide accumulation has not been observed in *Staphylococcus albus* and does not occur in all strains of *S. aureus.*[58] These data might suggest that the accumulation observed in *S. aureus,* strains H, Copenhagen, Duncan, and several other strains is a secondary reflection of the effect of penicillin on the organism. The method employed, however, is *not* a critical test of whether penicillin inhibits a reaction in bacterial cell wall synthesis. For example, the method depends upon the intracellular accumulation of the nucleotides. In some microorganisms accumulating nucleotides might leak into the culture medium. In *E. coli*, the leakage of ultraviolet-absorbing substances is one of the earliest observable effects of penicillin,[105] and extensive attempts to prevent it in ten separate strains with hypertonic sucrose, NaCl, or other salts have been unsuccessful. Therefore, the failure to detect nucleotide accumulation could have a simple explanation. In the medium, the ultraviolet-absorbing material is diluted in large amounts of buffer salts and other substances, and attempts to recover it have not been made. Isotopic experiments, presented below, provide some information relating to the possibility that penicillin is a specific inhibitor of cell wall synthesis in *E. coli*.

Other explanations of the failure to detect nucleotide accumulation must also be considered: (*a*) Accumulating nucleotides might be degraded by intracellular enzymes in some cases. (*b*) The kinetics of accumulation of compounds behind the point of a block are complex. Feedback mechanisms, differences in the relative amounts of various enzymes, or the existence of metabolic sidepaths in various organisms may prevent a demonstration of nucleotide accumulation. It is possible that those strains of *S. aureus* in which large amounts of nucleotide can be accumulated are unusual nonrepressible strains. (*c*) The organism might not be lysed in the procedure usually employed for extraction of nucleotides from *S. aureus*. (*d*) Nucleotide accumulation might not be directly related to the primary mechanism of action of penicillin.

3. Cytidine Nucleotide Intermediates

It was observed very early that a very small but variable amount of CMP accumulated in penicillin-treated *S. aureus*.[70, 106] CDP-ribitol and CDP-glycerol were first isolated from *Lactobacillus arabinosus*, strain 17/5.[82, 107-109] They have also been isolated from a number of other microorganisms,[110] some containing one or the other of these compounds, others containing both. The absence of an isolatable quantity of one of these compounds in a particular organism does not imply that it does not occur as an important metabolic intermediate, since a very low concentration might result from a rapid intracellular utilization of the compound. Reference has already been made to accumulation of CDP-ribitol in gentian violet-treated *S. aureus*[71] and to isolation of CMP-acetylneuraminic acid.[97]

IV. Enzymic Synthesis of the Nucleotide Cell Wall Precursors

A. Enzymic Synthesis of the Uridine Nucleotide Intermediates

The reaction sequence which leads to synthesis of UDP-GNAc-lactyl-peptide is summarized in Fig. 15. This representation, similar to that which describes glucuronide synthesis, serves to emphasize that uridine nucleotides act catalytically as carriers of the acetylamino sugar for modification and, presumably, eventual transfer in one of the reactions leading to cell wall synthesis.

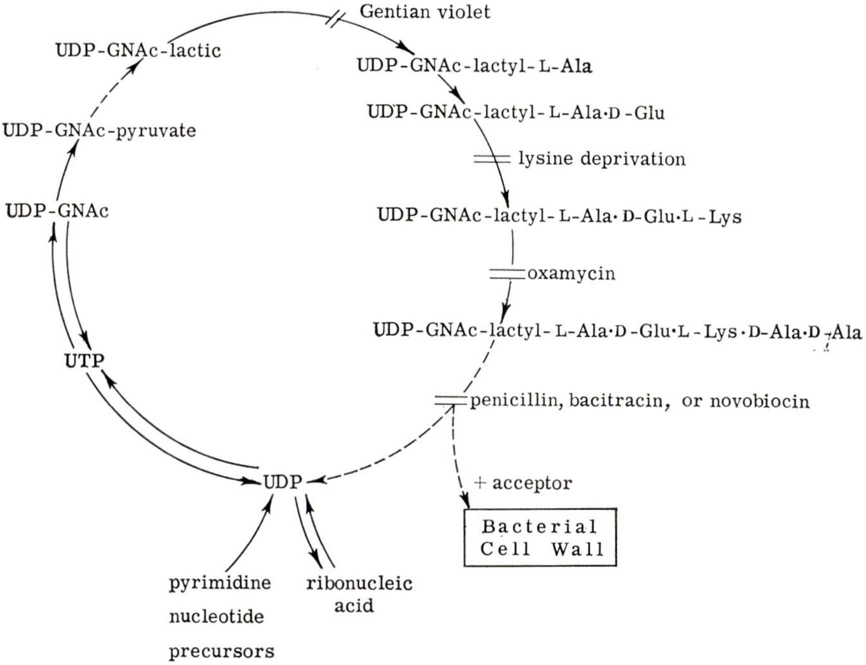

FIG. 15. Cycle that leads to synthesis of part of the glycopeptide in the cell wall of *Staphylococcus aureus*.

1. Transphosphorylation

The phosphorylation of UDP by ATP is catalyzed by an enzyme which is ubiquitous in nature.

$$\text{UDP} + \text{ATP} \leftrightharpoons \text{UTP} + \text{ADP}$$

The enzyme is termed nucleoside diphosphokinase.[111] Several other mechanisms for phosphorylation of UDP also exist (cf. ref. 2).

2. Synthesis of UDP-GNAc

The activation of sugars by reaction of the sugar-1-phosphate with UTP is catalyzed by a group of enzymes, termed pyrophosphorylases. Although these reactions are sometimes measured most conveniently in the direction of pyrophosphorolysis, their primary function in the cell is presumably the synthesis of the UDP-sugar derivative. UDP-GNAc pyrophosphorylase has been detected in liver, yeast, and bacteria and has been purified from liver and from *S. aureus*.[112-114]

$$\text{UTP} + \text{GNAc-1-P} \rightleftharpoons \text{UDP-GNAc} + \text{PP}$$

Using UDP-GNAc and PP as substrates, the reaction is conveniently measured by coupling the reaction to a phosphomonoesterase. Acetylglucosamine formed from GNAc-1-P is then determined colorimetrically. The amount of this enzyme in extracts of *S. aureus* is sufficient to catalyze the synthesis of an amount of UDP-GNAc equivalent to the acetylglucosamine and acetylmuramic acid in the cell wall in 5 min. at 37°C. under optimum *in vitro* conditions.[114]

3. UDP-GNAc-Pyruvate Transferase

The initial report by Strange[115] of the probable structure of muramic acid suggested that UDP-acetylmuramic acid might be synthesized by transfer of a 3-carbon fragment to UDP-GNAc. An enzyme was found which catalyzed transfer of pyruvic acid from 2-phosphoenolpyruvic acid, an intermediate in glycolysis, to UDP-GNAc.[116]

$$\text{UDP-GNAc} + \text{P-enolpyruvate} \rightarrow \text{UDP-GNAc-enolpyruvate-ether} + \text{Pi}$$

In crude preparations the reaction is measured by a colorimetric method which is relatively specific for 3-substituted acetylamino sugars. In purified preparations it can also be measured by the release of inorganic phosphate. This enzyme was purified about 10-fold. The amount present in extracts of *S. aureus* is sufficient to account for the rate of synthesis of acetylmuramic acid during cell growth and division. The enzyme has also been detected in *E. coli* and in *Aerobacter aerogenes*.

4. Reduction of UDP-GNAc-Pyruvate to UDP-GNAc-Lactic

The hypothesis that UDP-GNAc-pyruvate is a precursor of UDP-GNAc-lactic (UDP-acetylmuramic acid) was strengthened by the isolation of both of these compounds from gentian violet-inhibited *S. aureus*.[71] With C^{14}-UDP-GNAc-pyruvate and reduced DPNH as substrates and crude enzyme from *S. aureus*, the synthesis of a small amount of a radioactive compound with the electrophoretic mobility of UDP-GNAc-lactic was observed. The product, however, was not definitely identified, and this reaction should *not* be regarded as established. Isotopic experiments have supported a mechanism in which 2-phosphoenolpyruvate is the precursor of GNAc-lactic (acetylmuramic acid).[117]

5. Synthesis of the Peptide Bonds

Since uridine nucleotides containing incomplete peptides were isolated under various circumstances, the peptide appeared to be synthesized by stepwise addition of amino acids, in contrast to the synthesis of proteins which may take place by "simultaneous" assembly of a number of amino acids on a template.

TABLE IV

Uridine Nucleotide Requirement in Synthesis of Peptide Bonds[118]

Uridine nucleotide added	Substrate		
	C^{14}-L-Alanine	C^{14}-DL-Glutamic acid[a]	C^{14}-L-Lysine
None	0	0	0
UDP-GNAc-lactic	1364	0	0
UDP-GNAc-lactyl·L-alanine	—	3720	0
UDP-GNAc-lactyl·L-Ala·D-Glu	8	0	3530
UDP-GNAc-lactyl·L-Ala·D-Glu·L-Lys	2200[b]	0	20
UDP-GNAc-lactyl·L-Ala·D-Glu·L-Lys·D-Ala·D-Ala	100	0	20

Note: Reaction mixtures contained ATP as well as C^{14}-amino acids and uridine nucleotides indicated below. Data are recorded as c.p.m. incorporated into a charcoal adsorbable form.

[a] Although C^{14}-DL-glutamic acid was the substrate, it could be shown that only C^{14}-D-glutamic acid was enzymically active.

[b] This result is known to be due to the occurrence of alanine racemase and the enzymes catalyzing reactions 4 and 5 in the preparation.

The following reactions, which lead to the synthesis of the peptide, have been identified in extracts of *S. aureus*.[118, 119]

1. UDP-GNAc-lactic + L-Ala $\xrightarrow[Mn^{++}]{ATP}$ UDP-GNAc-lactyl·L-Ala

2. UDP-GNAc-lactyl·L-Ala + D-Glu $\xrightarrow[Mn^{++}]{ATP}$ UDP-GNAc-lactyl·L-Ala·D-Glu

3. UDP-GNAc-lactyl·L-Ala·D-Glu $\xrightarrow[Mn^{++}]{ATP}$ UDP-GNAc-lactyl·L-Ala·D-Glu·L-Lys

4. 2 D-Ala $\xrightarrow[Mn^{++}]{ATP}$ D-Ala·D-Ala

5. UDP-GNAc-lactyl·L-Ala·D-Glu·L-Lys + D-Ala·D-Ala $\xrightarrow[Mn^{++}]{ATP}$

   UDP-GNAc-lactyl·L-Ala·D-Glu·L-Lys·D-Ala·D-Ala

Each of these reactions is catalyzed by a separate enzyme. Reactions 1–3 are measured, after incorporation of a radioactive amino acid into the nucleotide, as charcoal adsorbable radioactivity (Table IV). Reaction 4 is measured in a similar manner by coupling the formation of C^{14}-D-Ala·D-Ala to reaction 5. Reaction 5 itself is measured most conveniently using C^{14}-D-Ala·D-Ala, prepared as a product of reaction 4, as a substrate, but can also be measured using enzymically synthesized UDP-GNAc-lactyl·L-Ala·D-Glu·C^{14}-L-Lys (Table V). The previous isolation of D-alanyl·D-alanine from *S. faecalis*[120] was an important clue to this mechanism.

Each of these five enzymes has been purified at least 20-fold and each of the products has been isolated and identified. The purified enzymes have absolute requirements for each of the substrates, including a specific uridine nucleotide, a specific amino acid with a particular optical configuration and

TABLE V

ENZYMIC ADDITION OF D-ALA·D-ALA TO UDP-GNAC-LACTYL·L-ALA·D-GLU·L-LYS[118]

Additions	C.p.m. in UDP-GNAc-lactyl ·L-Ala·D-Glu ·L-Lys·D-Ala·D-Ala
None	105
D-Ala·D-Ala (synthetic)	4790
D-Ala·D-Ala (isolated)	6600
L-Ala·L-Ala	20
D-Ala·L-Ala	56
D Alanine	81
L-Alanine	37

NOTE: The substrates were UDP-GNAc-lactyl·L-Ala·D-Glu·C^{14}-L-Lys (enzymically synthesized), ATP, and dipeptides indicated. Radioactivity in the uridine nucleotide product was measured after separation of the products by paper chromatography.

ATP (Table VI; see also Table IV). No other nucleoside triphosphate will substitute for ATP. The adenine nucleotide products have been identified as ADP and inorganic phosphate for the reactions catalyzed by the L-lysine-adding enzyme (400-fold purified)[119] and D-glutamic acid-adding enzyme (60-fold purified) from *S. aureus*.[121] In addition, Neuhaus[122] has identified ADP and inorganic phosphate as the products of an alanyl-alanine synthetase, about 250-fold purified, from *S. faecalis*.

A few preliminary studies of the mechanisms of these reactions by means of isotope exchange have been carried out. In the cases examined, the free amino acids do not catalyze an exchange of P^{32}-inorganic phosphate or of P^{32}-inorganic pyrophosphate with ATP. Presumably no amino-activated form of the amino acid occurs as an intermediate. Some early observations, suggesting that an ATP-inorganic phosphate exchange occurred in the presence of a uridine nucleotide substrate containing an amino acid with a free carboxyl group, have not been substantiated with later preparations of the enzyme. The detailed mechanisms of these reactions require more extensive study.

One interesting observation in this context is the fact that alanyl-alanine synthetase from *S. aureus* can be inactivated by incubation at pH 7 in the absence of substrates. The inactivated enzyme can be reactivated by incubation with ATP and Mn^{++} at pH 7 *prior* to assay at pH 9, the pH optimum.[123] The reactivation, however, requires a heat stable cofactor which can be demonstrated with dialyzed preparations or with enzyme passed through a column of Sephadex G-25 (Figs. 16, 17). Neither the crude nor

TABLE VI

REQUIREMENTS FOR THE ENZYMIC SYNTHESIS OF UDP-GNAc-LACTYL·L-ALA·D-GLU·L-LYS[119]

Experiment 1: System	C.p.m.	Experiment 2: Nucleotide added	C.p.m.
Complete	2620	UDP-GNAc-lactyl·Ala·Glu	5080
-UDP-GNAc-lactyl·Ala·Glu	0	UDP-GNAc-lactyl·Ala	0
-ATP	0	UDP-GNAc-lactyl·Ala·Glu·Lys	6
$-Mg^{++}$	0	UDP-GNAc-lactyl·Ala·Glu·Lys·Ala·Ala	0

NOTE: A purified lysine-adding enzyme was employed. In experiment 1, various components were omitted from the complete incubation mixture, which contained buffer, $MgCl_2$, C^{14}-lysine, ATP, UDP-GNAc-lactyl·L-Ala·D-Glu, and purified enzyme. In experiment 2, other uridine nucleotides were substituted for UDP-GNAc-lactyl·L-Ala·D-Glu. Data are expressed as counts per minute of C^{14}-lysine incorporated into nucleotide, measured as charcoal-adsorbable radioactivity.

the purified enzyme from *S. faecalis* has these properties. It is possible that identification of the heat-stable factor and an explanation of inactivation and reactivation of alanyl-alanine synthetase from *S. aureus* will shed some light on the mechanisms of these peptide bond syntheses.

The occurrence of a dipeptide, D-Ala·D-Ala, as an intermediate in the synthesis represents the first definitive evidence that a small peptide can occur as an intermediate in the synthesis of a larger peptide chain. Studies of the isotopic labeling of proteins by amino acids have indicated that the same amino acid residue at different points in a protein may have different specific activities (unequal labeling), and have, therefore, suggested the occurrence of peptide intermediates in protein synthesis.[124*]

* Recent data suggest, however, that unequal labeling can be explained by sequential addition of amino acids along a template already containing incomplete protein in various stages of assembly.

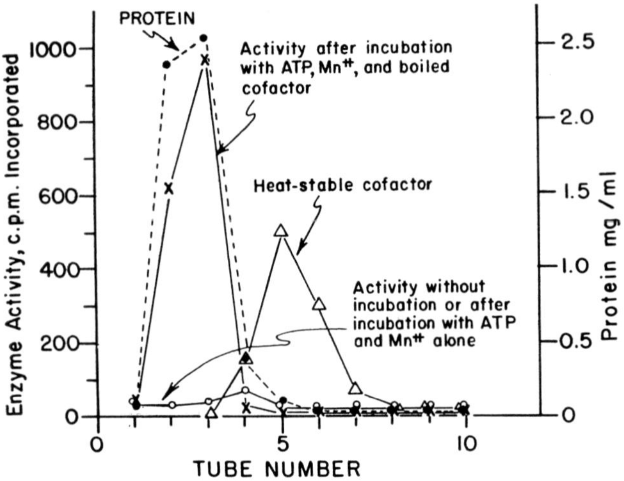

FIG. 16. Cofactor requirement for reactivation of D-alanyl·D-alanine synthetase from *Staphylococcus aureus* after chromatography on a column of Sephadex G-25.[123]

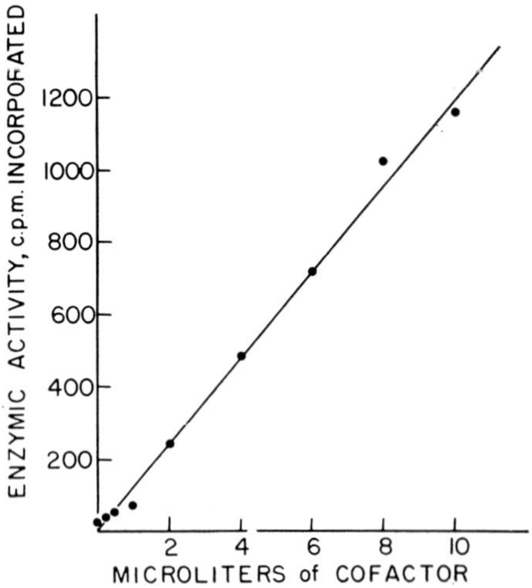

FIG. 17. Assay of the heat-stable cofactor required for reactivation of D-alanyl·D-alanine synthetase from *Staphylococcus aureus*.[123] Enzyme passed through Sephadex G-25 was used for this assay. Activity is proportional to the amount of cofactor added during preincubation with ATP and Mn^{++} at pH 7, prior to assay at pH 9.

These reactions are analogous to reactions which lead to synthesis of the peptide bond in glutamine,[125, 126]

$$\text{glutamic acid} + NH_3 + \text{ATP} \rightarrow \text{glutamine} + \text{ADP} + \text{Pi}$$

or to the synthesis of glutathione from glutamic acid, cysteine, and glycine.[127, 128]

$$\text{glutamic acid} + \text{cysteine} + \text{ATP} \rightarrow \gamma\text{-glutamyl}\cdot\text{cysteine} + \text{ADP} + \text{Pi}$$

$$\gamma\text{-glutamyl}\cdot\text{cysteine} + \text{glycine} + \text{ATP} \rightarrow \text{glutathione} + \text{ADP} + \text{Pi}$$

However, these reactions are mechanistically different from the reactions which lead to the synthesis of amino acyl-RNA and, presumably, of peptide bonds in protein via amino acyl-adenylates (see page 414). The participation of an amino acyl-adenylate has also been demonstrated in the synthesis of the dipeptide, carnosine (β-alanyl·histidine).[129] Moreover, a specific D-alanine activating enzyme has been found in several bacteria.[130]

$$\text{D-alanine} + \text{ATP} \leftrightharpoons \text{D-alanyl}\cdot\text{AMP} + \text{PP}$$

This enzyme has been separated from alanyl-alanine synthetase, both in *S. aureus* and in *S. faecalis*.[122, 131] No synthetic reaction involving D-alanyl-AMP has been found. It is possible that this enzyme is involved in the addition of D-alanine to the ribitol phosphate polymer, or conceivably in the activation of D-alanine at the end of the acetylmuramyl-peptide for some other reaction.

The peptide in the uridine nucleotide is, therefore, synthesized by stepwise addition of amino acids to a nucleotide of increasing complexity. Peptide bond synthesis proceeds without evidence of amino acid activation; and it is clear, in this case, at least, that techniques which depend upon the activation of isotope exchange reactions by amino acids or the formation of activated forms of amino acids would not have led to detection of these reactions. The specificity of peptide synthesis is determined by the specificity of individual enzymes without participation of a template (e.g., the individual reactions are insensitive to ribonuclease). One wonders if other peptides (e.g., the peptide antibiotics and peptide hormones) may be synthesized in this manner, and what the limitation in size and complexity of the product might be. Indeed, several mechanisms could be operative in the synthesis of very large peptide chains.

6. Comparative Biochemistry of Peptide Formation

It was of interest to examine the distribution of enzymes catalyzing peptide synthesis in view of the universal occurrence of the glycopeptide in bacterial cell walls and of the possibility that peptides with different structures might exist in some species. Enzymes catalyzing cell wall peptide synthesis as uridine nucleotide precursor were found in nine representative microorganisms.[132] The most interesting aspect of this study was the finding that those microorganisms which contain diaminopimelic acid in the basal structure are lacking the L-lysine-adding enzyme. Instead, they contain an enzyme which specifically catalyzes the addition of diaminopimelic acid to the nucleotide (Table VII).

$$\text{UDP-GNAc-lactyl}\cdot\text{L-Ala}\cdot\text{D-Glu} + meso\text{-DAP} \xrightarrow[\text{Mn}^{++}]{\text{ATP}} \text{UDP-GNAc-lactyl}\cdot\text{L-Ala}\cdot\text{D-Glu}\cdot meso\text{-DAP}$$

The DAP-adding enzymes from *Corynebacterium xerosis, Salmonella gallinarium, E. coli,* and *B. cereus* have been shown to have a strict specificity

TABLE VII

ADDITION OF LYSINE OR DIAMINOPIMELIC ACID TO UDP-GNAC-LACTYL·L-ALA·D-GLU[132]

Extract from	Substrate		Extract from	Substrate	
	C^{14}-Lys	H^3-DAP		C^{14}-Lys	H^3-DAP
Staphylococcus aureus	0.95	0.01	*Salmonella gallinarum*	0	0.60
Micrococcus lysodeikticus	0.49	0	*Bacillus cereus*	0	0.42
Staphylococcus albus	0.36	0	*Escherichia coli*	0	0.20
Sarcina lutea	0.90	0.02	*Corynebacterium xerosis*	0	5.12
Streptococcus faecalis	1.25	0			

NOTE: Extracts were prepared from logarithmic phase cells of the organisms listed. C^{14}-L-lysine and H^3-α,ϵ-diaminopimelic acid (mixture of LL- and *meso*-isomers) were used as substrates. Radioactivity converted to a charcoal-adsorbable form was measured, and data are expressed as mμmoles of amino acid added to the uridine nucleotide under the conditions of assay.

for the uridine nucleotide, UDP-GNAc-lactyl·L-Ala·D-Glu. The specificity of these enzymes is apparently the basis for the occurrence of DAP in some microbial cell walls and of lysine in others. The DAP-adding enzyme from *C. xerosis* is particularly active and has been purified to some extent. As in the case of the lysine-adding enzyme, ADP and inorganic phosphate are formed during the reaction. *meso*-Diaminopimelic acid is the substrate for this enzyme; both the LL- and the DD-isomers are inactive.

7. SYNTHESIS OF CELL WALL AMINO ACIDS

The reactions which lead to synthesis of the unusual amino acids in the cell wall are uniquely found in microorganisms. The synthesis of D-alanine from L-alanine is catalyzed by alanine racemase.[133]

$$\text{L-alanine} \xrightleftharpoons{\text{pyridoxal phosphate}} \text{D-alanine}$$

Several mechanisms may occur for D-glutamic acid synthesis. Early work in *B. subtilis* had suggested that D-glutamic acid is synthesized from D-alanine by a stereospecific transaminase.[134, 135]

D-alanine + α-ketoglutaric acid ⇌ D-glutamic acid + pyruvic acid

More recently, however, a microbial glutamic racemase has also been found (although the participation of pyridoxal phosphate in this reaction could not be demonstrated).[136]

L-glutamic acid ⇌ D-glutamic acid

meso-Diaminopimelic acid is synthesized from glutamic acid, aspartic acid, and pyruvic acid in several steps, all of which have not yet been characterized.[137, 138, 138a]

1. pyruvic acid + aspartic acid → → → *N*-succinyl-α-amino-ε-keto-pimelic acid
2. *N*-succinyl-α-amino-ε-keto-pimelic acid + glutamic acid → *N*-succinyl-LL-DAP + α-ketoglutaric acid
3. *N*-succinyl-LL-DAP → succinate + LL-DAP
4. LL-DAP ⇌ *meso*-DAP

B. Mechanism of Action of Oxamycin

The antibiotic oxamycin (D-cycloserine, D-4-amino-3-isoxazolidone) owes its selective toxicity to interference with the metabolism of D-alanine, a substrate which is uniquely important in microorganisms.[76, 139, 140*] As indicated above, the principal nucleotide accumulated in the presence of this antibiotic is distinct from the compounds which accumulate in the presence of other inhibitors of cell wall synthesis (e.g., penicillin) and may be represented as UDP-GNAc-lactyl·L-Ala·D-Glu·L-Lys.[76] This nucleotide is missing the two terminal D-alanine residues in the peptide sequence. This fact, as well as the occurrence of the D-configuration in oxamycin, suggested that the antibiotic might be a specific antagonist of D-alanine incorporation into the nucleotide. D-Alanine was shown to prevent or reverse nucleotide accumulation in whole cells of *S. aureus* in a manner which was competitively related to oxamycin incorporation (Table VIII). This phenomenon is similar to the competitive relationship between *p*-aminobenzoic acid and sulfonamides, demonstrated 20 years ago by Woods[144] and Fildes.[145]

It is apparent from the structures of these compounds that the antibiotic, oxamycin, is a structural analog of D-alanine (Fig. 18a). From molecular models (Fig. 18b and c) the structural representation of D-alanine in Fig.

* It is also of interest that L-cycloserine interferes with the metabolism of L-alanine.[141–143]

18a is clearly a closer two-dimensional representation of its structure than the usual representation

$$\begin{array}{c} CH_3\text{—}CH\text{—}COOH \\ | \\ NH_2 \end{array}$$

When these two models are viewed looking down on the amino group on the asymmetric carbon atoms (Fig. 18c), they are indistinguishable.

The enzymic basis for the antagonism has also been defined.[146] Oxamycin is a competitive antagonist both of alanine racemase and of D-alanyl·D-alanine synthetase (Figs. 19 and 20). It has no effect on the addi-

TABLE VIII

ANTAGONISM BY D-ALANINE OF URIDINE NUCLEOTIDE ACCUMULATION INDUCED BY OXAMYCIN[76]

Antagonist added	Experiment 1	Experimennt 2
None	41.4	30.0
D-Alanine (500 μg./ml.)	17.0	12.1
D-Alanine (5000 μg./ml.)	4.5	6.9
L-Alanine (5000 μg./ml.)	41.5	32.2
DL-Alanyl·DL-alanine (5000 μg./ml.)	—	33.5
D-Serine (5000 μg./ml.)	—	34.2

NOTE: In experiment 1, oxamycin (75 μg./ml.) and possible antagonists were added together at zero time. In experiment 2, oxamycin (75 μg./ml.) was added at zero time. At 45 min., 20.4 μmoles of nucleotide had accumulated. At this time possible antagonists were added and incubation was continued for 45 min. longer. Data are expressed as μmoles of uridine nucleotide per liter of culture at half-maximal growth.

tion of D-alanyl·D-alanine to the nucleotide. In the case of alanine racemase from *S. aureus*, K_m for either D-alanine or L-alanine is $6 \times 10^{-3}\ M$ while K_i for oxamycin is $6 \times 10^{-5}\ M$. For the dipeptide-synthesizing enzyme, K_m is $4 \times 10^{-3}\ M$ while K_i is $3 \times 10^{-5}\ M$. The ration, K_m/K_i, is for each of these enzymes about 100, emphasizing the efficiency of the antibiotic as competitor for the substrate.

The fact that the antibiotic has a much higher affinity for these enzymes than their natural substrates is probably related to the fact that the aminoxy bridge between the methylene and carbonyl carbons of oxamycin holds the molecule in a relatively fixed position. The possibilities for rotation are, therefore, far more restricted than those for D-alanine. The structure of the antibiotic may then be an excellent model of the configuration which the substrate assumes during its enzymic transformations. Oxamycin

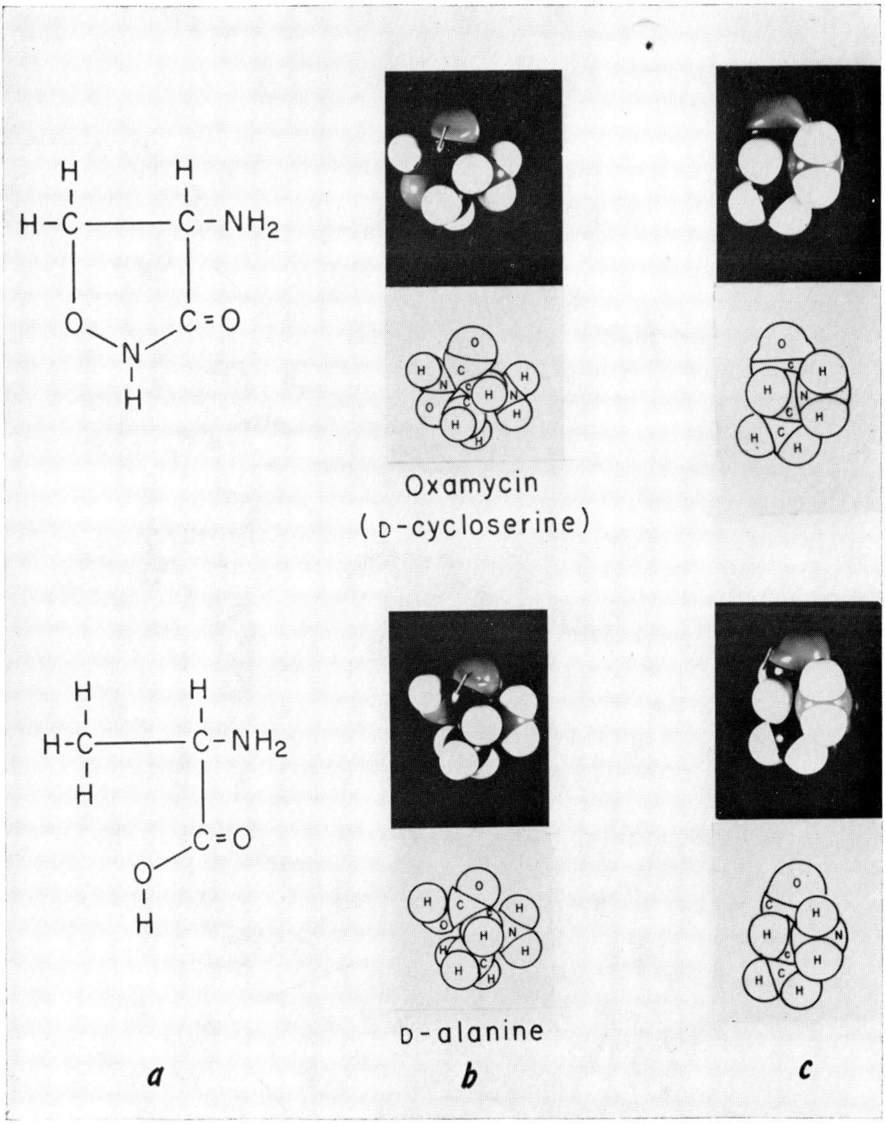

FIG. 18. Structures (*a*) and molecular models (*b* and *c*) of D-alanine and oxamycin. The models are viewed in (*b*) looking down on the hydrogen atoms on the asymmetric carbon atom, and in (*c*) looking down on the amino groups.

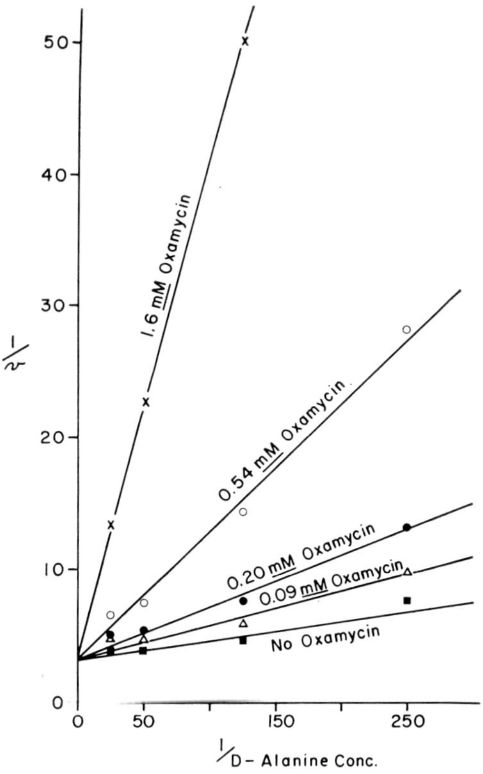

FIG. 19. Competitive inhibition of alanine racemase by oxamycin.[146]

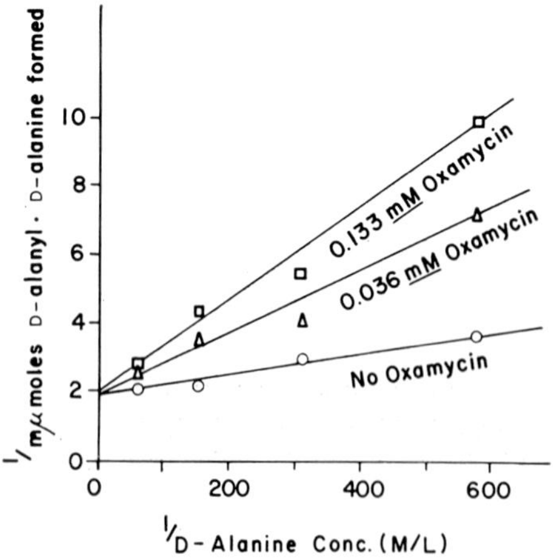

FIG. 20. Competitive inhibition of D-alanyl·D-alanine synthetase by oxamycin.[146]

will be an extremely interesting tool with which to explore the mechanisms of action of these two enzymes.

C. Enzymic Synthesis of Cytidine Nucleotide Intermediates

The syntheses of CDP-ribitol and CDP-glycerol are catalyzed by specific pyrophosphorylases.[147]

CTP + D-ribitol-5-phosphate ⇌ CDP-ribitol + PP

CTP + L-glycerol-1-phosphate ⇌ CDP-glycerol + PP

α-Glycerophosphate is presumably derived from glyceraldehyde-3-phosphate, but the mechanism of synthesis of D-ribitol-5-phosphate is still unknown. CMP-acetylneuraminic acid is synthesized in a similar reaction.[147a, 147b]

CTP + acetylneuraminic acid ⇌ CMP-acetylneuraminic acid + PP

V. Biosynthesis of the Cell Wall

A. Participation of Nucleotide Intermediates

No direct demonstration of the participation of the uridine nucleotides in cell wall synthesis has so far been made. The hypothesis that these compounds are cell wall precursors rests mainly on the striking similarity in composition of these nucleotides and of the glycopeptide component of the cell wall.[44, 60] The same amino acids are present with the same unusual optical rotations. Both contain the novel sugar, acetylmuramic acid.

Furthermore, transglycosylation of the sugar-peptide fragment from nucleotide to some cell wall precursor would be a typical reaction of a uridine nucleotide, for which many models exist.[1, 2]

UDP-X + acceptor ⇆ acceptor-X + UDP

In addition to the probable participation of UDP-GNAc-lactyl·L-Ala·D-Glu·L-Lys·D-Ala·D-Ala in such a reaction, it seems likely that UDP-GNAc and UDP-GNAc-lactic are substrates for transglycosylation. Such reactions could also occur with nucleotides containing acetylmuramic acid substituted by smaller or different peptides. Thus, in the cycle of cell wall intermediates in *S. aureus* (Fig. 15), sugar fragments may be transferred to cell wall from several different points in the cycle.

Similar structural and functional considerations apply to the possible participation of cytidine nucleotides in cell wall synthesis. Here the model reaction in phospholipid synthesis involves transfer of the alcohol *phosphate* to an acceptor.[1, 2] One would, therefore, anticipate that CDP-ribitol and CDP-glycerol are the activated intermediates for synthesis of the polyol phosphate component of the cell wall.

CDP-X + acceptor ⇌ acceptor-P-X + CMP

Attempts to measure these transfer reactions using radioactive intermediates of high specific activity have not yet yielded unequivocal results. Several methodological problems, however, complicate these studies, particularly the preparation of suitable acceptors (cf. ref. 2) and the probable location of the synthetic enzymes in the lipid cytoplasmic membrane. One may conceive of the enzyme, located in the membrane, as serving a dual function, that of transporting the activated group through the permeability barrier and that of catalyzing the synthetic reaction.[60]*

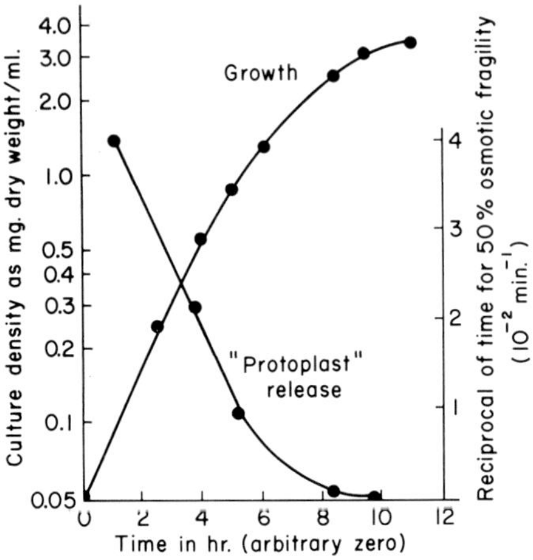

FIG. 21. Release of "protoplasts" of *Staphylococcus aureus*, strain Duncan, due to the action of autolysin at various times in the growth cycle (Mitchell and Moyle[148]).

B. AUTOLYSINS AND CELL WALL SYNTHESIS

Many microorganisms contain enzymes which are capable of hydrolyzing their own cell wall preparations.[65, 148-153] Physiologically, these autolysins are kept in check. With some aerobes, for example, only under conditions of relative anaerobiosis is the lytic activity evident. Although it might be presumed that such enzymes have a degradative function, in one case the autolysin is present in highest concentration during logarithmic growth and disappears in stationary phase[148] (Fig. 21). One wonders, then, if its physiological function might relate to cell growth, e.g., the preparation

* Very recently, Dr. S. Nathenson has demonstrated transfer of acetylglucosamine residues from C^{14}-UDP-GNAc to form acetylglucosaminyl-ribitol linkages, catalyzed by an enzyme preparation from *S. aureus*, strain Copenhagen.[147c] Both α and β-linked acetylglucosamine residues are formed. The acceptor in this reaction is the limit product of action of β-acetylglucosaminidase on teichoic acid.[34] Teichoic acid itself will not accept additional acetylglucosamine residues. The enzyme preparation which catalyzes this reaction is particulate and may be derived from the cell membrane.

of acceptor sites for the addition of new groups during wall synthesis. One manner of accomplishing the enlargement of a rigid sphere (as must occur in the cell wall of a coccus during growth) would be to open a series of holes in the sphere, which could then be enlarged by the addition of new pieces in the holes.

C. Cell Wall Syntheses and a Mechanism of Antibiotic Action

Clearly, penicillin, novobiocin, and bacitracin interfere in some manner in a penultimate stage of cell wall synthesis. This inference is supported both by isolation of accumulated cell wall intermediates and by demonstration of a physiological lesion in cell wall synthesis. It had been observed that several bacilli under the influence of lysozyme (which digests cell walls of sensitive bacteria) would undergo a transformation into spherical forms (protoplasts) which could be stabilized in hypertonic media.[154] These protoplasts are organisms denuded of cell walls, which are responsible for the shape and rigidity of bacteria. A similar spherical transformation takes place in the presence of penicillin,[155-160] oxamycin,[75] and bacitracin,[72] thus demonstrating at a physiological level a lesion in cell wall synthesis or maintenance. Further, when washed free of penicillin, spheroplasts of *E. coli* will revert to their normal bacillary form,[157] indicating that in hypertonic sucrose broths, little, if any, irreversible damage occurs. Again, spheroplasts of a strain of *Alcaligenes faecalis,* induced by penicillin, are able to grow and divide as spheroplasts, although at a slightly reduced rate[161-163] (Fig. 22). This experiment also demonstrates that damage by penicillin to the spheroplast is relatively minor.

Finally, isotopic measurements of cell wall synthesis support the argument that the lesion in cell wall synthesis induced by penicillin is a relatively specific one.[164-169] Thus, in *S. aureus* 70 to 90% inhibition of incorporation of C^{14}-lysine and P^{32}-inorganic phosphate into cell wall was observed under conditions where no inhibition of incorporation of these isotopes into cell protein or nucleic acid could be detected (Table IX). In *E. coli* isotopic measurements of cell wall synthesis are more complex because the special structure represents 80% of the wall. Here, one must be careful to discriminate between isotopes which are incorporated into basal structure and those which are incorporated predominantly into special structure. It had been observed that the incorporation of C^{14}-glucose into the cell wall of *E. coli* (which labels predominantly the special structure) was not inhibited by penicillin,[170] and it was incorrectly inferred that penicillin has no effect on cell wall synthesis in this organism. However, when DAP (which labels the basal structure uniquely) was used as the isotopic precursor,[169] an inhibition of cell wall synthesis by penicillin could be readily measured in this organism (Table IX). Similarly, it has been found that the absolute amount of DAP in the cell wall of *E. coli* is diminished during

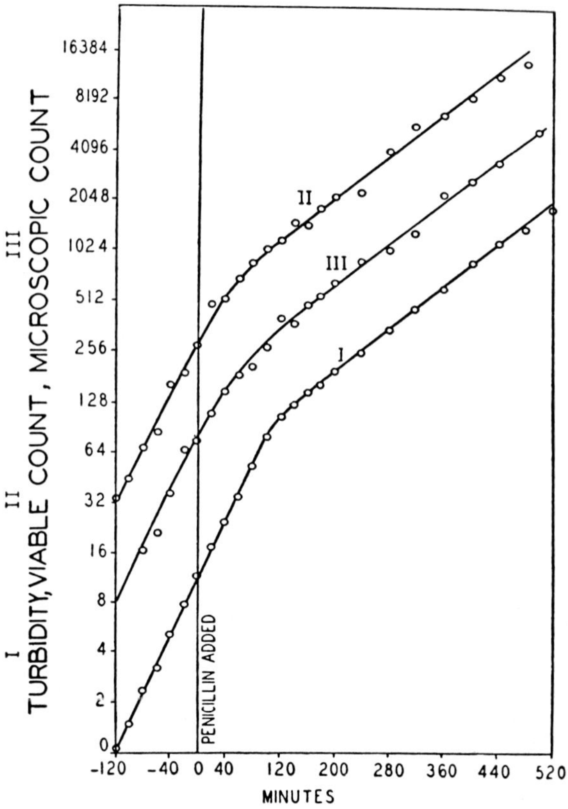

FIG. 22. Multiplication of spheroplasts of *Alcaligenes faecalis* in the presence of penicillin in a medium containing tryptone and NaCl (Lark[161]).

TABLE IX

EFFECTS OF PENICILLIN ON INCORPORATION OF ISOTOPES INTO CELL WALL OR INTO CELL PROTEIN AND NUCLEIC ACID IN *Staphylococcus aureus* AND IN *Escherichia coli*[169]

	Staphylococcus aureus				*Escherichia coli*	
	C^{14}-lysine		P^{32}-inorganic phosphate		H^3-DAP	C^{14}-glucose
	Cell wall	Protein	Cell wall	Nucleic acid	Cell wall	Cell wall
Control	34,800	5,100	155,000	11,600	1,040,000	389,000
Plus penicillin	3,290	4,960	48,900	11,600	297,000	334,000
Inhibition, %	91	2	68	0	72	14

NOTE: Data are expressed as specific activities (c.p.m./mg.).

penicillin treatment.[171] These experiments support the view that penicillin is an inhibitor of cell wall synthesis in *E. coli* as well as in *S. aureus.*

The question may well be asked, however, as to whether the effect of penicillin on cell wall synthesis is a direct one, or is the reflection of some more distant effect, such as interference in some general function of the cell membrane (where penicillin is fixed[74] and where the ultimate reactions in cell wall synthesis almost certainly occur) or interference with some reaction leading to synthesis of the acceptor in the uridine nucleotide transglycosylation reaction.* To this question there is still no experimental answer.

D. Mutants Blocked in Cell Wall Synthesis

Although mutants have been of great importance in the elucidation of several metabolic pathways, this approach has not yet been exploited greatly in studies of cell wall synthesis. Only recently have specific substances been known, lesions in the formation or utilization of which might result in specific blocks in cell wall synthesis (e.g., D-alanine and D-glutamic acid). In addition, not all cell wall mutants could be isolated by some procedures employed for concentration of mutants (e.g., the penicillin selection technique) because many of these mutants would themselves be killed under the conditions of selection. Successful isolation of cell wall mutants has, however, been made by a novel procedure employing sucrose-penicillin gradient plates[159] (Fig. 23).

Several classes of cell wall mutants are known at the present time. Those blocked in the synthesis of DAP have been studied for some time,[91-94, 138, 159] and D-glutamic acid-requiring mutants have been found more recently.[173, 174] Each of these types of bacteria lyses in the absence of the essential amino acid and forms protoplasts if the deprivation is carried out in a hypertonic medium. Some lactic acid bacteria require D-alanine in the absence of pyridoxal,[175, 176] but no bacteria have been isolated so far which have an absolute requirement for D-alanine. More recently, *Lactobacillus bifidus*, var. *Pennsylvanicus*, which grows as a bifid organism in the absence of an acetylglucosamine-containing oligosaccharide and as a rod in its presence, has been found to represent an interesting example of a cell wall mutant.[177-179] Precise definition of the lesion has, however, not yet been made.

VI. Some Additional Problems: The Role of Thymidine and Guanosine Nucleotides in Cell Wall Synthesis

Deoxyhexoses and heptoses are characteristic components of the lipopolysaccharides (see above). Recent isolations of nucleotide derivatives of

* Indeed, it has recently been indicated that viability may be inhibited earlier than inhibition of incorporation of DAP.[172]

some of these compounds have opened new avenues to study the mechanisms of synthesis of these unusual sugars and the means of their incorporation into cell wall lipopolysaccharide and into other compounds.

GDP-L-fucose (GDP-6-deoxy-L-galactose) from *Aerobacter aerogenes*[180] (also found in sheep's milk[181]) was the first of these compounds to

FIG. 23. Growth of *Escherichia coli,* strain Y-10, on sucrose agar containing 0 to 700 units/ml. of penicillin. Normal growth is seen above and growth as protoplasts is seen below. Only a few colonies grow in the intermediate zone; many of these are cell wall mutants (Lederberg and St. Clair[159]).

be isolated. This compound is synthesized from GDP-D-mannose by a complex transformation including dehydration, hydride shift, epimerizations and reduction[182-184] (Fig. 24). The recognition by Ginsburg[184] that 4-keto-6-deoxy-D-mannose was an intermediate, and that reduction took place at C-4 rather than C-6 in the over-all transformation has been an important contribution. GDP-colitose (GDP-3,6-dideoxy-L-galactose) from *E. coli,* strain O111, is also synthesized from GDP-D-mannose.[185]

Colitose is the antigenic determinant in the lipopolysaccharide of this microorganism. Finally, a GDP-heptose (probably GDP-D-glycero-D-mannoheptose) has been isolated from yeast,[186] where it occurs together with GDP-mannose.

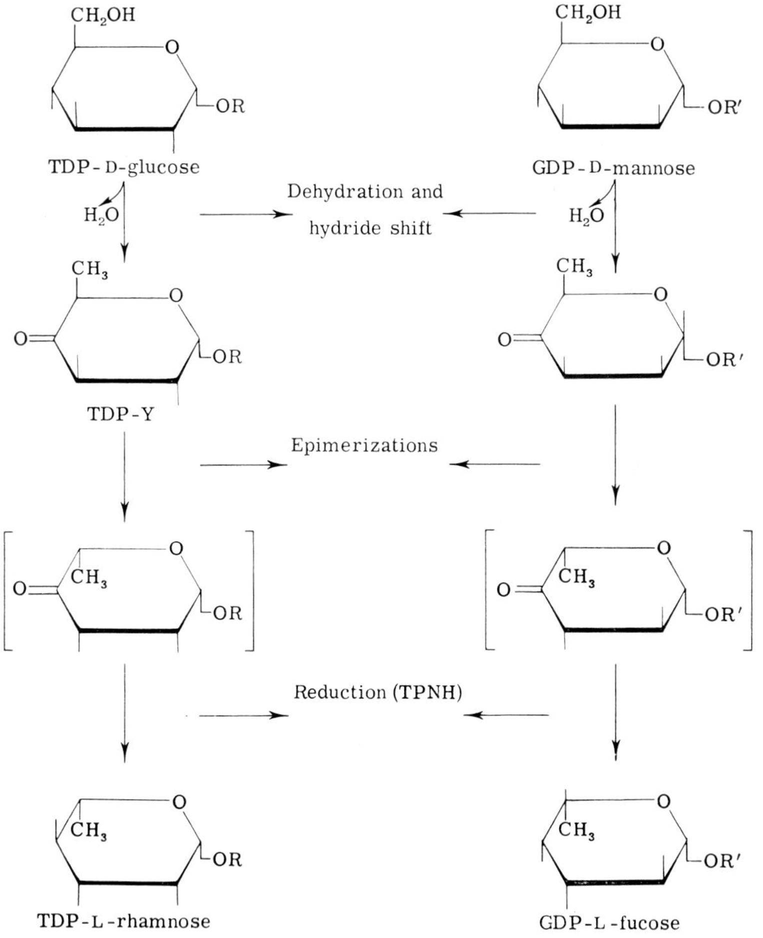

FIG. 24. Schemes for the biosynthesis of L-fucose and L-rhamnose. The mechanism for synthesis of GDP-L-fucose (right) was originally suggested by Ginsburg, and recent work suggests that a similar mechanism is employed in the biosynthesis of TDP-L-rhamnose (left); R = TDP, R′ = GDP.

The TDP-sugar compounds are another group of recently isolated nucleotides which is likely to contain lipopolysaccharide precursors. TDP-L-rhamnose (TDP-6-deoxy-L-mannose) from *Lactobacillus acidophi-*

lus,[187, 188] and several compounds, called TDP-X, TDP-Y, and TDP-Z, from *E. coli*, strain Y-10,[189] were the first of these to be isolated. TDP-Y has been identified as TDP-4-keto-6-deoxy-D-glucose, an intermediate in the synthesis of TDP-L-rhamnose,[190] and TDP-Z as a mixture of TDP-6-deoxy-D-galactose (TDP-D-fucose) and TDP-6-deoxy-D-glucose.[191] *E. coli*, strain Y-10, is probably a mutant blocked at a step in rhamnose biosynthesis and consequently accumulates a large amount of the intermediate, TDP-Y. TDP-D-fucose and TDP-6-deoxy-D-glucose are the reduction products of this intermediate, and probably arise from a nonspecific reduction, although the possibility that they are actual metabolic intermediates is not excluded. TDP-D-glucose is the precursor of the intermediate and of TDP-L-rhamnose.[192, 193] This pathway (Fig. 24), which is analogous to that proposed for L-fucose synthesis by Ginsburg,[182-184] has been studied by several investigators.[190-195]

In addition to these compounds, TDP-X[189] (from *E. coli*, strain Y-10) and TDP-X2 (from *E. coli*, strain B) are still unidentified. *E. coli*, strain B, also contains TDP-L-rhamnose, as do several other strains of *E. coli**[197] and *Streptomyces griseus*.[198] TDP-mannose has also been found in *S. griseus*,[198] an organism which produces mannosidostreptomycin. The occurrence of sugars, activated as derivatives of TDP, is a most unusual phenomenon, and one wonders if such deoxynucleotides are functionally related in some way to deoxyribonucleic acid to which they are so closely related chemically. They could, for example, represent part of a "code" for the synthesis of complex polysaccharides.†

These and other problems of bacterial cell wall synthesis will tax the skill and imagination of organic chemists, biochemists, microbiologists, and pharmacologists for some time to come.

* The interesting observation has been made that the steady state level of TDP-rhamnose in *E. coli*, strain O18, which contains rhamnose in its lipopolysaccharide, is so low as to be not detectable by ordinary methods. On the other hand, a rough variant of this strain, which lacks rhamnose in its lipopolysaccharide, has been found to contain large amounts of this nucleotide.[197] These strains have approximately the same amounts of enzymes which catalyze synthesis of TDP-rhamnose. Rough strains of *E. coli*,[48, 199] *Salmonella*,[49, 200] and *P. pseudotuberculosis*[52] in general lack the deoxysugars and dideoxysugars which are characteristic of the smooth strains. These organisms may, therefore, be important tools for the isolation of nucleotide derivatives of novel sugars and for study of the mechanism of lipopolysaccharide synthesis.

† Since this chapter was originally written, a number of novel sugar-containing nucleotides have been isolated from various bacteria. These include CDP-tyvelose (CDP-3,6-dideoxy-D-mannose),[201] CDP-abequose (CDP-3,6-dideoxy-D-galactose),[201] other incompletely identified CDP-sugar compounds,[202] GDP-D-rhamnose,[203] and GDP-6-deoxy-D-talose.[203] Moreover, the transformation of TDP-D-glucose to TDP-D-galactose[204, 205, 206] and the enzymic syntheses of CDP-glucose,[207] TDP-acetylglucosamine, and TDP-acetylgalactosamine[208] have been demonstrated. TDP-X and TDP-X_2 are unusual TDP-4-acetylamino-4,6-dideoxy-sugar compounds.[209, 210] The complexity of the problem of mechanisms in lipopolysaccharide synthesis adds to its fascination.

ACKNOWLEDGMENT

I am very grateful to the authors and publishers of the journals cited for permission to reproduce Figs. 1, 5, 21, 22, and 23.

REFERENCES

[1] Cf. R. M. Bock, *in* "The Enzymes" (P. D. Boyer, H. Lardy, and K. Myrback, eds.), Vol. 2, Chapter 1. Academic Press, New York, 1960; L. F. Leloir and C. E. Cardini, *ibid.*, Chapter 2; E. P. Kennedy, *ibid.*, Chapter 3; M. F. Utter, *ibid.*, Chapter 4.
[2] J. L. Strominger, *Physiol. Revs.* **40,** 55 (1960).
[3] M. R. J. Salton, *in* "The Bacteria" (I. C. Gunsalus and R. Y. Stanier, eds.), Vol. I. Academic Press, New York, 1960.
[4] M. R. J. Salton, *in* "Bacterial Anatomy." *Symposium Soc. Gen. Microbiol.*, **6,** 81 (1956).
[5] C. S. Cummins and H. Harris, *J. Gen. Microbiol.* **14,** 583 (1956).
[6] E. Work, *Nature* **179,** 841 (1957).
[7] J. L. Strominger, *Compt. rend. trav. lab. Carlsberg, Sér. chim.* **31,** 181 (1959).
[8] J. L. Strominger, S. S. Scott, and R. H. Threnn, *Federation Proc.* **18,** 334 (1959).
[9] M. R. J. Salton and J. M. Ghuysen, *Biochim. et Biophys. Acta* **36,** 552 (1959).
[10] J. M. Ghuysen and M. R. J. Salton, *Biochim. et Biophys. Acta* **40,** 462 (1960).
[11] J. M. Ghuysen, *Biochim. et Biophys. Acta* **40,** 473 (1960).
[12] M. R. J. Salton and J. M. Ghuysen, *Biochim. et Biophys. Acta* **45,** 355 (1960).
[13] J. M. Ghuysen, *Biochim. et Biophys. Acta* **47,** 561 (1961).
[14] H. R. Perkins and H. J. Rogers, *Biochem. J.* **72,** 647 (1951).
[15] H. R. Perkins, *Biochem. J.* **74,** 182 (1960).
[16] W. Brumfitt, A. C. Wardlaw, and J. T. Park, *Nature* **181,** 1783 (1958).
[17] M. R. J. Salton and J. M. Ghuysen, *Biochim. et Biophys. Acta* **24,** 160 (1957).
[18] W. Weidel and J. Primosigh, *J. Gen. Microbiol.* **18,** 513 (1958).
[19] J. Primosigh, H. Pelzer, D. Maass, and W. Weidel, *Biochim. et Biophys. Acta* **46,** 68 (1961).
[20] W. Weidel, H. Frank, and H. H. Martin, *J. Gen. Microbiol.* **22,** 158 (1960).
[21] D. G. Comb, W. Chin, and S. Roseman, *Biochim. et Biophys. Acta* **46,** 394 (1961).
[22] M. H. Mandelstam and J. L. Strominger, *Biochem. Biophys. Research Communs.* **5,** 466 (1961).
[23] M. R. J. Salton, *Biochim. et Biophys. Acta* **52,** 329 (1961).
[24] E. Work, *J. Gen. Microbiol.* **25,** 167 (1961).
[25] C. S. Cummins and H. Harris, *Intern. Bull. Bacteriol. Nomenclature and Taxonomy* **6,** 111 (1956).
[26] C. S. Cummins and H. Harris, *J. Gen. Microbiol.* **18,** 173 (1958).
[27] O. Westphal, *Ann. inst. Pasteur* **98,** 789 (1960).
[28] E. Lederer, *Angew. Chem.* **72,** 372 (1960).
[29] P. D. Mitchell and J. Moyle, *J. Gen. Microbiol.* **5,** 966 (1951).
[30] J. J. Armstrong, J. Baddiley, J. G. Buchanan, B. Carss, and G. R. Greenberg, *J. Chem. Soc.* p. 4344 (1958).
[31] J. Baddiley, *Proc. Chem. Soc.* p. 177 (1959).
[32] J. J. Armstrong, J. Baddiley, and J. G. Buchanan, *Biochem. J.* **76,** 610 (1960).
[33] J. J. Armstrong, J. Baddiley, and J. G. Buchanan, *Biochem. J.* **80,** 254 (1961).
[34] A. R. Sanderson, W. G. Juergens, and J. L. Strominger, *Biochem. Biophys. Research Communs.* **5,** 472 (1961).

[35] J. Baddiley, J. G. Buchanan, F. E. Hardy, R. D. Martin, U. L. RajBhandary, and A. R. Sanderson *Biochim. et Biophys. Acta* **52,** 406 (1961).
[36] W. G. Juergens, A. R. Sanderson, and J. L. Strominger, *Bull. soc. chim. biol.* **42,** 1669 (1960).
[37] M. McCarty, *J. Exptl. Med.* **104,** 629 (1956).
[38] M. McCarty, *J. Exptl. Med.* **108,** 311 (1958).
[38a] G. Haukenes, D. C. Ellwood, J. Baddiley, and P. Oeding, *Biochim. et Biophys. Acta* **53,** 425 (1961).
[38b] S. I. Morse, *J. Exptl. Med.* **115,** 295 (1962).
[39] L. A. Julianelle and C. W. Wieghard, *J. Exptl. Med.* **62,** 11, 23, 31 (1935).
[40] L. A. Julianelle and A. F. Hartmann, *J. Exptl. Med.* **64,** 149 (1936).
[41] J. L. Strominger, unpublished observations.
[42] R. M. Krause and M. McCarty, *J. Exptl. Med.* **114,** 127 (1961).
[43] H. Tauber and H. Russell, *J. Biol. Chem.* **235,** 961 (1960).
[44] J. L. Strominger, J. T. Park, and R. E. Thompson, *J. Biol. Chem.* **234,** 3263 (1959).
[44a] A. R. Archibald, J. J. Armstrong, J. Baddiley, and J. B. Hay, *Nature* **191,** 570 (1961).
[44b] J. M. Ghuysen, *Biochim. et Biophys. Acta* **50,** 413 (1961).
[45] A. Closse, O. Lüderitz, I. Fromme, and O. Westphal, personal communication.
[46] A. Burton and H. E. Carter, personal communication.
[47] F. Kauffmann, O. Lüderitz, H. Stierlin, and O. Westphal, *Zentr. Bakteriol Abt. I. Orig.* **178,** 442 (1960).
[48] F. Kauffmann, O. H. Braun, O. Lüderitz, H. Stierlin, and O. Westphal, *Zentr. Bakteriol. Abt. I. Orig.* **180,** 180 (1960).
[49] F. Kauffmann, L. Kruger, O. Lüderitz, and O. Westphal, *Zentr. Bakteriol. Abt. I. Orig.* **182,** 57 (1961).
[50] A. M. Staub and R. Tinelli, *Bull. soc. chim. biol.* **42,** 1637 (1960).
[51] R. Tinelli and A. M. Staub, *Bull. soc. chim. biol.* **42,** 583 (1960).
[51a] P. W. Robbins and T. Uchida, *Biochemistry* **1,** 323 (1962).
[52] D. A. L. Davies, *J. Gen. Microbiol.* **18,** 118 (1958).
[53] A. P. MacLennan and D. A. L. Davies, *Bull. soc. chim. biol.* **42,** 1373 (1960).
[54] D. A. L. Davies, *Advances in Carbohydrate Chem.* **15,** 271 (1960).
[55] O. Westphal and O. Lüderitz, *Angew Chem.* **72,** 881 (1960).
[56] O. Lüderitz, *Bull. soc. chim. biol.* **42,** 1349 (1960).
[57] A. P. MacLennan, *Biochim. et Biophys. Acta* **48,** 600 (1961).
[58] J. T. Park and M. J. Johnson, *J. Biol Chem.* **179,** 585 (1949).
[59] J. T. Park, *J. Biol. Chem.* **194,** 877, 885, 897 (1952).
[60] J. T. Park and J. L. Strominger, *Science* **125,** 99 (1957).
[61] R. Caputto, L. F. Leloir, C. E. Cardini, and A. C. Paladini, *J. Biol. Chem.* **184,** 333 (1950).
[62] R. E. Strange and J. F. Powell, *Biochem. J.* **58,** 80 (1954).
[63] R. E. Strange and F. A. Dark, *Nature* **177,** 186 (1956).
[64] R. E. Strange and L. H. Kent, *Biochem. J.* **71,** 333 (1959).
[65] R. E. Strange, *Bacteriol. Revs.* **23,** 1 (1959).
[66] L. H. Kent, *Biochem. J.* **67,** 5P (1957).
[67] Y. Matsushima and J. T. Park, *Federation Proc.* **20,** 78 (1961).
[68] J. L. Strominger and R. H. Threnn, *Biochim. et Biophys. Acta* **33,** 280 (1959).
[69] A. Berger and K. Linderstrøm-Lang, *Arch. Biochem. Biophys.* **30,** 106 (1957).
[70] J. L. Strominger, *J. Biol. Chem.* **224,** 509 (1957).
[71] J. L. Strominger, E. Ito, and R. H. Threnn, unpublished observations.
[72] E. P. Abraham and G. G. F. Newton, *Ciba Foundation Symposium, Amino Acids and Peptides with Antimetabolic Activity, 1958*, p. 205.
[73] H. Eagle and A. D. Musselman, *J. Exptl. Med.* **88,** 99 (1948).

[74] P. D. Cooper, *Bacteriol. Revs.* **20,** 28 (1956).
[75] J. Ciak and F. E. Hahn, *Antibiotics & Chemotherapy,* **9,** 47 (1959).
[76] J. L. Strominger, R. H. Threnn, and S. S. Scott, *J. Am. Chem. Soc.* **81,** 3803 (1959).
[76a] P. E. Reynolds, *Biochim. et Biophys. Acta* **52,** 403 (1961).
[76b] D. C. Jordan, *Biochem. Biophys. Research Communs.* **6,** 167 (1961).
[76c] C. Wallas and J. L. Strominger, submitted for publication.
[77] J. L. Strominger and R. H. Threnn, *Biochim. et Biophys. Acta* **36,** 83 (1959).
[78] G. D. Shockman, J. J. Kolb, and G. Toennies, *J. Biol. Chem.* **230,** 961 (1958).
[79] G. D. Shockman, M. J. Conover, J. J. Kolb, P. M. Phillips, L. S. Riley, and G. Toennies, *J. Bacteriol.* **81,** 36 (1961).
[80] J. T. Park, *Federation Proc.* **13,** 271 (1954).
[80a] J. J. Saukonnen, *Nature* **192,** 816 (1961).
[81] M. H. Mandelstam, Ph. D. Thesis, Dept. of Pharmacology, Washington University School of Medicine, St. Louis, Missouri, 1962.
[82] J. Baddiley, J. G. Buchanan, B. Carss, and A. P. Mathias, *J. Chem. Soc.* p. 4583 (1956).
[83] H. J. Rogers and H. R. Perkins, *Biochem. J.* **77,** 448 (1960).
[84] A. Tomasz and E. Borek, *Proc. Natl. Acad. Sci. U.S.* **45,** 929 (1959).
[85] A. Tomasz and E. Borek, *Proc. Natl. Acad. Sci. U. S.* **46,** 324 (1960).
[86] Y. Takagi and N. Otsuji, *Biochim. et Biophys. Acta* **29,** 227 (1958).
[87] N. Otsuji and Y. Takagi, *J. Biochem.* (*Tokyo*) **46,** 791 (1959).
[87a] M. Welsch, *Z. Schweiz. allgem. Pathol.* **21,** 741 (1958).
[88] N. Ishimoto, M. Saito, and E. Ito, *Nature* **182,** 959 (1958).
[89] E. Ito, N. Ishimoto, and M. Saito, *Arch. Biochem.* **80,** 431 (1958).
[90] E. Ito, N. Ishimoto, and M. Saito, *Nature* **181,** 906 (1958).
[91] N. Bauman and B. D. Davis, *Science* **126,** 170 (1957).
[92] K. McQuillen, *Biochim. et Biophys. Acta* **27,** 410 (1958).
[93] P. Meadow, D. S. Hoare, and E. Work, *Biochem. J.* **66,** 270 (1957).
[94] L. E. Rhuland, *J. Bacteriol.* **73,** 778 (1957).
[95] G. T. Barry, *J. Exptl. Med.* **107,** 507 (1958).
[96] D. G. Comb and S. Roseman, *J. Am. Chem. Soc.* **80,** 497 (1958).
[97] D. G. Comb, F. Shimizu, and S. Roseman, *J. Am. Chem. Soc.* **81,** 5513 (1959).
[98] P. J. O'Brien and F. Zilliken, *Biochim. et Biophys. Acta* **31,** 543 (1959).
[99] A. Rosenberg and E. Chargaff, *J. Biol. Chem.* **232,** 1031 (1958).
[100] P. Mandelstam, R. Loercher, and J. L. Strominger, *J. Biol. Chem.* **237,** in press (1962).
[101] G. Toennies, B. Bakay, and G. D. Shockman, *J. Biol. Chem.* **234,** 3269 (1959).
[102] M. Ikawa and E. E. Snell, *J. Biol. Chem.* **235,** 1376 (1960).
[103] J. A. Cifonelli and A. Dorfman, *J. Biol. Chem.* **228,** 547 (1957).
[104] R. H. McCluer, J. Van Eys, and O. Touster, *Amer. Chem. Soc., Abstr. Div. Biol. Chem.,* p 65c (1955).
[105] S. B. Binkley, *Arch. Biochem. Biophys.* **61,** 84 (1956).
[106] J. L. Strominger, *Federation Proc.* **12,** 277 (1953).
[107] J. Baddiley, J. G. Buchanan, B. Carss, A. P. Mathias, and A. R. Sanderson, *Biochem. J.* **64,** 599 (1956).
[108] J. Baddiley, J. G. Buchanan, A. P. Mathias, and A. R. Sanderson, *J. Chem. Soc.* p. 4186 (1956).
[109] J. Baddiley, J. G. Buchanan, and B. Carss, *J. Chem. Soc.* p. 1869 (1957).
[110] P. H. Clarke, P. Glover, and A. P. Mathias, *J. Gen. Microbiol.* **20,** 156 (1959).
[111] P. Berg and W. K. Joklik, *J. Biol. Chem.* **210,** 657 (1954).
[112] E. E. B. Smith and G. T. Mills, *Biochim. et Biophys. Acta* **13,** 386 (1954).
[113] L. Glaser and D. H. Brown, *Proc. Natl. Acad. Sci. U. S.* **41,** 253 (1955).
[114] J. L. Strominger and M. S. Smith, *J. Biol. Chem.* **234,** 1822 (1959).
[115] R. E. Strange, *Biochem. J.* **64,** 23P (1956).
[116] J. L. Strominger, *Biochim. et Biophys. Acta* **30,** 645 (1958).

[117] M. H. Richmond and H. R. Perkins, *Biochem. J.* **76,** 1P (1960).
[118] E. Ito and J. L. Strominger, *J. Biol Chem.* **235,** PC5 (1960); *J. Biol. Chem.* **237,** in press (1962).
[119] E. Ito and J. L. Strominger, *J. Biol. Chem.* **235,** PC7 (1960).
[120] M. Ikawa and E. E. Snell, *Arch. Biochem. Biophys.* **78,** 338 (1958).
[121] S. G. Nathenson and J. L. Strominger, unpublished observations.
[122] F. C. Neuhaus, *Biochem. Biophys. Research Communs.* **3,** 401 (1960).
[123] E. Ito, N. Ishimoto, and J. L. Strominger, unpublished observations.
[124] C. B. Anfinsen and D. Steinberg, *J. Biol. Chem.* **189,** 739 (1951).
[125] A. Meister, *Physiol. Revs.* **36,** 103 (1956).
[126] A. Meister, P. R. Krishnaswamy, V. Pamiljans, and G. Dumville, *Federation Proc.* **20,** 229 (1961).
[127] J. E. Snoke and K. Bloch, *J. Biochem.* **213,** 825 (1955).
[128] D. H. Strumeyer and K. Bloch, *J. Biol. Chem.* **235,** PC27 (1960).
[129] G. D. Kalyankar and A. Meister, *J. Biol. Chem.* **234,** 3210 (1959).
[130] J. Baddiley and F. C. Neuhaus, *Biochem. J.* **75,** 579 (1960).
[131] E. Ito and J. L. Strominger, unpublished observations.
[132] J. L. Strominger, E. Ito, and R. H. Threnn, *Federation Proc.* **20,** 380 (1961).
[133] W. A. Wood and I. C. Gunsalus, *J. Biol. Chem.* **190,** 403 (1951).
[134] C. B. Thorne, C. G. Gomez, and R. D. Housewright, *J. Bacteriol.* **69,** 357 (1955).
[135] R. D. Housewright, Chapter 9, this volume.
[136] L. Glaser, *J. Biol. Chem.* **235,** 2095 (1960).
[137] M. Antia, D. S. Hoare, and E. Work, *Biochem. J.* **65,** 448 (1957).
[138] C. Gilvarg, *Federation Proc.* **19,** 948 (1960).
[138a] C. Gilvarg, W. Farkas, and Y. Gugari, *Federation Proc.* **21,** 9–10 (1962).
[139] A. Bondi, J. Kornblum, and C. Forte, *Proc. Soc. Exptl. Biol. Med.* **96,** 270 (1957).
[140] G. D. Shockman, *Proc. Soc. Exptl. Biol. Med.* **191,** 693 (1959).
[141] A. E. Braunshtein, *in* "The Enzymes" (P. D. Boyer, H. Lardy, and K. Myrbäck, eds.), Vol. 2, p. 283. Academic Press, New York, 1960.
[142] R. M. Azarkh, A. E. Braunshtein, T. S. Paskhina, and H. T'ing-sen, *Biokhimiya* **25,** 954 (1960) [Consultants Bureau translation: *Biochemistry* **25,** 741, (1961)].
[143] P. Barbieri, A. Di Marco, L. Fuoco, P. Julita, A. Migliacci, and A. Ruxconi, *Biochem. Pharmacol.* **3,** 264 (1960).
[144] D. D. Woods, *Brit. J. Exptl. Pathol.* **21,** 74 (1940).
[145] P. Fildes, *Lancet* **1,** 955 (1940).
[146] J. L. Strominger, E. Ito, and R. H. Threnn, *J. Am. Chem. Soc.* **82,** 998 (1960).
[147] D. R. D. Shaw, *Biochem. J.* **66,** 56P (1957); *Biochem. J.* **82,** 297 (1962).
[147a] S. Roseman, *Proc. Natl. Acad. Sci. U. S.* **48,** 437 (1962).
[147b] L. Warren and R. S. Blacklow, *Biochem. Biophys. Research Communs.* **7,** 433 (1962).
[147c] S. G. Nathenson and J. L. Strominger, *Federation Proc.* **21,** 84 (1962).
[148] P. Mitchell and J. Moyle, *J. Gen. Microbiol.* **16,** 184 (1957).
[149] M. H. Richmond, *Biochim. et Biophys. Acta* **31,** 564 (1959).
[149a] M. H. Richmond, *Biochim. et Biophys. Acta* **33,** 92 (1959).
[150] G. D. Shockman, M. J. Conover, J. J. Kolb, L. S. Riley, and G. Toennies, *J. Bacteriol.* **81,** 36 (1961).
[150a] G. D. Shockman, J. J. Kolb, and G. Toennies, *J. Biol. Chem.* **230,** 961 (1958).
[151] R. E. Strange and F. A Dark, *J. Gen. Microbiol.* **16,** 236 (1957).
[152] J. R. Norris, *J. Gen. Microbiol.* **16,** 1 (1957).
[153] D. J. Mason and D. Powelson, *J. Gen. Microbiol.* **19,** 65 (1958).
[154] C. Weibull, *J. Bacteriol.* **66,** 688 (1953).
[155] J. P. Duguid, *Edinburgh Med. J.* **53,** 401 (1946).
[156] K. Liebermeister and E. Kellenberger, *Z. Naturforsch.* **11G,** 200 (1956).
[157] J. Lederberg, *Proc. Natl. Acad. Sci. U. S.* **42,** 574 (1956).
[158] J. Lederberg, *J. Bacteriol.* **73,** 144 (1957).

[159] J. Lederberg and J. St. Clair, *J. Bacteriol.* **75,** 143 (1958).
[160] F. E. Hahn and J. Ciak, *Science* **125,** 119 (1957).
[161] K. G. Lark, *Can. J. Microbiol.* **4,** 165, 179 (1958).
[162] C. Lark and K. G. Lark *Can. J. Microbiol.* **5,** 369, 381 (1959).
[163] C. Lark and K. G. Lark, *Biochim. et Biophys. Acta* **49,** 308 (1961).
[164] J. Mandelstam and H. J. Rogers, *Nature* **181,** 956 (1958).
[165] J. Mandelstam and H. J. Rogers, *Biochem. J.* **72,** 654 (1959).
[166] E. F. Gale, C. J. Shepherd, and J. P. Folkes, *Nature* **182,** 592 (1958).
[167] J. T. Park, *Biochem. J.* **70,** 2P (1958).
[168] R. Hancock and J. T. Park, *Nature* **181,** 1050 (1958).
[169] S. G. Nathenson and J. L. Strominger, *J. Pharm. Exptl. Therap.* **131,** 1 (1961).
[170] R. E. Trucco and A. B. Pardee, *J. Biol. Chem.* **230,** 435 (1958).
[171] M. R. J. Salton and F. Shafa, *Nature* **181,** 1321 (1958).
[172] P. Meadow, *Biochem. J.* **76,** 8P (1960).
[173] H. Momose and Y. Ikeda, *Nature* **186,** 567 (1960).
[174] H. Momose and Y. Ikeda, *Nature* **186,** 818 (1960).
[175] E. E. Snell, *J. Biol. Chem.* **158,** 497 (1945).
[176] J. T. Holden and E. E. Snell, *J. Biol. Chem.* **178,** 799 (1949).
[177] P. J. O'Brien, M. C. Glick, and F. Zilliken, *Biochim. et Biophys. Acta* **37,** 357 (1960).
[178] M. C. Glick, T. Sall, F. Zilliken, and S. Mudd, *Biochim. et Biophys. Acta* **37,** 361 (1960).
[179] F. Zilliken, *Federation Proc.* **18,** 966 (1959).
[180] V. Ginsburg and H. N. Kirkman, *J. Am. Chem. Soc.* **80,** 3481 (1958).
[181] R. Denaumur, G. Fauconneau, and G. Guntz, *Compt. rend. acad. sci.* **246,** 2820 (1958).
[182] V. Ginsburg, *J. Am. Chem. Soc.* **80,** 4426 (1958).
[183] V. Ginsburg, *J. Biol. Chem.* **235,** 2196 (1960).
[184] V. Ginsburg, *Federation Proc.* **20,** 903 (1961); *J. Biol. Chem.* **236,** 2389 (1961).
[185] E. C. Heath, *Biochim. et Biophys. Acta* **39,** 377 (1960).
[186] V. Ginsburg, *Biochem. Biophys. Research Communs.* **3,** 187 (1960).
[187] R. Okazaki, *Biochem. Biophys. Research Communs.* **1,** 34 (1959).
[188] R. Okazaki, *Biochim. et Biophys. Acta,* **44,** 478 (1960).
[189] J. L. Strominger and S. S. Scott, *Biochim. et Biophys. Acta* **35,** 582 (1959).
[190] R. Okazaki, T. Okazaki, and J. L. Strominger, *Federation Proc.* **20,** 906 (1961).
[191] R. Okazaki, T. Okazaki, J. L. Strominger, and A. M. Michelson, *J. Biol. Chem.* **237** (in press) (1962).
[192] J. H. Pazur and E. W. Shuey, *J. Am. Chem. Soc.* **82,** 5009 (1960).
[193] S. Kornfeld and L. Glaser, *Biochim. et Biophys. Acta* **42,** 548 (1960).
[194] J. H. Pazur and E. W. Shuey, *J. Biol. Chem.* **236,** 1780 (1961).
[195] S. Kornfeld and L. Glaser, *J. Biol. Chem.* **236,** 1791, 1795 (1961).
[196] R. Okazaki, T. Okazaki, and Y. Kuriki, *Biochim. et Biophys. Acta* **38,** 384 (1960).
[197] T. Okazaki, R. Okazaki, and J. L. Strominger, unpublished observations.
[198] J. Baddiley and N. L. Blumson, *Biochim. et Biophys. Acta* **39,** 376 (1960).
[199] E. Kröger, O. Lüderitz, and O. Westphal, *Naturwissenschaften* **13,** 1 (1959).
[200] O. Lüderitz, F. Kauffman, H. Stierlin, and O. Westphal, *Zentr. Bakteriol. Abt. I. Orig.* **179,** 180 (1960).
[201] H. Nikaido and K. Jokura, *Biochem. Biophys. Research Communs.* **6,** 304 (1961).
[202] S. Okuda, N. Suzuki, and S. Suzuki, Abstracts of 34th General Meeting of the Japanese Biochemical Society, November, 1961.
[203] A. Markovitz, *Biochem. Biophys. Research Communs.* **6,** 250 (1961).
[204] R. Tinelli, T. Okazaki, R. Okazaki, and J. L. Strominger, *Proc. Intern. Congr. Microbiol., 8th Congr., Montreal, 1962,* in press.

[205] J. H. Pazur, K. Kleppe, and A. Cepura, *Biochem. Biophys. Research Communs.* **7,** 157 (1962).
[206] E. F. Neufeld, *Biochem. Biophys. Research Communs.* **7,** 461 (1962).
[207] V. Ginsburg, P. J. O'Brien, and C. W. Hall, *Biochem. Biophys. Research Communs.* **7,** 1 (1962).
[208] S. Kornfeld and L. Glaser, *Biochim. et Biophys. Acta* **86,** 184 (1962).
[209] T. Okazaki, R. Okazaki, J. L., Strominger, and S. Suzuki, *Biochem. Biophys. Research Communs.* **7,** 300 (1962).
[210] R. W. Wheat and E. L. Rollins, *Federation Proc.* **21,** 80 (1962).

CHAPTER 11

The Synthesis of Proteins and Nucleic Acids*

ERNEST F. GALE

I. The Chemical Nature of Proteins 472
A. The Protein Content of Bacterial Cells 472
B. True Proteins 473
C. Some of the Problems of Protein Synthesis 477
D. Specific Peptides of Microbial Origin 478
II. The Chemical Nature of Nucleic Acids 479
A. The Composition of Nucleic Acids 479
B. Distribution 483
C. Problems of Nucleic Acid Synthesis 485
III. Evidence for a Relationship between the Synthesis of Proteins and the Presence of Nucleic Acids 485
A. Early Investigations: Growth Rate and Nucleic Acid Content 485
B. Dependence of Enzyme Formation on Ribonucleic Acid Synthesis 486
C. Dependence of Enzyme-Forming Potential on Deoxyribonucleic Acid 489
IV. Practical Considerations 489
A. The Estimation of Protein 489
B. Biological Preparations Used for Studies of Protein Synthesis 492
C. Contamination of Preparations 498
V. The Site of Protein Synthesis in the Cell 499
A. Mammalian Cells 499
B. Bacterial Cells 500
VI. Components of the Protein-Synthesizing Mechanism 502
A. Energy Source 502
B. Amino Acids 502
C. Amino Acid-Activating Enzymes 505
D. Nucleic Acids and Their Precursors 509
E. Other Soluble Factors Involved in Incorporation of Amino Acids into Ribosomes 513
F. Factors Involved in Release of Soluble Protein from Ribosomes 515
G. Incorporation Factors 516
H. Amino Acid–Lipid Complexes 517
I. Biotin 518
J. B_{12} 518
K. Incorporation Enzyme 519
L. Other Components 519

* The material for this chapter was first collected, organized, and written up in March 1959. The considerable advances in the field since then have given this chapter an old-fashioned look; an attempt has been made in the proof stage to bring the subject matter up to date by insertion of summaries and references to some of the main contributions appearing in the interim period.

Editors' footnote: This chapter in our opinion is an important contribution to the development of macromolecular synthesis. More recent genetic and chemical studies will be considered in Volume V on Heredity.

VII. Protein Synthesis ... 519
A. Amino Acid Incorporation by Microsomes: The Soluble Ribonucleic Acid Carrier System ... 519
B. Incorporation by the Disrupted Staphylococcal Cell Preparation ... 527
C. Incorporation by Preparations of Membrane Fragments ... 530
VIII. Peptides as Intermediates in Protein Synthesis ... 531
A. Do Peptide-Requiring Mutants Exist? ... 532
B. Can Peptides Be Made to Accumulate by Inhibition of Protein Synthesis? ... 534
C. Can Direct Labeling of Peptides Be Demonstrated during Protein Synthesis? ... 535
D. Is Glutathione Involved as an Intermediate in Protein Synthesis? .. 536
E. Do Nucleotide Complexes Carry Peptide Structures? ... 536
IX. The Synthesis of Specific Microbial Peptides ... 537
A. Capsular Polypeptides ... 537
B. Cell Wall Peptides ... 537
X. Ribonucleic Acid Synthesis ... 540
A. Synthesis by Intact Cells ... 540
B. Synthesis in the Disrupted Staphylococcal Cell ... 544
C. Synthesis by Membrane Fragments ... 545
D. Synthesis by Cell-Free Systems ... 545
XI. Deoxyribonucleic Acid Synthesis ... 549
A. Synthesis by Intact Cells ... 549
B. Synthesis by Cell-Free Systems ... 552
XII. Chloramphenicol ... 554
XIII. Discussion (1959) ... 557
Discussion (1961) ... 559
Addendum (1962) ... 562
References ... 565

I. The Chemical Nature of Proteins

A. The Protein Content of Bacterial Cells

The moisture content of bacteria varies, according to data collected by Porter,[1] from 75–90%, and nitrogen forms 8–15% (average 9.9%) of the residual dry matter. The main nitrogenous constituents of the cells are the proteins and the nucleic acids; the nitrogen content of both is about 15%, so we can calculate that an "average" bacterial cell contains about 65% of its dry matter as protein and nucleic acid.

A protein is a macromolecule constructed from amino acids joined together through peptide bonds. The peptide bond can be hydrolyzed by treatment with acid or alkali; if a protein is heated at 120°C. in 6 *N* hydrochloric acid for 18 hours or longer, all the peptide bonds are hydrolyzed and the macromolecule is broken down to a mixture of the constituent amino acids. Proteins obtained from the cytoplasm of animal, plant, and microbial cells can be hydrolyzed in this manner and yield a mixture of some twenty different amino acids. Some racemization occurs during hydrolysis and, if account is taken of this, it is then found that all of the amino acids in the original protein were in the L-form.

These statements are not necessarily true of hydrolyzates of intact microbial cells or of certain proteinlike products of some microorganisms. Cell walls and capsular substances of bacteria may contain peptides in which some of the amino acid residues are in the D-form; the cell walls may contain amino acids, such as diaminopimelic acid,[2] which are not present in the cytoplasmic proteins. Extracellular polypeptides, such as those found in tyrothricin and other peptide antibiotics excreted by bacteria, may contain unusual amino acids and in addition the D-isomers of "natural" amino acids.[3] Substances such as these specific polypeptides have chemical properties relating them to proteins but, as will be seen below, they appear to play different roles in the cells in which they are found and evidence is accumulating that, in some cases, their synthesis may proceed by mechanisms different from those involved in the formation of cytoplasmic proteins. Any discussion of protein synthesis must take account of these peptide materials and we must consequently divide our considerations into two: those concerned with the "true" proteins common to all tissues, and those concerned with the special peptides of microbial origin.

B. True Proteins

For the purposes of this chapter, we shall consider the true proteins to be those substances which, on hydrolysis, yield a mixture of amino acids all of which are in the L-form; these are listed in Table I, which gives the composition of some typical proteins. Most proteins contain some residues of all the amino acids listed but certain amino acids may be missing from specific proteins; e.g., insulin contains no tryptophan or methionine,[4] hemoglobin is free from tryptophan and isoleucine,[4] and flagellin from *Salmonella typhimurium* contains no tryptophan, cysteine, or hydroxyproline.[5] The protein molecule is large and may contain many molecules of each amino acid. The molecular weight of proteins varies from 13,000 for ribonuclease, as an example of a small molecule, to approximately 300,000 for edestin, as an example of a typical large protein molecule; values as high as 7×10^6 have been recorded for hemocyanin but this is probably an aggregate formed from similar units of smaller molecular weight. The molecular weight of tobacco mosaic virus[6] is 10^6. Svedberg proposed that proteins might be built up from units of molecular weight 17,600; this is almost certainly an oversimplification but it provides us with a convenient dimension. The average molecular weight of the naturally occurring amino acids is 145; consequently, in considering the problem of protein synthesis, we have to provide an explanation for the formation of a structural unit containing approximately 120 amino acid residues.

Proteins can play at least three roles in the cell: they may, by virtue of their physical properties, act as supporting or containing structures; they may act as catalysts or enzymes; they may act as carriers of nonprotein substances. The greatest variety of proteins is found in the catalyst or enzyme class. It is generally believed that all enzymes are proteins and that every enzyme is a specific protein; consequently a list of the enzymes that can be found within a cell provides an indication of the number of

TABLE I

AMINO ACID COMPOSITION[a] OF SOME TYPICAL PROTEINS

Amino acid	Hemo-globin[b]	Insulin[b]	Egg albumen[b]	Flagellin[c]
Alanine	7.4	4.4	6.7	10.8
Aspartic acid	10.4	7.5	9.3	18.4
Arginine	3.6	3.4	5.7	3.8
Cystine	1.4	12.5	2.4	—
Glutamic acid	8.5	18.6	16.5	12.8
Glycine	5.6	4.5	3.1	4.6
Histidine	7.9	4.9	2.4	0.5
Isoleucine	—	2.8	7.0	5.8
Leucine	16.0	12.9	9.2	8.8
Lysine	8.6	2.5	6.3	8.1
Methionine	1.5	—	5.2	0.5
Phenylalanine	7.9	8.0	7.7	2.1
Proline	3.9	2.6	3.6	1.3
Serine	5.8	5.3	8.2	5.3
Threonine	4.4	2.1	4.0	10.5
Tryptophan	—	—	1.2	—
Tyrosine	3.0	12.2	3.7	3.4
Valine	8.4	7.5	7.1	17.9

[a] Expressed in grams amino acid/100 g. protein.
[b] Data from Fruton and Simmonds.[4]
[c] Data from Ambler and Rees.[5]

different proteins in that cell. Dixon and Webb,[7] in their book "The Enzymes," have compiled a list of 659 enzymes properly characterized in the literature; since it is probable that at least an equal number remains to be discovered or properly described, we are led to the conclusion that there must be well over 1000 different proteins in a living cell.

Physical measurements on proteins show that they differ in molecular weight and in molecular shape. They are all built from the same twenty amino acids in different permutations and combinations and it is clear that the polypeptide chains must differ in length and folding in different proteins. The sequence of the amino acid residues within a chain is believed to be specific for a particular protein. The naturally occurring amino acids all possess an α-amino group next to the carboxyl group; the peptide bond between adjacent residues in a chain is formed between the α-amino group of one amino acid and the carboxyl group of the next. Many of the amino acids possess other polar groups; thus aspartic and glutamic acids possess a second terminal carboxyl; the basic amino acids, lysine, arginine, and ornithine, possess terminal amino; tyrosine, serine, and threonine possess hydroxyl groups, etc. When a series of amino acids forms a polypeptide chain, these polar groups will project from the chain and provide reactive or ionization centers. Clearly, the over-all properties of the molecule will be governed by the distribution and relative positions of these polar groups; it is probable that the active center of an enzyme is similarly determined and that its specificity will be decided by the nature and position of the groups capable of reacting with a potential substrate.

1. Amino Acid Sequences

The determination of amino acid sequences in a protein is a matter of great technical difficulty and has, so far, been completed in very few cases. The outstanding work of Sanger and his colleagues[8, 9, 10] first led to the determination of the sequences of the amino acid residues in the chains of insulin (see Fig. 1) and this now serves as a model for our considerations of protein structure.

The first question is whether the sequence in Fig. 1 is unique and invariable. Sanger's investigations showed that there are minor modifications in the sequence in insulins from different species: the sequence between the positions marked in Fig. 1 differs in insulin from pig, sheep, or ox. No modifications in the sequence for the insulin from any one of these species have ever been observed and the arguments used for the final piecing together of the smaller peptide structures, released by digestion of the molecule, depend upon the constancy of the structure. More recent investigations (Dr. L. F. Smith, personal communication) on the structure of insulins from a variety of species have shown that modifications occur in at least eight positions. Apart from those shown in Fig. 1, the other modifications are found near the ends of the polypeptide chains and "outside" the disulfide bridges. Human insulin differs from that of pig in that the terminal alanine is replaced by threonine. For any one species no amino acid residue is substituted by any other residue; within the limits of experimental detection this does not necessarily mean that mistakes are not made occasionally during the biosynthesis of insulin, but it does mean that if such mistakes occur, and the product is still insulin, then the frequency of their occurrence is too small for detection. The fact that an alteration of one residue in an amino acid sequence can result in a major change in the properties of the protein

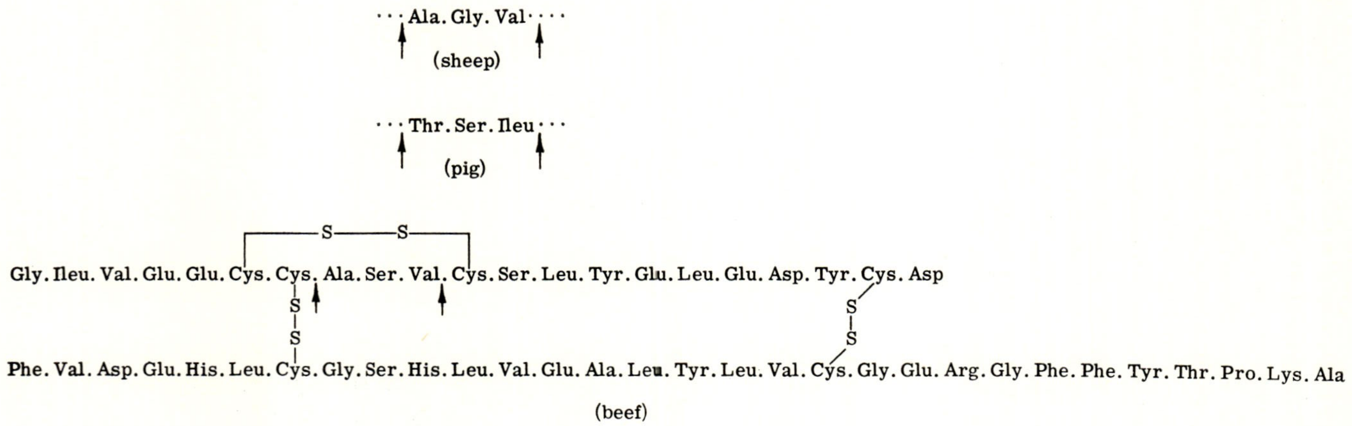

FIG. 1. Structure of insulin. Insulin molecules from beef, sheep, and pig differ only in the sequence between the arrows.[8, 9, 10]

is dramatically demonstrated by the difference between normal and sickle-cell hemoglobin where one glutamyl residue is replaced by a valyl residue.[11]

The sequences of ribonuclease[214, 215] and several fragments of other proteins are now known[12, 13, 216] and sufficient evidence is already available for it to be certain that no amino acid is ever constantly associated with any other amino acid, nor is there any evidence of any sort of consistent repeating pattern in sequences. Certain protein hormones of related activity possess part of their sequences in common: thus both the adrenocorticotropic and the melanophore-stimulating hormones possess the following sequence: Meth-Glut-His-Phen-Arg-Try-Gly- in otherwise dissimilar polypeptide chains. Similarly a number of hydrolytic enzymes (trypsin, chymotrypsin, elastase, and thrombin) are known to possess a common fragment with the sequence -Gly-Asp-Ser-Gly-.[217-222]

2. Cross Linkage

Figure 1 demonstrates a further property of the side chains: their ability to form cross linkages between separate, or folded, polypeptide chains. The constancy of the properties of a protein indicates that the folding and relative positions of polypeptide chains within the molecule must be constant and rigid. The chains are held in such constant relative positions by chemical bonds between side groups. The most important cross linkage is that formed, as in the insulin molecule, by disulfide bridges formed between two cysteine residues. A further possibility would be peptide bonds between the secondary amino groups of lysine residues and the secondary carboxyl groups of glutamic or aspartic acids; there is, however, no firm evidence that such bonds occur and it has been shown in some cases that the amount of lysine with free ω-NH_2 groups is equal to the lysine content of the protein. The presence of secondary carboxyl groups in glutamic and aspartic acids and of —OH in tyrosine gives rise to the possibility of hydrogen bonding between chains; such bonds undoubtedly play a part in determining the molecular shape in a native protein. X-ray analysis has shown that the polypeptide chains of many proteins fall into stable configurations which may be helical in form.[14, 15]

C. Some of the Problems of Protein Synthesis

This brief outline of protein structure indicates the problems which confront the biochemist who wishes to explain the mechanism of protein synthesis. These include: (*a*) How is the specific sequence for any one protein determined? (*b*) How is that sequence imposed upon the condensation of amino acids in the course of protein biosynthesis? (*c*) What imposes a limit to the length of a polypeptide chain? (*d*) How are polypeptide chains folded and brought into position so that the formation of disulfide bridges, etc., produces the rigidly defined structural relationships characteristic of native proteins? (*e*) What determines which proteins will be made by any particular cell?

D. Specific Peptides of Microbial Origin

These can be considered in three main groups:

1. Bacterial Capsules

A number of species in the genus *Bacillus* possess capsules consisting largely or wholly of peptidelike material. The capsules of *Bacillus anthracis* and *B. subtilis* have been studied in detail and prove to consist of a poly-

TABLE II

Amino Acid Components of the Cell Walls of Certain Gram-Positive Bacteria[a]

Organism	Aspartic acid	Alanine	Glutamic acid	Lysine	Diamino-pimelic acid	Glycine
Streptococcus pyogenes (A)	—	+	+	+	—	—
Streptococcus pyogenes (B)	—	+	+	+	—	—
Streptococcus faecalis	—	+	+	+	—	—
Corynebacterium diphtheriae	—	+	+	—	+	—
Corynebacterium pyogenes	—	+	+	+	—	—
Staphylococcus aureus	—	+	+	+	—	+
Sarcina lutea	—	+	+	+	—	+
Actinomyces viridans	—	+	+	+	—	—
Micrococcus cinnabareus	—	+	+	—	+	—
Lactobacillus casei	+	+	+	+	—	—
Lactobacillus plantarum	—	+	+	—	+	—

[a] From Cummins and Harris.[2]

glutamic acid peptide in which the residues are D-glutamic acid joined through γ-glutamyl links. Strains of other aerobic spore-bearing organisms produce polyglutamyl peptide either in the form of a capsule or released into the medium during growth.[16, 17]

2. Cell Wall Peptides

The studies of Salton,[18] Cummins and Harris,[2] and others have shown that cell walls contain amino acids but that the composition varies with the organism concerned. There is a major difference between walls from Gram-positive and Gram-negative cells in that the former are devoid of aromatic amino acids, whereas hydrolysis of the latter gives rise to the complete range of amino acids characteristic of true proteins. Hydrolysis of walls from Gram-positive bacteria may yield as few as four amino acids; Table II shows that many such species yield glutamic acid, alanine, lysine or diaminopimelic acid, and glycine or aspartic acid. Diaminopimelic acid is not found as a constituent of true proteins and examination of the optical specificity of the other amino acids shows that part, if not all, of the glutamic acid and the alanine is present as the D-form. The amino acid complex of the Gram-positive wall is therefore markedly different from true protein. It is probable that the wall of the Gram-negative cell contains both true protein and material similar to that found in the Gram-positive cell wall.

The structure of cell walls is discussed in Chapter 10 but it is relevant to the present discussion to consider the nature of the amino acid-containing portion of such walls. The work of Park[19] and Park and Strominger[20] has shown that the amino acid-containing portion of the Gram-positive cell wall is probably an *N*-acetylmuramic acid polymer in which the amino acids, D-glutamic acid, D-alanine, and lysine, form a peptide linked to muramic acid through a L-alanyl residue. The amino acids exist in a specific sequence[223] in the peptide so that synthesis of that peptide will involve sequence determination and so may be associated with a mechanism similar to that involved in sequence determination for proteins. Armstrong[21] and his co-workers have recently identified another major component of the walls of Gram-positive bacteria. This is "teichoic acid" and forms 40–60% of the wall substance, accounting for the phosphorus therein. Teichoic acid is a polymer of ribitol phosphate associated with other substances including alanine. Teichoic acid, isolated from *Lactobacillus arabinosus* and *B. subtilis*, contains α-glucosyl residues and *O*-alanyl groups probably linked to the ribitol; teichoic acid isolated from a strain of *Micrococcus pyogenes* var. *aureus* contains *N*-acylglucosamine in place of glucose.

3. Antibiotic Peptides

A number of *Bacillus* spp. excrete peptidelike substances into the medium during growth and these substances may possess antibiotic properties.[3] In some cases, e.g., gramicidin, gramicidin S, and tyrocidin, hydrolysis of the antibiotics yields a mixture of amino acids whereas in other cases, e.g., bacitracin[22] and polymyxin,[23] the molecules contain other structures in addition to amino acids. Antibiotic peptides are frequently cyclic in structure and contain D-isomers of amino acids.

II. The Chemical Nature of Nucleic Acids

A. The Composition of Nucleic Acids

Nucleic acids are macromolecules built up from units consisting of nucleotides linked through phosphate bridges. All free-living, self-reproducing cells contain two types of nucleic acid: ribonucleic acid (RNA) and deoxyribonucleic acid (DNA). On hydrolysis with mild alkali, ribonucleic acid is broken down to a mixture of nucleotides; 95% of the mixture is composed of four nucleotides present in roughly equal amounts. In these four nucleotides, the sugar moiety is ribose and the bases are

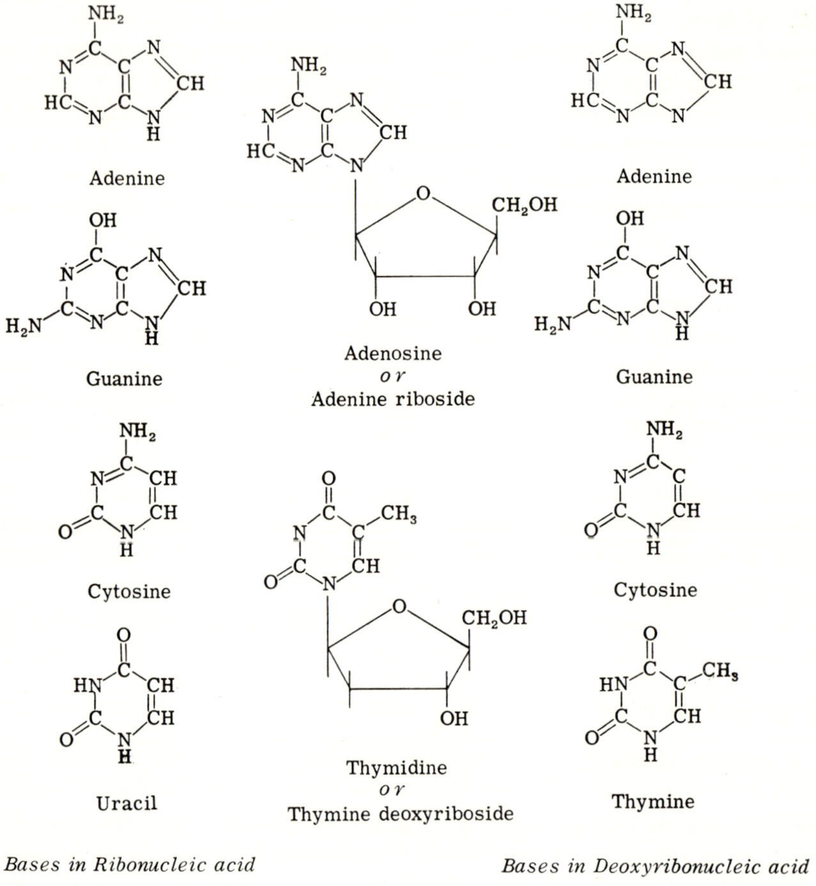

FIG. 2. Structures of bases and sugars in ribonucleic acid and deoxyribonucleic acid.

adenine, guanine, cytosine, and uracil (see Fig. 2). Deoxyribonucleic acid is also composed mainly of four nucleotides in which the sugar is deoxyribose and the bases, adenine, guanine, cytosine, and thymine. In the DNA obtained from certain bacteriophages (e.g. the T-even coliphages), cytosine is replaced by hydroxymethylcytosine.[24] The application of sensitive methods of paper chromatography has shown the existence of other bases present in trace amounts in nucleic acids: thus RNA has been shown to contain 6-methylaminopurine, 6-dimethylaminopurine, 1-methyladenine, 2-methyladenine, and thymine in traces while the presence of dimethylguanine, monomethylguanine, 6-methylaminopurine, methylcytosine, and hydroxymethylcytosine has been shown in DNA.[25-30] Cohn[31, 225] and Yu and Allen[226] have shown the presence in RNA of a new nucleotide identified as the phosphate of 5′-ribosyluracil which has properties similar to those of uridine and is frequently referred to as "pseudouridine." Smith and Dunn[32] have also found that RNA contains a small proportion of dinucleotide

TABLE III

MOLAR RATIOS OF BASES IN THE NUCLEIC ACIDS OF CERTAIN VIRUSES[a]

Virus	Molar ratios of nucleotides					
	Adenylic acid	Guanylic acid	Cytidylic acid	Hydroxymethylcytidylic acid	Uridylic acid	Thymidylic acid
Tobacco mosaic virus	1.20	1.01	0.74	—	1.06	—
Cucumber virus 3	1.03	1.03	0.74	—	1.21	—
Tomato bushy stunt virus	1.10	1.11	0.82	—	0.98	—
Tobacco necrosis virus	1.12	0.98	0.88	—	0.88	—
Bacteriophage T2, T4, T6	1.32	0.72	—	0.54	—	1.42
Bacteriophage T3	0.91	0.94	1.05	—	—	1.11
Bacteriophage T5	1.35	0.84	0.42	—	—	1.41
Polyhedral virus, gypsy moth	0.85	1.23	1.13	—	—	0.80
Polyhedral virus, silkworm	1.17	0.90	0.81	—	—	1.12
Capsule virus, budworm	1.29	0.78	0.70	—	—	1.24

[a] From C. A. Knight.[35]

structures which are resistant to alkaline hydrolysis; they differ from normal ribonucleotides in that they do not form complexes with borate and contain an abnormal sugar. Abnormal nucleotides corresponding to the four normal bases have been isolated and the sugar moiety tentatively identified as 2′ (or 3′)-*O*-methylribose.

In both RNA and DNA, nucleotides are jointed by 3′,5′-phosphodiester linkages to form polynucleotide chains. In DNA the polynucleotide may contain thousands of nucleotide residues and reach a length of several microns. In many analyses of base ratios in DNA preparations, it can be seen that there is a molar equivalence of thymine and adenine on one hand, and of cytosine and guanine on the other (see Table III). From this fact and the results of X-ray analysis, Watson and Crick[33] put forward the suggestion that the DNA molecule consists of a double helix, the two polynucleotide chains being held together by hydrogen bonds between thymine and adenine, or guanine and cytosine.

RNA exists in a variety of molecular species which are usually grouped into two classes: the "soluble RNA" (frequently abbreviated to "sRNA") and the ribosomal RNA. The two classes differ in the order of magnitude of their molecular weights, that of sRNA being approximately 20,000 and that of ribosomal RNA varying from $5–15 \times 10^5$. On centrifuging cell extracts at 100,000 *g* for 1–2 hours, ribosomal RNA is sedimented while sRNA remains in the supernatant. sRNA itself can be prepared from the supernatant solution by treatment with phenol followed by precipitation with ethanol.[227, 228] Table IV shows that, in the case of rat liver, sRNA contains a much higher proportion of "unusual," or trace, bases than does ribosomal RNA, and that 5′-ribosyluracil is the most abundant of these bases in both rat liver and *E. coli* sRNA.[229, 230]

TABLE IV

OCCURRENCE OF ADDITIONAL COMPONENTS IN sRNA[a]

Component	Moles per 100 moles uridylic acid		
	Rat liver ribosomal RNA	Rat liver sRNA	*E. coli* sRNA
5′-Ribosyluracil	7.5	25	14
5-Methylcytosine	0.4	10	—
6-Methylaminopurine	0.5	8.1	0.6
Thymine	—	—	7
1-Methylguanine	0.1	3.3	0.9
2-Dimethylaminopurine	0.1	3.0	—
2-Methylamino-6-hydroxypurine	0.1	2.3	—
2-Methyladenine	—	—	1.7
6-Dimethylaminopurine	0.1	0.1	—

[a] Data from Dunn.[229, 230]

sRNA extracted from *E. coli* has a molecular weight of 25,000 ± 10%, a sedimentation constant of 4 *S*, probably has a helical structure, and is not bound in the cell to other macromolecules.[231, 232] The material separated as sRNA is undoubtedly heterogeneous and a number of methods of fractionation are being worked out, especially with the object of separating molecules which act as acceptors for specific amino acids (see below). The ribosomal RNA exists as part of the ribonucleoprotein (RNP) material of the cell. This material appears to exist in the form of particles of varying size. In some mammalian cells, RNP particles are found in association with membranes and the cell fraction containing both fragments of membrane and RNP particles is frequently called the "microsome fraction." In the course of a recent symposium of the Biophysical Society[82] it became clear that the term "microsome" did not mean the same thing to all workers and a suggestion was made that the ribonucleoprotein particle should be called a "ribosome" to distinguish it from the more complex materials frequently used as microsomal fractions. The same symposium contains articles on the preparation and properties of microsomal fractions from a variety of cells and tissues. In the bactera so far investigated there are no intracellular membranes to contaminate the RNP ribosomes, although contamination with fragments of protoplast membrane is possible and is discussed below. Bacterial ribosomes appear to exist in varying degrees of aggregation, the main controlling factor being the concentration of magnesium ions in the medium. Examination by ultracentrifuge shows that RNP can be found in various conditions with sedimentation boundaries of 100, 70, 50, and 30 *S*. The 100 *S* particle appears to be a dimer of two 70 *S* particles while the 70 *S* particle appears to consist of an association of two unequal units corresponding to the 50 and 30 *S* particles[233, 234] (see Fig. 3). Examination of the RNA released from the particles shows that RNA from the 30 *S* particle has a sedimentation constant of 16 *S* and molecular weight 5.6×10^5 whereas the larger particles release RNA with sedimentation constants of 16 and 23 *S*.[234-236] There is evidence that this RNA consists of still smaller subunits since heating to 85°C. releases material with sedimentation constant 8 *S*, while the Carnegie group[237] has obtained disaggregation to 4 *S* in very low concentrations of magnesium, and proposes that such 4 *S* structures form the basic units from which larger particles are formed. Spahr and Tissières[240] claim that there is no essential difference in base composition between the 30, 50, and 70 *S* particles whereas Bolton *et al.*[238] believe that significant differences, especially in guanine content, can be shown between the 30 and 50 *S* particles whether these are separated in the native state or derived by the breakdown of 70 *S* ribosomes (see Table V). Bonner[241] has pointed out that ribosomes from various sources are very similar in composition, especially with respect to the size of units, magnesium content, and amino acid composition of the protein moiety. Bacterial ribosomes differ from those of higher organisms in possessing a higher RNA content and a somewhat different base composition.

B. Distribution

In cells showing internal differentiation into nucleus and cytoplasm, DNA is found exclusively within the nuclear membrane whereas RNA is distributed throughout nucleus and cytoplasm. Soluble RNA forms 10–20% of the total RNA of the cell in the case of *E. coli*.[230, 242] In bacterial cells, where no membrane has been described between the nucleus and the cytoplasm, both RNA and DNA are found in the cell contents. Spiegelman, Aronson and Fitz-James[34] have described the separation of DNA-containing structures from lysates of protoplasts of *B. megaterium*; these structures appear to consist of DNA gel wrapped around a ribonucleoprotein core and may represent the bacterial nucleus. In viruses only one type of nucleic acid is found; as a result the viruses can be classified as either RNA- or DNA-viruses. Table III shows the distribution of bases in the nucleic acids of a number of plant, insect, and bacterial viruses.[35] The four plant viruses are RNA-viruses, as shown by the nature of the sugar and the absence of thymine; the bacteriophages and insect viruses quoted are DNA-viruses and are free from uracil.

TABLE V

Base Composition of Ribonucleic Acids from *E. coli*

	Mole per cent			
	Adenine	Guanine	Cytosine	Uracil
Soluble RNA				
sRNA[a]	19.7	34.2	29.1	17.2
sRNA[b]	20.3	32.1	28.9	15.0
sRNA[c]	21.0	31.5	27.2	19.3
sRNA[d]	20.2	28.5	30.5	20.5
sRNA from cells grown in chloramphenicol[a]	19.9	35.6	27.2	17.7
Specific sRNA from T2 infected cells[d]	27.1	21.1	16.2	35.1
Ribosomal RNA				
30 *S* ribosome[e]	24.6	31.6	22.8	21.0
50 *S* ribosome[e]	25.6	31.4	20.9	22.1
70 *S* ribosome[a]	25.0	31.5	22.1	21.4
30 *S* ribosome[a]	24.2	30.4	22.2	23.1
30 *S* (from 70 *S*)[a]	24.3	31.6	23.6	20.5
50 *S* ribosome[a]	26.4	34.1	21.0	18.5
50 *S* (from 70 *S*)[a]	26.4	34.8	20.5	18.3
From cells grown in chloramphenicol[a]	28.6	32.3	20.2	19.0
30 *S* ribosome[d]	23.9	33.5	20.3	22.3
Specific RNA from T2 infected cells[d]	28.0	21.0	15.5	35.5

[a] Data from Bolton *et al.*[238]
[b] Data from Dunn.[230]
[c] Data from Osawa.[239]
[d] Data from Nomura, Hall, and Spiegelman.[410]
[e] Data from Spahr and Tissières.[240]

C. Problems of Nucleic Acid Synthesis

It can be seen from this outline that many of the problems arising in the field of protein synthesis apply in principle to nucleic acid synthesis, since in both cases we are concerned with the production of macromolecules of long chains of different units in specific sequences. It is possible that, instead of dealing with two similar series of problems, we are concerned with one underlying problem only in that the syntheses of proteins and nucleic acids may be intimately interrelated in the cell.

III. Evidence for a Relationship between the Synthesis of Proteins and the Presence of Nucleic Acids

A. Early Investigations: Growth Rate and Nucleic Acid Content

The first suggestion that there is a connection between nucleic acids and protein synthesis came from the studies of Caspersson[36] and Brachet[37] and their co-workers. Both groups produced evidence that there is a general correlation between the rate of growth of various tissues and the nucleic acid content of those tissues, and that there is a similar correlation between nucleic acid content and the rate of protein formation by secretory cells such as those of the pancreas. The work of Caspersson and his colleagues was based upon a photographic method in which the nucleic acid content of cells was deduced from their absorption of ultraviolet light of wavelength 258 mμ. The method was applied to bacteria by Malmgren and Hedèn,[38] who photographed large numbers of bacteria in this wavelength and estimated the density of the images of 200 or more cells on plates exposed under standard conditions. They showed that cells in the resting stage of the inoculum had little 258 mμ-absorbing material but that such material increased rapidly after the cells were placed in a nutrient medium and reached a maximum during the lag and early logarithmic phases of growth. The optical density at 258 mμ then decreased as growth continued and attained a low value again by the time the maximum stationary phase was reached. The cells thus contained the largest content of 258 mμ-absorbing material at the stage of most rapid growth. Evidence that the 258 mμ-absorbing material was nucleic acid was (*1*) that the ultraviolet spectrum of the cells during the lag phase was that of nucleoprotein, and (*2*) that treatment of the cells with ribonuclease removed the greater part of the 258-mμ absorption.

The general conclusion was confirmed at a later date by Caldwell and Hinshelwood,[39,40] who studied the rate of growth of bacteria under a variety of conditions and showed that there is a direct correlation between the ribonucleic acid content of the cells and their rate of growth, while the DNA content is approximately constant and does not vary with the growth rate. When more direct methods were available for studying the rate of protein synthesis in bacteria, it was further confirmed that there is a linear correlation between that rate and the RNA content of the cells.[41]

A correlation between rate of growth and RNA content has been confirmed for cells growing exponentially under constant conditions.[242-244,445,446] When organisms are transferred from a condition of high rate of growth to one of low rate, there is an immediate decrease in the rate of synthesis of RNA relative to that of protein or DNA and this holds until the RNA: protein ratio has arrived at the value characteristic of the new growth rate. Likewise when the rate of growth is increased, the rate of synthesis of RNA increases relative to that of protein or DNA until the characteristic ratio is achieved; this means that the rate of RNA synthesis during the transition period is greater than that occurring after the characteristic ratio has been established. Maaløe and co-workers[445,446] conclude that the rate of protein synthesis per unit RNA is approximately the same at all growth rates. Magasanik and his colleagues[243,244] have shown that the correlation may break down when the rate of growth is slow and when the rate of growth is changing. There is a direct correlation between the RNA: protein ratio when the rate of growth is greater than 0.6 doublings per hour and it seems possible that RNA may be the limiting factor under these conditions. At slower rates of growth, however, some other factor such as the supply of substrates may impose control.

Gros[245] has found that the RNA of intact *E. coli* cells is sensitive to the action of ribonuclease if incubation takes place in the absence of salts. This has enabled him to test the effect of partial removal of RNA on subsequent protein synthesis in a suitable medium. Ribonuclease treatment in this way results in a rapid drop in protein synthesizing capacity to about 50% of the normal value. On further treatment with ribonuclease, the rate of protein synthesis diminishes in proportion to the amount of RNA destroyed. Gros suggests that the initial rapid drop might be due to interaction of the enzyme with negatively charged components of the cell such as sRNA.

B. Dependence of Enzyme Formation on Ribonucleic Acid Synthesis

Pardee[42] studied the formation of β-galactosidase in a uracil-requiring mutant of *Escherichia coli* and found that the formation of enzyme was dependent upon the presence of uracil. If the uracil in the medium became exhausted in the course of the experiment, enzyme formation ceased; Pardee concluded that RNA synthesis is essential for enzyme synthesis since the RNA that is synthesized prior to the exhaustion of uracil has no action in the absence of uracil. Similar conclusions were reached by Spiegelman and co-workers,[43] who found that the formation of α-glucosidase in yeast is dependent upon a supply of purines and pyrimidines in the "pool" of the cell, and that the synthesis of β-galactosidase in *E. coli* is inhibited by the presence of 5-hydroxyuridine. Creaser[44] found that the formation of β-galactosidase by *Staphylococcus aureus* is inhibited by the presence of 8-azaguanine which is incorporated into the RNA of the organism in place of guanine. Creaser[44] found that β-galactosidase synthesis ceases very soon after the addition of 8-azaguanine to the incubation medium but the development of catalase by the same organism is much less sensitive to inhibition in this way. He suggested that enzyme synthesis is dependent upon the presence of RNA and that the functionally active form of RNA differs in stability for different enzymes. On this hypothesis the synthesis of β-galactosidase would depend upon the presence of unstable RNA which must be constantly synthesized in the system, whereas the synthesis of other enzymes such as catalase depends upon the presence of stable RNA and does not therefore show a requirement for RNA synthesis. Richmond[45] has demonstrated that 8-azaguanine inhibits the synthesis of an extracellular lysozymelike enzyme by *B. subtilis*. 8-Azaguanine becomes incorporated into the RNA of a number of bacteria[46] but little, if any, appears in the DNA; Smith and Matthews[47] have shown that most of the analog that is incorporated occurs as end groups of polyribonucleotides. The inhibition of growth produced by 8-azaguanine is not correlated with the amount of analog taken up by the cells and Mandel[48] has suggested that the action on enzyme formation may be associated with some function other than incorporation into RNA.

8-Azaguanine also inhibits the synthesis of protein in *B. cereus*, where it has little effect on DNA but increases the synthesis of RNA.[246] Again the presence of the analog inhibits the formation of some enzymes more than others.[247] The inhibitory effect can be released by the subsequent addition of guanosine but the longer the delay before the guanosine is added, the less effective it is in releasing inhibition. The analog has no effect on the synthesis of cell wall material.[45, 248] Chantrenne[249] suggests that 8-azaguanine affects the activity of a component of the protein synthesizing mechanism which undergoes rapid renewal and that such a component could well be sRNA. Similar effects are reported for 5-fluorouracil[250] which is partially able to replace uracil for RNA and protein synthesis in uracil-dependent *E. coli* incubated in the absence of exogenous uracil. Under such conditions the amounts of RNA and protein in the cells double but specific enzymes show varying effects: the syntheses of catalase and succinic dehydrogenase increase whereas the induction of β-galactosidase or serine dehydrase is blocked. Gros[245] finds that 5-fluorouracil increases the rate of incorporation of arginine, decreases the rate of incorporation of proline and tyrosine, but has no effect on the incorporation of other amino acids in *E. coli*. Protein newly synthesized under these conditions contains 20–30% less proline and tyrosine than normal, and the suggestion is again made that the effect must be on factors specific for amino acids, such as the sRNA amino acid-accepting components or "messenger RNA" (see also pp. 541, 561).

Direct evidence that RNA is necessary for enzyme formation was obtained by Gale and Folkes[49] with disrupted staphylococcal cells where development of catalase activity was decreased by removal of nucleic acid from the cell preparation and then restored by addition of staphylococcal RNA to the incubation mixture. The same preparation was capable of developing β-galactosidase activity but required the presence of purines and pyrimidines in the incubation mixture; no response to extracted RNA could be demonstrated, emphasizing previous conclusions that development of this enzyme is correlated with a synthesis of RNA rather than the presence of preformed nucleic acid. Landman and Spiegelman[50] studied the formation of β-galactosidase in protoplasts of *B. megaterium* and showed that enzyme development was impaired by removal of RNA from the preparation.

All these effects can be interpreted to mean that there is some connection or relationship between RNA and the protein synthesizing capacity of the cell. It is difficult to make more detailed or quantitative deductions especially since we have learned of the existence of several types of RNA within the cell. The analog experiments described above place emphasis on sRNA components, but, since these form only 10–20% of the cellular RNA, it is difficult to see how alterations in the sRNA content or composition could account for some of the grosser effects described. Mendelsohn and Tissières[251] find that changes in the RNA content of bacterial cells with growth phase can be related to the amount of ribosomal particles in the cells. The recent description of specific RNA and related ideas about "messenger RNA" (see Section XIII) provide a third site of control where the action of analogs might be especially effective. Peabody and Hurwitz[252] find that the induction of β-galactosidase in the presence of lactose in *E. coli* spheroplasts is accompanied by incorporation of radioactivity from C^{14}-labeled glycine into both protein and nucleic acid; however, when induction is produced by thiomethylgalactoside instead of lactose, then incorporation takes place into protein but not into nucleic acid. This would appear to be a case where the synthesis of new protein is not mandatorily coupled to nucleic acid synthesis. The whole concept has recently been rendered far more complicated by developments (dealt with elsewhere in this treatise) which suggest that induction of specific enzyme synthesis arises from a release of repression and that repressors may themselves be nucleic acids. Nucleic acids may therefore exert a more subtle form of control than can be indicated by the relatively crude type of experiment described in this section. Whatever the ultimate interpretation, such experiments have served their purpose in drawing the attention of biochemists to the fact that there is a relationship between RNA and protein synthesis and so leading, maybe indirectly and inaccurately, to the developments described below.

C. Dependence of Enzyme-Forming Potential on Deoxyribonucleic Acid

The studies outlined above have been interpreted to mean that RNA is in some way involved in the synthesis of proteins. Evidence that DNA is also concerned comes from rather different approaches; these will be mentioned briefly as they are discussed in detail elsewhere in this treatise. Three main lines of evidence can be cited: the importance of DNA as a component of chromosomes, the DNA-nature of transforming principles, and the transfer of DNA as the operative factor in transduction by bacteriophage. In all cases it can be shown that alteration of the DNA constitution of the cell is followed by an alteration in the enzyme make-up of that cell or, alternatively, that a change in the enzymic properties of a cell brought about by a biological agency can be shown to be accompanied by a transfer of DNA in, or as a result of the action of, that agent. The deduction drawn from such studies is that DNA is concerned in the determination of the specific proteins which can be made by a cell.

IV. Practical Considerations

A. The Estimation of Protein

Protein can be estimated by its nitrogen content, and studies of general protein synthesis can be carried out by determination of the nitrogen content of the protein fraction of the cells. Difficulties are, however, encountered in separating the true protein fraction of bacterial cells. The usual method of obtaining a protein fraction is to treat the experimental system with trichloracetic acid at a final concentration of 5% (w./v.) which precipitates proteins and, if the treatment is carried out in the cold, nucleic acids. Nucleic acids are hydrolyzed by this concentration of trichloracetic acid at 90°C. so the precipitate can be freed from nucleic acid simply by holding it in the trichloracetic acid solution at 90°–100°C. for 10–15 minutes. The protein is then contained in the hot trichloracetic acid precipitate. However, bacterial cell walls are insoluble in aqueous solutions and are also included in such hot trichloracetic acid precipitates made from cells or cell fragments. Consequently the nitrogen content of such precipitates includes the nitrogen of wall peptides and amino sugars in addition to true protein. The difficulty can be overcome by estimation of an amino acid which is not present in the wall peptides or by procedures which result in the selective removal of wall substances from the precipitates. In selected cases, such as *Bacillus megaterium* or *Micrococcus lysodeikticus*, wall material can be digested by the use of lysozyme. Hancock and Park[51] have devised a method whereby the true protein of the hot trichloracetic acid precipitate is digested by trypsin, leaving only the wall structures in the precipitate (although there is also a possibility that trypsin-resistant protein from the protoplast membrane may remain in the precipitate).

Determinations of protein nitrogen or amino acids give a measure of the total protein in a cell and can be used for studies of general protein formation. It does not follow that the synthesis of all proteins takes place by the same mechanism; possibly the most intriguing aspect of protein synthesis relates to the production of specific proteins and the biosynthesis of specific amino acid sequences. Enzymes are specific proteins and, under carefully controlled conditions, the activity of an enzyme can be used as a measure of its quantity. Quantitative estimations of enzyme activity are frequently used as measures of the synthesis of the corresponding specific protein, although care has to be taken that an increase in activity is due to an actual increase in enzyme and not to an unmasking, activation, or decrease in inhibition of an enzyme already formed. In cases of constitutive enzymes, synthesis can be followed by an increase over an initial value and such increases may not be spectacular. Consequently, studies of inducible enzyme formation, where the initial activity may be zero or negligible, offer what appears to be an excellent method of following specific enzyme synthesis. Again it does not follow that findings made with the formation of a single inducible enzyme can be taken as characteristic of the synthesis of all proteins. In fact, there are indications to the contrary[52] and the division of microbial enzymes into constitutive and inducible may itself be based upon a difference in synthetic mechanisms. The problem of whether induction is peculiar to certain types of enzyme or is a general phenomenon made apparent in certain cases by the condition of the biological material used is itself ancillary to the general one of the mechanism of protein synthesis and will be dealt with in another chapter.

The magnitude of the increase in protein content of a biological preparation, especially a subcellular one, may be so small that the increase comes within the experimental error of many chemical estimations. This difficulty can be overcome, with deceptive ease, by the use of isotopes where minute amounts of reaction can be readily estimated, especially in the case of radioactive isotopes. The sensitivity of this method is, however, so great that results may be obtained which are due to unsuspected reactions rather than to the process under investigation. It has been known for many years that incubation of an isotope-labeled amino acid with a biological preparation is frequently accompanied by fixation or incorporation of the isotope into the preparation. By appropriate techniques it is possible to separate the protein fraction and demonstrate that the incorporated radioactivity relates to amino acid residues within the polypeptides of the cell proteins. Many pioneer investigations of this nature have been carried out by Borsook[53] and his colleagues. Questions that arise are: (*a*) Is all the incorporated radioactivity truly associated with, and part of, the protein of the cell, and (*b*) do the incorporated residues enter the protein structure by the normal mechanism of protein synthesis?

The first question is of importance from a methodological standpoint since, if incorporation of radioactivity can relate to substances other than protein, then methods must be worked out whereby the nonprotein contamination can be estimated. Gale and Folkes[54] obtained incorporation of single amino acids, such as glutamic acid or glycine, when staphylococci were incubated with glucose and one amino acid; incorporation was associated with the hot trichloracetic acid precipitate and was only partially inhibited by chloramphenicol at concentrations which completely inhibit protein synthesis as measured by increase in bound L-glutamic acid or in enzyme activity. It was then found that the precipitate included cell wall peptides (which include glycine, L-lysine, D-glutamic acid, and DL-alanine in the staphylococcus) and that incorporation into such peptides takes place under simpler conditions than those required for protein synthesis.[51,55] The incorporation of amino acids into wall peptides is not inhibited by chloramphenicol. If the wall and protein fractions are separated, then it can be shown that optimal incorporation into protein requires a full complement of amino acids and is 90% inhibited by chloramphenicol at growth-inhibitory concentrations.[56] Optimal incorporation into wall peptides takes place when the incubation mixture contains the four amino acids of the wall substance together with glucose. When only one amino acid is present in the incubation mixture, incorporation may involve precursors of both protein and wall peptides.[56, 57] The studies of Hoagland[58, 59] and his colleagues have shown that the soluble RNA of mammalian cells will react with, and also incorporate, amino acids to form complexes precipitated by cold trichloracetic acid but decomposed by hot trichloracetic acid; consequently the procedure adopted for the removal of nucleic acid from trichloracetic acid precipitates should also result in removal of amino acyl complexes of soluble RNA. A less readily controlled experimental artifact has recently been indicated by Castelfranco[60, 61] and his colleagues, who have shown that amino acyladenylates (believed to be the products of the amino acid activation enzymes, see below) are highly reactive substances and can react nonenzymically and nonspecifically with both proteins and nucleic acids to give rise to "incorporation" reactions very similar to those described in biological studies. From these considerations it would appear that an amino acid may be fixed by a cell preparation in a number of ways including: (*1*) incorporation into proteins; (*2*) incorporation into nonprotein peptides (as cell walls); (*3*) nonspecific adsorption by proteins; (*4*) binding by soluble RNA and related substances; (*5*) nonspecific binding by RNA and related substances; (*6*) binding by precursors of protein and nucleic acids. Consequently, interpretation of a simple incorporation experiment may present many difficulties. Interpretation can be greatly helped if the true chemical nature of the substance estimated is known (e.g., protein, nucleic acid, polynucleotide, peptide). In investigations of protein synthesis, it is highly desirable that the incorporation of amino acids should be accompanied by an increase in protein content measured by some other method. Confirmatory evidence is supplied if (*a*) the incorporation of a single amino acid is dependent upon (or at least greatly increased by) the presence of the other amino acids necessary for protein synthesis, and (*b*) the incorporation is inhibited by chloramphenicol at growth-inhibitory concentrations. Ideally, studies of protein synthesis should be based upon estimation of a specific protein isolated in a pure condition and demonstration of the incorporation of labeled residues into that protein. A beginning along such lines has been made in the case of β-galactosidase of *Escherichia coli*[62, 63] and of hemoglobin synthesis by reticulocyte preparations,[133] but any general investigation on such a basis is scarcely feasible at present.

These warnings concerning the interpretation of experiments involving the use of isotopes do not, of course, mean that the technique is useless. We tend to concentrate upon the nature of the end product, the pure protein, and endeavor to plan experiments which will give us clear results in terms of that end product. There are undoubtedly many intermediate stages between free amino acids and finished proteins and the proper interpretation of experiments using isotopes may prove instrumental in tracing and elucidating these stages.

B. Biological Preparations Used for Studies of Protein Synthesis

1. Intact Cells

Washed suspensions of bacteria can synthesize proteins when incubated under suitable conditions, although it may be difficult to distinguish between an increasing protein content of the cells and "growth" in the sense of a general increase in all the cellular constituents leading eventually to cell division. Many investigations have been made on the synthesis of enzymes, especially inducible enzymes, in intact cell suspensions and much of the information quoted in this chapter on the use of inhibitors and mutant blocks has been obtained in studies with such suspensions. Difficulties in interpretation arise in cases where an activity may be "cryptic"[64, 65] and an apparent development of activity be due to loss of crypticity rather than to enzyme synthesis. Most investigators recognize that the crux of the protein synthesis problem lies in the relationship between proteins and nucleic acids. The existence of the relationship has been established, as described above, largely with studies on intact cells, but it is doubtful whether the effect of modifying the nucleic acid components can be studied in such cells. Nucleic acid is usually removed from biological preparations either by extraction with suitable salt solutions or by digestion with nucleases; the presence of an osmotic barrier in the intact cell renders these approaches ineffective. Brachet[66] has presented evidence that some cells are permeable to ribonuclease, and Gros[245] finds that intact cells of *E. coli* can be depleted of RNA by treatment with ribonuclease in the absence of salt. The necessity of removing and replacing nucleic acid has led investigators to search for subcellular preparations which are able to synthesize proteins.

One specific synthesis accomplished by intact cells of certain species is flagellin synthesis. The flagella of organisms such as *Proteus vulgaris*, *Salmonella typhimurium*, and *Bacillus subtilis* are composed of protein, usually called flagellin.[67] If flagellated organisms are submitted to the action of a rotary blender, the flagella are stripped off and can be separated from the cell bodies.[68] If the latter are now incubated in a suitable medium, they will regenerate flagella[69, 70] and the process can be used to investigate the synthesis of flagellin.

2. Protoplasts or Spheroplasts

The subcellular preparation most closely related to the intact cell is the protoplast in which the cell wall has been removed, or the spheroplast in which the wall structure has been weakened ar attenuated.[71] Such preparations possess most, if not all, of the metabolic and biosynthetic mechanisms of the intact cell and can, under appropriate conditions, synthesize protein and nucleic acid.[72] Since they possess an intact membrane, they do not appear to offer any advantages over intact cells from the standpoint of investigation of protein-nucleic acid relationships. Protoplasts are unstable and easily rupture if placed in hypotonic solutions or treated with enzymes such as trypsin, lipase, or ribonuclease.[73] The finding that protein synthesis in protoplasts is stopped by treatment with ribonuclease[74, 75] may mean nothing more than that the structure of the protoplast is destroyed and that lysates are inactive;[76] before it can be claimed that RNA is essential to the process it is necessary to show that the structure has not been destroyed and that synthesis can be restored by adding back the extracted RNA. Spiegelman[77] discovered that protoplasts can be stabilized so that they are not damaged by incubation with ribonuclease and has thus been able to study the effect of nuclease treatment on metabolic processes.

3. Disrupted Cells

The processes of protein and nucleic acid synthesis seem to require a degree of organization in the biological preparation and many methods of breaking bacterial cells result in loss of their ability to synthesize protein. Gale and Folkes[49] have described preparations of *Staphylococcus aureus* consisting of cells torn open by exposure to supersonic vibration at 25 kc./sec. These disrupted cells retain the ability to synthesize protein as measured by the development of enzymes, increase in protein nitrogen, and incorporation of labeled amino acids. Electron micrographs of the active units show that they consist of cells in which the outer membrane and wall structures have been torn and a proportion of the cell contents escaped. If the supersound treatment is continued until the cells have been totally broken, activity is lost.

4. Cell Fragments

Beljanski and Ochoa[78] have described a preparation of broken fragments of *Alcaligenes faecalis* which is capable of incorporating amino acids and is said to consist of pieces of cell membrane. It is not clear whether the fragments also contain wall material but, in view of the close structural relationship between wall and membrane in the Gram-negative organisms, it seems highly probable that this is so. Spiegelman[79] has overcome the difficulty of the presence of wall material in such preparations by preparing protoplasts (or spheroplasts) of *Escherichia coli*, disrupting the protoplasts, and obtaining preparations of membrane fragments by differential centrifugation. Nisman[80, 81] and his colleagues have made a somewhat similar preparation but have treated the fragments with digitonin and aqueous phenol to effect further purification. Rogers and Novelli[253] have investigated methods of breaking protoplasts which will give minimal contamination with whole protoplasts at the same time as significantly active extracts. Of the methods tested—water lysis, Hughes' press,[254] and French pressure cell[255]—they find that the Hughes' press used at −196°C. is the most effective.

5. Ribosomes

The cells of higher organisms display internal differentiation into organelles that have been described as nuclei, mitochondria, microsomes, and, in some cases, a system of reticular membranes. Methods have been worked out whereby such cells can be broken and the various organelles separated in a reasonably pure condition. Investigation of the biochemistry of the separated cell components shows localization of certain metabolic processes. Thus, in the liver cell, enzymes of the tricarboxylic acid cycle are localized in the mitochondria, enzymes of the glycolytic system are found in the soluble fraction of the cell contents, and amino acid incorporation is found to relate to the microsome fraction, although it is now known that mitochondrial preparations also possess limited ability to incorporate amino acids. Very considerable advance in our knowledge of protein synthesis has come from studies with the microsome fraction of mammalian cells. This fraction consists of particulate material sedimented at high speed and composed of ribonucleoprotein together with a variable amount of lipid. The lipid content may be as high as 40% but probably depends upon the source and method of preparation of the microsome fraction.

Electron micrographs of sections of bacteria show that the cytoplasm contains closely packed granules of diameter 10–30 mμ. If cells are broken, granular material can be separated by high speed centrifugation from the cytoplasmic contents. Schachman *et al.*[83] first showed that particulate material could be separated from disintegrated bacteria of various species and that the bulk of the RNA of the cell was associated with granules, 10–15 mμ in diameter, sedimenting at 40 Svedberg units. Later investigations by Tissières and Watson[84] showed that the state of aggregation of the granular material is markedly affected by the concentration of Mg ions in the extracting fluid. In high Mg concentrations, particles are obtained which sediment at 100 *S*; as the Mg concentration is decreased, disaggregation occurs to give particles separating at 75, 50, 35, and lower *S* values. The 100 *S* particle appears to consist of a dimer of 70 *S* particles while the 70 *S* structure is built up from two dissimilar particles of 30 and 50 *S*, respectively.[233, 251] The association of the two smaller particles to form the 70 *S* structure can be seen in the electron micrograph studies of ribosomes by Huxley and Zubay[234] (see Fig. 3). All the ribosomal particles from *E. coli* consist of approximately 63% RNA and 37% protein[233]; *in vivo* the particles are highly hydrated and approximately 80% of their volume consists of water.[256] A number of other substances have been described as trace components of ribosomes and these include the polyamines putrescine, spermine, spermidine, cadaverine, and 1,3-diaminopropane.[257] Cohen and Lichtenstein[258] find that some 15% of the cellular putrescine and spermidine is present in the ribosomes in a condition which does not allow exchange with these bases when added to the medium. The authors suggest that spermidine and magnesium are responsible for maintaining the structure of the particle.

The protein moiety of ribosomes appears to be a specific structure which

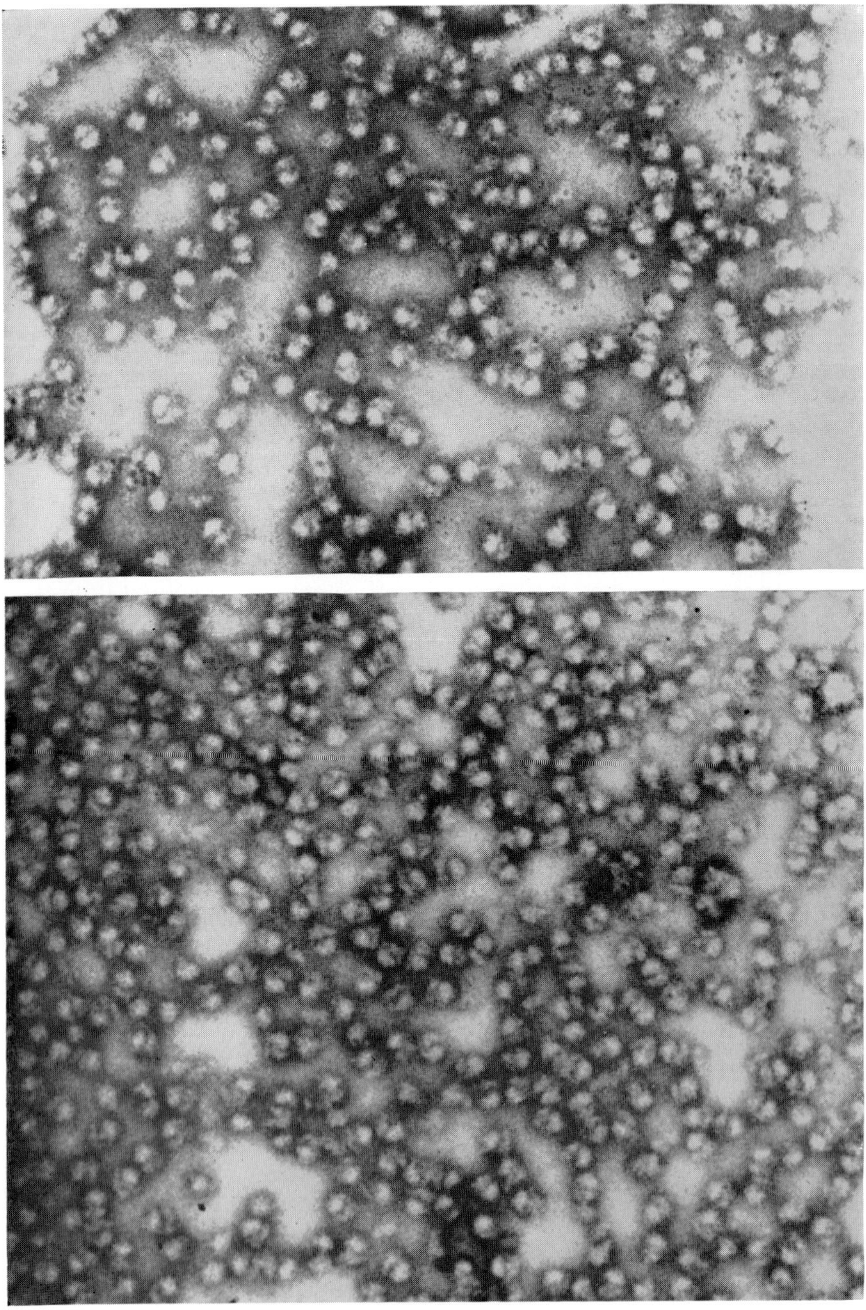

FIG. 3. Electron micrograph of preparation of 70 S particles from *E. coli*; negative staining with phosphotungstic acid. Magnification: × 220,000. (From Huxley and Zubay.[234])

is similar for ribosomes from different species.[241, 259] The protein is basic and somewhat similar in composition to histone.[260] Waller and Harris[261] have investigated the proteins from 30, 50, and 70 *S* ribosomes of *E. coli* and find that they fall into two major groups with N-terminal residues of either alanine or methionine. These two amino acids account for 85% of all the N-terminal groups (methionine 46–49%, alanine 36–40%) but smaller amounts of serine, threonine, aspartic acid, and glutamic acid were also found in the N-terminal position. The distribution of N-terminal groups was not significantly different in particles of different size. These results suggest that ribosomal protein is not a random sample from total cellular protein but forms a class of specific basic proteins. Fractionation of protein from 70 *S* ribosomes showed that each of the major groups (with either alanine or methionine as N-terminal residue) contained a number of different polypeptide structures. It was calculated that the average molecular weight of the constituent protein chains is 25,000.

Early preparations of ribosomes were found to possess dehydrogenase and electron transport enzyme systems but the occurrence of many of these activities decreased as preparative procedures improved, suggesting that the enzymic activities might have been due to contamination of the ribosomes with membrane fragments. Elson[262, 263] finds that essentially all of the cellular ribonuclease of *E. coli* is present in a latent form in the ribosomes and that the enzyme becomes active when the structure of the ribosome is destroyed. Evidence was obtained that the enzyme is associated exclusively with 30 *S* ribosomes.[264] A peptidase splitting prolylglycine was also found in *E. coli* ribosomes. If ribosomes are the site of synthesis of enzymes (see below), then traces of other enzymes in the course of synthesis might be expected in ribosomal preparations and could account for reports of small amounts of other enzymes in latent or partially latent form.[265, 447]

Controversy exists concerning the true nature of the 30, 50, and 70 *S* ribosomes and the extent to which these could be artefacts produced by the methods of preparation. Hanzon[266] and Sjöstrand[267] suggest that the particles result from precipitation of nucleoprotein gels during preparation and do not exist as such *in vivo*. Bowen *et al.*[268, 269] have shown that the size of the particles extracted from bacteria can be markedly affected by the salt concentration in the growth medium, the method of breaking the cells, the metabolic state of the cells, and the extraction medium used during separation of the particles. They suggest that the use of magnesium acetate to stabilize the particles during extraction may give rise to aggregates which do not exist in the cells before extraction. They have investigated this possibility by examining the whole crush obtained by breaking either *E. coli* or *B. cereus* in the Hughes' press in the presence and absence of added magnesium acetate. They find no evidence for the presence of other than a very small amount of 70 S particles in cell juice in the absence of added magnesium although 30 and 50 S boundaries can be clearly identified. Treatment of yeast cells under the same conditions shows the presence of 80 S boundaries in the cell juice, indicating that the two cells have a different "particle organization" and that the method will show the presence of boundaries greater than 50 S if they exist in the cells. Their conclusion is that the use of Tris buffer containing magnesium acetate as extraction and dilution fluid may give misleading results.

C. Contamination of Preparations

Bacteria are able to incorporate amino acids at such a high rate that contamination of subcellular preparations with intact cells can form a major experimental hazard, especially in those cases where preparations are made from suspensions of intact cells initially at high cell density. Where preparative methods involve breaking such cell suspensions and then separation of fractions by differential centrifugation, some contamination of fractions by intact cells is inevitable while measures which involve resuspension and refractionation may be undesirable if the fraction required is itself unstable or of short useful life. Use is frequently made of the fact that undamaged cells are unable to utilize ATP as a source of energy and attempts are made to demonstrate, first, that the process studied is ATP-dependent and, second, that the amount of intact cell contamination does not substantially contribute to the results recorded for the "cell free preparation." Some of the earlier attempts along these lines were those of Gale and Folkes[49] using preparations of *Staphylococcus aureus* disrupted by exposure to supersound. They reported a drop in viable count from 10^{10} to 10^6 cells per milligram dry weight of preparation and showed that this proportion of intact cells could contribute 0.01 % of the ATP-dependent incorporation of amino acids. With improved disintegration technique, the intact cell contamination was later reduced to 10^4 to 10^5 cells per milligram dry weight. Rogers and Novelli[270, 271] studied the synthesis of ornithine transcarbamylase by protoplasts of *Bacillus megaterium* and obtained significant synthesis after water lysis of the protoplasts. Such lysates were found to contain intact protoplasts and investigation showed that these were about seventimes more active in the lysate than in normal suspension; consequently the full activity of the water lysates was attributable to contamination with intact protoplasts. More satisfactory preparations were obtained by breaking protoplasts in the Hughes' press at −196°C. Protoplasts were still present in the preparations but the enzyme synthesis was ATP-dependent and calculation showed that quantitatively it was in excess of the possible contribution of intact protoplasts. The authors suggested that experiments of the nature described could be satisfactorily carried out if the intact protoplast count was not more than 10^6 per milligram protein. The same considerations apply to bacterial ribosome preparations and Lamborg and Zamecnik[272,273] state that the whole cell contamination must be less than 10^6 cells per milliliter if results of cell-free activity are to be assessed. The fact that activity does reside in the ribosome preparation can be shown by dilution of the preparation to a level where whole cell contamination is negligible.

Intercontamination of cell fragments further complicates experiments on localization of activity and seems to be particularly difficult in bacterial preparations. Mention was made above of the enzymic activities described in early preparations of bacterial ribosomes, activities now ascribed to contamination of these preparations with membrane fragments. Likewise it seems possible that some of the confusion regarding the roles of bacterial membranes and ribosomes in protein synthesis is due to contamination of preparations of membranes with ribosomes and, since it has been reported that membranes can break up into particulate material,[274] of preparations of ribosomes with membrane particles.

V. The Site of Protein Synthesis in the Cell

A. Mammalian Cells

Administration of radioactive amino acids to animal tissues, whether in the form of whole animals, tissue slices, or cell homogenates, is followed by incorporation of radioactive residues into the protein of the preparation. Cells can be broken and separated into fractions containing nuclei, mitochondria, microsomes, and soluble protein. Attempts to decide in which of these fractions amino acid incorporation first takes place have been made by "pulse" experiments, i.e. a small dose of highly radioactive amino acid is given to the preparation and then, starting as soon as practicable, samples of the preparation are taken, broken, separated into the various fractions, and the specific activity of each fraction determined. In such experiments with liver tissue and with reticulocytes,[85, 86] it has been shown that the amino acid is rapidly taken up by the microsome fraction and more slowly into the soluble protein. If the initial dose of radioactivity is small, it is found that the specific activity of the microsome fraction rises rapidly to a maximum and then decreases as activity appears in the soluble protein fraction, suggesting that incorporation first takes place in the microsome and that the radioactive residues are then transferred or released in the soluble protein fraction. From such evidence, it has been deduced that incorporation into protein takes place in the microsomes and this deduction has been supported by the demonstration that isolated microsomes can, under appropriate conditions, incorporate amino acids *in vitro*.[85, 87] The microsomes from liver contain lipid material but the ribonucleoprotein particles can be separated by treatment with digitonin and such ribosomes retain the ability to incorporate amino acids.

B. Bacterial Cells

The situation is not so clear-cut in bacteria. Until 1959 most of the preparations used for the investigation of protein synthesis in bacteria had been crude and complex compared with the microsomal preparations used in mammalian studies. Broken protoplasts and disrupted cells undoubtedly contain ribosomes as well as membranes. Pulse experiments with *B. megaterium* and *E. coli*, followed by separation of wall, membrane, and cytoplasmic fractions, suggested that the membrane might be the primary site of amino acid incorporation in this organism.[70, 88] Beljanski and Ochoa[78] obtained incorporation by "membrane fragments" of *Alcaligenes faecalis*. Spiegelman[79] fractionated protoplasts of *Escherichia coli* into membranes, ribosomes, and soluble protein, and Connell *et al.*[89] separated membranes and ribosomes from *Azotobacter*; both reported that the membrane fractions were highly active in incorporating amino acids whereas the ribosomes showed comparatively little activity. Spiegelman therefore suggested that only the membranes are capable of protein synthesis. However, protoplast membranes appear to be devoid of nucleic acid,[90, 91] so if they do play a role in protein synthesis in the bacterial cell, they must presumably do so as a result of some relationship with cytoplasmic or ribosomal components.

One of the reasons for the early inability to demonstrate a primary role of ribosomes in bacterial protein synthesis appears to rest in the time scale of the experiments. McQuillen, Roberts, and Britten[275] found that, if experiments were carried out for seconds rather than minutes, incorporation of radioactivity from $S^{35}O_4^{=}$ into the ribosomal fraction could be demonstrated and that the highest specific activity resided in the 70–85 *S* ribosomes. The specific activity of these 70 *S* particles rose rapidly during the first 20 seconds at 37°C. and thereafter remained approximately constant. The specific activity of 50 *S* particles rose more slowly and eventually reached a higher value than that of the 70 *S* particles. Experiments in which a short incubation (15 seconds) was followed by rapid dilution with unlabeled substrate showed that the radioactivity acquired by the 70 *S* particles during the initial 15 seconds was subsequently transferred to the soluble protein fraction. Clear cut results of this nature were obtained with $S^{35}O_4$ as radioactive source and the main conclusions were confirmed using C^{14}-labeled amino acids. The workers pointed out that, in dealing with bacterial systems, two types of synthesis are involved: synthesis of the ribosomal nucleoprotein itself, and synthesis of soluble proteins in the ribosome. The short time scale required to demonstrate the latter is a reflection of the difference in rate of the former between bacterial and animal tissues. Mc-

Quillen *et al.*[275] calculated that a single cell of *E. coli* contains approximately 6000 70–85 *S* ribosomes and 6×10^6 units of soluble protein of molecular weight 10,000. Since the growth rate at 37°C. under the conditions of their experiments was 0.02% per second, this corresponds to the synthesis of 1.2 ribosomes and 1200 protein units per second. Further, on the assumption that the 6000 ribosomes represent 6000 sites of protein synthesis, the time required for the synthesis of a complete protein unit works out at 5 seconds.

Once the ability of bacterial ribosomes to incorporate amino acids had been demonstrated in this fashion, other workers quickly followed with experiments showing that isolated ribosomes can incorporate amino acids in a manner similar to that reported for mammalian and plant RNP particles.[272, 273, 276, 277] The emphasis lies on the 70 *S* particles although Tissières *et al.*[277] find that it is only a proportion of these, which they call "active ribosomes," that are effective *in vitro* and can incorporate 15–40 times as much amino acid as the 30 and 50 *S* particles. It now seems that a system, analogous to that found in animal cells, exists in bacteria (although experiments have so far been carried out with very few species) and brings about the incorporation of amino acids into protein present in ribosomes. Consequently all the investigations on mammalian and plant systems are relevant to discussions of protein synthesis in bacteria and there would seem to be little point, at this stage in our knowledge, in treating the systems separately.

McQuillen *et al.*[275] were unable to demonstrate a release *in vitro* of radioactivity from particles labeled *in vivo*. Tissières *et al.*[277] found that radioactivity associated with "active ribosomes" could be transferred to the soluble protein after dilution of the ribosome preparation in unlabeled amino acids but that the cell-free synthesis obtained accounted for about 1% of the protein synthesizing activity of the intact cell. They suggested that a factor, not present in the *in vitro* system, acts within the cell to maintain the activity of the 70 *S* particles and to mediate the release of soluble protein from them. The question whether membranes play a role in protein synthesis is relevant in this connection. Lamborg and Zamecnik[273] state that fragments larger than ribosomes are not necessary for amino acid incorporation by bacterial extracts while McQuillen *et al.*[275] obtained some indication that ribosomes associated with membranous material (after lysozyme treatment followed by freezing and thawing) attained a higher specific activity than those released in a free condition.

Possible confusion may arise in this connection if the membrane is also the site of wall synthesis, as suggested by Crathorn and Hunter.[92] The syntheses of cell protein and wall peptide can be distinguished in the Gram-positive cell by the simpler nature of the wall substance together with its content of D-amino acids and the insensitivity of its synthesis to chloramphenicol. The Gram-negative cell wall appears to contain "complete" protein (see Section I, C, 2) and, so far, no method has been devised which will distinguish between the syntheses of cell protein and wall protein in the Gram-negative cell; it is not known, for example, whether chloramphenicol inhibits the synthesis of the Gram-negative cell wall. Consequently, results involving the incorporation of amino acids by membrane preparations from Gram-negative cells are, at present, difficult to interpret and require, more than ever, confirmatory evidence that true proteins (and/or enzymes) are being formed.

VI. Components of the Protein-Synthesizing Mechanism

In the preceding section, evidence was given which indicates that the ribosomes of mammalian and bacterial cells appear to provide the sites at which amino acids are incorporated into peptide bonds. The incorporation process requires the presence of components other than the amino acid and the ribosome preparation. It is not yet known whether the same components are required by all ribosome preparations. It seems best, therefore, to list the various requirements as separate components and then to discuss the hypotheses that are put forward to explain the nature of the mechanisms involving the components; later work may well disprove the hypotheses and alter the interpretations placed upon the roles of the components.

A. Energy Source

The synthesis of proteins, development of inducible enzymes, and the incorporation of amino acids into protein all require the provision of energy. With intact cell suspensions, energy is usually obtained by metabolism of utilizable carbohydrate; subcellular preparations may not be able to attack carbohydrate but can utilize adenosinetriphosphate (ATP) as energy source. ATP may be supplied direct or, since many preparations contain phosphatases destroying ATP, may be generated within the system from hexosediphosphate or other glycolytic substrates. Protoplasts may sometimes form an exception since Beljanski[75] has reported that amino acid incorporation by protoplasts is inhibited by the addition of ATP to the incubation mixture. If energy metabolism is inhibited or "uncoupled" by the presence of agents such as 2,4-dinitrophenol, then protein synthesis stops—this also applies to the protoplast system.

B. Amino Acids

Proteins are composed of some twenty different amino acids, all of which must be available for the synthetic process. The synthetic abilities of bacteria toward amino acids vary widely and an organism such as wild-type *Escherichia coli* is able to effect a rapid synthesis of all the amino acids if it is supplied with ammonia and glucose as nitrogen and carbon source, respectively. Organisms, such as strains of *Lactobacillus*, *Streptococcus*, *Staphylococcus*, etc., which have lost the ability to synthesize many of the amino acids, will require the addition of these amino acids to the medium before protein synthesis can proceed. In the same way, mutants of *E. coli* which are blocked in the synthesis of an amino acid are unable to synthesize protein unless this amino acid is supplied. Consequently, organisms blocked in the synthesis of one or more amino acids can be used to test whether an activity—such as the development of an enzyme activity or the incorporation of an amino acid—can be ascribed to protein synthesis or not.

If an organism is not supplied with an amino acid which it cannot synthesize, there arises the possibility that it could obtain that amino acid by the breakdown of protein already formed and that new proteins could be formed at the expense of the breakdown of old ones—the process usually called "turnover." A number of investigations have been designed to test the degree to which turnover takes place in bacterial cells. Hogness and co-workers[62] and also Rotman and Spiegelman[63] have studied the synthesis of the inducible enzyme, β-galactosidase, from this standpoint. The principle of their experiments was as follows: cells were grown in a medium which resulted in a labeling of the protein of those cells; the cells were then transferred to an unlabeled medium and induced to form β-galactosidase; conversely, cells were grown with unlabeled protein and then induced to form enzyme in a medium containing labeled amino acids. In both types of experiment, the β-galactosidase was isolated in a highly purified condition and its specific activity determined. The results indicated that less than 1% of the newly formed enzyme could be formed at the expense of cellular protein formed before induction took place, and that β-galactosidase was synthesized *de novo* from amino acids. Other workers have labeled the protein of cells and studied the release of label during subsequent incubation under various conditions; results vary with the organism and the conditions but agree in showing that the rate of protein breakdown is small compared with its rate of synthesis, and is less than 3% per hour in exponentially growing cells.[93a, 93b, 278, 279] An interesting exception to this general rule is the massive breakdown of cellular protein that precedes sporulation in *Bacillus thuringiensis* and provides the amino acids from which the crystalline protein inclusions are formed.[94] Mandelstam and Halvorson[280] have investigated the turnover in the ribosomes and soluble fraction of *E. coli*. They find a balanced degradation and resynthesis of ribosomal protein at a rate of 5% per hour. The ribosomal RNA is degraded at 5% per hour and resynthesized at 2½% per hour whereas RNA synthesis in the soluble fraction proceeds more rapidly than degradation so that there is an apparent transfer of RNA from the ribosomes to the soluble fraction.

As pointed out in Section I, B, some proteins are incomplete is that they do not contain the full range of twenty amino acids, and the question arises whether such incomplete proteins can be synthesized under more restricted conditions than those required for general protein synthesis. The regeneration of flagella by *Salmonella typhimurium* provides a system in which this can be tested. Flagellin (see Table I) does not contain tryptophan or cysteine. Kerridge[70, 281] has investigated the synthesis of protein and the regeneration of flagella by *Salmonella* mutants blocked in the synthesis of amino acids. A leucineless mutant is unable to synthesize protein or flagellin in the absence of added leucine but tryptophanless or cysteineless mutants can regenerate their flagella in the absence of the blocked amino acid, although no synthesis of cellular protein takes place. Flagellin contains amounts of histidine and methionine that correspond to only one residue per protein molecule of weight 30,000. Mutants blocked in the synthesis of either histidine or methionine can, however, regenerate flagella but cannot synthesize protein in the absence of these amino acids. Kerridge suggests that the small amounts of histidine or methionine necessary for the formation of flagellin can be provided by protein turnover.

Amino Acid Analogs

Since the synthesis of a protein can take place only if all the amino acids in that protein are supplied to the synthesizing system, it would seem probable that the process would be stopped by the addition of an amino acid analog which would effectively compete with the natural substrate. Halvorson and co-workers[95-96] demonstrated that the formation of α-glucosidase by yeast could be inhibited by the presence of analogs, such as *p*-fluorophenylalanine or tryptazan, and the inhibition specifically released by the corresponding amino acid only. The addition of such analogs to the growth medium apparently prevents the growth of small inocula but Munier and Cohen[97, 98] found that if the inoculum is large enough for growth to be measured, then the presence of *p*-fluorophenylalanine does not prevent the growth of *E. coli* but renders it linear instead of exponential. They were able to show that the analog does not prevent protein synthesis as such but gives rise to proteins containing *p*-fluorophenylalanine residues. Such proteins are devoid of enzymic activity. Analysis of the protein from cells grown in the presence of *p*-fluorophenylalanine showed that it contained 23% less phenylalanine, 47% less tyrosine but the same amounts of valine and leucine/isoleucine as those in "normal" protein. It is possible that an analog such as *p*-fluorophenylalanine may not become incorporated into all proteins to the same extent; thus Munier and Cohen[97] state that a given concentration of the analog prevents adaptation to lactose, maltose, or xylose, but does not affect the formation of β-galactosidase. No incorporation of the analog took place in the presence of chloramphenicol.

Kidder and Dewey[99] had earlier suggested that the inhibition of growth of *Tetrahymena pyriformis* by 7-azatryptophan might be due to displacement of tryptophan in the protein by the analog. Gross and Tarver[100] demonstrated that ethionine can be found in the proteins of the organism after growth in the presence of this analog. Pardee and co-workers[101] found that azatryptophan will support the growth of a tryptophanless mutant of *E. coli* although growth would continue only to the extent that the protein and RNA and cell count doubled. Similar results were obtained with tryptazan but 5-methyltryptophan would not support growth. Azatryptophan was found in the proteins of cells grown in its presence and the amount present was compatible with the analog having replaced tryptophan in newly formed protein. There was no increase in the activity of a number of enzyme systems during growth in the presence of the analog.

Cohen[102] has extended these studies to the special case of valine-sensitive strains of *E. coli*. Again, the presence of valine at a toxic level results in linear instead of exponential growth. Although protein synthesis continues under these conditions, inducible enzyme formation ceases and there is no increase in the activity of respiratory and other enzymes of the cells. Analysis of the protein from the cells shows that it contains an unusually high content of valine. It appears that valine, presumably by virtue of its structural similarity to leucine, acts as an analog and gives rise to the formation of inactive proteins of abnormal constitution.

C. Amino Acid-Activating Enzymes*

The incorporation of amino acids by biological preparations requires the presence of ATP or an ATP-generating system and it seems probable that the amino acids must undergo some metabolic modification before they can enter the protein synthesizing system. The incorporation of amino acids by mammalian microsomes requires, in addition to ATP, the presence of factors which can be precipitated at pH 5 from the soluble protein fraction of the cell.[87] Hoagland *et al.*[103, 104] found that this "pH 5 precipitate" contained enzymes which would bring about a reaction between amino acids and ATP. If hydroxylamine was added to the reaction mixture, amino acylhydroxamates were formed and could be estimated colorimetrically. If labeled inorganic pyrophosphate was added to the

* According to the recommendations of the Enzyme Commission of the International Union of Biochemistry, these enzymes should, in the future, be called the amino acid:sRNA ligases or amino acid:sRNA synthetases.

reaction mixture, an exchange reaction took place which (a) resulted in labeling of the ATP and (b) was dependent upon the presence of amino acids. Incubation of the enzyme system with ATP and an amino acid labeled with O^{18} in the carboxyl group, followed by recovery of the amino acyl product as the hydroxamate, showed that the O^{18} appeared in the phosphate group of AMP and not in the pyrophosphate.[282, 283] This indicates that the reaction takes place through phosphorylation of the carboxyl group by the ribose-bound phosphate of ATP. The exchange reaction was then formulated as:

$$\alpha\text{-CHNH}_2\text{—COOH} + \text{PP} \begin{matrix} A \\ | \\ P \\ \backslash \end{matrix} \rightleftharpoons \alpha\text{-CHNH}_2\text{—COO} \begin{matrix} A \\ | \\ P \\ \backslash \end{matrix} + \text{PP}$$

Amino acid   ATP   Amino acyladenylate   Pyrophosphate

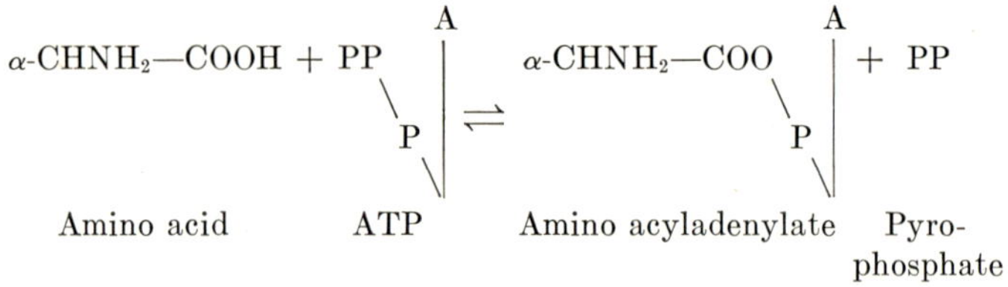

Amino acyl adenylates are highly reactive substances and so the enzymes responsible for the reaction were designated "amino acid-activating enzymes."

The amino acyl-AMP appears to exist in a form bound to the enzyme and proof of its existence has been obtained in two cases by the use of high concentrations of purified enzyme: tryptophan-adenylate[284-286] and serine adenylate.[287] Incubation of the tryptophan-activating enzyme at substrate concentration with tryptophan and ATP also gave a tryptophan ester of ATP itself (probably involving an ester link with the 2'- or 3'-hydroxyl of the ribose) but it was felt that this substance was of doubtful significance *in vivo*. The distribution of activity toward different amino acids suggests that a series of enzymes is concerned and it may be that a specific enzyme exists for the activation of each amino acid.[105, 106] The specificity of a number of activating enzymes has been confirmed by their isolation in highly purified condition; this applies so far to the enzymes activating tryptophan,[107] valine, methionine,[108] tyrosine,[288, 289] leucine,[290] alanine,[291, 292] serine,[293] and threonine.[294, 295] The reversibility of the last enzyme has been demonstrated and the equilibrium constant determined as 0.37 at 37°C. and pH 7; this indicates that the amino acyl adenylate bond is to be regarded as an "energy rich" bond. The tryptophan-activating enzyme will activate 7-azatryptophan and tryptazan but not 5-methyltryptophan[109] and it is of importance that the first two but not the last substance can be incorporated into proteins in biological systems.

The discovery[61] that synthetic amino acid adenylates can give rise to nonenzymic incorporation of amino acid residues into protein casts doubt on the interpretation of experiments involving "activated amino acids" and the importance of "activating enzymes" in protein biosynthesis. A series of investigations[297-300] has shown that enzyme-bound amino acid adenylates are transferred specifically and more efficiently to RNA than are free adenylates. Thus incubation of the tryptophan adenylate-enzyme complex with ribosomal RNA or sRNA is accompanied by a significant transfer of the tryptophan residue to RNA and this transfer is 100 times more efficient than the transfer of glycyl residues from glycyl-adenylate in the presence of the tryptophan-activating enzyme. If the enzyme complex is formed with labeled ATP, there is no transfer of the adenylate moiety to RNA. The non-enzymic incorporation of amino acyl adenylates into protein results in the transfer of the amino acid residue to N-terminal positions whereas incubation of tryptophan adenylate-enzyme complex with ribosomal particles results in incorporation of tryptophan into interpeptide bonds. Nevertheless, this latter incorporation was obtained in the absence of other amino acids or activating enzymes although it is, of course, possible that these might have been present in sufficient amonts in the ribosome preparations.

The tryptophan adenylate-enyme complex can be prepared by incubation of the specific activating enzyme with tryptophan and ATP, or with synthetic tryptophan adenylate. Catalysis of pyrophosphate exchange and hydroxamate formation by the enzyme is specific for tryptophan but ATP can be formed from some 16 other amino acid adenylates, including D-tryptophan adenylate. The enzyme is thus specific for the transfer of an amino acid residue but capable of transferring adenylyl residues from any amino acyl adenylate to pyrophosphate to yield ATP.

Table VI shows some typical results for the distribution of amino acid-activating enzymes in a variety of cells.[80, 110, 112, 113, 295] DeMoss and Novelli[111] examined the ability of extracts of bacteria to bring about the amino acid-dependent ATP-pyrophosphate exchange reaction and obtained active preparations from *Micrococcus*, *Aerobacter*, *Staphylococcus*, *Clostridia*, *Desulphovibrio*, *Proteus*, *Streptococcus*, *Serratia*, *Rhodospirilla*, *Azotobacter*, and *Escherichia* species. Detailed investigations were made in the case of extracts from *S. haemolyticus* and *E. coli*. These early results with bacteria were typical in that clear activation was obtained in the case of some eight of the naturally occurring amino acids only. There are two possible explanations of such results: (*a*) some amino acids are activated by a different mechanism or (*b*) that all amino acids are activated in this way but that the activating enzymes differ in stability so that the enzymes corresponding to some substrates are inactivated during preparation of the extract.

The latter explanation is borne out by the results quoted for pigeon pancreas,[110] guinea pig liver,[295] pea seedling,[112] and *E. coli* membrane[80] preparations where significant activation is obtained for all the amino acids tested (see Table VI). The results in some instances—such as glutamic acid activation in the mammalian systems—show an increase in pyrophosphate exchange only 20% greater than that in the control without added amino acids. In the case of the disrupted staphylococcus preparation, no significant increase in ATP-pyrophosphate exchange can be demonstrated in the presence of either glutamic or aspartic acids, yet both amino acids are rapidly incorporated by the disrupted cell preparation in the presence of ATP.[56, 113] Activating enzymes of the type described have been found in all cells and tissues so far examined for them but this does not rule out the possibility that other methods of preparing amino acids for later incorpora-

TABLE VI

ACTIVATION OF AMINO ACIDS BY PREPARATIONS OF ANIMAL, PLANT, AND MICROBIAL CELLS

Amino acid	Pigeon pancreas extract[a]	Guinea pig liver[b]	Pea seed extract[c]	Disrupted staphylococcus[d]	Membrane fraction of *E. coli*[e]	Yeast extract[c]
Alanine	2.9	0.53	5.9	0.81	2.6	0.3
Arginine	0.1	0.5	0.9	—	2.4	0.9
Aspartic acid	2.8	2.1	7.2	0	2.8	0.2
Cysteine	2.16	0.23	1.0	7.75	6.4	0.4
Glutamic acid	0.2	0.23	2.3	0	1.4	0.6
Glycine	0.6	2.1	0.3	1.14	0.9	0.1
Histidine	7.5	2.3	1.0	0.38	5.2	0.9
Isoleucine	1.37	16.1	2.2	11.4	13.6	3.0
Leucine	1.84	16.9	4.7	18.25	10.9	4.7
Lysine	2.3	3.5	12.0	0.4	7.6	1.9
Methionine	1.0	6.7	10.0	8.69	6.5	3.7
Phenylalanine	1.37	0.37	8.3	0.05	4.2	2.4
Proline	9.4	4.8	7.3		3.05	1.5
Serine	1.7	2.5	3.2	—	2.4	1.2
Threonine	5.7	3.2	5.4	—	4.9	0.8
Tryptophan	5.53	7.2	5.0	1.1	12.4	4.6
Tyrosine	2.8	1.8	7.6	3.23	8.5	8.8
Valine	3.43	1.3	6.5	14.23	29.7	2.6
Control	1.0	1.0	1.0	1.0	1.0	1.0

NOTE: In the original papers, results are expressed in somewhat different ways and in different units. To obtain some degree of comparison, the values have been recalculated and expressed in the table as the rate of activation (pyrophosphate exchange) relative to the control (without added amino acids) = 1.0.

[a] Data from Lipmann.[110]

[b] Data from Allen *et al.*[295].

[c] Data from Webster.[112]

[d] Data from Gale.[113]

[e] Data from Nisman.[80]

tion exist. There is, for example, a suggestion from the work of Ogata and Nohara[301-303] on the effect of ribonuclease on hydroxamate formation by the activating system of liver, that pyrophosphorylation of sRNA might sometimes be the first step, followed by substitution of an amino acid residue for pyrophosphate.

The activating enzymes so far described act on the free forms of L-amino acids as substrate. Baddiley and Neuhaus[304] have demonstrated the activation of D-alanine in *Lactobacillus arabinosus*, *Staphylococcus aureus*, and *Bacillus subtilis*. Relatively little attention has so far been paid to the activation of peptides. A rat liver preparation has been shown to activate di- and triglycine to the same extent as glycine[305] and also to activate glycylmethionine and alanylmethionine.[306] A pigeon liver preparation has been found to form hydroxamates of certain peptides in the presence of ATP and the activation of leucyltyrosine and glycyl-leucine so demonstrated. The formation of ATP by incubation of the preparation with glycyl-leucyl-AMP and pyrophosphate was also shown.[307]

D. Nucleic Acids and Their Precursors

1. Purines and Pyrimidines

Figure 4 shows the effect of purines and pyrimidines in an incubation mixture containing glucose and a complete amino acid mixture on the rate of synthesis of protein in washed suspensions of *Staphylococcus aureus*.[41] If the purines or pyrimidines are labeled, it can be shown that their addition to the medium leads to the formation of labeled nucleic acid in the cells and it can be deduced that their effect on protein synthesis is due to the increased nucleic acid content. Reference has already been made, in Section III, to the various investigations which have established a correlation between the rate of protein synthesis and the RNA content of the cells.

2. Nucleic Acids

Direct dependence of amino acid incorporation and protein synthesis on the presence of nucleic acids has been demonstrated in the disrupted staphylococcal cell preparation.[49] Nucleic acid can be removed from the preparation by incubation with nucleases or extraction with *M* sodium chloride; such extraction results in a loss of ability to synthesize protein or to incorporate amino acids. The ability to develop enzyme activity is restored by adding nucleic acid and/or purines and pyrimidines to the incubation medium; the components which must be added depend upon the enzyme whose activity is being studied and on the degree of nucleic acid depletion of the disrupted cell preparation. In partially depleted preparations, the development of catalase or glucozymase activity can be promoted by the addition of staphylococcal RNA; β-galactosidase formation requires the addition of purines and pyrimidines and will not respond to any RNA preparation so far tested. If depletion has resulted in the nucleic acid content of the preparation being reduced to less than 5% of the original value, the measures so far described often do not suffice to

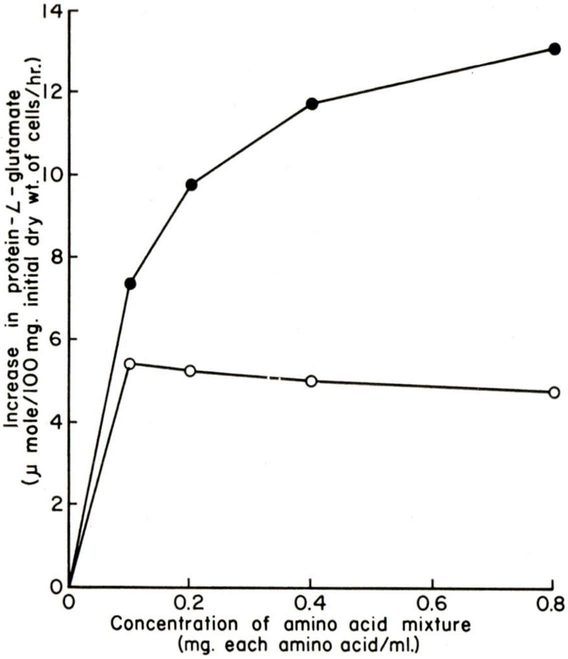

FIG. 4. Effect of purines and pyrimidines on the rate of synthesis of protein by *Staphylococcus aureus*.[41] Washed cells are incubated at 37°C. in the presence of glucose, a mixture of the 18 naturally occurring amino acids necessary for protein synthesis with (●) and without (○) the addition of a mixture of purines and pyrimidines at a concentration of 0.02 mg. each component/ml.

restore enzyme development and the further addition of DNA becomes necessary. The restoration of the ability to incorporate amino acids is accomplished by addition of either RNA or DNA to the incubation mixture, DNA being approximately twice as effective as RNA on a basis of the optical density at 260 mμ.

3. Soluble RNA

The pH 5 precipitate of the soluble fraction from liver cells (Section VI, C) contains ribonucleic acid of small molecular weight; Hoagland *et al.*[59, 116]

observed that when the fraction acts on leucine in the presence of ATP, the leucyl residue becomes attached to the RNA of the fraction to form an amino acyl ribonucleic acid complex. The reaction can be formulated in the following manner:

$$\alpha + \text{ATP} = \alpha\text{-AMP} + \text{PP}$$

$$\alpha\text{-AMP} + \text{sRNA} = \alpha\text{-sRNA} + \text{AMP}$$

where sRNA = soluble ribonucleic acid and α = an amino acid.† The action of the activating enzyme is to form an amino acyl adenylate which then reacts with the solùble ribonucleic acid and the amino acyl residue is transferred to the ribonucleic acid complex. The amino acyl residue is not combined through peptide bonds and the linkage is markedly more labile than that which binds the amino acids after incorporation by the microsomal particles. The amino acyl-sRNA complex is very labile to alkali, is 50% destroyed in 8 min. at pH 10 at 37°C., and is most stable over the pH range 3–6.[117] The soluble ribonucleic acid fraction can be prepared from the pH 5 enzyme fraction by shaking with aqueous phenol which precipitates the proteins.[59] The nature of sRNA is described in Section II, A. Incubation of the soluble RNA in the absence of amino acids or nucleoside triphosphates reduces its ability to accept amino acyl residues; this ability is restored by a second incubation in the presence of ATP and cytidine triphosphate (CTP). Hecht and others[118, 119] have obtained evidence that the polyribonucleotide chain must have a specific end group before it can accept amino acids; this end group can be represented sRNA—C—C—A where C = cytidine monophosphate and A = adenylic acid. Zachau *et al.*[120] have shown that ribonuclease digestion releases 2′ (or 3′)-leucyladenosine from leucyl-sRNA complex and it appears that the leucyl residue is attached through the 2′- or 3′-hydroxyl of the terminal adenylic acid residue†:

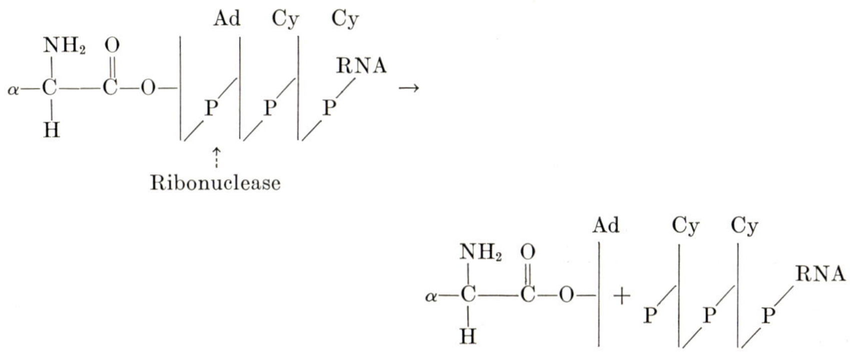

† The symbol α is used to represent an amino acid; where the nature of the bond is important α-$CHNH_2 \cdot COOH$ is used.

Similar results have been obtained for the leucyl-sRNA complex from *E. coli*.[121]

The question now arises whether there is any specificity between the polynucleotide acceptor and the amino acyl residue. The —C—C—A terminal group is apparently essential for the attachment of any amino acyl residue so that specificity, if it exists, must be determined by the structure of the polynucleotide further from the terminal group. Evidence for a specific relationship is: (*1*) If two or more radioactive amino acids are supplied at saturation concentrations, then the total incorporation corresponds to the sum of the radioactivities obtained when the amino acids are supplied one at a time.[117] (*2*) The acceptor activity of soluble RNA is lost if the terminal nucleotide is attacked by periodate; however, if the sRNA reacts with an amino acid, then the binding site for that amino acid is protected from destruction by periodate but the ability to react with other amino acids is destroyed.[121]

All tissues so far examined contain material which corresponds to the soluble RNA fraction, although the molecular size and degree of dissociation from other cellular fractions may vary with the tissue investigated. Thus Webster[122] finds that the acceptor material in pea seedling homogenates is associated with the microsomal fraction, whereas Hoagland,[123] working with pigeon pancreas, finds that the acceptor is a polynucleotide left in the supernatant after precipitation of the pH 5 enzyme fraction.

It seems probable that a specific sRNA acceptor exists for each amino acid. Column chromatography or prolonged centrifugation result in the removal of inactive material from sRNA[116, 308]; the active polynucleotides are referred to as "transfer RNA" to distinguish them from the inactive components. A number of attempts are currently being made to separate transfer RNA into components with acceptors for specific amino acids.[309-311] Zamecnik and co-workers[312-314] have adopted a trick which is dependent on the protection of the terminal adenosine of transfer RNA by combination with an amino acid. Transfer RNA is combined with valine and the unprotected polynucleotides then attacked with periodate. The dialdehydes so formed are then converted to hydrazones and coupled to a dye. The valyl-sRNA can now be separated from the polynucleotides coupled to the dye by precipitation with *n*-propanol buffer mixtures. In this way a twelvefold enhancement of the specific activity of the valine acceptor has been achieved. Brown *et al.*[315] have achieved partial purification of the tyrosine and histidine acceptors by the use of a diazotized polydiazostyrene column. Holley and his colleagues[316-319, 448] have employed countercurrent separation to purify and enrich acceptors for alanine, threonine, tyrosine, valine, histidine, and leucine from rat liver and yeast sRNA. The success, though partial at present, of these methods makes it clear that specific acceptors are present in the sRNA fraction. The various acceptors all have approximately the same molecular weight corresponding to a chain length of the order of 200 nucleotides.[320]

In our present state of knowledge it is not possible to say what are the chemical differences in acceptors that determine their amino acid specificity. In Section II, A, it was pointed out that sRNA differs from ribosomal RNA in possessing a high content of "unusual" bases, particularly 5′-ribosyluracil. Osawa[239, 321] has investigated the polynucleotides from sRNA of yeast, *E. coli*, and liver and found that the ability to bind leucine is proportional to the content of 5′-ribosyluracil.

Transfer RNA accepts amino acids after the latter have been activated by the specific activating enzymes. The enzymes themselves have been prepared free from RNA[322] and there seems to be a lack of species specificity in that transfer RNA from a variety of sources will accept amino acids from, say, the activating enzymes of liver.[323] There are minor quantitative differences in that, for example, sRNA from *E. coli* is a relatively poor acceptor of amino acids from liver enzymes, and the methionine-activating enzyme from *E. coli* will bind twice as much methionine to sRNA from *E. coli* as to the sRNA from yeast.[324] Transfer of amino acid adenylates from the surface of the activating enzyme to transfer RNA does not require the presence of any enzyme other than the activating enzyme itself. Labeling of the AMP moiety of the amino acid adenylate shows that only the amino acyl residue is transferred to sRNA.[300]

The sRNA fraction freshly isolated from rat liver and *E. coli* is naturally charged with amino acids, and hydrolysis has shown the presence in the fraction of most of the naturally occurring amino acids.[120, 129, 325] The amino acyl-sRNA complex thus provides a pool of activated amino acids available for synthetic purposes. In *E. coli* the amino acids bound to sRNA represent about 5% of the total amino acid "pool."[129] Whether the binding sites of transfer RNA are always occupied *in vivo* is not certain. sRNA isolated from *E. coli* grown in a medium free from added amino acids was found to have all its binding sites open to inactivation by periodate[121]; this might mean either that the binding sites were left free under such conditions of growth or that the amino acyl residues were easily split off during isolation of the sRNA. Lacks and Gros[129] have shown that specific binding sites can be rendered free in auxotrophs of *E. coli* starved of their specific amino acid requirements. Fixation by such free sites is very rapid, saturation taking place in less than 1 minute, and the process is independent of protein synthesis.

E. Other Soluble Factors Involved in Incorporation of Amino Acids into Ribosomes

Incorporation of amino acids into the protein of liver microsome preparations requires the presence of ATP and components present in the supernatant fraction of the homogenate in addition to the microsomes themselves.

The early studies with isolated microsomes showed that guanosine triphosphate (GTP) is an essential requirement and cannot be replaced by triphosphates of the other nucleosides.[116] Investigation of the supernatant components revealed the presence of the activating enzymes and sRNA as described above. In many tissue extracts, the activating enzymes and sRNA can be precipitated at pH 5 from the supernatant fraction and the supernatant left after removal of the precipitate is then referred to as fraction S_4. In the presence of ATP, activating enzymes, and sRNA, the amino acyl-sRNA complexes described above are formed but optimal incorporation of amino acids into the protein of ribosomes requires the further addition of GTP and factors present in the S_4 fraction. Since sRNA appears to be an essential component in the over-all transfer of amino acids to ribosomal protein[326, 327] it is a reasonable hypothesis that amino acyl-sRNA complexes are intermediate stages in the incorporation process. Hoagland *et al.*[116] first showed that isolated leucyl-sRNA can transfer some 20 % of the leucyl residue to microsomes under appropriate conditions while pulse experiments[326] show that valyl-sRNA of ascites tumor cells possesses the properties of a precursor of ribosomal-bound valine. As knowledge of conditions has improved, so has it been possible to demonstrate a more quantitative transfer of amino acyl residues from sRNA to ribosomal RNA. Grossi and Moldave[328, 329] showed that the transfer of leucyl residues requires at least two factors, one dialyzable and one nondialyzable, present in the supernatant. The dialyzable factor could be replaced by nucleoside triphosphates and was equivalent to GTP. A number of recent investigations have concentrated on the nondialyzable factor(s) and shown that heat-labile components, probably enzymes, are involved.[330-336] These components can be tentatively termed "transferring enzymes" and are characteristically activated by reducing agents including glutathione and mercaptoethylamine. The transferring enzymes can be separated from the amino acid-activating enzymes and there is some evidence that there may be a range of transferring enzymes each of which is specific for a different amino acid. Again there appears to be no marked species specificity about the sRNA component as amino acid donor since liver ribosomes will accept amino acids from *E. coli* sRNA as readily as from liver sRNA.[337] Nathans and Lipmann[367] report that there is a species specificity for the transferring factor or peptide-linking enzyme, since transfer to ribosomes of *E. coli* or liver requires factor from *E. coli* or liver, respectively. They were unable to demonstrate that different factors were required for different amino acids and so regard the peptide-linking enzyme as being nonspecific for amino acids. They found that the sRNA released after transferring its amino acid retains its terminal adenylic acid and can be recharged with amino acid. Consequently it acts in a cyclic fashion and can be regarded as a coenzyme for amino acid transfer.

An important question is whether the polynucleotide portion of transfer RNA (or any part thereof) accompanies the amino acid portion when the latter is transferred to the ribosome. Hoagland and Comly[338] have labeled liver amino acyl-sRNA with P^{32} in the nucleotide and C^{14} in the amino acid. Incubation of the doubly labeled substrate with ribosomes led to the simultaneous incorporation of both labels into the RNP particles. If the ribosomes were pre-incubated with P^{32}-labeled sRNA and then incubated with unlabeled sRNA under conditions suitable for protein synthesis, the label was released during the second incubation suggesting that the sRNA fixed during pre-incubation was later displaced. No detectable fragmentation of sRNA took place. The kinetics of the transfer process were consistent with the idea that the incorporation of amino acid into peptide bonds in the ribosome requires a large fraction of the sRNA to become bound firmly but transitorily to the ribosome.

F. Factors Involved in Release of Soluble Protein from Ribosomes

The components described so far in this section are concerned in the incorporation of labeled amino acids into peptide bonds associated with the ribosomal particles. Evidence that this incorporation involves protein synthesis will be discussed in the next section. Pulse experiments, described in Section V, A, suggest that incorporation of amino acids into ribosomal protein is a stage in the synthesis of soluble protein. It has so far been exceptionally difficult to prove that this is the case with *in vitro* experiments. In a few cases, incubation of ribosomes, labeled amino acids, and supernatant fractions have given a release of a small amount of labeled protein,[134] but in many other tests it has not been possible to demonstrate significant release[275] although an increase in specific proteins bound to ribosomes has been reported in a number of instances.[265, 339-341] Studies on the ability of rabbit reticulocyte ribosomes to form and release hemoglobin have suggested that ATP and an enzyme present in the S_4 fraction are necessary.[342, 343] Ribosomes from reticulocytes and *E. coli* have apparently lost the ability to turnover in the course of their isolation.[277, 342] Experiments on the net synthesis of protein by isolated RNP particles have been most successful in the case of the pea seedling particles studied extensively by Webster. These particles show an increase in net protein relatively readily[344, 345] when incubated in the presence of ATP, GTP, Mn and Mg ions, phosphoglyceric acid, sRNA, and a complete mixture of amino acids. The molecular weight of the proteins synthesized ranges from 7000–53,000 and Webster[344, 346] has identified adenosinetriphosphatase, aldolase, phosphoglucomutase, and pyruvic kinase among the proteins formed. Release of newly synthesized protein from the particles is dependent on the presence of ATP, Mg, and a highly labile "release enzyme."

G. Incorporation Factors

Gale and Folkes[114] studied the action of nucleic acids in restoring amino acid incorporation in nucleic acid-depleted staphylococci. They found that the activity of RNA was not decreased by digestion by ribonuclease but was released in dialyzable form. Fractionation of the digest led to the separation of a trace component or components provisionally called "incorporation factor(s)." It is not known whether the incorporation factors are true components of nucleic acid or whether they are contaminants adsorbed on the nucleic acid preparation and only released by breakdown of the polynucleotide chain. There is evidence that there are several incorporation factors which differ in their ability to replace nucleic acid for the incorporation of specific amino acids. The incorporation factor preparation described by Gale and Folkes[114] will replace nucleic acid completely for the incorporation of glycine, aspartic acid, glutamic acid, leucine, phenylalanine, lysine, and arginine, but will give only partial replacement for the incorporation of valine, isoleucine, proline, and tyrosine. The incorporation factor preparation thus appears to substitute for nucleic acid in the incorporation process but (see Section XI, A) the factor also promotes the synthesis of polyribonucleotides and RNA in the preparation; it may be that its action on amino acid incorporation is a consequence of the formation of nucleic acid components which act as acceptors for the amino acyl residues.[56, 115]

The activity of the factor preparation is unaffected by heating to 100°C. for 60 min. in N NaOH but is destroyed by heating for a similar period in N HCl; preparations slowly lose activity on prolonged storage. The active material has no electrophoretic mobility on paper in formate at pH 3.5, in acetate at pH 4.5, in phosphate at pH 2.0 or 7.0, or in borate at pH 9.0, but moves slowly toward the cathode in carbonate buffer at pH 10.5. The factor is soluble in ethanol and moves on paper chromatograms at $R_f = 0.8$ with isopropanol/ammonia as solvent or at $R_f = 0.7$ with ether/ethanol/ammonia as solvent. Aqueous solutions of the factor are colorless; they display no marked absorption in the ultraviolet at 260 mμ.

Investigation of a 5-month-old preparation of factor showed that the main component was glycerol.[347] Glycerol itself has little or no effect in promoting amino acid incorporation in the disrupted staphylococcal preparation and is markedly less active than freshly prepared factor.[348] However, incubation of the cell preparation with glycerol gives rise to active material after a lag period of 60–90 minutes. A concentration of 3×10^{-5} M glycerol, after such a lag period, is as active as the factor preparation although some disrupted cell preparations are unable to reach full activity in the presence of glycerol instead of factor. None of the breakdown products of glycerol that have been tested have had any effect but a "formin" preparation, made by heating glycerol with formic acid followed by fractional distillation, is three to five times more effective than glycerol itself in promoting amino acid incorporation. If disrupted staphylococcal cells are incubated, under conditions optimal for protein synthesis, with DL-glycerol-1-C^{14}, 50–60% of the radioactivity incorporated is found in the "nucleic acid fraction" (i.e., insoluble in hot ethanol or cold trichloracetic acid but soluble in hot trichloracetic acid). The radioactivity in the nucleic acid fraction is located in a number of components (not the normal nucleotides) of which at least one is an unstable substance readily breaking down to yield free glycerol. The nature of these substances is still under investigation at the time of writing (1961).

H. Amino Acid–Lipid Complexes

Hendler,[349] studying the short-term incorporation of alanine, phenylalanine, and valine by hen oviduct tissue, found a rapid binding by the lipid fraction of the cells. The lipid fraction took up radioactivity more rapidly than the nucleic acid, and lipid-bound amino acid residues then passed to the protein fraction. The lipid-bound amino acids acted as precursor to protein in preference to free amino acids. The amino acid-lipid bond was highly labile and the amino acids could be recovered by acid hydrolysis. In later studies[350, 351] Hendler separated the lipid complexes by countercurrent distribution and chromatography on silicic acid and showed that they were not produced simply by addition of amino acids to the lipid extracts of tissues before incubation and isolation. The amount of lipid-bound amino acid and the pattern obtained on chromatography thereof were markedly affected by addition of 2,4-dinitrophenol to the incubation mixture. The formation of lipid complexes with eleven amino acids was observed and similar materials isolated from ribosomes and cell wall preparations of *E. coli*. Hunter and colleagues[352, 353] have also observed the formation of amino acid-lipid complexes when protoplasts of *B. megaterium* are incubated with labeled amino acids. Chromatography on DEAE-cellulose separated both phospholipid and phosphorus-free amino acid complexes. The formation of the complexes was inhibited by the presence of chloramphenicol and, again, the labeled amino acids could be transferred from the lipid complexes to the protein fraction.

A different type of lipid-binding of amino acids has been described by Haining *et al.*[354] for the microsomal and supernatant fractions of rat liver. In this case the binding is unaffected by fluoride, *p*-chloromercuribenzoate, ribonuclease, or lipoxidase and is apparently independent of added energy sources. Amino acid residues are firmly bound to nonphosphorylated lipid fractions and are liberated only by prolonged acid hydrolysis. There would appear to be no reason to suppose that these complexes bear any simple relationship to the process of protein synthesis.

I. Biotin

Biotin is a growth factor whose mode of action has not yet been fully elucidated. In a number of investigations biotin has appeared to play a role in the synthesis of protein. Thus biotin-deficient cultures of *Lactobacillus arabinosus* are deficient in malic enzyme activity whether growth takes place in the presence or absence of malate.[124] The addition of biotin alone to suspensions of deficient organisms has no effect on the malic enzyme activity, but if the cells are incubated with biotin, glucose, and a mixture of amino acids, then a slow increase in malic enzyme activity sets in. This effect of biotin can only be demonstrated under conditions in which protein synthesis takes place and it has therefore been suggested that biotin mediates the synthesis of the apoenzyme moiety of malic enzyme. Similar results have been reported for enzymes involved in the synthesis of citrulline by *Streptococcus lactis*.[125] Recent studies by Lynen[355] have shown that biotin forms a prosthetic group of enzymes involved in carboxylation reactions. Some of the effects which have been attributed to a role of biotin in apoenzyme formation may, in fact, be due to conditions necessary for the synthesis of the active prosthetic group from biotin. Seaman[126] has studied the incorporation of glutamic acid by isolated kinetosomes from *Tetrahymena pyriformis* and finds that the process displays an absolute requirement for biotin despite the fact that the organism itself has no such nutritional requirement.

J. B_{12}

A series of papers by Wagle *et al.*[127] appeared in 1958 describing a requirement for B_{12} for the incorporation of amino acids by microsomes isolated from the livers of B_{12}-deficient rats. The incorporation in such B_{12}-deficient preparations is less than that in normal control preparations and is stimulated by the addition of B_{12} to the incubation medium. The incorporation is inhibited by antimetabolites of B_{12} such as the corresponding anilide although the inhibition is not released by B_{12}. The administration of Cobalt-labeled B_{12} to the animals shows that radioactivity is distributed between microsomes and supernatant cytoplasmic fluid, and much of the latter radioactivity is precipitated in the pH 5 enzyme fraction. The workers have therefore suggested that the activating enzyme preparation should be referred to as the "B_{12}-enzyme fraction" but any assumption that B_{12} is involved in the activation process would not appear to be justified by the present evidence. Attempts by other investigators to repeat the effects of B_{12}-deficiency in rats or chickens have so far failed.[356, 357] Incorporation of glutamic acid by the isolated kinetosome preparation of Seaman[126] is also dependent upon the presence of B_{12}, which is itself inactive in the absence of biotin. A further report from Wagle *et al.*[358] claims that the incorporation factor preparation (Section VI, G) of Gale replaces the pH 5 precipitate for amino acid incorporation in liver microsomes. The incorporation factor preparation does not contain B_{12} or biotin as determined by biological assay nor can its action be reproduced by either of the vitamins.

K. Incorporation Enzyme

Amino acid incorporation by the membrane fragment preparation from *Alcaligenes faecalis*[78] is stimulated by a heat-labile factor, called the "incorporation enzyme," present in the soluble extract of the cells. The enzyme has been purified and is free from amino acid-activating enzymes or polynucleotide phosphorylase activity. *A. faecalis* fragments can be separated into soluble and insoluble fractions by treatment with perfluorooctanoate. Both fractions are necessary for amino acid incorporation and the soluble fraction can be replaced by purified "incorporation enzyme." Highly purified preparations of the enzyme completely replace the pH 5 enzyme fraction in stimulating the incorporation of amino acids into the protein of rat liver microsomes.[128] The preparation catalyzes an Mg-dependent exchange of radioactive ADP with ATP and it is suggested that such a reaction may be involved in the process of amino acid incorporation.[359, 360]

L. Other Components

Specific requirements for metal ions and deoxyribonucleotides have been reported in particular investigations and will be discussed in later sections.

VII. Protein Synthesis

A. Amino Acid Incorporation by Microsomes: The Soluble Ribonucleic Acid Carrier System

In the preceding section, components have been described which are concerned in the following series of reactions:

(*a*) *Amino acid activation*

$$\alpha + \text{ATP} \rightleftharpoons \alpha\text{-AMP} + \text{PP} \tag{1}$$

A number of enzymes has been described, each carrying out the activation of a specific amino acid; it appears probable that there is a specific enzyme for each amino acid activated.

(*b*) *Binding of the activated amino acid by a component of the sRNA fraction*

$$\alpha\text{-AMP} + \text{sRNA} \rightleftharpoons \alpha\text{-sRNA} + \text{AMP} \tag{2}$$

Fractionation of sRNA indicates that there is probably a specific polynucleotide acceptor for each amino acid. If this acceptor is abbreviated as *pna* then reaction (2) is a general reaction including some 20 or more specific reactions:

$$\alpha\text{-AMP} + pna \rightleftharpoons \alpha\text{-}pna + \text{AMP}$$

(*c*) *Incorporation of the amino acid residue into peptide bonds in the protein of the ribosome*

$$n(\alpha\text{-sRNA}) + \text{RNP} \xrightleftharpoons{\text{GTP}} (\alpha)_n\text{-RNP} + n(\text{sRNA}) \quad (3)$$

In this way RNP-x-α^1-α^2-α^3-α^4-. . . can be built up and the amino acid residues built into peptide bonds. GTP is necessary for this reaction or reactions and is broken down in the course of the transfer. A transfer enzyme or series of transferring enzymes is required.

(*d*) *Release of the newly synthesized protein from the ribosome*

$$(\alpha)_n\text{-RNP} \xrightarrow{\text{ATP}} (\alpha^1\text{-}\alpha^2\text{-}\alpha^3\text{-}\alpha^4\ldots) + \text{RNP} \quad (4)$$

This reaction requires the presence of ATP and a highly labile "release enzyme" and results in the liberation of newly synthesized protein, leaving the ribosome ready to take up new amino acyl polyribonucleotides and to recycle the process. It is implicit in the above scheme that the residues α^1-α^2-α^3-α^4- of the newly synthesized protein have been inserted in a specific sequence but there is no evidence as yet for the mechanism controlling the sequential addition (see Section XIII).

There is ample evidence in the literature already cited that each of the above reactions can be demonstrated for certain amino acids and for a wide variety of cell preparations. The sRNA component appears to be surprisingly non-species-specific in that preparations from sources as widely different as liver, bacteria, tumor cells, and yeast will function both as amino acid acceptors for activating enzymes on the one hand and as donors for ribosomes on the other, whatever their sources. It is possible to demonstrate the incorporation of amino acids into, say, the ribosomes of ascites tumor cells by incubating them with sRNA from *E. coli* and activating enzymes from liver—although it is probable that crude preparations containing contaminating amounts of S_4 will work better than relatively clean materials. This would appear to mean that the chemical structure of amino acid-accepting polynucleotides must be the same in these various cells.

General acceptance of the scheme represented by reactions (1)–(4) above waits upon the answer to a number of questions:

Does the incorporation of amino acids into ribosomal protein, under the *in vitro* conditions described, represent true protein synthesis?

Is the course of amino acid incorporation as set out in reactions (1)–(4) a true reflection of the course of protein biosynthesis *in vivo*?

If so, is it the *only* course?

Is sRNA an obligatory intermediate in the process of protein biosynthesis or can it act simply as a reserve pool of activated amino acids?

These and other questions are inevitably in the minds of all who work in this field and, at the present time, few would be dogmatic about the answers. There is a formidable and daily growing mass of facts which can be readily interpreted in terms of the scheme set out above and no one doubts that there are further factors yet to be discovered as components of the over-all scheme. Two main difficulties present themselves at the moment: amino acid incorporation into ribosomes *in vitro* represents at best about 1% of the synthesis of protein *in vivo*, and only in rare cases it is possible to demonstrate anything more than a very small release of new protein from ribosomes *in vitro.* These difficulties may be resolved when all the necessary factors have been found and can be supplied to the *in vitro* system, and all the conditions and concentrations have been made optimal. Let us examine some of the questions in more detail.

The nature of the protein synthesized in the ribosome and its relation to amino acid incorporation. It is now established that the greater part of the protein of the ribosome is structural and does not consist of protein in the course of synthesis. This was first indicated when it was found that the composition of the protein of reticulocyte ribosomes did not correspond to that of hemoglobin[134] and has now been proved by detailed investigation of the structure and composition of *E. coli* ribosomes (see Section IV, B, 5). Consequently, protein in the course of synthesis must represent a very small part of the total protein of the ribosome. Schweet *et al.*[133] studied the incorporation of leucine, isoleucine, and valine by ribosomes from rabbit reticulocytes and showed that these three amino acids were incorporated in a ratio which was characteristic of hemoglobin but not of the ribosomal protein itself. This particular system does release a small amount of labeled protein in the soluble form and the investigators were able to show that about 82% of the protein released was radioactive hemoglobin. In this case, therefore, there is good evidence that incorporation in the ribosome represents the formation of a new specific protein. Further evidence was obtained by Webster[344] for the synthesis of soluble protein by pea seedling particles where the specific activity of the protein released was equal to that of the free amino acid added to the incubation mixture.

In Section IV, A, it was suggested that confirmatory evidence that amino acid incorporation can be equated with protein synthesis would be (*a*) that the incorporation of a labeled amino acid is dependent upon the presence of the other naturally occurring amino acids, and (*b*) that the incorporation is inhibited by chloramphenicol at growth inhibitory concentrations. In many studies of incorporation by *in vitro* systems, the presence of a mixture of unlabeled amino acids has not been necessary for, or had any effect upon, the incorporation of a single labeled amino acid. This may, however, merely be a reflection of the fact that incorporation of radioactivity represents a minute amount of synthesis of protein and sufficient amino acids, other than the added one, may be present in the preparations of sRNA and activating enzymes used. A further factor is that the process quickly comes to an end due to the inability of isolated ribosomes to release protein after synthesis. In those cases where release of protein does take place, e.g. hemoglobin synthesis by reticulocyte ribosomes[133] and soluble protein formation by pea seedling particles,[344] then the process is markedly stimulated by addition of an amino acid mixture. Webster[346] reports that the formation of soluble protein by the pea seedling system stops if any one of the naturally occurring amino acids is omitted from the incubation mixture. In the reticulocyte system, optimal stimulation is obtained if the mixture added contains amino acids in the proportions present in hemoglobin. The action of chloramphenicol cannot readily be used as a criterion in studies with mammalian and plant preparations where a growth inhibitory concentration is not known. However, Tissières *et al.*[277] find that amino acid incorporation by *E. coli* ribosomes is inhibited to the extent of 82% by a growth inhibitory concentration, while Webster reports inhibition of the pea seedling system by 100 μg. chloramphenicol per milliliter. It would seem therefore that, in specific cases, incorporation of amino acids by ribosomal preparations can be correlated with the synthesis of protein by those preparations. In bacterial systems, the problem may well be complicated by reactions concerned in the synthesis of the structural protein of ribosomes in addition to those leading to synthesis of soluble protein within the ribosome.

Although evidence can thus be obtained which relates amino acid incorporation to protein synthesis in specific examples, it does not follow that all investigations dealing with the incorporation of a single amino acid by ribosomes necessarily involve protein synthesis. Webster,[112] working with the pea seedling system, has shown that activation of all the natural amino acids will take place (see Table VI) and that the activated amino acids will form polynucleotide complexes. A proportion (up to 28% in early studies but considerably increased in later work) of the amino acid bound to sRNA can be transferred to protein in the presence of GTP. When a single amino acid was added to the system, the incorporated residues could be displaced by subsequent incubation with the corresponding unlabeled amino acyl polynucleotide but not by the corresponding free amino acid (see Table VII). This experiment demonstrates that reaction (3) of the scheme on p. 520 is reversible. However, investigation revealed that the residues incorporated under these conditions occupied N-terminal positions in the protein. If, on the other hand, the labeled amino acid were incorporated in the presence of a complete mixture of amino acids, then incorporation was no longer reversible and the residues were distributed throughout the protein in interpeptide bonds.[344] In the case of alanine incorporation, the proportion of residues present as N-terminal groups after 10 minutes' incubation was 95% when alanine was the only amino acid added to the incubation mixture or 5% if 18 amino acids, including alanine, were present. This might mean that the incorporation process is different when one amino acid is present to that occurring when 18 are added or, alternatively, that synthesis proceeds by N-terminal addition in sequence and necessarily stops when other amino acids required for continuing the sequence are not available.

TABLE VII

REVERSIBLE INCORPORATION BY RIBONUCLEOPROTEIN PARTICLES FROM PEA SEEDLINGS[a]

Additions	Radioactivity in 5-mg. particles	
	C^{14}-labeled glutamic acid	C^{14}-labeled methionine
No addition	359	168
C^{12}-glutamyl-polynucleotide	16	165
C^{12}-methionyl-polynucleotide	353	10
C^{12}-glutamic acid	356	163
C^{12}-methionine	355	160

NOTE: Incubation mixture contains 5 mg. ribonucleoprotein particles labeled either with C^{14}-glutamic acid or C^{14}-methionine, 0.5 mM guanosine triphosphate and additions as above, amino acids at 3 mM and amino acyl polynucleotides at 10 mg. Incubation at 38°C. for 15 minutes.

[a] From Webster.[112]

If this latter explanation is correct and we consider it in connection with the work of Hoagland and Comly,[338] it would appear that reaction (3) of the scheme on p. 520 is really the summation of a series of sequential steps, each involving the incorporation of a specific amino acyl polynucleotide as follows:

$$\alpha^1\text{-}pna^1 + \text{RNP-}x \rightleftharpoons \text{RNP-}x\text{-}\alpha^1\text{-}pna^1 \tag{3a}$$

$$\alpha^2\text{-}pna^2 + \text{RNP-}x\text{-}\alpha^1\text{-}pna^1 \rightleftharpoons \text{RNP-}x\text{-}\alpha^1\text{-}\alpha^2\text{-}pna^2 + pna^1 \tag{3b}$$

$$\alpha^3\text{-}pna^3 + \text{RNP-}x\text{-}\alpha^1\text{-}\alpha^2\text{-}pna^2 \rightleftharpoons \text{RNP-}x\text{-}\alpha^1\text{-}\alpha^2\text{-}\alpha^3\text{-}pna^3 + pna^2 \tag{3c}$$

summing to

$$n(\alpha\text{-sRNA}) + \text{RNP} = (\alpha)_n\text{-RNP} + n(\text{sRNA}) \tag{3}$$

The role of activating enzymes and soluble RNA. Activating enzymes and soluble RNA can be demonstrated in all cells so far tested but some investigators have thrown doubt on the suggestion that these fractions are consistently or *necessarily* components of the protein synthesizing mechanism. Two objections can be summarized as follows:

1. Some preparations are able to incorporate amino acids but appear to be devoid of activating enzymes and/or soluble RNA. Such preparations include the membrane fragments of Beljanski and Ochoa[78] which incorporate amino acids but are unable to effect an amino acid-dependent ATP-pyrophosphate exchange; preparations of mitochondria[130] and perfluoro-octanoate-treated microsomes[131] which incorporate amino acids by a process not dependent on the pH 5 enzyme fraction; and the disrupted staphylococcal preparation which possesses a range of activating enzymes (see Table VI) but no activity toward either glutamic or aspartic acids which are, nevertheless, rapidly incorporated in the presence of ATP. In the same connection, Heller *et al.*[132] report that there is no relationship between the activating enzymes of the silk gland of the silkworm and the concentration of the corresponding amino acids in silk protein; thus tryptophan and glycine give the highest and lowest "activation" values, respectively, but are present in the lowest and highest concentrations, respectively, in the silk protein.

2. Studies of the properties of the amino acyl adenylates believed to be the immediate products of the reactions catalyzed by the activating enzymes: Castelfranco and his colleagues[60,61] have shown that these substances are highly reactive and will combine with soluble RNA or with proteins to produce nonspecific incorporation reactions superficially similar to those described for microsome fractions, etc. These nonenzymic reactions are produced by concentrations of adenylates higher than those concerned in the biological reactions and the authors point out that enzymes may direct and organize reactions which will occur nonenzymically in nonspecific fashion. Nevertheless these studies indicate that considerable caution must be used in interpreting results of experiments which involve the production of adenylates in the presence of biological materials. Castelfranco *et al.*[60,61] emphasize the desirability of studying the synthesis of separable specific proteins and the incorporation of amino acids into known positions in the molecules of those proteins. More recent work on the relation between aminoacyl adenylates and enzymic reactions is discussed in Section VI, C.

The role of sRNA as an obligatory intermediate is still a matter of debate. The amino acid-sRNA complex could conceivably play two roles in the over-all passage of amino acids from the free state to peptide bond formation in the ribosome. These can be represented as

$$\begin{array}{c} \text{Free amino acids} \rightarrow \text{Amino acid Pool} \xrightarrow[1]{\text{sRNA}} \text{Protein} \\ 2 \upharpoonleft\downharpoonright \\ \text{sRNA} \end{array}$$

Position *1* is that assumed in the discussions so far and presented in the scheme on p. 520 whereas position *2* implies that sRNA provides a pool of activated but bound amino acids. Such a pool could protect amino acids from anabolic removal and could also hold a given amino acid in reserve should others not be immediately available at the right concentrations. Most kinetic studies that have so far been provided for *in vitro* studies do not enable a clear differentiation to be made between these possible roles. Aronson *et al.* point out that the sRNA-bound amino acids of *E. coli* could not supply the requirements of protein synthesis for more than 1 second.[361] They carried out an experiment in which cells were incubated with $S^{35}O_4^{=}$ and then, at a later stage, diluted with nonradioactive cystine, methionine, and $SO_4^{=}$. The radioactivity of the sRNA fraction continued to increase after the time at which the maximum rate of incorporation into protein was reached, and radioactivity persisted in the sRNA fraction after incorporation into protein had ceased. They suggest that these findings are more in accord with position *2* than *1*. Other experiments relevant to this point will be discussed later (p. 527). Webster[362] finds that pea seedling particles can synthesize soluble protein and incorporate radioactivity more efficiently when supplied with a mixture of radioactive amino acid-sRNA complexes than with a mixture of the free radioactive amino acids. Nevertheless the concentrations of the sRNA complexes that have to be added in such experiments are high and the results do not necessarily throw light on the function of sRNA as an intermediate carrier rather than as substrate.

There is, at present, a formidable mass of evidence which can be interpreted as supporting the general scheme of protein synthesis set out on p. 520. Nevertheless it certainly cannot be stated that this is necessarily the correct series of reactions or the only path of protein biosynthesis. There are the minority reports, some of which have been mentioned above and in Section VI, which throw doubt on the absolute necessity for activating enzymes or sRNA in synthetic systems or indicate alternative mechanisms involving lipid materials. It is still most desirable to investigate the details of protein synthesis in as many different types of biological material as possible.

The activating enzyme-sRNA-ribosome system in bacteria. The scheme set out on pp. 519–520 above has been developed by reference to investigations with many different sorts of cell. In Section V, B, it was pointed out that attention was not directed to the ribosome as a site of protein synthesis in bacteria until much of the scheme had been worked out by the use of mammalian and plant tissue extracts. It will be useful therefore to review here the extent to which the scheme can now be applied to bacterial preparations.

Activating enzymes were demonstrated in extracts of a wide variety of bacteria at an early stage in studies of amino acid incorporation *in vitro*,[111] and, similarly, extracts of *E. coli* were shown to contain sRNA, and the formation of amino acyl polyribonucleotides to follow the same course as in liver extracts.[117, 121] Lamborg and Zamecnik[273] have now isolated ribosomal

TABLE VIII

Amino Acid Incorporation by Extracts of *E. coli*[a, b]

Fraction number and description		Counts/ min.	Counts/min./ mg. protein
(1)	10,000 *g* (10 min.) supernatant	2870	757
(2)	30,000 *g* (20 min.) pellet	0	0
(3)	30,000 g (20 min.) supernatant	2795	777
(4)	100,000 *g* (2 hr.) pellet	714	340
(5)	100,000 *g* (2 hr.) supernatant	3	3
(2 + 5)	—	14	6
(4 + 5)	Ratio 1:1	1734	468
	Ratio 1:2	3000	527
	Ratio 1:3	3419	444

		Specific activity of complete system, %
(1)	Complete incubation system	100
	Mg absent	0
	KCl absent	41
	ATP absent	64
	GTP absent	76
	PEP, PEP kinase absent	47
	ATP, GTP, PEP, PEP kinase absent	0.2
	Amino acid mixture absent	38–80
	Complete + 1 μg. ribonuclease	4
	Complete + 30 μg. chloramphenicol	31

[a] Data from Lamborg and Zamecnik.[273]

[b] Incubation medium: 1.0 ml. contains 7×10^{-3} *M* $MgCl_2$, 6×10^{-2} *M* KCl, 1 μmole ATP, 0.25 μmole GTP, 5 μmoles phosphoenolpyruvate (PEP), 50 μg. PEP kinase, 0.2 μmole each of 18 L-amino acids, 1×10^{-2} *M* Tris buffer pH 7.9, 0.1 μmole (1.8×10^5 counts/minute) C^{14}-L-leucine. Incubation 15 minutes at 37°C.

particles from *E. coli* and shown (see Table VIII) that incorporation into the 100,000 *g* pellet is stimulated by the presence of the supernatant fraction and requires ATP, an energy-generating system, GTP, a mixture of amino acids, KCl and Mg ions. Incorporation is inhibited by ribonuclease and chloramphenicol. Tissières *et al.*[277] have confirmed the rapidity with which incorporation takes place in the bacterial system and found that the process is located primarily in a proportion of the 70 *S* particles forming less than 10% of the total ribosomes. These "active ribosomes" are neither formed nor broken down in the system during incorporation. The process was very sensitive to magnesium concentration, the optimal concentration varying between 0.007 and 0.011 *M* in different experiments. All the naturally occurring amino acids were incorporated and addition of the complete mixture increased by three times the incorporation of alanine by washed ribosomes. Calculation showed that the amount of protein synthesized was about 3.5 μg. per milligram ribosomes and 25–30% of this material was released into the soluble fraction. Chromatography on DEAE-cellulose indicated that radioactive amino acids were incorporated into a wide range of different proteins in the soluble fraction. Incorporation was inhibited by ribonuclease, deoxyribonuclease, and chloramphenicol.

Lacks and Gros[129] have shown that the fixation of amino acyl residues by the sRNA fraction is independent of protein synthesis but that the turnover of the fraction is reduced by more than 75% if protein synthesis is reduced by the presence of chloramphenicol or omission of an essential amino acid. They have demonstrated a transfer of S^{35}-labeled methionine from the sRNA fraction to the protein fraction and have obtained evidence that at least a part of the incorporation of methionine into protein can be ascribed to the following mechanism:

$$\alpha + \text{sRNA} \rightarrow \alpha\text{-sRNA} \rightarrow \text{sRNA} + \text{protein}$$

However, the turnover of methionine-labeling of the sRNA fraction that occurs in the absence of protein synthesis indicates that there is probably also reversible fixation of activated amino acids by the sRNA. At low temperatures, the maximum rate of turnover of the sRNA fraction is significantly lower than the rate of incorporation into protein and this finding cannot, at present, be reconciled with a direct and simple transfer of amino acyl residues from the sRNA fraction to protein.

B. Incorporation by the Disrupted Staphylococcal Cell Preparation

When the disrupted cell preparation is incubated with a labeled amino acid and ATP, incorporation of the radioactivity into both trichloracetic acid-soluble and -insoluble material takes place. If the incubation mixture contains C^{14}-labeled glutamic acid, then the radioactivity that is incorporated is associated with glutamyl residues only and can therefore be taken as a measure of glutamic acid incorporation.[54] The course of the incorporation varies with the conditions of the incubation: if glutamic acid is present as one component of a mixture of the eighteen naturally occurring amino acids required for protein synthesis, then incorporation proceeds in a linear fashion for several hours. If glutamic acid is the only amino acid present in the incubation mixture, incorporation starts off rapidly but decreases in rate and ceases after 60–90 min. at 37°C. Gale *et al.*[57] investigated

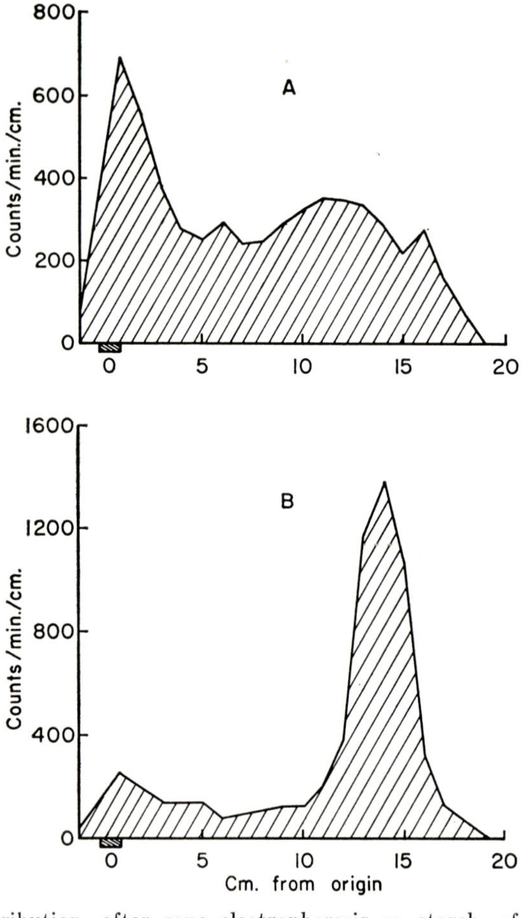

FIG. 5. Distribution, after zone electrophoresis on starch, of radioactivity in nucleoprotein extracts from disrupted staphylococcal cells incubated with an energy source and C^{14}-labeled glutamic acid (B) alone and (A) in the presence of 17 other unlabeled amino acids.[56, 57] Ordinate = distance moved from origin in $4\frac{1}{2}$ hr. at 20 v./cm. Abscissa = radioactivity eluted from 1 cm. section of starch column.

the nature of the incorporation under these conditions by application of zone electrophoresis on starch to nucleoprotein extracts from the disrupted cells after incorporation had taken place. Figure 5 shows the distribution of radioactivity along the starch column after electrophoresis. When the full amino acid mixture was present during incubation (Fig. 5A), glutamic acid is incorporated into material which either remains at the origin or moves slowly toward the anode; when glutamic acid only is present during incubation (Fig. 5B), radioactivity is incorporated into a fast-moving fraction running just behind the nucleic acid front.

The material, which remains at the origin or moves slowly towards the anode, is precipitated with 5% trichloracetic acid and has the properties of nucleoprotein. The fast-moving material, which is labeled by glutamic acid when that amino acid is the only one added to the incubation system, is not precipitated by trichloracetic acid and is clearly not protein in nature. The binding of glutamic acid to the protein-containing fractions is irreversible and stable; the binding by the fast-moving component is reversible and labile. Glutamic acid is released from the fast-moving component by alkali (90% release after 5 min. at pH 10 and 37°C.), by acid hydrolysis (70% release by treatment at pH 2 for 8 hr. at 37°C.; complete release in 60 min. at 100°C. in *N* HCl), and 30-40% released by incubation with ribonuclease. The fast-moving component, "glutamyl-X," has properties similar to those of an amino acyl-soluble RNA complex but differs in being soluble in cold 5% trichloracetic acid, perchloric acid, or 60% ethanol.

If the disrupted staphylococcal preparation is incubated with C^{14}-labeled glutamic acid and ATP to form glutamyl-X and then reincubated with the other seventeen amino acids required for protein synthesis, electrophoresis of the nucleoprotein extract shows that there has been a redistribution of the glutamyl residues. Approximately half of the radioactivity in the glutamyl-X fraction is transferred during the second incubation to the slow-moving fractions where it becomes incorporated into protein. Glutamyl-X can act, therefore, as an intermediate in the incorporation of glutamic acid residues into the protein of the preparation.

Investigation of the optical specificity of the glutamic acid released from glutamyl-X by acid hydrolysis shows that it consists of approximately equal amounts of D- and L-isomers. The material which is transferred to the protein fractions during incubation with the other amino acids (above) is the L-isomer, the D-glutamic acid remaining in the fast-moving fraction. *Staphylococcus aureus* contains D-glutamic acid as a component of its cell wall (see Section I, C, 2) but the disrupted cell preparation has lost most of its ability to synthesize wall peptide; it may be that the portion of glutamyl-X labeled with the D-isomer represents a precursor of the wall D-glutamic acid peptide.

Removal of nucleic acid from the disrupted cell preparation results in a marked decrease in the ability of the preparation to incorporate amino acids; this ability can be restored by adding nucleic acid or the incorporation factor preparation (see Section VI, G) to the incubation mixture.

C. Incorporation by Preparations of Membrane Fragments

The fragments of wall and membrane from *Alcaligenes faecalis*[78] incorporate all of the natural amino acids on incubation; the addition of ATP has little effect but incorporation is inhibited by 2,4-dinitrophenol or removal of oxygen. Washing the fragments with *M* NaCl reduces the incorporation and this can be stimulated by the addition of the soluble portion of the bacterial extract. The stimulation is due to two fractions, one thermolabile and the other thermostable. Beljanski and Ochoa[78] have investigated the thermolabile factor and have purified the incorporation enzyme involved; the purified preparation is devoid of amino acid activating enzymes nor does it appear to play any role in oxidative phosphorylation. Other properties of the preparation are described in Section VI, K.

Spiegelman[79] has studied the incorporation of amino acids into hot and cold trichloracetic acid stable linkages in various fractions obtained by disintegration of protoplasts of *E. coli*. He found that the membrane fragment fraction will incorporate amino acids into hot trichloracetic acid stable linkages and that this ability exceeded by a factor of 100 the ability of other fractions (soluble fraction, ribosomes, etc.) to incorporate amino acids. There was no significant difference between the incorporation into hot and cold acid stable linkage, so there was no clear evidence for incorporation into an RNA moiety. After a short lag period, amino acid incorporation continued in a linear fashion for more than 3 hours. Optimal incorporation is dependent upon the presence in the incubation medium of a number of components. These are listed in Table IX which shows the effect of their omission, one at a time, on the incorporation of leucine. A complete mixture of amino acids is necessary for incorporation by the membrane fragments, and the process is abolished by chloramphenicol. Other essential components are ATP, Mn ions, and mixtures of both ribonucleotides and deoxyribonucleotides. Investigation of the lag period that occurs during the onset of incorporation showed that it could be eliminated by addition to the medium of the triphosphates, but not the diphosphates, of guanosine, cytidine, and uridine in addition to that of adenosine. It seems probable (see Section X, C) that many of these components are necessary for the synthesis of ribonucleic acid as a factor in the incorporation process.

Nisman[80] has studied incorporation by membrane fragments of *E. coli* and has obtained similar results to those quoted above. He noted that incorporation, studied in the presence of Tris buffer, is inhibited if the preparation is preincubated with 0.05 *M* phosphate buffer. Nisman and Hirsch[81] lysed protoplasts in the presence of digitonin and separated three fractions: (A) material sedimented at 30,000 *g*; (B) material sedimented at 105,000 *g*; and (C) soluble fraction not sedimented at 105,000 *g*. All fractions activated and incorporated amino acids when incubated in the presence of ATP,

Mg and Mn ions, a complete amino acid mixture, and RNA from the same organism. Fraction B incorporated amino acids into both hot trichloracetic acid stable and extractable fractions, whereas incorporation into fraction C was all in the extractable fraction. A recent communication[365] indicates that fraction B consists of ribosomes whereas fraction A contains DNA, RNA, and protein and is broken down by the action of deoxyribonuclease to liberate ribosomes. Wachsmann *et al.*[366] have prepared a membrane fraction from *B. megaterium* which contains 17% of the protein, 12% of the RNA, and 37% of the DNA of the cell. Incorporation of amino acids by

TABLE IX

COMPONENTS REQUIRED FOR THE INCORPORATION OF LEUCINE INTO THE PROTEIN OF MEMBRANE FRAGMENTS FROM *Escherichia coli*[a]

Omissions from incubation mixture	Incorporation mμmole amino acid per mg. protein	Complete reaction, %
None	600	100
ATP	120	20
5′-Ribonucleotides	220	37
5′-Deoxyribonucleotides	200	33
(Ribo- and deoxyribonucleotides)	100	17
Other amino acids	20	3
Mn	0	0
Addition of 200 μg. chloramphenicol/ml.	40	7

NOTE: The complete incubation mixture contains per milliliter: 20 μmole KCl; 50 μmole Tris; 50 μmole maleate; 5 μmole Mn; 1 μmole Mg; 0.1 μmole of each of the four 5′-ribonucleotides; 0.1 μmole of each of the four 5′-deoxyribonucleotides; 2 mg. amino acid mixture containing all the amino acids in the proportions found in *E. coli* protein; 2 μmole ATP and 0.005 μg. membrane fraction.

[a] From Spiegelman.[79]

this fraction is greater than by other cell fractions and is stimulated by ribonucleoside diphosphates and sRNA from either *E. coli* or *B. megaterium* but is insensitive to the action of ribonuclease. Butler *et al.*[352] have shown that their membrane fraction from *B. megaterium*, which they claim is the initial site of amino acid incorporation in this organism, liberates amino acid activating enzymes and adhering ribosomes when washed with dilute buffers.

VIII. Peptides as Intermediates in Protein Synthesis

In the considerations set out above, we have discussed the transfer of amino acids from the free state to polypeptide structures associated with microsomal or membraneous material with the assumption that these polypeptide structures possess the full complexity of the specific proteins whose (immediate) precursors they are. The question arises whether peptides formed from a small number of amino acid residues are formed as primary condensation products or whether the first amino acid complex is that possessing the full specific sequence. In other words, in the assembly of a specific sequence ABCDEFGH from the individual residues A, B, C, etc., is there an intermediate production of AB, DCE, GH, etc.? No final answer can be given at present despite a number of attempts to tackle the problem from an experimental point of view.

A. Do Peptide-Requiring Mutants Exist?

If specific oligopeptides are intermediates in protein synthesis, then these must presumably be synthesized through the action of specific enzymes which must, in turn, be genetically controlled. If so, then mutants blocked in the activity of these enzymes should occur and should display nutritional requirements for the peptide products of the blocked enzymes. Consequently there has been an intensive search for microbial mutants which will require the presence of peptides for growth and which will incorporate those peptides, without prior hydrolysis, into their proteins.

There are many examples in the literature of organisms whose growth is accelerated by the presence of peptides.[135-142] In nearly all of the cases that have been examined in detail the organism has been found capable of hydrolyzing the peptides involved and incorporating the amino acid residues liberated. Thus Simmonds and Fruton[143] have described an organism which will grow more rapidly on leucyl-glycine than on glycine + leucine, but growth is not dependent on the presence of the peptide. It is of interest that this organism becomes markedly more dependent upon peptide when penicillin is added to the growth medium.

The outstanding example of a peptide requirement is the dependence of various Lactobacteriaceae on the presence of strepogenin. Investigations on this substance are mainly due to Woolley and his collaborators,[144,145,146] who first demonstrated that a peptidelike substance present in protein digests was essential for the growth of certain bacteria. The active factor was obtained by incomplete digestion of crystalline insulin and so must be of peptide structure and, furthermore, one that is an integral part of native protein. Fractionation of insulin digests has led to the separation of a number of oligopeptides with strepogenin activity. The most active of these contained cystine, glutamic acid, glycine, serine, valine, leucine, and isoleucine. The natural peptides oxytocin and vasopressin also possess activity and a most important finding is that synthetic oxytocin has a high activity. This means that the activity must reside in a peptide structure and not in an impurity associated with biological preparations. There are considerable differences in the composition of the various active structures that have been isolated—synthetic oxytocin differs from the most active peptide from insulin in that it contains tyrosine, proline, and aspartic acid, but no serine—so that it would appear that strepogenin activity cannot reside in any one specific amino acid sequence.

Kihara and his colleagues[147,148] have investigated a number of instances where peptides appear to be required for growth. *L. casei* can grow in a medium free from vitamin B_6 if a suitable mixture of amino acids is supplied, but the growth then becomes dependent upon the presence of incompletely digested protein. Fractionation of the active material in a digest of soybean protein showed that it consisted of a mixture of simple peptides containing either alanine or tyrosine; synthesis of a number of such dipeptides showed that the activity resided in alanyl peptides. Analysis of the situation revealed that growth in the absence of B_6 is dependent upon the addition of D-alanine to the medium and that the D-alanine inhibits the utilization of L-alanine but not of L-alanyl peptides. Growth is dependent upon the supply of both isomers of alanine but the L-isomer can only be supplied effectively in the form of a peptide. It seems probable that D-alanine inhibits the transport of the L-isomer across the cell membranes (inhibits the alanine permease) but is without effect on the transport of alanyl peptides. The peptides must penetrate by a different mechanism and then act, after hydrolysis, as a source of L-alanine within the cell. Recent studies[368, 369] with C^{14}-labeled glycine, alanine, and their peptides have shown the presence of three separate systems in *L. casei* which are responsible for the uptake of (1) glycine, (2) D- or L-alanine, and (3) glycyl-L-alanine or L-alanylglycine. The uptake of the peptide is markedly more efficient than that of the free amino acid but utilization of the peptide residues follows only after hydrolysis to the free amino acids within the cells. Similarly, peptides of D-alanine are taken up by *S. faecalis* but the D-alanine can only be used for growth purposes in those cases where rapid hydrolysis liberates the free amino acid after or during uptake.

Snell and his co-workers[149,150] then widened their investigation to cover a number of other instances where utilization of an amino acid for growth was inhibited by an analog of that amino acid and the inhibition released by the presence of a peptide containing the same amino acid. In none of these cases was there a specific requirement for the peptide as such, and the peptide was hydrolyzed before utilization within the cell so that the requirement related to events controlling the passage of substances into the cell. In the case of an organism responding to histidine peptides, it was found that peptides containing histidine and an unnatural amino acid were as effective as normal peptides so that, again, it is improbable that the peptide structure is required as such.

In some instances organisms possess enzymes which destroy essential amino acids. Thus *Streptococcus faecalis* strains possess arginine dihydrolase which attacks essential arginine and tyrosine decarboxylase which attacks essential tyrosine. With these strains it is found that growth is promoted by the addition of arginyl or tyrosyl peptides to the medium.[148,151] The peptides are not attacked by the enzymes but nevertheless will act as sources of the essential amino acids for growth—presumably the growth mechanisms have a higher affinity for the amino acids as they are liberated by hydrolysis within the cell than the enzymes which catabolize the free amino acids.

Experiments of this nature indicate that the nutrition of an organism may result from considerations more complex than enzyme blocks in amino acid synthesis or protein formation. It is therefore difficult, at present, to assess whether the strepogenin effect results from a synthetic disability relating to protein structure or to an indirect effect arising from transport phenomena. Kihara and Snell[370] suggest that the strepogenin effect is not a specific one. Single peptides supply limiting amino acids in a form more readily utilized than the corresponding free amino acids and complex peptides can be replaced by appropriate mixtures of simple peptides. The answer requires a demonstration either that a strepogenin-like substance is required for synthesis of a protein in a cell-free system or, less directly, that the amino acids in strepogenin are incorporated into protein in the same sequence without prior hydrolysis.

B. Can Peptides Be Made to Accumulate by Inhibition of Protein Synthesis?

Halvorson and his colleagues[94,95,96] attempted to use analogs of specific amino acids to solve this problem. The argument was as follows: if *p*-fluorophenylalanine (for example) inhibits protein synthesis by blocking the incorporation of phenylalanine residues, the presence of the analog should not affect the intermediate formation of structures containing amino acids other than phenylalanine and these should, consequently, accumulate. *p*-Fluorophenylalanine was therefore added to yeast in sufficient concentration to prevent enzyme formation and the amino acids of the pool examined quantitatively and qualitatively for conversion to combined forms. No new substances could be demonstrated and, furthermore, the disappearance of glutamic acid from the pool was inhibited by the presence of the unrelated fluorophenylalanine. It therefore appeared as though the presence of an analog to one amino acid inhibited the utilization of all other amino acids and did not lead to the accumulation of intermediate peptides. However, this work was carried out before the demonstration that such analogs become incorporated into inactive proteins and do not prevent protein synthesis as such.[97,98] Reinvestigation of the effect on pool amino acids has now shown[152] that *p*-fluorophenylalanine accelerates the rate of breakdown of preformed protein as well as becoming part of the newly formed protein, so that the apparent inhibition of utilization of pool glutamic acid is due to replenishment of the pool by protein breakdown, this replacement balancing the disappearance due to synthesis of analog-containing protein. The experiments do not therefore throw any light on the peptide intermediate hypothesis and should be repeated with the use of an amino acid analog which inhibits the utilization of an amino acid but is not itself incorporated in place of that amino acid.

Gale and van Halteren[153] found that washed suspensions of *Staphylococcus aureus* incubated with simple mixtures of amino acids could, in certain cases, give rise to the formation of peptides. Peptides were produced when the cells were incubated with glucose, glutamic acid, and either cysteine, alanine, or glycine; the combined forms were isolated by chromatography and the following peptides identified: (*1*) giving glutamic acid and cysteine on hydrolysis; (*2*) giving glutamic acid and alanine; (*3*) giving glutamic acid, alanine, and glycine; (*4*) glutamic acid and glycine; and (*5*) polyglycine. It was thought that these structures were formed in the presence of unbalanced mixtures that were unable to support protein synthesis. However, this study was made before the nature of the staphylococcal cell wall was known and before it had been demonstrated that incubation of such cells with glucose and mixtures containing glutamic acid, glycine, and alanine leads to synthesis of cell wall material. In the light of this modern knowledge, it seems probable that the peptide materials formed in the investigations of Gale and van Halteren[153] were related to wall peptides rather than to protein synthesis.

C. Can Direct Labeling of Peptides Be Demonstrated during Protein Synthesis?

Turba and co-workers[154, 371, 372] and McManus[155] have shown that a progressive labeling of pool amino acids and cell protein takes place during the growth of yeast in the presence of C^{14}-labeled acetate. The pool can, however, be separated into free and combined amino acids, the latter being broken down by acid hydrolysis and designated the "peptide fraction." If the course of labeling is followed with time, it is found that the peptide fraction becomes labeled before the protein; thus label can be shown in the peptide fraction within 3 minutes of the start of incubation, whereas significant label appears in the protein only after 30 minutes. Electrophoresis of the peptide fraction shows that it contains many different substances (Turba records 42 different peptide bands) which release amino acids on hydrolysis. Many of the peptides that have been examined in greater detail contain glutamic acid or cysteine and their relation to glutathione, with the possibility that some of them are γ-glutamyl peptides, has yet to be elucidated. McManus[155] separated a peptide containing leucine and showed that the specific activity of the leucyl residue was of the same order as that of the free leucine of the pool and markedly higher than that of the leucyl residues of the protein. These studies provide the best evidence at present available for a role of peptides as intermediates in protein synthesis but a conclusive proof of this role must await full and proper characterization of the peptide fraction.

D. Is Glutathione Involved as an Intermediate in Protein Synthesis?

One peptide whose occurrence in many tissues is well-authenticated is glutathione. This is also needed for the maintenance in synthetic media of certain strains of *Neisseria gonorrhoeae* although the requirement is not shown on first isolation.[156] Roberts and co-workers[157] found that glutathione is rapidly labeled if S-starved cells are incubated with labeled sulfate and that about half of the glutathione-S was transferred to protein during subsequent growth of the cells. Such results could, however, only be demonstrated in S-starved cells. Hanes *et al.*[158,159] discovered an enzyme in sheep kidney which will bring about the transfer of the glutamyl residue from glutathione to another amino acid to yield a γ-glutamyl peptide and cysteinylglycine:

$$\gamma\text{-Glutamylcysteinylglycine} + \text{R} = \gamma\text{-glutamyl-R} + \text{cysteinylglycine}$$

and suggested that this reaction might be a source of peptide bonds in protein synthesis. It was later found that glutathione is not a specific substrate for the transpeptidation, any γ-glutamyl peptide acting as substrate. No evidence has been forthcoming to show that glutamyl peptides are incorporated as such into protein; where incorporation of the amino acid residues has been obtained, it has been preceded by hydrolysis of the peptide to the free amino acids. Samuels[160] has shown the presence of the transpeptidase in extracts from *Proteus vulgaris* but the role of the enzyme is obscure, other than in the synthesis of γ-glutamyl peptides occurring in capsular substances (see Section IX, A).

E. Do Nucleotide Complexes Carry Peptide Structures?

It has been suggested (see Section IX, B) that the uridine diphosphate-muramic acid peptides that accumulate in penicillin-treated cells are precursors of cell wall substance, including the cell wall peptides, and the question therefore arises whether similar structures, involving a nucleotide and a peptide, play a part in protein synthesis. In Section VI, C, mention was made of the activation of certain peptides by enzyme systems akin to the amino acid activating enzymes.[305-307, 363, 364] It would therefore be reasonable to expect that peptide nucleotides would be found in cell extracts. In general, however, reports of the existence of such substances indicate that they are more complex than simple peptide adenylates. Rieth[373] has described an aspartyl uridylate in ascites tumor cells while Brown[374, 375] has isolated nucleotide-bound peptides from *S. faecalis* and these yield a range of amino acids and adenine on hydrolysis; it appears, however, that these substances are probably amino acid derivatives of polyadenylic acid. Koningsberger and co-workers[376, 377] have isolated carboxyl-activated peptide oligonucleotides from yeast. Somewhat similar complexes have been obtained from other sources,[378, 379] and their nature and importance await the proper characterization and elucidation of the soluble RNA fraction in general.

IX. The Synthesis of Specific Microbial Peptides

A. Capsular Polypeptides

The capsule of certain *Bacillus* species contains poly-D-glutamyl peptide (see Section I, C, 1). Glutamic acid racemase is, apparently, missing in *B. subtilis* or *B. anthracis* and D-glutamic acid is formed in these cells as a result of the action of two enzymes:[161,162]

1. Alanine racemase

$$\text{L-Alanine} \rightleftharpoons \text{DL-alanine} \rightleftharpoons \text{D-alanine}$$

2. D-Alanine, α-ketoglutaric acid amino transferase

$$\text{D-alanine} + \alpha\text{-ketoglutaric acid} \rightleftharpoons \text{pyruvic acid} + \text{D-glutamic acid}$$

Williams and others[163,164] isolated an enzyme from culture filtrates of *B. subtilis* which hydrolyzed the capsular polypeptide; the same enzyme brings about the synthesis of glutamyl dipeptides by transferring the γ-glutamyl residue from glutamine to glutamic acid with the formation of γ-glutamyl-glutamic acid. Dipeptide is synthesized when either D- or L-glutamine is the substrate and the most rapid reaction is obtained when L-glutamine is present together with D-glutamic acid. When supplied with the appropriate substrates, the enzyme catalyzes the formation of γ-D-glutamyl-D-glutamic acid and this can act as substrate for a transpeptidase which will bring about the formation of poly-D-glutamyl peptides. Thorne[165] suggests that the capsular polypeptide of *B. subtilis* may be synthesized by a combination of transamidation and transpeptidation reactions similar to those catalyzed by these cell-free preparations.

B. Cell Wall Peptides

The cell walls of Gram-positive bacteria contain peptides containing a limited variety of amino acids, e.g., the walls of *Staphylococcus aureus* contain glycine, D-glutamic acid, DL-alanine, and L-lysine (see Section I, C, 2). Park and Strominger[20] have isolated a uridine diphosphate-muramic acid peptide complex from cells growing in the presence of penicillin and have shown that the ratio of muramic acid to D-glutamic acid, alanine, and lysine is the same in the nucleotide complex as in the cell wall peptide. They have suggested therefore that the nucleotide complex may be a precursor of the wall muramic acid peptide structure. The concept has received further support from Ito and others,[166] who have isolated muramic acid complexes containing glycine and aspartic acid from another strain of *S. aureus* that possesses aspartic acid as a cell wall component. Although *S. aureus* requires a full complement of amino acids before it can synthesize true protein, wall peptide synthesis can occur if only glucose and the wall amino acids are present in the incubation mixture. The wall peptide synthesis further differs from protein synthesis in that it is not inhibited by chloramphenicol.[51,55] Since the wall peptide contains several amino acids in a determined sequence,[167, 223] it might be expected that its synthesis would be related in mechanism to that of protein synthesis, and that the initial stages (amino acid activation, fixation by a nucleotide acceptor) would be the same for both processes. In Section VII, B, it was shown that the incorporation of glutamic acid by disrupted staphylococcal cells leads to the formation of a glutamyl-X which contains both L- and D-glutamyl residues. The formation of the uridine complex may be a stage in wall peptide synthesis similar to that occurring at the soluble RNA stage in protein synthesis, and it is possible that investigations of wall synthesis may throw light on the more complex process of protein synthesis.

The most complex of the nucleotides isolated in the presence of penicillin has the structure uridine diphosphate-muramic acid-L-alanyl-D-glutamyl-L-lysyl-D-alanyl-D-alanine[223] (U-M-peptide). The course of the synthesis of the peptide chain has been studied in detail by Strominger and colleagues.[380-382] Deprivation of lysine leads to the accumulation of U-M-alanyl-glutamic acid while the presence of gentian violet in the incubation mixture leads to the accumulation of U-M itself. From these findings and the nature of the original nucleotides discovered by Park,[19] it appears that the peptide is built up from U-M in a stepwise fashion. Ito and Strominger[382] have confirmed that this is so by the use of labeled amino acids and extracts of *S. aureus*. Their results are summarized in Table X where it can be seen that the addition of each amino acid takes place in the presence of a specific nucleotide. Thus U-M cannot incorporate D-glutamic acid unless and until L-alanine is already in position while the L-lysyl residue is not attached unless both the L-alanyl and D-glutamyl residues are in place and in the right sequence. The incorporation of radioactivity from L-alanine by U-M-alanylglutamyllysine is due to prior synthesis of D-alanyl-D-alanine by the action of alanine racemase and a dipeptide-synthesizing enzyme present in the crude extracts used. During purification of the extracts, it was found that the ability to incorporate activity from alanine in this last stage was lost but could be restored by addition of material accumulated by the organism when deprived of lysine. This material was found to contain D-alanyl-D-alanine in addition to U-M-alanylglutamic acid (as above). The natural material and also synthetic dipeptide are attached directly to U-M-alanylglutamyllysine; D-alanyl-D-alanine thus becomes the first properly established case of a dipeptide being incorporated in the course of synthesis of a larger polypeptide. The enzyme responsible for the synthesis of the dipeptide has been studied by Neuhaus.[383] As a result of these investi-

TABLE X

STEPWISE SYNTHESIS OF MURAMIC ACID PEPTIDE BY EXTRACTS OF *Staphylococcus aureus*[a, b]

Substrate added	Radio-activity incorporated from:			
	L-Alanine-C^{14}	D-Glutamic-C^{14}	L-Lysine-C^{14}	D-Alanyl-D-alanine-C^{14}
U-M	1364	0	0	0
U-M-ala	0	3720	0	—
U-M-ala-glu	8	0	3530	—
U-M-ala-glu-lys	(2200)	0	20	4790

[a] Data from Ito and Strominger.[382]

[b] Incubation mixture: in 0.05 ml. 0.1 *M* Tris buffer pH 8.4, 200 μmoles ATP, 400 μmoles $MgCl_2$, 7 μmoles muramic acid nucleotide (U-M-x) as above, 0.07–0.5 mg. enzyme protein prepared by precipitating with 3 volumes saturated $(NH_4)_2SO_4$ solution the sonicate of *S. aureus*.

gations, Strominger has put forward the following stepwise reactions for the synthesis of the peptide:

$$\text{U-M} + \text{L-alanine} \xrightarrow[\text{Mn}]{\text{ATP}} \text{U-M-L-alanine} \tag{1}$$

$$\text{U-M-ala} + \text{D-glutamic acid} \xrightarrow[\text{Mn}]{\text{ATP}} \text{U-M-L-alanyl-D-glutamic acid} \tag{2}$$

$$\text{U-M-ala-glu} + \text{L-lysine} \xrightarrow[\text{Mn}]{\text{ATP}} \text{U-M-L-ala-D-glu-L-lysine} \tag{3}$$

$$2\ \text{D-alanine} \xrightarrow[\text{Mn}]{\text{ATP}} \text{D-alanyl-D-alanine} \tag{4}$$

$$\text{U-M-ala-glu-lys} + \text{D-ala-D-ala} \xrightarrow[\text{Mn}]{\text{ATP}} \text{U-M-L-ala-D-glu-L-lys-D-ala-D-alanine} \tag{5}$$

The enzyme responsible for reaction (3) has been purified some 450 times and the following reversible reaction established:

$$\text{U-M-ala-glu} + \text{L-lysine} + \text{ATP} \rightleftharpoons \text{U-M-ala-glu-lysine} + \text{ADP} + \text{P}$$

The enzyme is specific for the nucleotide dipeptide as substrate. Strominger points out that this is the longest sequential biosynthesis of a peptide chain known and wonders how long a polypeptide can be fabricated by stepwise addition in this manner. The process is extremely interesting as it depends upon stepwise addition by specific enzymes, and the inability to comprehend how such a mechanism could give rise to the specific sequences of protein chains has been one of the reasons for speculation concerning the need for templates in protein synthesis. No requirement for nucleic acid has been observed so far. Reaction [4] is inhibited by oxamycin.[363]

X. Ribonucleic Acid Synthesis

A. Synthesis by Intact Cells

The processes of protein and nucleic acid synthesis appear to be closely linked so that many, if not all, of the components of the protein-synthesizing mechanism are also necessary for the formation of nucleic acid. These, listed in Section VI, will not be set out again except where particular comment is necessary.

1. Purines and Pyrimidines: Incorporation of Analogs

Synthesis of nucleic acids requires the supply of the five natural bases, adenine, guanine, uracil, cytosine, and thymine. Just as certain amino acid analogs may become incorporated into protein in place of the corresponding natural amino acids, so certain analogs of purine or pyrimidine can be incorporated into nucleic acids. Matthews[168] found that 8-azaguanine becomes incorporated into the RNA of tobacco mosaic virus and can be isolated therefrom as the corresponding nucleotide. It was later shown[46,47] that the same analog becomes incorporated into the RNA of a wide variety of bacteria but, with the possible exception of *Bacillus cereus*, incorporation does not take place into DNA. The base is incorporated almost exclusively as 8-azaguanylic acid but there is little or no correlation between the amount of base incorporated and the degree of growth inhibition produced. This absence of correlation has led Mandel[48,169] to suggest that the growth-inhibitory action of 8-azaguanine lies in some action other than its substitution for guanine in nucleic acid. Examination of the distribution of 8-azaguanine after incorporation shows that it is largely associated with small molecular weight RNA material and is present as end groups on polyribonucleotides.[47]

A number of other purines and pyrimidine analogs have been examined from this point of view. Dunn and colleagues[170] have investigated the

5-halogen-substituted uracils and shown that the inhibition of growth they produce may be antagonized by thymine but not by uracil. The bromo and iodo derivatives are incorporated into DNA in place of thymine in both *E. coli*[170] and *Streptococcus faecalis*[171] and the corresponding deoxyribosides and deoxyribotides have been separated from the DNA concerned. 5-Chlorouracil appears to be more toxic than the bromo or iodo derivatives and to antagonize both thymine and uracil. 5-Fluorouracil, on the other hand, acts specifically as a uracil antagonist.[245, 384] When fluorouracil is added to an exponentially growing culture of *E. coli*, growth becomes linear and changes take place in the composition of the proteins formed and in the activity of some of the enzymes produced (see pp. 487–8). In the presence of the analog, *E. coli* synthesizes a protein which is antigenically related to β-galactosidase but is biologically inactive; the rate of synthesis of this protein corresponds to the rate of synthesis of β-galactosidase by the organism in the absence of analog. Alkaline phosphatase formed in the presence of analog has a greater thermal sensitivity than the corresponding normal enzyme. Consequently the replacement of uracil in some nucleic acid component by fluorouracil brings about a change in the synthesis of certain proteins. Whether the nucleic acid component concerned is a transfer RNA whose function is altered, or a template or "messenger RNA" (see Discussion 1961) is not yet known.

2. Amino Acids

Cells supplied with a source of energy, pentose, purines, and pyrimidines cannot synthesize nucleic acids unless amino acids are also present. Figure 6 shows that optimal synthesis of nucleic acid by washed *S. aureus* requires the presence of the eighteen amino acids necessary for protein synthesis; omission of a single essential amino acid results in almost complete loss of ability to form nucleic acid.[41] Figure 7 shows the time course of nucleic acid synthesis, measured by the incorporation of C^{14}-labeled adenine into the nucleic acid fraction, in disrupted staphylococci.[115] In the absence of amino acids or in the presence of a single amino acid such as glutamic acid, there is a small incorporation complete in about 30 minutes. In the presence of a complete mixture of eighteen amino acids, incorporation of adenine continues in a linear fashion for 2–3 hours and is accompanied by a parallel increase in the optical density at 260 mμ of the nucleic acid fraction. This synthesis in the presence of amino acids is markedly stimulated by the further addition of the other purines and pyrimidines. If incomplete mixtures of amino acids are used, adenine incorporation takes place during the first 30–60 minutes only, to an extent which varies with the number and nature of the amino acids present. These results have been obtained with a strain of *S. aureus* which is nutritionally exacting to a large number of amino acids, and the same general principle applies to other organisms.[172-174] Thus, mutants of *E. coli* which are blocked in the synthesis of an amino acid cannot synthesize RNA unless that amino acid is supplied in the

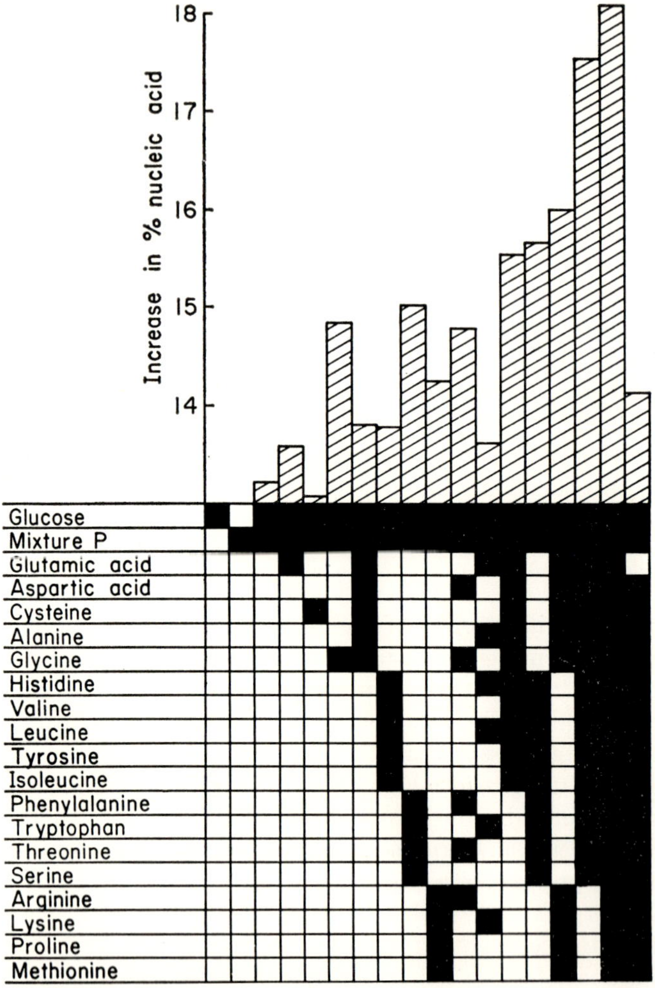

FIG. 6. Dependence of nucleic acid synthesis in *Staphylococcus aureus* on the presence of amino acids.[41] Washed cells were incubated in the presence of glucose, mixture P (= purines and pyrimidines), and amino acids as indicated by black blocks.

medium. It would seem that nucleic acid synthesis can take place only under conditions in which protein synthesis is also occurring. However, the addition of chloramphenicol at a concentration which gives 98% inhibition of protein synthesis has no inhibitory action on nucleic acid synthesis and may give rise to an enhanced rate of RNA synthesis for a limited period. Recent investigations indicate that the RNA which is formed under these conditions differs in a number of ways from normal RNA (see Section XII).

Certain amino acid analogs are able to replace the corresponding amino acids during protein synthesis; Gros and Gros[175] have investigated the ability of *p*-fluorophenylalanine and ethionine to replace phenylalanine

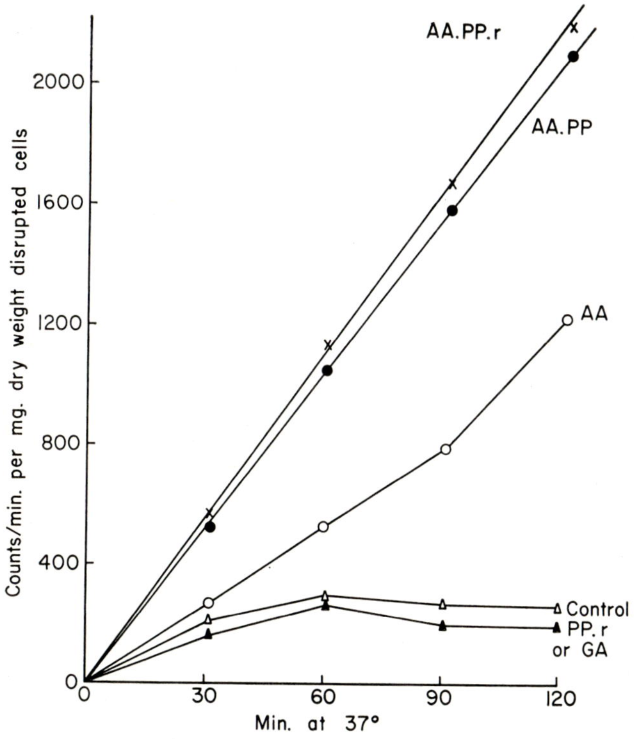

FIG. 7. Course of C^{14}-labeled adenine incorporation by disrupted staphylococcal cell preparation.[115] Disrupted cell preparation incubated with C^{14}-labeled adenine and ATP (control) in the presence of a mixture of purines and pyrimidines (PP) other than adenine, ribose (r), glutamic acid (GA) and a mixture of the 18 amino acids required for protein synthesis (AA).

and methionine for the promotion of RNA synthesis. The analogs promote a much smaller RNA synthesis than do the natural amino acids but are relatively more effective if chloramphenicol is also present.

Goldstein and Brown[385] have observed that incorporation of P^{32} into RNA continues during starvation of leucine- and proline-less mutants of *E. coli* but no net synthesis of RNA or protein can be measured under these conditions. Turnover of both protein and RNA occurs to the extent of about 10% per hour and is equally distributed in cell debris, ribosomal and supernatant fractions. The question arises why there should be no net synthesis of RNA stimulated by the amino acids liberated during protein turnover. The authors suggest either that protein turnover does not yield free amino acids or, alternatively, that RNA and protein turnover are so closely coupled that protein breakdown is accompanied by an equivalent depolymerization of RNA and no net synthesis is possible.

B. Synthesis in the Disrupted Staphylococcal Cell

The disrupted staphylococcal cell preparation accomplishes a linear synthesis of nucleic acid if incubated in the presence of ATP and a complete mixture of purines, pyrimidines, and the amino acids necessary for protein synthesis (see Fig. 7). If nucleic acid synthesis is measured by the incorporation of C^{14}-labeled adenine, approximately 88% of the radioactivity is associated with the RNA fraction. Removal of nucleic acid from the disrupted cells results in almost complete loss of the ability to incorporate adenine, and this ability can be fully restored by addition to the incubation medium of the incorporation factor preparation (see Section VI, G) at the same concentration as that necessary for optimal restoration of amino acid incorporation.[115]

When chloramphenicol was added to the medium, it was found that the incorporation factor preparation was less effective than nucleic acid in restoring adenine incorporation. When adenine incorporation is promoted by incorporation factor in place of nucleic acid, approximately 6% of the radioactivity relates to the DNA fraction and this is further suppressed by the presence of chloramphenicol; under these conditions, therefore, there is little or no synthesis of DNA. Consequently if the disrupted cell has been efficiently depleted of nucleic acid, no source of DNA is present in the system. This led to the finding that DNA is necessary for the full action of the incorporation factor in promoting RNA synthesis. The full requirements for RNA synthesis by the disrupted staphylococcal cell can now be stated: a source of energy, purines, pyrimidines, ribose, the full complement of amino acids needed for protein synthesis, DNA, and the incorporation factor preparation.

In Section VII, B it was shown that the incorporation factor preparation can apparently replace nucleic acid in promoting amino acid incorporation in the disrupted staphylococcal cell. The results now available show that the factor preparation is also active in promoting RNA synthesis. It seems probable therefore that the action of the factor on amino acid incorporation may be a secondary one, its primary effect relating to the synthesis of RNA or polyribonucleotide derivatives which then act as acceptors for the amino acid residues.[56]

C. Synthesis by Membrane Fragments

Spiegelman[79] has shown by pulse experiments that polyribonucleotide synthesis takes place in the membrane fraction of *E. coli* and that preparations of isolated membrane fragments can bring about polyribonucleotide formation under conditions similar to those required for amino acid incorporation (see Table IX, Section VII, C). Initial inconsistent results were found to be due to destruction of the polyribonucleotide product but linear synthesis was obtained when spermine was added as a stabilizing agent. The incubation mixture contained the four ribotides, four deoxyribotides, ATP, Mn, and the riboside triphosphates of guanine, uracil, and cytosine. No reaction took place in the absence of either ATP or Mn while omission of the riboside triphosphates (other than ATP) reduced the synthesis by 80 %. The triphosphates could not be replaced by the corresponding diphosphates. The reaction was inhibited by either ribonuclease or deoxyribonuclease and was dependent upon the presence of the four deoxyribotides. The product was attacked by ribonuclease but not by deoxyribonuclease and Spiegelman suggests that further examination of this system might reveal clues pertinent to the relationship between RNA and DNA. It is interesting that, as in the case of the disrupted staphylococcal cells, optimal conditions for the incorporation of amino acids are again those which are optimal for the production of polyribonucleotides.

D. Synthesis by Cell-Free Systems

1. Addition of Terminal Nucleotides to sRNA

In Section VI, D, 3, it was shown that polynucleotides of transfer RNA must possess the terminal sequence cytidyl-cytidyl-adenylic before they can act as amino acid acceptors. Incubation of sRNA in the absence of amino acids and ATP leads to inactivation of the transfer system. The ability of the transfer RNA to act as amino acid acceptor can then be restored by further incubation in the presence of CTP and ATP.[118, 119] Extracts of many different tissues, including bacterial cells, have now been found to contain an enzyme or enzymes which bring about the terminal addition of nucleotide residues to polynucleotides according to the general equation[386-395]:

$$\text{XTP} + \text{sRNA} \rightleftharpoons \text{XMP-sRNA} + \text{PP}$$

Terminal attachment of this nature is specific for ATP and CTP, little activity being shown with either GTP or UTP. When ATP alone is provided, the incorporated AMP is found in the terminal position but adjacent to CMP. It is possible that the terminal addition of CMP and AMP from triphosphate donors is brought about by the same enzyme. Terminal AMP originates in this way from ATP and not from amino acyl-AMP, and the terminal-nucleotide-adding enzyme can be completely separated from amino acid activating enzymes. Consequently the terminal addition in cell-free systems does not display an amino acid dependence of the type described above for RNA synthesis in intact cell systems. Canellakis and Herbert[389, 390] have purified the enzyme from rat liver and shown that the terminal addition is reversible. Fractionation of the enzyme into protein and nucleic acid components showed that both fractions were necessary for activity. Evidence was obtained for the probable existence of sRNA with terminal sequences other than cytidyl-cytidyl-adenylic.

2. RNA Synthesis with Interpolynucleotide Incorporation of Ribonucleoside Triphosphates

The enzyme systems described in the last section bring about terminal addition of CMP or AMP to sRNA when incubated in the presence of the corresponding triphosphates. Moldave,[396] studying the incorporation of AMP, separated active material from amino acid activating enzymes and found that it could be further separated into two fractions, one of which brought about terminal addition of AMP whereas the other gave incorporation into interpolynucleotide linkages. Similarly, investigations of incorporation of UTP into RNA-like material by extracts of rat liver nuclei[397] and *E. coli*[398, 399] have led to the description of a new enzyme system which requires the presence of the four triphosphates (ATP, GTP, CTP, and UTP) and brings about incorporation of any of the corresponding nucleosides into interpolynucleotide linkages. In the *E. coli* system, Stevens[399] found that the product is sedimentable at 115,000 *g* for 2 hours and, on centrifuging through a sucrose density gradient, is distributed in the region of small ribonucleoprotein particles. The enzyme system was present in the 105,000 *g* supernatant fraction but was pelleted from this fraction by centrifuging 5 hours at 115,000 *g*. Seventy-five per cent of the activity of the pellet was precipitated at pH 5 but this precipitate carried down less than 10% of the polyribonucleotide phosphorylase of the preparation (see below). Hurwitz *et al.*[398] found that the UMP incorporating system from *E. coli* required the four triphosphates and a nucleic acid fraction while the reaction was prevented by the presence of either ribonuclease or deoxyribonuclease. The nucleic acid fraction was prepared by lanthanum precipitation of ribosome-free supernatant of *E. coli* and could be replaced by DNA from thymus gland, rat liver nuclei, or bacteriophage T2; RNA from *E. coli* would not replace the nucleic acid fraction and all the active preparations were inactivated by pretreatment with deoxyribonuclease. The rat liver nuclei system studied by Weiss[397] is also inactivated by "extremely small quantities of deoxyribonuclease" whereas similar treatment with ribonuclease is not as effective. The implication of these findings is that a synthesis of polyribonucleotide can arise from mixtures of the four nucleoside triphosphates in the presence of a specific enzyme system and a DNA primer.

A DNA-dependent synthesis of RNA, requiring the presence of the four nucleoside triphosphates as substrates, has now been demonstrated for extracts from *Micrococcus lysodeitkicus*,[449] *Lactobacillus arabinosus*,[450] and *Azotobacter vinelandii*[450] in addition to *E. coli*. The *A. vinelandii* enzyme has been partially purified and can be primed by DNA from *A. vinelandii*, spleen, or bacteriophage T4, whereas RNA from *A. vinelandii*, yeast, or tobacco mosaic virus is ineffective.[451] DNA is inactivated by heat, and the heat-treated material reduces the effect obtained with native DNA. Weiss

TABLE XI

BASE COMPOSITION OF SYNTHESIZED RNA COMPARED WITH THAT OF PRIMER DNA[a]

Base	Base proportions (moles per cent)							
	Pseudomonas DNA	Synthesized RNA	Salmon DNA	Synthesized RNA	*Serratia* DNA	Synthesized RNA	*E. coli* DNA	Synthesized RNA
Guanine	33.0	35.0	20.8	21.0	29.0	32.6	24.1	23.4
Adenine	18.2	16.5	29.7	29.0	21.1	18.6	25.4	23.4
Thymine (uracil)	18.8	19.0	29.1	27.0	20.9	20.8	24.8	28.4
Cytosine	30.0	29.5	20.4	23.0	29.0	28.0	25.7	24.8
G + C	63.0	64.5	41.2	44.0	58.0	60.6	49.8	48.2
$\frac{A + T(U)}{G + C}$	0.59	0.55	1.43	1.27	1.72	0.65	1.01	1.07

[a] From Weiss and Nakamoto.[452]

and Nakamoto[449] have demonstrated a net synthesis of RNA with the enzyme from *M. lysodeikticus* and investigated the effect on the composition of the product of the nature of the DNA used as primer.[452] Primers from a variety of bacteria, calf thymus, rat liver, and salmon were used, and it was found that the products differed in base composition and in the position of CMP residues within the RNA chain. It can be seen from Table XI that there are wide alterations in the base composition of the products and that there is a similarity between the composition of the product and of the DNA used as primer. It seems possible that the enzyme assembles ribopolynucleotides in a sequence complementary to that present in the DNA.

3. Polyribonucleotide Phosphorylase

Grunberg-Manago, Ochoa and Ortiz[176, 177, 400] have described the isolation of an enzyme from extracts of *Azotobacter vinelandii*; the enzyme will carry out the reaction:

$$n\text{X-r-P-P} \rightleftharpoons (\text{X-r-P})_n + n\text{P}$$

where r = ribose, P = orthophosphate, P-P = diphosphate, and X = one or more of the bases adenine, hypoxanthine, guanine, uracil, and cytosine. If the enzyme is incubated with adenosine diphosphate, an adenylic polymer is formed with the liberation of inorganic phosphate. If a mixture of diphosphates is used, a mixed polymer is formed which has the characteristics of ribonucleic acid. The molecular weight of the polynucleotides formed varies from about 30,000 to 1,000,000. The polymers can be digested by ribonuclease and the nucleotides are linked through 3′-phosphoribose ester bonds as in RNA. The composition of the synthetic polymer appears to be largely determined by the relative concentrations of the diphosphates used as substrate. The polymer synthesized from a mixture of the four diphosphates present in equimolar concentrations is highly active in stimulating streptolysin S formation by *Streptococcus pyogenes*, whereas polymers formed from mixtures deficient in guanosine diphosphate were inactive.

Purified preparations of the enzyme require the presence of an oligonucleotide primer before they can accomplish polymer synthesis. The synthesis of polyadenylate can be primed by the addition of di-, tri- or tetra-adenylate; the primers all give rise to the same final rate of polymer synthesis but the concentration required to give that rate decreases as the chain length of the primer increases. The use of P^{32}-labeled primers has shown that the primer is built into the polymerized product and it appears that the action of the enzyme is to add nucleotide residues to the primer chain. The action of a diadenylate primer on the formation of a polyuridylic acid can be represented:

$$(\text{A-r-P})(\text{A-r-P}) + n\text{U-r-P-P} = (\text{A-r-P})(\text{A-r-P})(\text{U-r-P})(\text{U-r-P})\cdots$$

Polymers will themselves act as primers but a degree of specificity is displayed; thus polyadenylate (as opposed to oligoadenylate) will prime the synthesis of polyadenylate but not other polymers, whereas polycytidylate will prime the formation of all other polymers including the mixed ones. The formation of the mixed, RNA-like polymer can be primed by polycytidylate, the mixed polymer itself, or RNA. The mechanism of priming may be different in the two cases; oligonucleotides undoubtedly act as primer chains which are elongated by the enzymic addition of further residues but the polynucleotides may prime by providing a basis for replication. Thus incubation of the enzyme with adenosine 5′-P^{32}-pyrophosphate or the corresponding P^{32}-labeled UDP, together with the other three diphosphates in unlabeled condition, results in incorporation of the labeled nucleotide in positions distributed throughout the polynucleotide chain.[401]

Polyribonucleotide phosphorylase has been demonstrated in extracts from a wide variety of aerobic, anaerobic, Gram-positive and -negative bacteria, and also in yeast and spinach. The polymerase action is reversible. In the presence of Mg^+ and orthophosphate, polynucleotides and ribonucleic acids are degraded by the enzyme. Phosphorolysis of short polynucleotide chains takes place in stepwise fashion starting at the terminal nucleotide with an unesterified 3′-hydroxyl group.[402] The amino acid acceptor activity of sRNA is not decreased after 20–30% of the RNA of the fraction has been digested by incubation with the phosphorylase,[403] which suggests that transfer RNA is protected in some way (possibly by hydrogen bonding) while inactive components are rapidly destroyed. Grunberg-Manago[404] has investigated the relationship between phosphorolysis and the macromolecular configuration of polyribonucleotides. The phosphorylase readily attacks monopolymers in the presence of magnesium and inorganic phosphate; the rate of breakdown of polyguanylic acid is about one-quarter that of the other monopolymers while the mixed polymer is attacked at about one-third of the rate for polyadenylate. Most ribonucleic acids are attacked slowly, the rate corresponding to 10–15% of that of polyadenylate; tobacco mosaic virus RNA is exceptional in that it is attacked as rapidly as polyadenylate. When polyadenylate and polyuridylate are mixed, a two-stranded helix is formed which is broken down at 30% of the rate of breakdown of polyadenylate itself. The author concludes that the enzyme attacks single-stranded polymers much more rapidly than multistranded RNA and that, as a consequence, RNA within the cell may well escape phosphorolysis even in the presence of a high concentration of inorganic phosphate. Skoda *et al.*[405] have observed that 6-azauridine pyrophosphate inhibits the phosphorylase from *E. coli*.

Ochoa *et al.*[450, 451] investigated the possibility that polyribonucleotide phosphorylase might be concerned in the DNA-dependent synthesis of RNA described in Section X, D, 2 above. The possibility arises since cell extracts usually bring about some degradation of nucleoside triphosphates to the diphosphates, so providing substrates for phosphorylase, while DNA might act by removing inhibitors or ribonucleases. The suggestion was abandoned, however, when it was found that the extract from *L. arabinosus* which brought about DNA-dependent synthesis of RNA was free from phosphorylase, while the similar activity in *A. vinelandii* was not affected by concentrations of inorganic phosphate which completely inhibited phosphorylase.

XI. Deoxyribonucleic Acid Synthesis

A. Synthesis by Intact Cells

Less detailed evidence is available concerning DNA synthesis in bacteria although many investigations have dealt with the synthesis of DNA in bacteriophage-infected bacteria and are dealt with elsewhere in this treatise. Braun[180, 181] have obtained evidence that bacterial DNA synthesis can be markedly stimulated by a factor present in digests of DNA; the factor is not replaced by deoxyribonucleotides. The formation of DNA by amino acid-requiring mutants of *E. coli* indicates that amino acids are stimulatory to synthesis, but this is not as dependent upon amino acids as RNA synthesis.[174, 406] Okazaki and Okazaki[407] have studied the synthesis of DNA in a strain of *L. acidophilus* requiring deoxyribosides for growth. In the absence of added deoxyriboside, DNA synthesis ceases but RNA and protein synthesis continue. If uracil is omitted, then synthesis of both RNA and protein ceases whereas that of DNA continues. However the further omission of amino acids in addition to uracil results in cessation of DNA synthesis.

A survey of the action of antibiotics on protein and nucleic acid synthesis[182] shows that those substances which inhibit protein synthesis also bring about an inhibition of DNA synthesis. The question therefore arises whether DNA synthesis is linked to prior protein synthesis; this problem has received particular attention with relation to the synthesis of DNA occurring in *E. coli* after infection with specific bacteriophage. The T-even series of coliphages contain hydroxymethylcytosine in place of cytosine in the nucleic acid (see Section II, B), so that the synthesis of phage nucleic acid as opposed to host nucleic acid can be relatively easily investigated. After infection, the DNA formed by the cell appears to be almost completely that characteristic of the phage. If 5-methyltryptophan or chloramphenicol is added to the system at the time of infection, or within a few minutes after infection, synthesis of DNA ceases and no DNA synthesis takes place in amino acid-requiring mutants unless the essential amino acid is added.[183-186] If chloramphenicol is added some time after infection has been established and DNA synthesis has started, then the antibiotic has little or no effect on that synthesis. These findings have been adduced as evidence for a need for protein synthesis to take place before DNA synthesis can occur. The phage DNA contains the new base, hydroxymethylcytosine, and this presumably means that an enzyme or enzymes must be produced to form the new base before phage DNA can be produced. However, this is not the whole answer since repetition of the same type of experiment with other phages, not containing hydroxymethylcytosine, shows that DNA synthesis is inhibited in some cases but not in others.[187-189] The fact that chloramphenicol in concentrations sufficient to prevent protein synthesis does not inhibit DNA formation in some instances would seem to abolish the argument that such DNA synthesis necessarily requires prior protein synthesis. The reasoning depends, however, upon chloramphenicol having a direct and primary action on protein synthesis *per se* and it is not possible, in the present state of knowledge, to be certain that this is in fact the case.

Hershey[190,191] found that if *E. coli* was infected with phage in the absence of P^{32} and then transferred to peptone containing labeled phosphate, there was a rapid incorporation of P^{32} into the RNA. Since there was little or no increase in RNA under these circumstances, it appeared that the incorporation must be due to turnover of RNA following infection; no turnover could be detected in uninfected cells. This finding has been followed up by Volkin and Astrachan[192-194] who have shown that there is a rapid turnover of a small fraction of the RNA after infection with phage. If the RNA is labeled prior to infection, and the system then diluted with unlabeled phosphate after infection, a rapid loss of label from RNA takes place but label continues to appear in DNA, suggesting that P may be transferred from RNA to DNA. In the presence of chloramphenicol, there is no loss of label from RNA after dilution, nor is there any transfer of P to DNA; on removal of the chloramphenicol, labeled P once more passes from RNA to DNA. Investigation of the composition of the RNA fraction which undergoes rapid turnover shows a striking correlation between the ratio of the specific activities of the nucleotides in the fraction, and the ratio of the analogous nucleotides in the phage DNA that is being synthesized. The authors suggested that this may, again, be an indication of a role of protein synthesis in DNA formation but further work, since 1959, has modified the interpretation of these findings[408, 409] (see Discussion). If chloramphenicol is added to the system prior to infection, the composition of the RNA nucleotides after infection is the same as that of uninfected cells. If the antibiotic is added immediately before or after infection, both RNA turnover and DNA synthesis are inhibited. If the antibiotic is added some 9 minutes after infection, then P^{32} incorporation into DNA increases, and RNA turnover takes place. Consequently RNA turnover stops when DNA synthesis is inhibited, but DNA synthesis proceeds when RNA turnover is occurring. One explanation put forward by Astrachan and Volkin[408] at this stage was that, in the presence of chloramphenicol, RNA is synthesized from precursors which are also used for the synthesis of DNA. Experiments were next carried out with a mutant of *E. coli* which required both adenine and arginine for growth and phage reproduction. Deoxyadenosine would not substitute for adenine for phage growth unless infected cells were first preincubated in the presence of adenine. In the latter condition the subsequent yield of phage could be correlated with the amount of specific RNA formed during the preincubation period. Uninfected cells, put through the same procedure, incorporated adenine into the nucleotide pool, DNA, and RNA (the labeling of RNA being different from that occurring in infected cells) but were unable to support phage growth on subsequent infection and incubation in the presence of deoxyadenosine and arginine. The conclusion drawn rom these experiments was that a synthesis of specific RNA is necessary before synthesis of phage protein and DNA can take place. It is possible that formation of this specific RNA is a prerequisite of the synthesis of either phage protein or phage DNA or both. Nomura *et al.*[410] have investigated the nature of the RNA newly synthesized by *E. coli* after T2 infection and found that it consists of two fractions: a specific ribosomal RNA associated with 30 *S* particles but readily dissociated therefrom, and a sRNA fraction with a base composition equivalent to that of the specific ribosomal RNA (see Table V). When the 30 *S* ribosomes are suspended in low magnesium concentrations, the new RNA is released in a form with sedimentation constant 8 *S*.

B. Synthesis by Cell-Free Systems

A system accomplishing the formation of polydeoxyribonucleotides has been described by Kornberg and his colleagues.[178, 179] The enzyme has been extracted from *E. coli* and carries out a reaction which can be represented as:

$$\begin{array}{l} n\text{dAPPP} \\ n\text{dGPPP} \\ n\text{dCPPP} \\ n\text{TPPP} \end{array} + \text{DNA} = \text{DNA}\!-\!\left[\begin{array}{c} \text{dAP} \\ \text{dGP} \\ \text{dCP} \\ \text{TP} \end{array}\right]_n + 4n\text{PP}$$

where dAPPP = deoxyadenosinetriphosphate, dAP = deoxyadenosinemonophosphate, PP = pyrophosphate, G = guanosine, C = cytosine, T = thymine, etc. The four nucleoside triphosphates are necessary for the reaction, omission of any one of the four reducing the action to about 1% of that obtained in the presence of the four. Diphosphates do not substitute for the triphosphates and a DNA primer is essential. Under optimal conditions a considerable synthesis of polymer occurs to give a product with molecular weight which may be as high as 5×10^6. When only one nucleoside triphosphate is provided as substrate, a small but significant reaction takes place which involves a new residue being attached to the end of the primer. If nucleoside triphosphates containing certain analogs of natural bases are supplied, the enzyme brings about incorporation of the analogs into the polymer; thus 5-bromouracil becomes incorporated in place of thymine and 5-methylcytosine in place of cytosine.

In the presence of the four deoxyriboside triphosphates, the rate of polymerization is many times greater than with a single triphosphate and the composition of the product is related to that of the primer. When calf thymus DNA is used as primer, the molecular weight of the product formed is approximately equal to that of the primer.[411] The number of residues of adenine incorporated is equivalent to those of thymine, while guanine and cytosine are incorporated in approximately equivalent amounts (see Section II, A) but when the primers are altered so that the ratio of the pairs, i.e. ratio of adenine (= thymine) to guanine (= cytosine), varies between 0.5 and 40, then the base ratios of the products reflect the composition of the primers.[412] Bollum[413-417] has obtained a polymerase with similar properties from calf thymus and found that native DNA preparations which did not act as primers became active after "denaturation" at 99° for 10 minutes. Sonicated DNA preparations that had lost their activity[418] regained their function as primers after such heat denaturation. It seems probable that "denaturation" in this sense involves separation of the strands of the DNA and that the single-stranded DNA then acts as a primer by being replicated in the presence of the polymerase. "Denatured" DNA has also been implicated as the primer in the bacterial polymerase system[419] where one of the most effective primers is the DNA from bacteriophage ϕX174[420] in which the DNA exists as a single stranded structure of molecular weight 1.7×10^6.

The bacterial polymerase is not primed by DNA preparations that have been digested by deoxyribonuclease I but such digests will stimulate the calf thymus enzyme. Bollum[416] finds that the latter enzyme will incorporate single deoxyriboside triphosphates in reactions primed with homogeneous polydeoxyribonucleotides. Schachman *et al.*[421] have recently found that the bacterial polymerase incubated with the four deoxyriboside triphosphates in the absence of primer will give rise, after a long lag period, to a polymer of deoxyadenylic and thymidylic acids. dCMP and dGMP do not occur in the product nor does the presence of dCTP and dGTP in the incubation mixture have any effect on the reaction. The polymer product consists of alternating units of deoxyadenylic acid and thymidylic acid and appears to be double stranded. It seems probable that a dinucleotide of dAMP and TMP is produced during the lag period and this then acts as a primer for synthesis of a copolymer. If this interpretation is correct, the mechanism of the initial formation of the dinucleotide should throw considerable light on the nature of the whole polymerase system. Hurwitz[422] has found that incubation of the bacterial polymerase with an extract from *E. coli* and a mixture of ribonucleotide triphosphates results in incorporation of ribonucleotides into DNA-like material; degradation of the product releases dinucleotides containing both ribo- and deoxyribonucleotides.

The enzyme is reversible and will bring about the degradation of polymers in the presence of excess pyrophosphate. The degradation reaction is again dependent upon the presence of DNA but, when measured by exchange of labeled PP with added triphosphate, is relatively little affected by omission of three out of four triphosphates. Thus, whereas the polymerization reaction with one triphosphate proceeds at 1% of the rate with four, the pyrophosphate exchange with one triphosphate present takes place at 45% of the rate obtained when four are present.

XII. Chloramphenicol

Mention has been made on several occasions of the use of chloramphenicol as an inhibitor of protein synthesis. Any substance inhibiting an unknown process is worthy of investigation since it may well throw light on the process itself, e.g., inhibitors have been of considerable use in studies of biosynthesis as they frequently lead to accumulation of intermediates whose further utilization is blocked by the inhibitor. The evidence that chloramphenicol inhibits protein synthesis, rather than any of the ancillary processes necessary for that process, rests on a number of studies in which it has been shown that the formation of proteins or enzymes in sensitive cells is prevented by growth-inhibitory concentrations, whereas other metabolic reactions, such as glycolysis, respiration, membrane transport, nucleic acid synthesis, glutathione synthesis, wall peptide synthesis, etc., are either insensitive or attacked at concentrations much higher than those needed to prevent growth.[41,49,51,55,195-202]

In the staphylococcus, concentrations of chloramphenicol which limit protein synthesis give rise to an enhanced rate of RNA synthesis; in other organisms RNA synthesis is unaffected by concentrations of the antibiotic giving complete inhibition of protein synthesis. Consequently the presence of chloramphenicol leads to complete dissociation of two processes—protein and nucleic acid syntheses—which appear to be intimately connected in all other conditions. This has led workers to turn their attention to the nature of the RNA that is formed in the presence of the antibiotic. Gros and Gros,[175] and Pardee and Prestidge[174] reported that its formation resembled that of normal RNA in that the full complement of amino acids was still necessary for significant synthesis. Neidhardt and Gros,[203] however, found that it differed from normal RNA in that it proved to be unstable in cells from which chloramphenicol had been removed; the breakdown could be prevented by chloramphenicol and, if breakdown had taken place, addition of the antibiotic led to resynthesis of RNA-like material. Horowitz and colleagues[205] confirmed this instability of the RNA and showed that it was equally unstable in the presence of chloramphenicol but that the presence of the antibiotic promoted synthesis at a rate exceeding the rate of breakdown.

Much of our thinking about RNA has been altered during recent years by the discovery that there are a number of different molecular species of RNA, and aggregates of ribonucleoprotein, in cells and that, in particular, there is a small molecular weight species, the soluble RNA, which may play a role as a carrier of activated amino acyl residues (see Section VII, A). Pardee *et al.*[204] have investigated the nature of the RNA extracted from *E. coli* by zone electrophoresis on starch and shown that it consists of a number of components of differing electrophoretic mobility. The faster moving components are of smaller molecular weight and different base composition from the major slow-moving components. Extracts from cells incubated in the presence of chloramphenicol showed an increase in the amount of fast moving, small molecular weight components. If P^{32} was included in the medium, radioactivity normally became incorporated into all the electrophoretic fractions but, when chloramphenicol was present, incorporation took place almost exclusively in the fast-moving fractions. Dagley and Sykes[206] have also shown that the presence of chloramphenicol leads to the formation of ribonucleoprotein aggregates with smaller *S* values in the ultracentrifuge than those obtained in normal extracts under otherwise similar conditions. Nomura and Watson[423] compared the sedimentation patterns of ribonucleoprotein extracts from *E. coli* grown in the presence and absence of the antibiotic. They found the normal pattern of 30 and 50 *S* particles in extracts from normal cells but a new component at 15 *S* appeared in extracts from chloramphenicol-grown cells. Since in the latter case, the 30 *S* component was missing and the 50 *S* particle greatly reduced, it seemed that there might be breakdown of the normal organization of the ribonucleoprotein in addition to formation of new material. Table V gives figures for the base analysis of RNA extracted from *E. coli* grown in the presence of chloramphenicol.[238] It is clear from these studies that chloramphenicol, in concentrations which inhibit protein synthesis, has an effect on RNA synthesis which leads to the accumulation of "small" molecular weight materials.

These results have been confirmed and extended in *Staphylococcus*.[56] If labeled glutamic acid is added to the incubation mixture, incorporation normally occurs into all the electrophoretic fractions of the nucleoprotein, with some 60% of the total label being associated with slow-moving components. If chloramphenicol is present during incubation, glutamic acid becomes fixed by fast-moving components but not by the slow-moving fractions and a distribution is obtained which is essentially similar to that occurring when disrupted staphylococci are incubated with ATP and glutamic acid alone (see Section VII, B). Chloramphenicol does not affect the activating enzymes nor does it affect the binding of glutamic acid by the primary acceptors in the fast-moving electrophoretic fractions. Lacks and Gros[129] have studied the fixation of amino acids by the soluble RNA fraction of *E. coli* and find that neither fixation nor release is affected by chloramphenicol, although the rate of turnover is decreased as a result of the inhibition of protein synthesis by the antibiotic. In the recent studies by Tissières *et al.*[277] on amino acid incorporation by ribosomes isolated from *E. coli*, the over-all process was 84% inhibited by 30 μg. chloramphenicol per milliliter, the concentration required to inhibit protein synthesis in intact cells. Nathans and Lipmann[367] find that 60 μg. chloramphenicol per milliliter gives 65% inhibition of the transfer of amino acyl residues from sRNA to ribosomes in the *E. coli* system. It appears that chloramphenicol inhibits a stage in nucleoprotein synthesis which occurs later than the formation of polyribonucleotides and the fixation of amino acyl residues by the soluble RNA fraction but before the formation of peptide bonds in the ribosome.

The question arises of the nature of the selective action of chloramphenicol. Synthesis of protein by intact bacteria is completely inhibited by concentrations of the order 1–10 μg./ml. An amount of 30 μg./ml. gives 84% inhibition of amino acid incorporation by ribosomes isolated from *E. coli*, but investigations on incorporation in isolated mammalian ribosomes show insignificant inhibition by concentrations of 300 μg./ml. Table XII provides data collected from studies with various types of tissue and subcellular preparation; unfortunately few of the papers quoted give the effects of a wide range of antibiotic concentration. It is a general finding that chloramphenicol, although able to inhibit the production of specific proteins by bacteria to the extent of 100%, does not give complete inhibition of amino acid incorporation in subcellular systems even at high concentrations, the inhibition approaching a plateau value at 80–90%. Consequently, a single finding that 3000 μg. chloramphenicol per milliliter gives 81% inhibition in any system does not provide all the information necessary, since lower concentrations of antibiotic might well give inhibition of the same order. Nevertheless, it does seem clear from the data in Table XII that incorporation of amino acids by ribosomes from mammalian and plant sources is significantly less sensitive to inhibition by chloramphenicol than is the process in ribosomes from *E. coli*. There are a number of possible explanations: the incorporation process in bacterial ribosomes may differ in some respect from the process in extracts from higher organisms; the process of synthesis of soluble protein (which is the main function of ribosomes from mammalian sources such as liver) is less sensitive than the process of synthesis of the structural protein of ribosomes themselves which may also occur in ribosomes from bacteria harvested in exponential growth; "ribosomes" from bacteria may differ from those in higher organisms. The last possibility is supported by the work of Mager[428] who was puzzled by the insensitivity of incorporation in "microsomes" from *Tetrahymena pyriformis* despite the sensitivity of the growth of the whole organism to chloramphenicol. He found that amino acid incorporation took place in two cell fractions resembling, on grounds of particle size, "microsomes" and "mitochondria," respectively. The process in the latter, larger particles was sensitive to chloramphenicol whereas that in the "microsomes" was relatively insensitive (see Table XII).

TABLE XII

EFFECT OF CHLORAMPHENICOL ON REACTIONS ASSOCIATED WITH PROTEIN SYNTHESIS IN PREPARATION FROM CELLS OF DIFFERENT SPECIES

Preparation	Source	Type of experiment	Chloramphenicol (μg./ml.)	% Inhibition	Reference
Intact cells	*S. aureus*	Protein synthesis	10	100	197
Disrupted cells	*S. aureus*	Leucine incorporation	30	82	49
Ribosomes	*E. coli*	Amino acid incorporation	30	84	277
			100	84	277
Ribosomes	Maize endosperm	Amino acid incorporation	200	27	424
			400	62	424
Ribosomes	Pea seedling	AMP-ATP exchange	3000	81	122
		Soluble protein synthesis	300	83	344
Nuclei	Calf thymus	Amino acid incorporation	100	10	425
			2200	"almost complete"	426
Microsomes	Rat liver	Amino acid incorporation	250	negligible	428
"Microsomes"	*Tetrahymena pyriformis*	Amino acid incorporation	50	1	428
"Mitochondria"	*Tetrahymena pyriformis*	Amino acid incorporation	50	78	428
Soluble extract	Pigeon pancreas	Amylase formation	1	"complete"	427
Ribosomes	*E. coli*	Transfer leucyl-sRNA to ribosome	60	67	367

XIII. Discussion (1959)

The mechanism of protein and nucleic acid synthesis has for years been the subject of much speculation and device. This was inevitable as long as experimental evidence was scanty and difficult to obtain, but the present situation in which experimental development is rapid means that theories can now await the discovery of facts. The present is not a good time to attempt a synthesis of knowledge or a forecast of mechanism as we are only now beginning to learn some of the facts and to see the nature of past misinterpretations. For this reason the account in this chapter has been presented with emphasis on practical findings. The discovery which is exercising many minds in this field at the moment is the nature and role of the soluble RNA fraction. We can no longer refer simply to the RNA of a cell but must specify what sort of RNA—this may be a beginning of a still more elaborate specification in which we must characterize the particular polyribonucleotide structure associated with a specific amino acid. On the other hand, there are the experiments of recent date in which soluble RNA fractions and activating enzymes are not definitely involved although a process of protein synthesis can be demonstrated. Then there is the peculiar position of the cell-free enzymes which produce polynucleotides apparently akin to ribonucleic acid and deoxyribonucleic acid but under conditions so much simpler than those required for the processes *in vivo*. It is a well-known psychological trait in man that he will attempt to place facts and events in simple, linear, "logical" series and it may well be that our present confusion over matters of mechanism is due to our attempt to impose order on a few isolated events which are really components of a highly complex continuum.

The big problem—and one which is not touched by our present knowledge—is that of imposing specific sequences on amino acids in a polypeptide and on nucleotides in a polynucleotide. The impressive though indirect evidence associating DNA, RNA, and proteins has led to a belief that specificity of structural synthesis lies in the relationship between these macromolecules.[207,208] The part played by DNA in genetic control suggests that the structure of the DNA molecule carries the information needed for ordering amino acid sequences into specific proteins. The variables along a DNA molecule consist of the bases and the only way we can conceive of information being carried is in the sequence of these bases. If base sequences determine amino acid sequences, then the bases must function in groups of two or three if they are to determine the position of twenty different amino acids and the suggestion has been made that amino acids are coded by specific sequences of bases taken two or three at a time. A mutation affecting a single base pair would thus alter the coding and might lead to the insertion of a "wrong" amino acid in a given polypeptide sequence. This situation has been realized in the case of sickle-cell anemia in which Ingram[11] has shown that one glutamic acid residue of normal hemoglobin has been replaced by a valine residue; whether this is produced by a change in the corresponding nucleic acid has, of course, yet to be investigated.

On this general hypothesis, a deoxyribonucleotide chain carries the coding for a specific protein. If each protein had to be made on the surface of the DNA, steric difficulties might well arise and control of the relative amounts of different proteins would introduce mechanical problems. This would be overcome if the information carried by the DNA were translated into amino acid sequences by the mediation of an independent catalyst: this is the role envisaged for RNA. Specific RNA molecules would be made in conformation with a DNA template and would carry the same coding as the DNA; this specific RNA would then leave the DNA and become a focus for the condensation of amino acids in a sequence determined by the coded information. It is generally assumed that the RNA carrying the information for protein synthesis is that which is associated with the ribonucleoprotein of the microsome (in mammalian systems) and the role, if any, of the soluble RNA fraction in this scheme has yet to be explained.

It will be clear that experimental support for such hypotheses has yet to be obtained. The evidence so far available appears to relate to events concerned in the preparation of amino acids and nucleotides for their later condensation into specific macromolecules. In Section X evidence was presented which shows that DNA or its precursors are required for the synthesis of RNA. Goldstein and Plaut,[209] working with amoebae, have shown that RNA synthesis takes place inside the nucleus and that the RNA then passes into the cytoplasm. They further demonstrated that amino acid incorporation can take place into the protein of the cytoplasm in enucleated organisms. That RNA can carry information and control the synthesis of specific protein is also shown by the ability of the nucleic acid moiety of tobacco mosaic virus to establish infection and give rise to complete virus.[210-213] While these findings are in general agreement with the hypothesis set out above, there is clearly very much more work to be done before the mechanism of protein synthesis is understood.

XIII. Discussion (1961)

Two years have elapsed since material was originally collected for this chapter. The period between finishing the script and correcting the galley proofs has seen the publication of over 300 papers relating to the general field which this chapter is supposed to cover, and references to some 200 of these have been darned in. A number of reviews and symposia have also appeared.[429-440]

Experimental facts relating to the mechanism of protein and nucleic acid synthesis are accumulating at increasing pace although understanding of the field as a whole is still in its infancy. It is inevitable that major changes in thought and comprehension will take place in a period as long as two years, making the task of bringing an article such as this up-to-date both interesting and frustrating. One of the major changes in thinking about protein synthesis in bacteria relates to the site where that process takes place within the cell. Until 1959 experiments had suggested that bacteria differed from other cells in that the site of peptide bond formation in bacteria was associated with membrane fractions whereas in cells of higher organisms this site was to be found in the ribosomes. The work of McQuillen *et al.*[275] showed that pulse experiments of sufficiently short duration led to the conclusion that ribosomes were also the primary site of peptide bond formation in *E. coli*. Amino acid activating enzymes and sRNA had already been demonstrated in bacteria and experiments were quickly designed to show that the events of what might be called the Hoagland-Zamecnik system also take place in bacterial preparations. There are still doubts, indicated in the relevant sections above, that the ribosomes *alone* are responsible for peptide bond formation and synthesis of soluble protein in bacteria, or that the activating enzyme-sRNA-ribosome pathway is the *only* pathway of protein synthesis in bacteria or other tissues. Certainly the contributions of the last two years have added many details to the pathway and increased its verisimilitude. The demonstration that bacterial cells (assuming that facts shown for one or two species apply to all) fall into line has hardened the belief of many workers that this pathway is the true route of protein biosynthesis. However, hardening of the shell does not mean that the egg will hatch and there are quantitative considerations that are, at present, difficult to reconcile with the present theory (see, for example, the excellent review by Raacke[431]). The time is nevertheless early, and further work may well abolish many of the present difficulties without necessarily leading to drastic reconstruction of the biochemical framework at present erected.

Important developments have taken place in the field of RNA synthesis with the separation of processes involved in terminal addition of nucleotides to sRNA from those concerned in what would appear to be polynucleotide synthesis *de novo* which is dependent upon the presence of DNA. The role of polyribonucleotide phosphorylase is still obscure and it may be that it is primarily concerned with breakdown, especially since the recent studies described on p. 549 indicate that functional forms of RNA and sRNA are not susceptible to phosphorolysis. Considerable progress has been made in the characterization of sRNA and the separation of specific amino acid-accepting polynucleotides. Glimmerings of mechanism are appearing in the reversible addition of amino acyl-sRNA residues in N-terminal positions in peptide chains,[112] in the transient binding of the sRNA moiety during incorporation in the ribosome,[338] and in the re-cycling of the discharged sRNA[367]. Deductions are drawn at present from experiments conducted with preparations from widely different tissues and may be justified on the belief that the general mechanism of a common and fundamental process such as protein biosynthesis may well be the same in all tissues. The species non-specificity of sRNA in incorporation systems *in vitro* is in accord with this belief. If the general chemical mechanism is so conveniently nonspecific, we must be prepared for a very fine control over those mechanisms.

Protein synthesis is still a happy hunting ground for speculation and a treatise such as this should attempt to give some indication of the trend of speculative thought at the time of writing (which is not the same as the time of publication). In 1961, a change is taking place in the attitude toward ribosomes. Pulse experiments indicate that the ribosome is the site within the cell where amino acids are bound in specific peptide sequences. Early calculations of the number of ribosomes in a cell and of the amount of amino acid taken up per ribosome suggested that each ribosome possessed the equipment for making one protein and that, consequently, different ribosomes were specific for the fabrication of different proteins. Attempts have indeed been made to use serological methods to locate specific proteins and separate the ribosomes involved in their synthesis.[265] The basic idea, or "central dogma,"[208] underlying speculation on protein synthesis still involves the template and transmission of information by coding systems as outlined in 1959. If the ribosome is the site of specific sequence formation, then the template must be in the ribosome and presumably lies in the ribosomal RNA. The protein moiety of the ribosome is now known to be mainly structural while protein-in-the-course-of-synthesis represents only a very small fraction of the ribosomal protein. The primary source of coding information is believed to be DNA.[437] If the code is constructed of nucleotide sequences in the DNA and this is transmitted through the medium of RNA, then (1) RNA synthesis should be modeled on DNA and (2) there should be correlation between nucleotide sequences of specific RNA fragments and the DNA supplying the code for their formation. The recent investigations[397, 398, 449, 452] on RNA synthesis which requires priming with DNA provide evidence for the existence of the former process while the work of Volkin and Astrachan[192-194, 408, 409] first revealed the existence of RNA with a high rate of turnover and base composition akin to that of DNA in bacteriophage infected *E. coli*. This "new" RNA represents only a small proportion of the total RNA of the cell.[410, 442] RNA with a high rate of turnover and composition related to DNA has now been described in yeast,[443] and Gros[384, 453] has shown that P^{32} is incorporated, within a few seconds of administration to *E. coli*, into RNA of relatively small molecular weight (12–15 *S*) which, at a later stage, becomes associated with ribosomal particles but forms only a very small proportion of the ribosomal RNA. Brenner *et al.*[442, 453] have obtained evidence that there is no significant increase in the ribosomes of *E. coli* after infection with bacteriophage despite the production of new enzymes by the infected cell. These findings have been brought together by the development of a new theory—and the reader is also referred to Jacob and Monod[441] and Lwoff[444]—based on the assumption that the ribosome itself is a nonspecific site of synthesis that is made functional and specific by association with specific or "messenger RNA." "Messenger RNA" is synthesized with DNA as primer and carries the code of the

DNA to the ribosome so that the synthetic capacity of the ribosome is then determined and controlled by the presence of this specific RNA. The labile nature of "messenger RNA" conveniently explains the instability of the isolated ribosome as an *in vitro* system for protein synthesis and also the dependence of inducible enzyme synthesis on continued RNA synthesis (see Section III, B). On this basis we must consider that at least three types of RNA are involved in protein synthesis: transfer RNA with 20 or more components, each specific for the transfer of an amino acid (or peptide?); messenger RNA determining amino acid sequences at the site of peptide bond formation; and the RNA moiety of ribosomal nucleoprotein providing the site where transfer RNA and messenger RNA meet to accomplish the forging of specific peptide bonds. This is an exciting idea and provides a new basis for experimental investigation.

There have been many and major advances since 1959 but, in the main, the same problems exist and some of the conclusions still hold. Especially "the present is not a good time to attempt a synthesis of knowledge or a forecast of mechanism" and "there is clearly very much more work to be done before the mechanism of protein synthesis is understood."

Addendum (1962)

The 9 months that have passed since correction of galley proofs have seen changes and advances that are of funadmental importance to the understanding and interpretation of much that has been written in this chapter.

Reference is made, immediately above, to the hypothesis that the ribosome is a relatively nonspecific center of protein synthesis and that the activity of this center, together with the specificity of its synthetic activity, might be determined by the presence of an unstable "messenger RNA" synthesised in conjunction with, and carrying the code of, DNA.[442, 453] Matthaei and Nirenberg[454] have now obtained extracts of *E. coli* which can be dialyzed and stored without marked loss of activity. Incorporation of amino acids into acid-precipitable material by these preparations requires the presence of ribosomes, supernatant fractions, ATP, and ATP-generating system; the process is stimulated by the presence of an amino acid mixture, and is inhibited by chloramphenicol or puromycin. Incorporation was found to be sensitive to ribonuclease, whereas deoxyribonuclease had little effect on the initial rate of incorporation but markedly decreased the amount of amino acid incorporated after the first 10–15 minutes incubation. Incorporation by deoxyribonuclease-treated preparations was markedly stimulated by the addition of ribosomal RNA but not by sRNA. Ribosomal RNA from a number of difference sources (yeast, tobacco mosaic, *E. coli*) was effective and there appeared to be a quantitative relationship between the amount of RNA added and the amount of amino acid incorporated.

This general finding has been confirmed by other workers and attention turned to the nature of the stimulatory RNA. Tissières and Hopkins[455] found that stimulation could be produced by addition of DNA and the four nucleoside triphosphates under conditions in which the DNA-dependent RNA polymerase (see Section X, D, 2) was active, and that the stimulatory material separated with properties similar to those of the "messenger RNA" described by Gros *et al.*[453] Further investigation of the RNA polymerase from *E. coli* has shown that it can be primed by a variety of DNA preparations including both single-stranded and double-stranded DNA from ϕX174 bacteriophage.[456] Wood and Berg[457] have carried the investigation a step further and studied the effect on amino acid incorporation in the *E. coli* system of the addition of purified RNA polymerase together with a variety of DNA primers. The unprimed polymerase was without effect but RNA synthesized by the enzyme in the presence of primer DNA was active; however, the degree of activity varied with the source of the primer DNA. DNA from T6, T2, or T5 phage increased the rate of incorporation, and increased its extent by as much as 20 times. DNA from *E. coli*, λ phage, salmon sperm, or calf thymus was much less effective and ϕX174 DNA was inactive.

It is clear from these experiments that amino acid incorporation in the *in vitro* system is limited by the amount of specific RNA present, and that stimulatory material can be synthesized by DNA-primed RNA polymerase. Since the polymerase can be primed by a wide variety of different DNA preparations (but the products are not equally effective in promoting amino acid incorporation by the ribosomes), it would seem that there may be some further site of specificity, possibly in the ribosome itself. Nevertheless, this recent work provides impressive evidence in favor of the "messenger" theory. An odd finding ("odd" in the sense that it does not yet fit readily into the theory) is that purified polymerase incubated with ATP alone synthesizes polyadenylate; the synthesis still requires a DNA primer.[456] The other nucleoside triphosphates do not give rise, under similar conditions, to production of homopolymers but their addition to ATP reduces the amount of polyadenylate formed

The fact that amino acid incorporation could be stimulated by addition of RNA led Nirenberg and Matthaei[458] to test the effect of addition of polynucleotides synthesized by means of polyribonucleotide phosphorylase (see Section X, D, 3). Addition of polyuridylic acid (polyU)—but not other homopolymers—gave a startling increase in the incorporation of phenylalanine but not of other amino acids. The process yielded material with properties resembling those of poly-L-phenylalanine and involved phenylalanyl-sRNA as an intermediate.[459] It seemed from this result that polyU can "code" phenylalanine and that the polynucleotide was acting as a "messenger" or template promoting the synthesis by the ribosomes of the corresponding amino acid polymer. Polyribonucleotide phosphorylase provides a means for the preparation of mixed polymers in which the relative proportions of the various nucleotides, but not their sequences, can be calculated, and the experiments of Nirenberg and Matthaei provide a system in which the possible activity of "synthetic" polymers as templates for amino acid incorporation can be tested. A series of such investigations is now in progress.[460-462] No results with the definiteness of the relationship between polyU and phenylalanine have so far emerged, and many polymer preparations are found to exert some stimulatory effect on the incorporation of several amino acids. Genetic analysis of deletion and addition mutants within a single cistron of bacteriophage T4 by Crick *et al.*[463] suggests that the DNA code consists of non-overlapping triplets, read in one direction from a fixed point, with the probability that the code may be degenerate in the sense that one amino acid may be coded by one of several triplets. The explanation put forward[461] for the spread of results in incorporation experiments with synthetic polymers is that, for example, a polymer of uridylic (U) and cytidylic acids (C) will contain all triplet combinations of U and C and will consequently promote, to some extent, the incorporation of any amino acid coded by any one of those triplets. One approach is to determine the incorporation, relative to that of phenylalanine, of a given amino acid and compare this with the probable amount, relative to UUU, of a given triplet in the polynucleotide; in this way it should be possible to deduce which triplet codes the amino acid. This method, as at present applied, involves a number of assumptions and an uncertain decision concerning the significance of incorporated radioactivity; nevertheless, two groups of workers have published lists of "codes" for 13 amino acids, and these lists are in reasonable agreement.[461, 462] Papers are appearing in rapid profusion often supported by (sometimes conflicting) footnotes while the clearest expositions, with the farthest flung deductions, appear in the daily newspapers.[464]

Template or messenger RNA forms only a small part of the total RNA of the cell, and the synthesis and nature of the presumably nonspecific ribosomal RNA remains to be elucidated. Britten, McCarthy, and Roberts[465] have produced evidence that ribosomes are assembled from rapidly labeled "eosomes" sedimenting at 14 *S*; there would appear to be some confusion at present in the interpretation of data involving ribonucleic acid sedimenting at about 14 *S* in that such material may be template RNA, ribosomal RNA precursor, or possibly a mixture of both.

There is, without doubt, a great advance inherent in these latest discoveries but much careful and detailed work will be required before the exact nature of the processes, primers, promoters, and products are known and the true significance of the advance assessed. Of immediate importance is that a new experimental approach has been opened up, an understanding of sequence determination seems to be within reach, and an understanding of the relative inefficiency and instability of the "old" *in vitro* system provided. And this is a very pleasant note on which to end what has almost become a "history of protein synthesis."

References

[1] J. R. Porter, "Bacterial Chemistry and Physiology," p. 360. Wiley, New York, 1946.
[2] C. S. Cummins and H. Harris, *J. Gen. Microbiol.* **14,** 583 (1956).
[3] E. Bricas and C. Fromageot, *Advances in Protein Chem.* **8,** 1 (1953).
[4] J. S. Fruton and S. Simmonds, "General Biochemistry," p. 126. Wiley, New York, 1953.
[5] R. P. Ambler and M. W. Rees, *Nature* **184,** 56 (1959).
[6] S. S. Cohen and W. M. Stanley *J. Biol. Chem.* **144,** 589 (1942).
[7] M. Dixon and E. Webb, "The Enzymes," p. 185. Longmans, Green, New York, 1958.
[8] F. Sanger and H. Tuppy, *Biochem. J.* **49,** 463 (1951).
[9] F. Sanger and E. D. P. Thompson, *Biochem. J.* **53,** 353 (1953).
[10] H. Brown, F. Sanger and R. Kitai, *Biochem. J.* **60,** 556 (1955).
[11] V. M. Ingram, *Nature* **180,** 326 (1957).
[12] J. T. Edsall, *J. Cellular Comp. Physiol.* **47,** 163 (1956).
[13] C. B. Anfinsen and R. R. Redfield, *Advances in Protein Chem.* **11,** 1 (1956).
[14] L. Pauling, R. B. Corey, and H. R. Branson, *Proc. Natl. Acad. Sci. U. S.* **37,** 205 (1951).
[15] L. Pauling, *in* "Les proteines: rapports et discussions" (R. Stoops, ed.), p. 63. Inst. Intern. de Chimie Solvay, Brussels, 1953.
[16] G. Ivanovics, *Zentr. Bakteriol. Parasitenk. Abt. 1 Origi.* **138,** 449 (1937).
[17] G. Ivanovics and V. Bruckner, *Z. Immunitätsforsch.* **90,** 304 (1937).
[18] M. R. J. Salton, *Symposium Soc. Gen. Microbiol.* **6,** 81 (1956).
[19] J. T. Park, *J. Biol. Chem.* **194,** 877, 885, 897 (1952).
[20] J. T. Park and J. L. Strominger, *Science* **125,** 99 (1957).
[21] J. J. Armstrong, J. Baddiley, J. C. Buchanan, B. Carss, and G. R. Greenberg, *J. Chem. Soc.* p. 4344 (1958).
[22] L. C. Craig, W. Konigsberg, and R. T. Hill, *in* "Amino Acids and Peptides with Antimetabolic Activity" (G. E. W. Wolstenholme and C. M. O'Connor, eds.), p. 226. Churchill, London, 1958.
[23] P. H. Bell, J. F. Bone, J. P. English, C. E. Fellows, K. S. Howard, M. M. Rogers, R. G. Shepherd, R. Winterbottom, A. C. Dormbush, S. Kishner, and Y. Subba-Row, *Ann. N. Y. Acad. Sci.* **51,** 897 (1949).
[24] G. R. Wyatt and S. S. Cohen, *Biochem. J.* **55,** 774 (1953).
[25] F. F. Davis and F. W. Allen, *J. Biol. Chem.* **227,** 907 (1957).
[26] M. Adler, B. Weissmann, and A. B. Gutman, *J. Biol. Chem.* **230,** 717 (1958).
[27] H. Amos and M. Korn, *Biochim. et Biophys. Acta* **29,** 444 (1958).
[28] D. B. Dunn and J. D. Smith, *Biochem. J.* **68,** 627 (1958).
[29] D. B. Dunn and J. D. Smith, *Proc. Intern. Congr. Biochem., 4th Congr., Vienna* **7,** 72 (1958).
[30] J. W. Littlefield and D. B. Dunn, *Nature* **181,** 254 (1958).

[31] W. E. Cohn, *Federation Proc.* **16,** 166 (1958).
[32] J. D. Smith and D. B. Dunn, *Biochim. et Biophys. Acta* **31,** 573 (1959).
[33] J. D. Watson and F. H. C. Crick, *Nature* **171,** 737 (1953).
[34] S. Spiegelman, A. I. Aronson, and P. C. Fitz-James, *J. Bacteriol.* **75,** 102 (1958).
[35] C. A. Knight, *Advances in Virus Research* **2,** 168 (1954).
[36] T. Caspersson, *Symposia Soc. Exptl. Biol.* **1,** 127 (1947).
[37] J. Brachet, *Arch. biol. (Liège)* **53,** 207 (1941).
[38] B. Malmgren and C. G. Heden, *Acta Pathol. Microbiol. Scand.* **24,** 417, 437, 448 (1947).
[39] P. C. Caldwell and and C. Hinshelwood, *J. Chem. Soc.* 3156 (1950).
[40] P. C. Caldwell, E. L. Mackor, and C. Hinshelwood, *J. Chem. Soc.* 3151 (1950).
[41] E. F. Gale and J. P. Folkes, *Biochem. J.* **53,** 483 (1953).
[42] A. B. Pardee, *Proc. Natl. Acad. Sci. U. S.* **40,** 263 (1954).
[43] S. Spiegelman, H. O. Halvorson, and R. Ben-Ishai, *in* "Amino Acid Metabolism" (W. D. McElroy and B. Glass, eds.), p. 124. Johns Hopkins Press, Baltimore, Maryland, 1955.
[44] E. H. Creaser, *Biochem. J.* **64,** 539 (1956).
[45] M. H. Richmond, *Biochim. et Biophys. Acta* **34,** 325 (1959).
[46] R. E. F. Matthews and J. D. Smith, *Nature* **177,** 271 (1956).
[47] J. D. Smith and R. E. F. Matthews, *Biochem. J.* **66,** 323 (1957).
[48] H. G. Mandel, *J. Biol. Chem.* **225,** 137 (1957).
[49] E. F. Gale and J. P. Folkes, *Biochem. J.* **59,** 661, 675 (1955).
[50] O. E. Landman and S. Spiegelman, *Proc. Natl. Acad. Sci. U. S.* **41,** 698 (1955).
[51] R. Hancock and J. T. Park, *Nature* **181,** 1050 (1958).
[52] E. F. Gale, *in* "Enzymes: Units of Biological Structure and Function" (O. H. Gaebler, ed.), p. 49. Academic Press, New York, 1956.
[53] H. Borsook *in* "Chemical Pathways of Metabolism" (D. M. Greenberg, ed.), Vol. 2, p. 173. Academic Press, New York, 1954.
[54] E. F. Gale and J. P. Folkes, *Biochem. J.* **55,** 721 (1953).
[55] J. Mandelstam and H. J. Rogers, *Nature* **181,** 956 (1958).
[56] E. F. Gale, "Synthesis and Organization in the Bacterial Cell." Wiley, New York, 1959.
[57] E. F. Gale, C. J. Shepherd, and J. P. Folkes, *Nature* **182,** 592 (1958).
[58] M. B. Hoagland, *Intern. Congr. Biochem., 4th Congr., Vienna* **8,** 199 (1958).
[59] M. B. Hoagland, P. C. Zamecnik, and M. L. Stephenson, *Biochim. et Biophys. Acta* **24,** 215 (1957).
[60] P. Castelfranco, A. Meister, and K. Moldave, *in* "Microsomal Particles and Protein Synthesis" (R. B. Roberts, ed.), p. 115. Pergamon Press, London, 1958.
[61] K. Moldave, P. Castelfranco, and A. Meister, *J. Biol. Chem.* **234,** 841 (1959).
[62] D. S. Hogness, M. Cohn, and J. Monod, *Biochim. et Biophys. Acta* **16,** 99 (1955).
[63] B. Rotman and S. Spiegelman, *J. Bacteriol.* **68,** 419 (1954).
[64] G. N. Cohen and J. Monod, *Bacteriol. Revs.* **21,** 169 (1957).
[65] J. Monod, "*The Bacteria*"
[66] J. Brachet, *Nature* **174,** 876 (1954).
[67] H. Koffler, *Bacteriol. Revs.* **21,** 227 (1957).
[68] B. A. D. Stocker, *J. Pathol. Bacteriol.* **73,** 314 (1957).
[69] D. Kerridge, *Biochim. et Biophys. Acta* **31,** 579 (1959).
[70] D. Kerridge, *J. Gen. Microbiol.* **21,** 168 (1959).
[71] C. Weibull, *Symposium Soc. Gen. Microbiol.* **6,** 111 (1956).
[72] K. McQuillen *Symposium Soc. Gen. Microbiol.* **6,** 127 (1956).

[73] S. Spiegelman *in* "Enzymes: Units of Biological Structure and Function" (O. H. Gaebler and E. B. Ford, eds.), p. 67. Academic Press, New York, 1956.
[74] R. L. Lester, *J. Am. Chem. Soc.* **75,** 5448 (1953).
[75] Mirko Beljanski, *Biochim. et Biophys. Acta* **15,** 425 (1954).
[76] S. Brenner, *Biochim. et Biophys. Acta* **18,** 531 (1955).
[77] S. Spiegelman, *in* "The Chemical Basis of Heredity" (W. D. McElroy and B. Glass, ed.), p. 232. Johns Hopkins Press, Baltimore, Maryland, 1956.
[78] Mirko Beljanski and S. Ochoa, *Proc. Natl. Acad. Sci. U. S.* **44,** 496 (1958).
[79] S. Spiegelman *in* "Recent Progress in Microbiology" (G. Tunevall, ed.), p. 81. Almquist and Wiksell, Stockholm, Sweden, 1959.
[80] B. Nisman, *Biochim. et Biophys. Acta* **32,** 18 (1959).
[81] B. Nisman and M. L. Hirsch, *Ann. inst. Pasteur* **95,** 615 (1958).
[82] R. B. Roberts, "Microsomal Particles and Protein Synthesis." Pergamon Press, London, 1958.
[83] H. K. Schachman, A. B. Pardee, and R. Y. Stanier, *Arch. Biochem. Biophys.* **38,** 245 (1952).
[84] A. Tissières and J. D. Watson, *Nature* **182,** 778 (1958).
[85] J. W. Littlefield, E. B. Keller, J. Gross, and P. C. Zamecnik, *J. Biol. Chem.* **217,** 111 (1955).
[86] M. Rabinowitz and M. E. Olson, *Exptl. Cell Research* **10,** 747 (1956).
[87] P. C. Zamecnik and E. B. Keller, *J. Biol. Chem.* **209,** 337 (1954).
[88] J. A. V. Butler, A. R. Crathorn, and G. D. Hunter, *Biochem. J.* **69,** 544 (1958).
[89] G. E. Connell, P. Lengyel, and R. C. Warner, *Biochim. et Biophys. Acta* **31,** 391 (1959).
[90] A. R. Gilby, A. V. Few, and K. McQuillen, *Biochim. et Biophys. Acta* **29,** 21 (1958).
[91] C. Weibull and L. Bergstrom, *Biochim. et Biophys. Acta* **30,** 340 (1958).
[92] A. R. Crathorn and G. D. Hunter, *Biochem. J.* **69,** 47P (1958).
[93a] R. J. Podolsky, *Arch. Biochem. Biophys.* **45,** 327 (1953).
[93b] J. Mandelstam, *Biochem. J.* **64,** 55P (1956).
[94] R. E. Monro, *Biochem. J.* **81,** 225 (1961).
[95] H. O. Halvorson and S. Spiegelman, *J. Bacteriol.* **64,** 207 (1952).
[96] H. O. Halvorson, S. Spiegelman, and R. L. Hinman, *Arch. Biochem. Biophys.* **55,** 512 (1955).
[97] R. Munier and G. N. Cohen, *Biochim. et Biophys. Acta* **21,** 592 (1956).
[98] G. N. Cohen and R. Munier, *Biochim. et Biophys. Acta* **31,** 347, 375 (1959).
[99] G. W. Kidder and V. C. Dewey, *Biochim. et Biophys. Acta* **17,** 288 (1955).
[100] D. Gross and H. Tarver, *J. Biol. Chem.* **217,** 169 (1955).
[101] A. B. Pardee, V. G. Shore, and L. S. Prestidge, *Biochim. et Biophys. Acta* **21,** 406 (1956).
[102] G. N. Cohen, *Ann. inst. Pasteur* **94,** 15 (1958).
[103] M. B. Hoagland, *Biochim. et Biophys. Acta* **16,** 288 (1955).
[104] M. B. Hoagland, E. B. Keller, and P. C. Zamecnik, *J. Biol. Chem.* **218,** 345 (1956).
[105] R. D. Cole, J. Coote, and T. S. Work, *Nature* **179,** 199 (1957).
[106] V. V. Koningsberger, A. M. van de Ven, and J. T. G. Overbeek, *Koninkl. Ned. Akad. Wetenschap., Proc. Ser. B* **60,** 141 (1957).
[107] E. W. Davie, V. V. Koningsberger, and F. Lipmann, *Arch. Biochem. Biophys.* **65,** 21 (1956).
[108] P. Berg, *J. Biol. Chem.* **222,** 1025 (1956).
[109] N. Sharon and F. Lipmann, *Arch. Biochem. Biophys.* **69,** 219 (1957).
[110] F. Lipmann, *Proc. Natl. Acad. Sci. U. S.* **44,** 70 (1958).

[111] J. A. DeMoss and G. D. Novelli, *Biochim. et Biophys. Acta* **22,** 49 (1956).
[112] G. C. Webster, *Arch. Biochem. Biophys.* **82,** 125 (1959).
[113] E. F. Gale, *in* "Recent Progress in Microbiology" (G. Tunevall, ed.), p. 104. Almqvist & Wiksell, Stockholm, Sweden, 1959.
[114] E. F. Gale and J. P. Folkes, *Biochem. J.* **69,** 611 (1958).
[115] E. F. Gale and J. P. Folkes, *Biochem. J.* **69,** 620 (1958).
[116] M. B. Hoagland, M. L. Stephenson, J. F. Scott, L. I. Hecht, and P. C. Zamecnik, *J. Biol. Chem.* **231,** 241 (1958).
[117] P. Berg and E. J. Ofengand. *Proc. Natl. Acad. Sci. U. S.* **44,** 78 (1958).
[118] L. I. Hecht, M. L. Stephenson, and P. C. Zamecnik, *Proc. Natl. Acad. Sci. U. S.* **45,** 505 (1959).
[119] L. I. Hecht, P. C. Zamecnik, M. L. Stephenson, and J. F. Scott, *J. Biol. Chem.* **233,** 954 (1958).
[120] H. G. Zachau, G. Acs, and F. Lipmann, *Proc. Natl. Acad. Sci. U. S.* **44,** 885 (1958).
[121] J. Preiss, P. Berg, E. J. Ofengand, F. H. Bergmann, and M. Dieckmann, *Proc. Natl. Acad. Sci. U. S.* **45,** 319 (1959).
[122] G. C. Webster, *J. Biol. Chem.* **229,** 535 (1957).
[123] M. B. Hoagland, personal communication.
[124] M. L. Blanchard, S. Korkes, A. D. Campillo, and S. Ochoa, *J. Biol. Chem.* **187,** 875 (1950).
[125] R. F. Sund, J. M. Ravel, and W. Shive, *J. Biol. Chem.* **231,** 807 (1958).
[126] G. Seaman, *Proc. Intern. Congr. Biochem., 7th Congr., Abstr.* p. 142 (1958).
[127] S. R. Wagle, R. Mehta, and B. C. Johnson, *J. Biol. Chem.* **230,** 137, 917; **233,** 619 (1958).
[128] Mirko Beljanski and S. Ochoa, *Proc. Natl. Acad. Sci. U. S.* **44,** 1157 (1958).
[129] S. Lacks and Francois Gros, *J. Mol. Biol.* **1,** 301 (1959).
[130] M. V. Simpson, J. R. McLean, G. L. Cohn, and I. K. Brandt, *Federation Proc.* **16,** 249 (1957).
[131] P. Cohn, *Biochim. et Biophys. Acta* **33,** 284 (1959).
[132] J. Heller, P. Szafranski, and E. Sulkowski, *Nature* **183,** 397 (1959).
[133] R. S. Schweet, H. Lamfron, and E. H. Allen, *Proc. Natl. Acad. Sci. U. S.* **44,** 1029 (1958).
[134] H. M. Dintzis, H. Borsook, and J. Vinograd *in* "Microsomal Particles and Protein Synthesis" (R. B. Roberts, ed.), p. 95. Pergamon Press, London, 1958.
[135] S. Simmonds, E. L. Tatum, and J. S. Fruton, *J. Biol. Chem.* **169,** 91 (1947).
[136] S. Simmonds and J. S. Fruton, *J. Biol. Chem.* **174,** 705 (1948).
[137] S. Simmonds and J. S. Fruton, *J. Biol. Chem.* **180,** 635 (1949).
[138] S. Simmonds and J. S. Fruton, *Science* **111,** 329 (1950).
[139] S. P. Taylor, S. Simmonds, and J. S. Fruton, *J. Biol. Chem.* **187,** 613 (1950).
[140] A. I. Virtanen and V. Murmikko, *Acta Chem. Skand.* **5,** 681 (1951).
[141] R. B. Malin, M. N. Camien, and M. S. Dunn, *Arch. Biochem. Biophys.* **32,** 106 (1951).
[142] H. K. Miller and H. Waelsch, *Arch. Biochem. Biophys.* **35,** 184 (1952).
[143] S. Simmonds and J. S. Fruton, *Science* **109,** 561 (1949).
[144] H. Sprince and D. W. Woolley, *J. Am. Chem. Soc.* **67,** 1734 (1945).
[145] D. W. Woolley and R. B. Merrifield, *J. Am. Chem. Soc.* **76,** 316 (1954).
[146] R. B. Merrifield and D. W. Woolley, *J. Am. Chem. Soc.* **78,** 358, 4646 (1956).
[147] H. Kihara, W. G. McCullough, and E. E. Snell, *J. Biol. Chem.* **197,** 783 (1952).
[148] H. Kihara and E. E. Snell, *J. Biol. Chem.* **197,** 791 (1952).
[149] H. Kihara and E. E. Snell, *J. Biol. Chem.* **212,** 83 (1955).
[150] V. J. Peters, J. M. Prescott, and E. E. Snell, *J. Biol. Chem.* **202,** 521 (1953).
[151] E. F. Gale, *Brit. J. Exptl. Pathol.* **26,** 255 (1945).

[152] G. N. Cohen, H. O. Halvorson, and S. Spiegelman, *in* "Microsomal Particles and Protein Synthesis" (R. B. Roberts, ed.), p. 100. Pergamon Press, London, 1958.
[153] E. F. Gale and M. B. van Halteren, *Biochem. J.* **50**, 34 (1951).
[154] F. Turba and H. Esser, *Biochem. Z.* **327**, 93 (1955).
[155] I. R. McManus, *J. Biol. Chem.* **231**, 777 (1958).
[156] R. G. Gould, *J. Biol. Chem.* **153**, 143 (1943).
[157] R. B. Roberts, P. H. Abelson, D. B. Cowie, E. T. Bolton, and R. J. Britten, Studies of Biosynthesis in *Escherichia coli*. *Carnegie Inst. Wash. Publ. No* **607**, 1955.
[158] C. S. Hanes, F. J. R. Hird, and F. A. Isherwood, *Nature* **166**, 288 (1950).
[159] C. S. Hanes, F. J. R. Hird, and F. A. Isherwood, *Biochem. J.* **51**, 25 (1952).
[160] P. J. Samuels, *J. Gen. Microbiol. Proc.* **8**, viii (1953).
[161] C. B. Thorne, C. G. Gomez, and R. D. Housewright, *J. Bacteriol.* **69**, 357 (1953).
[162] C. B. Thorne and D. M. Molnar, *J. Bacteriol.* **70**, 420 (1955).
[163] W. J. Williams and C. B. Thorne, *J. Biol. Chem.* **210**, 203; **211**, 631 (1954).
[164] W. J. Williams, J. Litwin, and C. B. Thorne, *J. Biol. Chem.* **212**, 427 (1955).
[165] C. B. Thorne, *Symposium Soc. Gen. Microbiol.* **6**, 68 (1956).
[166] E. Ito, N. Ishimoto, and M. Saito, *Arch. Biochem. Biophys.* **80**, 431 (1959).
[167] J. L. Strominger and R. H. Threnn, *Biochim. et Biophys. Acta* **33**, 280 (1959).
[168] R. E. F. Matthews, *Virology* **1**, 165 (1955).
[169] H. G. Mandel, G. I. Sugarman, and R. A. Apter, *J. Biol. Chem.* **225**, 151 (1957).
[170] D. B. Dunn, J. D. Smith, S. Zamenhof, and G. Griboff, *Nature* **174**, 305 (1954).
[171] F. Weygand, A. Wacker, and H. Dellweg, *Z. Naturforsch.* **7b**, 19 (1952).
[172] E. Borek, A. Ryan, and J. Rickenbach, *J. Bacteriol.* **69**, 460 (1955).
[173] Francois Gros and Francoise Gros, *Biochim. et Biophys. Acta* **22**, 200 (1956).
[174] A. B. Pardee and L. S. Prestidge, *J. Bacteriol.* **71**, 677 (1956).
[175] F. Gros and F. Gros, *Exptl. Cell Research* **14**, 104 (1958).
[176] M. Grunberg-Manago, P. J. Ortiz, and S. Ochoa, *Biochim. et Biophys. Acta* **20**, 269 (1956).
[177] S. Ochoa, *in* "Recent Progress in Microbiology" (G. Tunevall, ed.), p. 122. Almqvist & Wiksells, Uppsala, Sweden, 1958.
[178] M. J. Besseman, I. R. Lehman, E. S. Simms, and A. Kornberg, *J. Biol. Chem.* **233**, 171 (1958).
[179] M. J. Besseman, I. R. Lehman, J. Adler, S. B. Zimmerman, E. S. Simms, and A. Kornberg, *Proc. Natl. Acad. Sci. U. S.* **44**, 633, 641 (1958).
[180] W. Firshein and W. Braun, *Proc. Natl. Acad. Sci. U. S.* **44**, 918 (1958).
[181] W. Braun, *J. Cellular Comp. Physiol.* **52**, 337 (1958).
[182] E. H. Creaser, quoted in ref. 52.
[183] S. S. Cohen and C. B. Fowler, *J. Exptl. Med.* **85**, 771 (1947).
[184] K. Burton, *Biochem. J.* **61**, 473 (1955).
[185] N. E. Melechen, *Genetics* **40**, 584 (1955).
[186] J. Tomizawa and S. Sunakawa, *J. Gen. Physiol.* **39**, 553 (1956).
[187] L. V. Crawford, *Biochem. J.* **65**, 17P (1957).
[188] L. V. Crawford, *Biochim. et Biophys. Acta* **28**, 208 (1958).
[189] J. G. Flaks and S. S. Cohen, *Biochim. et Biophys. Acta* **25**, 667 (1957).
[190] A. Hershey, *J. Gen. Physiol.* **37**, 1 (1954).
[191] A. Hershey, *J. Gen. Physiol.* **38**, 145 (1955).
[192] E. Volkin and L. Astrachan, *Virology* **2**, 149, 433 (1956).
[193] E. Volkin and L. Astrachan, *in* "Chemical Basis of Heredity" (W. D. McElroy and B. Glass, eds.), p. 686. Johns Hopkins Press, Baltimore, Maryland, 1957.
[194] E. Volkin, L. Astrachan, and J. L. Countryman, *Virology* **6**, 545 (1958).
[195] F. E. Hahn and C. L. Wisseman, Jr., *Proc. Soc. Exptl. Biol. Med.* **76**, 533 (1951).

[196] G. N. Smith, C. S. Worrel, and A. L. Swanson, *J. Bacteriol.* **58,** 803 (1949).
[197] E. F. Gale and J. P. Folkes, *Biochem. J.* **53,** 493 (1953).
[198] G. N. Smith, *Bacteriol. Revs.* **17,** 19 (1953).
[199] C. L. Wisseman, Jr., J. E. Smadel, F. E. Hahn, and H. E. Hopps, *J. Bacteriol.* **67,** 663 (1954).
[200] F. E. Hahn, C. L. Wisseman, Jr., and H. E. Hopps. *J. Bacteriol.* **67,** 674 (1954).
[201] F. E. Hahn, C. L. Wisseman, Jr., and H. E. Hopps, *J. Bacteriol.* **69,** 215 (1955).
[202] P. J. Samuels, *Biochem. J.* **55,** 441 (1953).
[203] F. C. Neidhardt and Francois Gros, *Biochim. et Biophys. Acta* **25,** 513 (1951).
[204] A. B. Pardee, K. Paigen, and L. S. Prestidge, *Biochim. et Biophys. Acta* **23,** 162 (1957).
[205] J. Horowitz, A. Lombard, and E. Chargaff, *J. Biol. Chem.* **233,** 1517 (1958).
[206] S. Dagley and J. Sykes, *Nature* **183,** 1608 (1959).
[207] A. L. Dounce, *Enzymologia* **15,** 251 (1952).
[208] F. H. C. Crick, *Symposium Soc. Exptl. Biol.* **12,** 138 (1958).
[209] L. Goldstein and W. Plaut, *Proc. Natl. Acad. Sci. U. S.* **41,** 874 (1955).
[210] A. Gierer and G. Schramm, *Nature* **177,** 702 (1956).
[211] A. Siegel, W. Ginoza, and S. G. Wildman, *Virology* **3,** 554 (1957).
[212] H. Fraenkel-Conrat and B. Singer, *Biochim. et Biophys. Acta* **24,** 540 (1957).
[213] H. Fraenkel-Conrat, B. Singer, and R. C. Williams *Biochim. et Biophys. Acta* **25,** 87 (1957).
[214] C. H. W. Hirs, S. Moore, and W. H. Stein, *J. Biol. Chem.* **235,** 633 (1960).
[215] C. H. W. Hirs, *J. Biol. Chem.* **235,** 625 (1960).
[216] G. E. Perlmann and R. Diringer, *Ann. Rev. Biochem.* **29,** 151 (1960).
[217] G. H. Dixon, D. L. Kauffman, and H. Neurath, *J. Amer. Chem. Soc.* **80,** 1260 (1958).
[218] R. A. Oosterbaan, P. Kunst, J. Van Rotterdam, and J. A. Cohen, *Biochim. et Biophys. Acta* **27,** 549 (1958).
[219] N. K. Schaffer, L. Simet, S. Harshman, R. R. Engle, and R. W. Drisko, *J. Biol. Chem.* **225,** 197 (1957).
[220] B. S. Hartley, M. A. Naughton, and F. Sanger, *Biochim. et Biophys. Acta* **34,** 214 (1959).
[221] M. A. Naughton, F. Sanger, B. S. Hartley, and D. C. Shaw, *Biochem. J.* **77,** 149 (1960).
[222] J. A. Gladner and K. Laki, *J. Amer. Chem. Soc.* **80,** 1264 (1958).
[223] J. L. Strominger and R. H. Threnn, *Biochim. et Biophys Acta* **33,** 281 (1959).
[224] D. B. Dunn, *Biochim. et Biophys. Acta* **46,** 198 (1961).
[225] W. E. Cohn, *Biochim. et Biophys. Acta* **32,** 569 (1959).
[226] C. Yu and F. W. Allen, *Biochim. et Biophys. Acta* **32,** 393 (1959).
[227] A. Gierer and G. Schramm, *Nature* **177,** 702 (1956).
[228] K. S. Kirby, *Biochem. J.* **64,** 405 (1956).
[229] D. B. Dunn, *Biochim. et Biophys. Acta* **34,** 286 (1959).
[230] D. B. Dunn, J. D. Smith, and P. F. Spahr, *J. Mol. Biol.* **2,** 113 (1960).
[231] A. Tissières, *J. Mol. Biol.* **1,** 365 (1959).
[232] G. L. Brown and G. Zubay, *J. Mol. Biol.* **2,** 287 (1960).
[233] A. Tissières, J. D. Watson, D. Schlessinger, and B. R. Hollingworth, *J. Mol. Biol.* **1,** 221 (1959).
[234] H. E. Huxley and G. Zubay, *J. Mol. Biol.* **2,** 10 (1960).
[235] B. D. Hall and P. Doty, *J. Mol. Biol.* **1,** 111 (1959).
[236] C. G. Kurland, *J. Mol. Biol.* **2,** 83 (1960).
[237] A. I. Aronson, E. T. Bolton, R. J. Britten, D. B. Cowie, J. D. Duerksen, B. J.

McCarthy, K. McQuillen, and R. B. Roberts, *Carnegie Inst. Wash. Yearbook* **59,** 228 (1960).
[238] E. T. Bolton, R. J. Britten, D. B. Cowie, B. J. McCarthy, K. McQuillen, and R. B. Roberts, *Carnegie Inst. Wash. Yearbook* **58,** 275 (1959).
[239] S. Osawa, *Biochim. et Biophys. Acta* **42,** 244 (1960).
[240] P. F. Spahr and A. Tissières, *J. Mol. Biol.* **1,** 237 (1959).
[241] J. Bonner, *in* "Protein Biosynthesis" (R. J. C. Harris, ed.), p. 323. Academic Press, New York, 1961.
[242] D. Herbert, *Symposium Soc. Gen. Microbiol.* **11,** 391 (1961).
[243] F. C. Neidhardt and B. Magasanik, *Biochim. et Biophys. Acta* **42,** 99 (1960).
[244] B. Magasanik, A. K. Magasanik, and F. C. Neidhardt, *Ciba Foundation Symposium. The Regulation of Cell Metabolism* p. 334 (1959).
[245] Francois Gros, *in* "The Nucleic Acids" (E. Chargaff and J. N. Davidson, eds.), Vol. 3, p. 409. Academic Press, New York, 1960.
[246] H. Chantrenne and S. Devreux, *Biochim. et Biophys. Acta* **39,** 486 (1960).
[247] H. Chantrenne and S. Devreux, *Biochim. et Biophys. Acta* **41,** 239 (1960).
[248] D. B. Roodyn and H. G. Mandel, *J. Biol. Chem.* **235,** 2036 (1960).
[249] H. Chantrenne, *Biochem. Pharmacol.* **1,** 233 (1959).
[250] J. Horowitz, J. J. Saukkonen, and E. Chargaff, *J. Biol. Chem.* **235,** 3266 (1960).
[251] J. Mendelsohn and A. Tissières, *Biochim. et Biophys. Acta* **35,** 248 (1959).
[252] R. A. Peabody and C. Hurwitz, *Biochim. et Biophys. Acta* **39,** 184 (1960).
[253] P. Rogers and G. D. Novelli, *Biochim. et Biophys. Acta* **44,** 298 (1960).
[254] D. E. Hughes, *Brit. J. Exptl. Pathol.* **32,** 97 (1951).
[255] C. S. French and H. W. Milner, *in* "Methods in Enzymology" (S. P. Colowick and N. O. Kaplan, eds.), Vol. I., p. 64. Academic Press, New York (1955).
[256] W. E. Dibble and H. M. Dintzis, *Biochim. et Biophys. Acta* **37,** 152 (1960).
[257] W. Zillig, W. Krone, and M. Albers, *Z. physiol. Chemie, Hoppe-Seyler's* **317,** 131 (1959).
[258] S. S. Cohen and J. Lichtenstein, *J. Biol. Chem.* **235,** 2112 (1960).
[259] C. F. Crampton and M. L. Peterman, *J. Biol. Chem.* **234,** 2642 (1959).
[260] J. A. V. Butler, A. R. Crathorn, and G. D. Hunter, *Biochim. et Biophys. Acta* **38,** 386 (1960).
[261] J-P. Waller and J. I. Harris, *Proc. Natl. Acad. Sci. U.S.* **47,** 18 (1961).
[262] D. Elson, *Biochim. et Biophys. Acta* **27,** 216 (1958).
[263] D. Elson, *Biochim. et Biophys. Acta* **36,** 372 (1959).
[264] D. Elson and M. Tal, *Biochim. et Biophys. Acta* **36,** 281 (1959).
[265] D. B. Cowie, S. Spiegelman, R. B. Roberts, and J. D. Duerksen, *Proc. Natl. Acad. Sci. U.S.* **47,** 114 (1961).
[266] V. Hanzon, *Progr. in Biophys. and Biophys. Chem.* **11,** 242 (1961).
[267] F. S. Sjöstrand, *in* "Biophysical Science, A Study Program" (J. L. Oncley, ed.), p. 306. Wiley, New York, 1959.
[268] T. J. Bowen, S. Dagley, and J. Sykes, *Biochem. J.* **72,** 419 (1959).
[269] T. J. Bowen, S. Dagley, J. Sykes, and D. G. Wild, *Nature* **189,** 638 (1961).
[270] P. Rogers and G. D. Novelli, *Federation Proc.* **18,** 311 (1959).
[271] P. Rogers and G. D. Novelli, *Biochim. et Biophys. Acta* **44,** 298 (1960).
[272] M. R. Lamborg, *Federation Proc.* **19,** 346 (1960).
[273] M. R. Lamborg and P. C. Zamecnik, *Biochim. et Biophys. Acta* **42,** 206 (1960).
[274] P. D. Mitchell and J. Moyle, *Symposium Soc. Gen. Microbiol.* **6,** 150 (1956).
[275] K. McQuillen, R. B. Roberts, and R. J. Britten, *Proc. Natl. Acad. Sci. U.S.* **45,** 1437 (1959).

[276] T. Kameyama and G. D. Novelli, *Biochem. Biophys. Research Communs.* **2,** 393 (1960).
[277] A. Tissières, D. Schlessinger, and Francois Gross, *Proc. Natl. Acad. Sci. U.S.* **46,** 1450 (1960).
[278] G. Fox and J. W. Brown, *Biochim. et Biophys. Acta* **46,** 387 (1961).
[279] J. Mandelstam, *Bacteriol. Revs.* **24,** 289 (1960).
[280] J. Mandelstam and H. O. Halvorson, *Biochim. et Biophys. Acta* **40,** 43 (1960).
[281] D. Kerridge, *Symposium Soc. Gen. Microbiol.* **11,** 41 (1961).
[282] M. B. Hoagland, P. C. Zamecnik, N. Sharon, F. Lipmann, M. P. Stulberg, and P. D. Boyer, *Biochim. et Biophys. Acta* **26,** 215 (1957).
[283] R. W. Bernlohr and G. C. Webster, *Arch. Biochem. Biophys.* **72,** 276 (1958).
[284] M. Karasek, P. Castelfranco, P. R. Krishnaswamy, and A. Meister, *J. Amer. Chem. Soc.* **80,** 2335 (1958).
[285] H. S. Kingdon, L. T. Webster, and E. W. Davie, *Proc. Natl. Acad. Sci. U.S.* **44,** 757 (1958).
[286] S. B. Weiss, H. G. Zachau, and F. Lipmann, *Arch. Biochem. Biophys.* **83,** 101 (1959).
[287] L. T. Webster and E. W. Davie, *Biochim. et Biophys. Acta* **35,** 559 (1959).
[288] R. S. Schweet and E. H. Allen, *J. Biol. Chem.* **233,** 1104 (1958).
[289] A. M. van de Ven, V. V. Koningsberger, and J. Th. G. Overbeek, *Biochim. et Biophys. Acta* **28,** 134 (1958).
[290] F. H. Bergmann, P. Berg, J. Preiss, E. J. Ofengand, and M. Dieckmann, *Federation Proc.* **18,** 751 (1959).
[291] R. W. Holley and J. Goldstein, *J. Biol. Chem.* **234,** 1765 (1959).
[292] G. Webster, *Biochem. Biophys. Research Communs.* **2,** 56 (1960).
[293] L. T. Webster and E. W. Davie, *Federation Proc.* **18,** 348 (1959).
[294] G. Acs, G. Hartmann, H. G. Boman, and F. Lipmann, *Federation Proc.* **18,** 178 (1959).
[295] E. H. Allen, E. Glassman, and R. S. Schweet, *J. Biol. Chem.* **235,** 1061 (1960).
[296] J. Leahy, E. Glassman, and R. S. Schweet, *J. Biol. Chem.* **235,** 3209 (1960).
[297] M. Karasek, P. Castelfranco, P. R. Krishnaswamy, and A. Meister, *J. Amer. Chem. Soc.* **80,** 2335 (1958).
[298] K. K. Wong, A. Meister, and K. Moldave, *Biochim. et Biophys. Acta* **36,** 531 (1959).
[299] P. R. Krishnaswamy and A. Meister, *J. Biol. Chem.* **235,** 408 (1960).
[300] K. K. Wong and K. Moldave, *J. Biol. Chem.* **235,** 694 (1960).
[301] H. Nohara and K. Ogata, *Biochim. et Biophys. Acta* **31,** 142 (1959).
[302] K. Ogata and H. Nohara, *Biochim. et Biophys. Acta* **31,** 149 (1959).
[303] K. Ogata, H. Nohara, and S. Miyazaki, *Biochim. et Biophys. Acta* **32,** 287 (1959).
[304] J. Baddiley and F. C. Neuhaus, *Biochim. et Biophys. Acta* **33,** 277 (1959).
[305] R. Rendi, *Boll. soc. ital. biol. sper.* **31,** 1410 (1955).
[306] R. Rendi, A. Di Milia, and C. Fronticelli, *Biochem. J.* **70,** 62 (1958).
[307] S. Tuboi and A. Huzino, *Arch. Biochem. Biophys.* **86,** 309 (1959).
[308] D. A. Goldthwait and J. L. Starr, *J. Biol. Chem.* **235,** 2025 (1960).
[309] L. Bosch, H. Bloemendal, and M. Sluyser, *Biochim. et Biophys. Acta* **41,** 444 (1960).
[310] L. Bosch, H. Bloemendal, M. Sluyser, and P. H. Pouwels, *in* "Protein Biosynthesis" (R. J. C. Harris, ed.), p. 133. Academic Press, New York, 1961.
[311] K. C. Smith, E. Cordes, and R. S. Schweet, *Biochim. et Biophys. Acta* **33,** 386 (1959).
[312] P. C. Zamecnik and M. L. Stephenson, *Ann. N.Y. Acad. Sci.* **88,** 708 (1960).
[313] P. C. Zamecnik, M. L. Stephenson, and J. F. Scott, *Proc. Natl. Acad. Sci. U.S.* **46,** 811 (1960).
[314] P. C. Zamecnik, M. L. Stephenson and C.-T. Yu, *in* "Protein Biosynthesis" (R. J. C. Harris, ed.), p. 125. Academic Press, New York, 1961.

[315] G. Brown, A. V. W. Brown, and J. Gordon, *Brookhaven Symposia in Biol.* **12,** 47 (1959).
[316] R. W. Holley and S. H. Merrill, *Federation Proc.* **18,** 982 (1959).
[317] R. W. Holley, J. Apgar, and B. P. Doctor, *Ann. N.Y. Acad. Sci.* **88,** 745 (1960).
[318] R. W. Holley, E. F. Brunngraber, F. Saad, and H. H. Williams, *J. Biol. Chem.* **236,** 200 (1961).
[319] R. W. Holley and B. P. Doctor, *Federation Proc.* **19,** 348 (1960).
[320] W. A. Klee and G. L. Cantoni, *Proc. Natl. Acad. Sci. U.S.* **46,** 322 (1960).
[321] S. Osawa and E. Otaka, *Biochim. et Biophys. Acta* **36,** 549 (1959).
[322] E. Herbert, *Ann. N.Y. Acad. Sci.* **81,** 679 (1959).
[323] E. H. Allen, E. Glassman, E. Cordes, and R. S. Schweet, *J. Biol. Chem.* **235,** 1068 (1960).
[324] F. H. Bergmann, P. Berg, J. Preiss, E. J. Ofengand, and M. Dieckmann, *Federation Proc.* **18,** 191 (1959).
[325] G. Acs, G. Hartmann, H. G. Boman, and F. Lipmann, *Federation Proc.* **8,** 700 (1959).
[326] P. C. Zamecnik, *Harvey Lectures, 1958–1959* **54,** 256 (1960).
[327] E. H. Allen and R. S. Schweet, *Biochim. et Biophys. Acta* **39,** 185 (1960).
[328] L. G. Grossi and K. Moldave, *Biochim. et Biophys. Acta* **35,** 275 (1959).
[329] L. G. Grossi and K. Moldave, *J. Biol. Chem.* **235,** 2370 (1960).
[330] W. C. Hulsmann and F. Lipmann, *Biochim. et Biophys. Acta* **43,** 123 (1960).
[331] D. Nathans and F. Lipmann, *Biochim. et Biophhys. Acta* **43,** 126 (1960).
[332] D. Nathans, *Ann. N.Y. Acad. Sci.* **88,** 718 (1960).
[333] M. Takanami and T. Okamoto, *Biochim. et Biophys. Acta* **44,** 379 (1960).
[334] A. Von der Decken and T. Hultin, *Biochim. et Biophys. Acta* **40,** 189 (1960).
[335] A. Von der Decken and T. Hultin, *Biochim. et Biophys. Acta* **45,** 139 (1960).
[336] T. Hultin and A. Von der Decken, *in* "Protein Biosynthesis" (R. J. C. Harris, ed.), p. 83. Academic Press, New York, 1961.
[337] D. Nathans and W. C. Hulsmann, *Federation Proc.* **19,** 347 (1960).
[338] M. B. Hoagland and L. T. Comly, *Proc. Natl. Acad. Sci. U.S.* **46,** 1563 (1960).
[339] P. N. Campbell, *in* "Protein Biosynthesis" (R. J. C. Harris, ed.), p. 19. Academic Press, New York, 1961.
[340] D. Elson, *in* "Protein Biosynthesis" (R. J. C. Harris, ed.), p. 291. Academic Press, New York, 1961.
[341] P. Siekevitz and G. E. Palade, *J. Biophys. Biochem. Cytol.* **7,** 619, 631, (1960).
[342] J. Bishop, E. H. Allen, J. Leahy, A. Morris, and R. S. Schweet, *Federation Proc.* **19,** 346 (1960).
[343] H. Lamfron, *Federation Proc.* **19,** 350 (1960).
[344] G. C. Webster, *Arch. Biochem. Biophys.* **85,** 159 (1959).
[345] I. D. Raacke, *Biochem. et Biophys. Acta* **34,** 1 (1959).
[346] G. C. Webster and T. B. Lingrel, *in* "Protein Biosynthesis" (R. J. C. Harris, ed.), p. 301. Academic Press, New York, 1961.
[347] F. A. Kuehl, A. L. Demain, and E. L. Rickes, *J. Amer. Chem. Soc.* **82,** 2079 (1960).
[348] E. F. Gale and J. P. Folkes, *Biochem. J.* **83,** 430 (1962).
[349] R. W. Hendler, *J. Biol. Chem.* **234,** 1466 (1959).
[350] R. W. Hendler, *Progr. in Biophys. and Biophys. Chem.* **11,** 249 (1961).
[351] R. W. Hendler, *Federation Proc.* **19,** 346 (1960).
[352] J. A. V. Butler, G. N. Godson, and G. D. Hunter, *in* "Protein Biosynthesis" (R. J. C. Harris, ed.), p. 349. Academic Press, New York, 1961.
[353] G. D. Hunter and R. A. Goodsall, *Biochem. J.* **78,** 564 (1961).
[354] J. L. Haining, T. Fukin, and B. Axelrod, *J. Biol. Chem.* **235,** 160 (1960).

[355] F. Lynen, *J. Cellular Comp. Physiol.* **54,** Suppl. 1, 33 (1959).
[356] M. J. Fraser and E. S. Holdsworth, *Nature* **183,** 519 (1959).
[357] H. R. V. Arnstein and J. L. Simkin, *Nature* **183,** 523 (1959).
[358] S. R. Wagle, R. Mehta, and B. C. Johnson, *Biochim. et Biophys. Acta* **39,** 500 (1960).
[359] Mirko Beljanski, *Biochim. et Biophys. Acta* **41,** 104, 111 (1960).
[360] Mirko Beljanski, *Progr. in Biophys. and Biophys. Chem.* **11,** 238 (1961).
[361] A. I. Aronson *et al.*, *Carnegie Inst. Wash. Yearbook* **59,** 269 (1960).
[362] G. C. Webster, *Arch. Biochem. Biophys.* **89,** 53 (1960).
[363] J. L. Strominger, E. Ito, and R. H. Threnn, *J. Amer. Chem. Soc.* **82,** 998 (1960).
[364] F. Lipmann and H. M. Bates, *in* "Protein Biosynthesis" (R. J. C. Harris, ed.), p. 5. Academic Press, New York, 1961.
[365] R. Cohen and B. Nisman, *Compt. Rend. acad. sci.* **252,** 1062)1961).
[366] J. T. Wachsmann, H. Fukuhara, and B. Nisman, *Biochim. et Biophys Acta* **42,** 388 (1960).
[367] D. Nathans and F. Lipmann, *Proc. Natl. Acad. Sci. U.S.* **47,** 497 (1961).
[368] F. R. Leach and E. E. Snell, *J. Biol. Chem.* **235,** 3523 (1960).
[369] H. Kihara, M. Ikawa, and E. E. Snell, *J. Biol. Chem.* **236,** 172 (1961).
[370] H. Kihara and E. E. Snell, *J. Biol. Chem.* **235,** 1409 (1960).
[371] F. Turba, G. Hüskins, L. Büscher-Daubembüchel, and H. Pelzer, *Biochem. Z.* **327,** 410 (1955).
[372] F. Turba, A. Leismann, and G. Keinbenz, *Biochem. Z.* **329,** 97 (1957).
[373] W. S. Rieth, *Nature* **178,** 1393 (1956).
[374] A. D. Brown, *Biochim. et Biophys. Acta* **30,** 447 (1958).
[375] A. D. Brown, *Biochem. J.* **71,** 5P (1959).
[376] V. V. Koningsberger, C. O. van der Grinten, and J. T. G. Overbeek, *Biochim. et Biophys. Acta* **26,** 483 (1957).
[377] V. V. Koningsberger, *in* "Protein Biosynthesis" (R. J. C. Harris, ed.), p. 207. Academic Press, New York, 1961.
[378] V. Habermann, *Biochim. et Biophys. Acta* **32,** 297 (1959).
[379] E. Hase, S. Mihara, H. Otsuka, and H. Tamiya, *Arch. Biochem. Biophys.* **83,** 170 (1959).
[380] J. L. Strominger and R. H. Threnn, *Biochim. et Biophys. Acta* **36,** 83 (1959).
[381] J. L. Strominger, *J. Biol. Chem.* **234,** 1520 (1959).
[382] E. Ito and J. L. Strominger, *J. Biol. Chem.* **235,** PC5, PC7 (1959).
[383] F. C. Neuhaus, *Biophys. Biochem. Research Communs.* **3,** 401 (1960).
[384] Francois Gros and S. Naono, *in* "Protein Biosynthesis" (R. J. C. Harris, ed.), p. 195. Academic Press, New York, 1961.
[385] A. Goldstein and B. J. Brown, *Biochim. et Biophys. Acta* **44,** 491 (1960).
[386] C. Heidelberger, E. Harbers, K. C. Leibman, Y. Takagi, and V. R. Potter, *Biochim. et Biophys. Acta* **20,** 455 (1957).
[387] E. Herbert, *J. Biol. Chem.* **231,** 975 (1958).
[388] L. I. Hecht, M. L. Stephenson, and P. C. Zamecnik, *Proc. Natl. Acad. Sci. U.S.* **45,** 505 (1959).
[389] E. S. Canellakis and E. Herbert, *Biochem. et Biophys. Acta* **45,** 121 (1960).
[390] E. S. Canellakis and E. Herbert, *Proc. Natl. Acad. Sci. U.S.* **46,** 170 (1960).
[391] M. Edmonds and R. Abrams, *Federation Proc.* **19,** 317 (1960).
[392] D. A. Goldthwait, *J. Biol. Chem.* **234,** 3251 (1959).
[393] E. Harbers and C. Heidelberger, *Biochim. et Biophys. Acta* **35,** 381 (1959).
[394] J. Preiss and P. Berg, *Federation Proc.* **19,** 317 (1960).
[395] C. Coutsogeorgopoulos, *Biochim. et Biophys. Acta* **44,** 189 (1960).
[396] K. Moldave, *Biochim. et Biophys. Acts* **43,** 188 (1960).

[397] S. B. Weiss, *Proc. Nat.. Acad. Sci. U.S.* **46,** 1020 (1960).
[398] J. Hurwitz, A. Bresler, and R. Diringer, *Biochem. Biophys. Research Communs.* **3,** 15 (1960).
[399] A. Stevens, *Biochem. Biophys. Research Communs* **3,** 92 (1960).
[400] L. A. Heppel, P. J. Oritz, and S. Ochoa, *J. Biol. Chem.* **229,** 679, 695 (1957).
[401] P. J. Ortiz and S. Ochoa, *J. Biol. Chem.* **234,** 1208 (1959).
[402] M. F. Singer, R. J. Hilmoe, and M. Grunberg-Manago, *J. Biol. Chem.* **235,** 2705 (1960).
[403] M. F. Singer, S. Luborsky, R. A. Morrison, and G. L. Cantoni, *Biochim. et Biophys. Acta* **38,** 568 (1960).
[404] M. Grunberg-Manago, *J. Mol. Biol.* **1,** 240 (1959).
[405] J. Skoda, J. Kara, Z. Sormova, and F. Sorm, *Biochim. et Biophys. Acta* **33,** 579 (1959).
[406] A. Goldstein, D. B. Goldstein, B. J. Brown, and S-C. Chou, *Biochim. et Biophys. Acta* **36,** 163 (1959).
[407] T. Okazaki and R. Okazaki, *Biochim. et Biophys. Acta* **35,** 434 (1959).
[408] L. Astrachan and E. Volkin, *Biochim. et Biophys. Acta* **32,** 449 (1959).
[409] E. Volkin, *Proc. Natl. Acad. Sci. U.S.* **46,** 1336 (1960).
[410] M. Nomura, B. D. Hall, and S. Spiegelman, *J. Mol. Biol.* **2,** 306 (1960).
[411] H. K. Schachman, I. R. Lehman, M. J. Bessman, J. Adler, E. S. Simms, and A. Kornberg, *Federation Proc.* **17,** 304 (1958).
[412] I. R. Lehman, S. B. Zimmerman, J. Adler, M. J. Bessman, E. S. Simms, and A. Kornberg, *Proc. Natl. Acad. Sci. U.S.* **44,** 1191 (1958).
[413] F. J. Bollum, *Federation Proc.* **17,** 193 (1958).
[414] F. J. Bollum, *Federation Proc.* **18,** 194 (1959).
[415] F. J. Bollum, *J. Biol. Chem.* **234,** 2733 (1959).
[416] F. J. Bollum, *J. Biol. Chem.* **235,** 2399, PC18 (1960).
[417] F. J. Bollum, *Federation Proc.* **19,** 305 (1960).
[418] P. Doty, B. B. McGill, and S. A. Rice, *Proc. Natl. Acad. Sci. U.S.* **44,** 432 (1958).
[419] I. R. Lehman, *Ann. N.Y. Acad. Sci.* **81,** 745 (1959).
[420] R. L. Sinsheimer, *J. Mol. Biol.* **1,** 43 (1959).
[421] H. K. Schachman, J. Adler, C. M. Radding, I. R. Lehman, and A. Kornberg, *J. Biol. Chem.* **235,** 3242 (1960).
[422] J. Hurwitz, *J. Biol. Chem.* **234,** 2351 (1959).
[423] M. Nomura and J. D. Watson, *J. Mol. Biol.* **1,** 204 (1959).
[424] R. Rabson and G. D. Novelli, *Proc. Natl. Acad. Sci. U.S.* **46,** 484 (1960).
[425] V. G. Allfrey, A. E. Mirsky, and S. Osawa, *J. Gen. Physiol.* **40,** 451 (1957).
[426] J. W. Hopkins, *Proc. Natl. Acad. Sci. U.S.* **45,** 1461 (1959).
[427] F. B. Straub and A. Ullman, *Biochim. et Biophys Acta* **23,** 665 (1957).
[428] J. Mager, *Biochim. et Biophys. Acta* **38,** 150 (1960).
[429] M. B. Hoagland, *in* "The Nucleic Acids" (E. Chargaff and J. N. Davidson, eds.), Vol. III, p. 349. Academic Press, New York, 1960.
[430] Francois Gros, *in* "The Nucleic Acids" (E. Chargaff and J. N. Davidson, eds.), Vol. III, p. 409. Academic Press, New York, 1960.
[431] I. D. Raacke, *in* "Metabolic Pathways" (D. M. Greenberg, ed.), Vol. II, p. 263. Academic Press, New York, 1961.
[432] P. N. Campbell, *Biol. Revs.* **35,** 413 (1960).
[433] K. McQuillen, *Progr. in Biophys. and Biophys. Chem.* **11,** 294 (1961).
[434] R. B. Roberts, K. McQuillen, and I. Z. Roberts, *Ann. Rev. Microbiol.* **13,** 1 (1959).
[435] J. L. Simkin, *Ann. Rev. Biochem.* **28,** 145 (1959).
[436] G. N. Cohen and Francois Gros, *Ann. Rev. Biochem.* **29,** 525 (1960).

[437] W. Hayes and R. C. Clowes (eds.) *Symposium Soc. Gen. Microbiol.* **10,** 1 (1960).
[438] *Faraday Society Discussion on Cytoplasmic Particles, Progr. in Biophys. and Biophys. Chem.* (1961).
[439] *Ann. N.Y. Acad. Sci.* **88,** 708 (1960).
[440] "Protein Biosynthesis" (R. J. C. Harris, ed.). Academic Press, New York, 1961.
[441] F. Jacob and J. Monod, *J. Mol. Biol.* **5,** 318 (1961).
[442] S. Brenner, F. Jacob, and M. Meselson, *Nature* **190,** 576 (1961).
[443] M. Ycas and W. S. Vincent, *Proc. Natl. Acad. Sci. U.S.* **46,** 804 (1960).
[444] A. Lwoff, *Proc. Roy. Soc. B* **154,** 1 (1961).
[445] N. O. Kjeldgaard, O. Maaløe, and M. Schaechter, *J. Gen. Microbiol.* **19,** 607 (1958).
[446] M. Schaechter, O. Malløe, and N. O. Kjeldgaard, *J. Gen. Microbiol.* **19,** 592 (1958).
[447] H. Kihara, A. S. L. Hu, and H. O. Halvorson, *Proc. Natl. Acad. Sci. U.S.* **47,** 489 (1961).
[448] B. P. Doctor, J. Apgar, and R. W. Holley, *J. Biol. Chem.* **236,** 1117 (1961).
[449] S. B. Weiss and T. Nakamoto, *J. Biol. Chem.* **236,** PC18 (1961).
[450] S. Ochoa, D. P. Burma, H. Krøger, and J. D. Weill, *Proc. Natl. Acad. Sci. U.S.* **47,** 670 (1961).
[451] D. P. Burma, H. Krøger, S. Ochoa, R. C. Warner, and J. D. Weill, *Proc. Natl. Acad. Sci. U.S.* **47,** 749 (1961).
[452] S. B. Weiss and T. Nakamoto, *Proc. Natl. Acad. Sci. U.S.* **47,** 694 (1961).
[453] Francois Gros, H. Hiatt, W. Gilbert, C. G. Kurland, R. W. Risebrough, and J. D. Watson, *Nature,* **190,** 581 (1961).
[454] J. H. Matthaei and M. W. Nirenberg, *Proc. Natl. Acad. Sci. U.S.* **47,** 1580 (1961).
[455] A. Tissières and J. W. Hopkins, *Proc. Natl. Acad. Sci. U.S.* **47,** 2015 (1961).
[456] M. Chamberlin and P. Berg, *Proc. Natl. Acad. Sci. U.S.* **48,** 81 (1962).
[457] W. B. Wood and P. Berg, *Proc. Natl. Acad.Sci. U.S.* **48,** 94 (1962).
[458] M. W. Nirenberg and J. H. Matthaei, *Proc. Natl. Acad. Sci. U.S.* **47,** 1588 (1961).
[459] M. W. Nirenberg, J. H. Matthaei, and O. W. Jones, *Proc. Natl. Acad. Sci. U.S.* **48,** 104 (1962).
[460] P. Lengyel, J. F. Speyer, and S. Ochoa, *Proc. Natl. Acad. Sci. U.S.* **47,** 1936 (1961).
[461] J. F. Speyer, P. Lengyel, C. Basilio, and S. Ochoa, *Proc. Natl. Acad. Sci. U.S.* **48,** 63, 282 (1962).
[462] R. G. Martin, J. H. Matthaei, O. W. Jones, and M. W. Nirenberg, *Biochem. Biophys. Research Communs* **6,** 410 (1962).
[463] F. H. C. Crick, L. Barnett, S. Brenner, and R. J. Watts-Tobin, *Nature* **192,** 1227 (1961).
[464] J. A. Osmundsen, *The New York Times,* February 2 (1962).
[465] B. J. McCarthy, R. J. Britten, and R. B. Roberts, *Biophys. J.* **2,** 57, 83 (1962).

CHAPTER 12

The Synthesis of Enzymes

ARTHUR B. PARDEE

I. The Problems of Enzyme Formation 577
II. The Kinds of Enzymes Synthesized by Bacteria 578
A. Genes and Enzymes 578
B. Deoxyribonucleic Acid and Enzyme Synthesis 585
C. Other Effects of Genes on Enzyme Production 590
III. The Quantities of Enzymes Synthesized by a Bacterium 592
A. Genetic Modifiers of Enzyme Amounts 592
B. Inducible Enzymes 594
C. Constitutive and Repressible Enzymes 604
D. Miscellaneous Specific Modifiers of Enzyme Synthesis 610
IV. Metabolic Control and the Regulation of Enzyme Synthesis 615
A. Relative Merits of Types of Enzyme Production 615
B. Functions of Inductions and Repressions 616
V. Summary and Current Problems 618
A. Genes 619
B. Ribonucleic Acids and Enzyme Synthesis 620
C. Inducers and Repressors 620
References 621

I. The Problems of Enzyme Formation

The synthesis of enzymes by bacteria will be discussed in this chapter. Since mechanisms by which peptide chains can be assembled from amino acids have been dealt with in the preceding chapter and in recent reviews,[1, 2] we shall concern ourselves here with questions regarding qualitative and quantitative aspects of enzyme synthesis. Which proteins are formed, and how many molecules of each? An attempt will be made to state the problems of enzyme synthesis and to present the currently accepted views concerning the parts of the enzyme-synthesizing machinery and their interactions, together with the major evidence in support of these interpretations. Alternative views, with evidence supporting them, will be brought forward. Also there will be some discussion of the principal unexplained observations and discrepancies in our understanding of enzyme formation. References to the literature will be somewhat selective, and reviews will be referred to wherever possible to provide a broader coverage. The literature to January, 1961, has been surveyed. For literature through February 1962 see M. Riley and A. B. Pardee, Ann. Rev. Microbiol. 1962.

One may first wonder how it is that only a few proteins are formed, of the vast number that theoretically could be constructed from the common amino acids. Fundamental to this problem is an understanding of how the specific sequences of amino acids and foldings of peptide chains necessary for production of active enzymes are determined. Evidence from genetic studies and from chemistry support our present concepts regarding this problem, which can be referred to as the "qualitative" question.

The "quantitative" question—how the amount of each enzyme is determined—must next be considered. Different enzymes normally can be present as several per cent of the total bacterial protein or as only a few molecules per bacterium. Furthermore, the amount of a particular enzyme in an organism can change by a thousandfold or more depending on nutritional conditions. What variables determine the enzymic composition of bacteria? How is each enzyme level brought about? These problems provide a fascinating study of biological specificity at a fundamental level.

A third, related problem is that of metabolic control in the living organism. Is the enzyme content correlated with growth requirements? There are indications that this is the case.[3] Whether or not a given enzyme can be made and how much of it is produced under different circumstances would seem vital to the survival of a bacterial species in competition with other organisms. Production of enzymes is also vital in protecting microorganisms against antibiotics.[4, 5]

Enzyme formation impinges on other areas. For commercial production organisms accumulate vast amounts of specific metabolites; these overproductions depend on enzyme-forming abilities or deficiencies of the organisms.[6] For enzyme isolation from bacteria, a large amount of enzyme per cell is most important.[7] Bacterial classification depends in part on the characteristic enzymic reactions of taxonomic groups.

II. The Kinds of Enzymes Synthesized by Bacteria

A. Genes and Enzymes

1. Limitations on Kinds of Enzymes

The qualitative problem of enzyme synthesis can be stated as follows: what mechanism limits a bacterium* to synthesizing only a relatively few kinds of proteins compared to the enormous variety which could be formed from its amino acids? The total number of proteins that could be built from the 20 common amino acids is inconceivably large. For example, if only proteins (molecular weight about 10^5) composed of exactly

* Data obtained with other organisms will be referred to when pertinent to these questions and when similar information has not been obtained with bacteria.

830 amino acids are considered, there are 20^{830} possible structures. A sphere constructed from one of each of these molecules would have a radius of 10^{345} light-years. By contrast, a bacterium of a few microns radius contains about a million protein molecules (assuming an average molecular weight of 10^5). A much smaller number of different kinds of enzyme molecules, perhaps none, would have exactly 830 amino acids. Clearly, bacteria must possess extraordinarily effective selection mechanisms for determining what kinds of proteins can be synthesized.

Bacteria seemingly have unlimited metabolic abilities, suggesting vast flexibility in enzyme formation. This idea is basic to the enrichment culture technique[7] by means of which an organism capable of metabolizing almost any organic substance can be isolated. Bacteria are known, for example, which can degrade hydrocarbons, aromatic acids, or tobacco mosaic virus. In accord with the current notion that every biochemical reaction is catalyzed by an enzyme, one concludes that bacteria can possess a very large number of different enzymes. The number of different enzymes found in all bacteria cannot be calculated with any precision, but a rough estimate of 3000 to 10,000 is suggested.

This versatility does not mean that every bacterium possesses every known enzyme. In fact, the evidence is extensive that a given organism can make only a fraction of the known kinds of enzymes. In the enrichment culture technique, only a few of the numerous kinds of bacteria present in the original population are selected, by virtue of their unique ability to grow under the conditions provided. Different species of bacteria are found in different environments, presumably because they possess enzymes suitable for growth in that set of conditions. Also, bacterial classification schemes which depend in part on ability or inability of the species to perform certain reactions (such as acid, gas, or indole production, or fermentation of a given carbon source) show that not all bacteria possess the same enzymes. By direct assays, one finds that *Escherichia coli* can form β-galactosidase but *Salmonella* strains lack this enzyme. *Aerobacter aerogenes* lacks tryptophanase and possesses inositol dehydrogenase, whereas *E. coli* makes the former and not the latter enzyme.

Each bacterium seems to be able to produce a unique, limited set of enzymes, perhaps a few thousand different kinds. Apparently the enzyme-synthesizing mechanism of a bacterium permits only a limited set of proteins to be made. This implies that protein structures are specified by the mechanism, and leads to the question of how accurate these specifications are. What fraction of a given kind of protein molecule does not have the same structure as the majority? Is there a predominant structure for each enzyme? This problem has been reviewed recently; the conclusion reached is that the structure is highly specific and that errors are rare.[8]

One approach to the problem of homogeneity is to isolate a protein in as pure a state as possible and then determine by amino acid sequence analysis and separation methods whether the preparation is chemically, physically, and enzymically homogeneous. Disagreement exists as to how homogeneous such preparations are, but in the best cases most of the molecules appear identical or fall into a few classes which differ at one amino acid position.[8] One objection to this method is that the process of purification eliminates those proteins that differ markedly from the norm: this is more likely the better the purification. Conversely, changes in structure introduced during purification must be avoided.

A second method for the study of homogeneity is to fractionate unpurified extracts by a general separation technique such as electrophoresis and then to determine enzyme activities in the fractions. By such means, several proteins with apparently the same enzyme activities have been separated from a single source.[9, 10, 11] However, one must be cautious in interpreting such results because complexes between the enzyme and other materials in crude preparations could produce several active fractions and because several different proteins with the same sort of enzyme activity can occur in one tissue.[12]

Another approach is to try to force an error in protein synthesis. Mistakes in protein structure could be made in the folding of the polypeptide chains, in the loss of terminal amino acids, or in the sequence of the amino acids. Methods for detecting or causing errors of the first sort are not yet at hand,[8] but one can alter the availability of one amino acid and determine whether another can substitute for it with a measurable frequency. Thus, if amino acid-requiring mutants are deprived of their essential amino acids, they are not able to form proteins, and therefore the naturally occurring amino acids do not readily replace one another. Theoretical calculations have been made on the probability of errors in the synthesis of protein molecules.[13] In agreement with the above observations, substitution of one natural amino acid for another should be rare, with the exceptions of substitution of glycine for alanine or valine for isoleucine (each of which should occur to the extent of about 5%). It seems unlikely, then, that a bacterium supplied with all of the amino acids would make the mistake of incorporating the "wrong" amino acid. As an exceptional case several per cent extra valine was incorporated in place of other amino acids into proteins of *E. coli* K12.[14]

The specificity of protein synthesis can be tested with analogs, some of which have structures more similar to naturally occurring amino acids than do other natural amino acids. A number of these analogs are incorporated into proteins (see ref. 8). For example, the incorporation of *p*-fluorophenylalanine into the α-amylase of *Bacillus subtilis* has recently been demonstrated by isolation of the crystalline enzyme containing the analog.[15] The penicillinase of *Bacillus cereus* is changed in both activity and immunological specificity when the bacteria are exposed to canavanine or *p*-fluorophenylalanine.[16] In the extreme case, an analog can be built into the structure of an enzyme so effectively that enzymic activity is retained, as with selenomethionine in the β-galactosidase of *E. coli*,[17] or ethionine in the α-amylase of *B. subtilis*.[18] The presence of analogs often prevents appearance of active enzymes; usually because analogs seem to prevent all protein synthesis, rather than being incorporated into the proteins. These results suggest that the structural requirements for amino acid incorporation are severe.

Although the chance of building the wrong amino acid into any specific site in the protein may be slight, the chance of incorporating the wrong amino acid into some place in the entire protein is greater, increasing in proportion to the number of amino acids in the protein. If the above estimates are correct one or two errors might be built into several per cent of the protein molecules.

2. The One-Gene One-Enzyme Hypothesis

The current interpretation of this limited, specific ability to synthesize enzymes is that organisms contain one (or more) units of a sort which permit the cell to make each enzyme. These units are the genes; the ability of the cell to produce its particular set of enzymes and no others is determined by its hereditary material. If the active form (allelle) of a gene is not present in the organism, the corresponding enzyme cannot be produced. This concept is known as the one-gene one-enzyme hypothesis,[19] and postulates a one-to-one correspondence between genes and enzymes. The hypothesis was originally based on extensive genetic studies with nutritional mutants of *Neurospora crassa*: the dissimilar nutritional abilities of mated parental strains were passed on to progeny in accord with classical laws of genetics. The hereditary units which controlled metabolic abilities of the mold behaved as classical genes.

Although genes were originally defined and studied as units which carry hereditary properties in definite ways (for a brief review, see refs. 20, 21), they are now thought of as actual material pieces of the chromosome in which are built the specificity that controls a function. The mass of information that supports this view cannot be presented here, but it has been amply discussed in recent reviews,[22-24] and books.[21, 25, 26]

In the great majority of cases, metabolic abilities seem to be carried by genes in the chromosomes. Mutations to nutritional requirements of virtually every known sort are located in the chromosomes.[19] The great majority of these mutations are highly specific; the mutants require only one supplementary nutrient to permit growth. Thus, of 612 mutants of the mold *Aspergillus nidulans* that grew on a complex medium but not on a simple medium, 95% grew on the simple medium plus a single supplement.[27] Similarly, with the mold *Neurospora crassa*, 84% of the nutritional mutants responded to one supplement.[28]

Cases have been described where hereditary properties do not seem to be carried by classical genes.[29, 30] For instance, the inheritance of respiratory systems in baker's yeast[31] and in *Neurospora*[32] does not obey the usual laws of genetics, but rather behaves as if the ability to make the enzymes and pigments resides in cytoplasmic particles. Cytoplasmic control of frequency of division and of morphological properties has been studied by techniques of nuclear transplantation in amoeba.[33] A self-reproducing system has been postulated to operate in the formation of chloroplasts of *Euglena gracilis*.[34, 35] The heritable conversion of *Salmonella paratyphi* to an L-form also has been postulated to be controlled by a nongenetic mechanism.[36]

The mechanism for hereditary specificity not carried in the chromosomes might be of several sorts. It might be carried by ribonucleic acid of the cytoplasm, since specificity of plant[37, 38] and animal[39, 40] viruses is contained in this material. Specificity can also be carried by cytoplasmic genelike structures named episomes[41]; the production of certain phages, colicines, enzyme-forming abilities, and mating factors has been attributed to them. A third possibility is that the hereditary property might be carried by self-perpetuating metabolic systems.[42-44] A well-documented case of this sort has been described with regard to the formation of β-galactosidase in *E. coli*. Ordinarily, the enzyme is not formed in the presence of low concentrations of certain β-galactosides (inducers, see Section III, B), nor in the presence of glucose. However, after temporary exposure to a high concentration of the inducer, β-galactosidase continues to be formed under the originally inadequate conditions (low inducer or in the presence of glucose) for hundreds of generations.[45, 46] The explanation lies in the creation at high inducer concentrations of a mechanism (permease) which accumulates the inducer inside the bacteria; thereafter the low extracellular concentration of inducer creates a sufficiently high intracellular inducer concentration to permit enzyme and further permease formation.

3. Mutation and Enzyme Synthesis in Bacteria

The set of enzymes found in a strain of bacteria remains constant from one generation to the next, suggesting that the controlling factors must be inherited. Like other inheritable properties, the enzyme-forming abilities can undergo mutation, and mutant strains of bacteria can be isolated from an original (wild-type) strain (see review[47]). These mutated strains are very often deficient in the ability to produce one enzyme; a summary of such mutants has been compiled.[22]

The hereditary loss of enzyme-forming ability strongly suggests that the one-gene one-enzyme hypothesis is valid for bacteria. Stronger support is provided by experiments on the recombination of genetic material of bacteria. Bacteria cannot be mated in the same way as can higher organisms. However, hereditary material from genetically different parents can be recombined by the processes of transformation, transduction, and recombination.[48, 49] Such experiments are basic to the area of bacterial genetics, now summarized in books,[21, 50] and numerous articles.[51] The genetic determinants of enzyme formation fall on a linear genetic map, on which independently obtained mutations for a given enzyme-forming ability are found closely linked[48] (see for example, results obtained with tryptophan-requiring mutants[52] and β-galactosidase-negative mutants[53]).

Perhaps more direct evidence for the role of specific genes in enzyme formation is obtained by injecting the genetic material governing synthesis of the enzyme into bacteria lacking enzyme-forming ability. When this is done by conjugation,[54] transformation,[55] or transduction[56-58] the synthesis of enzyme in the receptor cell commences almost at once. Therefore, the added gene provides a necessary component for enzyme synthesis.

Further discussion of the gene-enzyme relation requires a more precise definition of functional genes. When a number of independent mutations cause the loss of ability to produce one enzyme, and these mutations are all located in one region of the genetic map, how is one to tell whether these sets of mutations are in one functional genetic unit or in two or more units? The *cis-trans* test was devised to distinguish between these possibilities. Consider two mutants, each deficient in the reaction in question. If a cell that contains genetic material from both mutants (a zygote) is able to perform the reaction, the two mutations are in separate functional units, whereas if it cannot, they are in the same unit (and both inactivate the same enzyme). As a control, the reaction should occur in a zygote containing the doubly mutated genetic material and the corresponding normal chromosome. The functional gene defined by this test is called a cistron.[59]

Few groups of bacterial mutants have been analyzed for cistron relationships. All ten of a set of β-galactosidase-negative mutants of *E. coli* were defective in the same cistron, as shown by inability of zygotes made from pairs of the mutants to produce the enzyme.[53]

Several cases are known in which two or more cistrons must cooperate to permit an enzyme to be formed. Several enzymes of the mold *Neurospora* are each under the control of a group of adjacent cistrons. When the mold carries nuclei of two mutants (each unable to form the enzyme) in the same cytoplasm, the enzyme is made in reduced amounts compared to the wild-type organism. How this phenomenon, known as complementation, operates is not well understood.[22, 23, 60] However, since extracts of two complementary mutants deficient in adenylosuccinase showed enzyme activity when mixed,[61] a combination of cytoplasmic gene products seems likely. Since adenylosuccinase produced by complementing *Neurospora* is in some cases altered in heat stability and other properties, the enzyme cannot be formed by combination of normal subunits.[61a]

Complementation has also been found in bacteria, for example, among the loci that govern the synthesis of histidine[62] or of tryptophan.[52] Often complementation between nutritional mutants is between genes that govern the synthesis of different enzymes in the same biosynthetic pathway. Surprisingly, in some cases such as the histidine and tryptophan pathways in *Salmonella typhimurium,* the genes are located on the genetic map in the same order as the steps of the pathway they regulate.[62] Therefore, adjacent mutants that lack separate enzymes in the same pathway can complement each other when in the same cell. In one instance, two adjacent cistrons are needed to permit formation of tryptophan synthetase in *E. coli*.[52] Each cistron is involved in synthesis of a different protein, and the two proteins combine to produce the active enzyme. Complementation between mutations which affect the formation of the same enzyme, imidazole glycerol dehydrogenase, of *S. typhimurium* has also been reported.[62]

4. Fine Structure Alterations of Genes and Enzymes

A functional gene is thought of as a linear structure composed of many parts, each capable of an alteration which produces a mutation. These changes can be mapped by genetic methods and provide a fine structure map of the gene.[62-64] Genes governing the production of certain enzymes—tryptophan synthetase,[52] histidine enzymes,[62] alkaline phosphatase[65]—have been mapped intensively. The most complete mapping has been done on the rII region of coliphage T4, where hundreds of mutant sites have been placed in linear order on two cistrons.[59] One therefore thinks of a cistron as composed of a linear sequence of many mutable parts, and of each part as important for the functioning of the entire cistron.

The linear substructure of a cistron suggests a one-to-one correspondence with the linear structure of the enzyme whose synthesis it controls. An alteration in a gene should then bring about a change in the enzyme at a specific corresponding position. Although a complete loss of enzyme protein is noted in many mutants, in other instances mutants do indeed form altered proteins.[22, 23]

Most convincing would be an instance of alteration of a single amino acid in an enzyme, in response to a limited (point) mutation in the corresponding gene. This result has not yet been obtained with bacteria. However, in each of several hereditary disorders of hemoglobin synthesis in humans, of which sickle cell anemia is one,[66] the hemoglobin is changed in a single amino acid at a specific place (for summaries of these results, see refs. 22, 23).

Other evidence is available for modified proteins produced by mutated organisms.[22, 23] Some of these enzymes are unusually heat-labile. These include the coupling enzyme of pantothenate synthesis of *E. coli*[67] and the tyrosinase of *Neurospora*.[68] A glutamic dehydrogenase of *Neurospora* that is heat-activated has also been reported.[69] A *Neurospora* mutant produces a tryptophan synthetase that is especially strongly inhibited by heavy metals.[70]

As a result of mutation, *E. coli* and *Neurospora* can produce materials (CRM) that cross react with antibodies to tryptophan synthetase.[52] CRM to β-galactosidase has been found in eight of sixteen β-galactosidase-negative *E. coli*.[71] These CRM must be proteins only slightly different from the normal enzymes.

Results such as these are explained by the "template hypothesis," according to which one or more genes act as carriers of specificity for each enzyme: subunits of the gene determine the sequential order of amino acids in the enzyme. The details of this correspondence between gene parts and enzyme parts remain completely vague, and there is reason to believe that the enzyme is not built directly on the gene (see Section II, B, 3). However, considerable information consistent with the template hypothesis has been obtained in the last few years, and this hypothesis is the most practical one at present.

In summary, one or a few genes seem to carry the structural specificity for the formation of each enzyme. It is sometimes said that genes contain the "information" required to determine the structure of the protein, although of course this more picturesque way of presenting the idea does not aid one to understand the connection between genes and enzymes. A gene alteration can result in inability to make an enzyme, or can alter the enzyme structure. This is not the same as saying that all genes act to determine structure. There are genes that do not determine the structures of proteins but rather control the structural genes (see Sections III, A and III, C, 3).

B. Deoxyribonucleic Acid and Enzyme Synthesis

1. Deoxyribonucleic Acid (DNA) as a Carrier of Genetic Specificity

The idea that the genetic fine structure determines the fine structure of enzymes leads one immediately to inquire into the chemical composition of genetic material. Can the correspondence between parts of the genetic material and parts of the related enzyme be determined?

DNA carries genetic information in bacteria.[72, 73] Of a variety of data supporting this view, the most direct evidence comes from the ability of purified DNA to act as transforming principle,[74] i.e., to convert bacteria from one genetic type to another. With regard to enzyme production, DNA from mannitol dehydrogenase-[75] or amylomaltase-[55] positive strains of pneumococcus can transform individuals lacking the ability to form these enzymes into bacteria that can synthesize them. Two enzymes involved in glucuronic acid metabolism were found in a capsulated strain of pneumococcus,[76, 77] and since this strain can be created by transformation from a noncapsulated strain which lacks these enzymes, the ability to form the enzymes must be carried by the transforming principle. Transformation of strains of *Bacillus subtilis* which require various nutrients (nicotinic acid, indole, tryptophan) into nutritionally independent strains has been accomplished.[78] The ability of the bacteria to make the corresponding enzymes appears to have been transformed.

These observations, and also a long series of cytological, genetic, and biochemical studies, point to a major role of DNA in genetic material (see review[72]). One should not, however, conclude that the functional gene is composed only of DNA. Quite probably DNA in the cell is associated with other material such as protein and RNA.[79, 80] A unique protein has been reported to be combined with the DNA of *E. coli*.[81] Purified DNA, such as is added in transformation experiments, apparently must be integrated into the genetic apparatus of the receptor bacteria.[55]

Other evidence that DNA carries the genetic information necessary for enzyme synthesis derives from work with bacteriophages. Most of the DNA and little of the protein of phage T2 enter *E. coli* upon infection.[82] Therefore the injected DNA must provide the specificity for formation of all parts of the progeny phage. The possibility that the DNA simply activates specific genes in the host bacteria would seem to be eliminated by the observation that phage can reproduce in bacteria which have received doses of ultraviolet light sufficient to abolish enzyme synthesis.[83] Infection produces antigenic proteins related to the phage; irradiation of the phage prevents formation of the antigenic material.[84] Many new enzyme activities appear after phage infection.[85-89] UV-irradiated phage can cause such enzyme formation.[90] Either the phage DNA acting as a carrier of specificity might be responsible for enzyme formation, or injection might cause a physiological change which releases an inhibition of enzyme activity that existed in the uninfected bacteria, as found with DNase activation.[91]

Transducing phages can carry the specificity for formation of several enzymes.[48, 92] By analogy with phage T2, presumably the phage DNA enters the bacteria and therefore is the specific agent. Direct evidence for this DNA transfer is not available, but since the transducing ability can be inactivated, after infection of the bacteria, by decay of incorporated radiophosphorus P^{32} (a component of DNA rather than of protein), DNA would seem to be involved.[93] Similarly, the material involved in transmission of genetic information by bacterial conjugation is sensitive to P^{32} decay. Therefore, it is likely that in the conjugation process specificity is transferred via DNA.[54, 94]

A final piece of evidence for the genetic role of DNA derives from studies of chemically induced mutations. Bromouracil can be incorporated in place of thymine into the DNA of bacteria and phages (see review[95]). Biochemical evidence shows a strong relationship between the mutagenic action of bromouracil toward phages and DNA metabolism;[96] genetic mapping and studies of back mutation[97] reveal that bromouracil brings about highly localized mutations, as if the analog altered the structure of DNA at the point of its incorporation. Nitrous acid, a mutagen in living cells, causes mutation of transforming principle.[97a]

2. DNA as a Template Material

Granting that DNA carries the information for synthesis of enzymes, one next seeks an explanation in terms of structures of the two macromolecules. As is now well known[98, 99] DNA is built of four deoxynucleotides, linked by 3′,5′-phosphodiester bonds into long polynucleotide strands. The molecules of DNA all seem to be about the same size,[100] containing about 20,000 bases (molecular weight at least 6,000,000). Two of these strands, running in opposite directions, are combined to form a double helix around a central axis. The strands are held together by specific hydrogen bonds between adenine and thymine, and also between guanine and cytosine.

The structure of DNA seems to be a satisfactory one for a template capable of directing the formation of proteins.[101] According to current views,[98] groups of bases in sequence on the DNA each are supposed to determine the position of a specific amino acid in the protein. Quite likely, the DNA does not itself act as the template; rather, in some mysterious way, it guides the synthesis of specific RNA, the template on specific sites of which activated amino acids are attached and combine to form proteins (see the preceding chapter). According to these views, the sequence of bases of DNA determines, via the base sequence of RNA, the sequence of amino acids in a protein.[102, 103]

A DNA molecule would seem amply large to specify the structure of one enzyme. In fact, one DNA molecule might contain several such templates. This would seem almost essential in the case of a minute pleuropneumonia-like organism which contains only about 10 molecules of DNA,[104] and even in *E. coli* which has about 300 DNA molecules per nucleus. Viruses carry genetic specificity, but contain even less nucleic acid: the Shope papilloma virus and phage ϕX174[105] contain only about 2,000,000 molecular weight units of DNA, and tobacco mosaic virus contains a similar amount of RNA.[106]

A few estimates of the size of functional genes are available, all considerably smaller than the size of a DNA molecule. The A and B cistrons of the rII region of phage T2 each contain roughly 3000 nucleotides.[59, 73] Transducing phages can carry at least 4 cistrons and contain about 100,000 bases, which sets an upper limit of 25,000 bases per cistron. The transforming principle for streptomycin resistance is estimated to have a molecular weight of about 500,000.[107]

The long, stranded structure of DNA has led to several attempts to devise a "code" by which a one-to-one correspondence can be made between groups of bases in DNA and individual amino acids in the proteins.[109, 110] The composition of DNA provides a puzzle in this respect because, although all molecules of DNA in any one species seem to have nearly the same base composition,[100, 111] the ratio of guanine to adenine in the total DNA of different species varies from 0.45 to 2.7.[112] Since the compositions of proteins of different organisms do not differ very much, the highly variable DNA base ratios should not be found if certain groups of the four bases specify each amino acid. As one resolution of this problem, it has been suggested that there are only two symbols of the nucleic acid code.[113] Most schemes have been based on the assumption that there are four independent "letters" in the DNA "alphabet," corresponding to the four bases. Comparisons of viral RNA and protein compositions have been made.[114] Attempts at coding have so far not been successful, and coding presents one of the major problems of genetic biochemistry.

Direct proof of the hypothesis that there is a one-to-one correspondence between each amino acid and a group of bases will depend on determinations of both the base sequence in a specific cistron and the amino acid sequence in the corresponding protein. This is not experimentally feasible at present, but it is possible to locate mutational changes in a gene by fine structure genetic mapping and to compare the locations of the different defects with the locations of alterations in structure of the corresponding protein. Several laboratories are busy with this difficult problem.[52, 65, 115] Results are awaited with interest, since a correspondence of this sort would greatly strengthen the code hypothesis.

3. Functioning of DNA

How DNA is involved in the synthesis of bacterial enzymes is not yet clear. Either DNA could exert a *direct* action—it could, for example, retain the specificity by functioning as a template.[116, 117] Or, it could function by *indirect* action; it could control the synthesis of a substance, such as RNA, directly involved in enzyme formation.[98, 118] DNA synthesis need not occur for enzyme synthesis to take place, as shown by numerous experiments in which DNA synthesis is inhibited (by mitomycin c, for example[119]) without affecting enzyme production.

In the more highly organized cells of animals and plants, the weight of evidence indicates that DNA plays an indirect role, and that RNA is the material directly involved in enzyme synthesis.[120] Enucleation of cells and study of their fractions show that protein synthesis is possible without a nucleus.[121, 122] Enzymes are produced by enucleated *Acetabularia* (alga),[123] at least for a short time. The synthesis of RNA is stopped by enucleation of human cells[121] or *Acetabularia*[122]; treatment of the latter with ribonuclease stops protein synthesis permanently (although nucleated cells recover).[124]

Direct evidence is not available on the role of the nucleus in synthesis of enzymes by bacteria. It is commonly supposed that the genes guide formation of RNA which serves as a template for enzyme production.[118, 125] In *Neurospora*, RNA is made by the nuclei.[126] One difficulty with supposing that the ribonucleoproteins (ribosomes) carry the information originally present in the genes is that the base compositions of DNA and ribosomes are not the same; therefore it is hard to see how the two structures can carry the same specificity.[127]

Other data are also difficult to interpret on the basis of an indirect action hypothesis in which specificity is transferred to a stable ribosomal RNA. The presence of intact DNA seems to be necessary for enzyme formation. When P^{32} is incorporated into bacteria and allowed to decay, the bacteria rapidly lose the ability to form several inducible and constitutive enzymes.[128] The P^{32} damage is in DNA rather than in RNA. Even when the P^{32} is located only in specific genetic material, introduced by bacterial conjugation into a nonradioactive receptor, decay causes loss of β-galactosidase forming ability.[54] The prevention of synthesis of certain enzymes by infection with virulent phages[129] may be a consequence of destruction of bacterial DNA by phage-activated deoxyribonuclease.[91]

Experiments on the rapidity of action of genetic material suggest a close connection between genes and enzymes. Genetic material introduced by either conjugation or transformation expresses itself very rapidly, within a few minutes of entry into the receptor bacteria. The same is true of expression of phage genes after their entry into the bacteria (for references see Section II, B, 1). Thus, introduction of new genetic material causes a rapid appearance of enzymes, and destruction of DNA causes a rapid disappearance of enzyme-forming ability. These results suggest a close connection between DNA and enzyme synthesis; the bacterial cytoplasm alone does not seem adequate to form enzymes.

Currently, the genetic material is thought to produce a special specific RNA (messenger RNA) which becomes attached to the ribosomes where it completes the template. Evidence favoring this view is accumulating rapidly. Thus, infection with T2 phage stops synthesis of bacterial ribosomes, but a new sort of unstable RNA with base composition similar to that of phage DNA is produced.[130] A similar RNA has been reported in yeast[131] and in bacteria.[132] RNA synthesis is essential for the production of the new phage.[130] In the phage case the new RNA can combine with the ribosomes.[133] This messenger RNA is reported to be so similar to phage DNA that it can form double strands with the latter under appropriate conditions.[134]

Another sort of evidence for implication of a rapidly formed RNA in enzyme synthesis comes from work with nucleic acid base analogs which permit the formation of protein but not of normal enzymes.[135] For example, 2-thiouracil causes *Escherichia coli* to produce an immunologically cross-reacting material rather than normal β-galactosidase.[136] Similarly, 5-fluorouracil, which permits protein synthesis but not active β-galactosidase formation,[137] causes a CRM to appear.[135] 5-Fluorouracil does not inhibit formation of the alkaline phosphatase of *E. coli* but the enzyme produced is thermolabile.[135] These analogs are thought to exert their effect by altering the "code" in the messenger RNA, and thereby causing a sort of phenotypic mutation.

The final step in the sequence from gene to enzyme is thought to be the assembly of amino acids on the ribosomes. Newly formed bacterial protein first appears in these particles,[138] and protein can be formed on them in broken cell preparations.[139] A special β-galactosidase associated with ribosomes has been reported in induced and constitutive *E. coli*, but not in uninduced cells or negative mutants.[140] Thus, the sequence of steps by which genes provide specificity for enzyme synthesis seems to be taking shape rapidly.

C. Other Effects of Genes on Enzyme Production

Genes of the sort discussed above may be called structural genes because they are thought to provide the information for the structure of corresponding enzymes. Mutation can prevent enzyme formation in other ways than by damage to the structural gene.[28, 141] Various possibilities for indirect effects of mutation on enzyme formation, although not often thoroughly explored from a biochemical point of view, often seem to be due to metabolic interactions.[20, 142] These effects are not basic to our central theme, which is that the ability to produce enzymes requires the proper structural genes.

Numerous cases have been reported in which mutation at several sites causes loss of ability to produce an enzyme. For example, production of tyrosinase by the fungi *Glomerella*[143] or *Neurospora*[144] is prevented by several mutations at different loci. In the latter case, one gene carries the structure of the enzyme and the others affect the functioning of the structural gene. Several widely separated loci drastically reduce the lactase of *Neurospora*.[145] Four genes are involved in utilization of raffinose by *E. coli*.[146] At least seven loci in *E. coli* can be mutated to prevent growth on lactose.[147] However, only one of these appears to prevent production of β-galactosidase; another prevents formation of the galactoside permease.[53] Another instance of a nutritional requirement without loss of the corresponding enzyme is found with a pantothenate-requiring mutant of *Neurospora*.[148] An inhibition of enzyme activity in crude extracts seems to be responsible. Absence of the enzyme is only one cause of a nutritional requirement.

A mutation can affect enzyme formation by reversing the effect of an earlier mutation. In some cases the second event is a back mutation, i.e., it occurs at the same locus as the first. But in other cases the second mutation occurs at a different genetic locus and is called a supressor mutation.[23] Supressor mutations have been found for the tryptophan synthetase of *Neurospora* and *E. coli*,[52] and in the purine pathway of *E. coli*.[149]

If the primary mutation directly prevents the enzyme from being made, the supressor mutation can act in other ways than by restoring the lost structural specificity. As one possibility, an alternate pathway can be established, as seems to be the case for reversal of mutation from thymine requirement in *E. coli*.[150] Or, a suppressor mutation might act by complementation, to replace the damaged part of the altered enzyme with a differently changed piece. This might be the case with tryptophan synthetase of *E. coli*,[52] or adenylosuccinase of *Neurospora*.[61a]

If the primary mutation does not damage the structure of an enzyme, but rather causes loss of enzyme activity owing to inhibition, other possibilities for supressor gene action exist. Complexities of the supressor gene action suggest a metabolic basis[20]; for example, production of acetate in *Neurospora* appears to be inhibited by acetaldehyde in a mutant which cannot directly convert pyruvate to acetate; a supressor mutation alters metabolism so as to reduce the inhibitor level.[142] One such case of supression is well understood. The inability of one *Neurospora* mutant to form tryptophan synthetase is attributed to a modification of the enzyme which makes it strongly susceptible to inhibition by metal ions (probably Zn^{++}). The supressor mutation reverses this inhibition by reducing the metal ion concentration.[70]

III. The Quantities of Enzymes Synthesized by a Bacterium

How the amount of each enzyme is determined, the quantitative question, will be discussed in this section. Although genetic composition determines the kinds of enzymes that can be formed by an organism, some enzymes are present in large amounts, others in small, and still others can be virtually absent. The differentiated tissues of higher organisms, all of the same genetic origin, contain different amounts of enzymes. For instance, mammalian liver is high in arginase and low in cholinesterase, whereas the opposite is true of heart. (For an extensive listing of relative activities of this sort see ref. 151.) Whether the genetic composition of different tissues is actually the same is open to some question: in some organisms, such as the fly *Oligerces paradoxus,* many chromosomes of the germ cell are eliminated during development.[152] Also, the properties of the nucleus change with development.[153] As other examples of variable enzyme amounts, differences in enzyme levels have been observed[154] in the vegetative and spore stages of bacilli, and a striking increase in the activity of amylase is observed in *B. subtilis* when the resting stage is reached.[155]

Enzyme concentrations can vary enormously in bacteria depending on the physical and nutritional conditions under which growth takes place —a phenomenon especially readily investigated (see below). All of these results lead to the conclusion that factors in addition to the genes are of great importance in determining the quantities of those enzymes that an organism can form. In this article interest will center upon the more readily interpreted effects of nutrients and metabolites upon enzyme formation.

A. Genetic Modifiers of Enzyme Amounts

Genes exert quantitative as well as qualitative influences on enzyme formation. The quantity of an enzyme per cell probably varies in proportion to the number of corresponding genes per cell, other conditions being equal. Polyploid yeast and animal cells, which contain several times the normal number of each chromosome, seem structurally and metabolically like normal ones, except larger.[156-158] The amounts of their enzymes are greater per cell but remain in the same proportions to one another. If the genes do not remain in the same proportion, relative amounts of enzymes probably differ since striking morphological changes are observed when the ratios of the chromosomes are not normal[20]; measurements of enzyme levels in these organisms would be of interest.

A relation between numbers of genes and quantities of enzymes is obtained by comparing heterozygous with homozygous organisms. In several cases, the former organisms, which have only one positive gene for a given function, are less active than the latter, which have two.[20] For example, rabbits heterozygous for atropine esterase make an average of 107 units of enzyme per milliliter of serum as compared to 271 units for the homozygous animals.[159] However, hybrids of respiration-sufficient and -deficient yeasts behave like the respiration-sufficient yeast of the same ploidy[156]; possibly the factor controlling respiration does not depend on gene numbers. The specific activity of an enzyme is determined more by the ratio of the specific gene to the total genes than the absolute number of genes of a certain type. Data on this point have not yet been obtained with bacteria.

Certain genes determine how nutritional conditions affect the rates at which enzymes are produced[160, 161] under the direction of other, structural genes. Enzyme formation by bacteria may or may not depend on nutritional conditions, depending on the enzyme and organism. Some enzymes are produced in a relatively constant amount, irrespective of the conditions under which bacteria are grown, e.g., with some rate-limiting enzymes of glucose oxidation. This kind of enzyme is said to be produced constitutively.[162] Other enzymes, which are called inducible, are made in relatively small amounts unless some specific compound (the inducer) is present.[163] Others, repressible enzymes,[164] are made in large amounts whenever a compound (a repressor) is absent.

Bacteria are converted by mutation from the inducible or repressible form to the constitutive, or vice versa.[165] Mutations to constitutivity have been described for the β-galactosidase and amylomaltase of *E. coli*.[166] Different penicillinase-constitutive *B. cereus* produce amounts of this enzyme that vary by factors of up to 5000-fold.[167] Strains of yeast capable of inducible or constitutive synthesis of β-glucosidase have also been described.[168] Constitutive mutants for tryptophan synthetase,[169] ornithine transcarbamylase,[170] and aspartate transcarbamylase[171] have been obtained from repressible organisms. How such mutants are isolated will be outlined in Section III, C, 1.

The genes governing the inducibility or constitutivity of β-galactosidase[161, 172] and tryptophan synthetase[169] have been mapped. The ways in which these affect the rate of functioning of structural genes will be discussed in Section III, C, 3.

B. Inducible Enzymes

1. The Phenomenon of Induction

Concentrations of some enzymes in bacteria can vary strikingly, by factors of hundreds or thousands, depending on nutritional conditions. Consequently these enzymes have encouraged investigation during the last 20 years because they offer hope of obtaining information regarding several central problems of biology—mechanism of protein synthesis, basis of specificity, and regulation of metabolism.

Enzyme induction, the specific increase in the amount of an enzyme in response to a nutrient or metabolite, has been known for almost 80 years.[173] The rather diffuse historical data will be referred to only when they are useful in the light of present-day concepts. Interesting reviews of this early work exist.[162, 174] Curiously, as late as 1946, it was argued that enzyme induction does not take place.[175] A thought-provoking historical summary was written in 1947.[176] Modern developments in enzyme induction may be dated from about this time. Information on enzyme induction has been collected in numerous reviews,[177, 178] some very recent.[167, 168] Several articles in a symposium on "Adaptation in Microorganisms"[179] are still of great value.

A property implicit in the definition of enzyme induction[163] is that induction is a physiological change (phenotypic change) in all of the cells rather than a genetic change in part of the population of bacteria. Considerable confusion existed between these two modes by which a population of bacteria can change its enzyme content (see, for example, refs. 180–182). It is now clear that in many cases enzyme concentrations are greatly altered without a corresponding population change, although, of course, genetic changes can also cause enzyme concentrations to vary in a population of bacteria.[183, 184]

The first evidence that induction does not involve selection of mutants capable of forming the enzyme came from experiments with nongrowing cultures. Thus, "galactozymase" activity (a measure of the galactose-oxidizing group of enzymes and coenzymes) of yeast was shown to increase manyfold upon exposure to galactose in a medium in which the cells did not increase in number.[185, 186] As another example, enzymes involved in degradation of aromatic compounds by *Pseudomonas fluorescens* were formed within an hour in the absence of an added nitrogen source.[187]

In growing cultures of *E. coli,* individual bacteria are all induced equally to form β-galactosidase when exposed to lactose under appropriate conditions.[188] This was shown as follows: The bacteria are induced and then infected with a bacteriophage, with lactose as the sole carbon source. The time of lysis of each bacterium depends on how extensively it was induced: bacteria with the most enzyme metabolize most rapidly and lyse first. (Infection with a phage prevents enzyme induction,[129, 189] and therefore the level of enzyme in an infected bacterium is that which existed at the time of infection.) A comparison of the amount of enzyme released by the lysing bacteria with the number of bacteria lysed at any moment showed that under appropriate conditions each bacterium that lysed released nearly the same amount of enzyme, and therefore each bacterium was equally induced (after several intervals of induction).

Another proof that all of the bacteria in a culture can gain the ability to form β-galactosidase was reached by experiments in which the state of induction of an individual bacterium was transmitted to all of its descendants, so that the enzyme contents of cultures each consisting of descendants of one bacterium showed whether or not all the original bacteria were induced.[45, 46]

A second essential property of induction is that it represents a formation *de novo* of enzyme molecules rather than activation of preexisting precursors. A requirement of amino acids and energy for the production of certain inducible enzymes[190-192] strongly suggested that at least a part of each enzyme molecule had to be synthesized from simple components. The rapidity with which incorporable amino acid analogs stop active enzyme synthesis also supports this view.[193] In *Escherichia coli,* a protein (Pz) immunologically similar to β-galactosidase was found. It was at first thought to be a precursor of the enzyme, but this view was later abandoned.[194] It has now been shown by experiments in which radioactive protein precursors were incorporated into β-galactosidase over short times that essentially all (more than 99%) of the carbon or sulfur incorporated into the enzyme comes from the medium and very little from preformed cell components.[195, 196] However, a possible objection to these experiments is that the bacteria were starved before the radioactive compound was added, so that any bacterial precursors might have disappeared before the induced enzyme was formed. With penicillinase, too, the enzyme formed in the presence of radioactive precursors has been isolated and shown to be formed almost entirely from amino acids in the medium.[197] The bacteria do not contain precursors sufficient to make enzyme for more than one-half minute.

One can conclude, then, that enzyme induction exists: cases are known in which an enzyme is formed at an altered rate in response to a nutritional stimulus, and this enzyme is formed *de novo* in bacteria of constant genotype. One cannot conclude that the increase in enzyme activity of a culture is due to induction in all cases. For example, enzymically inactive molecules that behave similarly to citredesomolase in the ultracentrifuge are thought to be precursors of the enzyme.[198] The α-amylase of *Pseudomonas saccharophila* is built in part from precursors in growing bacteria, and completely in resting cells.[199] The α-amylase of *B. subtilis* is made from a well-defined precursor.[200] As another mode of activity change, deoxyribonuclease activity increases due to release of inhibition.[91, 201]

Numerous studies on enzyme induction have been performed with yeast. Since this work has been summarized recently,[168] it will not be discussed here.

A principal aim of studies on enzyme induction is to discover the mechanism by which low molecular weight compounds specifically stimulate the synthesis of enzymes. But does *the* mechanism of enzyme induction exist? Perhaps there are several mechanisms by which small molecules specifically influence enzyme formation. Only very few cases of enzyme induction have been at all thoroughly studied; one cannot yet decide if there is a general mechanism, especially since two cases do not behave similarly in regard to a number of important aspects. It seems advisable at this time when the unity of mechanism is not at all apparent to summarize briefly the information on β-galactosidase and penicillinase induction and then to compare these findings with results obtained on less completely studied systems.

2. β-Galactosidase as a Model Inducible Enzyme

Probably more has been done and written regarding the induction of β-galactosidase in *E. coli* than on all other inductions in bacteria combined. What goes on here at the nutritional level is so much more clearly understood than in any other case that it has become a standard to which other inductions can be compared. Most of the original confusion, based largely on inadequate methods (see review[194]) has been swept away, and a remarkably clear picture remains. Therefore, it is worthwhile to outline the main features of β-galactosidase induction prior to extending the general picture beyond that given in the preceding section. There are excellent reviews primarily on the induction of this enzyme[194, 202] which provide references to the hundred-odd publications on the subject.

The major points discussed above—enzyme change in each bacterium and *de novo* synthesis of the enzyme molecule—were established for β-galactosidase. The enzyme has been crystallized.[203, 204] It seems to be an aggregate of 5 molecules each of molecular weight about 133,000 and has a turnover number per molecule of 48,000 per minute.[194] The enzyme does not differ in any obvious way from most proteins, except for this ability to aggregate. It is activated by alkali metal ions, K^+ or Na^+ being the most active depending on the substrate.[205]

The kinetics of induction are very simple; if to a culture of exponentially growing *E. coli* an inducer is added (under conditions to be discussed below) enzyme formation commences within a very few minutes[206, 207] and continues in an amount proportional to the amount of total bacterial protein produced.[191] The actual rate of enzyme formation increases with increasing inducer concentration.[206] Eventually the enzyme reaches a maximum of about 5% of the total protein.[194] As has been pointed out above, the enzyme is formed from low molecular weight compounds, and all evidence is against appreciable amounts of a precursor in the bacteria. If the inducer is removed, enzyme formation ceases almost at once[194, 207]; the enzyme already formed is stable and is diluted out among the bacterial progeny.[188, 196, 208] β-Galactosidase is found inside the growing bacteria and not in the medium.

The induction is quite specific. A variety of β-D-galactosides or thio-β-D-galactosides, possessing small aglycone groups, serve as inducers.[209] Of the α-galactosides, melibiose is an excellent inducer, and methyl-α-D-galactoside is a weak inducer. Alterations in the galactose moiety abolish inducing activity. The inducers are substrates or at least inhibitors of the extracted enzyme; but this affinity is not at all proportional to the inducing ability in intact cells. Thus, methyl-β-D-galactoside is an excellent inducer of low affinity for the enzyme, whereas phenyl-β-D-thiogalactoside is an extremely weak inducer and inhibits induction by other galactosides,[206] but it has a high affinity for the enzyme. Neolactose is an excellent substrate although not an inducer.[165] Such results suggest that enzyme induction is independent of enzyme action, although this conclusion is open to question (see Section III, B, 4 below).

3. Induction of Penicillinase

The second most extensively studied inducible enzyme of bacteria is the penicillinase of *B. cereus*.[167] This enzyme catalyzes the hydrolysis of penicillin to penicilloic acid. Penicillinase is excreted into the medium, although enzyme activity is also found on the cell surface of the bacteria.[210] It has been crystallized and found to have a molecular weight of 31,500 and a turnover number of 160,000 per minute.[167]

Penicillinase is induced in essentially each bacterium of a culture at the same time, as shown by increased penicillin resistance of all the bacteria after induction.[167] Therefore, the enzyme concentration increases because of induction rather than selection of constitutive penicillin-producing mutants. It is formed *de novo* from amino acids in the medium, as shown by measurements of specific activity of the isolated enzyme after growth in the presence of radioactive amino acids.[197]

Induction is quite specific: only penicillin or a few closely related compounds such as cephalosporin induce the enzyme. Alterations of the ring system inactivate penicillin as an inducer.[211]

Penicillinase differs from β-galactosidase most strikingly in the kinetics of its induction. Upon addition of penicillin, penicillinase increases after a lag of about 15 minutes.[212] This lag is considerably longer than the one found in induction of β-galactosidase. It can be separated into two parts. First, there is a brief period during which penicillin becomes fixed to the bacteria; after less than a minute penicillin can be removed from the medium and is no longer essential for induction.[213] The enzyme continues to be formed at a constant rate (with time) after the inducer is removed. This fixation process does not require metabolism by the cell. During the remainder of the lag period, aerobic metabolism is necessary for subsequent enzyme production.[212] The sensitivity to ultraviolet light of penicillinase formation becomes smaller after the lag period is over.[214] All evidence points to the formation of some intermediate or intermediary mechanism after the addition of the inducer and before the production of enzyme. One explanation for this continued formation of penicillinase is that the inducer is stably bound to the bacteria and can continue to act when it is the bound form, whereas the inducers of β-galactosidase are only loosely bound and are easily removed. Actually, β-galactosidase induction also can continue after the extracellular inducer (galactose) is removed; however, galactose remains in the bacteria.[215] *E. coli* exposed briefly to high lactose concentrations are reported to form β-galactosidase after resuspension in lactose-free medium.[216]

Measurements of the binding of penicillin-S^{35} to *B. cereus* reveal two sorts of attachment, probably to the cell membrane. One correlates with the rate of penicillinase formation, reaching a saturating value at the concentration of penicillin which also saturates enzyme formation. The second sort of binding is relatively nonspecific in that penicilloic acid (not an inducer) can substitute for penicillin, and this binding does not show any correlation with enzyme induction.[217] The specific binding is suggested to represent the attachment of inducer to the sites important for induction. Relatively few molecules of penicillin are specifically bound: approximately 200 penicillin molecules per bacterium permit full induction. Since each bacterium produces about 30 times this number of enzyme molecules per hour, the inducer is thought to have a catalytic role in induction, i.e., a molecule of the inducer is not used up each time an enzyme molecule is formed. Therefore, the inducer cannot be an integral part of the enzyme structure.[167, 218]

4. Specificity of Induction and Enzyme Activity

In numerous cases induction has been shown to be a specific process (for references, see Pollock[167] and Halvorson[168]). Only compounds in a limited group structurally related to the substrate of the enzyme serve as inducers. Certain portions of the molecule cannot be altered without abolishing induction, whereas changes in the structure of other portions have only a quantitative effect on the rate of enzyme production. In this sense, specificity of induction resembles specificity of enzyme activity.

As a consequence of these specificity studies, all current theories of enzyme induction presuppose as a first step some site to which the inducer attaches specifically. The most obvious sites are on the enzyme molecules themselves. Alternatively, some other large molecules that serve as templates for formation of the enzyme might bind the inducer. The kind of molecule to which the inducer is bound determines the sort of model that can be constructed to explain the mechanism of induction. For example, the idea that the inducer forms a complex with the enzyme is basic to a "mass action" hypothesis[174, 219] in which the inducer binds the free enzyme, and thereby displaces an equilibrium between enzyme and inactive precursor molecules with resultant production of more enzyme. If this is so, inducers should act either as substrates or as inhibitors of enzyme action. Conversely, compounds that are bound to the enzyme should act as inducers or as inhibitors of induction. Classically, substrates are the compounds tested for inducting ability, and hence affinities for both the enzyme and the induction site are commonly found. However, to decide the question of whether the inducer and enzyme must combine for induction to occur, one would like to find whether compounds exist that are inducers but have no affinity for the enzyme, or at least whose affinities for the two sites are very different.

Relative affinities of inducers for enzyme and induction site have been studied extensively in the case of β-galactosidase induction, and correlation between the two affinities is not found in *E. coli*[209] or *B. megaterium*.[220] In general, D-galactosides or thiogalactosides were the only compounds active in either sense; but some compounds were excellent inducers and showed little affinity for the enzyme. Other compounds with no inducing ability were excellent inhibitors or substrates of the enzyme. The results led to the conclusion that a combination between inducer and enzyme is not involved in induction.[209] Induction of a CRM with no affinity for the inducer seems conclusively to dispose of the mass action hypothesis.[220a]

Compounds are known which serve in other systems as inducers but not as substrates or enzyme inhibitors. Methyl-α-glucoside is an inducer for α-glucosidase of yeast but has little affinity for the enzyme.[221] Similarly, methyl-β-glucoside is a very good inducer of β-glucosidase of yeast but has almost no affinity for the enzyme.[168] L-Leucine stimulates production of L-serine deaminase of *E. coli* but does not seem to combine with the enzyme; conversely, L-serine, the substrate, does not stimulate enzyme production.[222] Cephalosporin C is an inducer of penicillinase of *B. cereus* but not a substrate.[211] β-Thioglucuronides are inducers for β-glucuronidase in *E. coli* and are inhibitors but not substrates of the enzyme.[223] Conversely, D-threonine is a substrate but not an inducer for D-serine deaminase of *E. coli*; however, it does inhibit enzyme formation.[222]

Complications in interpretation of data of the above sort are now apparent, arising from difficulties in comparison of measurements of enzyme activity in extracts and induction in intact bacteria. For instance, melibiose might not itself be an inducer of β-galactosidase, but could be converted to an active compound inside the cell; the latter might have a high affinity for the enzyme although melibiose itself does not. Also, choline esters appear to induce cholinesterase of *P. fluorescens* after they are hydrolyzed.[224] Or, a substrate might not be able to induce because it cannot penetrate the bacterial membrane. Furthermore, mechanisms for the active transport of inducers into the bacteria are known (see Section III, B, 5), and owing to these the intracellular concentrations of inducers can be very different from their concentrations in the medium. Such effects could invalidate a quantitative comparison between inducer and substrate actions.[225] In spite of such objections, specificity data provide the principal evidence in favor of a binding site distinct from the enzyme molecules. There is a great need for extended experiments in this area.

Sometimes an enzyme is induced under conditions where it is not active. Examples include production of the maltozymase of yeast at high pH,[226] the lysine decarboxylase of *B. cadaveris* under aerobic conditions,[227] and the catalase of yeast in the presence of its inhibitor.[228] These examples suggest that the active enzyme is not a part of the enzyme forming system. However, it cannot be ruled out that a very few molecules of enzyme per bacterium are involved in induction.

5. Kinetics of Induction

The kinetics of enzyme formation provide a second group of data that must be satisfied by any theory of induction. Kinetic experiments are full of pitfalls and must be interpreted very carefully, since the rate of enzyme formation is highly dependent on the way an experiment is performed.[178, 194] For example, with succinic dehydrogenase and cytochrome oxidase of yeast, linear or "autocatalytic" curves could be obtained at will[229]; of course each of these curves by itself would lead to a different model of induction.

Kinetic studies of induction were until recently confused by at least three complications. The first of these involved the use of "nongratuitous" conditions under which the development of the inducible enzyme was necessary for the production of energy, which in turn was essential for enzyme production. For example, when maltase induction is studied with maltose as both the inducer and the sole source of carbon, the enzyme has to be formed before the maltose can be utilized, and yet maltose metabolism is necessary to provide energy and building blocks for enzyme synthesis.[230] Under these conditions complex kinetics of an autocatalytic type were observed; and these kinetics led to various hypotheses (such as the "plasmagene theory") regarding enzyme-inducer interaction and autocatalytic enzyme production that are now of only historical interest. For a summation of these hypotheses, see ref. 231.

To eliminate complications of other roles of the inducer it is necessary to use conditions that are as "gratuitous"[178] as possible: the added inducer ideally should have no effect on metabolism except to permit enzyme formation, nor should it be metabolized. Reasonably gratuitous conditions for β-galactosidase can be achieved by using a carbon source such as maltose and as an inducer an alkyl-β-thiogalactoside which is not metabolized. Even in this case, the thiogalactoside causes a small change in respiration[232] and is acetylated.[233] Other inducers which seem to have little or no effect on the over-all metabolism of the bacteria and are not themselves altered by the bacteria are β-thioglucuronides (for β-glucuronidase)[223] and α-methyl glucoside of yeast.[221]

Permeability provides a second complication in the study of kinetics of induction. The intracellular concentration of the inducer can be very different from the concentration in the medium, owing to the presence of enzymelike mechanisms (permeases) for accumulation of the inducer inside the cell (for a review see ref. 234). These permeases can themselves be induced,[172, 215, 235] usually but not always[215] by the same inducer as the enzyme itself. The process of enzyme induction can proceed as follows (as it does in the induction of β-galactosidase with lactose): The uninduced bacteria on exposure to inducer form a small amount of permease, the permease concentrates the inducer intracellularly, and this higher concentration of inducer rapidly brings about formation of more permease and of β-galactosidase. If the extracellular concentration of inducer is low, considerable time can elapse before the initial permease is formed; after this step induction proceeds rapidly in that particular bacterium. Once permease is formed, the enzyme is produced by extracellular concentrations of inducer that originally would not have been sufficient to cause induction.[45, 46] Therefore, in the population as a whole, low inducer concentrations initiate enzyme formation over a considerable period.[45, 236] Kinetics of enzyme induction can be interpreted easily only when the functioning of permeases is eliminated.

A final set of errors in the study of kinetics comes under the heading of improper ways to assay the enzyme activity. In the past, assays often have been carried out on whole bacteria, where problems of introduction of the substrate into the cells and disappearance of reaction products through secondary reactions exist. Therefore, such methods often do not measure the actual enzyme content of the bacteria (for comparisons of whole and disrupted cells, see, for example, refs. 237, 238). Sometimes a whole set of reactions is measured, as with the determination by oxygen uptake of the metabolism of a sugar or other substrate. Only under unusual circumstances would the data obtained in this way reflect the change in a single enzyme of the sequence.[239, 240] Also, enzymes are sometimes measured after intervals of hours or even days of induction. It would appear only reasonable to study enzyme formation over times[207] comparable to the times required for protein synthesis or detectable growth (seconds or minutes).[241]

From these remarks it should be apparent that most of the data on kinetics of enzyme formation are worthless for setting up models of induction. Whether at present the conditions are sufficiently rigorous to permit valid conclusions to be drawn remains to be seen.

The kinetics of induction of certain enzymes appears remarkably simple when activities are measured properly, and induction is carried out under approximately gratuitous conditions and in the absence of permease effects. The quantity of enzyme increases in proportion to new cell mass from approximately the moment an inducer is added. (This comparison is made with cell mass rather than with time to eliminate the effect of exponential growth of the bacterial culture.) Such kinetics have been shown most rigorously with induction of β-galactosidase by isopropyl-β-D-thiogalactoside in permease-negative *E. coli*.[206] An S-shaped dependence of rate of enzyme formation upon inducer concentration was found. The less-than-proportional increase of enzyme at low levels of inducer is not understood; the ultimate maximal rate of induction is explained in terms of saturation of the site for binding the inducer. Although the maximal rates of induction by two inducers were nearly the same, the concentrations of inducer required to half saturate the system were different: 2×10^{-4} *M* for isopropyl-β-thiogalactoside and 4×10^{-3} *M* for methyl-β-thiogalactoside.

With other enzymes, such as D- and L-serine deaminases[222] and tryptophanase[207] of *E. coli*, enzyme production is proportional to mass increase from a few minutes after addition of inducer, under conditions where the inducer is in excess and does not appear appreciably to influence the growth rate of the bacteria. The β-glucosidase of yeast is also formed in proportion to mass increase upon induction.[168]

Does the inducer permit creation of a template rather than its activation? These simple kinetics would appear to eliminate the former possibility. However, closer examination of the initial few minutes of induction of β-galactosidase and five other enzymes shows about three minutes to elapse before enzyme formation commences.[207] The results are not consistent with a lag before entry of inducer into the bacteria. Three minutes is a very appreciable interval on the time scale of protein synthesis, and may be comparable to the time needed to synthesize a ribonucleoprotein particle.[242] The time required to synthesize a protein molecule is estimated as considerably less than one minute.[138, 197, 241] Therefore, this short lag before enzyme formation commences is available for synthesizing an enzyme-forming template subsequent to the addition of inducer. The data suggest a requirement for formation of a nucleic acid prior to enzyme formation. This process would have to be complete (or at least not rate-limiting) within a few minutes.

A particularly striking case of a lag between addition of inducer and appearance of enzyme is seen in the induction of penicillinase by penicillin (see above). Numerous other enzymes are induced only after very appreciable lags; however, the induction of permeases or development of an energy supply are probably involved in the latter cases.

The loss of enzyme after removal of inducer has also been studied; "enzymic reversion" has been suggested to designate this event.[243] With some enzymes, for example, nitratase and β-galactosidase of *E. coli* and histidase and glycerol dehydrogenase of *A. aerogenes*, enzyme formation ceases almost at once when inducer is removed. The enzyme is simply diluted out among the progeny as the bacteria grow. In a second group of enzymes, penicillinase, succinic dehydrogenase,[244] or myoinositol dehydrogenase (see Pollock[243] for references), the removal of inducer does not stop enzyme formation.

A third group of enzymes disappear from the culture after the inducer is removed. These results do not necessarily prove that the protein part of the enzyme is inactivated or degraded, although this has been suggested. The loss of lysine decarboxylase of *B. cadaveris* is attributed to removal of the coenzyme by the reaction product,[227] and other examples of coenzyme loss have been noted. Loss of permeases is involved in the deadaptations of tartrate-utilizing enzymes of *Pseudomonas*,[245] and of maltozymase of yeast.[240] The disappearance of enzyme activity is not related to induction of enzyme in any known case.

In summary, kinetic data suggest that the sites for enzyme formation either are present in the bacteria before the inducer is added, or are made very quickly thereafter. They do not permit one to decide whether the inducer combines with genetic material or with some enzyme-forming system produced by the gene. Any components which increase after induction (possibly including the enzyme molecules themselves) cannot be rate-limiting for long, because the rate of enzyme synthesis does not increase rapidly with time once started. In spite of the great quantity of kinetic data, the conclusions that can be drawn are rather meager.

6. Other Inducible Enzymes

Numerous reports of changes in enzyme activities of bacteria under various conditions have appeared.[176] Some of these have not been described clearly enough to permit one to conclude that a true induction was observed; many of the others do not add anything to our present knowledge of induction beyond the idea that inductions must be widespread. Some of the more interesting references to induction, not discussed above, that have come to the author's attention include tryptophanase,[207, 246] uronic acid enzymes,[247-249] levansucrase,[250] amylomaltase,[251] 2-ketogluconoreductase,[252] adenosine deaminase,[253] chlorophyll-forming enzymes,[254] cyanase,[255] thymidine phosphorylase,[256] and hydrogenase.[257]

C. Constitutive and Repressible Enzymes

1. Basal Enzyme and Constitutive Enzyme

An inducible enzyme is generally present even when bacteria are grown without inducer. The actual level of this basal enzyme can be quite low: for β-galactosidase it is estimated at 20 active sites per bacterium,[258] and for penicillinase 75 molecules per bacterium.[167] Actually, these are not minute levels, being comparable to the numbers of molecules per cell of at least one biosynthetic enzyme, aspartate transcarbamylase.[171]

How can one account for the basal enzyme? Traces of inducers present in the medium or synthesized by the bacteria might be responsible. Perhaps this is why *E. coli* has a higher basal nitratase[259] or β-galactosidase in a rich medium than in a synthetic medium. Among sources of basal enzyme are the few constitutive mutants, producers of high enzyme levels in the absence of inducer, which are present in every inducible culture.[166] There is no reason to believe that basal enzymes are formed by a mechanism different than those for inducible or constitutive enzymes.

A constitutive enzyme[162] is, ideally, one that comprises a constant fraction of the bacterial protein, irrespective of conditions of growth.[260] The utilization of glucose by many bacteria is mediated by constitutive enzymes. However, even before constitutivity was defined, the ability of *E. coli* to ferment glucose was shown to vary by a factor of at least four in different media.[261] More recent experiments have revealed marked changes in individual enzymes of glucose utilization[262] and of respiration.[263] Other constitutive enzymes also can vary considerably in amount.[264]

Such similarities between constitutive and inducible enzyme raise the question of whether the difference between them is simply a quantitative one, all enzymes actually being inducible. Certain enzymes whose concentrations increase by factors of hundreds when their inducer is added might conveniently be called inducible, while others for which there is no known inducer could be called constitutive if they are easily measured, or basal if they are not readily measured. However, in the intermediate range of five- to tenfold variations in activity the distinction is not so simple, since some constitutive enzymes and also some inducible ones (e.g., penicillinase in some *B. cereus*[167]) vary by this factor. An attempt to separate constitutive, weakly inducible, and basal enzymes according to a scale of absolute activities would appear useless.

Fortunately, mutational data show a clear difference between inducible and constitutive enzyme production. Mutant bacteria produce penicillinase,[167] β-galactosidase,[194] or β-glucosidase of yeast[168] without inducer and relatively independent of nutrients, but at rates similar to those at which the enzymes are produced by fully induced inducible bacteria. Therefore, a distinct, qualitative difference between constitutivity and inducibility exists.

Constitutive mutant strains can be selected from the "wild-type" inducible bacteria by several procedures. One depends on the ability of the constitutive bacteria to utilize a nutrient that the inducible cells cannot use. For example, neolactose is a substrate but not an inducer of β-galactosidase, and therefore is not utilized by inducible strains; but the constitutive mutants will grow on neolactose.[165] Constitutive mutants for β-galactosidase have also been selected after numerous transfers of the culture from a lactose to a glucose medium and back again.[166] During each period in glucose the inducible bacteria lose, by dilution, the enzyme that was induced during growth on lactose. On resuspension in lactose medium, the constitutive cells commence growth at once, whereas the inducible ones grow only after a period required for formation of the enzyme. Therefore, at each cycle the culture is progressively enriched in constitutive bacteria, and eventually a single colony of constitutive organisms can be selected. The same method was used to obtain constitutive amylomaltase mutants.[166] A third technique involves growth on a very low concentration of lactose in a chemostat (a device for the continuous culture of bacteria under constant conditions[265]). Constitutive bacteria outgrow the inducible ones, because the former are better able to utilize lactose as their carbon source, since lactose does not induce at the low concentration used.[45]

Constitutive mutants can also be selected by subjecting bacteria to conditions where only organisms capable of forming a surplus of a metabolite will grow. Thus when antimetabolites structurally related to amino acids or to nucleic acid bases inhibit the growth of bacteria, mutants resistant to the analog appear. Some of these mutants overproduce the normal compound corresponding to the analog.[266] When *E. coli* is inhibited by 5-methyltryptophan, selective growth of mutants constitutive for tryptophan synthesis is obtained.[169] When *E. coli* is inhibited by 6-azauracil, some mutants which can grow are richer in aspartate transcarbamylase than the wild type.[171]

At present there is no reason to believe that any fundamental difference exists between the mechanisms for formation of inducible and constitutive enzymes (but 8-azaguanine is stated to inhibit formation of certain constitutive enzymes less than it does inductions[267]). Rather a secondary mechanism that controls the rate of the enzyme-forming apparatus is thought to be responsible.[268] Evidence for this view will be presented below.

2. Repression of Enzyme Synthesis

Synthesis of a repressible enzyme is blocked by a specific metabolite named the repressor. The subject of enzyme repression has recently been extensively reviewed,[269] and in the present article we shall only illustrate the phenomenon with a few of the better characterized examples before proceeding to a discussion of how repression, induction, and constitutivity are related.

Several brief reports of prevention of enzyme formation by products of enzyme action appeared in 1953 (see Cohn and Monod[268] and Vogel[269] for references). The specificity of these effects with respect to both the repressor and the enzyme were demonstrated. In 1957, several other cases were studied more thoroughly; the basic properties of repression became clearer, and the similarity to induction was soon recognized.

Two of these cases of repression were in the arginine pathway—for acetylornithinase[270] and ornithine transcarbamylase.[271] The latter enzyme has been studied fairly extensively. Arginine reduces its concentration well below the level found in repressible bacteria growing in a medium lacking arginine.[271] When an arginine- and histidine-requiring mutant of *E. coli* is grown in the chemostat[265] on limiting amounts of arginine, the enzyme concentration rises to at least 100 times the value found in the wild-type bacteria. But when the growth rate is equally restricted by histidine in the presence of excess arginine, the enzyme concentration is very small. If arginine cannot be concentrated inside the bacteria, repression is much diminished.[272, 273] Therefore arginine, or some compound made in its presence, is the repressor. Formation of the repressed enzyme requires nitrogen and carbon sources and is inhibited by chloramphenicol; hence it seems to require protein synthesis and is not merely activation of a precursor.[272]

In the other of these cases, the three primary enzymes of pyrimidine synthesis—aspartate transcarbamylase, dihydroorotase, and dihydroorotic dehydrogenase—all are repressed by uracil.[274] The first of these enzymes was studied most extensively. Its specific activity increased up to 1000-fold upon release of repression by depriving a mutant of uracil or poisoning the wild type with an inhibitor of pyrimidine synthesis. Repression was specific in the sense that these enzymes increased, and not others, nor protein synthesis in general; repression depended specifically on the pyrimidine concentration. The substrates or products of the enzymes were not required as inducers; mutants incapable of making these compounds formed the enzymes. The rise in enzyme activity was not a result of metabolic changes due to alterations in growth rate. The increase in activity of one of these enzymes (aspartate transcarbamylase) represented *de novo* synthesis of protein as shown by tracer incorporation and inhibitor studies. This enzyme has been highly purified and crystallized;[171] it can comprise about 7% of the bacterial protein. Repression of aspartate transcarbamylase is released within about 4 minutes of removal of uracil, and is reestablished quickly on addition of uracil.[207] Other enzymes are also reported to be repressed and derepressed quickly, although precise kinetic data have not yet appeared.

About two dozen repressible enzymes have been reported to date.[269] Some of those that contribute to our understanding of the phenomenon will be mentioned in the next section.

3. The Similar Basis of Induction, Repression, and Constitutivity

Induction and repression are in many ways similar phenomena. They are both related to constitutivity since constitutive mutants can be derived for either inducible or repressible enzymes. A "unitary hypothesis" was early put forward to relate these phenomena,[268] and other more detailed hypotheses have more recently been presented.[164, 269, 275] In particular, a lucid and thorough organization of the evidence has been compiled.[276]

Perhaps the most fundamental thought underlying this problem is that since a single mutation can change a bacterium from an inducible to a constitutive state, some single and perhaps simple difference must be responsible. If one could understand how the mutation eliminated the requirement for inducer, one would have an insight into what the inducer does. The original hypothesis was that the mutation permitted the constitutive bacteria to produce their own inducer, perhaps by causing the production of a new, inducer-synthesizing enzyme.[268] A second hypothesis, at present in favor, is that the inducible (i^+) allele produces a repressor which prevents formation of an enzyme in the inducible bacteria, and that, furthermore, an externally added inducer can overcome the repressor.

These hypotheses were tested[53] by bringing together in one bacterium the inducible and constitutive alleles of the *i* gene, for β-galactosidase in the cytoplasm formed in the presence of one allele or the other. When these combinations were accomplished, by means of bacterial conjugation the nature of the cytoplasm was found to determine whether production of the enzyme was inducible or constitutive. Only when the genes were mixed in the cytoplasm of a constitutive cell was there a rapid production of β-galactosidase in the absence of an inducer. This result rules out interaction of *i* and *z* (structural) genes within the chromosome because enzyme was produced when the z^+ and i^- (constitutive) genes were on different chromosomes. The inducible allele was dominant over the constitutive allele, for when both were present the zygote soon became inducible. But according to the hypothesis of an intracellular inducer, the constitutive genetic material should be dominant since bacteria which contain both alleles would form the intracellular inducer. The results are consistent with the hypothesis that an inducible cell contains a repressor of β-galactosidase synthesis which is absent in constitutive bacteria and which can be overcome by extracellular inducers.

Somewhat similar results have been reported for the repressible and constitutive formation of some enzymes governing tryptophan synthesis.[169] These enzymes are repressed by tryptophan in the wild-type organism. A mutant of *E. coli* that produced excess tryptophan was isolated as a consequence of its resistance to 5-methyl tryptophan. In the presence or absence of tryptophan, it was found to perform several reactions of tryptophan synthesis at rates up to 20 times the rates in the wild-type organism. The genetic marker that controlled the repressible-constitutive property was far removed from the loci of genes important for tryptophan synthesis and the repressor allele was dominant over the constitutive allele. The concept that emerges from these two studies and one with tyrosinase of *Neurospora*,[144] is that a repressor, formed under the control of a separate gene in inducible cells, regulates the rate of functioning of other genes.

With most repressible enzymes, there is no evidence for participation of an inducer (a compound that competes with the repressor and blocks its action). Data obtained on the repression of the set of three enzymes of pyrimidine synthesis[274] and the four enzymes of histidine synthesis[277] show that the participation of an intracellular inducer is not essential for enzyme production by repressible bacteria. However, in the case of β-galactosidase the induction seems to be just such a competition of inducer and repressor. The repression of ornithine transcarbamylase of *E. coli* is counteracted by ornithine under conditions where ornithine is not further metabolized (except to citrulline). Ornithine, the substrate of the enzyme, plays the role of an inducer in counteracting arginine, the repressor.[170] How often both an inducer and a repressor are involved in the regulation of synthesis of an enzyme remains to be determined.

The next question is what site the repressor combines with. Recalling the discussion of the functioning of DNA (Section II, B, 3), one might imagine that the repressor blocks production of the messenger RNA or the functioning of the ribosomal template.[202] Evidence favoring the former view is mostly of a genetic nature (see ref. 276). Certain constitutive β-galactosidase mutants of *E. coli,* different from the ones discussed above, are still able to make the repressor but are altered in their ability to bind the repressor to its site of action.[161] This site appears to be a genetic locus, called the operator gene, adjacent to the structural gene which controls β-galactosidase formation, which in turn is very close to the structural gene for galactoside permease formation. A modification of the operator locus permits constitutive production of *both* β-galactosidase and permease. This is so only if all three loci are on the same chromosome, because heterozygotes in which the constitutive form of the operator is on one chromosome and the set of active structural genes (the "operon") are on the other do not form the enzyme and permease constitutively. The operator locus is thought of as a sort of switch for turning both of the structural genes off and on. In the constitutively mutated operator, the switch is always on. In the inducible strain the repressor combines with the operator and prevents enzyme synthesis, which otherwise proceeds. Another sort of mutated operator is also found in which the "switch" is always off, and neither enzyme nor permease is formed. The details of these extraordinary experiments, and the reasoning that leads to the above ideas is best appreciated by reading further.[276]

Another case supports the operator concept. Four enzymes in the pathway of histidine synthesis are repressible. These enzymes increase about 15-fold when the histidine supply is limited, and all the enzymes increase in proportion.[277] This proportional change, called coordinate repression, suggests that the structural genes are controlled as a group. The genes do indeed lie next to one another on the genetic map, and mutations at certain locations affect all of the activities.[278] Not all sets of biosynthetic enzymes seem to undergo coordinate repression. Three enzymes of pyrimidine biosynthesis[274] are repressed to different extents, as are three proteins of tryptophan synthesis[279] and two of purine synthesis.[280]

In summary, one currently thinks of repression and induction as being due to the action of regulatory genes that affect the rate of functioning of the structural genes. Evidence for this view comes from a very few systems. The nature of the actual repressor in the cell is unknown; it is thought not to be identical with the corepressor, the added or endogenously synthesized metabolite that represses,[275] and there is a suggestion that it contains RNA, although the evidence for this is most tenuous.[281] Finally, how the inducer interferes with the repressor is completely unknown.

D. Miscellaneous Specific Modifiers of Enzyme Synthesis

1. Specific Inhibitors of Enzyme Synthesis

Processes that are not readily shown to involve induction and repression influence the quantities of enzymes that bacteria synthesize. Sometimes the effect is on the metabolism as a whole rather than on enzyme synthesis specifically; then it is not of great interest in view of present knowledge. Thus, inhibition by dinitrophenol which blocks energy-producing reactions or chloramphenicol which inhibits the synthesis of proteins,[282, 283] merely shows that enzyme formation requires energy and protein synthesis. Likewise, the close connection between nucleic acid precursors and enzyme synthesis,[284] and numerous data relating synthesis of proteins with RNA[118, 168] are rapidly being replaced by more detailed information (see Section II, B, 3). Unlike RNA, neither the amount of DNA per bacterium (which can be made to vary by addition of β-2-thienylalanine to the culture[285]) nor the synthesis of DNA[119, 284] is directly related to protein synthesis. The significance of effects of amino acid and nucleic acid base analogs have also been discussed in earlier sections. The mechanism of a unique stimulation of β-galactosidase and D-serine deaminase production by 5-amino-2,4-bis (substituted amino)-pyrimidines is obscure.[286]

Glucose has long been known to inhibit the formation of many enzymes.[287, 288] For example, glucose prevents the appearance of β-galactosidase in *E. coli*. If the bacteria are provided with limited amounts of glucose and lactose, a pattern of growth named "diauxie" is observed[289]: first growth on glucose is seen until this sugar disappears; then there is a short lag during which induction of β-galactosidase takes place; following this a second phase of growth occurs as the lactose is metabolized.

The mechanism or mechanisms by which glucose prevents enzyme synthesis are not clearly understood. The hypothesis most favored at present is that surplus metabolites produced from glucose repress formation of enzymes specifically related to them.[290, 291] Evidence in agreement with this hypothesis is that glucose permits more rapid growth than do other carbon sources, and it causes the accumulation of metabolic intermediates to a greater extent.[292] Furthermore, histidase and urocanase,[290] and also serine deaminase[293] are not formed in the presence of glucose and ammonia, but are produced when the corresponding amino acid is the sole available source of nitrogen. These results suggest that nitrogen-containing metabolites formed when glucose is present are the direct repressors of the synthesis of these enzymes. Objections to the repressor hypothesis of glucose action have been raised.[294]

Myoinositol dehydrogenase is induced by inositol and continues to be formed after the inducer is removed; it ceases to be made when glucose is present and reappears in the absence of inositol when the glucose is removed.[292] These results suggest that the glucose-repressor inhibits a stable enzyme forming system made under the influence of inositol.

Ingenious experiments with a mutant of *S. typhimurium* which does not show diauxic growth reveal that the mutant grows slowly on glucose and contains 3 times the normal amount of glucose-6-phosphatase.[295] These observations can be interpreted according to the above hypothesis since the phosphatase could diminish the rate of glycolysis and thereby reduce the pool of repressing metabolites. Alternatively, diauxie could be related to a lowering of the intracellular phosphate concentration by glucose metabolism. A mutant of *A. aerogenes* that does not show glucose repression also grows more slowly on glucose.[291]

Several other explanations have been given for glucose inhibition, and arguments against these have been presented.[288, 292, 296] Glucose can, for example, prevent entry of substrates into bacteria by inhibiting the formation of permeases.[296, 297] However, glucose inhibition is still found when permease effects are eliminated.[206] Furthermore, one is left with the question of how glucose inhibits permease formation. According to a second hypothesis, enzymes of glucose metabolism remove nutrients such as phosphate, mentioned above,[295] which would otherwise be used for the formation of the inhibited enzymes.[292] But surpluses of known metabolites, provided in rich media, do not eliminate the glucose inhibition,[292] and there are more involved objections to this hypothesis.[296] Glucose appears to inhibit β-galactosidase induction by two mechanisms, one strong and competitive with the inducer, the other weak and noncompetitive.[296] The mode of glucose inhibition remains obscure.

2. Other Conditions Which Cause Differential Effects

Under certain nutritional conditions a single enzyme is made at a rapid rate relative to other enzymes or net protein synthesis, as when a culture is starved with respect to some essential component. As examples, *E. coli* form β-galactosidase rapidly with little net protein synthesis provided no carbon source other than lactose is available.[298-300] In diauxie, and generally under nongratuitous conditions, β-galactosidase synthesis takes place while net protein synthesis is slow.[301] In the absence of a nitrogen source and with lactose or maltose as carbon sources, *E. coli* synthesize nitrate reductase and tetrathionase but not β-galactosidase.[302] Aspartate transcarbamylase[274] and ornithine transcarbamylase[272] are formed in the absence of uracil but β-galactosidase is not.

This synthesis of enzymes without net protein synthesis has been called a preferential synthesis of the enzyme. However, in view of the relatively rapid breakdown and resynthesis of proteins in starved bacteria, net protein formation is not a measure of total protein synthesis, and it has been suggested that in reality the ratio of the rates of enzyme synthesis and protein synthesis is the same under these conditions as in growing cultures.[303] But there might be an effect beyond this, related to the mechanism by which glucose inhibits enzyme production; e.g., a repression could be released in the starved bacteria[288]; the α-amylase of *P. saccharophila*[304] and tryptophanase[305] of *E. coli* are made equally rapidly with or without a good carbon source.

Modifiers of enzyme formation often are less obviously related to the enzyme than are inducers. Nutritional requirements specific for an enzyme induction have been reported frequently. For the synthesis of formic hydrogenlyase in growing cultures of *E. coli* several amino acids not needed for growth must be present,[306] and another set are required for the synthesis of this enzyme in resting bacteria.[307] Arginine decarboxylase is formed only in a rich medium, largely owing to an iron requirement for formation of this enzyme.[308] Zinc and biotin affect formation of DPNase and alcohol dehydrogenase in *Neurospora*.[309] Most amino acid decarboxylases are formed only at low pH, whereas deaminases are formed at high pH.[310] β-Fructofuranosidase synthesis also depends on pH.[311] Oxygen stimulates the appearance of some enzymes,[312-314] and diminishes the amounts of others.[315, 316] Instability of the enzyme to oxygen could be responsible in some cases.[317] L-Serine deaminase activity in *E. coli* increases when glycine and leucine are supplied, but L-serine does not affect the formation of this enzyme.[222] The age of a culture also affects its enzyme-forming ability.[216, 318] These peculiar nutritional requirements, numerous in the literature, possibly foreshadow a new sort of regulation of enzyme amounts based on the idea that enzyme-forming systems compete with different avidities for the metabolites in a common pool,[192] although inductionlike specific effects which interact in a not yet obvious way could be responsible.

At different temperatures enzyme formation usually parallels growth,[319] although few careful studies of this sort have been made. Some enzymes do not appear to be made at higher temperatures owing to their instability.[68, 310] Formation of the tetrathionase of a coliform organism is actually more temperature-sensitive than growth or enzyme activity.[320] The bacteria grow well at 44°C., and tetrathionase is active and stable at this temperature, but formation of the enzyme is greatly depressed at temperatures above 40°C. The results suggest a specific, highly temperature-sensitive step in the synthesis of this enzyme. Whether the induction becomes more temperature-resistant after it has proceeded for a short time is not known.

3. Irradiation

Irradiation of bacteria[83, 321] or yeast[322, 323] with ultraviolet light (UV) inhibits enzyme formation. UV does not simply destroy the integrity of the entire synthetic apparatus of the cell, since bacteria subjected to doses of UV that severely damage enzyme formation support almost normal replication of bacteriophage T2.[83, 324] Further evidence for a selective damage is that formation of β-galactosidase, tryptophanase, and D-serine deaminase is more strongly inhibited by UV than is total protein synthesis.[324a] The irradiated bacteria might produce protein in the form of inactive enzymes or these enzyme-forming systems might be more sensitive than most.

The effects of UV on induction of β-galactosidase or tryptophanase in *E. coli* are very simply described. In a salts-glycerol medium the bacteria cease production of the enzyme within a few minutes after irradiation (at 260 mμ) with about 70,000 ergs/cm^2. Smaller doses partially inactivate enzyme formation, according to an approximately "three-hit" curve. The efficiency of inactivation is the same before or after the inducer is added, provided the inducer can act independently of permease. However, if permease must be induced to permit induction of the enzyme, the UV sensitivity decreases after induction has commenced. This is because initially two essential processes existed which could be damaged by UV, but later permease formation is not required and only the sensitivity directly connected to enzyme formation remains. The results argue against a UV-stable enzyme-forming system.[324a]

The inhibition by UV of penicillinase and protein formation in *B. cereus* provides a different result.[214] Penicillin is added to induce the bacteria and is destroyed after one minute; when after another minute the bacteria are irradiated, enzyme formation is inhibited about 2.5 times as strongly as is protein synthesis. Enzyme formation becomes progressively more resistant to UV until at 21 minutes it is only 0.6 times as strongly inhibited as is protein synthesis. During this 19-minute interval, corresponding approximately to the lag before enzyme formation starts, some process occurs which makes induction relatively independent of the original UV-sensitive material.

The system is not understood well enough to permit the conclusion that the UV-sensitive site is involved in production of an enzyme-forming system of relatively high UV resistance. An alternative interpretation, for instance, is based on a two-step induction. In the first step, penicillin would induce an enzyme, not penicillinase, which converts some naturally occurring metabolite into a compound X not otherwise found in the bacteria. In the second step, X would induce penicillinase. The increased resistance to UV after induction could be explained as a reduction in the number of UV-sensitive targets from two to one when the first enzyme is formed, as with β-galactosidase-permease induction. A "sequential" mechanism of this sort would also explain why formation of penicillinase continues long after penicillin is removed—it would continue until the first enzyme was diluted out by growth. This model would be consistent with the metabolism-dependent lag before penicillinase appears.

Similar but less striking instances of increased resistance to UV after induction have been observed with galactozymase[322] and α-glucosidase of yeast.[323] Whether these changes are due to a two-step induction process, involving a permease, for example, or whether they represent some stabilization connected with production of an active enzyme-forming system is not clear.

Action spectrum studies show that the site of UV damage to enzyme formation is in nucleic acid. This has been demonstrated for the galactozymase of yeast,[322] the lysine decarboxylase of *Clostridium cadaveris*,[325] and the β-galactosidase and tryptophanase of *E. coli*.[108]

Inactivation of enzyme formation shows a multi-hit curve, suggesting nuclear damage, as has been suggested for loss of viability.[326]

Far fewer data are available on the effects of other radiations or radiationlike agents on enzyme formation. Irradiation with X-rays causes a much more rapid loss of colony-forming ability than of inducibility, both in the case of β-galactosidase formation in *E. coli*,[327] and of maltozymase formation in yeast.[328] The X-ray target size for inactivation of lysine decarboxylase formation is calculated as 20,000,000.[329] X-rays are stated to stimulate enzyme formation in *E. coli* under special conditions[330]; this may be a permeability effect. Mustard gas, which in many ways affects bacteria like X-rays, also inhibits colony formation much more strongly than enzyme formation or total protein synthesis, and the latter two are about equally inhibited.[284]

When radiophosphorus P^{32} is incorporated into bacterial components and allowed to decay, enzyme-forming ability and colony-forming ability are lost at about the same rate; total protein synthesis is somewhat more resistant.[128] The damaging P^{32} decays are in the bacterial DNA because bacteria labeled preferentially in RNA do not rapidly lose their ability to form enzymes.[128, 285] Decay of tritium incorporated into *E. coli* DNA as H^3-thymidine is effective in stopping enzyme formation; tritium decay in other parts of the cell is far less effective.[331]

These data in general point to a role of DNA in the synthesis of inducible enzymes. This necessity for intact DNA would seem to persist in the induction of β-galactosidase even after enzyme formation is well under way.[54] However, in other cases as with penicillinase, the system created after induction might be relatively stable.

IV. Metabolic Control and the Regulation of Enzyme Synthesis

A. Relative Merits of Types of Enzyme Production

Bacteria seem to possess enzymes that are often utilized and lack enzymes that they do not normally use. This suggests evolutionary advantages both for ability and inability to form enzymes. An enzyme would seem advantageous for the growth of the organism when its substrate is a useful and available nutrient.[332, 333] Enzymes that confer resistance to drugs[4, 334] also give an advantage to an organism whenever the drug is present. In spite of these considerations, individual bacteria can metabolize only a limited number of compounds, and are in general not resistant to drugs. One is inclined to believe that advantages in possessing an enzyme used only occasionally must commonly be outweighed by some selective benefit in not being able to synthesize the enzyme. The most evident benefit would lie in more efficient use of energy and metabolites for synthesis of other parts of the bacterium, and hence more rapid growth. Evolutionary selection, quite rapid in bacteria, should lead to the elimination of slower growing strains: a mutant with a growth rate of 60 minutes, appearing with a frequency of 10^{-8}, would in theory overtake its parent strain with growth rate 60.5 minutes in about 3 months.

Examples of auxotrophic mutants with growth rates (in the presence of the nutrient required by the mutant) more rapid than wild-type bacteria have been reported. These include five amino acid-requiring mutants of *E. coli* which grew 0.8 to 4% faster than the wild type,[335] and a methionine-requiring strain which grew 40% faster.[336] Mutants selected in the chemostat for ability to use lactose efficiently were "superconstitutive" producers of β-galactosidase; they grew more slowly than the wild type.[337] But a histidine-requiring *E. coli* mutant grew at the same rate as its parent.[338] More data are needed to test the hypothesis that inability to form a single enzyme gives an appreciable advantage in growth rate.

Inducibility must sometimes provide a selective advantage over constitutivity or inactivity, because wild-type bacteria possess inducible enzymes. This advantage is probably based in part on the utility of the enzyme under some circumstances and the saving of energy and metabolites at times when the enzyme molecules are not synthesized; also this benefit of economy appears to be fortified in several ways. Induction is not an all-or-none phenomenon, the level of an inducible enzyme is regulated by the amount of available substrate. A consequence of this regulated enzyme synthesis is that toxic concentrations of enzymes are avoided. Thus, sulfanilamide-requiring strains of *Neurospora* appear to be overactive in one-carbon transfer reaction to such an extent that they will not grow well unless sulfanilamide is present.[339] If enzymes such as histidase or tryptophanase were not inducible, they could prevent growth by destroying histidine or tryptophan, respectively, as fast as these amino acids were formed,[340] and the same can be said for arginine dihydrolase.[341] A third advantage to inducibility is that bacteria could not provide space or nutrients for the full complement of enzymes they are capable of forming (some as much as 5% of the total protein) if all were made at once. Therefore an organism with many inducible enzymes can be more versatile than one with only constitutive enzymes. Induction or repression would seem to furnish a very advantageous compromise between enzyme deficiency and constitutivity. Inducibility has its price in regulatory genes and repressors, but this must be small in many cases.

B. Functions of Inductions and Repressions

Growth, based on an enormously complex set of enzymically catalyzed reactions, would hardly be possible if there were not some ways by which living cells automatically coordinated the velocities of their individual reactions with their whole metabolism. Induction and repression, among others, are means by which the amounts of enzymes can be adjusted to meet the needs of the cell. The subject of control of metabolism has been reviewed at length recently,[3] and symposia have been held on the subject,[342, 342a] therefore, only a brief summary dealing specifically with aspects related to enzyme synthesis will be incorporated into this chapter.

The most obvious role of induction in metabolism is to provide enzymes for the utilization of temperarily available nutrients. This role can be extended to an entire sequence of reaction by the process of sequential induction which operates as follows[343]: when a nutrient A is fed to the bacteria, an inducible enzyme E_a is formed, attacks A and converts it to B; then B serves as an inducer for E_b, which converts B to C, following which C induces E_c, and so forth. Even cyclical, self-perpetuating sequences have been suggested.[44] Several pathways of sequential induction have been explored, including those for degradation of histidine,[344] tryptophan,[345] and nicotinic acid[346]; the technique has indeed aided in the elucidation of these pathways.

It has been assumed in sequential induction that each metabolite serves as an inducer for the enzyme which directly metabolizes it. However, an alternative possibility, similar to coordinate repression,[277] is that one metabolite induces the whole series of enzymes. Evidence in favor of the first view derives from the fact that if an intermediate such as *C* in the scheme above is fed, only the compounds that follow (*D*, *E*, *F*, etc.) are readily metabolized and not the preceding compounds (*A* and *B*). An exception, supporting the second hypothesis, is that galactose, substrate of the first enzyme, induces the second enzyme, galactose-1-phosphate transferase in an organism which cannot convert galactose to the substrate of the second enzyme, galactose-1-phosphate.[347]

Inductions and repressions are probably responsible for regulating the levels of enzymes in the main pathways of energy production. Shifting from anaerobic to aerobic conditions causes increases of up to 70-fold in the quantities of Krebs cycle enzymes in *Pasteurella pestis*.[348] In *Aspergillus niger* the condensing enzyme increases 10-fold under conditions of citrate accumulation, and at the same time isocitric dehydrogenase and aconitase decrease to very low levels.[349] The citrate forming reaction in *E. coli* is much less active in bacteria grown in rich medium than in those grown on synthetic medium.[350] Enzyme changes in *Micrococcus denitrificans*,[351] *Pseudomonas ovalis*,[352] and *Rhodopseudomonas spheroides*[254] cause shifts between several metabolic pathways under different nutritional conditions. Induction can produce an enzyme to break down an excess of an intracellular metabolite: a tryptophan-requiring mutant of *Neurospora* accumulates protocatechuic acid, and when this occurs a protocatechuic oxidase is induced.[353] Presumably other instances in which induction aids in removal of surplus metabolites and in detoxifications will be discovered.

Glucose inhibition of induction serves to block the synthesis of numerous inducible enzymes which would otherwise provide the same metabolites as does glucose. The inhibition thereby saves the energy that would otherwise be spent on synthesis of the extra enzymes.

Whereas inductions by nutrients seem to be of importance in catabolism, repressions are found more often in biosynthetic processes. An early indication that bacteria grown in rich medium do not possess many of the enzymes required for amino acid synthesis was seen when these cells, upon transfer to synthetic medium, had to be supplemented with amino acids before they could produce bacteriophage.[354] Repression can assure the production of adequate amounts of anabolic enzymes; furthermore, by limiting the amounts of these enzymes the overproduction of end products can be prevented. Control by repression should be self-regulating because the excess or shortage of the repressing end product determines whether enzyme formation is arrested or continues. The regulation of purine synthesis[280, 355] illustrates the role of repression in metabolic regulation.

Regulation by repression should adjust to change rather slowly, especially if a surplus of an enzyme has been formed and then can be only slowly diluted out by growth of bacteria. A second mechanism of regulation has been found which often supplements repression. This is feedback inhibition of enzyme activity in which an end product of the reaction sequence inhibits the *activity* of an early enzyme of the pathway. By this means, surplus enzymes are prevented from catalyzing their reactions at rates greater than required to provide a low concentration of end product. A few of about a dozen pathways in which both repression and feedback inhibition function include: tryptophan synthesis,[356] pyrimidine synthesis,[357] threonine-isoleucine synthesis,[358, 359] purine synthesis,[355] and arginine synthesis.[360] The relationship between the two modes of control is shown in a particularly interesting way in the last article.

Antimetabolites and analogs may affect growth through modified enzyme patterns. Sometimes these compounds simply inhibit protein synthesis, and sometimes they replace the natural metabolite to produce enzymes of altered properties. They might also function as corepressors; or, as "false feedback inhibitors" they could block the formation of the natural metabolite and permit derepression of an enzyme-forming system.[274, 361, 362]

It is not too hard to imagine how, by evolution, each repressible or inducible step permitting an acceleration of the growth rate would be selected so that a balanced set of mechanisms for enzyme formation is brought into existence.[363] In some cases, however, the expenditure involved in synthesizing a control mechanism could be greater than that required to permit constitutive synthesis. Thus, for trace constituents such as coenzymes, the most efficient procedure might be to permit a constitutive synthesis in micro amounts, the control of the amount somehow residing in the ability of genetic material to produce enzymes at a relatively constant rate, in excess of the maximum required to provide the metabolite.[364]

V. Summary and Current Problems

The processes that confer specificity on enzyme molecules and regulate the amounts of enzymes are evidently not all understood at the molecular level. A reasonable description of them is possible in a very few cases, in terms of genetic units, RNA of different sorts, nutrients, and metabolites. The main problems at present, aside from generalizing the observations already presented, lie in devising interpretations of these phenomena in terms of interactions between molecules.

A. Genes

The structure of each part of an enzyme appears to be uniquely de-determined by the structure of a corresponding part of a gene. Strong correlations are found between locations and sizes of mutations, measured genetically, and effects on synthesis and properties of proteins. More elaborate efforts, in progress, on correlation of genetic fine structure mapping with chemical mapping of changes in the amino acid sequences of the protein corresponding to the gene should provide a powerful test of this concept. A more detailed understanding of the structures of enzymes and, even more so, of the structures of genes would seem essential for the clarification of the idea of correspondence between parts of genes and enzymes. So, too, would knowledge concerning the binding affinities of structures in DNA for other molecules, and of activated amino acids for nucleic acid groupings. With information of this sort, ultimately one might describe templates, with known structures arranged in a sequence, which direct the activated amino acids into the pattern of an enzyme molecule. In spite of elegant work on the coding of specificity for amino acid sequences into genetic information, no satisfactory model is forthcoming; but restrictions on the possibilities have been established.

Mutations obtained with agents such as 5-bromouracil or nitrous acid suggest that a change in a single base in a definite position in one strand[365] of DNA is sufficient to bring about a mutation. Other mutagens might act in informative ways. Possibly, the change in DNA caused by agents such as these can be correlated with changes in protein structure, and thereby the relation between gene and enzyme structure can be clarified.

The simple picture of the gene as a linear structure that uniquely determines an amino acid sequence in the protein is complicated by the recently discovered phenomenon of complementation in which many genetically independent units cooperate in the production of one active enzyme. These data can perhaps be interpreted in terms of an interaction of the products of the separate genetic units in some way, after they have been formed. How big is a functional gene?

In the actual functioning of genes, other questions arise. Is the rate of a gene-catalyzed reaction proportional to the number of genes of a certain type? What fraction of genes have other roles than to guide specificity of protein synthesis, e.g., for control of function of other genes.[276] Do suppressor genes always function through enzymes they produce? What of the genes for inducibility or repressibility—do they function solely to control the activities of other genes? What controls them?

Are there units other than nuclear genes which provide specificity for enzyme formation? What examples of cytoplasmic inheritance can be found in bacteria? Do genes determine the entire structures of enzymes;

for example, are certain structures common to most proteins, and are they made by a common, possibly nongenic, mechanism? How great is the probability of an error occurring at the step where the gene acts, and at other steps in enzyme synthesis?[366]

B. Ribonucleic Acids and Enzymes Synthesis

The mechanism of transfer of specificity from genes to enzymes presents major problems. Is an intermediate information-carrying structure such as ribonucleic acid (RNA) interposed between gene and enzyme? Since bacterial or viral genetic material can function within a few minutes after it is introduced into a new cytoplasmic environment, intermediate enzyme-forming systems must be made very rapidly. Furthermore, the intermediate systems cannot function for very long after DNA is destroyed by radiation or by P^{32} decay: these soon bring about a halt in enzyme synthesis. Possibly the intermediate systems are not metabolically stable (at least those involved in formation of inducible enzymes).

Considerable evidence connects RNA with protein synthesis. At least four types of RNA are postulated to exist in bacteria: amino acid transport RNA, RNA on the cell membrane, ribonucleoprotein particles (ribosomes), and messenger RNA. The first is clearly important in the production of active amino acids. Membranes of *B. megaterium* and *E. coli* are quite active in the uptake of amino acids into protein, but whether the membranes form enzymes, and whether the associated RNA is vital and whether it differs from ribosomal RNA is not known. Recent data suggest an important role of the ribosomes, since a very rapid incorporation and removal of amino acids occurs on these particles.[125] Most recently, the importance of the messenger RNA, supposed to transmit specificity from the genes to the ribosomes, has been put forward.[132, 366a] Work now in progress should soon define the activity of this material. The alterations in enzyme structure, like phenotypic mutations, brought about by fluorouracil and other base analogs should provide further information regarding functioning of the RNA.

C. Inducers and Repressors

Much concerning the stimulatory or inhibitory actions on enzyme synthesis (induction and repression) remains a mystery; it seems useless at present to add another "model" to the large collection now available. Do the small molecules act directly on the genes so as to stop gene-product formation or do they combine with secondary products of the genes such as RNA so as to inhibit enzyme formation by the latter? Neither kinetic nor specificity data at present permit a decision between these possibilities though genetic data favor the former model.[276] The rapid cessation (in some cases) of enzyme formation upon removal of the inducer requires that any active or activated enzyme-forming system created in the presence of the inducer disappear when the inducer is removed.

The idea, derived principally from studies with one enzyme (β-galactosidase), that inducible and constitutive bacteria differ owing to a repressor present in the former and absent in the latter needs verification with other enzymes. Systems for forming biosynthetic enzymes can be repressed, and in general no compounds are known to reverse the repression. In these cases, a simple removal of the repressor could control enzyme formation *in vivo* according to a mass action principle. For other enzymes, the original concept that the enzyme-forming system is not active unless combined with an inducer remains to be tested.

It would seem very desirable to determine the structure of the postulated repressor. What relation does the low molecular weight end product of biosynthesis (the corepressor) have to the actual repressor molecule that prevents enzyme formation? Enzyme formation studies with broken cell preparations, many of which have been attempted, may answer some of these questions; the difficulties of experimentation and of interpretation appear great.

Numerous problems exist in the area of regulation of enzyme synthesis and growth, in the alterations of enzyme activity as the bacteria go through their division cycle,[367-369] in the enzyme changes that take place in bacteriophage-infected cells, and on sporulation.[154, 370] A general mathematical formulation of the rate of enzyme synthesis in terms of number of genes, supply of nutrients, and concentrations of inducers and repressors might be hoped for some day (for a start, see refs. 371, 372).

Possibly as this chapter is written we approach the end of a period in the study of enzyme synthesis which could be called *The Nutritional Period.* Questions have been posed clearly enough now so that we can look forward to a new stage, *The Chemical Period.* Chemical techniques for separation of enzymes and nucleic acids and study of their structures, protein and nucleic acid synthesis in extracts, and genetic fine structure-mapping techniques should permit us to pose, and perhaps before too long to answer, questions of enzyme synthesis in precise chemical terms.

ACKNOWLEDGMENT

The author is happy to acknowledge the very considerable debt he owes Dr. Jacques Monod for ideas gained in many discussions.

REFERENCES

1 M. B. Hoagland, *in* "The Nucleic Acids" (E. Chargaff and J. N. Davidson, eds.), Vol. III, p. 349. Academic Press, New York, 1960.

2 I. D. Raacke, *in* "Metabolic Pathways" (D. M. Greenberg, ed.), Vol. II, p. 263. Academic Press, New York, 1961.

[3] A. B. Pardee, *in* "The Enzymes" (P. D. Boyer, H. A. Lardy, and K. Myrbäck, eds.), p. 681. Academic Press, New York, 1959.
[4] G. E. W. Wolstenholme and C. M. O'Conner, eds., "Drug Resistance in Microorganisms." Little, Brown, Boston, Massachusetts, 1957.
[5] B. B. Brodie, R. P. Maickel, and W. R. Jondorf, *Federation Proc.* **17,** 1163 (1958).
[6] S. Kinoshita, *Advances in Appl. Microbiol.* **1,** 201 (1959).
[7] O. Hayaishi, *in* "Methods in Enzymology" (S. P. Colowick and N. O. Kaplan, eds.), Vol. I, p. 126. Academic Press, New York, 1955.
[8] M. Vaughan and D. Steinberg, *Advances in Protein Chem.* **14,** 115 (1959).
[9] J. M. Gillespie, M. A. Jermyn, and E. F. Woods, *Nature* **169,** 487 (1952).
[10] A. P. Nygaard, *Biochim. et Biophys. Acta* **35,** 212 (1959).
[11] G. Terui, H. Okada, and Y. Oshima, *Osaka Univ. Technol. Repts.* **9,** 237 (1959).
[12] C. L. Markert and F. Møller, *Proc. Natl. Acad. Sci. U.S.* **45,** 753 (1959).
[13] L. Pauling, "Festschrift Arthur Stoll," p. 597. Birkhauser, Basel, Switzerland, 1957.
[14] G. N. Cohen, *Ann. inst. Pasteur* **94,** 15 (1958).
[15] A. Yoshida, *Biochim. et Biophys. Acta* **41,** 98 (1960).
[16] M. H. Richmond, *Biochem. J.* **77,** 112, 121 (1960).
[17] D. B. Cowie and G. N. Cohen, *Biochim. et Biophys. Acta* **26,** 252 (1957).
[18] A. Yoshida and M. Yamasaki, *Biochim. et Biophys. Acta* **34,** 158 (1959).
[19] G. W. Beadle, *Chem. Revs.* **37,** 15 (1945).
[20] R. P. Wagner and H. K. Mitchell, "Genetics and Metabolism." Wiley, New York, 1955.
[21] B. S. Strauss, "An Outline of Chemical Genetics." Saunders, Philadelphia, Pennsylvania, 1960.
[22] J. R. S. Fincham, *Ann. Rev. Biochem.* **28,** 343 (1959).
[23] C. Yanofsky and P. St. Lawrence, *Ann. Rev. Microbiol.* **14,** 311 (1960).
[24] G. W. Beadle, *Ann. Rev. Physiol.* **22,** 45 (1960).
[25] S. Zamenhof, "The Chemistry of Heredity." Charles C Thomas, Springfield, Illinois, 1959.
[26] C. B. Anfinsen, "The Molecular Basis of Evolution." Wiley, New York, 1959.
[27] G. Pontecorvo, *Advances in Enzymol.* **13,** 121 (1952).
[28] N. H. Horowitz and U. Leupold, *Cold Spring Harbor Symposia Quant. Biol.* **16,** 65 (1951).
[29] D. L. Nanney, *in* "The Chemical Basis of Heredity" (W. D. McElroy and B. Glass, eds.), p. 134. Johns Hopkins Press, Baltimore, Maryland, 1957.
[30] D. G. Catcheside, *Proc. Roy. Soc.* **B148,** 285 (1958).
[31] B. Ephrussi, "Nucleo-cytoplasmic Relations in Microorganisms." Oxford Univ. Press, London and New York, 1953.
[32] M. B. Mitchell, H. K. Mitchell, and A. Tissieres, *Proc. Natl. Acad. Sci. U.S.* **39,** 606 (1953).
[33] J. F. Danielli, *in* "Microbial Genetics." *Symposium Soc. Gen. Microbiol.* **10,** 294 (1960).
[34] M. de Deken-Grenson, *Exptl. Cell Research* **18,** 185 (1959).
[35] G. Brawerman and E. Chargaff, *Biochim. et Biophys. Acta* **37,** 221 (1960).
[36] O. E. Landman and H. S. Ginoza, *J. Bacteriol.* **81,** 875 (1961).
[37] H. Fraenkel-Conrat, *J. Am. Chem. Soc.* **78,** 882 (1956).
[38] A. Gierer and G. Schramm, *Z. Naturforsch.* **11b,** 138 (1956).
[39] J. S. Colter, H. H. Bird, A. W. Moyer, and R. A. Brown, *Virology* **4,** 522 (1957).
[40] I. M. Mountain and H. E. Alexander, *Proc. Soc. Exptl. Biol. Med.* **101,** 527 (1959).

[41] F. Jacob, P. Schaeffer, and E. L. Wollman, *in* "Microbial Genetics." *Symposium Soc. Gen. Microbiol.* **10,** 67 (1960).
[42] S. Wright, *Am. Naturalist* **79,** 289 (1945).
[43] M. Delbrück, *in* "Unités Biologiques Douées de Continuité Genetique," p. 33. Centre National de la Recherche Scientifique, Paris, 1949.
[44] M. R. Pollock, *in* "Adaptation in Micro-organisms." *Symposium Soc. Gen. Microbiol.* **3,** 150 (1953).
[45] A. Novick and M. Weiner, *Proc. Natl. Acad. Sci. U.S.* **43,** 553 (1957).
[46] M. Cohn and K. Horibata, *J. Bacteriol.* **78,** 613 (1959).
[47] J. Lederberg, *Methods in Med. Research* **3,** 5 (1950).
[48] P. E. Hartman and S. H. Goodal, *Ann. Rev. Microbiol.* **13,** 465 (1959).
[49] F. Jacob and E. L. Wollman, *in* "The Biological Replication of Macromolecules." *Symposia Soc. Exptl. Biol.* **No. 12,** 75 (1958).
[50] H. Hayes and R. C. Clowes, eds., "Microbial Genetics." *Symposium Soc. Gen. Microbiol.* **10** (1960).
[51] E. A. Adelberg, "Bacterial Genetics." Little, Brown, Boston, Massachusetts, 1960.
[52] C. Yanofsky, *Bacteriol. Revs.* **24,** 221 (1960).
[53] A. B. Pardee, F. Jacob, and J. Monod, *J. Mol. Biol.* **1,** 165 (1959).
[54] M. Riley, A. B. Pardee, F. Jacob, and J. Monod, *J. Mol. Biol.* **2,** 216 (1960).
[55] S. Lacks and R. D. Hotchkiss, *Biochim. et Biophys. Acta* **45,** 155 (1960).
[56] P. Starlinger, personal communication (1960).
[57] S. Luria and J. Monod, personal communication (1960).
[58] A. Matsushiro and K. Mizobuchi, *Japan J. Genetics* **34,** 282 (1959).
[59] S. Benzer, *in* "The Chemical Basis of Heredity" (W. D. McElroy and B. Glass, eds.), p. 70. Johns Hopkins Press, Baltimore, Maryland, 1957.
[60] M. E. Case and N. H. Giles, *Proc. Natl. Acad. Sci. U.S.* **46,** 659 (1960).
[61] D. O. Woodward, *Proc. Natl. Acad. Sci. U.S.* **45,** 846 (1959).
[61a] C. W. H. Partridge, *Biochem. Biophys. Research Communs.* **3,** 613 (1960).
[62] M. Demerec and P. E. Hartman, *Ann. Rev. Microbiol.* **13,** 377 (1959).
[63] G. Pontecorvo, "Trends in Genetic Analysis." Columbia Univ. Press, New York, 1958.
[63a] E. W. Nester and J. Lederberg, *Proc. Natl. Acad Sci. U.S.* **47,** 52 (1961).
[64] E. Ephrati-Elizur, P. R. Srinivasan, and S. Zamenhof, *Proc. Natl. Acad. Sci. U.S.* **47,** 56 (1961).
[65] A. Garen, *in* "Microbial Genetics." *Symposium Soc. Gen. Microbiol.* **10,** 239 (1960).
[66] V. M. Ingram, *Nature* **178,** 792 (1956).
[67] W. K. Maas and B. D. Davis, *Proc. Natl. Acad. Sci. U.S.* **38,** 785 (1952).
[68] N. H. Horowitz and S. C. Shen, *J. Biol. Chem.* **197,** 513 (1952).
[69] J. R. S. Fincham, *Biochem. J.* **65,** 721 (1957).
[70] S. R. Suskind and L. I. Kurek, *Proc. Natl. Acad. Sci. U.S.* **45,** 193 (1959).
[71] D. Perrin, A. Bussard, and J. Monod, *Compt. rend. acad. sci.* **249,** 778 (1959).
[72] R. D. Hotchkiss, *in* "The Nucleic Acids" (E. Chargaff and J. N. Davidson, eds.), Vol. II, p. 435. Academic Press, New York, 1955.
[73] R. L. Sinsheimer, *Ann. Rev. Biochem.* **29,** 503 (1960).
[74] O. T. Avery, C. M. Macleod, and M. McCarty, *J. Exptl. Med.* **79,** 137 (1944).
[75] J. Marmur and R. D. Hotchkiss, *J. Biol. Chem.* **214,** 383 (1955).
[76] E. E. B. Smith, G. T. Mills, H. P. Bernheimer, and R. Austrian, *Biochim. et Biophys. Acta* **28,** 211 (1958).
[77] E. E. B. Smith, G. T. Mills, H. P. Bernheimer, and R. Austrian, *Biochim. et Biophys. Acta* **29,** 640 (1958).

[78] J. Spizizen, *Federation Proc.* **18**, 957 (1959).
[79] C. D. Darlington, *Nature* **176**, 1139 (1955).
[80] J. A. V. Butler and P. F. Davison, *Advances in Enzymol.* **18**, 161 (1957).
[81] G. Zubay and M. R. Watson, *J. Biophys. Biochem. Cytol.* **5**, 51 (1959).
[82] A. D. Hershey and M. Chase, *J. Gen. Physiol.* **36**, 39 (1952).
[83] F. Jacob, A. M. Torriani, and J. Monod, *Compt. rend. acad. sci.* **233**, 1230 (1951).
[84] I. Watanabe, *J. Gen. Physiol.* **40**, 521 (1957).
[85] D. J. Ralston, B. S. Baer, M. Lieberman, and A. P. Krueger, *Proc. Soc. Exptl. Biol Med.* **89**, 502 (1955).
[86] M. H. Adams and B. H. Park, *Virology* **2**, 719 (1956).
[87] J. G. Flaks and S. S. Cohen, *J. Biol. Chem.* **234**, 1501 (1959).
[88] A. Kornberg, S. B. Zimmerman, S. R. Kornberg, and J. Josse, *Proc., Natl. Acad. Sci. U.S.* **45**, 772 (1959).
[89] R. Somerville, K. Ebisuzaki, and G. R. Greenberg, *Proc. Natl. Acad. Sci. U.S.* **45**, 1240 (1959).
[90] M. L. Dirksen, J. S. Wiberg, J. F. Koerner, and J. M. Buchanan, *Proc. Natl. Acad. Sci. U.S.* **46**, 1425 (1960).
[91] L. M. Kozloff, *Cold Spring Harbor Symposia Quant. Biol.* **18**, 209 (1953).
[92] S. E. Luria, D. K. Fraser, J. N. Adams, and J. W. Burrows, *Cold Spring Harbor Symposia Quant. Biol.* **23**, 71 (1958).
[93] P. Starlinger, *Z. Naturforsch.* **14b**, 523 (1959).
[94] G. S. Stent and C. R. Fuerst, *Advances in Biol. and Med. Phys.* **7**, 1 (1959).
[95] S. Zamenhof, *in* "Recent Progress in Microbiology" (G. Tunevall, ed.), p. 139. Almqvist and Wiksell, Stockholm, 1959.
[96] R. M. Litman and A. B. Pardee, *Biochim. et Biophys. Acta* **42**, 131 (1960).
[97] E. Freese, *J. Mol. Biol.* **1**, 87 (1959).
[97a] R. M. Litman and H. Ephrussi-Taylor, *Compt. rend. acad. sci.* **249**, 838 (1959).
[98] F. H. C. Crick, *in* "The Biological Replication of Macromolecules." *Symposia Soc. Exptl. Biol.* **No. 12**, 138 (1958).
[99] J. D. Watson and F. H. C. Crick, *Cold Spring Harbor Symposia Quant. Biol.* **18**, 123 (1953).
[100] R. Rolfe and M. Meselson, *Proc. Natl. Acad. Sci. U.S.* **45**, 1039 (1959).
[101] J. D. Watson and F. H. C. Crick, *Nature* **171**, 964 (1953).
[102] G. Zubay, *Nature* **182**, 112 (1958).
[03] G. S. Stent, *Advances in Virus Research* **5**, 95 (1958).
[104] H. J. Morowitz and R. C. Cleverdon, *Biochim. et Biophys. Acta* **34**, 578 (1959).
[105] R. L. Sinsheimer, *J. Mol. Biol.* **1**, 43 (1959).
[106] W. Frisch-Niggemeyer, *Nature* **178**, 307 (1956).
[107] L. J. Lerman and L. S. Tolmach, *Biochim. et Biophys. Acta* **33**, 371 (1959).
[108] G. Rushizky, M. Riley, L. S. Prestidge, and A. B. Pardee, *Biochim. et Biophys. Acta* **45**, 70 (1960).
[109] C. Levinthal, *Rev. Modern Physics* **31**, 249 (1959).
[110] M. Yčas, *in* "Symposium on Information Theory in Biology" (H. P. Yockey, ed.), p. 70. Pergamon Press, New York, 1956.
[111] J. Marmur, P. Doty, and N. Sueoka, *Nature* **183**, 1429 (1959).
[112] A. N. Belozersky and A. S. Spirin, *in* "The Nucleic Acids" (E. Chargaff and J. N. Davidson, eds.), Vol. III, p. 147. Academic Press, New York, 1960.
[113] R. L. Sinsheimer, *J. Mol. Biol.* **1**, 218 (1959).
[114] M. Yčas, *Nature*, **188**, 209 (1960).
[115] W. Dreyer, *Brookhaven Symposia in Biol.* **13**, 243 (1960).
[116] S. Emerson, *Ann. Missouri Botan. Garden* **32**, 243 (1945).

[117] H. McIlwain, *Proc. Roy. Soc.* **B136,** 12 (1949).

[118] S. Spiegelman, *in* "The Chemical Basis of Heredity" (W. D. McElroy and B. Glass, eds.), p. 232. Johns Hopkins Press, Baltimore, Maryland, 1957.

[119] S. Shiba, A. Terawaki, T. Taguchi, and J. Kawamata, *Biken's J.* **1,** 179 (1958).

[120] D. M. Prescott, *Ann. Rev. Physiol.* **22,** 17 (1960).

[121] L. Goldstein, J. Micou, and T. J. Crocker, *Biochim. et Biophys. Acta* **45,** 82 (1960).

[122] G. Richter, *Biochim. et Biophys. Acta* **34,** 407 (1959).

[123] K. Keck, *Biochem. Biophys. Research Communs.* **3,** 56 (1960).

[124] H. Stich and W. Plaut, *J. Biophys. Biochem. Cytol.* **4,** 119 (1958).

[125] F. Gros, *in* "The Nucleic Acids" (E. Chargaff and J. N. Davidson, eds.), Vol. III, p. 409. Academic Press, New York, 1960.

[126] M. Zalokar, *Exptl. Cell Research* **19,** 559 (1960).

[127] P. F. Spahr and A. Tissières, *J. Mol. Biol.* **1,** 237 (1959).

[128] E. McFall, A. B. Pardee, and G. S. Stent, *Biochim. et Biophys. Acta* **27,** 282 (1958).

[129] J. Monod and E. Wollman, *Ann. inst. Pasteur* **73,** 937 (1947).

[130] E. Volkin, *Proc. Natl. Acad. Sci. U.S.* **46,** 1336 (1960).

[131] M. Yčas and W. S. Vincent, *Proc. Natl. Acad. Sci. U.S.* **46,** 804 (1960).

[132] F. Gros, H. H. Hiatt, W. Gilbert, C. G. Kurlard, R. W. Risebrough, and J. D. Watson, *Nature,* **190,** 581 (1961).

[133] M. Nomura, B. D. Hall, and S. Spiegelman, *J. Mol. Biol.* **2,** 306 (1960).

[134] B. D. Hall and S. Spiegelman, *Proc. Natl. Acad. Sci. U.S.* **47,** 137 (1961).

[135] G. N. Cohen and F. Gros, *Ann. Rev. Biochem.* **29,** 525 (1960).

[136] R. Hamers and C. Hamers-Casterman, *Biochim. et Biophys. Acta* **33,** 269 (1959).

[137] J. Horowitz, J. J. Saukkonen, and E. Chargaff, *J. Biol. Chem.* **235,** 3266 (1960).

[138] K. McQuillan, R. B. Roberts, and R. J. Britten, *Proc. Natl. Acad. Sci. U.S.* **45,** 1437 (1959).

[139] A. Tissières, D. Schlessinger, and F. Gros, *Proc. Natl. Acad. Sci. U.S.* **46,** 1450 (1960).

[140] D. B. Cowie, S. Spiegelman, R. B. Roberts, and J. D. Duerksen, *Proc. Natl. Acad. Sci. U.S.* **47,** 114 (1961).

[141] K. C. Atwood and F. Mukai, *Proc. Natl. Acad. Sci. U.S.* **39,** 1027 (1953).

[142] B. S. Strauss, *Am. Naturalist* **89,** 141 (1955).

[143] C. L. Markert and R. D. Owen, *Genetics* **39,** 818 (1954).

[144] N. H. Horowitz, M. Fling, H. L. Macleod, and N. Sueoka, *J. Mol. Biol.* **2,** 96 (1960).

[145] O. E. Landman and D. M. Bonner, *Arch. Biochem. Biophys.* **41,** 253 (1952).

[146] G. Lester and D. M. Bonner, *J. Bacteriol.* **73,** 544 (1957).

[147] J. Lederberg, E. M. Lederberg, N. D. Zinder, and E. R. Lively, *Cold Spring Harbor Symposia Quant. Biol.* **16,** 413 (1951).

[148] R. P. Wagner and C. H. Haddox, *Am. Naturalist* **85,** 319 (1951).

[149] J. S. Gots, *Carnegie Inst. Wash. Publ.* **No. 612,** 87 (1956).

[150] R. Mantsavinos and S. Zamenhof, *J. Biol. Chem.* **236,** 876 (1961).

[151] M. Dixon and E. C. Webb, "Enzymes," p. 642. Academic Press, New York, 1958.

[152] T. S. Painter, *Proc. Natl. Acad. Sci. U.S.* **45,** 897 (1959).

[153] R. Briggs and T. J. King, *J. Morphol.* **100,** 269 (1957).

[154] H. Halvorson and B. Church, *Bacteriol. Revs.* **21,** 112 (1957).

[155] M. Nomura, G. Hosoda, H. Yoshikawa, and S. Nichimura, *Proc. Intern. Symposium Enzyme Chem., Tokyo and Kyoto,* p. 359 (1958).

[156] M. Ogur, *Arch. Biochem. Biophys.* **53,** 484 (1954).

[157] S. Kit and A. L. Gross, *Biochim. et Biophys. Acta* **36,** 185 (1959).

[158] W. Laskowski, E. Lochmann, A. Wecker, and W. Stern, *Z. Naturforsch.* **15b,** 734 (1960).

[159] P. B. Sawin and D. Glick, *Proc. Natl. Acad. Sci. U.S.* **29,** 55 (1943).
[160] F. Jacob and J. Monod, *Compt. rend. acad. sci.* **249,** 1282 (1959).
[161] F. Jacob, D. Perrin, C. Sanchez, and J. Monod, *Compt. rend. acad. sci.* **250,** 1727 (1960).
[162] H. Karström, *Ergeb. Enzymforsch.* **7,** 350 (1938).
[163] M. Cohn, J. Monod, M. R. Pollock, S. Spiegelman, and R. Y. Stanier, *Nature* **172,** 1096 (1953).
[164] H. J. Vogel, *Proc. Natl. Acad. Sci. U.S.* **43,** 491 (1957).
[165] J. Lederberg, *in* "Genetics in the 20th Century" (L. C. Dunn, ed.), p. 263. Macmillan, New York, 1951.
[166] G. Cohen-Bazire and M. Jolit, *Ann. inst. Pasteur* **84,** 937 (1953).
[167] M. R. Pollock, *in* "The Enzymes" (P. D. Boyer, H. Lardy, and K. Myrbäck, eds.), Vol. I, 2nd ed., p. 619. Academic Press, New York, 1959.
[168] H. O. Halvorson, *Advances in Enzymol.* **22,** 99 (1960).
[169] G. Cohen and F. Jacob, *Compt. rend. acad. sci.* **248,** 3490 (1959).
[170] L. Gorini, *Proc. Natl. Acad. Sci. U.S.* **46,** 682 (1960).
[171] M. Shepherdson and A. B. Pardee, *J. Biol. Chem.* **235,** 3233 (1960).
[172] H. V. Rickenberg, G. N. Cohen, G. Buttin, and J. Monod, *Ann. inst. Pasteur* **91,** 829 (1956).
[173] J. Wortman, *Z. physiol. Chem. Hoppe Seyler's* **6,** 287 (1882).
[174] J. Yudkin, *Biol. Revs.* **13,** 93 (1938).
[175] M. G. Sevag, *Advances in Enzymol.* **6,** 33 (1946).
[176] J. Monod, *Growth* **11,** 223 (1947).
[177] R. Y. Stanier, *Ann. Rev. Microbiol.* **5,** 35 (1951).
[178] J. Monod and M. Cohn, *Advances in Enzymol.* **13,** 67 (1952).
[179] R. Davies and E. F. Gale, eds., "Adaptation in Micro-organsisms." *Symposium Soc. Gen. Microbiol.* **3** (1953).
[180] A. C. R. Dean and Sir C. Hinshelwood, *in* "Adaptation in Micro-organisms." *Symposium Soc. Gen. Microbiol.* **3,** 21 (1953).
[181] A. W. Ravin, *in* "Adaptation in Micro-organisms." *Symposium Soc. Gen. Microbiol.* **3,** 46 (1953).
[182] M. J. Thornley and J. Yudkin, *J. Gen. Microbiol.* **20,** 355 (1959).
[183] H. P. Klein and M. Doudoroff, *J. Bacteriol.* **59,** 739 (1950).
[184] F. J. Ryan, *J. Gen. Microbiol.* **7,** 69 (1952).
[185] F. Dienert, *Ann. inst. Pasteur* **14,** 139 (1900).
[186] M. Stephenson and J. Yudkin, *Biochem. J.* **30,** 506 (1936).
[187] R. Y. Stanier, *J. Bacteriol.* **54,** 339 (1947).
[188] S. Benzer, *Biochim. et Biophys. Acta* **11,** 383 (1953).
[189] I. H. Sher and M. F. Mallette, *Arch. Biochem. Biophys.* **53,** 354 (1954).
[190] D. Ushiba and B. Magasanik, *Proc. Soc. Exptl. Biol. Med.* **80,** 626 (1952).
[191] J. Monod, A. M. Pappenheimer, Jr., and G. Cohen-Bazire, *Biochim. et Biophys. Acta* **9,** 648 (1952).
[192] S. Spiegelman, H. O. Halvorson, and R. Ben-Ishai, *in* "Amino Acid Metabolism" (W. D. McElroy and H. B. Glass, eds.), p. 124. Johns Hopkins Press, Baltimore, Maryland, 1955.
[193] A. B. Pardee and L. S. Prestidge, *Biochim. et Biophys. Acta* **27,** 330 (1958).
[194] M. Cohn, *Bacteriol. Revs.* **21,** 140 (1957).
[195] B. Rotman and S. Spiegelman, *J. Bacteriol.* **68,** 419 (1954).
[196] D. S. Hogness, M. Cohn, and J. Monod, *Biochim. et Biophys. Acta* **16,** 99 (1955).
[197] M. R. Pollock and M. Kramer, *Biochem. J.* **70,** 665 (1958).
[198] S. Dagley and J. Sykes, *Arch. Biochem. Biophys.* **62,** 338 (1956).

[199] J. M. Eisenstadt and H. P. Klein, *J. Bacteriol.* **77,** 661 (1959).
[200] A. Yoshida, T. Tobita, and J. Koyawa, *Biochim. et Biophys. Acta* **44,** 388 (1960).
[201] S. Zamenhof and E. Chargaff, *J. Biol. Chem.* **180,** 727 (1949).
[202] J. Monod, *Rec. trav. chim.* **77,** 569 (1958).
[203] K. Wallenfels, M. L. Zarnitz, G. Laule, H. Bender, and M. Keser, *Biochem. Z.* **331,** 459 (1959).
[204] A. S. L. Hu, R. G. Wolfe, and F. J. Reithel, *Arch. Biochem. Biophys.* **81,** 500 (1959).
[205] M. Cohn and J. Monod, *Biochim. et Biophys. Acta* **7,** 153 (1951).
[206] L. A. Herzenberg, *Biochim. et Biophys. Acta* **31,** 525 (1959).
[207] A. B. Pardee and L. S. Prestidge, *Biochim. et Biophys. Acta* **49,** 77 (1961).
[208] H. V. Rickenberg, C. Yanofsky, and D. M. Bonner, *J. Bacteriol.* **66,** 683 (1953).
[209] J. Monod, G. Cohen-Bazire, and M. Cohn, *Biochim. et Biophys. Acta* **7,** 585 (1951).
[210] R. Sheinin, *J. Gen. Microbiol.* **21,** 124 (1959).
[211] M. R. Pollock, *Biochem. J.* **66,** 419 (1957).
[212] M. R. Pollock, *Brit. J. Exptl. Pathol.* **33,** 587 (1952).
[213] M. R. Pollock, *Brit. J. Exptl. Pathol.* **31,** 739 (1950).
[214] A. M. Torriani, *Biochim. et Biophys. Acta* **19,** 224 (1956).
[215] A. B. Pardee, *J. Bacteriol.* **73,** 376 (1957).
[216] C. J. Porter, R. Holmes, and B. F. Crocker, *J. Gen. Physiol.* **37,** 271 (1953).
[217] M. R. Pollock and C. J. Perret, *Brit. J. Exptl. Pathol.* **32,** 387 (1951).
[218] M. Kogut, M. R. Pollock, and E. J. Tridgell, *Biochem. J.* **62,** 391 (1956).
[219] J. Mandelstam and J. Yudkin, *Biochem. J.* **51,** 686 (1952).
[220] O. E. Landman, *Biochim. et Biophys. Acta* **23,** 558 (1957).
[220a] D. Perrin, F. Jacob, and J. Monod, *Compt. rend. acad. sci.* **251,** 155 (1960).
[221] S. Spiegelman and H. O. Halvorson, *J. Bacteriol.* **68,** 265 (1954).
[222] A. B. Pardee and L. S. Prestidge, *J. Bacteriol.* **70,** 667 (1955).
[223] F. Stoeber, *Compt. rend. acad. sci.* **244,** 950 (1957).
[224] D. B. Goldstein, *J. Bacteriol.* **78,** 695 (1959).
[225] H. V. Rickenberg, *Nature* **185,** 240 (1960).
[226] S. Spiegelman, J. M. Reiner, and M. Sussman, *Federation Proc.* **6,** 209 (1947).
[227] J. Mandelstam, *J. Gen. Microbiol.* **11,** 426 (1954).
[228] H. Chantrenne, *Biochim. et Biophys. Acta* **16,** 410 (1955).
[229] C. R. Hebb and J. Slebodnik, *Exptl. Cell Research* **14,** 286 (1958).
[230] S. Spiegelman, *Cold Spring Harbor Symposia Quant. Biol.* **11,** 256 (1946).
[231] J. Mandelstam, *Intern. Rev. Cytol.* **5,** 51 (1956).
[232] A. Kepes, *Compt. rend. acad. sci.* **244,** 1550 (1957).
[233] I. Zabin, A. Kepes, and J. Monod, *Biochem. Biophys. Research Communs.* **1,** 289 (1959).
[234] G. N. Cohen and J. Monod, *Bacteriol. Revs.* **21,** 169 (1957).
[235] F. Stoeber, *Compt. rend. acad. sci.* **244,** 1091 (1957).
[236] A. Novick and M. Weiner, *in* "A Symposium on Molecular Biology" (R. E. Zirkle, ed.), p. 78. Univ. Chicago Press, Chicago, 1959.
[237] B. P. Sleeper, M. Tsuchida, and R. Y. Stanier, *J. Bacteriol.* **59,** 129 (1950).
[238] B. Rotman, *J. Bacteriol.* **76,** 1 (1958).
[239] A. L. Sheffner and D. O. McClary, *Arch. Biochem. Biophys.* **52,** 74 (1954).
[240] J. J. Robertson and H. O. Halvorson, *J. Bacteriol.* **73,** 186 (1957).
[241] M. Zalokar, *Federation Proc.* **18,** 358 (1959).
[242] R. B. Roberts, ed., "Microsomal Particles and Protein Synthesis." Washington Academy of Sciences Press, Washington D.C., 1958.
[243] M. R. Pollock, *Proc. Roy. Soc.* **B148,** 340 (1958).
[244] K. P. Jacobsohn and M. D. Azevado, *Bull. soc. chim. biol.* **37,** 139 (1955).

[245] M. Shilo and R. Y. Stanier, *J. Gen. Microbiol.* **16,** 482 (1957).
[246] P. Fildes, *Biochem. J.* **32,** 1600 (1938).
[247] S. S. Cohen, *J. Biol. Chem.* **177,** 607 (1949).
[248] G. Ashwell, A. J. Wahba, and J. Hickman, *J. Biol. Chem.* **235,** 1559 (1960).
[249] S. Hollmann and E. Thofern, *Z. physiol. Chem. Hoppe Seyler's* **312,** 111 (1958).
[250] R. Dedonder, *Intern. Congr. Biochem. Abstr. 4th Congr.* p. 122 (1958).
[251] H. Wiesmeyer and M. Cohn, *Biochim. et Biophys. Acta* **39,** 417 (1960).
[252] J. de Ley and J. Defloor, *Biochim. et Biophys. Acta* **33,** 47 (1959).
[253] A. L. Koch and G. Vallee, *J. Biol. Chem.* **234,** 1213 (1959).
[254] J. Lascelles, *J. Gen. Microbiol.* **23,** 487, 499, 511 (1960).
[255] A. Taussig, *Biochim. et Biophys. Acta* **44,** 510 (1960).
[256] M. Rachmeler, J. Gerhart, and J. Rosner, *Biochem. et Biophys. Acta,* **49,** 222 (1961).
[257] E. M. Linday and P. J. Syrett, *J. Gen. Microbiol.* **19,** 223 (1958).
[258] D. S. Hogness, *in* "Biophysical Science—A Study Program" (J. L. Oncley, ed.), p. 256. Wiley, New York, 1959.
[259] S. D. Wainwright, *Brit. J. Exptl. Pathol.* **31,** 495 (1950).
[260] E. F. Gale, *Biochem. J.* **36,** 64 (1942).
[261] M. Stephenson and E. F. Gale, *Biochem. J.* **31,** 1311 (1937).
[262] N. D. Gary, R. E. Klausmeier, and R. C. Bard, *J. Bacteriol.* **68,** 437 (1954).
[263] E. Englesberg and J. B. Levy, *J. Bacteriol.* **69,** 418 (1955).
[264] J. H. Quastel, *Enzymologia* **2,** 37 (1937).
[265] A. Novick, *Ann. Rev. Microbiol.* **9,** 97 (1955).
[266] E. A. Adelberg, *J. Bacteriol.* **76,** 326 (1958).
[267] E. H. Creaser, *Biochem. J.* **64,** 539 (1956).
[268] M. Cohn and J. Monod, *in* "Adaptation in Micro-organisms." *Symposium Soc. Gen. Microbiol.* **3,** 132 (1953).
[269] H. J. Vogel, *in* "Control Mechanisms in Cellular Processes" (D. M. Bonner, ed.), Ronald Press. New York, 1961.
[270] H. J. Vogel, *in* "The Chemical Basis of Heredity" (W. D. McElroy and B. Glass, eds.), p. 276. Johns Hopkins Press, Baltimore, Maryland, 1957.
[271] L. Gorini and W. K. Maas, *Biochim. et Biophys. Acta* **25,** 208 (1957).
[272] P. Rogers and G. D. Novelli, *Biochim. et Biophys. Acta* **33,** 423 (1959).
[273] W. K. Maas, *Biochem. Biophys. Research Communs.* **1,** 13 (1959).
[274] R. A. Yates and A. B. Pardee, *J. Biol. Chem.* **227,** 677 (1957).
[275] L. Szilard, *Proc. Natl. Acad. Sci. U.S.* **46,** 277 (1960).
[276] F. Jacob and J. Monod, *J. Mol. Biol.* **3,** 318 (1961).
[277] B. N. Ames and B. Garry, *Proc. Natl. Acad. Sci. U.S.* **45,** 1453 (1959).
[278] B. N. Ames, B. Garry, and L. A. Herzenberg, *J. Gen. Microbiol.* **22,** 369 (1960).
[279] F. Gibson and C. Yanofsky, *Biochim. et Biophys. Acta* **43,** 489 (1960).
[280] A. P. Levin and B. Magasanik, *J. Biol. Chem.* **236,** 184 (1961).
[281] A. B. Pardee and L. S. Prestidge, *Biochim. et Biophys. Acta* **36,** 545 (1960).
[282] E. F. Gale and J. P. Folkes, *Biochem. J.* **53,** 493 (1953).
[283] C. L. Wisseman, Jr., J. E. Smadel, F. E. Hahn, and H. E. Hopps, *J. Bacteriol.* **67,** 662 (1954).
[284] A. B. Pardee, *Proc. Natl. Acad. Sci. U.S.* **40,** 263 (1954).
[285] A. B. Pardee and L. S. Prestidge, *Biochim. et Biophys. Acta* **27,** 412 (1958).
[286] R. E. Kunkee, *J. Bacteriol.* **79,** 43 (1960).
[287] M. Stephenson, "Bacterial Metabolism," p. 309. Longmans, Green, New York, 1949.
[288] B. Magasanik, *Ann. Rev. Microbiol.* **11,** 221 (1957).

[289] J. Monod, "Recherches sur la Croissance des Cultures Bacteriennes." Hermann, Paris, 1941.
[290] F. C. Neidhardt and B. Magasanik, *J. Bacteriol.* **73,** 253 (1957).
[291] F. C. Neidhardt, *J. Bacteriol.* **80,** 536 (1960).
[292] F. C. Neidhardt and B. Magasanik, *Biochim. et Biophys. Acta* **21,** 324 (1956).
[293] E. A. Dawes, *J. Bacteriol.* **63,** 647 (1952).
[294] A. M. MacQuillan, S. Winderman, and H. O. Halvorson, *Biochem. Biophys. Research Communs.* **3,** 77 (1960).
[295] E. Englesberg, *Proc. Natl. Acad. Sci. U.S.* **45,** 1494 (1959).
[296] M. Cohn and K. Horibata, *J. Bacteriol.* **78,** 624 (1959).
[297] J. Mandelstam, *Biochim. et Biophys. Acta* **22,** 313 (1956).
[298] A. B. Pardee, *J. Bacteriol.* **69,** 233 (1955).
[299] S. Løvtrup, *Biochim. et Biophys. Acta* **19,** 433 (1956).
[300] G. Weinbaum and M. F. Mallette, *J. Gen. Physiol.* **42,** 1207 (1959).
[301] H. V. Rickenberg and G. Lester, *J. Gen. Microbiol.* **13,** 279 (1955).
[302] S. D. Wainwright and A. Neville, *J. Gen. Microbiol.* **14,** 47 (1956).
[303] J. Mandelstam, *Bacteriol. Revs.* **24,** 289 (1960).
[304] J. M. Eisenstadt and H. P. Klein, *Biochim. et Biophys. Acta* **44,** 206 (1960).
[305] A. B. Pardee and L. S. Prestidge, unpublished.
[306] D. Billen and H. C. Lichstein, *J. Bacteriol.* **61,** 515 (1951).
[307] M. J. Pinsky and J. L. Stokes, *J. Bacteriol.* **64,** 151 (1952).
[308] G. Melnykovych and E. E. Snell, *J. Bacteriol.* **76,** 518 (1958).
[309] A. Nason, N. O. Kaplan, and H. A. Oldewurtel, *J. Biol. Chem.* **201,** 435 (1953).
[310] E. F. Gale, *Bacteriol. Revs.* **7,** 139 (1943).
[311] R. Davies, *Biochem. J.* **55,** 484 (1953).
[312] P. P. Slonimski, *Proc. Intern. Congr. Biochem. 3rd Congr., Brussels, 1955,* p. 226.
[313] F. O. Moss, *Australian J. Exptl. Biol. Med. Sci.* **34,** 395 (1956).
[314] R. K. Clayton, *Biochim. et Biophys. Acta* **37,** 503 (1960).
[315] H. M. Lenhoff and N. O. Kaplan, *Nature* **172,** 730 (1953).
[316] H. E. Umbarger and B. Brown, *J. Bacteriol.* **73,** 105 (1957).
[317] E. C. C. Lin, A. P. Levin, and B. Magasanik, *J. Biol. Chem.* **235,** 1824 (1960).
[318] M. J. Pinsky and J. L. Stokes, *J. Bacteriol.* **64,** 337 (1952).
[319] R. Knox, *in* "Adaptation in Micro-organisms." *Symposium Soc. Gen. Microbiol.* **3,** 184 (1953).
[320] R. Knox, *J. Gen. Microbiol.* **4,** 388 (1950).
[321] N. Entner and R. Y. Stanier, *J. Bacteriol.* **62,** 181 (1951).
[322] P. A. Swenson and A. C. Giese, *J. Cellular Comp. Physiol.* **36,** 369 (1950).
[323] H. Halvorson and L. Jackson, *J. Gen. Microbiol.* **14,** 26 (1956).
[324] T. F. Anderson, *J. Bacteriol.* **56,** 403 (1948).
[324a] M. Masters and A. B. Pardee, *Biochim. et. Biophys. Acta,* **56,** 609 (1962).
[325] B. Rajewsky, H. Bücker, and H. Pauly, *Arch. Biochem. Biophys.* **82,** 229 (1959).
[326] K. C. Atwood and A. Norman, *Proc. Natl. Acad. Sci. U.S.* **35,** 696 (1949).
[327] C. Yanofsky, *J. Bacteriol.* **65,** 383 (1953).
[328] L. S. Baron, S. Spiegelman, and H. Quastler, *J. Gen. Physiol.* **36,** 631 (1953).
[329] H. Pauly, *Nature* **184,** 1570 (1959).
[330] H. Laser and M. Thornley, *Proc. Roy. Soc.* **B150,** 539 (1959).
[331] M. Rachmeler and A. B. Pardee, *Biochim. et. Biophys. Acta,* in press (1962).
[332] R. Y. Stanier, *in* "Adaptation in Micro-organisms." *Symposium Soc. Gen. Microbiol.* **3,** 1 (1953).
[333] R. E. O. Williams and C. C. Spicer, eds., "Microbial Ecology." *Symposium Soc. Gen. Microbiol.* **7** (1957).

[334] P. Rogers and G. D. Novelli, *Biochim. et Biophys. Acta* **33,** 423 (1959).
[335] R. R. Roepke, R. L. Libby, and M. H. Small, *J. Bacteriol.* **48,** 401 (1944).
[336] J. Monod, *Ann. inst. Pasteur* **72,** 879 (1946).
[337] A. Novick, *in* "Growth in Living Systems" (M. X. Zarrow, ed.), p. 93. Basic Books, New York, 1961.
[338] F. J. Ryan and L. K. Schneider, *J. Bacteriol.* **56,** 699 (1948).
[339] S. Emerson, *Cold Spring Harbor Symposia Quant. Biol.* **14,** 40 (1950).
[340] B. Magasanik, *J. Biol. Chem.* **213,** 557 (1955).
[341] R. E. Hartman and L. N. Zimmerman, *J. Bacteriol.* **80,** 753 (1960).
[342] G. E. W. Wolstenholme and C. M. O'Conner, eds., "Regulation of Cell Metabolism." Little, Brown, Boston, Massachusetts, 1959.
[342a] *Cold Spring Harbor Symposia Quant. Biol.* **26** (1961).
[343] R. Y. Stanier, *Bacteriol. Revs.* **14,** 179 (1950).
[344] H. Tristram, *J. Gen. Microbiol.* **23,** 425 (1960).
[345] O. Hayaishi and R. Y. Stanier, *J. Bacteriol.* **62,** 691 (1951).
[346] S. D. Wainwright and D. M. Bonner, *Can. J. Biochem. and Physiol.* **37,** 741 (1959).
[347] H. M. Kalckar, *Advances in Enzymol.* **20,** 111 (1958).
[348] E. Englesberg and J. B. Levy, *J. Bacteriol.* **69,** 418 (1955).
[349] C. V. Ramakrishnan, R. Steel, and C. P. Lentz, *Arch. Biochem. Biophys.* **55,** 270 (1955).
[350] H. E. Umbarger, *J. Bacteriol.* **68,** 140 (1954).
[351] H. L. Kornberg, J. F. Collins, and D. Bigley, *Biochim. et Biophys. Acta* **39,** 9 (1960).
[352] H. L. Kornberg, A. M. Gotto, and P. Lund, *Nature* **182,** 1430 (1958).
[353] S. R. Gross, *J. Biol. Chem.* **233,** 1146 (1958).
[354] S. S. Cohen, *Bacteriol. Revs.* **13,** 1 (1949).
[355] B. Magasanik and D. Karibian, *J. Biol. Chem.* **235,** 2672 (1960).
[356] A. Novick and L. Szilard, *in* "Dynamics of Growth Processes" (E. J. Boell, ed.), p. 21. Princeton Univ. Press, Princeton, New Jersey, 1954.
[357] R. A. Yates and A. B. Pardee, *J. Biol. Chem.* **221,** 757 (1956).
[358] H. E. Umbarger and B. Brown, *J. Biol. Chem.* **233,** 415 (1958).
[359] E. H. Wormser and A. B. Pardee, *Arch. Biochem. Biophys.* **78,** 416 (1958).
[360] L. Gorini, *Bull. soc. chim. biol.* **40,** 1939 (1958).
[361] J. S. Gots and E. G. Gollub, *Proc. Soc. Exptl. Biol. Med.* **101,** 641 (1959).
[362] H. S. Moyed, *J. Biol. Chem.* **235,** 1098 (1960).
[363] A. B. Pardee, *in* "Microbial Reaction to Environment." *Symposium Soc. Gen. Microbiol.* **12,** 19 (1961).
[364] A. C. Wilson and A. B. Pardee, *J. Gen. Microbiol.*, in press (1962).
[365] D. Pratt and G. S. Stent, *Proc. Natl. Acad. Sci. U.S.* **45,** 1507 (1959).
[366] R. B. Loftfield, L. I. Hecht, and E. A. Eigner, *Federation Proc.* **18,** 276 (1959).
[366a] S. Brenner, F. Jacob, and M. Meselson, *Nature* **190,** 576 (1961).
[367] Y. Maruyama and H. Mitsui, *J. Biochem.* **45,** 169 (1958).
[368] F. E. Abbo and A. B. Pardee, *Biochim. et Biophys. Acta* **39,** 478 (1960).
[369] K. G. Lark, *Biochim. et Biophys. Acta* **45,** 121 (1960).
[370] H. M. Nakata and H. O. Halvorson, *J. Bacteriol.* **80,** 801 (1960).
[371] C. Heinmets and A. Herschman, *Physics in Med. and Biol.* **4,** 238 (1960).
[372] C. J. Perret, *J. Gen. Microbiol.* **22,** 589 (1960).

AUTHOR INDEX

Numbers in parentheses are reference numbers and indicate that an author's work is referred to although his name is not cited in the text. Numbers in italic show the page on which the complete reference is listed.

A

Aaronson, S., 276(178), *291*
Abbo, F. E., 621(368), *630*
Abdel Kader, M. M., 266(94), *289*
Abderhalden, R., 262(63), *288*
Abelson, P. H., 18(70), *38*, 173(16), 188(100), 195(132, 133), 196(132, 133), 202(16), 205, 207(202), 208(202), 212(202), 215(202), *244*, *245*, *246*, *248*, 311(63), 313(63), 327(63), *332*, 536(157), *569*
Abraham, E. P., 433, 434(72), 459(72), *466*
Abrams, R., 301(42), *332*, 545(391), *574*
Abramsky, T., 349(84), *370*
Abramson, C., 308(52), 310(52), *332*
Ackermann, W. W., 299(25), *331*
Acs, G., 506(294), 511(120), 513(120, 325), *568*, *572*, *573*
Adams, E., 219(253, 254), 229(292), *249*, *250*
Adams, J. N., 587(92), *624*
Adams, M. H., 586(86), *624*
Adelberg, E. A., 171, 172(9, 10), 210, 212(225), 214(9, 10, 237, 238), 238 (337), *243*, *248*, *249*, *251*, 258(36), *287*, 583(51), 606(266), *623*, *628*
Adler, E., 153(117, 118, 119, 120), 154(117, 118, 119, 120), *164*, *165*
Adler, J., 72, 74(117), *114*, 552(179), 553(411, 412, 421), *569*, *575*
Adler, M., 296(3), *331*, 481(26), *565*
Ajl, S. J., 36(126), *40*
Akita, S., 237(332), *251*
Albers, M., 495(257), *571*
Albert, A., 279(197), *292*
Albrecht, A., 284(240), *294*
Aldous, E., 195(132), *246*
Aldrich, R. A., 350(98), *370*
Aleem, M. I. H., 31, *39*, 123(11), *162*
Alexander, B. H., 374(9), *386*
Alexander, H. E., 582(40), *622*
Alexander, M., 123(9, 11), 144(99), *162*, *164*
Allen, B. K., 330(167), *334*
Allen, E. H., 492(133), 506(288), 507(295), 508(295), 513(323), 514(327), 515(342), 521(133), 522(133), *568*, *572*, *573*
Allen, F. W., 296(7), *331*, 481, *565*, *570*
Allen, M. B., 29(106, 107, 108), *39*
Allfrey, V. G., 557(425), *575*
Allison, R. M., 136(75, 76), *163*
Almasy, F., 357(117), *371*
Almquist, H. J., 285(260, 263), *294*
Altenbern, R. A., 181(58), 203(58), *245*, 271(131b), *290*
Amano, T., 390(11), 403(11), 404(11), 405(11, 69), 408(11), *410*, *411*
Ambler, R. P., 473(5), 474, *565*
Ames, B. N., 172(14), 184(79a), 217(14, 245, 246, 247, 248, 249, 250, 251, 252), 219(14), 221, *243*, *245*, *249*, 309(60), 317(60), *332*, 609(277), 610(278), 617(277), *628*
Amos, H., 296(6), 327(157), *331*, *334*, 481(27), *565*
Anderson, M. L., 279(204, 205), *292*
Anderson, R. J., 286(274), *294*
Anderson, T. F., 613(324), *629*
Andrejew, A., 341(30), *368*
Anfinsen, C. B., 154(124, 125), *165*, 449(124), *468*, 477(13), *565*, 581(26), *622*
Angier, R. B., 279(195), *292*
Anslow, W. P., Jr., 196(141), *247*
Antia, M., 202(184), *247*, 453(137), *468*
Apgar, J., 512(317, 448), *573*, *576*
Aprison, M. H., 145(108), *164*
Apter, R. A., 540(169), *569*
Aramaki, Y., 262(64), *288*
Archibald, A. R., 423(44a), *466*
Armstrong, J. J., 419(30, 32, 33), 420(32), 423(44a), *465*, *466*, 479, *565*
Arnon, D. I., 11(50), 17(65), 29(50, 106, 107, 108, 109), 34, 35, *38*, *39*, *40*, 127(49), *163*

Arnstein, H. R. V., 205(204), *248*, 260(50), *288*, 518(357), *574*
Aronson, A. I., 483(237), 484, 525, *566*, *570*, *571*, *574*
Asai, M., 263(79), 265(79, 84), *289*
Ashton, D. M., 314(78), *332*
Ashwell, G., 604(248), *628*
Astrachan, L., 551, 561(192, 193, 194, 408), *569*, *575*
Atkinson, D. E., 122(6a), *162*
Atwood, K. C., 590(141), 614(326), *625*, *629*
Aubert, J. P., 16, 19, 23(77, 78), 31(114), *39*, 189(110), *246*
Austrian, R., 586(76, 77), *623*
Avery, O. T., 340, *368*, 586(74), *623*
Avigad, G., 376(22), 383(66, 76, 77), 384(66, 76, 77), 385(77), *387*, *388*
Avineri-Shapiro, S., 383(80), *388*
Axelrod, A. E., 257(32), 275(169, 172), *287*, *291*
Axelrod, B., 60(77), *114*, 517(354), *573*
Axelrod, D. R., 401(60), 408(60), *411*
Ayengar, P., 183(70), 236(325), *245*, *251*, 398(42), *411*
Azarkh, R. M., 453(142), *468*
Azevado, M. D., 603(244), *627*

B

Baalsrud, K., 4(18, 23), 5, 19, *37*
Baalsrud, K. S., 4(18, 23), 5, 19, *37*
Baas-Becking, L. G. M., *37*
Bach, M. K., 140(87), 150, *164*
Bachhawat, B. K., 75, 76, 77(132), *115*
Bachmann, B. J., 353(108), *370*
Bachrach, U., 272(134), *290*
Baddiley, J., 273(146), 274(146), *291*, 419, 420(31, 32, 35), 421(38a), 423(44a), 438(82), 444(107, 108, 109), 451(130), 464(198) *465*, *466*, *467*, *468*, *469*, 479(21), 509, *565*, *572*
Baer, B. S., 586(85), *624*
Bagatell, F. K., 297(14), 298(14), 317(14), *331*
Bail, O., 407, *412*
Bailey, R. W., 374(11), 377(26, 29, 30), 378(30), *386*, *387*
Bakay, B., 443(101), *467*
Balis, M. E., 306(49), 308(49), 311(49), 312(49, 66, 67), 313(49), *332*
Ballentine, R., 257(30), 270(30), *287*
Ballio, A., 268(105c, 107), *290*
Bandurski, R. S., 60(77), 61, 62(83), 63, 64, *114*, 207(213a), *248*
Banfi, R. F., 50(43), 51(43), 54(43), 65(43), *113*
Barbieri, P., 453(143), *468*
Barchelli, R., 283(227a), *293*
Bard, R. C., 360(149), *371*, 605(262), *628*
Bariéty, M., 345(60), 346(60), *369*
Barkemyer H., 278(193a), *292*
Barker, H. A., 23(93), 24, *39*, 43(11), 47(27), 85, 98, 99, 101, 103, 104(201, 208), 105, 106, 108, 109, *112*, *116*, *117*, 156,(140, 141), *165*, 198, *247*, 280, 281(274a, 214b, 214c, 214d, 215a, 215b), 283(299a), 284(236), *292*, *293*
Barker, S. A., 374(7, 10), 377(26, 27, 29), 378(7), 379(40), *386*, *387*
Barner, H. D., 191(115), *246*, 323(125), 324(125), 325(144), 328, *333*, *334*
Barnett, L., 409(87), *412*, 564(463), *576*
Baron, L. S., 614(328), *629*
Barrett, J. M., 282(216a), *293*
Barrett, R., 254(4), *287*
Barron, E. S. G., 47, *112*
Barry G. T., 442(95), *467*
Barsha, J., 381(51), *387*
Barton, L. S., 284(243), *294*
Bashford, M., 236(318), *251*
Basilio, C., 409(95), *412*, 564(461), *576*
Bassham, J. A., 13(51, 52, 53), 14, 16 17(56), *38*
Bateman, J. B., 405(71), *411*
Bates, H. M., 536(364), *574*
Bauer, N., 144(94), *164*
Bauman, N., 199(161), *247*, 440(91), 461(91), *467*
Baumann-Grace, J. B., 391(18), 393(23), *410*
Beadle, G. W., 171, 229(295), *243*, *250*, 581(19, 24), *622*
Beaucamp, K., 131(62b), *163*
Becher, E., 283(225, 229), *293*
Beck, E. S., 408(89), *412*
Beck, J. V., 4(19), 31(19, 115), *37*, *39*, 43(11), 98(11), 99(199), 105, 106, 107, *112*, *117*
Beck, W. S., 72, 85, *114*
Beechey, R. B., 270(118), *290*
Beerstecher, E., 260(48), *288*

Beijerinck, M. W., 3(11), *37*, 373, 383(2), *386*
Beljanski, Mirko, 341(39, 40), 361(40), *369*, 493(75), 494, 500, 502, 519(78, 128, 359, 360), 524, 530, *576*, *568*, *574*
Beljanski, Monique, 341(40), 361(40), *369*
Bell, D. J., 383(65), *388*
Bell, P. H., 479(23), *565*
Bell, T. T., 260(45), *288*
Bellamy, W. D., 270(117), *290*
Belozersky, A. N., 588(112), *624*
Belt, M., 284(234), *293*
Benante, C., 401(59), 408(59), *411*
Bender, H., 596(203), *627*
Ben-Ishai, R., 300(28), 306(28), *331*, 486(43), *566*, 595(192), 612(192), *626*
Benson, A., 350(94), 351(100), *370*
Benson, A. A., 13(52, 53), 14, 16(56), 17(56, 58), *38*
Bentley, M., 301(42), *332*
Benzer, S., 583(59), 584(59), 588(59), 594(188), 597(188), *623*, *626*
Berg, P., 322(123), *333*, 445(111), *467* 506(108, 290), 511(117), 512(117, 121), 513(121, 324), 526(117, 121), 545(394), 563, *567*, *568*, *572*, *573*, *574*, *576*
Berger, A., 429(69), *466*
Bergersen, F. J., 142, *164*
Bergmann, E. D., 300(28), 306(28), *331*
Bergmann, F. H., 22, 23(83), *39*, 506(290), 512(121), 513(121, 324), 526(121), *568*, *572*, *573*
Bergstroem, S., 300(26), *331*
Bergstrom, L., 500(91), *567*
Bernath, P., 359(142b), *371*
Bernhauer, K., 283(225, 226, 229), *293*
Bernheimer, H. P., 586(76, 77), *623*
Bernlohr, R. W., 506(283), *572*
Bessman, M. J., 318(87, 88), 323(87, 88), 324(87), 325(87, 88), *333*, 552(178, 179), 553(411, 412), *569*, *575*
Bettex-Galland, M., 91(188), *116*
Betz, R. F., 318(86), 328(86), *333*
Bhat, J. V., 23(94), 24, *39*
Bicking, J. B., 89(183), *116*
Bigley, D., 123(86), *39*, 617(351), *630*
Billen, D., 272(131), *290*, 612(306), *629*
Binkley, F., 196(141), *247*
Binkley, S. B., 285(254), *294*, 444(105), *467*
Binns, F., 267(102), *289*
Bird, H. H., 582(39), *622*
Bird, O. D., 282(217), *293*
Bishop, J., 515(342), *573*
Black, S., 194(126, 128), 195(134), *246*
Blakley, R. L., 324(143), *334*
Blanchard, M. L., 91(189), *117*, 390(1), *409*, 518(124), *568*
Blind, G. R., 390(9), 396(9), 398(9), *410*
Bloch, K., 185(84, 85), 224, 227(274, 275), *245*, *250*, 348(78), *370*, 451(127, 128), *468*
Bloemendal, H., 512(309, 310), *572*
Bloom, W. L., 390(10), 402(10), 407(10, 75, 76, 77), 408(10), *410*, *411*
Blumson, N. L., 464(198), *469*
Bock, R. M., 414(1), 457(1), *465*
Bönicke, R., 341(29), *368*
Boezi, J. A., 177(36), *244*
Bogard, M. O., 376(17), *386*
Bogorad, L., 350(95, 96), 351, 352(95), 366(173b), *370*, *372*
Bohonos, M., 342(44, 45), *369*
Bokman, A. H., 267(100), *289*
Bollum, F. J., 553, *575*
Bolton, E. T., 18(70), *38*, 173(16), 188(100), 195(132, 133), 196(132, 133), 200(133), 202(16), 207(213), *244*, *245*, *246*, *248*, 311(63), 313(63), 327(63), *332*, 483, 484, 536(157), 555(238), *569*, *570*, *571*
Boman, H. G., 506(294), 513(325), *572*, *573*
Bond, G., 125, *162*
Bondi, A., 453(139), *468*
Bone, J. F., 479(23), *565*
Bonner, D., 229(235), *250*
Bonner, D. M., 172(9), 188(104), 189, 231(297), 234(312), *243*, *246*, *250*, *251*, 265(89), *289*, 591(145, 146), 597(208), 617(346), *625*, *627*, *630*
Bonner, J., 260(51), *288*, 483, 497(241), *571*
Booij, H. L., 351(99, 104), 352(104), *370*
Boothe, J. H., 279(195), *292*
Borek, B. A., 401(55), 402(55), *411*
Borek, E., 158(158), *166*, 399(47), *411*, 439(84, 85), *467*, 542(172), *569*

Boretti, G., 283(227, 227a, 227b), *293*
Borkenhagen, L. F., 206(208), *248*
Borsook, H., 491, 515(134), 521(134), *566*, *568*
Bosch, L., 512(309, 310), *572*
Boss, M. L., 265(92), *289*
Bourgois, S., 179(53), *244*
Bourne, E. J., 374(7, 10), 377(26, 27, 29), 378(7), 379(40), *386*, *387*
Bovarnick, M., 394(25), 401, 402(55), 404, 406(25), 408(57, 58, 59, 60), *410*, *411*
Bovarnick, M. R., 401(57), 408(57), *411*
Bovey, F. A., 376(20), 377(24), *386*, *387*
Bowden, J. P., 275(171), *291*
Bowen, T. J., 497, *571*
Boyd, J. M., 329(164, 165), *334*
Boyer, P. D., 54, *113*, 158(160), *166*, 506(282), *572*
Boylen, J. B., 323(134), *334*
Brachet, J., 485, 493, *566*
Bradbeer, C., 127(46), *163*
Bradley, R. M., 79(145), *115*
Brady, R. O., 77, 78, 79, *115*, 283(229a), *293*
Brady, W. T., 284(252a), *294*
Brandt, I. K., 524(130), *568*
Branson, H. R., 477(14), *565*
Braun, O. H., 425(48), 464(48, 49), *466*
Braun, W., 550, *569*
Braunstein, A. E., 154(127, 128), *165*, 179(49, 50, 51, 52), 180, *244*, 397(36), *410*, 453(141, 142), *468*
Brawerman, G., 328(160), *334*, 582(35), *622*
Bray, R., 282(223), *293*
Breed, R. S., 390(12), *410*
Bregoff, H. M., 144(101), 145, *164*, 272(132), *290*
Brenner, S., 409(97), *412*, 493(76), 561, 562(442), 564(463), *567*, *576*
Brenoer, S., 620(366a), *630*
Bresler, A., 546(398), 561(398), *575*
Breslow, R., 89, *116*, 131(62a), *163*
Brewer, C. R., 396, *410*
Bricas, E., 390, *410*, 473(3), 479(3), *565*
Briggs, R., 592(153), *625*
Britten, R. J., 18(70), *38*, 173(16), 195(133), 196(133), 200(133), 202(16), *244*, *246*, 311(63), 327(63), *332*, 483(237, 238), 484(238), 500, 501(275), 515 (275), 536(157), 555(238), 560(275), 564, *569*, *570*, *571*, *576*, 590(138), 603 (138), *625*
Broberg, P. L., 265(81b), *289*
Brockman, J. A., Jr., 284(238), *293*
Brockman, R. W., 308(55), 310(55), 327 (152), *332*, *334*
Brodie, B. B., 578(5), *622*
Brooke, M. S., 306(48, 49), 308(49), 311 (49, 64), 312(49, 64), 313(49), 326 (148), *332*, *334*
Broquist, H. P., 219(255), *249*, 284(237, 240), *293*, *294*, 319(97), *333*
Brown, A., 367(181b), *372*
Brown, A. D., 537, *574*
Brown, A. M., 382(60), *387*
Brown, A. V. W., 512(315), *573*
Brown, B., 174(23, 24), 210(218), 211(24), 213(23, 226), 214(230), 243(226), *244*, *248*, *249*, 612(316), 618(358), *629*, *630*
Brown, B. J., 543, 550(406), *574*, *575*
Brown, D. H., 446(113), *467*
Brown, E. G., 263(71), 265(71), 279(71), *288*, 349(88a), *370*
Brown, G., 512(315), *573*
Brown G. B., 306(49), 308(49), 311(49), 312(49, 66, 67, 68), 313(49), *332*
Brown, G. L., 482(232), *570*
Brown, G. M., 211(223), 222(265), *248*, *250*, 260(48c, 48f), 261(48c), 262(61a), 266(95a, 95b), 272(143), 273(143, 144, 145), 274, 278(194a, 194b), 279(194b), *288*, *289*, *291*, *292*
Brown, G. W., Jr., 155(130), *165*
Brown, H., 475(10), 476(10), *565*
Brown, J. M. A., 63, *114*
Brown, J. W., 503(278), *572*
Brown, M. E., 125(23, 24), *162*
Brown, R. A., 582(38), *622*
Bruce, G. T., 374(7), 378(7), *386*
Bruckner, V., 390(5), 394(5), 404(5), 405 (5), 407(5), *410*, *411*, 478(17), *565*
Brumfitt, W., 416(16), *465*
Brummond, D. O., 17(62), *38*
Brunngraber, E. F., 512(318), *573*
Bryant, G., 376(17), *386*
Buchanan, J. M., 18(72), *38*, 43(16), 70, *112*, *114*, 160, *166*, 184(75), 198(150,

155), 204(150), *245*, *247*, 298, 299, 300, 301(23), 303, 304, 308(32), 318(89), 319(32), *331*, *333*, 419(30, 32, 33), 420(32, 35), 438(82), 444(107, 108, 109), *465*, *466*, *467*, 479(21), *565*, 586 (90), *624*
Buckley, S., 186(92), *245*
Bücker, H., 614(325), *629*
Büscher-Daubembüchel, L., 535(371), *574*
Burdon, K. L., 407(85), *412*
Burkhardt H. J., 232(303), *250*
Burkholder, P. R., 259(42), *288*
Burma D. P., 547(450, 451), 549(450, 451), *576*
Burnet, F. M., 407, *412*
Burris, R. H., 22(83), 23(83), *39*, 89(185), *116*, 121(13), 122(5), 124(13, 20), 125 (29), 127(43, 46), 129(20), 130(20), 135(67), 136(20, 68, 69, 70, 71, 72, 73, 75, 76), 138(20), 139, 141(20), 82) 145(105, 107, 108, 109), 146(110, 111, 112), 149(5), 150(5), 151(5), 157(72), *162*, *163*, *164*
Burrows, J. W., 587(92), *624*
Burton, A., 425(46), *466*
Burton, K., 49, *113*, 384(86), *388*, 550 (184), *569*
Burton, M. O., 342(46a), *369*
Bussard, A., 585(71), *623*
Buswell, A. M., 109(226), *117*
Butkevich, V. S., 43(9), *112*
Butler, J. A. V., 497(260), 500(88), 517 (352), 531, *567*, *571*, *573*, 586(80), *624*
Buttin, G., 593(172), *626*
Byrne, W. L., 206(209), *248*

C

Caldwell, P. C., 485, *566*
Calvin, M., 13(51, 52, 53), 14, 16(56), 17, 22(85), 27(55), *38*, 39, 51, 60(78), 61, *113*, *114*
Camien, M. N., 183(73), *245*, 532(141), *568*
Camiener, G. W., 260(48c, 48f), 261(48c), *288*
Campbell, L. L., Jr., 272(135), 274(154), *290*, *291*, 329(166), *334*
Campbell, P. N., 515(339), 559(432), *573*, *575*
Campbell Smith, S., 345(69), *370*
Campbell, W. P., 285(257), *294*
Campillo, A. D., 518(124), *568*
Canellakis, E. S., 327(153), *334*, 545(389, 390), 546, *574*
Cannata, J., 65(97), *114*
Cantoni, G. L., 513(320), 549(403), *573*, *575*
Caputto, R., 426(61), *466*
Cardini, C. E., 414(1), 426(61), 457(1), *465*, *466*
Cardon, B. P., 99(201, 202), 104(201), 105(202), *117*
Carlson, A. S., 379(34), *387*
Carlson, G. L., 211(223), *248*, 262(61a), *288*
Carnahan, J. E., 121, 124(17), 125(36), 126(36, 40), 127(16, 17), 128(16, 17), 130, 131(17), 132(17), 134(65), 136 (17, 18), 137(17), 138(17), 141(17, 90), 142(17), 143(17), 149(17), 150(36), *162*, *163*, *164*
Carsiotis, M., 58(69), *113*
Carson, S. F., 43(13), 47(27), 82(149, 152), 85, 88, *112*, *115*, *116*, 357(120), *371*
Carss, B., 419(30), 438(82), 444(107, 109), *465*, *467*, 479(21), *565*
Carter, C. E., 308(51), *332*
Carter, H. E., 435(46), *466*
Case, M. E., 584(60), *623*
Casida, L. E., 199(167), 237(167), *247*
Caspersson, T., 485, *566*
Castelfranco, P., 491, 506(61, 284), 507 (297), 524, *566*, *572*
Castle, J. E., 121(16, 17), 124(17), 125 (36), 126(36, 40), 127(16, 17), 128(16, 17), 130(17), 131(17), 132(17), 134(65), 136(17, 18), 137(17), 138(17), 141(17, 90), 142(17), 143(17), 149(17), 150(36), *162*, *163*, *164*
Castro-Mendoza, H., 71(109), 73(120), *114*, *115*
Caswell, M. C., 270(123), *290*
Catcheside, D. G., 582(30), *622*
Cathou, R. E., 198(155), *247*
Cauthen, S. E., 276(181), *291*
Ceithaml, J., 58, *113*
Chaikoff, I. L., 71, *114*

Chaix, P., 357(119), 359(134, 135b, 141b), *371*

Chakraborty, K. P., 323(127), *334*

Chamberlain, N., 319(93), *333*

Chamberlin, M., 563(457), *576*

Chance, B., 33, *40*

Chantrenne, H., 359(143, 144, 145b), *371*, 487, *571*, 600(228), *627*

Chargaff, E., 327(158), 328(160), *334*, 442(99), *467*, 487(250), 554(205), *570*, *570*, *571*, 582(35), 590(137), 596(201), *622*, *625*, *627*

Chase, M., 586(82), *624*

Chattaway, F. W., 320(104), *333*

Cheldelin, V. H., 255(23), 262(60), *287*, *288*

Chen, J., 72(112), 74(112), *114*

Cheney, L. C., 285(254), *294*

Chernigoy, F., 312(69), *332*

Chin, C. H., 359(141a), *371*

Chin, W., 417(21), 442(21), *465*

Chopard dit Jean, L. H., 285(255), *294*

Chou, S-C., 550(406), *575*

Chu, H. P., 391(17), 407(17), *410*

Church, B., 592(154), 620(154), *625*

Ciak, J., 434(75), 459(75, 160), *467*, *468*

Cifonelli, J. A., 443(103), *467*

Clark, F. E., 391(14), *410*

Clarke, G. D., 346(72), *370*

Clarke, P. H., 444(110), *467*

Clayton, R. K., 360(151b, 151c), 361 (151b), *371*, 612(314), *629*

Cleverdon, R. C., 588(104), *624*

Closse, A., 425, *466*

Clowes, R. C., 559(437), 561(437), *576*, 583(50), *623*

Coffee, W. B., 145(105), *164*

Cohen, A., 270(111), *290*

Cohen, G., 593(169), 606(169), *626*

Cohen, G. N., 194(127, 128a), 196(135, 136), *246*, 314(79), *332*, 493(64), 504, 505, 534(97, 98), 535(152), 559(436), *566*, *567*, *569*, *575*, 580(14), 581(17), 590(135), 593(172), 601(234), *622*, *625*, *626*, *627*

Cohen, J. A., 477(218), *570*

Cohen, P. P., 155(130), 160(167, 168, 169, 170), *165*, *166*, 190(111), *246*

Cohen, R., 531(365), *574*

Cohen, S. S., 60(76), *114*, 191(115), *246*, 296(2), 297(13), 298(13), 317(13), 323 (125, 135), 324(125, 140), 325(144, 145, 146), 328, *331*, *333*, *334*, 473(6), 480(24), 495, 550(183, 189), *565*, *569*, *571*, 586(87), 604(247), 617(354), *624*, *628*, *630*

Cohen-Bazire, G., 344(50), 362(163), 365 (50), *369*, *372*, 593(166), 595(191), 597(191, 209), 604(166), 605(166), *626*, *627*

Cohn, G. L., 524(130), *568*

Cohn, M., 492(62), 503(62), *566*, 582(46), 593(163), 594(163, 178), 595(46, 194, 196), 596(194), 597(194, 196, 205, 209), 600(178, 194), 601(178), 604 (251), 605(194), 606, 607(268), 608 (268), 611(296), *623*, *626*, *627*, *628*, *629*

Cohn, P., 524(131), *568*

Cohn, W. E., 296(8), *331*, 481, *566*, *570*

Cole, R. D., 506(105), *567*

Collette, J., 179(53), *244*

Collins, J. F., 23(86), *39*, 617(351), *630*

Colter, J. S., 582(39), *622*

Colvin, J. R., 382(59), *387*

Comb, D. G., 417(21), 442, 444(97), *465*, *467*

Comly, L. T., 515, 523, 560(338), *573*

Comstock, E., 407(85), *412*

Connell, G. E., 500, *567*

Conover, M. J., 437(79), 444(79), *467*

Contopoulos, R., 7(38), 10(38), 35(38), 36(38), *37*

Cook, R. P., 155(131), 156(131), *165*, 178(37), *244*

Cookson, G. H., 350(92), *370*

Coon, M. J., 75, 76, 77, *115*, 219(256), *249*, 272(133), *290*

Cooper, P. D., 434(74), 461(74), *467*

Cooper, R., 344(52), 345(52), 363(52), *369*

Coote, J., 506(105), *567*

Corcoran, J. W., 282(221), *293*

Cordes, E., 512(311), 513(323), *572*, *573*

Corey, R. B., 477(14), *565*

Cori, C. F., 373(3), *386*

Cori, G. T., 373(3), *386*

Corman, J., 376(17), *386*

Corsey, M. E., 214(231, 233), *249*

Coulter, C. B., 345(70), *370*

Countryman, J. L., 551(194), 561(194), *569*
Courtois, C., 359(144), *371*
Couslich, D. B., 279(195), *292*
Coutsogeorgopoulos, C., 545(395), *574*
Cowie, D. B., 18(70), *38*, 173(16), 195 (133), 196(133), 200(133), 202(16), 207 (213) *244*, *246*, *248*, 311(63), 313(63), 327(63), *332*, 483(237, 238), 484(238), 497(265), 515(265), 536(157), 555 (238), 561(265), *569*, *570*, *571*, 581 (17), 590(140), *622*, *625*
Craig, L. C., 479(22), *565*
Crampton, C. F., 497(259), *571*
Crane, F. L., 266(97), 286(272), *289*, *294*
Crathorn, A. R., 497(260), 500(88), 501, *567*, *571*
Crawford, I., 327(151), *334*
Crawford, L. V., 328(161), *334*, 550(187, 188), *569*
Creaser, E. H., 486(44), 487, 550(182), *566*, *569*, 606(267), *628*
Cremer, M., 373, *386*
Cresson, E. L., 199(168), *247*, 274(156), 276(175, 179), *291*
Crick, F. H. C., 409, *412*, 481, 559(208), 561(208), 564, *566*, *570*, *576*, 587(98, 99, 101), 589(98), *624*
Crocker, B. F., 598(216), 612(216), *627*
Crocker, T. J., 589(121), *625*
Cromartie, W. J., 390(10), 402(10), 407 (10, 75, 76, 77), 408(10), *410*, *411*
Crosbie, G. W., 320(103), 321(103), 329 (103), 330(103), *333*
Cross, M. J., 279(199), *292*
Cummins, C. S., 199(159), *247*, 415(5), 418(5, 25, 26), *465*, 473(2), 478, *565*
Cutinelli, C., 8, 10, *38*, 89, *116*
Cutts, N. S., 319(93), *333*

D

Dabrowska, W., 401(56), *411*
Dagley, S., 239(338), *251*, 497(268, 269), 555, *570*, *571*, 596(198), *626*
Dalby, A., 313(75), *332*
Dalgliesh, C. E., 265(90), 267(90), 270, *289*
Dam, H., 285(261), *294*
Danielli, J. F., 582(33), *622*
Dark, F. A., 427(63), 458(151), *466*, *468*
Darlington, C. D., 586(79), *624*
Das, N. B., 153(119), 154(119), *164*
Davie, E. W., 506(107, 285, 287, 293), *567*, *572*
Davies, D. A. L., 425(52, 53, 54), 464(52), *466*
Davies, D. D., 205(207), *248*
Davies, R., 594(179), 612(311), *626*, *629*
Davies, R. E., 32, 33, *40*
Davis B. D., 169(1a) 170(2, 3), 171(6), 172(11, 13, 15), 176(3, 30, 31, 32), 183(67), 185(89, 90), 186(89, 90), 187(97, 98), 189(31), 191(31, 116), 194(125), 196(31), 197(31), 198(149), 199(30, 161), 204(31, 197), 207(210), 212(89), 217(244), 222(263, 266, 268), 233(6, 269, 270, 272, 273), 224(15, 278), 225(279), 226(11, 15, 282), 227 (13, 284, 288), 289(291), 237(30, 89), 240(341), *243*, *244*, *245*, *246*, *247*, *248*, *249*, *250*, *251*, 258(34, 37), 271 (124), 276(183), *287*, *290*, *291*, 298 (22), *331*, 440(91), 461(91), *467*, 585 (67), *623*
Davis, D., 265(93), *289*
Davis, F. F., 296(7), *331*, 481(25), *565*
Davis, J. M., 327(152), *334*
Davison, P. F., 586(80), *624*
Dawes, E. A., 611(293), *628*
Day, P. L., 330(167), *334*
Dean, A. C. R., 594(180), *626*
DeBusk, B. G., 284(249, 250), *294*
de Deken-Grenson, M., 582(34), *622*
Dedonder, R., 60(73), *114*, 383(65, 75, 78), 384(75, 79, 81, 83), *387*, 604(250), *628*
Defloor, J., 604(252), *629*
de Haan, P. G., 197(148), 204(148), *247*
Deitzel, E., 365(166b), *372*
Dekker, E. E., 76(129), 77(129), *115*
Delavier-Klutchko, C., 131(62c), *163*
Delbruck, M., 582(43), *623*
del Campillo, A., 53(62, 63), 55(62, 63), 56(62, 63), 76(129), 77(129), *113*, *115*
de Ley, J., 604(252), *628*
Dellweg, H., 283(225, 226, 229), *293*, 312 (74), *332*, 541(171), *569*
Delwiche, C. C., 141, *164*
Delwiche, E. A., 82(149), 85(149, 161), *115*, *116*, 357(120), *371*

Demain, A. L., 342(46b, 46c), *369*, 516 (347), *573*
Demerec, M., 584(62), *623*
DeMoss, J. A., 507, 526(111), *568*
DeMoss, R. D., 60(75), *114*, 177(36), *244*
Denaumur, R., 462(181), *469*
den Boer, D. H. W., 394(29, 30), 398(29, 30), *410*
den Dooren de Jong, L. E., 23(91), *39*
Denes, G., 158(159), *166*, 404(66), *411*
Denman, R. F., 200(176), *247*
de Robichon-Szulmajster, H., 194(128a), *246*, 314(79), *332*
Devreux, S., 487(246, 247), *571*
Dewar, N. E., 22(84), *39*
Dewey, D. L., 199(165), 200(175), *247*
Dewey, V. C., 312(73), 313(73), *332*, 505, *567*
Dibble, W. E., 495(256), *571*
Dickel, D. F., 275(172), *291*
Dickens, F., 60(72), *114*
Dieckmann, M., 506(290), 512(121), 513 (121, 324), 526(121), *568*, *572*, *573*
Dienert, F., 594(185), *626*
DiGrado, C. J., 401(55, 59), 402(55), 408 (59), *411*
Di Marco, A., 283(227, 227a, 227b), *293* 453 (143), *468*
Di Milia, A., 509(306), 536(306), *572*
Dinning, J. S., 298(21), 330(167, 168), *331*, *334*
Dintzis, H. M., 495(256), 515(134), 521 (134), *568*, *571*
Diringer, R., 477(216), 546(398), 561 (398), *570*, *575*
Dirksen, M. L., 586(90), *624*
Disraely, M. N., 278(194c), *292*
Dittmer, K., 274(153), 275(161), *291*
Dixon, G. H., 477(217), *570*
Dixon, M., 58(68), *113*, 474, *565*, 592(151), *625*
Doctor, B. P., 512(317, 319, 448), *573*, *576*
Doisy, E. A., 285(254), *294*
Donker, H. J. L., 175, *244*
Dorfman, A., 443(103), *467*
Dormbush, A. C., 479(23), *565*
Doty, P., 483(235), 553(418), *570*, *575*, 588(111), *624*
Doudoroff, M., 7(38), 10(38), 35(38), 36 (38), *37*, 379(38), *387*, 594(183), *626*
Douglas, H. C., 233(307), *250*
Dounce, A. L., 559(207), *570*
Downing, M., 298(20), *331*
Doy, C. H., 232, 233(305, 306, 308a), *250*, *251*
Drabkin, D. L., 354(116), 359(116), *371*
Drell, W., 322(121), *333*
Dresel, E. I. B., 349(90), 350(93, 94, 97), *370*
Dreyer, W., 588(115), *624*
Driscoll, C. A., 274(156, 157), 276(157), *291*
Drisko, R. W., 477(219), *570*
Dubuc, J., 58(69), *113*
Duerksen, J. D., 483(237), 497(265), 515 (265), 561(265), *570*, *571*, 590(140), *625*
Duguid, J. P., 459(155), *468*
Dumville, G., 451(126), *468*
Dunn, D. B., 296(4, 5, 9), *331*, 481, 482, 484, 540, 541(170), *565*, *566*, *569*, *570*
Dunn, M. S., 183(73), *245*, 532(141), *568*
du Vigneaud, V., 196(141), *247*, 274(153), *291*
Dworkin, M., 23(88), *39*

E

Eagle, H., 433(73), *466*
Eakin, E. A., 274(155), *291*, 299(25), *331*
Eakin, R. E., 260(48), 274(155), *288*, *291*
Earl, J M., 157(144), *165*, 272(131c), *290*
Ebert, M., 277(188), *292*
Ebisuzaki, K., 325(147), *334*, 586(89), *624*
Edelhock, H., 405(71), *411*
Edmonds, M., 545(391), *574*
Edsall, J. T., 477(12), *565*
Eggerer, H., 82(151), 85(151, 160), *115*, *116*
Ehrensvärd, G., 8(41, 42, 43), 10(41, 42, 43), *38*, 89(179, 180, 181), *116*, 208 (215), 212, 222(267), 224, 236(318, 320), *248*, *250*, *251*
Eigner, E. A., 620(366), *630*
Eisenberg, F., Jr., 376(21), 384(21), *386*, 401(58), 408(58), *411*
Eisenstadt, J. M., 596(199), 612(304), *627*, *629*
Eisgruber, H., 345(65), 346(65), 347(65), *369*
Elion, G. B., 312(66, 67), *332*

Ellfolk, N., 156(134, 136), *165*
Ellinger, P., 266(94), *289*
Elliot, W. H., 157, 158(152, 157), *165*
Ellwood, D. C., 421(38a), *466*
Elsden, S. R., 10, 31(118), 35, *38*, *39*, *40*, 99(200), 102, *117*
Elson, D., 497, 515(340), *571*, *573*
Emerson, S., 589(116), 616(329), *624*, *630*
Emmart, E. W., 262(69), *288*
Emmelot, P., 394(29, 30), 398(29, 30), *410*
Englard, S., 265(87), *289*
Engle, H., 24, *39*
Engle, M. S., 123(9), *162*
Engle, R. R., 477(219), *570*
Englesberg, E., 359(146, 147, 148), *371*, 605(263), 611(295), 617(348), *628*, *629*, *630*
English, J. P., 479(23), *565*
Ennis, H. L., 192(121), *246*
Entner, N., 613(321), *629*
Ephrati-Elizur, E., 584(64), *623*
Ephrussi, B., 359(138), *371*, 582(31), *622*
Ephrussi-Taylor, H., 587(97a), *624*
Eppling, F. J., 127(43), *163*
Epps, H. M. R., 178(41), *244*
Erdos, L. Z., 390(6), 406(6), *410*
Esposito, R. G., 125(34), 126(34, 39), *162*
Esser, H., 535(154), *569*
Etingof, R. N., 271(131a), *290*
Evans, E. A., Jr., 43(15), 47, 49, 64, *112*, *113*, *114*
Evans, R. J., 341(31b), *369*
Everett, J. E., 153(120), *165*
Eyring, E. J., 210(218), 214(230), *248*, *249*

F

Fager, E. W., 14(54), 16, *38*
Fahrenbach, M. J., 279(195), *292*
Fairhurst, A. S., 154(126), *165*
Fairley, J. L., 329(164), *334*
Falk, J. E., 336(4), 338(10), 349, 350(93, 94, 97), 367(181b), *368*, *370*, *372*
Fauconneau, G., 462(181), *469*
Faulkner, P., 53(60), *113*
Feeney, R. E., 394(31), *410*
Feger, V. H., 376(17), *386*
Feingold, D. S., 376(22), 383(66, 67, 68, 76), 384(66, 76), *387*, *388*
Feldman, L. I., 180, 203(55), *244*, 397 (37a), *410*
Fellows, C. E., 479(23), *565*
Few, A. V., 500(90), *567*
Fewson, C. A., 122(3a, 4, 4a), 151(4), *161*
Fewster M. E., 239(338), *251*
Fieber, S., 401(57), 408(57), *411*
Fieger, E. A., 257(31), *287*
Fieser, L. F., 285(257), *294*
Fildes, P., 183(69), 229(293), *245*, *250*, 340(14), *368*, 453, *468*, 604(246), *628*
Fincham, J. R. S., 153(121, 122), *165*, 178, 189(105), *244*, *246*, 581(22), 583 (22), 584(22), 585(22, 69), *622*, *623*
Fink, H., 345(58), 346(58), *369*
Firshein, W., 550(180), *569*
Fischer, G. A., 197(143), *247*
Fischer, H., 336, 345(58), 346(58), 362(7), *368*, *369*
Fisher, D. J., 127(47, 48), 128(48), 152 (47), *163*
Fisher, M. W., 339(26), 340(26), *368*
Fitz-James, P. C., 484, *566*
Flaks, J. G., 191(115), *246*, 324(140), 325 (145, 146), *334*, 550(189), *569*, 586 (87), *624*
Flavin, M., 71, 72(113), 73(120), *114*, *115*, 196(140), *246*, 312(72, 72), 313(71, 72), *332*
Fling, M., 195(129, 130), *246*, 591(144), 608(144), *625*
Flinn, B. C., 275(169), *291*
Foda, I. O., 357(126), *371*
Foldes, J., 390(13), *410*
Folkers, K., 214(237), *249*, 276(175, 176, 179), *291*
Folkes, J. P., 459(166), *469*, 486(41), 488, 491, 494, 498, 509(41, 49), 510(41), 516, 527(54), 528(57), 541(41, 115), 542(41), 543(115), 544(115), 554(41, 49, 197, 557(49, 197), *566*, *568*, *570*, 610(282), *628*
Fontaine, F. E., 99, *117*
Ford, J. E., 280(213), 282(213, 215, 216), *292*, *293*
Forgione, P., 365(168a), *372*
Formica, J. V., 878, *115*
Forrest, H. S., 265(83, 86, 86a), 279(196), *289*, *292*
Forsander, O., 261(57), *288*

Forte, C., 453(139), *468*
Foster, J. W., 8(39), 23(88, 89), 24, *37*, *39*, 43(13), 47(27), *112* 200(171, 172), *247*
Fowler, C. B., 550(183), *569*
Fowler, J. F., 130(54), *163*
Fox, C. L., Jr., 299(24), *331*
Fox, G., 503(278), *572*
Fraenkel-Conrat, H., 559(212, 213), *570*, 582(37), *622*
Francis, J., 285(258), *294*
Frank, H., 417(20), *465*
Frank, I. F., 284(244), *294*
Fraser, D., 324(139), *334*
Fraser, D. K., 587(92), *624*
Fraser, M. J., 518(356), *574*
Frazer, P. E., 185(87, 88), *245*
Fred, E. B., 44(20), *112*
Freed, M., 407(77), *411*
Freeman, M., 279(210), *292*
Freese, E., 587(97), *624*
Frei, W., 357(117), *371*
French, C. S., 494(255), *571*
French, D., 380, *387*
French, T. C., 301(36), 305(46), 306(46), *331*, *332*
Frenkel, A. W., 30(110, 111), *39*
Fresco, J., 191(116), *246*
Friedkin, M., 324(136, 142), 330(136), *334*
Friedman, D. L., 79(147), 95, 96, *115*, *117*
Friedman, H., 319(94, 95), *333*
Friedman, M., 175, 238(25), *244*
Friedman, S., 311(65), *332*
Friedmann, B., 204(196), *248*
Fries, N., 300(26), *331*
Frisch-Niggemeyer, W., 588(106), *624*
Fromageot, C., 357(119), *371*, 390, *410*, 473(3), 479(3), *565*
Fromme, I., 425, *466*
Fronticelli, C., 509(306), 536(306), *572*
Fruton, J. S., 473(4), 474, 532, *565*, *568*
Fry, E. M., 285(257), *294*
Fuchs, A., 383(63), *387*
Fuerst, C. R., 587(94), *624*
Fukin, T., 517(354), *573*
Fukuhara, H., 531(366), *574*
Fuller, R. C., 17(58), 22(85), 27(85, 103), *38*, *39*, *49*, *114*
Fuoco, L., 283(227b), *293*, 453(143), *468*
Fuson, R. C., 71(106), *114*
Futterman, S., 279(206), *292*

G

Gaevskaya, M. S., 43(9), *112*
Gaffron, H., 6(30), 7, 8(37), 14, 28(37), 35, *37*, *38*
Gajdos, A., 345(60), 346(60), *369*
Gajdos-Torok, M., 345(60), 346(60), *369*
Gale, E. F., 156(135), 158(157), *145*, 178 (39, 40, 41), 211(220), *244*, *248*, 359(137), *371*, 459(166), *469*, 486(41), 488, 490(52), 491 494, 498, 507(113), 508, 509(41, 49), 510(41), 516, 527(54), 528, 534(151), 535, 541(41, 115), 542(44), 543(115), 544(56, 115), 554(41, 49, 197), 555(56), 557(49, 197), *566*, *568*, *569*, *570*, *573*, 594(179), 604 (260), 605(261), 610(282), 612(310), 613(310), *626*, *628*, *629*
Gallop, R. C., 407, *412*
Ganguly, J., 78, *115*
Gardner, I. C., 125, *162*
Garen, A., 584(65), 588(65), *623*
Garibaldi, J. A., 342(43), 345(62), 346 (62), *369*, 394(31), *410*
Garnjobst, L., 222(267), *250*
Garry, B., 609(277), 610(278), 617(277), *628*
Garry, B. J., 184(79a), 217(252), 221 (79a), *245*, *249*, 309(60), 317(60), *332*
Gary, N. D., 360(149), *371*, 605(262), *628*
Gascoigne, J. A., 382(60), *387*
Gehatia, M., 383(67), *388*
Gehring, L. B., 306(47), 308(47), 319(47), *332*
Genghof, D. S., 274(158), 275(158), 276 (180), *291*
Georg, L. K., 265(91), *289*
George, P., 62, *114*
Gerhart, J., 604(256), *628*
Gernez-Rieux, C., 341(30), *368*
Gery, I., 272(134), *290*
Gest, H., 36(127), *40*, 144(100, 101, 103), 145(101), *164*
Getzendaner, M. E., 299(25), *331*
Ghuysen, J. M., 416, 418(13), *465*
Gibbons, R. J., 339(31a), 341(31a), *369*
Gibbs, M., 17, 27(103), *38*, *39*, 60(75), *114*
Gibeon, F., 610(279), *628*
Gibor, A., 359(146, 147), *371*
Gibson, D. M., 74(123), 77, *115*

Gibson, F., 232, 233(305), 234(315), *250*, *251*
Gibson, J., 73, *115*
Gibson, K. D., 349(86, 89), 353(110), 354, *370*, *371*
Gierer, A., 482(227), 559(210), *570*, 582 (38), *622*
Giese, A. C., 613(322), 614(322), *629*
Gilbert, W., 561(453), 562(453), 563(453), *576*, 590(132), 620(132), *625*
Gilby, A. R., 500(90), *567*
Gilder, H., 335(1), 336(1), 339(16, 17), 340, 361(16), *368*
Giles, N. H., 584(60), *623*
Gillespie, J. M., 580(9), *622*
Gilvarg, C., 170(3), 176(3), 200, 201(178, 182), 202(124, 183), 223(271), 224, 227(274, 275, 287, 291), 229(291), *243*, *246*, *247*, *250*, 453(138), 461(138), *468*
Ginoza, H. S., 271(131b), *290*
Ginoza, W., 559(211), *570*, 532(36), *622*
Ginsburg, V., 462, 463, 464(182, 183, 184), *469*
Gladner, J. A., 477(222), *570*
Gladstone, G. P., 183(69), *245*, 396, *410*
Glaser, L., 381(56), *387*, 398(43), *411*, 446(113), 453(136), 464(193, 195), *467*, *468*, *469*
Glassman, E., 507(295), 508(295), 513 (323), *572*, *573*
Glavind, J., 285(261), *294*
Glick, D., 593(159), *626*
Glick, M. C., 461(177, 178), *469*
Glock, G. E., 60(72), *114*, 179(46), 240 (46), *244*
Gloor, U., 286(273), *294*
Glover, J., 10(45), 26, *38*
Glover, P., 444(110), *467*
Godson, G. N., 517(352), 531(352), *573*
Goldberg, M., 88, *116*
Goldblum, J., 383(70), *388*
Golden, J. H., 362(162b), *372*
Goldman, D. S., 154, *165*
Goldstein, A., 543, 550(406), *574*, *575*
Goldstein, D. B., 550(406), *575*, 600 (224), *627*
Goldstein, J., 314, *332*, 506(291), *572*
Goldstein, L., 560, *570*, 589(121), *625*
Goldthwait, D. A., 184(74), *245*, 512 (308), 545(392), *572*, *574*
Gollub, E. A., 301(37), 307(37), 309(37), *331*
Gollub, E. G., 618(361), *630*
Gollub, M. C., 53(58), *113*
Gomez, C. G., 236(327), *251*, 390(9), 394 (26, 28), 396(9), 397(138), 398(9, 38), 400(26), 401(26), 406(26), 407(84), *410*, *411*, *412*, 453(134), *468*, 537(161), *569*
Goodal, S. H., 583(48), 587(48), *623*
Goodale, T. C., 13(52), *38*
Goodsall, R. A., 517(353), *573*
Goodwin, T. W., 255(14, 14a), 263(71, 74, 75), 264(74), 265(71), 279(71), *287*, *288*, *289*, 345(56), 346(56), 362(165a), *369*, *372*
Gordon, J., 512(315), *573*
Gordon, M., 299(25), *331*
Gordon, R., 391(14), *410*
Gorelick, M. K., 408(89), *412*
Gorini, L., 174(19), 188(101), 191(19, 101), 192, 207(210), 243(19), *244*, *246*, *248*, 593(170), 606(271), 609(170), 618(360), *626*, *628*, *630*
Gots, J. S., 232(304), *250*, 300(27), 301 (37), 305(27, 44, 45), 306(27, 45), 307 (37, 50), 308(52, 54), 309(37), 310 (52), 311(65), 313, 314, *331*, *332*, 591 (149), 619(361), *625*, *630*
Gotto, A. M., 617(352) *630*
Gould, R. G., 536(156), *569*
Graff, S., 312(71, 72), 313(71, 72), *332*
Grafflin, A. L., 51(50), *113*
Granick, S., 335(1, 2), 336, 337(1), 339 (16, 17), 340, 345(68), 348, 349(2, 87), 350(95), 351(101, 105), 352(95), 358, 361(16), 362, *368*, *370*
Grant, P. M., 377(28), *387*
Grau, F. H., 127(50), *163*
Graves, J. L., 65(94), *114*
Gray, C. H., 345(55), 346(55), 352(106), *369*, *370*
Greathouse, G. A., 381(54), 382(61), *387*
Green, D. E., 126(41), 128(41), 142(41), 151(41), *162*, 203(185), *247*, 390(1), *409*
Green, H., 176(32), *244*
Green, J. A., 365(170, 171), 366(176, 177), *372*

Green, M., 144(98, 99), *164*, 323(135), *334*, 402, *411*
Greenberg, D. M., 197(145), *247*, 279 (207, 208), *292*
Greenberg, G. R., 184(74), 189(194), 204 (194), *245*, *248*, 279(212), *292*, 300, 325(147), *331*, *334*, 419(30), *465*, 479 (21), *565*, 586(89), *624*
Greiner, C. M., 60(77), 61, 63, 64, *114*
Greull, G., 89(183), *116*, 211(222), *248*
Griboff, G., 540(170), 541(170), *569*
Griffiths, M., 344(53a), 363(53a), *369*
Grimshaw, J., 253(1), 266(1), *287*
Grisolia, S., 49(41), *113*
Gromet, Z., 381(50, 52 53), *387*
Gromet-Elhanan, Z., 381(49), *387*
Gros, Francois, 486, 487, 493, 501(277), 503 (203), 513, 515(277), 522(277), 526 (277), 527, 541(245, 384), 542(173), 543, 554, 555(129), 556(277), 557(277), 559(430, 436), 561, 562(453), 563, *568*, *569*, *570*, *571*, *572*, *574*, *575*, *576*, 589(125), 590(132, 135, 139), 620(125, 132), *625*
Gros, Francoise, 542(173), 543, 554, *569*
Gross, A. L., 592(157), *625*
Gross, D., 505, *567*
Gross, J., 499(85), *567*
Gross, S. R., 216(243a, 243b), 222(267), *249*, *250*, 617(353), *630*
Grossi, L. G., 514, *573*
Grossman, J. P., 357(122), *371*
Grossowicz, N., 158(158), *166*, 399(47), *411*
Gruber, W., 106(216), *117*, 284(243), *294*
Grunberg-Manago, M., 547, 549, *569*, *575*
Gryder, R. M., 184(76), *245*
Guarino, A. J., 298(18), *331*
Guest, J. R., 198(152a), *247*, 283(232), *293*
Guex-Holzer, S., 391(16), 393(21, 22), *410*
Guirard, B. M., 270, *290*
Gunness, M., 275(162), *291*
Gunsalus, C. F., 270(120), *290*
Gunsalus, I. C., 61, 88, 106(216), *114*, *117*, 180, 203(55, 188), 205(206), 206(206), 213(227), 235(317), 236(188), *244*, *248*, *249*, *251*, 270(113, 117), 284(236, 242, 243, 244), *290*, *293*, *294*, 397(37a, 37b), *410*, 452(133), *468*
Gunther, G., 153(118, 120), 154(118, 120), *164*, *165*
Guntz, G., 462(181), *469*
Gurin, S., 77, *115*
Gurnani, S., 85(158), *116*
Gutman, A. B., 296(3), *331*, 481(26), *565*
György, P., 274(152), 276(177), *291*

H

Haas, P., 390(2), *409*
Haas, V., 99(197, 198), 101(198), *117*
Haas, V. A., 13(52), *38*
Habermann, V., 537(378), *574*
Haddox, C. H., 591(148), *625*
Hague, E., 274(153), *291*
Hagy, H., 404(67), *411*
Hahn, F. E., 434(75), 459(75, 160), *467*, *468*, 554(195, 199, 200, 201), *569*, *570*, 610(283), *628*
Haining, J. L., 517, *573*
Hale, J. H., 352(106), 361(159), *370*, *372*
Halenz, D. R., 74, 75, 79, 95, *115*, *117*
Hall, B., 282(217), *293*,
Hall, B. D., 483(235), 484, 552(410), 561 (410), *570*, *575*, 590(133, 134), *625*
Hall, L. M., 160(167, 168, 169), *166*
Halpern, Y. S., 178(45), 183(68), 193(45), 211(221), 242(221), *244*, *245*, *248*
Halvorson, H. O., 486(43), 497(447), 503, 504, 534, 535(152), *566*, *567*, *569*, *572*, *576*, 592(154), 593(168), 594(168), 595 (192), 596(168), 599, 600(168), 601 (221), 602(240), 603(168, 240), 605 (168), 610(168), 611(294), 612(192), 613(323), 614(323), 620(154, 370), *625*, *626*, *627*, *629*, *630*
Hamers, R., 590(136), *625*
Hamers-Casterman, C., 590(136), *625*
Hamilton, D. M., 378(33), 379(35), *387*
Hamilton, P. B., 124(21), 127(21), 131 (55), 142(91, 92), 147(55), 151(55), *162*, *164*
Hamilton, P. G., 127(44), 131(55), *163*
Hanby, W. E., 390(8), 405, 406(8), *410*
Hancock, R., 459(168), *469*, 490, 491(51), 538(51), 554(51), *566*
Handler, P., 268(103, 105a), *289*, *290*
Hanes, C. S., 399(49, 50), *411*, 536, *569*

Hanford, J., 205(207), *248*
Hanshoff, G., 320(111), *333*
Hanzon, V., 497, *571*
Happold, F. C., 239(338), *251*, 270(118), *290*
Harary, I., 49, 50(43), 51(43), 54(43), 55 (34), 65(43), *113*
Harbers, E., 545(386, 393), *574*
Hardy, F. E., 420(35), *466*
Harington, C. R., 260(52), *288*
Harris, A. Z., 14(56), 16(56), 17(56), *38*
Harris, D. L., 260(46, 49), 262(66), *288*
Harris, H., 199(159), *247*, 415(5), 418(5, 25, 26), *465*, 473(2), *478*, *565*
Harris, J. I., 497, *571*
Harris, J. O., 267(102), *289*
Harris, J. S., 197(146), *247*
Harshman, S., 477(219), *570*
Hartley, B. S., 477(220, 221), *570*
Hartman, P. E., 583(48), 584(62), 587 (48), *623*
Hartman, R. E., 616(341), *630*
Hartman, S. C., 18(72), *38*, 160, *166*, 300, 301(34), 308(32), 318(89), 319(32), *331*, *333*
Hartmann, A. F., 421, *466*
Hartmann, G., 506(294), 513(325), *572*, *573*
Hase, E., 537(379), *574*
Hassid, W. Z., 379(38), *387*
Hastings, A. B., 43(16), *112*
Hatch, F. T., 198(150, 155), 204(150), *247*
Hatefi, Y., 285(269), *294*
Haukenes, G., 421(38a), *466*
Haworth, J. W., 270(111), *290*
Hay, J. B., 423(44a), *466*
Hayaishi, O., 108, *117*, 219(257), *249*, 578(7), 579(7), 617(345), *622*, *630*
Hayashi, K., 214(234), *249*
Hayes, W., 559(437), 561(437), *576*, 583 (50), *623*
Heath, E. C., 462(185), *469*
Heath, H., 354(113, 114), *371*
Hebb, C. R., 359(142b), *371*, 601(229), *627*
Hecht, L. I., 510(116), 511, 512(116), 514(116), 545(118, 119, 388), *568*, *574*, 620(366), *630*
Heckley, R. J., 390(10), 402(10), 407(10, 76), 408(10), *410*, *411*
Heden, C. G., 236(318), *251*, 485, *566*
Hegre, C. S., 75(125), 79(125), 95(125), *115*
Hehre, E. J., 373(4), 374(12), 375, 377(4, 16), 378(4, 33), 379(34, 35, 36), 383 (69), *386*, *387*, *388*
Heidelberger, C., 545(386, 393), *574*
Heimmets, C., 621(371), *630*
Heinrich, M. R., 312(73), 313(73), *332*
Helleiner, C. W., 198(152), 204(152), *247*
Heller, J., 524, *568*
Hellman, N. N., 376(17, 18), 377(18), *386*
Hellmann, H., 232(303), 233(308), *250*
Hellstrom, V., 153(118), 154(118), *164*
Hemingway, A., 43(12, 17), 47(17, 24, 26), *112*
Hendee, E. D., 346(71), 358(71), *370*
Henderson, L. M., 265(93), *289*
Hendler, R. W., 517, *573*
Hendlin, D., 270(123), *290*, 342(46b, 46c), *369*
Henis, Y., 383(74), *388*
Henseleit, K., 186, *245*
Heppel, L. A., 323(128), *334*, 547(400), *575*
Herbert, D., 48, 49, *113*, 339(24), 340 (24), 341, *368*, 483(242), 486(242), *571*
Herbert, E., 323(129), *334*, 513(322), 545 (387, 389, 390), 546, *573*, *574*
Herbst, E. T., 396(33), *410*
Herman, E. C., 279(205), *292*
Herrmann, R. L., 329(164, 165), *334*
Herschman, A., 621(371), *630*
Hershey, A., 551, *569*
Hershey, A. D., 586(82), *624*
Herzenberg, L. A., 217(252), *249*, 597 (206), 602(206), 610(278), 611(206), *627*, *628*
Hesseltine, C. W., 342(44, 45), *369*
Hestrin, S., 373(5), 376(21, 22), 380(43, 44, 46), 381(43, 46, 50, 52, 53), 382 (43), 383(43, 66, 70, 73, 76, 77, 80), 384 (21, 66, 76, 77, 82), 385(77), *386*, *387*, *388*
Heyman, U., 153(119), 154(119), *164*
Heyman-Blanchet, T., 359(141b), *371*
Hiatt, H., 561(453), 562(453), *576*
Hibbert, H., 381(51), *387*
Hickman, J., 604(248), *628*
Higashi, T., 361(156b), *372*

Hildinger, M., 233(308), *250*
Hill, A. G., 13(53), *38*
Hill, R. T., 479(22), *565*
Hill, T. G., 390(2), *409*
Hilmoe, R. J., 549(402), *575*
Hinman, R. L., 504(96), 534(96), *567*
Hino, S., 127(42), *162*
Hinshelwood, C., 485, *566*, 594(180), *626*
Hird, F. J. R., 399(49, 50), *411*, 536(158, 159), *569*
Hirs, C. H. W., 477(214, 215), *570*
Hirsch, H. E., 13(53), *38*
Hirsch, H. M., 359(142a), *371*
Hirsch, M. L., 194(127), 196(135, 136), *246*, 494(81), 530, *567*
Hitchings, G. H., 312(66, 67), *332*
Hoagland, M. B., 491, 505, 506(282), 510, 511(59), 512, 514, 515, 523, 559(429), 560(338), *566*, *567*, *568*, *572*, *573*, *575*, 577(1), *621*
Hoare, D. S., 199(162), 200(175, 176), 202 (184), *247*, 354(113, 114), *371*, 440 (93), 453(137), 461(93), *467*, *468*
Hoch, G. E., 122(5), 127(45), 139(5), 145, 149, 150, 151, *162*, *163*, *164*
Hodgkiss, W., 345(56), 346(56), *369*
Högström, G., 8(43), 10(43), *38*
Hoffman, C. A., 376(17, 18), 377(18), *386*
Hoffmann, C. E., 91(189), *117*, 284(234), *293*
Hofmann, J., 257(32), 275(168, 169, 172), *287*, *291*
Hofmann, T., 3, *37*
Hogenkamp, H. P. C., 281(214d), *293*
Hogness, D. S., 492(62), 503, *566*, 595 (196), 597(196), 604(258), *626*, *628*
Holcomb, W. F., 285(254), *294*
Holden, J. T., 461(176), *469*
Holdsworth, E. S., 282(216), *293*, 313 (75), *323*, 518(356), *574*
Holley, R. W., 506(291), 512, *572*, *573*, *576*
Hollingworth, B. R., 483(233), 495(233), *570*
Hollmann, S., 360(151a), *371*, 604(249), *628*
Hollunger, G., 33, *40*
Holm, K., 277(188), *292*
Holmes, R., 598(216), 612(216), *627*
Holm-Hansen, O., 60(78), *114*
Holt, A. S., 362(165b), *372*
Holt, L. B., 345(55), 346(55), 352(106), *369*, *370*
Holt, L. E., 259(43), *288*
Holzer, H., 131(62b), *163*
Hong, M. M., 154(127, 128), *165*, 179(49, 50, 51, 52), *244*
Hopkins, J. W., 557(426), 563, *575*, *576*
Hopps, H. E., 554(199, 200, 201), *570*, 610(283), *628*
Horecker, B. L., 15(68), 17(59, 60, 61, 68), 19(60, 61), 22(68), 24(68), *38*, 59, 60(71), 61(79), 63(89), 69(101), 88 (79), *113*, *114*, 217(250), *249*
Horibata, K., 582(46), 595(46), 611(296), *623*, *429*
Horowitz, J., 487(250), 554, *570*, *571*, 590 (137), *625*
Horowitz, N. H., 195(129, 130), 196(142), 197(142), *246*, *247*, 582(28), 585(68), 590(28), 591(144), 608(144), 613(68), *622*, *623*, *625*
Horvath, St., 393, 394, *410*
Hosoda, G., 592(155), *625*
Hotchkiss, R. D., 583(55), 586(55, 72, 75), *623*
Houlahan, M. B., 176(29), 199(29), *244*
Housewright, R. D., 181(58), 203(58), 236(327), *245*, *251*, 390(9), 394(26, 27), 395(27), 396(9), 397(37, 38), 398(9, 38, 44), 400(26), 401(26), 402(27), 406(26), 407(84), *410*, *411*, *412*, 453 (134, 135), *468*, 537(162), *569*
Howard, K. S., 479(23), *565*
Howe, A. F., 396(33), *410*
Hu, A. S. L., 497(447), *576*, 596(204), *627*
Hubbard, J. A., 53, 56(61), *113*
Hubbard, N., 358(128), *371*
Hubbard, R., 365(168b), 366(168b), *372*
Huennekens, F. M., 279(202, 203a, 209, 210), *292*
Hüskins, G., 535(371), *574*
Huff, J. W., 320(107, 108), *333*
Hughes, D. E., 20, *39*, 128 *163*, 278(104, 105), 272(142), 273(146), 274(146), *289*, *290*, *291*, 494, *571*
Hughes, E. G., 270(111), *290*
Hulsmann, W. C., 514(330, 327), *573*
Hultin, T., 514(334, 335, 336), *573*
Hultquist, M. E., 279(195), *292*
Hunter, G. D., 497(260), 500(88), 501, 517, 531(352), *567*, *571*, *573*

Hurlbert, R. B., 322(120), 323(126, 127), *333*, *334*
Hurwitz, C., 488, *571*
Hurwitz, J., 17(60, 61), 19(60, 61), *38*, 63(89), *114*, 226(283), *250*, 270(121, 122), *290*, 546, 553, 561(398), *575*
Hutchings, B. L., 279(195), *292*, 342(44, 45), *369*
Hutner, S. H., 280(213), 282(213), *292*
Huxley, H. E., 483(234), 495, 496, *570*
Huzino, A., 509(307), 536(307), *572*

I

Ichihara, K., 156, *165*
Ikawa, M., 181(61), 197(61), 236(323), *245*, *251*, 270(114), 274(147), *290*, *291*, 443(102), 448(120), *467*, *468*, 533(369), *574*
Ikeda, K., 282(222), *293*
Ikeda, Y., 237(330), *251*, 461(173, 174), *469*
Imai, K., 263(78), 264(81), 265(78, 81, 81d, 87a, 87b, 88), *289*
Imsande, J., 269(107a), *290*
Ingram, G. L. Y., 285(265), *294*
Ingram, V. M., 477(11), *565*, 585(66), *623*
Inoue, S., 352(104a), *370*
Isherwood, F. A., 399(49, 50), *411*, 536 (158, 159), *569*
Ishii, K., 278(193), *292*
Ishimoto, N., 439(88), 440(89, 90), 449 (123), 450(123), *467*, *468*, 538(166), *569*
Isler, O., 285(259a), *294*
Ito, E., 430(71), 437(71), 438(71), 439(71, 88), 440(89, 90), 444(71), 446(71), 447(118, 119), 448(118, 119), 449 (119, 123), 450(123), 451(131), 452(132), 454(146), 456(146), *466*, *467*, *468*, 536 (363), 538(166, 382), 539, 540(363), *569*, *574*
Itoh, K., 236(326), *251*
Ivanovics, G., 255(19), 271(19), *287*, 390(5, 6, 13), 393, 394, 397(35), 404(5), 405(5), 406(6), 407(5, 35), *410*, 478(16, 17), *565*

J

Jackson, C. P., 382(58), *387*
Jackson, L., 613(323), 614(323), *629*
Jackson, R. W., 376(17, 18), 377(18), *386*
Jacob, F., 561(442), 562(442), *576*, 582 (41), 583(49, 53, 54), 586(83), 587(54), 589(54), 591(53), 593(160, 161, 169), 599(220a), 606(169), 607(276), 608 (53), 609(161, 276), 613(83), 615(54), 619(276), 620(276, 366a), *623*, *624*, *626*, *627*, *628*, *630*
Jacobsohn, K. P., 603(244), *627*
Jacoby, W., 17(62), *38*
Jaenicke, L., 189(194), 204(194), *248*, 279 (212), *292*
James, A. E., 374(10), *386*
Jamikorn, M., 345(56), 346(56), *369*
Jang, R., 60(77), *114*
Jeanes, A., 374(8), *386*
Jensen, H. L., 125(31, 32, 33), 126(31, 32, 33, 38), *162*
Jensen, J., 339(34), 341, 361(34, 35, 160), *369*, *372*
Jermyn, M. A., 580(9), *622*
Johns, A. T., 85(162, 163), *116*
Johnson, B. C., 85(158), *116*, 518(127, 358), *568*, *574*
Johnson, M. J., 44(20), 49, 61(40), 62, 86, 99(195), *112*, *113*, *116*, *117*, 131(56), *163*, 426(58), 429(58), 443(58), *466*
Joklik, W. K., 322(123), *333*, 445(111), *467*
Jokura, K., 464(201), *469*
Jolit, M., 593(166), 604(166), 605(165), *626*
Jondorf, W. R., 578(5), *622*
Jones, K. M., 198(152a), *247*
Jones, M. E., 159, 160, *166*, 190(112), 191, *246*, 320(110), *333*
Jones, M. J., 207(211), 234(315), *248*, *251*
Jones, O. T. G., 123(8), *162*
Jones, O. W., 409(93, 96), *412*, 563(459), 564(462), *576*
Joslyn, M. A., 262(62), *288*
Josse, J., 323(132), 324(132), *334*, 586(88), *624*
Juergens, W. G., 420(34, 36), 421(34), 422 (34, 36), 458(34), *465*, *466*
Jütting, G., 74(124), 76(124), 92(124), 93 (124), 95(124), 97(124), 98(124), *115*
Juillard, M., 283(228), *293*
Jukes, T. H., 91(189), *117*, 284(240), *294*

Julianelle, L. A., 421, *466*
Julita, P., 283(227, 227a), *293*, 453(143), *468*
Jungwirth, C., 216(243a, 243b), *249*
Juni, E., 211, *248*

K

Kajiro, Y., 261(58a, 58b), *288*
Kajtar, M., 404(67), *411*
Kalan, E. B., 224(278), 225(279), 226(282), 227(284, 288), *250*
Kalckar, H. M., 298(16), *331*, 384(85), *388*, 617(347), *630*
Kalle, G. P., 308(52, 54), 310(52), *332*
Kalnitsky, G., 51, *113*
Kaltenbach, J. P., 51, *113*
Kalyankar, G. D., 451(129), *468*
Kamen, M. D., 8(40), 10, 26(45), 36(126, 127), *37*, *38*, *40*, 43(13, 14), 99, 101, *112*, *117*, 144(100, 101), 145, *164*, 343(49), *369*
Kameyama, T., 501(276), *572*
Kammen, H. O., 323(126), *333*
Kanagawa, H., 156(137), *165*
Kanazir, D., 191(115), *246*
Kandler, O., 17, *38*
Kaneda, T., 70, *114*
Kaplan, N. O., 358(132), 360(132), *371*, 612(309, 315), *629*
Kara, J., 549(405), *575*
Karasek, M., 506(284), 507(287), *572*
Karibian, D., 308(56), 314, 316(56), 317(56), 320(56), *332*, *333*, 618(355), *630*
Karlsson, J. L., 98, 106, *117*
Karström, H., *112*, 593(162), 594(162), 604(162), *626*
Katagiri, H., 263(78), 264, 265(78, 81, 81d, 87a, 87b, 88), *289*
Katagiri, M., 18(71), *38*, 108(220), *117*, 226(280), 227(290), *250*
Kattermann, R., 359(145a), *371*
Katunuma, N., 277(190, 191), 278(190, 191), *292*
Katz, J., 71, *114*
Katznelson, H., 261(53, 54), *288*
Kaudewitz, F., 232(303), *250*
Kauffman, D. L., 477(217), *570*
Kaufman, S., 53(63), 55(63), 56(63), *113*
Kauffmann, F., 425(47, 48, 49), 464(48, 49, 200), *466*, *469*
Kawaguchi, S., 13(53), *38*
Kawamata, J., 589(119), 610(119), *625*
Kawasaki, K., 282(224a), *293*
Kawasaki, T., 260(48e), *288*
Kay, L. D., 14(56), 16(56), 17(56), *38*
Kazenko, A., 376(18), 377(18), *386*, 401(56), *411*
Kaziro, Y., 72, 74, 75, 76(111), 84(111), 90(111), 91(111), 95, *114*, *117*
Kazlowski, J., 401(57), 408(57), *411*
Kearney, E. B., 265(87), 268(106), *289*, *290*
Keck, K., 589(123), *625*
Keech, D. B., 24, 25, 26, *39*, 65(98), 68, 69, 70(98, 103), 79, 80, 108, *114*, *117*
Keglevic, D., 205(204), *248*
Keinbenz, G., 535, *574*
Kellenberger, E., 459(156), *468*
Keller, E. B., 499(85, 87), 505(104), *567*
Kelley, H. J., 301(39), *332*
Kennedy, E. P., 206(208), *248*, 414(1), 457(1), *465*
Kent, L. H., 405(72), 406, 407(72), *411*, 427, *466*
Kepes, A., 601(232, 233), *627*
Keppie, J., 406(82), 407(82, 83), *412*
Kerr, S. E., 312(68, 69), *332*
Kerridge, D., 493(69, 70), 500(70), 504, *566*, *572*
Keser, M., 596(203), *627*
Kessler, B. J., 401(59), 408(59), *411*
Khambata, S. R., 23(94), 24(94), *39*
Khorana, H. G., 328(162), *334*
Kidder, G. W., 312(73), 313(73), *332*, 505, *567*
Kiese, M., 341(38), *369*
Kihara, H., 497(447), 533, 534, *568*, *574*, *576*
Kikuchi, G., 349(88b), 353(88b, 108, 109), *370*
Kikuoka, H., 156(138), *165*
King, H. K., 154(126), *165*
King, T. J., 592(153), *625*
Kingdon, H. S., 506(285), *572*
Kinoshita, S., 188, 189(103), 192(122), 237(331, 332), 238(333, 334, 336), *246*, *251*, 578(6), *622*
Kinoshita, T., 262(64), *288*
Kirby, K. S., 482(228), *570*
Kirkman, H. N., 462(180), *469*
Kishi, T., 263(79), 265(79, 84), *289*

Kishner, S., 479(23), *565*
Kisliuk, R. L., 198(151, 152), 204(151, 152, 193), *247*, *248*, 283, *293*
Kit, S., 592(157), *625*
Kitada, S., 237(331), *251*
Kitai, R., 475(10), 476(10), *565*
Kitay, E., 298(19), 328(19), *331*
Kjeldgaard, N. O., 486(445, 486), *576*
Klausmeier, R. E., 605(262), *628*
Kleck, K., 324(139), *334*
Klee, W. A., 513(320), *573*
Klein, H. P., 77, *115*, 594(183), 596(199), 612(304), *626*, *627*, *629*
Klemperer, F. W., 43(16), *112*
Kline, L., 284(236), *293*
Klingenberg, M., 33, *40*
Klotz, A. W., 257(29), 270(29), *287*
Klungsöyr, S., 382, *387*
Kluyver, A. J., 23(95), *39*, 43, *112*, 175, *244*, 361(157), *372*
Knappe, J., 74(124), 75, 76(124), 92(124), 93(124), 95(124), 97(124), 192a, 98 (124), *115*, *117*
Knight, B. C., 350(94), *370*
Knight, B. C. J. G., 183(69), *245*, 254 (7), 268(7), *287*
Knight, C. A., 481, 484(35), *566*
Knight, M., 73, *115*
Knox, R., 339(27), 340(27), *368*, 613(319, 320), *629*
Kobat, E. A., 375(13), *386*
Koch, A. L., 311(62), 317(82), 318(62), 319(90, 92), *332*, *333*, 604(253), *628*
Koeppe, O. J., 158(160), *166*
Koepsell, H. J., 86, *116*, 131(56), *163*, 376 (17, 18), 377(18, 27), *386*, *387*
Koerner, J. F., 323(133), *334*, 586(90), *624*
Koffler, H., 22(84), *39*, 493(67), *566*
Kögl, F., 394(29, 30), 398, *410*
Kogut, M., 360(152), *372*, 598(218), *627*
Kohn, H. I., 197(146), *247*
Koike, M., 262(65), 284(251, 252, 253), *288*, *294*
Koike, S., 352(104a), *370*
Kolb, J. J., 437(78, 79), 440(78, 79), 444 (79), *467*
Kon, S. K., 254(5), 280(214), *287*, *292*
Konigsberg, W., 479(22), *565*
Koningsberger, V. V., 506(106, 107, 289), 537(376), *567*, *572*
Konishi, S., 196(139), *246*
Korey, S. R., 49(34), 55(34), *113*
Korkes, S., 51(49), 53(62, 63), 55(62, 63), 56(62, 63), 88, *113*, *116*, 518(124), *568*
Korn, M., 296(6), *331*, 481(27), *565*
Kornberg, A., 49(41, 42), 51(42), 53(42), 57, *113*, 265(88a), 269(108), *289*, *290*, 297(15), 318(87), 320(113, 114, 115), 321(116), 322, 323(132), 324(87, 136), 325(87, 132), 327(151), 328(163), 330 (136), *331*, *333*, *334*, 552, 553(411, 412, 421), *569*, *575*, 586(88), *624*
Kornberg, H. L., 23, *39*, 49, 53(36), 55 (42), 66(36), *113*, 169(1, 1a), 205(201), *243*, *248*, 617(351, 352), *630*
Kornberg, S. R., 323(132), 324(132), *334*, 586(88), *624*
Kornblum, J., 453(139), *468*
Kornfeld, S., 464(193, 195), *469*
Korte, F., 278(192, 193a), *292*
Koshland, D. E., Jr., 158(161), *166*
Kosow, D. P., 75(125), 79(125), 95(125), *115*
Kovacs, J., 404, *411*
Kowalsky, A., 158(161), *166*
Koyama, M., 282(224a), *293*
Koyawa, J., 596(200), *627*
Kozloff, L. M. 586(91), 589(91), 596(91), *624*
Kramar, E., 390(3), *409*
Kramer, M., 595(197), 598(197), 603(197), *626*
Krampitz, L. O., 47, 48, 86(169), 87(169, 172), 89, 103, *113*, *116*, 211(222), *248*, 296(12), *331*
Krasna, A. I., 282(224), *293*
Krasnovskii, A. A., 342(48a), *369*
Krause, R. M., 423(42), *466*
Kream, J., 327(158), *334*, 401(55), 402(55), *411*
Krebs, H. A., 32, 33, *40*, 49, 53(36, 55), 66(36, 55), *113*, 157(148), *165*, 169(1), 186, 205(201), *243*, *245*, *248*, 384(86), *388*
Krishnaswamy, P. R., 158, *166*, 451(126), *468*, 506(284), 507(297, 299), *572*
Kritzmann, M. G., 180, *244*
Kröger, E., 464(199), *469*
Krøger, H., 547(450, 451), 549(450, 451), *576*

Krone, W., 495(257), *571*
Krueger, A. P., 586(85), *624*
Krueger, K. K., 275(170), *291*
Kruger, L., 425(49), 464(49), *466*
Kuehl, F. A., 516(347), *573*
Kuh, E., 279(195), *292*
Kumar, A., 349(88b), 353(88b, 109), *370*
Kunisawa, R., 7(38), 10(38), 35(38), 36 (38), *37*
Kunkee, R. E., 610(286), *628*
Kunst, P., 477(218), *570*
Kupiecki, F. P., 77, *115*, 272(133), *290*
Kupke, D. W., 362(162a), *372*
Kurahashi, K., 53(53), 64, 65, 66(53), 67, *113*, *114*
Kurek, L. I., 585(70), 592(70), *623*
Kuriki, Y., 464(196), *469*
Kurimura, O., 390(11), 403(11), 404(11), 405(11, 69), 408(11), *410*, *411*
Kurland, C. G., 283(236), 561(453), 562 (453), *570*, *576*, 590(132), *625*
Kurz, H., 341(38), *369*
Kusunose, E., 78(142), *115*
Kusunose, M., 78(142), *115*
Kuwada, S., 263(79), 265(79), *289*

L

Labbe, R. F., 350(98), 358(127, 128), *370*, *371*
Lacks, S., 513, 527, 555, *568*, 583(55), 586(55), *623*
Ladd, J. N., 85(157), *116*, 281(214b, 214c, 215b), *293*
Lagerkvist, U., 301(41), *332*
Laine, T., 272(130), *290*
Laki, K., 477(222), *570*
Lamborg, M. R., 499, 501, 526, *571*
Lamfron, H., 492(133), 515(343), 521 (133), *568*, *573*
Lampen, J. O., 207(211), *248*, 327(159), *334*
Landman, O. E., 488, *566*, 582(36), 591 (145), 599(220), *622*, *625*, *627*
Landy, M., 256(26), *287*
Lane, M. D., 74, 75, 79, 95, *115*, *117*
Lang, H. M., 17(66), 30(66), *38*
Langer, L., 158(161), *166*
Langham, W. H., 231(296), *250*
Langley, M., 284(245, 246, 247), *294*
Lankford, C. E., 262(61), *288*
Larabee, A. R., 198(155), *247*
Lardy, H. A., 50, 53, 54(57), 72, 74, 89 (185), *113*, *114*, *116*
Lark, C., 439(162, 163), 459(162, 163), *468*, *469*
Lark, K. G., 439(162, 163), 459(161, 162, 163), 460, *468*, *469*, 621(369), *630*
Larkum, N. W., 256(26), *287*
Larminie, H. E., 367(181b), *372*
Larsen, A., 277(189), *292*
Larsen, H., 5, 6, 7(27), *37*, 344, 362(164b), *369*, *372*
Larson, E. A., 336(3), *368*
Lasada, M., 34(124), *40*
Lascelles, J., 27, 28, 30, *39*, 257(28), 277 (28, 185), 278(28), *287*, *291*, 339(37), 341(37), 342(48), 343(48, 107), 344 (48), 345(67), 346(48, 67), 352(107), 353(112b), 354(67, 107, 112b), 355(37, 67), 356(67), 358(37), 360(37), 361 (37), 364(112b), 365(112b), 367(179b), 368(48, 179b), *369*, *370*, *371*, *372*, 604 (254), 617(254), *628*
Laser, H., 614(330), *629*
Laskowski, M., 401(56), *411*
Laskowski, W., 592(158), *625*
Laule, G., 596(203), *627*
Laver, W. G., 349(85, 86), *370*
Leach, F. R., 284(251, 252), *294*, 536(368), *574*
Leadbetter, E. R., 23(89), 24, *39*
Leahy, J., 515(342), *572*, *573*
Leaver, F. W., 82, *115*
Leavitt, R. J., 213(228), *249*
Lebedev, A. F., 42, *112*
Le Bras, G., 194(128a), *246*, 314(79), *332*
Leder, I. G., 260(48b, 48d), *288*
Lederberg, E. M., 591(147), *625*
Lederberg, J., 258(35), *287*, 379(38), *387*, 459(157, 158, 159), 461(159), 462, *468* 582(47), 584(63a), 591(147), 593(165), 597(165), 605(165), *623*, *625*, *626*
Lederer, E., 419(28), *465*
Lee, J. M., 276(180), *291*
Lee, S. B., 144(95, 97), *164*
Leeper, L. C., 239(339), *251*
Lees, H., 3, 31, *37*, *39*, 123(10), *162*
Legge, J. W., 336, *368*
Lehman, I. R., 318(87), 322(87), 324(87),

325(87), 328(163), *333*, *334*, 552(178, 179), 553(411, 412, 419, 421), *569*, *575*
Leibman, K. C., 545(386), *574*
Leifer, E., 231(296), *250*
Lein, J., 233(311), *251*
Leismann, A., 535(372), *574*
Leitner, F., 226(282), *250*
Leloir, L. F., 203(185), *247*, 414(1), 426, 457(1), *465*, *466*
Lemberg, R., 336, *368*
Lengyel, P., 72, 85(115), *114*, 409(94, 95), *412*, 500(89), 564(460, 461), *567*, *576*
Lenhoff, H. M., 358(132), 360(132), *371*, 612(315), *629*
Lentz, C. P., 617(349), *630*
Lentz, K. E., 102, 103, 104, 108, *117*
Leonard, C. G., 394(27), 395(27), 398(44), 402(27), 403, 405(70), 408, *410* *411*
Leone, E., 72(111), 74(111), 75(111), 76 (111), 84(111), 90(111), 91(111), 95 (111, 191), 96(111), *114*, *117*
Leonian, L. H., 275(163, 167), *291*
Lerman, L. J., 588(107), *624*
Lerner, P., 234(313), *251*
Lester, G., 591(146), 612(301), *625*, *629*
Lester, R. L., 286(272), *294*, 493(74), *567*
Leupold, U., 582(28), 590(28), *622*
Lev, M., 285(267, 268), *294*, 341(32), *369*
Levenberg, B., 184(75), *245*
Levin, A. P., 301(38), 305(38), 317(38), *331*, 610(280), 612(317), 618(280), *628*, *629*
Levin, J. G., 227, *250*
Levinthal, C., 588(109), *624*
Levintow, L., 159, *166*, 183(64, 65), *245*
Levitch, M. E., 284(252), *294*
Levy, H. R., 319(92), *333*
Levy, J. B., 359(146, 147, 148), *371*, 605 (263), 617(348), *628*, *630*
Levy, L., 219(256), *249*
Lewis, K. F., 214(231), *249*
Libby, R. L., 203(189), *248*, 615(325), *630*
Lichstein, H. C., 91(186, 187), *116*, 272 (131), 275(160, 173), *290*, *291*, 319(96), *333*, 612(306), *629*
Lichtenstein, J., 323(125), 324(125), 325(146), 328, *333*, *334*, 495, *571*
Lieberman, I., 160, *166*, 297(15), 320(113, 114, 115), 321(116), 322, 323, *333*
Lieberman, M., 586(85), *624*
Liebermeister, K., 459(156), *468*
Lieske, R., 3(12), 4(12), *37*
Lilly, V. G., 275(163, 167), *291*
Lin, E. C. C., 612(317), *629*
Linday, E. M., 604(257), *628*
Lindeberg, G., 383(64), *387*
Linderstrøm-Lang, K., 429(69), *466*
Lindstrom, E. S., 125(30), *162*
Lingens, F., 232(303), 233(308), *250*
Lingrel, T. B., 515(346), 522(346), *573*
Linstead, R. P., 362(162b), *372*
Lipmann, F., 86(166, 167), 87(166), *116*, 131(57), 158(156), 159(165, 166), *163*, *165*, *166*, 188(102), 190(112), 191(112, 113), 207(213b), *246*, *248*, 272(140), 274(140), *290*, 320(110), *333*, 506(107, 109, 282, 286, 294), 507(110), 508, 511(120), 513(120, 325), 514, 536(364), 556, 557(367), 560(367), *567*, *568*, *572*, *573*, *574*
Liston, J., 345(56), 346(56), *369*
Litman, R. M., 587(96, 97a), *624*
Little, H. N., 145(107), *164*
Littlefield, J. W., 296(5), *331*, 481(30), 499(85), *565*, *567*
Litwin, J., 184(82), *245*, 399(53), *411*, 537 (164), *569*
Lively, E. R., 591(147), *625*
Ljungdahl, L., 102, *117*
Lochhead, A. G., 261(54), *288*, 342(46a), *369*
Lochmann, E., 592(158), *625*
Locke, L. A., 212(224), 215(239), 216(239), *248*, *249*
Lockwood, W. H., 351(100), *370*
Loeb, M. R., 297(13), 298(13), 317(13), *331*
Loercher, R., 442(100), 443(100), *467*
Løvtrup, S., 612(299), *629*
Loewenberg, J. R., 383(72), *388*
Loewus, F. A., 61(81), 62(81), 63(81), 65 (81), 66(81), *114*
Loftfield, R. B., 620(366), *630*
Lohmar, R. L., Jr., 374(9), *386*
Lombard, A., 554(205), *570*
London, I. M., 348(80), 349(81), *370*
Long, B., 274(153), *291*
Long, M. V., 82(152), 85(152), *116*
Loomis, W. E., 399(48), *411*
Lorch, E., 74(124), 76(124), 92(124), 93 (124), 95(124), 97(124, 192a), 98(124), *115*, *117*

Loring, H. S., 320(106), *333*
Losada, M., 11(50), 29(50), *38*, 127(49), *163*
Loutit, J. S., 191(114), *246*, 322(118), *333*
Love, S. H., 305(45), 306(45), *332*
Luborsky, S., 549(403), *575*
Lucas, D. R., 341(33), *369*
Luchsinger, W. W., 158(160), *166*
Lüderitz, O., 425, 464(48, 199, 200), *466*, *469*
Lukens, L. N., 70, *114*
Lund, P., 617(352), *630*
Luria, S. E., 295(1), *331*, 583(57), 587(92), *623*, *624*
Lwoff, A., 259(39), *287*, 340(18, 21, 23), 361(18), *368*, 561, *576*
Lwoff, M., 259(39), *287*, 340(18, 19, 20, 22), 361(18), *368*
Lynch, V., 13(53), *38*
Lynen, F., 74, 75, 76, 78, 82(151), 85(151, 160), 91, 92, 93, 95, 97, *115*, *116*, *117*, 207(212), *248*, 518, *574*
Lytle, V. L., 284(235), *293*

M

Maaløe, O., 486, *576*
Maas, F., 217(247), *249*
Maas, W. K., 174(19), 188(102), 191(116), 192, 208(214), 243(19), *244*, *246*, *248*, 255(20), 257(27), 258(27), 259(27), 271(20, 27, 124, 125), 272(125, 137, 138, 139), *287*, *290*, 585(67), 606(271), 607(273), *623*, *628*
Maass, D., 416(19), 417(19), *465*
McCarter, J. R., 285(266), *294*
McCarthy, B. J., 483(237, 238), 484(238), 555(238), 564, *570*, *571*, *576*
McCarty, M., 421, 423(42), *466*, 586(74), *623*
Macchi, M. E., 284(238), *293*
McClary, D. O., 602(239), *627*
McCloy, E. W., 407, *412*
McCluer, R. H., 443(104), *467*
MacCorquodale, D. W., 285(254), *294*
McCoy, E., 99(195), *117*
McCullogh, W. G., 266(95), *289*, 396(33), *410*, 533(147), *568*
MacDonald, J. B., 339(31a), 341(31a), *369*
McDougall, B. M., 324(143), *334*
McEvoy, D., 255(14a), *287*
McFall, E., 589(128), 615(128), *625*
McGhee, W. J., 390(10), 402(10), 407(10), 408(10), *410*
McGill, B. B., 553(418), *575*
Machlis, L., 284(241), *294*
McIlwain, H., 183(69), *245*, 589(117), *625*
McIntosh, E. N., 271(128), *290*
McKee, R. W., 285(254), *294*
McKeon, J. E., 365(168a), *372*
Mackor, E. L., 485(40), *566*
MacLaren, J. A., 255(12), 263(12), *287*
McLean, J. R., 524(130), *568*
McLean, P., 179(46), 240(46), *244*
MacLennan, A. P., 425(53, 57), *466*
MacLeod, C. M., 586(74), *623*
Macleod, H. L., 591(144), 608(144), *625*
McManus, I. R., 51, *113*, 535(155), 536, *569*
McNall, E. G., 122(6a), *162*
McNary, J., 135(67), *163*
McNutt, W. S., 255(13), 265(83, 86, 86a), *287*, *289*
McNutt, W. S., Jr., 255(16), *287*, 298(19), 318(83, 85), 328(19, 85), *331*, *333*
Macow, J., 255(22), 271(22), 272(22), *287*
MacQuillan, A. M., 611(294), *629*
McQuillen, K., 199(164), *247*, 440(92), 461(92), *467*, 483(237, 238), 493(72), 500, 501, 515(275), 555(238), 559(433, 434), *566*, *567*, *570*, *571*, *575*, 590(138), 603(138), *625*
Macturk, H. M., 285(258), *294*
McVeigh, I., 259(42), *288*
Madinaveitia, J., 285(258), *294*
Magasanik, A. K., 486(244), *571*
Magasanik, B., 18(73), *38*, 161, *166*, 170 (4), 172(8), 177(4), 184(78, 79), 205 (198), 219, 220, 221(79, 262), *243*, *245*, *248*, *249*, 301(35, 38, 40), 305(38, 43) 306(40, 43, 47, 48, 49), 308(40, 47, 49, 56), 309(57, 58, 59, 61), 311(49, 61, 64), 312(49, 64), 313(49), 314, 316(56, 80), 317(38, 56, 57, 58, 59), 319(43, 47, 57, 91), 320(99, 100), 326(148), 327(157), *331*, *332*, *333*, *334*, 486, *571*, 595(190), 610(280, 288), 611(288, 290, 292), 612 (288, 317), 616(340), 618(280, 355), *626*, *628*, *629*, *630*
Magee, W. E., 127(44), *163*
Mager, J., 309(57), 317(57), 319(57), *332*, 556, 557(428), *575*

Mahler, H. R., 126(41), 128(41), 142(41), 151(41), *162*, 324(139), *334*
Maikel, R. P., 578(5), *622*
Makino, K., 262(64, 65), *288*
Maley, F., 323(130), 324(138), *334*
Maley, G. F., 263(77, 78a), 264(81a), 265 (85, 85a), *289*, 324(138), *334*
Malin, R. B., 532(141), *568*
Mallette, M. F., 595(189), 612(300), *626*, *629*
Malmgren, B., 485, *566*
Mandel, H. G., 487, 540, *566*, *569*, *571*
Mandeles, S., 185(85), *245*
Mandelstam, J., 459(164, 165), *469*, 491 (55), 503, 538(55), 554(55), *566*, *567*, *572*, 599(219), 600(227), 601(231), 603 (227), 611(297), 612(303), *627*, *629*
Mandelstam, M. H., 417(22), 422(22), 423(22), 424(22), 428(22), 437(81), 442(100), 443(100), *465*, *467*
Mangum, J. H., 31(115), *39*
Mano, Y., 261(58a), *288*
Mans, R. J., 317(82), *333*
Mantsavinos, R., 591(150), *625*
Mardeshev, S. R., 271(131a), *290*
Margalith, P., 400, 401(54), 403, 405, *411*
Margolin, P., 215(243), 216(243a, 243b), *249*
Margreiter, H., 260(50), *288*
Marion, L., 253(1), 266(1), *287*
Markert, C. L., 580(12), 591(143), *622*, *625*
Markovitz, A., 464(203), *469*
Marks, G. S., 366(173b), *372*
Marmur, J., 586(75), 588(111), *623*, *624*
Marnati, M. P., 283(227b), *293*
Marnay, C., 256(25), 266, *287*
Marshall, M., 160(170), *166*, 190(111), *246*
Martin, H. H., 200(172), *247*, 417(20), *465*
Martin, R. D., 420(35), *466*
Martin, R. G., 184(79a), 221(79a), *245*, 309(60), 317(60), *332*, 409(96), *412*, 564(462), *516*
Maruyama, Y., 621(367), *630*
Mason, D. J., 458(153), *468*
Mason, H. S., 139a(76a), *163*
Massini, P., 14, 27(55), *38*
Mastuo, Y., 197(145), *247*
Masuda, T., 263(76, 79), 264(80), 265(79, 82, 84), *289*
Mathias, A. P., 273(146), 274(146), *291*, 438(82), 444(107, 108, 110), *467*
Matsuda, N., 156(138), *165*
Matsushima, Y., 427(67), *466*
Matsushiro, A., 583(58), *623*
Matsuyama, A., 270(129), 271(129), *290*
Matthaei, J. H., 409, *412*, 562, 563, 564, *576*
Matthews, R. E. F., 487, 540, *566*, *569*
Mauer, P. H., 408(90), *412*
Maul, W., 274(148), *291*
Mauzerall, D., 351(101, 105), 358, *370*
Maxwell, E. S., 323(128), *334*
May, E. L., 267(99), *289*
May, O. E., 43(10), *112*
Mazumder, R., 72(115), 85(115), *114*
Meadow, P., 199, 200(177), 239(166), *247*, 440(93), 461(93, 172), *467*, *469*
Mecchi, F., 285(260), *294*
Medina, A., 122(6), *162*
Meek, J. S., 86(168), *116*
Mehler, A. H., 49(41, 42), 51(42), 53(42), 55(42), 91(189), *113*, *117*, 219(257), *249*, 267(98, 99), *289*
Mehta, R., 518(127, 358), *568*, *574*
Meister, A., 157, 158(161a), *165*, *166*, 180, 181, 183(64, 65), 185, 186(92), 189(106, 107), 192(56), 193, 202(56), 203(56, 59, 186, 187), *245*, *246*, *247*, *248*, 270 (119), *290*, 398, *411*, 451(125, 126, 129), *468*, 491(60, 61), 506(61, 284), 507 (297, 298, 299), 524(60, 61), *566*, *572*
Melechen, N. E., 550(185), *569*
Melnykovych, G., 612(308), *629*
Melville, D. B., 275(161), 276(180), *291*
Menard-Jeker, D., 91(188), *116*
Mendelsohn, J., 488, 495, *571*
Menon, G. K. K., 79(147), *115*
Merrifield, R. B., 532(145, 146), *568*
Merrill, S. H., 512(316), *573*
Meselson, M., 561(442), 562(442), *576*, 587 (100), 588(100), 620(366a), *624*, *630*
Metcalfe, G., 125(23, 24), *162*
Metzenberg, R. L., 160(167, 168, 169, 170), *166*, 190(111), *246*
Metzler, D. E., 181, 197(61), *245*
Mevius, M., 23(92), 24, *39*
Meyerhof, O., 3(9), *37*
Michelson, A. M., 322(121), *333*, 464 (191), *469*
Mickelsen, O., 254(6), 274(6), *287*
Micou, J., 589(121), *625*

Middlebrook, G., 341(28), *368*
Migliacci, A., 283(227, 227a, 227b), *293*, 453(143), *468*
Mihara, S., 537(379), *574*
Milhaud, G., 16(77), 19(77, 78), 23(77, 78), 31(114), *39*, 189(110), *246*
Miller, C. S., 89(183), *116*, 320(108), *333*
Miller, D. A., 205(199), *248*
Miller, H. K., 532(142), *568*
Miller, I. M., 282(222), *293*
Miller, P., 169(1a), *243*
Miller, P. A., 183(71), *245*
Millet, J., 16(77), 19(77, 78), 23(77, 78), 31(114), *39*, 189(110), *246*
Mills, G. T., 446(112), *467*, 586(76. 77), *623*
Mills, R. C., 396(33), *410*
Milner, H. W., 494(255), *571*
Minakami, S., 359(135a), *371*
Minghetti, A., 283(227, 227a), *293*
Mingioli, E., 169(1a), *243*
Mingioli, E. S., 172(11, 12), 198(149), 217(244), 223(269), 226(11, 12), 227(291), 229(291), *243*, *247*, *249*, *250*, 258(37), *287*, 298(22), *331*
Minor, F. W., 381(54), *387*
Mirsky, A. E., 557(425), *575*
Mistry, S. P., 85(158), *116*
Mitchell H., K., 176(29), 199(29), 217(245, 246, 247), 231(296), 233(311), *244*, *249*, *250*, *251*, 258(33, 38), *287*, 322(121), *333*, 581(20), 582(32), 591(20), 593(20), *622*
Mitchell, M. B., 217(247), *249*, 582(32), *622*
Mitchell, P. D., 419, 458, *465*, *468* 499(274), *571*
Mitoma, C., 219, 239(339), *249*, *251*, 319, *333*
Mitsuhashi, S., 223(270, 273), *250*
Mitsui, H., 359(135a), *371*, 621(367), *630*
Mitz, M., 257(32), *287*
Miyake, A., 267(100), *289*
Miyazaki, S., 509(303), *572*
Mizobuchi, K., 583(58), *623*
Moat, A. G., 319(94, 95), *333*
Möller, E. F., 274(148), *291*
Møller, F., 580(12), *622*
Moggridge, R. C. G., 260(52), *288*
Moldave, K., 491(60, 61), 506(61), 507(298, 300), 513(300), 514, 524(60, 61), 546, *566*, *572*, *573*, *574*
Molnar, D. M., 139, *163*, 236(328), *251*, 394(28), 397(39), 398(39), *410*, 537(162), *569*
Momose, H., 237(330), *251*, 461(173, 174), *469*
Moncel, C., 31(114), *39*
Mond, J., *576*
Monk, G. W., 365(172), *372*
Monod, J., 379(37, 39), *387*, 492(62), 493(64, 65), 503(62), *566*, 583(53, 54, 57), 585(71), 586(83), 587(54), 589(54, 129), 591(53), 593(160, 161, 163, 172), 594(163, 176, 178), 595(129, 191, 196), 596(202), 597(191, 196, 205, 209), 599(209, 220a), 600(178), 601(178, 233, 234), 604(176), 606, 607(268, 276), 608(53, 268), 609(161, 202, 276), 610(289), 613(83), 615(54, 336), 619(276), 620(276), *623*, *624*, *625*, *626*, *627*, *628*, *629*, *630*
Monro, R. E., 503(94), *567*
Monty, K. J., 85(159), *116*
Moore, A. M., 323(134), *334*
Moore, S., 477(214), *570*
Morley, A. V., 362(165b), *372*
Morowitz, H. J., 588(104) *624*
Morris, A., 515(342), *573*
Morris, J. G., 270(115, 116), *290*, 345(57), 346(57), *369*
Morris, L. R., 106(218), *117*
Morrison, R. A., 549(403), *575*
Morse, S. I., 421(38f), *466*
Mortenson, L. E., 121(16, 17), 122(86), 123(86), 124(17, 18), 126(40), 127(16, 17, 44), 128(16, 17, 52), 130(17, 53), 131(17, 18, 52), 132(17), 133(86), 134(64, 65), 136(17, 18, 52), 137(17), 138(17, 18, 53), 140(86), 141(17, 53), 142(17), 143(17), 146(112), 149(17, 18), 151(86), 152(18), *162*, *163*, *164*
Mortimer, R. G., 144(94), *164*
Mortlock, R. P., 87, *116*, 131(59, 60), *163*
Morton, R. A., 285(270), *294*
Moses, V., 60, *114*
Moses, W., 262(62), *288*
Moss, F. O., 360(150), 361(158), 367(181b), *371*, *372*, 612(313), *629*
Mountain, I. M., 582(40), *622*

Mowat, J. H., 279(195), *292*
Mower, H. F., 121(16, 17), 124(17, 17a, 19), 127(16, 17), 128(16, 17), 130(17), 131(17, 17a), 132(17), 134(65, 66), 136(17, 18), 137(17), 138(17), 141(17), 142(17, 19), 143(17), 149(17), *162*, *163*
Moyed, H. S., 18(73), *38*, 161, *166*, 175, 184(78), 220, 238(25, 26), *244*, *245*, 301(40), 306(40, 47), 308(40, 47), 309(58, 59), 317(58, 59), 319(47), 320 (99), *332*, *333*, 618(362), *630*
Moyer, A. J., 43(10), *112*
Moyer, A. W., 582(39), *622*
Moyle, J., 58(68), *113*, 419, 458, 465, *468*, 499(274), *571*
Mozen, M. M., 139, 141(81), *163*
Mudd, S., 461(178), *469*
Mühlethaler, K., 380(42), *387*
Müller, F. M., 7, 28(35), *37*
Mueller, J. H., 183(71), *245*, 257(29), 270 (29), *287*
Mukai, F., 590(141), *625*
Mumford, F. E., 141(90), *164*
Munch-Peterson, A., 156(140), *165*, 281 (214b), *293*
Munier, R., 504, 505(97), 534(97, 98), *567*
Munro Fox, H., 366(179a), *372*
Murmikko, V., 532(140), *568*
Murray, E. G. D., 390(12), *410*
Musselman, A. D., 433(73), *466*
Muto, T., 282(224a), *293*
Myers, J., 192(118), *246*
Myers, J. W., 214(238), *249*, 258(36), *287*

N

Nadel, H., 376(19), *386*
Näveke, R., 24, *39*
Nagler, A., 169(1a), *243*
Najjar, V. A., 254(4), 259(43), *287*, *288*
Nakada, H. I., 204(196), *248*
Nakagama, K., 237(331), *251*
Nakamoto, T., 547, 561(449, 452), *576*
Nakata, 621(370), *630*
Nakayama, H., 260(50a), *288*
Nanney, D. L., 582(29), *622*
Naono, S., 541(384), 561(384), *574*
Narrod, S. A., 398(41), *410*
Nason, A., 31, *39*, 122(2), 123(2), 140 (85), *161*, *164*, 360(156a), *372*, 612 (309), *629*
Nathans, D., 514, 556, 557(367), 560(367), *573*, *574*
Nathenson, S. G., 448(121), 458, 459(169), 460(169), *468*, *469*
Naughton, M. A., 477(220, 221), *570*
Nawa, H., 284(252a), *294*
Neely, W. B., 374, 376(6), 377(6, 23), 378 (31, 32), *386*, *387*
Neidhardt, F. C., 170(4), 177(4), *243*, 486(243, 244), 554, *570*, *571*, 611(290, 291, 292), *629*
Neidle, A., 183(66), 184(66), 219, 221, *245*, *249*, 320(101), *333*
Neilands, J. B., 342(41, 42, 43), 345(62, 63), 346(62, 63), 354, 356(63), 358 (63), *369*
Nelson, W. O., 271(126, 127), *290*
Nemeth, G., 125(25), *162*
Nester, W., 584(63a), *623*
Neuberger, A., 349(83, 85, 86, 89), 353 (112a), 357(112a), *370*, *371*
Neuhaus, F. C., 206(209), *248*, 448, 451 (122, 130), *468*, 509, 539, *572*, *574*
Neurath, H., 477(217), *570*
Neve, R. A., 350(98), *370*
Neville, A., 612(302), *629*
Newburgh, R. W., 4(25), 5, *37*
Newton, G. G. F., 433, 434(72), 459(72), *466*
Newton, J. W., 125(30), 136(70, 75), 144 (102), *162*, *163*, *164*
Nichimura, S., 592(155), *625*
Nicholas, D. J. D., 122(2a, 3a, 4, 4a, 4b, 6), 123(2a, 8), 127(47, 48), 128(48), 130(54), 151(4), 152(47), *161*, *162*, *163*, 358(132), 360(132), *371*
Nier, A. O., 43(12, 17), 47(17, 24, 26), *112*
Nierlich, D., 301(35), 314(35), *331*
Nikaido, H., 464(201), *469*
Nimmo-Smith, R. H., 277(185), *291*
Nirenberg, M. W., 409, *412*, 562, 563, 564, *576*
Nishida, G., 358(127), *371*
Nisman, B., 194(127), 196(135), *246*, 494, 507(80), 508, 530, 531(365, 366), *567*, *574*
Noblesse, C., 60(73), *114*
Noblesse, M., 384(81), *388*

Noboru, S., 588(111), *624*
Nocito, V., 203(185), *247*
Noda, H., 277(191), 278(191), *292*
Nohara, H., 509, *572*
Noll, H., 285(256), *294*
Nomura, M., 484, 552, 555, 561(410), *575*, 590(133), 592(155), *625*
Norman, A., 614(326), *629*
Norman, J. O., 286(271), *294*
Norris, J. R., 126(38), *162*, 458(152), *468*
Northey, E. H., 279(195), *292*
Nose, Y., 260(48e), *288*
Novak, R., 164(83), *164*
Novelli, G. D., 86, *116*, 188(102), *246*, 272(138, 141), *290*, 494, 498, 501(276), 507, 526(111), 557(424), *568*, *571*, *572*, *575*, 607(272), 615(334), *628*, *630*
Novick, A., 582(45), 595(45), 602(45, 236), 606(45, 265), 615(337), 618(356), *623*, *627*, *628*, *630*
Nowakowska, K., 282(219), *293*
Noyes, H. E., 394(26), 400(26), 401(26), 406(26), *410*
Nozaki, K., 127(49), *163*
Nutting, L. A., 88, *116*
Nye, J. F., 205(203), 231(296), *248*, *250*
Nygaard, A. P., 580(10), *622*

O

O'Brien, P. J., 442(98), 461(177), *467*, *469*
O'Callaghan, C. H., 361(159), *372*
Ochoa, S., 15(68), 17(62, 68), 22(68), 24 (68), *38*, 49, 50(33, 43), 51, 53, 54 (43), 55(34, 42, 62), 57, 58, 65(43), 69 (101), 71, 72, 73, 74(110, 111, 112, 121), 75(111), 76, 84(111), 85, 90(111), 91(111, 189), 95(111, 191), 96, *113*, *114*, *115*, *117*, 176(33), *244*, 409(94, 95), *412*, 494, 500, 518(124), 519(78, 128), 524, 530, 547, 548(401), 549, 564(460, 461), *567*, *568*, *569*, *575*, *576*
O'Connell, D. J., 401(58, 60), 408(58, 60), *411*
O'Conner, C. M., 578(4), 615(4), *622*, *630*
Oeding, P., 421(38a), *466*
Ofengand, E. J., 506(290), 511(117), 512 (117, 121), 513(121, 324), 526(117, 121), *568*, *572*, *573*
Ogata, K., 509, *572*
Ogata, S., 34(124), *40*
Ogur, M., 592(156), 593(156), *625*
Okada, H., 580(11), *622*
Okamoto, T., 514(333), *573*
O'Kane, D. J., 87(174), *116*, 131(58), *163*, 284(235), *293*
Okazaki, R., 464(187, 188, 190, 191, 196, 197, 204), *469*, 550, *575*
Okazaki, T., 464(190, 191, 196, 197, 204), *469*, 550, *575*
Okuda, S., 464(202), *469*
Oldewurtek, H. A., 612(309), *629*
Olivard, J., 181(60), *245*
Olsen, O., 340(15), *368*
Olson, J. A., 154(123, 124, 125), *165*
Olson, M. E., 499(86), *567*
Oosterbaan, R. A., 477(218), *570*
Oparin, A. I., 26, *39*
Ordal, E. J., 123(7), 140(7), 150(7), *162*
Orgel, G., 22(84), *39*
Orla-Jensen, M. D., 285(261), *294*
Orla-Jensen, S., 285(261), *294*
Ormerod, J. G., 9, 10, 27, 28, 35(44, 125), 36, *38*, *40*, 144(103), *164*
Ortega, M. V., 222(265), *250*, 266(95a, 95b), *289*
Orth, H., 336, 362(7), *368*
Ortiz, P. J., 71(107), *114*, 547, 548(401), *569*, *575*
Osawa, S., 484, 513, 557(425), *571*, *573*, *575*
Osborn, M. J., 279(202. 203a, 209, 210), *292*
Oshima, Y., 580(11), *622*
Osler, O., 285(255), *294*
Osmundsen, J. A., 564(464), *576*
Oswald, E. J., 255(26), *287*
Otaka, E., 513(321), *573*
Otsuji, N., 439(86, 87), *467*
Otsuka, H., 537(379), *574*
Overath, P., 82(151), 85(151, 160), *115*, *116*
Overbeek, J. T. G., 506(106, 289), 537 (376), *567*, *572*, *574*
Owades, P., 401(58), 408(58), *411*
Owen, R. D., 591(143), *625*

P

Packer, L., 4(21), 31(119), 32(21), *35*, *37*
Paege, L. M., 298(17), 327(156), *331*, *334*

Paigen, K., 555(204), *570*
Painter, T. S., 592(152), *625*
Palade, G. E., 515(341), *573*
Paladini, A. C., 426(61), *466*
Pamiljans, V., 158(161a), *166*, 451(126), *468*
Pappenheimer, A. M., Jr., 345, 346(54, 71), 358(71), 368(183), *369*, *370*, *372*, 595(191), 597(191), *626*
Pardee, A. B., 174(18, 21), 196(137), 243 (18), *244*, *246*, 320(112), 321, 322(112), 326, 327(150), *333*, *334*, 459(170), *469*, 486, 495(83), 505, 542(174), 550(174), 554, 555, *566*, *567*, *569*, *570*, 578(3), 583(53, 54), 587(54, 96), 589(54, 128), 591(53), 593(171), 595(193), 597(207), 598(215), 600(222), 601(215), 602 (207, 222), 603(207), 604(171, 207), 606(171), 607(171, 207, 274), 608(53), 609(274), 610(274, 281, 284, 285), 612 (222, 274, 298, 305), 613(324a), 614 (108, 284), 615 (54, 128, 285), 616(3), 618(274, 357, 359, 363), 621(368), *622*, *623*, *624*, *625*, *626*, *627*, *628*, *629*, *630*
Park, B. H., 586(86), *624*
Park, J. T., 416(16), 423(44), 426(44, 58, 59, 60), 427(59, 67), 428(44, 59), 429(58), 437(80), 439(167), 443(58), 457(44, 60), 458(60), 459(167, 168), *466*, *467*, *469*, 479, 490, 491(51), 538, 554(51), *565*, *566*
Parker, C. A., 152(114), *164*
Parks, G. S., 3, *37*
Parks, L. W., 233(307), *250*
Partridge, C. W. H., 231(297), *250*, 584 (61a), 591(61a), *623*
Paskhina, T. S., 453(142), *468*
Patterson, D. S. P., 345(60a), 346(60a), *369*
Patterson, E. L., 284(238, 240), *293*, *294*
Pauling, L., 477(14, 15), *565*, 580(13), *622*
Pauly, H., 614(325, 329), *629*
Pawelkiewicz, J., 254(11), 280(214), 282 (11, 218, 219, 220), 283(218), *287*, *292*, *293*, 345(64), 346(64), *369*
Pazur, J. H., 464(192, 194), *469*
Peabody, R. A., 184(74), *245*, 488, *571*
Peanasky, R., 72, 74(116), *114*
Peaud-Leonël, C., 383(78, 79), 384(79, 83, 84), 385(84), 386(79), *388*
Peck, R. L., 276(175, 176), *291*
Peel, J. L., 31(118), *39*
Pelzer, H., 416(19), 417(19), *465*, 535 (371), *574*
Pendlington, S., 255(14), 263(71), 265 (71), 279(71), *287*, *288*
Pengra, R. M., 125(35), 126(35), *162*
Pennington, R. J., 65(94), *114*
Pentler, L. F., 285(260), *294*
Perkins, H. R., 416(14, 15), 439(83), 446 (117), *465*, *467*, *468*
Perlman, D., 281(215c), 282(216a), *293*
Perlmann, G. E., 477(216), *570*
Perret, C. J., 598(217), *627*
Perrin, D., 585(71), 593(161), 599(220a), 609(161), *623*, *626*, *627*
Perrit, C. J., 621(372), *630*
Perry, J. J., 200(171), *247*
Peterkofsky, A., 283(229b, 229c), *293*
Peterkofsky, B., 201(182), *247*
Peterman, M. L., 497(259), *571*
Peters, H., 267(101), *289*
Peters, J. M., 279(207, 208), *292*
Peters, V. J., 533(150), *568*
Peters, V. P., 270(123), *290*
Peterson, B. H., 282(219), *293*
Peterson, E. A., 181(62), *245*
Peterson, W. H., 44(20), 99(195), *112*, *117*, 275(170, 171), 276(174), *291*
Petit, J. F., 359(135b), *371*
Pfeiffer, R., 340(13), *368*
Phares, E. F., 82, 85(149, 152, 165), *115*, *116*
Phelps, A. S., 144(96), *164*
Phillips, P. M., 437(79), 440(79), 444(79), *467*
Pidacks, C., 342(44, 45), *369*
Piérard, A., 153, 154, *164*, 179, 203(48), *244*
Pierce, J. G., 320(106), *333*
Pierce, J. V., 284(238), *293*
Pierpoint, W. S., 272(142), 273(146), 274 (146), *290*, *291*
Pine, L., 103, 108, *117*, 284(236), *293*
Pine, M. J., 109, *117*
Pineau, E., 189(110), *246*
Pinsky, M. J., 612(307, 318), *629*
Plaut, G. W. E., 50, 57(67), *113*, 255(15), 263(72, 73, 77, 78a), 264, 265(81b, 85, 85a), *287*, *288*, *289*
Plaut, W., 560, *570*, 589(124), *625*
Podolsky, R. J., 503(93a), *567*

Pogell, B. M., 184(76), *245*
Pollock M. R., 582(44), 593(163, 167), 594(163, 167), 595(197), 597(167), 598(167, 197, 211, 212, 213, 217, 218), 599, 600(211), 603, 604(167), 605(167), 616(44), *623*, *626*, *627*
Pomerantz, S. H., 82(150), *115*
Pon, N. G., 17(63), *38*, 51, 61, *113*
Pontecorvo, G., 253(2), 266(2), *287*, 582 (27), 584(63), *622*, *623*
Pope, H., 275(166), *291*
Porra, R. J., 367(181b), *372*
Porter, C. J., 598(216), 612(216), *627*
Porter, J. R., 472, *565*
Porter, J. W. G., 254(5), 282(215), *287*, *293*
Postgate, J. R., 357(121, 122), 358(130), *371*
Potter, A. L., 379(38), *387*
Potter, R. L., 89(185), *116*
Potter, V. R., 296(10), 305(10), 322(10), 323(129), *331*, *334*, 545(386), *574*
Pouwels, P. H., 512(310), *572*
Powell, D., 265(93), *289*
Powell, J. F., 200(173), *247*, 427(62), *466*
Powelson, D., 458(153), *468*
Pratt, D., 619(365), *630*
Preiss, H., 407(73, 74), *411*
Preiss, J., 268(105a), *290*, 506(290), 512 (121), 513(121, 324), 526(121), 545 (394), *568*, *572*, *573*, *574*
Prescott, D. M., 589(120), *625*
Prescott, J. M. 533(150), *568*
Prestige, L. S., 91(188), *116*, 505(101), 542(174), 550(174), 554, 555(204), *567*, *569*, *570*, 595(193), 597(207), 600 (222), 602(207, 222), 603(207), 604 (207), 607(207), 610(281, 285), 612 (222, 305), 613(324a), 614(108, 285), *624*, *626*, *627*, *628*, *629*
Price, L., 362(161), *372*
Pricer, W. E., Jr., 57, 106, 107, *113*, *117*, 204(195), *248*, 279(211), *292*
Primosigh, J., 416(18, 19), 417(19), *465*
Proctor, M. H., 125, 126(37), *162*
Pronyakova, G. V., 282(220a), *293*
Purko, M., 271(126, 127, 128), *290*
Putnam, E. W., 379(38), *387*

Q

Quastel, J. H., 153, 155, *164*, 178(38), *244*, 605(264), *628*
Quastler, H., 614(328), *629*
Quayle, J. R., 17(58), 24, 25, 26, *38*, *39*, 65(98), 69, 70(98, 103, 108, *114*, *117*)

R

Raacke, I. D., 515(345), 559(431, 560, *573*, *575*, 577(2), *621*
Rabinowitch, E. I., 336, 344(9), *368*
Rabinowitz, J. C., 106, 107, *117*, 131(61), *163*, 204(195), *248*, 279(211), *292*
Rabinowitz, M., 499(86), *567*
Rabson, R., 557(424), *575*
Rachmeler, M., 234(314), *251*, 604(256), 615(331), *628*, *629*
Radding, C. M., 553(421), *575*
Radhakrishnan, A. N., 186(92), 211, 214(235, 236), *245*, *248*, *249*
Radin, N. S., 270(114), *290*
Rafelson, M. E., 236(318, 319, 320, 321), *251*
Rafelson, M. E., Jr., 215(242), *249*
Raggio, M., 125(29), *162*
Raggio, N., 125, *162*
Ragland, J. B., 207(213c), *248*
Rainbow, C., 319(93), *333*
RajBhandary, U. L., 420(35), *466*
Rajewsky, B., 614(325), *629*
Ralston, D. J., 586(85), *624*
Ramakrishnan, C. V., 617(349), *630*
Ramamutri, K., 382(58), *387*
Ramasarma, G. B., 183(70), *245*
Ranby, B. G., 380(45), *387*
Randles, C. I., 376(19), *386*
Rankin, J. C., 374(8), *386*
Ranson, S. L., 36(126), *40*, 62(84), 63 (84), *114*
Rao, M. R. R., 357(124, 125), *371*
Rappoport, D. A., 365(170, 171), *372*
Ratner, S., 186, 191(117), *245*, *246*, 390 (1), *409*
Ravel, J. M., 255(21), 271(21), 272(21), *287*, 518(125), *568*
Ravin, A. W., 594(181), *626*
Rawlinson, W. A., 352(106), *370*
Razin, S., 272(134), *290*

Razzell, W. E., 328(162), *334*
Record, B. R., 405(72), 406(72), 407(72), *411*
Redfield, B., 283(229b, 229c), *293*
Redfield, R. R., 477(13), *565*
Reed, L. J., 283(233), 284(233, 249, 250, 251, 252, 252a), *293, 294*
Rees, M. W., 473(5), 474, *565*
Reese, R. T., 383(72), *388*
Regan, M., 91(189), *117*
Reichard, P., 160, *166*, 318(84), 320(102, 111), 322(120), 323(131), 324(141), 327(154, 155), 329(102), 330(102), *333, 334*
Reid, L., 8(41, 42, 43), 10(41, 42, 43), *38*
Reiner, J. M., 600(226), *627*
Reiner, L., 44(21), *112*
Reio, L., 89(179, 180, 181), *116*, 208(215), 212(215), 224(276, 277), 236(320), *248, 250, 251*
Reithel, F. J., 596(204), *627*
Remberger, V., 80, 97, *115*
Remy, C. N., 308(53), *332*
Rendi, R., 509(305, 306), 536(305, 306), *572*
Repaske, R., 139, 145(78), *163*
Revel, H. R. B., 319(91), *333*
Reynolds, E. S., 320(114), *333*
Rhuland, L. E., 199(163), 200, *247*, 440 (94), 461(94), *467*
Rice, S. A., 553(418), *575*
Richmond, M. H., 446(117), 458(149, 150), *468*, 487, *566*, 581(16), *622*
Richter, G., 589(122), *625*
Rickenbach, J., 542(172), *569*
Rickenberg, H. V., 593(172), 597(208), 600(225), 612(301), *626, 627, 629*
Rickes, E. L., 516(347), *573*
Riedmüller, L., 357(117), *371*
Rieth, W. S., 537, *574*
Rievra, A., Jr., 233(308a), *251*
Riley, L. S., 437(79), 440(79), 444(79), *467*
Riley, M., 583(54), 587(54), 589(54), 614 (108), 615(54), *623, 624*
Rilling, H. C., 76(129), 77, *115*
Rimington, C., 336(6), 341(33), 348(76), 349(90), 350(92), 351(99, 104), 352 (104, 106), 365(168b), 366(168b), *368, 369, 370, 372*
Ringelmann, E., 74(124), 76(124), 92 (124), 93(124), 95(124), 97(124), 98 (124), *115*
Risebrough, R. W., 561(453), 562(453), 563(453), *576*
Rist, C. E., 374(9), *386*
Ritt, E., 33(121), *70*
Rittenberg, D., 185(91), *245*, 348(78, 79, 80), 349(81), *370*
Ritter, G. J., 99(195), *117*
Rizki, M. T. M., 366(174, 175), *372*
Robbins, P. W., 207(213b), *248*
Roberts, E., 183(70), 236(325), *245, 251*, 272(132), *290*, 398(42), *411*
Roberts, E. C., 197(147), 204(147), *247*
Roberts, E. R., 139, 141, *163, 164*
Roberts, I. Z., 559(434), *575*
Roberts, R. B., 18(70), *38*, 173, 195(133), 196(133), 200(133), 202(16), *244, 246*, 311(63), 313(63), 327(63), *332*, 483 (82, 237, 238), 484(238), 497(265), 500, 501(275), 514(265, 275), 536, 555 (238), 559(434), 560(275), 561(265), 564, *567, 569, 571, 575, 576*, 590(138, 140), 603(138, 242), *625, 627*
Robertson, J. J., 602(240), 603(240), *627*
Robinson, F. A., 259(40, 44), 262(67), *288*
Robinson, W. G., 75(126, 127), 76(126, 127), *115*
Roelofsen, P. A., 6, 7(29), *37*
Roepke, R. R., 203(189), 207(211), *248*, 615(325), *630*
Roessler, W. G., 396(33), *410*
Rogers, H. J., 320(105), *333*, 416(14), 439(83), 459(164, 165), *465, 467, 469*, 491(55), 538(55), 554(55), *566*
Rogers, L. L., 275(165), *291*
Rogers, M. M., 479(23), *565*
Rogers, P., 494, 498, *571*, 607(272), 615 (334), *628, 630*
Rogovin, S. P., 376(17), *386*
Rolfe, R., 587(100), 588(100), *624*
Roncoli, G., 359(134), *371*
Roodyn, D. B., 487(248), *571*

Rose, I. A., 65(95), *114*
Roseman, S., 417(21), 442(21, 96, 97), 444(97), *465*, *467*
Rosenberg, A., 442(99), *467*
Rosenberg, J. L., 14(54), *38*
Rosenberger, R. F., 360(152), *372*
Rosenblum, C., 282(224), *293*
Rosenblum, E. D., 136(69, 73), 140(84), 146(110), *163*, *164*
Rosenthal, S. M., 192(123), *246*
Rosenthal, W. S., 401(60), 408(60), *411*
Rosner, J., 604(256), *628*
Ross, S. H., 232(304), *250*
Rothaas, A., 365, *372*
Rotman, B., 492(63), 503, *566*, 595(195), 602(238), *626*, *627*
Rottenberg, M., 300(26), *331*
Roughton, F. J. W., 49, *113*
Roush, A. H., 318(86), 328(86), *333*
Rowatt, E., 339(25), 340(25), *368*
Roxburgh, J. M., 70, *114*
Ruben, S., 43(11, 13, 14), 47(27), *112*
Rudman, D., 180, 181(56), 192(56), 202(56), 203(56), *245*
Rudzinska, M. A., 345(68), *370*
Rüegg, R., 285(255), *294*
Ruhland, W., 3(7), *37*
Rushizky, G., 614(108), *624*
Russell, C. S., 349(82, 84), *370*
Russell, H., 423(43), 425(43), *466*
Russi, S., 268(105c), *290*
Rutberg, L., 323(131), *334*
Rutman, R. J., 62, *114*
Rutter, W. J., 53, 54(57), *113*
Ruxconi, A., 453(143), *468*
Ryan, A., 542(172), *569*
Ryan, F. J., 257(30), 270(30), *287*, 594(184), 615(338), *626*, *630*
Rydon, H. N., 390(8), 405, 406(8), *410*

S

Saad, F., 512(318), *573*
Sable, H. Z., 297(14), 298(14, 18), 317(14), 327(159), *331*, *334*
Sagers, R. D., 106, 107, *117*
Sahashi, Y-., 282(224a), *293*
St. Clair, J., 459(159), 461(159), 462, *468*
St. Lawrence, P., 581(23), *622*
Saito, M., 439(88), 440(89, 90), *467*, 538(166), *569*
Saito, Y., 108(220), *117*
Sakami, W., 204(193), *248*
Salamon, I. I., 222(268), *250*
Sall, T., 461(178), *469*
Sallach, H. J., 205, *248*
Salles, J. B. V., 50, 51(43), 53(56), 54(43), 65(43), *113*
Salton, M. R. J., 199(158), 236(324), *247*, *251*, 415, 416, 418(4, 23), 461(171), *465*, *469*, 478, *565*
Saluste, E., 8(42, 43), 10(42, 43), *38*, 89(179, 180, 181), *116*, 208(215), 212(215), 236(318), *248*, *251*
Sampson, W. L., 285(259), *294*
Samuels, P. J., 185(83), *245*, 536, 554(202), *569*, *570*
Sanadi, D. R., 88, *116*, 284(245, 246, 247, 248), *294*
Sanchez, C., 593(161), 609(161), *626*
Sanderson, A. R., 420(34, 35, 36), 421(34), 422(34, 36), 444(107, 108), 458(34), *465*, *466*, *467*
Sandheimer, E., 183(72), *245*
Sands, M. K., 207(213), *248*
Sanger, F., 475, 476(8, 9, 10), 477(220, 221), *565*, *570*
Sano, S., 352(104a), *370*
San Pietro, A., 17(66, 67), 30(66, 67), *38*
Santer, M., 5, 20, 22, 23, *37*, *39*
Santer, U. V., 366(178, 178a), *372*
Sarett, H. P., 262(60, 68), *288*
Sato, R., 227(290), *250*, 360(155), *372*
Saukonnen, J. J., 437(80a), *467*, 487(250), *571*, 590 (137), *625*
Sawin, P. B., 593(159), *626*
Sawyer, E., 353(111), *371*
Saz, H. J., 53, 56(61), *113*
Scarano, F., 324(137), *334*
Schachman, H. K., 495, 553, *567*, *575*
Schaechter, M., 486(445, 446), *576*
Schaeffer, P., 345(61), 346(61), 347(73, 74), 360(73), *369*, *370*, 582(41), *623*
Schaffer, N. K., 477(219), *570*
Schatz, A., 4(20), 32(20), *37*
Schen, S. C., 154(127, 128), *165*
Scher, W. I., 189(108), *246*
Schicke, H. G., 278(192), *292*
Schink, C. A., 255(23), *287*
Schlenk, F., 298(17), 327(156), *331*, *334*
Schlessinger, D., 483(233), 495(233), 501(277), 515(277), 522(277), 526(277),

556(277), 557(277), *570*, *572*, 590(139), *625*
Schlossmann, K., 207(212), *248*
Schmidt, G., 373(3), *386*
Schneider, K. C., 122(5), 127(46), 135 (67), 139(5), 145(109), 149(5), 150 (5), 151(5), *162*, *163*, *164*
Schneider, L. K., 257(30), 270(30), *287*, 615(338), *630*
Schnellen, G. T. P., 23(90, 95), *39*
Schoenheimer, R., 182(63), *245*
Schopfer, W. H., 260(47), 265(92), *288*, *289*
Schramm, G., 482(227), 559(210), *570*
Schramm, M., 380(44, 46), 381(46, 50, 52, 53), *387*, 582(38), *622*
Schrecker, A. W., 265(88a), *289*
Schulman, M. P., 300(30), *331*
Schuster, C. W., 205(206), 206(206), *248*
Schwabacher, H., 341(33), *369*
Schwartz, S., 282(222), *293*
Schweet, R. S., 492(133), 506(288), 507 (295), 508(295), 512(311), 513(323), 514(327), 515(342), 521, 522(133), *568*, *572*, *573*
Schweigert, B. S., 267(100), *289*, 298(20), *331*
Schwerdt, R. F., 60(74), *114*
Schwinck, I., 229(292), *250*
Scott, D. B. M., 60(76), *114*
Scott, J. F., 510(116), 511(119), 512(116, 331), 514(116), 545(119), *568*, *572*
Scott, J. J., 349(83, 89), *370*
Scott, S. S., 416(8), 434(76), 440(8), 441 (8), 453(76), 454(76), 464(189), *465*, *467*, *469*
Scutt, P. B., 152(114), *164*
Seaman, G., 518, *568*
Seaman, G. R., 284(239, 253), *294*
Searle, D. S., 44(21), *112*
Searls, R. L., 284(246, 248), *294*
Seeger, D. R., 279(195), *292*
Semb, J., 279(195), *292*
Semenza, G., 91, *116*
Senez, J. C. 357(123), *371*
Senti, F. R., 376(17), *386*
Seraidarian, K., 312(68), *332*
Serlupi-Crescenzi, G., 268(107), *290*
Seubert, W., 80, 97, *115*
Sevag, M. G., 278(193), *292*, 594(175), 601(175), *626*
Sewell, C. E., 154(126), *165*
Shafa, F., 461(171), *469*
Sharon, N., 506(109, 282), *567*, *572*
Sharpe, E. S., 376(18), 377(18, 27), *386*
Shatton, J. B., 214(231, 232, 233), *249*
Shaw, D. C., 477(221), *570*
Shaw, D. R. D., 457(147), *468*
Shedlovsky, A., 316(80), *332*
Sheffner, A. L., 602(239), *627*
Sheinin, R., 597(210), *627*
Shemin, D., 185(91), *245*, 282(221, 223), *293*, 348, 349(75, 81, 82, 84, 88b), 353 (88b, 108, 109), *370*
Shen, S. C., 179, *244*, 585(68), 613(68), *623*
Shepherd, C. J., 459(166), *469*, 491(57), 528(57), *566*
Shepherd, R. G., 479(23), *565*
Shepherdson, M., 320(112), 322(112), *333*, 593(171), 604(171), 606(171), 607 (171), *626*
Sher, I., 595(189), *626*
Shiba, S., 589(119), 610(119), *625*
Shigura, H. T., 172(15), 224(15), 226 (15), *243*
Shilo, M., 603(245), *628*
Shimanuki, H., 408(89), *412*
Shimazono, H., 108(220), *117*
Shimazono, N., 261(58a, 58b), *288*
Shimizu, F., 442(97), 444(97), *467*
Shimono, M., 238(333), *251*
Shimura, K., 196(138, 139), 214(234), *246*, *249*
Shimwell, J. L., 380, *387*
Shintani, S., 262(64), *288*
Shiota, T., 278(194, 194c), *292*
Shirk, H. G., 381(54), *387*
Shive, W., 197(147), 204(147), *247*, 255 (21, 22), 260(48), 271(21, 22), 272(21, 22), 275(165), *287*, *288*, 291, 299(25), *331*, 518(125), *568*
Shockman, G. D., 437(78, 79), 440(78, 79), 443(101), 444(79), 453(140), *467*, *468*
Shoda, T., 277(190, 191), 278(190, 191), *292*
Shore, V. G., 505(101), *567*
Shuey, E. W., 464(192, 194), *469*
Shug, A. L., 126(41), 128(41), 131(55), 132(63), 133(63), 142, 147, 151(41, 55), *162*, *163*, *164*

Shuster, C. W., 91, *117*
Sickels, J. P., 279(195), *292*
Siebert, G., 58, *113*
Siegel, A., 559(211), *570*
Siegel, J. M., 8(40), 36(126), *37*, *40*
Siekevitz, P., 515(341), *573*
Sifferlen, J., 345(60), 346(60), *369*
Silber, R. H., 401(59), 408(59), *411*
Silverman, M., 189(194), 204(194), *248*, 261(59), *288*
Silvester, D. J., 130(54), *163*
Simet, L., 477(219), *570*
Simkin, J. L., 518(357), 559(435), *574*, *575*
Simmer, H., 267(101), *289*
Simmonds, S., 205(199), *248*, 473(4), 474, 532, *565*, *568*
Simms, E. S., 297(15), 318(87), 322, 324 (87), 325(87), 327(151), 328(163), *331*, *333*, *334*, 552(178, 179), 553(411, 412), *569*, *575*
Simon, H., 279(198), *292*
Simpson, J. R., 31, *39*
Simpson, M. S., 308(55), 310(55), *332*
Simpson, M. V., 524(130), *568*
Singer, B., 559(212, 213), *570*
Singer, M. F., 549(402, 403), *575*
Singer, T. P., 268(106), *290*, 359(142b), *371*
Singh, R. N., 127(46), *163*
Sinsheimer, R. L., 553(420), *575*, 586 (73), 588(73, 105), *623*, *624*
Sistrom, W. R., 344(50, 53a), 363(53a), 365(50), 367(50), *369*
Siu, P. M. L., 69, 82, 91(100), *114*
Sjöstrand, F. S., 497, *571*
Sjolander, J. R., 214(237), *249*
Skaggs, P. K., 262(61), *288*
Skarnes, R. C., 407(78), *412*
Skeggs, H. R., 89(183), *116*, 276(175), *291*, 320(107, 108), *333*
Skipper, H. E., 301(39), 308(55), 310(55), *332*
Skoda, J., 549, *575*
Sköld, O., 327(154, 155), *334*
Slade, H. D., 43(12), *112*
Slaughter, C., 196(140), *246*
Slebodnik, J., 359(142b), *371*, 601(229), *627*
Sleeper, B. P., 602(237), *627*
Slenczka, W., 33(121), *40*
Sloan, J. W., 374(9), *386*
Slocum, D. H., 157(155), *165*
Slonimsky, P. P., 345(66), 346(66), 347 (66), 359, *370*, *371* 612(312), *629*
Slotin, L., 43(15), 47, 49(38), 64(38), *112*, *113*
Sluyser, M., 512(309, 310), *572*
Smadel, J. E., 554(199), *570*, 610(283), *628*
Small, M. H., 203(189), *248*, 615(325), *630*
Smith, E. E. B., 446(112), *467*, 586(76, 77), *623*
Smith, G. N., 534(196, 198), *570*
Smith, H., 406(82), 407, *412*
Smith, J. D., 296(4), *331*, 481, 482(230), 483(230), 484(230), 487, 540(46, 47, 170), 541(170), *565*, *566*, *569*, *570*
Smith, J. H. C., 345(53b), 362(162a, 164a), 363(166a), *369*, *372*
Smith, J. M., 279(195), *292*
Smith, K. C., 512(311), *572*
Smith, L., 357(118), 365(168a), *371*, *372*
Smith, M. I., 262(69), *288*
Smith, M. S., 308(53), 323(133), *332*, *334*, 446(114), *467*
Smith, N. R., 390(12), 391(14), *410*
Smith, R. A., 205(206), 206(206), *248*, 284(244), *294*, 353(111), *371*
Smith, R. M., 85(159), *116*
Smith, W., 352(106), 361(159), *370*, *372*
Smyrniotis, P. Z., 17(59, 60), 19(60), *38*, 59, 60(71), *113*
Smyth, R. D., 85(157), *116*, 156(141), *165*, 198(154), *247*, 281(214a, 214b, 214c, 215b), *292*, *293*
Snell, E. E., 181(60, 61), 197(61), 211, 214(235, 236), 219, 229, 236, *245*, *248*, *249*, *250*, *251*, 269(109), 270, 272(143), 273(143, 145), 274(145, 147), 284(237), *290*, *291*, *293*, 298(19), 319, 328(19), *331*, *333*, *334* 443(102), 448(120), 461 (175, 176), *467*, *468*, *469*, 533(147, 148, 149, 150, 368, 369), 534, *568*, *574*, 612 (308), *629*
Snoke, J. E., 185(84), *245*, 451(127), *468*
Snow, G. A., 285(258, 262), *294*
Sober, H. A., 181(62), 185(87), 203(187), *245*, *248*, 270(119), *290*
Soda, J. A., 200, *247*
Sohngen, K., 3(8), 23(8), *37*

Sollo, F. W., Jr., 109(226), *117*
Solomon, A. K., 43(16), *112*
Somerville, R., 325(147), *334*, 586(89), *624*
Sorm, F., 549(405), *575*
Sormova, Z., 549(405), *575*
Sowden, F. J., 342(46a), *369*
Spahr, P. F., 482(230), 483, 484, *570*, *571*, 589(127), *625*
Spalla, C., 283(227a, 227b), *293*
Sparks, C., 308(55), 310(55), *332*
Speck, J. F., 53(58), *113*, 157, 158(150), *165*
Spector, L., 159(165, 166), 160, *166*, 190 (112), 191(112, 113), *246*, 320(110), *333*
Spell, W. H., Jr., 298(21), *331*
Spencer, D., 125(32), 126(32), 140(85), *162*, *164*
Speyer, J. F., 409(94, 95), *412*, 564(460, 461), *576*
Spicer, C. C., 615(333), *629*
Spicer, D. S., 320(107), *333*
Spiegelman, S., 484, 487, 488, 493(63, 73), 494, 497(265), 500, 503, 504(95, 96), 515(265), 530, 531, 534(95, 96), 535 (152), 545, 552(410), 561(265, 410), *566*, *567*, *569*, *571*, *575*, 589(118), 590 (133, 134, 140), 593(163), 594(163), 595(192, 195), 599(221), 600(226), 601 (221, 230), 610(118), 612(192), 614 (328), *625*, *626*, *627*, *629*
Spilman, E. L., 300(31), *331*
Spirin, A. S., 588(112), *624*
Spizizen, J., 586(78), *624*
Sprague, J. M., 89(183), *116*
Sprecher, M., 172(15), 224(15, 278), 226 (15), *243*, *250*
Sprince, H., 532(144), *568*
Sprinson, D. B., 18(71), *38*, 172(15), 224 (15, 278), 225(279), 226(15, 280, 281), 227, 236, *243*, *250*, 282(224), *293*
Squires, C., 78(141), *115*, 207(213a), *248*
Srinivasan, P. R., 18(71), *38*, 172(15), 224(15, 278), 225(279), 226(15, 280, 281), 227(285, 286), 233(308a), 235, 236(285), *243*, *250*, *251*, 276(183a), *291*, 584(64), *623*
Stacey, M., 374(7, 10), 377(26, 27, 29), 378(7), *386*, *387*
Stadtman, E. R., 82, 85(151, 160), *115*, *116*, 157, *165*, 194(128a), *246*, 255(17), 263(17), 265(17), 272(131c, 136), *287*, *280*, 314(79), *332*
Stadtman, T. C., 23(93), *39*, 109, *117*, 204 (191), *248*, 281(213a), *292*
Stahly, G. L., 376(19), *386*
Stahmann, M. A., 402, *411*
Stanier, R. Y., 7, 10, 35, 36, *37*, 239, *251*, 344(50, 53a), 345(53b), 362(163, 164a), 363(53a, 166a), 365(50), 366(50), *369*, *372*, 495(83), *567*, 593(163), 594(163, 177, 187), 602(237), 603(245), 613 (321), 615(332), 616(343), 617(345), *626*, *627*, *628*, *629*, *630*
Stanley, J. L., 407(83), *412*
Stanley, W. M., 473(6), *565*
Starkey, R. L., 3(10), 4(14, 15), *37*
Starlinger, P., 583(56), 587(93), *623*, *624*
Starr, J. L., 512(308), *572*
Starr, T. J., 354(115), *371*
Staub, A. M., 425(50, 51), *466*
Steel, R., 617(349), *630*
Steenholf-Eriksen, T., 153(117), 154(117), *164*
Steers, E., 312(70), *332*
Stein, W. H., 477(214), *570*
Steinberg, D., 449(124), *468*, 579(8), 580(8), *622*
Steinberger, R., 50, *113*
Stent, G. S., 587(94, 103), 589(128), 615 (128), 619(365), *624*, *625*, *630*
Stephenson, M., 594(186), 605(261), 610 (287), *626*, *628*
Stephenson, M. L., 491(59), 510(59, 116), 511(59, 118, 119), 512(116, 312, 313, 314), 514(116), 545(118, 119, 388), *566*, *568*, *572*, *574*
Stepka, W., 13(52), *38*
Stern, J. R., 79, 95, 96, *115*, *117*
Stern, W., 592(158), *625*
Sterne, M., 391(15), 396(15), 407(15), 408, *410*
Stetten, M. R., 182(63), *245*, 299(24), *331*
Stevens, A., 546(399), *575*
Stevenson, G., 125(28), *162*
Steward, F. C., 157(145), *165*
Steyn-Parvé, E. P., 261(56), *288*
Stich, H., 589(124), *625*
Stich, W., 345(65), 346(65), 347(65), *369*
Stickland, R. G., 63(88), *114*

Stierlin, H., 425(47, 48), 464(48, 200), *466*, *469*
Stjernholm, R., 8(42, 43), 10(42, 43), *38*, 82(150), 89(179, 180, 181), *115*, *116*, 208(215), 212(215), *248*
Stjernholm, R. L., 69(100), 82(100), 83, 84(155, 156), 85(154), 91(100, 156), 97(156), *114*, *116*
Stocker, B. A. D., 493(68), *566*
Stodola, F. H., 377(27), *387*
Stoeber, F., 600(223), 601(223, 235), *627*
Stokes, J. L., 275(162), 277(189), *291*, *292*, 357(125), *371*, 612(307, 318), *629*
Stokstad, E. L. R., 279(195), 284(234, 238, 240), *292*, *293*, *294*
Stone, F. M., 345(70), *370*
Stone, R. W., 47(23), *112*
Stoppani, A. O. M., 22, 27, *39*, 65(97), *114*
Strange, R. E., 199(160), 200(173), *247*, 427, 446, 458(65, 151), *466*, *467*, *468*
Strassman, M., 208, 210(216), 212, 213, 214(231, 232, 233), 215(239), 216, *248*, *249*
Straub, F. B., 557(427), *575*
Strauss, B. S., 581(21), 583(21), 591(142), *622*, *625*
Strecker, H. J., 87, *116*, 174(22), 185(22), *244*
Strehler, B. L., 203(190), *248*
Streightoff, F., 256(26), *287*
Stringer, C. S., 376(17), *386*
Strittmatter, C. F., 359(136), *371*
Strominger, J. L., 323(128), *334*, 414(2), 416(7, 8), 417(22), 420(34, 36), 421(34, 41), 422(22, 34, 36), 423(7, 22, 44), 424(22), 426(44, 60), 427(59), 428(7, 22, 44, 68), 429(70), 430(70), 431(70, 71), 433(70), 434(68, 76), 435(77), 436 (77), 437(71), 438(71), 439(71), 440(8), 441(8), 442(100), 443(41, 70, 100), 444 (70, 71, 106), 445(2), 446(71, 114, 116), 447(118, 119), 448(118, 119, 121), 449 (119, 123), 450(123), 451(131), 452(132), 453(76), 454(76, 146), 456(146), 457 (2, 44, 60), 458(2, 60), 458(34, 147a), 459(169), 460(169), 464(189, 190, 191, 197, 204), *465*, *466*, *467*, *468*, *469*, 479, 536(363), 538, 539, 540, *565*, *569*, *570*, *574*
Strumeyer, D. H., 451(128), *468*
Stulberg, M. P., 506(282), *572*
Stumpf, P. K., 78(141), *115*, 399(48), *411*
Stutts, P., 327(152), *334*
SubbaRow, Y., 279(195), *292*, 479(23), *565*
Sueoka, N., 591(144), 608(144), *625*
Sugarman, G. I., 540(169), *569*
Sugisaki, Z., 238(335), *251*
Sulkowski, E., 524(132), *568*
Sumiya, C., 352(104a), *370*
Sunakawa, S., 550(186), *569*
Sund, R. F., 518(125), *568*
Sung, S. C., 57(67), *113*
Suskind, S. R., 234(312), *251*, 585(70), 592(70), *623*
Sussman, M., 600(226), *627*
Suzuki, I., 21, 30(82), *39*, 62(86), 63, 65 (86), *114*, 211(222), *248*
Suzuki, N., 464(202), *469*
Suzuki, S., 464(202), *469*
Swanson, A. L., 554(196), *570*
Swenson, P. A., 613(322), 614(322), *629*
Swick, R. W., 73(119), 81, 82, 84(119), 85(160), *115*, *116*
Swoboda, O. P., 277(187), 279(187), *292*
Sykes, J., 497(268, 269), 555, *570*, *571*, 596(198), *626*
Synnatschke, G., 278(193a), *292*
Syrett, P. J., 604(257), *628*
Szafranski, P., 524(132), *568*
Szilard, L., 607(275), 610(275), 618(356), *628*, *630*
Szongott, H., 390(4), 407(4), *410*
Szulmajster, J., 198(153), *247*

T

Tabor, C. W., 192(123), *246*
Tabor, H., 192(123), 219(257), *246*, *249*
Tacquet, A., 341(30), *368*
Taguchi, T., 589(1119), 610(119), *625*
Takagi, Y., 439(86, 87), *467*, 545(38b), *574*
Takahashi, H., 122(2), 123(2), 140(85), *161*, *164*, 360(156a), *372*
Takanami, M., 514(333), *573*
Takeda, I., 263(78), 264(81), 265(78, 81, 81d), *289*
Takeda, Y., 360(154), 361(154), *372*
Takeyama, S., 198(150, 155), 204(150), *247*
Tal, M., 497(264), *571*

Talmage, P., 353(109), *370*
Tamiya, H., 537(379), *574*
Tanabe, Y., 352(104a), *370*
Tanaka, K., 237(332), *251*
Tanaka, R., 261(58a), *288*
Tanner, F. W., Jr., 254(9, 10), *287*
Tarnanen, J., 156, *165*
Tarver, H., 505, *567*
Tatum, E. L., 171, 172(9, 10), 214(9, 10, 237), 222(267), 229(295), *243*, *249*, *250*, 260(45), 275(164), *288*, *291*, 532 (135), *568*
Tauber, H., 423(43), 425(43), *466*
Taussig, A., 604(255), *628*
Taylor, S. P., 532(139), *568*
Tchen, T. T., 61, 62, 63, 65, 66, *114*
Teas, H. J., 195(129, 131), *246*
Teltscher, H., 234(315), *251*
Ten Hagen, M., 342(45), *369*
Terawaki, A., 589(119), 610(119), *625*
Terroine, T., 266(96), *289*
Terui, G., 580(11), *622*
Thayer, S. A., 285(254), *294*
Thjötta, T., 340, *368*
Thofern, E., 339(34), 341, 360(151a), 361 (34, 35), *369*, *371*, 604(249), *628*
Thoma, R. W., 276(174), *291*
Thomas, A. J., 208(216), 210(216), 212 (224), 215(239), 216(239), *248*, *249*
Thomas, J. O., 43(14), *112*
Thompson, E. D. P., 475(9), 476(9), *565*
Thompson, J. F., 157(145), *165*
Thompson, R. C., 259(41), 274(41), 286 (41), *288*
Thompson, R. E., 423(44), 426(44), 428 (44), 457(44), *466*
Thorne, C. B., 184(80, 81, 82), 236(327, 328, 329), *245*, *251*, 390(9), 394(26, 27, 28), 395(27), 396(9), 397(37, 38, 39, 40), 398(9, 38, 39, 40, 44), 399, 400(26), 401(26), 402(27), 403, 405(46, 70), 406 (26), 407, 408, *410*, *411*, *412*, 453(134), *468*, 537, *596*
Thornley, M. J., 594(182), 614(330), *626*, *629*
Threnn, R. H., 416(8), 428(68), 430(71), 434(68, 76), 435(77), 436(77), 437(71), 438(71), 439(71), 440(8), 441(8), 444 (71), 446(71), 452(132), 453(76), 454 (76, 146), 456(146), *466*, *467*, *468*, 479 (223), 536(363), 538(167, 223, 380), 540(363), *569*, *570*, *574*
Tice, S. V., 185(87), 203(186, 187), *245*, *247*, *248*
Tietz, A., 71, 72(110), 73(121), 74(110, 121), 76, *114*, *115*
Tinelli, R., 425(50, 51), 464(204), *466*, *469*
T'ing-sen, H., 453(142), *468*
Tishler, M., 285(259), *294*
Tissières, A., 358(133), *371*, 482(231), 483, 484, 488, 495, 501, 515(277), 522, 526, 556, 557(277), 563, *567*, *570*, *571*, *572*, *576*, 582(32), 589(127), 590(139), *622*, *625*
Titani, K., 359(135a), *371*
Titchener, E. B., 74(123), 77(136, 137), *115*
Tobita, T., 596(200), *627*
Todd, C. M., 345(59), 346(60) *369*
Toennies, G., 437(78, 79), 440(78, 79), 443 (101), 444(79), *466*
Tolbert, N. E., 13(53), *38*
Tolmach, L. S., 588(107), *624*
Tomasz, A., 439(84, 85), *467*
Tomcsik, J., 390(4), 391(16, 18), 393(20, 21, 22, 23), 407(4), *410*
Tomisek, A. J., 301(39), *332*
Tomizawa, J., 550(186), *569*
Tonzetich J., 270(120), *290*
Toohey, J. I., 85(157), *116*, 281(214b, 215a, 215b), *293*
Torii, M., 390(11), 402, 403(11), 404(11), 405(11, 69), 408(11), *410*, *411*
Torriani, A. M., 379(37, 39), *387*, 586(83), 598(214), 613(83, 214), *624*, *627*
Touster, O., 443(104), *467*
Towne, J. C., 22(83), 23(83), *39*
Townsley, P. M., 345(63), 346(63), 354, 356(63), 358(63), *369*
Trams, E. G., 79(145), *115*
Treble, D. H., 263(74, 75), 264(74), *289*
Trebst, A. V., 11, 29(50), 34(124), *38*, *40*, 277(187), 279(187), *292*
Trice, S. V., 270(119), *290*
Tridgell, E. J., 598(218), *627*
Tristram, H., 616(344), *630*
Trucco, R. E., 459(170), *469*
Trudinger, P. A., 19, 20, 21, 30, 31, *38*, *39*
Tsuchida, M., 602(237), *627*
Tsuchiya, H. M., 376(17, 18), 377(18, 25), *386*, *387*

Tuboi, S., 509(307), 536(307), *572*
Tuppy, H., 475(8), 476(8), *565*
Turba, F., 535, *569*, *574*
Tuttle, L. C., 131(57), 158(156), *163*, *165*
Twort, F. W., 285(265), *294*

U

Uchida, M., 156(137, 138), *165*
Udaka, S., 188, 189(103), 192(122), 237 (332), 238(333, 334, 336), *246*, *251*
Udenfriend, S., 349(85), *370*
Ueda, K., 260(48e), *288*
Ullman, A., 557(427), *575*
Umbarger, H. E., 172(8), 174(20, 23, 24), 178(45), 193(45), 205(20), 210(218), 211 (24, 221), 213(23, 226, 228), 214 229, 230), 216(243a, 243b), 242(221), 243(226), *243*, *244*, *248*, *249*, 612(316), 617(350), 618(358), *629*, *630*
Umbriet, W. W., 4, 5, 31(17), *37*, 270 (117), *290*
Ushiba, D., 326(148), *334*, 595(190), *626*
Utsumi, S., 390(11), 402, 403, 404, 405, 408(11), *410*, *411*
Utter, M. F., 50, 53(53, 54), 55(54), 63 (54), 64, 65, 66, 67, 68, 71, 79, 80, 86, 87(166), 88, 89, 91(45), 97(32), 106, *113*, *114*, *116*, 414(1), 457(1), *465*

V

Vagelos, P. R., 78, *115*, 157, *165*, 272 (131c), *290*
Valentik, K. A., 320(107), *333*
Valentine, R. C., 87(171), *116*, 131(59), *163*
Valiant, J., 276(179), *291*
Vallee, G., 311(62), 318(62), *332*, 604(253), *628*
van der Grinten, C. O., 537(376), *574*
Vanderwerff, H., 312(66, 67), *332*
van de Ven, A. M., 506(106, 289), 537 (376), *567*, *572*
Van Eys, J., 443(104), *467*
van Genederen, H., 10(45), 26(45), *38*
van Halteren, M. B., 535, *569*
van Heyningen, W. E., 358(131), *371*
Van Lanen, J. M., 254(9, 10), *287*
van Niel, C. B., 5, 7, 8, 24(97), 28(32), 34, 35(32), *37*, *39*, 43(14), *112*, 342, 343, 364(47), *369*
Van Rotterdam, J., 477(218), *570*
Varner, J. E., 157(154, 155), 159 *165*, *166*
Vaughan, M., 579(8), 580(8), *622*
Vaughn, R. H., 357(126), *371*
Vennesland, B., 43(16), 49(38), 53(58), 58, 61(81), 62(81, 85), 63(81), 64(38, 91), 65(81, 85, 94), 66(81), *113*, *114*
Verhoeven, W., 360(153, 154), 361(154, 157), *372*
Vernon, L. P., 10, 30, 31(115), *38*, *39*
Victor, J., 401(58), 408(58), *411*
Vincent, W. S., 561(443), *576*, 590(131), *625*
Vinograd, J., 515(134), 521(134), *568*
Virtanen, A. I., 43, 84, *112*, 121(14), 140 (14), 156, *162*, *165*, 272(130), *290*, 532 (140), *568*
Vishniac, W., 4(21), 5, 15, 17, 18, 20, 22, 23, 27, 31(119), 32(21), 35, *37*, *38*, *39*, 69, *114*
Vogel, H. J., 174(17), 176(28), 185(90), 186, 187(97, 99), 188(99, 100, 104), 189, 192, 199(170), 208(214), 217(244), 239(170), 243(17), *244*, *245*, *246*, *247*, *248*, *249*, 257(27), 258(27), 259(27), 271(20, 27), *287*, 366(178), 368(181a), *372*, 593(164), 606, 607(164, 269), *626*, *628*
Vogler, K. G., 4, 5, 31(17), *37*
Voinovskaya, K. K., 342(48a), *369*
Volcani, B. E., 85(157), 98(194), 99(202), *116*, *117*, 281(214b, 214c, 214d, 215b), *293*, 300(28), 306(28), *331*, 400, 401 (54), 403, 405, *411*
Volkin, E., 317(81), *333*, 551, 561(192, 193, 194, 408, 409, *569*, *575*, 590(130), *625*
Von der Decken, A., 514(334, 335, 336), *573*
von Euler, H., 153, 154(117, 118, 119), *164*
von Saltza, M. H., 284(240), *294*

W

Wachsmann, J. T., 156(139), *165*, 531, *574*
Wacker, A., 277(187, 188), 279(187, 198), *292*, 312(74), *332*, 541(171), *569*
Waelsch, H., 158(158), *166*, 183(66), 184 (66), 219, 221, *245*, *249*, 320(101), *333*, 399(47), *411*, 532(142), *568*

Wagenknecht, A. C., 136(70), *163*
Wagle, S. R., 518, *568*, *574*
Wagner, R. P., 214(235), *249*, 581(20), 591 (20, 148), 593(20), *622*, *625*
Wahba, A. J., 324(142), *334*, 604(248), *628*
Wahlin, H. P., 127(43), *163*
Wainfan, E., 158(158), *166*, 399(47), *411*
Wainright, S. D., 604(259), 612(302), 617 (346), *628*, *629*, *630*
Wakil, S. J., 74, 77(136, 137, 138), 78, *115*
Waksman, S., 4(14), *37*
Waley, S. G., 404, 405(68), 406(68), *411*
Walker, D. A., 53(59), 62(84), 63, *113*, *114*
Walker, G. C., 122(2a), 123(2a), *161*
Walker, J., 279(196), *292*
Walker, J. B., 192(118), *246*
Wall, J. S., 136(70), *163*
Wallenfels, K., 596(203), *627*
Waller, C. W., 279(195), *292*
Waller, J-P., 497, *571*
Wallis, R. G., 405(72), 406(72), 407(72), *411*
Walton, C. J., 282(222), *293*
Wang, L. C., 127(46), *163*
Wang, T. P., 327(159), *334*
Warburg, O., 10(47), *38*
Ward, G. B., 273(145), 274(145), *291*
Wardlaw, A. C., 416(16), *465*
Waring, W. S., 358(129), *371*
Warner, A. C. I., 178, *244*
Warner, R. C., 58(69), 72(112), 74(112), *113*, *114*, 500(89), 547(451), 549(451), *567*, *576*
Wasserman, H. H., 365(168a), 366(178a), *372*
Watanabe, I., 586(84), *624*
Watanabe, Y., 196(138, 139), 214(234), *246*, *249*
Watson, C. J., 336(3), *368*
Watson, D. W., 390(10), 402(10), 407, 408(10), *410*, *411*, *412*
Watson, J. D., 481, 483(233), 495, 555, 561(453), 562(453), 563(453), *566*, *567*, *570*, *575*, *576*, 587(99, 101), 590(132), 620(132), *624*, *625*
Watson, M. R., 586(81), *624*
Watts-Tobin, R. J., 409(97), *412*, 564 (463), *576*
Webb, E., 474, *565*
Webb, E. C., 592(151), *625*
Webb, M., 215, *249*
Webster, G., 506(292, 293), *572*
Webster, G. C., 157, 158(147), 159, *165*, *166*, 506(283), 507(112), 508, 512, 515 521, 522, 523, 525, 557(122, 344), 560 (112), *568*, *572*, *573*, *574*
Webster, L. D., 506(285, 287), *572*
Wecker, A., 592(158), *625*
Weed, L. L., 320(108, 109), *333*
Weibull, C., 459(154), *468*, 493(71), 500 (91), *566*, *567*
Weidel, W., 416(18, 19), 417(19, 20), *465*
Weigl, J., 381(55), *387*
Weil, E., 407(79), *412*
Weill, J. D., 547(450, 451), 549(450, 451), *576*
Weil-Malherbe, H., 261(55), *288*
Wein, J., 404(67), *411*
Weinbaum, G., 612(300), *629*
Weiner, M., 582(45), 595(45), 602(45, 236), 606(45), *623*, *627*
Weinhouse, S., 204(196), 208(216), 210 (216), 212(224), 214(231, 232, 233), 215(239), 216(239), *248*, *249*
Weiss, B., 227(286), *250*, 276(183a), *291*
Weiss, C. M., 365(169), *372*
Weiss, S. B., 546, 547, 561(397, 449, 452), 506(286), *572*, *575*, *576*
Weiss, U., 171(6), 172(12), 223(6, 269), 226(12), 227(291), 229(291), *243*, *250*
Weissbach, A., 17(59, 60, 61), 19(60, 61) *38*, 63(89), *114*, 226(283), *250*, 281 (214a, 214b, 214c, 215a, 215b), *292*, *293*
Weissbach, H., 85(157), *116*, 156(141), *165*, 198(154), *247*, 283(229b, 229c), *293*
Weissmann, B., 296(3), *331*, 481(26), *565*
Weissman, N., 390(10), 402(10), 407(10), 408(10), *410*
Weitkamp, H., 278(192), *292*
Wells, P. A., 43(10), *112*
Welsch, M., 439(87a), *467*
Wende, R. D., 407(85), *412*
Werkman, C. H., 21, 30(82), *39*, 42(3), 43(12, 17), 44(3, 18, 19), 45(18), 46 (19, 22), 47, 48(30), 51(48), 62(86), 63, 65(86), 86(166, 169), 87(166, 169), *112*, *113*, *114*, *116*, 261(59), *288*, 358(129), *371*
Westall, R. G., 350(91), *370*

Westenbrink, H. G. K., 261(58), *288*
Westheimer, F. H., 50, *113*
Westlake, D. W. S., 127(45), 146(111), *163*, *164*
Westphal, O., 419(27), 423(27), 425, 464 (48, 199, 200), *465*, *466*, *469*
Weygand, F., 277(187), 279(187), *292*, 312(74), *332*, 541(171), *569*
Whatley, F. R., 29(106, 107, 108), *39*
Wheat, R. W., 268(105b), *290*
White, F., 284(245, 247), *294*
White J., 219(257), *249*
Whitehill, A. R., 342(44, 45), *369*
Whiteley, H. R., 85, *116*, 123(7), 140(7), 150(7), *162*, 279(202), *292*
Whitham, G. M., 362(162b), *372*
Whittingham, C. P., 10(48), *38*
Wiame, J. M., 153, 154, *164*, 176(34), 177, 179, 203(48), *244*, 367(180), *372*
Wiberg, J. S., 586(90), *624*
Wieghard, C. W., 421, *466*
Wieland, T., 274(148), *291*
Wieringa, K. T., 98, *117*
Wiesendanger, S. B., 194(127), 196(135), *246*
Wiesmeyer, H., 604(251), *628*
Wijesundera, S., 197(144), *247*
Wild, D. G., 497(269), *571*
Wildman, S. G., 559(211), *570*
Wilkins, C. N., Jr., 319(94), *333*
Williams, A. E., 136(71), *163*
Williams, H. H., 512(318), *573*
Williams, J. H., 342(44, 45), *369*
Williams, O. B., 274(154), *291*
Williams, R. C., 559(213), *570*
Williams, R. E. O., 615(333), *629*
Williams, R. J., 260(48), *288*
Williams, R. P., 286(271), *294*, 365(170, 171, 173a), 366(176, 177), *372*
Williams, V. R., 257(31), 276(181), *287*, *291*
Williams, W. J., 184(80, 81, 82), *245*, 399, 405(46), *411*, 537, *569*
Williamson, D. H., 268(104, 105), *289*, *290*
Wilson, A. C., 618(364), *630*
Wilson, A. T., 14(56), 16(56), 17(56), *38*
Wilson, D. C., 183(72), *245*
Wilson D. W., 320(108, 109), *333*
Wilson, J., 86, 87(169), *116*
Wilson, L. G., 207(213a), *248*
Wilson, P. W., 121(12, 13, 15), 124(13, 15, 20, 21), 125, 126(34, 35, 37, 39, 41), 127(21, 42, 43, 46, 50), 128(41), 129 (20), 130(20), 131(15, 55) 132(63), 133(63), 136(20, 69, 72, 73, 75), 138 (20), 139, 140(83, 84), 141(20), 142, 144(95, 96, 97, 98, 99, 102), 145(78, 105, 106), 147(55), 151(41, 55), 157 (73), *162*, *163*, *164*
Wilson, R. M., 156(140, 141), *165*, 281 (214b), *293*
Wilson, T. G. G., 139, *163*
Wilson, T. H., 49, *113*
Winderman, S., 611(294), *628*
Windsor, E., 199(168), *247*
Winfield, M. E., 146(113), 147, 148, 149, 151, *164*
Winkler, K. C., 197(148), 204(148), *247*
Winnick, T., 275(168), *291*
Winogradsky, S., 2, *37*, 42, *112*
Winterbottom, R., 479(23), *565*
Winterstein, A., 285(255), *294*
Wiss, O., 267(101), 285(255, 259a), 286 (273), *289*, *294*
Wisseman, C. L., Jr., 554(195, 199, 200, 201), *569*, *570*, 610(283), *628*
Woessner, J. F., Jr., 76, *115*
Wolf, D. E., 276(175, 179), *291*
Wolfe, R. G., 596(204), *627*
Wolfe, R. S., 87, *116*, 131(58, 59, 60), *163*
Wolff, I. A., 374(9), *386*
Wolff, J. B., 362(161), *372*
Wollman, E., 589(129), 595(129), *625*
Wollman, E. L., 582(41), 583(49), *623*
Wolstenholme, G. E. W., 578(4), 615(4), *622*, *630*
Wong, K. K., 507(298, 300), 513(300), *572*
Wood, H. G., 42(3), 43(12, 17), 44(3, 18, 19), 45(18), 46(19, 22), 47, 48(30), 50 (33), 64(91), 65, 69(100), 71, 73(119), 81, 82, 83, 84(119, 155, 156) 85(154), 87(172), 88, 89, 91(100, 156), 97(32, 156), 100, 101, 102, 103, 104, 106, *112*, *113*, *114*, *115*, *116*, *117*
Wood, R. C., 312(70), *332*
Wood, T. R., 270(123), 276(175), *290*, *291*
Wood, W. A., 60(74), 61(79), 88(79), *114*, 203(188), 213(227), 235(317), 236(188), *248*, *249*, *251*, 270(113), 271(126, 127,

128), *290*, 296(11), *331*, 397(37b), 398 (41), *410*, 452(133), *468*
Wood, W. B., 563, *576*
Woods, D. D., 197(144), 198(151, 152, 153), 204(151, 152), *247*, 254(8), 256 (24), 257(28), 270(115), 276(182), 277 (28, 184, 185, 186), 278(28, 184), 279 (184, 203), 283, *287*, *290*, *291*, *292*, *293*, 453, *468*
Woods, E. F., 580(9), *622*
Woodward, D. O., 584(61), *623*
Woodward, R. B., 235(316), 242(316), *251*
Woolf, B., 153, 155, 156(131, 132), *164*, *165*, 178(37, 38), *244*
Woolfolk, C. A., 123(7), 140(7), 150, *162*
Woolley, D. W., 255(18), 262(18, 70), 274 (159), 282(70), 285(266), 286(275), *287*, *288*, *291*, *294*, 532, *568*
Work, E., 199, 200(174, 175, 176, 177), 202(184), 239(166), *247*, 415(6), 418(6, 24), 440(93), 453(137), 461(93), *465*, *467*, *468*
Work, T. S., 506(105), *567*
Wormser, E. H., 196(137), *246*, 618(359), *630*
Worrel, C. S., 554(196), *570*
Wortman, J., 594(173), *626*
Wrede, F., 365, *372*
Wright, B. E., 204(191, 192), 205(200), *248*, 279(200, 201, 204, 205), 280(201), *292*
Wright, D. E., 260(48a), *288*
Wright, E. M., 297(14), 298(14), 317(14), *331*
Wright, L. D., 199(168), *247*, 274(156, 157), 276(157, 175, 179), *291*, 320(107, 108), *333*
Wright, N. G., 194(126, 128), 195(134), *246*
Wright, S., 582(42), *623*
Wyatt, G. R., 296(2), *331*, 480(24), *565*
Wyngaarden, J. B., 314(78), *332*
Wyss, O., 144(95), *164*
Wyttenback, C., 158(161), *166*

Y

Yamada, H., 266(87a, 87b, 88), *289*
Yamagami, H., 282(224a), *293*
Yamamuro, H., 390(11), 403(11), 404(11), 405(11, 69), 408(11), *410*, *411*
Yamasaki, M., 581(18), *622*
Yamaura, Y., 78(142), *115*
Yaniv, H., 223(271), *250*
Yčnofsky, C., 18(74), *38*, 222(264), 230, 231, 232, 233(301, 309, 310), 234(309, 312, 313, 314), 235(309), *249*, *250*, *251*, 253(3), 265(89), 266, *287*, *289*, 581(23), 583(52), 584(23, 52), 585(23, 52), 588 (52), 59(52), 597(208), 610(279), 614 (327), *622*, *623*, *627*, *628*, *629*
Yates, R. A., 174(18, 21), 243(18), *244*, 321, 326, 327(150), *333*, *334*, 607(274), 609(274), 610(274), 612(274), 617 (274, 357), *628*, *630*
Yavit, J., 260(49), *288*
Yčas, M., 354(115, 116), 359(116), *371*, 561(443), *576*, 588(110, 114), 590(131), *624*, *625*
Yoneda, M., 368(183), *372*
Yoshida, A., 581(15, 18), 596(200), *622*, *627*
Yoshikawa, H., 592(155), *625*
Young, R. S., 330(167, 168), *334*
Yu, C., 481, *570*
Yu, C.-T., 512(314), *572*
Yudkin, J., 594(174, 182, 186), 599(174, 219), *626*, *627*
Yura, T., 186(93, 94), *245*

Z

Zabin, I., 205(203), *248*, 601(233), *627*
Zachau, H. G., 506(286), 511, 513(120), 513(120), 560(286), *568*, *572*
Zalockar, M., 589(126), 602(241), 603 (241), *625*, *627*
Zambito, A. J., 401(59), 408(59), *411*
Zamecnik, P. C., 491(59), 499, 501, 505 (87, 104), 506(282), 510(69, 116), 511 (59, 118, 119), 512, 514(116, 326), 526, 545(118, 119, 368), *566*, *567*, *568*, *571*, *572*, *573*, *574*
Zamenhof, S., 540(170), 541(170), *569*, 581(25), 584(64), 587(95), 591(150), 596(201), *622*, *623*, *624*, *625*, *627*
Zarnitz, M. L., 596(203), *627*
Zelikson, R., 383(71, 73), *388*
Zelitch, I., 136(69, 72, 73, 74), *163*

Ziegler-Gunder, I., 279(198), *292*
Zillig, W., 495(257), *571*
Zilliken, F., 442(98), 461(177, 178, 179), *467*, *469*
Zimmerman, E. F., 309(61), 311(61), *332*
Zimmerman, L. N., 616(341), *630*
Zimmerman, S. B., 205(206), 206(206), *248*, 323(132), 324(132), *334*, 552(179), 553(412), *569* *575*, 586(88), *624*
Zinder, N., 258(35), *287*, 591(147), *625*
Zodrow, K., 282(218, 220), 283(218), *293*, 345(64), 346(64), *369*
Zubay, G., 482(232), 483(234), 495, 496, *570*, 586(81), 587(102), *624*

SUBJECT INDEX

A

Acetabularia,
enucleate, enzyme synthesis in, 589
Acetaldehyde,
acetate mutant and, 591
riboflavin synthesis and, 264
valine biosynthesis and, 210–212
Acetate, 58
alanine formation from, 89
calcium requirement and, 126
corrin synthesis and, 282
deoxyribose formation and, 298
exchange, pyruvate and, 87
glucose fermentation and, 43
heme synthesis and, 348–349, 361
isoleucine biosynthesis and, 212
lactate formation from, 88–89
leucine biosynthesis and, 209, 215, 216
methane formation from, 109
photometabolism of, 7, 8–11, 35–36
photosynthesis and, 22, 23, 27, 29, 35
prodigiosin synthesis and, 366
protein labeling by, 535–536
purine fermentation and, 98, 105–107
riboflavin synthesis and, 263–265
ribose synthesis and, 297
succinate formation and, 43
total synthesis, carbon dioxide and, 98–108
tyrosine synthesis and, 227
valine biosynthesis and, 208
Acetic acid,
cell wall extraction and, 423
Acetoacetate, 48
leucine degradation and, 75
riboflavin synthesis and, 265
Acetoacetyl coenzyme A,
methylmalonyl-oxalacetic transcarboxylase and, 84
reduction of, 36
Acetobacter, dextran of, 374
Acetobacter capsulatum,
dextran formation by, 378
Acetobacter melanogenum,
hemes in, 357
Acetobacter rancens,
pyridoxal in, 270
Acetobacter suboxydans,
coenzyme A synthesis by, 274
hemes in, 357
pantetheine formation by, 272–273
Acetobacter viscosum,
dextran formation by, 378
Acetobacter xylinum,
cellulose synthesis by, 380–382
α-Aceto-α-hydroxybutyrate,
formation of, 213–214
isoleucine biosynthesis and, 209, 213–214
Acetoin,
formation of, 169, 211
riboflavin synthesis and, 264–265
Acetokinase, nitrogen fixation and, 131
Acetolactate,
biosynthetic pathways and, 169
formation of, 210–212, 242–253
leucine biosynthesis and, 209
valine biosynthesis and, 174, 209–212, 214
Acetolactate decarboxylase,
metabolic function of, 211
valine excretion and, 238
Acetone, carbon dioxide fixation and, 8
Acetyl carboxylase,
biotin and, 90
carbon dioxide fixation and, 52, 77–79
Acetyl coenzyme A,
biosynthetic pathways and, 169
exchange, pyruvate and, 87
leucine degradation and, 75
methylmalonyl-oxalacetic transcarboxylase and, 84
nitrogen fixation and, 131
pyruvic carboxylase and, 79–80
riboflavin synthesis and, 265
Acetylglucosamine,
antigenicity and, 421
cell walls and, 415, 417–420, 424, 479
Acetylglucosamine-1-phosphate,
cell wall synthesis and, 445–456
α-Acetylglucosaminidase,
cell walls and, 421
β-Acetylglucosaminidase,
cell walls and, 417, 420, 421, 424

β-1, 6-Acetylglucosaminyl-acetylmuramic acid,
cell walls and, 416
N-Acetylglutamate,
arginine synthesis and, 187–188
carbamylaspartate synthesis and, 320, 321
carbamyl phosphate synthetase and, 160
citrulline synthesis and, 190
N-Acetylglutamic semialdehyde,
ornithine synthesis and, 187, 188
transamination of, 188
N-Acetyl-γ-glutamylphosphate,
ornithine synthesis and, 188
Acetylmannosamine, cell walls and, 442
Acetylmethylcarbinol,
riboflavin synthesis and, 263–264
Acetylmuramic acid,
cell walls and, 415, 416, 417, 418, 479
isolation of, 427, 441
Acetylneuraminic acid,
cell walls and, 441, 442
Acetylornithinase,
activation of, 188
occurrence of, 188, 189
repression of, 606
N^{α}-Acetylornithine,
ornithine synthesis and, 187
Acetyl phosphate,
exchange, pyruvate and, 87
formation of, 86–87
nitrogen fixation and, 131–132, 134, 135
Achromobacter,
denitrification by, 122
nitrate reductase in, 122
nitrogen fixation by, 122, 125, 127
Aconitase,
citrate accumulation and, 617
isocitric dehydrogenase and, 58
Aconitate,
aspartase and, 156
glutamate synthesis and, 176
Acrylyl coenzyme A,
β-alanine formation from, 272
amination of, 157
Actinomyces viridans,
cell walls of, 478
Actinomycetes,
dipyrrylmethenes in, 365
Adenine,
cobamide coenzymes and, 281
incorporation of, 541–544
nicotinic acid biosynthesis and, 266
nucleotide formation from, 308
requirement for, 305
riboflavin synthesis and, 255
Adenine deaminase,
guanine formation and, 309–310
Adenosine,
aspartase and, 156
guanine formation from, 311
Adenosine deaminase, induction of, 604
Adenosine diphosphate,
formation of, 301
photosynthesis and, 12, 13, 30, 36
propionyl carboxylase and, 96
Adenosine monophosphate,
feed-back control and, 315–317
formation of, 301, 304, 307
from other purines, 308, 309
histidine synthesis from, 220–221, 242
imidazoleglycerol phosphate and, 184
protein synthesis and, 414
Adenosine triphosphatase,
synthesis of, 515
Adenosine triphosphate,
amino acid biosynthesis and, 239
asparagine synthetase and, 159
biotin-containing enzymes and, 90–91, 93–96
carbamyl phosphate synthetase and, 159–160
carbon dioxide fixation and, 5, 52, 71–81
cellulose synthesis and, 382
cell wall synthesis and, 445, 447–449
citrulline synthesis and, 190
formate utilization and, 25
formation of, 301
glutamine synthetase and, 157, 158
histidine biosynthesis and, 221
homoserine methylation and, 198
nitrogen fixation and, 131–132, 135
nucleic acid synthesis and, 543, 544, 545
phosphoenolpyruvic carboxykinase and, 65
photosynthesis and, 12, 13, 17, 18, 20–22, 29–37
polyglutamate synthesis and, 400
protein release and, 515

protein synthesis and, 498, 502, 505, 508, 511, 513–514, 526, 530, 531, 562
pyridine nucleotide reduction and, 32–33
riboflavin synthesis and, 264
soluble nucleic acid and, 545–546
thiamine synthesis and, 260
thiaminokinase and, 261
Adenosine triphosphate sulfurylase,
sulfur reduction and, 207
S-Aldenosylmethionine, 198
Adenyl carbon dioxide,
carbon dioxide fixation and, 76–77
Adenylcobamide coenzyme,
properties of, 281
Adenylic deaminase, 311
Adenylic polymer, formation of, 548
Adenylosuccinase,
complementation and, 584
guanosine triphosphate and, 317
mutants and, 307
suppressors and, 591
Adenylosuccinate, cleavage of, 301, 304
Adrenocorticotropic hormone,
amino acids of, 477
Aerobacter,
amino acid activation in, 507
aromatic ring biosynthesis by, 223
levan of, 383
tryptophan auxotrophs of, 232
Aerobacter aerogenes, 579
acetolactate formation by, 211
alanine formation by, 203
cell wall synthesis in, 446, 462
cytochromes in, 358, 361
glutamate synthesis by, 170, 176–177
glutamine auxotroph of, 183
glycerol dehydrogenase of, 603
histidase of, 603
histidine synthesis by, 184, 220–221
inositol and, 286
isoleucine auxotrophs of, 203
leucine auxotrophs of, 215
mutant, glucose repression and, 611
nitrogen fixation by, 122, 124, 125, 127
ornithine synthesis by, 189
orotic acid excretion by, 325–326
pantothenate synthesis by, 258
proline auxotrophs of, 189
proline synthesis by, 186
purine synthesis by, 319
thiamine synthesis by, 259
valine-isoleucine synthesis by, 214
xanthosine-5′-phosphate aminase of, 161
Aerobacter indologenes,
hemoprotein formation by, 358
Aerobacter levanicum,
levansucrase of, 383
Aerobiosis, enzyme induction and, 617
Alanine,
acceptor ribonucleic acid for, 512
acetate and, 9, 89
activation of, 506
biosynthesis of, 168, 182, 202–203, 208, 242
cell walls and, 415–417, 419, 420, 422–424, 428, 442, 443, 447–448, 478, 538–539
D-glutamate formation and, 398
incorporation, 523
lipid and, 517
glycine utilization and, 580
pyridoxal and, 270
transamination of, 180
Alanine dehydrogenase, 161
ammonia incorporation and, 154–155, 179
Alanine racemase, 397, 452
capsules and, 537
cell walls and, 539
occurrence of, 236, 398
oxamycin and, 454
D-Alanine,
activation of, 451, 509
cell walls and, 423–424, 428, 440, 443, 447–448, 461, 479
formation of, 397–398, 452
occurrence of, 236
oxamycin and, 453–457
polyglutamate formation and, 236–237
requirement for, 533
D-Alanine-α-ketoglutarate transaminase, 537
β-Alanine,
natural auxotrophs and, 257
origin of, 255, 256, 272
pantothenate synthesis and, 255–257, 271–272
pyrimidine biosynthesis and, 329
Alanylalanine synthetase,
cell wall synthesis and, 448–450

Alanylalanine synthetase—*Continued*
  oxamycin and, 454
D-Alanyl-D-alanine,
  cell wall synthesis and, 538–539
β-Alanyl coenzyme A synthetase,
  ammonia incorporation and, 157
Alanylmethionine, activation of, 509
Alanyl peptides, requirement for, 533
Albumin,
  egg, amino acids of, 474
Alcaligenes faecalis,
  cell wall synthesis in, 439
  protein synthesis in, 494, 500, 519, 530
  spheroplasts of, 459–460
  thiamine synthesis by, 259
Alcohol dehydrogenase, formation of, 612
Alder, nitrogen fixation in, 122
Aldoheptoses, cell walls and, 425
Aldolase,
  photosynthesis and, 12, 17, 21, 36
  synthesis of, 515
  levansucrase and, 384–385
Alkaline phosphatase,
  fluorouracil and, 541, 590
  gene mapping and, 584
Alkyl-β-thiogalactoside,
  β-galactosidase induction and, 601
Allomyces macrogynus,
  lipoic acid analog in, 284
Alloxan, riboflavin synthesis and, 262
Amethopterin, folic acid reduction and, 279
Amide(s),
  nitrogen fixation and, 136–138
  synthesis, ammonia incorporation and, 157–161
Amino acid(s),
  analogs,
    peptides and, 533
    protein specificity and, 580–581
    protein synthesis and, 504–505, 534–535
    resistance to, 238
  biosynthesis,
    cell walls and, 452–453
  comparative biochemistry and, 238–239
    control of, 242–243
    general features of, 239–240
    special features of, 240–242
  catabolic vs. biosynthetic pathways, 168–169
  deaminases, formation of, 612
  decarboxylases, formation of, 612
  deoxyribonucleic acid synthesis and, 550
  excretion of, 237–238
  formic hydrogenlyase formation and, 612
  incorporation, fluorouracil and, 488
  nucleic acid synthesis and, 541–544
  obligatory intermediates of, 169–171
  protein synthesis and, 502–504
  sequence of, 475–477
D-Amino acid(s),
  formation of, 236–237
  occurrence of, 473
  significance, cell walls and, 428–429
Amino acid-activating enzymes,
  distribution of, 507–508, 526
  protein synthesis and, 505–509, 523–527
Amino acid-lipid complexes,
  protein synthesis and, 517
Aminoacrolein fumaric acid,
  nicotinic acid biosynthesis and, 267
Aminoacroleinmaleic acid,
  nicotinic acid biosynthesis and, 267
Amino acyl adenylate(s),
  existence of, 506–507
  incorporation of, 491–492, 524
  peptide synthesis and, 451
Amino acyl ribonucleic acid,
  stability of, 511
α-Aminoadipic acid,
  lysine formation from, 199, 200, 239
p-Aminobenzoic acid, 227, 453
  anthranilate hydroxylation and, 266
  biosynthesis of, 276
  folic acid synthesis and, 277
  methionine formation and, 197
  pantothenate and, 271
  requirement for, 222, 223
  serine and, 204–205
  sulfonamides and, 256, 277
p-Aminobenzoylglutamic acid,
  folic acid synthesis and, 276–277
5-Amino-2,4-bis(substituted amino)-pymidines,
  enzyme synthesis and, 610
α-Aminobutyrate,
  alanine formation and, 203
  isoleucine biosynthesis and, 212
  transamination of, 180

pyrimidine biosynthesis and, 329
valine synthesis and, 215
5-Amino-4-carboxyimidazoleribonucleotide,
carbon dioxide and, 52
2-Amino-4,6-dihydroxypteridine,
folic acid synthesis and, 278
2-Amino-4,7-dihydroxypteridine-6-carboxylic acid,
folic acid synthesis and, 278
2-Amino-4,5-dimethyl-1′-ribitylaminobenzene,
riboflavin synthesis and, 262, 263
2-Amino-4-hydroxy-6-hydroxymethylpteridine,
folic acid synthesis and, 278
2-Amino-4-hydroxy-6-hydroxymethyltetrahydropteridine,
tetrahydrofolate synthesis and, 278
2-Amino-4-hydroxypteridine-6-aldehyde,
folic acid synthesis and, 278
2-Amino-4-hydroxypteridine-6-carboxylic acid,
folic acid synthesis and, 278
4-Aminoimidazole,
purine fermentation and, 107
4-Amino-5-imidazolecarboxamide,
excretion of, 299, 316
formylation of, 318
ribotide formation from, 308
utilization of, 306
4-Amino-5-imidazolecarboxamide ribotide, 184
cyclization of, 301, 303
histidine synthesis and, 220–221
4-Amino-5-imidazolecarboxylic acid,
purine fermentation and, 107
5-Amino-4-imidazolecarboxylic acid ribotide,
formation of, 303
Aminoimidazoleribonucleotide carboxylase,
carbon dioxide fixation and, 52, 70–71
Aminoimidazole ribotide,
formation of, 302–303
δ-Amino levulinic acid,
corrin synthesis and, 282
enzyme forming, 353
hemin requirement and, 341
porphyrin synthesis and, 348–350, 352–353, 355–356
prodigiosin synthesis and, 366
δ-Aminolevulinic acid dehydrase,
bacterial, 353–354
purification of, 349
synthesis of, 367
Aminolevulinic acid synthetase,
repression of, 368
synthesis of, 367
2-Amino-3-(4′-methylthiazole-5′-)propionic acid,
thiazole synthesis and, 260–261
α-Amino-*cis,trans*-muconic acid,
nicotinic acid biosynthesis and, 267
Aminopropane, spermine and, 192
1-Amino-2-propanol,
vitamin B_{12} synthesis and, 282
Aminopterin,
folic acid reduction and, 279
riboflavin synthesis and, 279
5-Amino-1-ribosylimidazole,
excretion of, 305–306, 319
5-Amino-1-ribosyl-4-imidazolecarboxamide,
excretion, 305
inhibition of, 313–314
guanine formation and, 309, 310
5-Amino-1-ribosyl-4-imidazolesuccinocarboxamide-5-phosphate,
cleavage of, 301, 303, 306–307
Ammonia,
alanine formation and, 203
aspartate formation and, 153
assimilation, 120, 123, 178–179
amide synthesis and, 157–161
conclusions regarding, 161
dehydrogenases and, 153–155
saturation amination and, 155–157
carbamic acid formation and, 191
carbon dioxide fixation and, 2–3
cystathionine cleavage and, 195, 197
glutamine synthesis and, 183
guanylic acid formation and, 301
histidine synthesis and, 184, 221
nicotinic acid biosynthesis and, 266, 269
nitrogen fixation and, 130, 134–145
oxidation of, 123
phosphoribosylamine formation from, 301
purine synthesis and, 319
pyridoxal biosynthesis and, 270

Ammonia—*Continued*
thermodynamic efficiency and, 3
uridine triphosphate aminase and, 322, 323
Ammonium salts,
polyglutamate formation and, 394
Amoebae,
cytoplasmic inheritance in, 582
ribonucleic acid synthesis in, 559
Amylase,
formation, chloramphenicol and, 557
resting stage and, 592
α-Amylase,
ethionine and, 581
p-fluorophenylalanine and, 580–581
formation of, 612
precursors of, 596
Amylodextrins,
dextran synthesis from, 374
Amylomaltase,
constitutive, 593, 605
induction of, 604
reaction catalyzed, 379
transformation and, 586
Amylopectin, dextrandextrinase and, 378
Amylose,
dextrandextrinase and, 378
synthesis of, 378–379
Amylosucrase, product of, 379
Anaerobiosis,
nitrogen fixation and, 126–127
porphyrin formation and, 347, 353
Analogs,
biosynthetic pathways and, 174–175
N^5, N^{10} - Anhydroformyltetrahydrofolic acid,
glycinamide and, 318
Anthranilic acid, 227
formation of, 235–236
hydroxylation of, 266
tryptophan biosynthesis and, 229–233
Anthranilic ribonucleotide,
tryptophan biosynthesis and, 232
Anthrax,
infection, polyglutamate and, 407–408
Antibiotics,
amino acids in, 236, 479
cell walls and, 415, 459–461
deoxyribonucleic acid synthesis and, 550
Antimetabolites,
modes of action, 618
mutant selection and, 606
vitamin synthesis and, 255–256
Arginase, tissue content of, 592
Arginine,
biosynthesis, 169, 182, 186–192, 239, 322, 330
feedback and, 618
enzyme repression and, 606–607, 609
glutamate formation and, 177
incorporation, 5-fluorouracil and, 488
Arginine decarboxylase,
formation of, 612
Arginine dihydrolase,
inducibility of, 616
peptide requirement and, 534
Argininosuccinate, formation of, 190, 191
Argon, nitrogen fixation and, 142–143
Aromatic rings,
enzyme induction and, 594
photosynthesis and, 18
Arsenate, nitrogen fixation and, 135
Arthrobacter globiformis,
porphyrin excretion by, 346
Arthrobacter histidinolvorans,
histidine formation by, 219
Arthrobacter terregens,
growth factors for, 342
Ascaris, malic enzyme of, 53, 56–57
Ascites cells,
aspartyl uridylate in, 537
Ashbya gossypii,
riboflavin synthesis by, 254, 255, 262–263, 265
Asparagine,
formation of, 193
nitrogen fixation and, 136
Asparagine synthetase,
ammonia incorporation and, 159
Aspartase,
ammonia incorporation and, 155–156, 178–179
function of, 192–193
Aspartate,
acetate metabolism and, 9
activating enzymes and, 508
β-alanine and, 255, 256, 271, 272
arginine synthesis and, 190, 191
biosynthesis of, 153, 155, 182, 192–193, 242

biotin requirement and, 89, 257
cell walls and, 440, 442, 478
cross linkages and, 477
diaminopimelate formation and, 200, 453
histidine synthesis and, 314
isoleucine biosynthesis and, 212
β-methylaspartase and, 157
photosynthesis and, 20, 22, 23
purine synthesis and, 299, 319
pyrimidine synthesis and, 320–322, 329
transamination of, 180
Aspartate kinase, inhibition of, 314
Aspartate transcarbamylase,
constitutive, 593, 606
enzyme amounts and, 604
feedback inhibition of, 326
purification of, 320
repression of, 326, 607
uracil and, 612
D-Aspartate, 397
Aspartic deaminase, biotin and, 276
Aspartic semialdehyde,
diaminopimelate formation from, 198–202
formation of, 185, 188, 193, 194
function as intermediate, 193–194
homoserine formation from, 195–196
lysine formation from, 200
Aspartic semialdehyde dehydrogenase,
mutant lacking, 194
Aspartokinase, feedback control of, 194
β-Aspartyl phosphate,
formation of, 193, 194
Aspartyl uridylate,
occurrence of, 537
Aspergillus nidulans,
mutants of, 582
nicotinic acid synthesis by, 265
Aspergillus niger,
biotin sulfoxide and, 274, 276
condensing enzyme of, 617
oxalacetate cleavage by, 108
pimelic acid and, 274
Athiorhodaceae,
carbon dioxide fixation by, 7–11
nitrogen fixation by, 144–145
porphyrin excretion by, 342–345
Atropine esterase, amounts formed, 593
Autolysins,
cell wall synthesis and, 458–459
Autolysis,
nitrogen fixing preparations and, 128
Autotrophs,
carbon dioxide fixation by, 111
Avidin,
biotin and, 274
biotin-containing enzymes and, 90–91
β-methylcrotonyl carboxylase and, 76
methylmalonyl-oxalacetic transcarboxylase and, 81, 91
porphyrin synthesis and, 353
propionate fermentation and, 84
propionyl carboxylase and, 74
pyruvic carboxylase and, 80–81
8-Azaguanine,
bacteriochlorophyll synthesis and, 364
enzyme synthesis and, 487, 606
incorporation of, 540
Azaserine, purine synthesis and, 301
7-Azatryptophan,
activation of, 506
protein synthesis and, 505
6-Azauracil,
mutant selection and, 606
nucleotide accumulation and, 439
6-Azauridine diphosphate,
polyribonucleotide phosphorylase and, 549
Azaxanthine,
riboflavin synthesis and, 263
Azide, nitrogen fixation and, 150
Azines, nitrogen fixation and, 140, 149
Azotobacter,
amino acid activation in, 507
protein synthesis in, 500
Azotobacter vinelandii,
coenzyme Q of, 286
copolymer synthesis by, 409
hydroxylamine reductase in, 123
mutant, nitrogen fixation by, 141
nitrogen fixation in, 122, 124–128, 136–140, 142, 144, 146, 150, 152
oxalacetic decarboxylase of, 50–51
ribonucleic acid synthesis by, 547–549

B

Bacillaceae,
ornithine synthesis in, 189
Bacillus,
D-amino acids in, 236

Bacillus—Continued
capsule of, 537
levan of, 383
Bacillus anthracis,
D-amino acids in, 236, 537
capsule of, 391–393, 478
polyglutamic acid, 390, 393
biosynthesis of, 396, 398
hydrolysis of, 400, 405
structure of, 402–405
taxonomic status of, 390
transaminases of, 397–398
Bacillus brevis, vitamin K of, 285
Bacillus cadaveris,
lysine decarboxylase of, 600, 603, 614
Bacillus cereus,
8-azaguanine in, 540
canavanine and, 581
cell wall synthesis in, 452
p-fluorophenylalanine and, 581
hemoprotein synthesis by, 360
penicillinase of, 593, 597–598, 600, 605, 613–614
porphyrin excretion by, 346, 347
protein synthesis by, 487
sporulation, amino sugar and, 427
Bacillus licheniformis,
denitrification by, 122
taxonomic status of, 390–391
Bacillus macerans, amylase of, 379–380
Bacillus megaterium,
amino acid-lipid complexes in, 517
cross-septa of, 393
deoxyribonucleic acid of, 484
β-galactosidase of, 599
membrane, protein synthesis by, 620
nucleotide accumulation and, 443
phosphogluconic dehydrogenase in, 60
polyglutamic acid in, 390, 393, 403–405
protein assay in, 490
protein synthesis by, 488, 498, 500, 531
N-succinylglutamate in, 189
riboflavin production by, 254
vitamin B_{12} in, 282, 283
Bacillus mesentericus,
taxonomic status of, 390–391
thiamine synthesis by, 259
Bacillus paraalvei,
thiamine synthesis by, 261
Bacillus polymyxa,
nitrogen fixation by, 122, 127
Bacillus pumilus,
hydroxylamine reductase in, 123
nitrite assimilation by, 122
taxonomic status of, 390
Bacillus pyocyaneus,
nitrite assimilation by, 122
Bacillus sphaericus,
diaminopimelate in, 200
Bacillus subtilis,
alanine dehydrogenase in, 154, 179
alanine formation by, 203
amino acid activation in, 509
D-amino acids in, 236–237
amylase of, 592, 596
capsule of, 478
cell walls of, 419, 420, 443, 479
cytochromes of, 359
enzyme synthesis in, 487
ethionine and, 581
flagella, synthesis of, 493
p-fluorophenylalanine and, 580–581
glutamate synthesis by, 177
D-glutamate synthesis by, 453, 537
glutamine synthesis by, 398
γ-glutamyl transfer in, 184
hemoprotein synthesis by, 360
homoserine in, 195
levansucrase of, 383
nicotinic acid synthesis by, 266
nitrate reductase in, 122
nitric oxide reductase in, 122
nitrite reduction by, 122
peptidase of, 400–401, 405
phosphogluconic dehydrogenase in, 60
polyglutamic acid, 390
biosynthesis of, 394–396, 399–400, 409
hydrolysis of, 400, 405
structure of, 402–405
porphyrin excretion by, 345, 346
taxonomic status of, 390–391
transaminases in, 180, 397
transformation in, 586
Bacillus thuringiensis,
sporulation of, 503
Bacitracin,
nature of, 479
nucleotide accumulation and, 433–434
spheroplasts and, 459
Bacteria,
chemolithotrophic, 19–26

halotolerant, hydroxylamine reductase in, 123
number of enzymes in, 579
photolithotrophic, 5–11
protein content of, 472–473
protein synthesis in, 500–502
Bacteriochlorophyll,
carboxydismutase and, 28
structure of, 338, 362
synthesis, 362–364
porphyrin excretion and, 343–344, 354–356
Bacteriophage,
capsules and, 407–408
deoxyadenosine utilization and, 317
deoxycytidylate kinase and, 323
deoxycytidylic deamination and, 324
deoxyguanylate kinase and, 318
deoxyribonucleic acid, 480
synthesis of, 550–551
enzymes,
induction of, 594–595
synthesis of, 586–587, 589–590
episomes and, 582
lysozyme of, 416, 417
nucleic acid of, 296
nucleotides of, 481
r II region, 584
mapping of, 584
size of, 588
thymidylate kinase and, 325
thymidylate synthetase and, 324, 325
ultraviolet and, 613
Bacteriophage ϕX 174,
deoxyribonucleic acid, 587
synthesis and, 553
Bacteriopheophorbide,
chlorophyll synthesis and, 364
Bacterium cadaveris,
aspartic deaminase of, 276
Bacterium linens,
pantothenate requirement of, 271
Benzimidazolecobamide coenzyme,
formation of, 281, 283
Betacoccus arabinosaceus,
dextran of, 374, 377
Biocytin,
biological activity of, 276
biotin binding and, 92
Biosynthetic pathways,
analysis,
extracted enzymes and, 174
isotopes and, 172–173
metabolite analogs and, 174–175
mutants and, 171–172
catabolic vs. biosynthetic, 168–169
obligatory intermediates, 169–171
unity of, 175–176
Biotin,
acetyl carboxylase and, 90
analogs of, 89
biosynthesis of, 274–276
bound form of, 275–276
carbon dioxide fixation and, 71, 74, 76, 89–98
cysteine and, 208
enzyme formation and, 612
mechanism of action, 92–93
β-methylcrotonyl carboxylase and, 90
methylmalonyl-oxalacetic transcarboxylase and, 81, 90
nitrogen fixation and, 122, 126, 129, 131
pimelic acid and, 274
porphyrin synthesis and, 348, 353, 357
propionyl carboxylase and, 90
protein synthesis and, 518
purine synthesis and, 319
pyruvic carboxylase and, 80–81, 90
replacement of, 257
Biotindase, bound biotin and, 276
Biotin-L-sulfoxide,
biological activity of, 276
ϵ-N-Biotinyl-L-lysine,
biological activity of, 276
Biuret test, polyglutamate and, 404
Bordetella pertussis,
hemin requirement of, 339, 440
Bromouracil,
incorporation of, 541, 552
mutagenesis and, 587, 619
Brucella abortus,
alanine formation by, 203
transaminases of, 181
Butyrate,
carbon dioxide exchange and, 78
carbon dioxide fixation and, 9–11
fermentation, methane and, 108–109
metabolism of, 36
Butyribacterium rettgeri,
acetate formation by, 99, 103–104, 108
lipoic acid in, 284

Butyryl coenzyme A
carboxylation of, 79
methylmalonyl-oxalacetic transcarboxylase and, 84
propionyl carboxylase and, 97

C

Cadaverine, ribosomes and, 495
Calcium,
nitrogen fixation and, 122, 126, 129
polyglutamate synthesis and, 395
Calothrix, nitrogen fixation by, 122
Canavanine, enzyme activity and, 581
Candida albicans,
folic acid synthesis by, 278
Candida flareri,
riboflavin synthesis by, 255
Caproic acid,
carbon dioxide exchange and, 78
Capsule(s),
composition of, 391–393
peptides of, 478
virulence and, 407–408
Capsule virus, nucleotides of, 481
Carbamate phosphokinase,
carbamic acid and, 191
Carbamic acid, formation of, 191
Carbamylaspartate,
pyrimidine synthesis and, 320–321
Carbamyl group, transfer of, 191
Carbamyl phosphate, 239, 242
citrulline synthesis and, 189–190
formation of, 191, 320–322
pyrimidine synthesis and, 160, 330
Carbamyl phosphate synthetase,
ammonia incorporation and, 159–160
Carbohydrate,
formation, pyruvate and, 66–68
N-Carbomethoxy-biotin methylester,
formation of, 92
Carbon dioxide,
assimilation,
acetate synthesis and, 98–108
adenosine triphosphate and, 71–81
biotin and, 89–98
decarboxylation and, 86–89
early studies on, 46–51
importance of, 110
methane and, 108–109
original proof of, 44–46
primary reactions of, 51–81
pyridine nucleotides and, 52–61
summary of, 109–112
transcarboxylation and, 81–86
without extra energy source, 61–71
citrate formation and, 43
coenzyme F synthesis and, 279
glutamate synthesis and, 176
leucine degradation and, 75
photosynthetic fixation,
adenosine triphosphate and, 5
autotrophic mechanism of, 11–19
chemolithotrophic, 19–26
cyanide and, 10–11, 28–29
energetics of, 29–37
mechanism of, 26–29
oxygen and, 4
thermodynamic efficiency of, 3–4
types of, 1–2
polyglutamate synthesis and, 396, 407–408
propionate formation and, 81–82
purine synthesis and, 299, 303
riboflavin synthesis and, 255
Carbon monoxide,
growth on, 23–24
nitrogen fixation and, 134–135, 142, 145–147, 150
Carbonyl phosphate,
propionyl carboxylase and, 73
Carboxy dismutase,
cyanide and, 11
formate utilization and, 25
heterotrophic growth and, 23
induction of, 27–28
lactate and, 22
occurrence of, 27
oxalate utilization and, 25
photosynthesis and, 12–13, 17, 19, 20, 21, 23
β-Carboxy-β-hydroxyisocaproate,
leucine synthesis and, 216
Carboxypeptidase,
polyglutamate hydrolysis by, 402
1-(*o*-Carboxyphenylamino-1-deoxyribulose 5-phosphate,
tryptophan biosynthesis and, 231, 232
N-*o*-Carboxyphenylribosylamine-5-phosphate,
tryptophan biosynthesis and, 230, 232

Carnosine, synthesis of, 451
Carotenoids, synthesis of, 364–365
Catalase, 338, 343, 346
  hemin requirement and, 341
  induction of, 600
  oxygen and, 359, 360
  synthesis, 354, 361–362
    iron and, 358
    ribonucleic acid and, 487, 488, 510
Cell(s),
  polyploid, enzymes in, 592
Cell membrane,
  nucleic acid synthesis by, 545
  protein synthesis and, 500–502, 524, 530–531, 559–560, 620
Cellobiose, dextransucrase and, 377
Cellulodextrins,
  cellulose formation and, 381
Cellulose,
  photosynthesis and, 18
  structure and occurrence of, 380
  synthesis of, 380–382
Cell walls, 473
  biosynthesis, 457–458
    additional problems, 461–464
    antibiotics and, 459–461
    autolysins and, 458—59
    intermediates and, 426–444
    mutants and, 461
    nucleotide precursors and, 444–457
  composition of, 414
  peptides, 478–479
    synthesis of, 537–450
  protein assay and, 489–491, 501–502
  structure of, 415–425
Cephalosporin,
  penicillinase and, 598, 600
Chemolithotrophy,
  definition of, 1, 24, 26
Chemostat,
  mutant selection and, 605–606
Chloral hydrate, polyglutamate and, 407
Chloramphenicol,
  amino acid-lipid complexes and, 517
  bacteriochlorophyll synthesis and, 364
  cell wall peptides and, 538, 554
  deoxyribonucleic acid synthesis and, 550–551
  nucleic acid composition and, 484
  nucleic acid synthesis and, 542–544, 554–555
  nucleotide accumulation and, 437
  polyglutamate synthesis and, 409
  protein synthesis and, 491, 492, 501–502, 521, 522, 526, 527, 530, 531, 554–557, 562, 610
  repressed enzymes and, 607
Chlorella,
  chlorophyll synthesis by, 362
  photosynthesis by, 13
  porphyrin formation by, 350, 352
Chlorobacteriaceae,
  carbon dioxide fixation by, 5–6
Chlorobium, nitrogen fixation by, 122
Chlorobium limacola,
  carbon dioxide fixation by, 5
Chlorobium thiosulfatophilum,
  carbon dioxide fixation by, 5–6
  chlorophylls of, 362
  porphyrin excretion by, 344
p-Chloromercuribenzoate,
  amino acid-lipid complexes and, 517
Chlorophyll(s),
  absorption spectra of, 337
  formation, 348
    algae and, 362
    factors affecting, 364–365
    induction of, 604
    photosynthetic bacteria and, 362–364
Chlorophyll *a*, structure of, 338
5-Chlorouracil, inhibition by, 541
Cholinesterase,
  induction of, 600
  tissue content of, 592
Chromatium,
  acetate utilization by, 35
  carboxydismutase in, 27
  coenzyme Q of, 286
  nitrogen fixation by, 122, 127
Chymotrypsin,
  amino acids of, 477
  polyglutamate and, 402
Cistron,
  definition of, 583
  size of, 588
Citratase, 58
Citrate,
  formation of, 43
  glutamate synthesis and, 176
  polyglutamate synthesis and, 394
Citridesmolase, precursors of, 595–596
Citrovorum factor, formation of, 279

Citrulline,
  arginine formation from, 191–192
  biosynthesis, 189–191, 242, 322
    biotin and, 518
  carbamyl phosphate synthetase and, 159
  degradation of, 191
Clostridia,
  amino acid activation in, 507
  hemes in, 357
  hydrogenase of, 151
  nitrogen fixation by, 122, 124, 127
  uracil degradation by, 329
Clostridium aceticum,
  acetate synthesis by, 98
Clostridium acetobutylicum,
  riboflavin synthesis by, 254, 263–264
Clostridium acidi-urici,
  acetate formation by, 98
  carbon dioxide fixation by, 105–108
Clostridium botulinum,
  thiamine and, 259
Clostridium butylicum,
  phosphoroclastic reaction in, 86
Clostridium butyricum,
  pyruvate cleavage by, 87
Clostridium cadaveris,
  lysine decarboxylase of, 614
Clostridium cylindrosporum,
  acetate formation by, 99
  carbon dioxide fixation by, 105–108
  tetrahydrofolate formylase of, 279
Clostridium HF,
  serine-glycine interconversion by, 204
Clostridium kluyveri,
  carbon dioxide exchange reaction in, 78
  carbon dioxide-pyruvate exchange in, 91
Clostridium pasteurianum,
  hydroxylamine reductase in, 123
  nitric oxide reductase in, 122
  nitrite assimilation by, 122
  nitrogen fixation by, 121, 124–126, 128, 129, 131–132, 134, 136, 140, 142, 144, 146, 147, 151, 152
Clostridium perfringens,
  coenzyme Q and, 286
Clostridium propionicum,
  β-alanine formation by, 272
Clostridium septicum,
  pantothenate requirement of, 257
Clostridium sporogenes,
  glutamic dehydrogenase of, 177
Clostridium sticklandii,
  adenylcobamide coenzyme of, 282
  folic acid reduction by, 279
Clostridium tetani,
  glutamine requirement of, 183
  thiamine and, 259
Clostridium tetanomorphum,
  cobamide coenzymes of, 281, 283
  glutamate fermentation by, 156
Clostridium thermoaceticum,
  acetate formation by, 99–103, 108
  glucose fermentation by, 102–103
  xylose fermentation by, 102
Clostridium uracilicum,
  β-alanine formation by, 272
Clostridium welchii,
  alanine formation by, 203
  pyridoxamine phosphate in, 270
Cobalt,
  aspartase and, 156
  glutamine synthetase and, 158
  nitrogen fixation and, 126
  polyglutamate synthesis and, 395–396
  porphyrin and, 282, 355
Coenzyme(s), synthesis of, 618
Coenzyme A, 367
  aspartokinase and, 194
  biosynthesis of, 272–274
  catalase formation and, 361
  nitrogen fixation and, 129, 131–132
  phosphoroclastic reaction and, 87
  porphyrin synthesis and, 353, 356
Coenzyme A transferase,
  propionate formation and, 82–83
Coenzyme F,
  biosynthesis of, 279–280
  nature of, 276
Coenzyme Q,
  function of, 285
  growth stimulation by, 286
Colicines, episomes and, 582
Coliminic acid, occurrence of, 442
Complementation, operation of, 584, 619
Compound Z1,
  anthranilate formation and, 236
  aromatic biosynthesis and, 224, 226–227
Compound III,
  histidine synthesis and, 220–221

Condensing enzyme,
glutamate synthesis and, 170, 176
induction of, 617
Conjugation,
enzyme formation and, 583, 587, 589
Copper,
nitrite assimilation and, 122
nitrogen fixation and, 126
Coprogen, isolation of, 342
Coproheme, formation of, 358
Coproporphyrin,
catalase synthesis and, 362
vitamin B_{12} and, 282
Coproporphyrin I,
excretion of, 345, 346, 368
Coproporphyrin III,
accumulation of, 335–336, 343–346, 352
formation of, 348, 353–355
structure of, 337
Coproporphyrinogen III,
formation of, 348, 351, 353
Corepressor, repressor and, 621
Corrin,
biosynthesis of, 282
vitamin B_{12} and, 280
Corynebacteria,
cell walls of, 418
diaminopimelate in, 199
levan of, 383
orotic acid formation by, 320–321
polysaccharide synthesis by, 378–379
Corynebacterium bovis,
lipoic acid and, 284
Corynebacterium diphtheriae,
biotin requirement of, 274
cell walls of, 478
cobamide synthesis by, 282, 283
nicotinic acid utilization by, 268
pantothenate requirement of, 257
porphyrin biosynthesis in, 352
porphyrin excretion by, 345–347, 368
purine utilization by, 313
Corynebacterium erythrogenes,
porphyrin excretion by, 345, 346
Corynebacterium pyogenes,
cell walls of, 478
Corynebacterium xerose,
biotin and, 274, 275
cell wall synthesis in, 452
Cross reacting material,
induction of, 599
mutations and, 585
tryptophan synthetase and, 234
uracil analogs and, 590
Crystal violet,
nucleotide accumulation and, 437–439
Cucumber virus 3, nucleotides of, 481
Cyanase, induction of, 604
Cyanide,
photosynthesis and, 10–11, 28–29
vitamin B_{12} and, 280
Cyanocobalamin, structure of, 280
D-Cycloserine, cell walls and, 415, 434
Cystathionine, 173
cleavage of, 195, 197, 239
formation of, 195–197
methionine formation and, 196
Cysteic acid,
pantothenate and, 255, 256
Cysteine,
biosynthesis of, 182, 206–208, 242
coenzyme A synthesis and, 272, 273
cross linkages and, 477
degradation of, 207
flagella synthesis and, 504
methionine formation and, 196
thiazole synthesis and, 260–261
Cysteine desulfhydrase,
cysteine auxotrophs and, 207
Cystine, metabolic function of, 207
Cytidine, utilization of, 327
Cytidine diphosphate,
phospholipid synthesis and, 414
Cytidine diphosphate abequose,
isolation of, 464
Cytidine diphosphate diglyceride,
formation of, 415
Cytidine diphosphate glycerol,
accumulation of, 444
cell wall synthesis and, 457
synthesis of, 457
Cytidine diphosphate ribitol,
accumulation of, 437, 444
cell wall synthesis and, 457
synthesis of, 457
Cytidine diphosphate tyvelose,
isolation of, 464
Cytidine monophosphate,
accumulation of, 437, 444
utilization of, 323

Cytidine monophosphate acetylneuraminic acid, 457
cell walls and, 442
Cytidine phosphates,
biosynthesis of, 322–323
Cytidine triphosphate,
ammonia incorporation and, 160–161
protein synthesis and, 511
soluble ribonucleic acid and, 545
Cytochrome(s), 338, 343
formation, iron and, 358
nitrate reductase and, 122
nitrite oxidase and, 123
photosynthesis and, 30–31
Cytochrome a,
oxygen and, 359, 360
porphyrin synthesis and, 347
Cytochrome a_1,
iron and, 358
oxygen and, 359
Cytochrome a_2,
iron and, 358
oxygen and, 361
Cytochrome a_3, oxygen and, 359
Cytochrome b,
nitrogen fixation and, 142
oxygen and, 359, 360
porphyrin excretion and, 346–347
Cytochrome b_1, oxygen and, 359
Cytochrome b_2, iron and, 358
Cytochrome c,
ammonia oxidation and, 123
iron-oxidizing bacteria and, 31
oxygen and, 359, 360
porphyrin synthesis and, 347, 354
Cytochrome oxidase,
induction, kinetics of, 600–601
Cytochrome peroxidase,
oxygen and, 359, 360
Cytoplasm,
inducible enzymes and, 608
inheritance and, 582, 619–620
Cytosine, utilization of, 327–328
Cytosine deaminase,
cytosine utilization and, 327

D

Decarboxylation,
reversal, carbon dioxide fixation and, 86–89
Dehydrogenases,
ammonia incorporation and, 153–155
Dehydroquinase, properties of, 223
5-Dehydroquinate,
aromatic ring synthesis and, 222–223
precursors of, 223–226
5-Dehydroshikimic acid,
aromatic ring synthesis and, 222–223
5-Dehydroshikimic reductase,
occurrence of, 223
Denitrification,
organisms performing, 122
Denitrobacillus,
denitrification by, 122
Deoxyadenosine,
phage growth and, 551–552
utilization of, 317
3-Deoxy-D-arabinoheptulosonic acid-7-phosphate,
dehydroshikimate formation from, 226
Deoxycytidine,
deamination of, 327, 328, 329
utilization of, 328, 329
Deoxycytidine phosphates,
biosynthesis of, 323
deamination of, 324
Deoxycytidylate hydroxymethylase,
formation of, 325
Deoxycytidylic acid,
utilization of, 328, 329
Deoxyguanylate kinase,
bacteriophage and, 318
Deoxyribonuclease,
deoxyribonucleic acid synthesis and, 553
inhibitor of, 596
nucleic acid synthesis and, 545, 546
protein synthesis and, 527, 562, 589
Deoxyribonucleic acid, *see also* Nucleic acids
amount per cell, 610
8-azaguanine and, 540
coding and, 558, 561, 564, 588
components of, 296
composition of, 480–481, 588
distribution of, 483–484
functioning of, 589–590
protein associated with, 586
protein synthesis and, 489, 510, 563, 585–590, 610, 615
repressors and, 609

ribonucleic acid synthesis and, 544, 546–547, 563
structure of, 481–482
synthesis,
cell-free systems and, 552–554
intact cells and, 549–552
templates and, 587–588
thymidine diphosphate sugars and, 464
Deoxyribonucleotides,
formation of, 317–318
kinases for, 318
protein synthesis and, 530, 531
ribonucleic acid synthesis and, 545
Deoxyribose, formation of, 298
6-Deoxytalose, cell walls and, 425
Deoxyuridine monophosphate,
formation of, 324
kinase for, 324
Dephospho-coenzyme A, formation of, 272
Desamido-diphosphopyridine nucleotide,
formation of, 268
glutamic dehydrogenase and, 154
Desamidonicotinamide mononucleotide,
formation of, 268
Desthiobiotin,
biological activity of, 274–275
formation of, 274
Desulfovibrio,
amino acid activation in, 507
Desulfovibrio desulfuricans,
hemes in, 357
hydroxylamine reductase in, 123
nitrite assimilation by, 122
nitrogen fixation by, 122
Deuterium,
nitrogen fixation and, 145–146, 149
phosphoenolpyruvic carboxykinase and, 65
phosphoenolpyruvic carboxylase and, 63–64
Deuteroporphyrin,
growth promotion by, 339, 440
Dextrans,
antigenicity of, 375
branching of, 377–378
structure and occurrence of, 374–375
synthesis,
dextrins and, 378
sucrose and, 375–377
Dextrandextrinase, substrates of, 378
Dextransucrase,
acceptors for, 377
reaction catalyzed by, 375
Dextrins, dextran formation from, 378
Diadenylate,
polyadenylate formation and, 548
α,γ-Diaminobutyrate,
polyglutamate and, 404
Diaminopimelate, 173, 473
accumulation of, 237
biosynthesis of, 168, 169, 192–194, 198–202, 453
cell walls and, 415–417, 440–442, 452, 459–461, 478–479
dipicolinic acid and, 200
distribution of, 199
isomers, interconversion of, 202
lysine formation from, 199, 200, 238–240, 242
Diaminopimelic decarboxylase,
distribution of, 200
Diaminopimelic racemase, 237
1,3-Diaminopropane, ribosomes and, 495
2,6-Diaminopurine,
resistance to, 308
utilization of, 311–312
Diaminopyrimidines,
pteridine synthesis and, 279
Diaminouracil,
riboflavin biosynthesis and, 263, 265
Diauxie, enzyme formation and, 610
6-Diazo-5-oxo-L-norleucine,
purine synthesis and, 301
Dicarboxylic acids, synthesis of, 53
ϵ-(2,4-Dichlorosulfanilido) caproic acid,
biotin and, 274
3,6-Dideoxyhexoses,
cell walls and, 425, 462–463
Digitonin,
protein synthesis and, 494, 500
Dihydrolipoamide, activity of, 284
Dihydrolipoic acid, utilization of, 284
Dihydroorotase, repression of, 326, 607
Dihydroorotate dehydrogenase,
cofactors of, 330
repression of, 326, 607
Dihydroorotic acid,
formation of, 320–321
Dihydroporphyrins,
excretion of, 344–345

3,4-Dihydropyridazinone-5-carboxylic acid,
formation of, 140
Dihydrouracil, utilization of, 329
Dihydroxyacetone-3-phosphate,
ribose formation and, 296, 297
α,β-Dihydroxyisovalerate,
leucine biosynthesis and, 209
valine biosynthesis and, 209, 214
Dihydroxymaleic acid, 48
β,γ-Dihydroxy-β-methylbutyric acid,
pantothenate synthesis and, 255, 256
α,β-Dihydroxy-β-methylglutaric acid,
pyridoxal biosynthesis and, 270
α,β-Dihydroxy-β-methylvalerate,
isoleucine biosynthesis and, 209
Dihydroxytartaric acid, 48
2-Dimethylaminopurine,
occurrence of, 482
6-Dimethylaminopurine, 296
occurrence of, 481, 482
5,6-Dimethylbenzimidazole,
formation of, 282
5,6-Dimethylbenzimidazolecobamide coenzyme,
formation of, 281, 282, 283
1,2-Dimethyl-4,5-diaminobenzene,
riboflavin synthesis and, 262
vitamin B_{12} synthesis and, 282
Dimethylguanine, occurrence of, 481
6,7-Dimethyl-8-(D-ribityl)lumazine,
riboflavin synthesis and, 263–265
Dinitrofluorobenzene,
cell wall fragments and, 418
2,4-Dinitrophenol,
protein synthesis and, 502, 517, 530, 610
Dinitrophenylation,
cell wall peptide and, 428
1,3-Diphosphoglyceric acid,
photosynthesis and, 12, 13, 29
Diphosphopyridine nucleotidase,
formation of, 612
Diphosphopyridine nucleotide, 240
alanine dehydrogenase and, 154, 179
biosynthesis of, 268–269
glutamic dehydrogenase and, 154, 177
histidine synthesis and, 219
homoserine methylation and, 198
hydroxylamine reductase and, 123
hyponitrite assimilation and, 122
isocitric dehydrogenase and, 57
malic enzyme and, 55–56
nitrate reductase and, 122
nitric oxide reductase and, 122
nitrite assimilation and, 122
nitrite reduction and, 122
orotic acid formation and, 330
phosphogluconic dehydrogenase and, 60
photosynthesis and, 20, 21, 30–35
porphyrin synthesis and, 353
proline synthesis and, 186
requirement for, 259
riboflavin synthesis and, 264
Diphosphothiamine,
acetolactate synthesis and, 211–212
Dipicolinic acid, formation of, 200
Diplococcus glycinophilus,
carbon dioxide fixation by, 99, 104–105, 108

E

Edestin, molecular weight of, 473
Elaeganus,
nitrogen fixation in, 122, 125
Elastase, amino acids of, 477
Electron transfer,
nitrogen fixation and, 139
Electrophoresis, enzymes and, 580
Endotoxins, cell walls and, 419
Energy, protein synthesis and, 502
Enoloxalacetate,
phosphoenolpyruvic carboxylase and, 63
Enoyl hydrase,
leucine degradation and, 76
Enterobacteriaceae,
purine interconversions by, 307–312
purine synthesis by, 301, 305–307
Enterococcus stei,
coenzyme F synthesis by, 279
folic acid synthesis by, 277
Enzymes,
assay, induction and, 602
constitutive, 604–610
amounts of, 593, 604
episomes and, 582
extracted, biosynthetic pathways and, 174
fine structure alterations in, 584–585
formation,

deoxyribonucleic acid and, 585–590
differential effects on, 611–613
genes and, 578–585, 590–592
irradiation and, 613-615
mathematical formulation and, 621
metabolic control of, 615-618
problems of, 577–578, 618–621
quantitative aspects of, 592–615
specific inhibitors of, 610–611
homogeneity of, 580
inducible, 490, 607–610
amounts of, 593
formation of, 594–598
kinetics of, 600–604
selective advantage of, 616
specificity of, 599-600
repressible, 606–610
amounts of, 593
Epimerase,
photosynthesis and, 12, 16, 17, 21
Episomes, inheritance and, 582
Eremothecium ashbyii,
pteridine synthesis and, 279
riboflavin synthesis by, 254, 255, 262–265
Erythrocytes,
heme synthesis by, 348–349, 351
Erythrose 4-phosphate,
aromatic biosynthesis and, 224–226
biosynthetic pathways and, 169
photosynthesis and, 12, 15, 18
ribose formation and, 297
Escherichia coli,
acetolactate formation by, 211
alanine biosynthesis by, 203
β-alanine formation by, 272
amino acid activation in, 507, 508
amino acid biosynthesis by, 176
p-aminobenzoate auxotroph of, 278
ammonia assimilation by, 178
amylomaltase of, 593
analog resistant, 238
arginine synthesis by, 191-192, 239
aromatic ring biosynthesis by, 222–223
aspartases of, 156, 178–179, 193
aspartate formation by, 153, 155, 192–193
aspartate transcarbamylase of, 606
aspartic semialdehyde dehydrogenase of, 194, 196
aspartokinase of, 194
biotin and, 275
carbamylaspartate synthesis by, 320
carbon dioxide fixation by, 47, 69
catalase synthesis by, 361–362
cell walls, 416–419, 425, 439, 462, 464
amino acid-lipid complexes and, 517
strain differences and, 440–442
synthesis of, 446, 452, 459–460
chloramphenicol and, 557
citrate forming reaction in, 617
citrulline auxotrophs of, 191
cobamide synthesis by, 282, 283
coenzyme A synthesis by, 274
coenzyme F synthesis by, 279
coenzyme Q of, 286
cystathionine cleavage by, 197
cysteine auxotroph of, 197, 207
cystine and, 207
cytidine deaminase in, 327
cytidine triphosphate aminase of, 161
deoxycytidine phosphate formation by, 323
deoxycytidine utilization by, 328, 329
deoxycytidylate deamination by, 324
deoxyribonucleic acid, 588
synthesis of, 553
deoxyribonucleoprotein in, 586
deoxyribonucleotide formation by, 317
deoxyribonucleotide kinases of, 318
deoxyribose synthesis by, 298
diaminopimelate auxotroph of, 198
diaminopimelic decarboxylase of, 200
enzymes of, 579, 614
induction of, 594–595
repression of, 607
p-fluorophenylalanine and, 504
fluorouracil effect on, 541, 590
fluorouracil resistant, 327
folic acid synthesis by, 257, 277–279
formate exchange by, 86–88
formic hydrogenlyase of, 612
β-galactosidase of, 583, 593, 595–597, 599, 603, 604, 609, 612–614
genetic code and, 409
glucose fermentation by, 43, 605
β-glucuronidase of, 600
glutamate auxotroph of, 186
glutamate synthesis by, 176, 177
glutamic dehydrogenase of, 153
glutathione synthesis in, 185
growth rate, mutants and, 615

Escherichia coli—Continued
halogenated uracils and, 541
hemin requiring, 341
hemoprotein synthesis by, 360
histidine auxotrophs of, 217, 219
histidine synthesis by, 184, 219–221, 320
homocysteine methylation by, 198
homoserine in, 195
α-hydroxyethyl thiamine in, 262
infected, enzyme synthesis by, 586
isoleucine auxotroph of, 180
isoleucine synthesis by, 213–214
ketopantoate synthesis by, 271
lactose utilization by, 591
leucine auxotrophs of, 215
lysine auxotrophs of, 199, 200
membrane, protein synthesis by, 620
methionine auxotrophs of, 196
methionine synthesis by, 198, 283
5-methyltryptophan and, 608
mutant, amino acid synthesis by, 169
mutant enzymes of, 585
nicotinic acid synthesis by, 256, 266, 269
nitratase of, 603, 604
nitrate reductase in, 122, 612
nucleic acid synthesis by, 542, 546–547, 561
nucleoside utilization by, 311
nucleotide accumulation by, 443–444
ornithine auxotrophs of, 186–187
ornithine synthesis by, 188
ornithine transcarbamylase of, 191, 609
oxalacetate formation by, 51
pantothenate synthesis by, 255, 257, 258, 585
permease negative, 602
phenylalanine auxotrophs of, 227, 229
phenylalanine synthesis by, 239
phosphogluconic dehydrogenase in, 60
polyribonucleotide phosphorylase of, 549
proline auxotrophs of, 185
proline synthesis by, 186
protein synthesis, 486–488, 492, 494, 500–503, 526–527, 530–531, 560, 562, 563
chloramphenicol and, 555–556
purine synthesis by, 300, 319
pyrimidine auxotrophs of, 321, 323, 326-327
raffinose utilization by, 591
riboflavin synthesis by, 264–265
ribonuclease and, 493
ribose synthesis by, 297
ribosomes, 497–498, 515, 521, 522, 526–527
amino acid-lipid complexes in, 517
nucleic acid of, 482, 484
peptidase of, 497
ribonuclease of, 497
turnover of, 503
chloramphenicol and, 555
rough, lipopolysaccharide of, 464
selenomethionine and, 581
D-serine deaminase of, 600, 602–603
L-serine deaminase of, 600, 602–603
serine/glycine auxotrophs of, 205
serine synthesis by, 205–206
soluble ribonucleic acid of, 482, 483, 484, 512, 513, 555
spheroplasts of, 459
succinyldiaminopimelate deacylase of, 202
sulfanilamide inhibited, 299
sulfathiazole resistant, 278
sulfur auxotrophs of, 207
supressor mutations in, 591
tetrathionase of, 612, 613
T-even bacteriophage of, 296
thiamine synthesis by, 259
thiazole auxotrophs of, 260
2-thiouracil and, 590
threonine synthesis by, 195, 196
threonine utilization by, 173
thymidine utilization by, 328, 329
thymidylate synthetase of, 324
thymine auxotroph of, 325, 328, 591
transaminases of, 180–181, 202, 216
trans-N-deoxyribosylase and, 318
tryptophan analogs and, 505
tryptophanase of, 602–603, 612–614
tryptophan auxotrophs of, 229, 232–235
tryptophan synthesis, 231–232, 591
repression of, 608
tryptophan synthetase of, 584, 591, 606
tyrosine biosynthesis by, 239
unusual nucleotides in, 482
uracil utilization by, 327
uridine monophosphate synthesis by, 322
uridine triphosphate aminase of, 323

valine auxotrophs of, 210–211
valine incorporation by, 580
valine-sensitive, 505
valine synthesis by, 174, 208–211, 214
vitamin B_6 auxotrophs of, 270, 298
vitamin B_{12} auxotrophs of, 258
vitamin K synthesis by, 285
xylose fermentation by, 88
Ethanol,
fermentation, methane and, 108–109
Ethionine,
α-amylase and, 581
nucleic acid synthesis and, 543, 545
protein synthesis and, 505
Ethylene, mass analysis of, 100
Ethylmalonyl coenzyme A,
formation of, 97
Euglena gracilis,
chloroplasts, reproduction of, 582

F

Factor B,
cobamide synthesis and, 282, 283
nitrogen fixation and, 129
phosphate esters of, 283
Fatty acids,
photometabolism of, 7
synthesis, carbon dioxide and, 77–79, 86, 110
Feedback,
acetohydroxybutyrate formation and, 214
acetolactate formation and, 211
analogs and, 175
arginine synthesis and, 192
aspartokinase and, 194
biosynthetic pathways and, 242–243
escape from, 237–238
histidine synthesis and, 221, 316
homoserine kinase and, 196
α-ketobutyrate formation and, 213
lysine and, 314
metabolic control and, 618
porphyrin synthesis and, 368
purine synthesis and, 312–317
pyrimidine nucleotide synthesis and, 325–327
threonine and, 314
Ferrichrome,
heme formation and, 358
source of, 342
Ferrous salts,
carbon dioxide fixation and, 2, 4
oxidation of, 31
Flagella, synthesis of, 493, 504
Flagellin, amino acids of, 473, 474
Flavin, nitrogen fixation and, 142
Flavin adenine dinucleotide,
formation of, 265
homoserine methylation and, 198
hydroxylamine reductase and, 123
nitrate reductase and, 122
nitric oxide reductase and, 122
nitrite assimilation and, 122
nitrite reduction and, 122
nitrogen fixation and, 129
orotic acid formation and, 330
Flavin mononucleotide,
nitrate reductase and, 122
nitric oxide reductase and, 122
nitrite reduction and, 122
Flavobacterium polyglutamicum,
polyglutamate hydrolysis by, 400–401, 405
Flavokinase, substrates of, 265
Flavonoids, 222
Flavoprotein,
nitrogen fixation and, 147
Fluoride,
amino acid-lipid complexes and, 517
carbon dioxide fixation and, 46
Fluorokinase, identity of, 73–74
p-Fluorophenylalanine,
bacteriochlorophyll synthesis and, 364
hemoprotein formation and, 359
incorporation of, 580–581
nucleic acid synthesis and, 543
protein synthesis and, 504–505, 534–535
Fluorouracil,
alkaline phosphatase and, 590
β-galactosidase and, 590
protein synthesis and, 487, 541, 620
nucleotide accumulation and, 439
resistance, uracil utilization and, 327
Folic acid,
aspartase and, 156
biosynthesis of, 257, 276–280
formylation of, 279
methionine formation and, 195, 196
pantothenate synthesis and, 271
purine synthesis and, 312
reduction of, 279

Folic Acid—*Continued*
serine-glycine interconversion and, 204
thymidylate synthesis and, 330
Formaldehyde,
acetate formation and, 104, 107–108
lactate formation and, 104
methionine biosynthesis and, 283
tetrahydrofolic acid and, 280
thymidylate synthesis and, 330
Formamide,
cell wall extraction and, 423
Formate,
acetate formation and, 103–104, 107
decomposition of, 87
exchange, pyruvate and, 86, 87, 89
growth on, 23–25, 70
histidine synthesis and, 219
photosynthesis and, 22
purine synthesis and, 299, 319
riboflavin synthesis and, 255
serine formation and, 204
Formic dehydrogenase, 87
formate utilization and, 25
Formic hydrogenlyase, 87
formation of, 612
Formiminoglycine,
purine fermentation and, 106–107
Formin,
amino acid incorporation and, 516–517
Formylacetic acid,
β-alanine formation and, 272
N-Formylglycinamidine ribotide,
formation of, 184, 301, 302
glutamine and, 184
Formyl groups,
purine synthesis and, 318-319
Formylkynurenine,
nicotinic acid biosynthesis and, 265
β-Formylpropionic acid,
polyglutamate and, 404
Formyltetrahydrofolic acid,
purine nucleotide cycle and, 314–315
carboxamide and, 318
β-Fructofuranosidase,
formation of, 612
Fructose,
cellulose formation from, 381
dextransucrase and, 375
formate utilization and, 25
Fructose-1,6-diphosphate,
photosynthesis and, 12, 27
Fructose-6-phosphate,
photosynthesis and, 12, 15, 16, 17, 21, 27
ribose formation from, 297
xylose fermentation and, 88
Fructosylsucrose, formation of, 384
Fucose, cell walls and, 425
Fumarate,
ammonia assimilation and, 178
argininosuccinase and, 192
aspartate and, 153, 155–156
biosynthetic pathways and, 169
carbon dioxide fixation and, 9–11, 47
photophosphorylation and, 30
Fusarium,
pyridine nucleotide synthesis in, 268
Fusiformis (*Bacteroides*) *melaninogenicus*,
hemin requirement of, 339, 441
Fusiformis nigrescens,
menadione and, 285

G

Galactose,
dextransucrase and, 377
sequential induction and, 617
Galactose-1-phosphate transferase,
induction of, 617
β-Galactosidase, 579, 621
basal amounts of, 604
cis-trans test of, 583
constitutive, 593, 604, 605, 609
enzymic reversion and, 603
p-fluorophenylalanine and, 505
fluorouracil and, 541, 590
glucose and, 610–611
induction of, 594–603, 608, 609, 615
mutants and, 583, 585
properties of, 596–597
protein turnover and, 503
selenomethionine and, 581
"superconstitutive", 615
synthesis, 492, 582, 591, 610, 612
radioactive decay and, 589
ribonucleic acid and, 486–488, 510
2-thiouracil and, 590
ultraviolet and, 613, 614
x-rays and, 614
Galactozymase,
induction of, 594
ultraviolet and, 614

Galsucrose,
dextransucrase and, 376
levansucrase and, 385
Gene(s),
enzyme amounts and, 592–593
enzymes and, 578–585, 590–592, 619
fine structure alterations in, 584–585, 619
size of, 588
Genetic code, nature of, 409
Gentian violet,
cell wall synthesis and, 538
nucleotide accumulation and, 437–439, 446
Glomerella, tyrosinase of, 591
Gluconate,
cellulose formation from, 381
Glucosamine phosphate,
formation of, 184
lipopolysaccharide and, 425
amino acid biosynthesis and, 169, 170
amylomaltase and, 379
carbon dioxide fixation and, 27
cellulose formation from, 380–382
cell walls and, 418, 419, 459, 479
citrate transport and, 176
enzyme formation and, 610–611, 617
fermentation, 43
constitutivity and, 604–605
formate utilization and, 25
glutamine synthetase and, 157
hemoprotein formation and, 358, 359
histidine and, 205
nitrogen fixation and, 141
polyglutamate synthesis and, 396
riboflavin synthesis and, 263
ribose synthesis from, 296–297
serine formation from, 205
shikimate synthesis from, 224, 226
Glucose,
acetate formation from, 99
Glucose-1-phosphate,
cellulose synthesis and, 381
dextrandextrinase and, 378
dextransucrase and, 377
polysaccharides and, 373, 378–379
riboflavin phosphate formation and, 265
Glucose-6-phosphatase, diauxie and, 611
Glucose-6-phosphate,
cellulose synthesis and, 381
5-dehydroshikimate formation from, 224
glutamine and, 184
ribose formation from, 297
Glucose-6-phosphate dehydrogenase,
isocitrate formation and, 58
malate synthesis and, 55
phosphogluconic dehydrogenase and, 59–60
propionyl carboxylase and, 74
α-Glucosidase,
amino acid analogs and, 504
induction of, 599
synthesis, ribonucleic acid and, 487
ultraviolet and, 614
β-Glucosidase,
constitutive, 605
induction of, 600, 603
types of, 593
Glucozymase,
formation, nucleic acid and, 510
Glucuronate,
metabolism, transformation and, 586
β-Glucuronidase, induction of, 600, 601
Glutamate,
activating enzymes and, 508
alanine formation and, 203
ammonia assimilation and, 178–179
arginine formation from, 182, 186–192
biosynthesis of, 169, 170, 176–178, 241
carboxydismutase and, 27
cell walls and, 415, 416, 417, 423, 441, 442, 478
cross linkages and, 477
diaminopimelate formation and, 201, 453
excretion of, 238
glutamine formation from, 182
hemoglobin and, 558
histidine and, 206, 217
incorporation of, 527–529, 538
nitrogen fixation and, 136, 138
nucleic acid synthesis and, 541–543
ornithine formation from, 186–189
photosynthesis and, 22, 23, 27
polyglutamate synthesis and, 394, 396
proline formation from, 182, 185–186
racemases for, 398
transamination and, 180–182

Glutamate-β-methylaspartate isomerase, 85
Glutamate-tryptophan transaminase, 177
D-Glutamate,
cell walls and, 423, 424, 428, 440, 447–448, 479, 538, 539
formation of, 397, 398, 453, 537
occurrence of, 236
requirement for, 435, 461
Glutamic dehydrogenase, 161
ammonia incorporation and, 153–154
glutamate synthesis and, 177–179
mutations and, 585
Glutamic racemase, 453, 537
occurrence of, 236
Glutamic γ-semialdehyde,
ornithine synthesis and, 189
proline biosynthesis and, 185, 187, 240
Glutamine,
p-aminobenzoate synthesis and, 276
anthranilate formation and, 236
asparagine synthetase and, 159
diphosphopyridine nucleotide synthesis and, 268
formation of, 182, 398, 451
histidine synthesis and, 184, 221
nitrogen fixation and, 136
polyglutamate synthesis and, 399
purine synthesis and, 160, 299, 301, 302, 304, 319
requirement for, 183
transfer reactions of, 183–185
uridine triphosphate aminase and, 323
Glutamine synthetase,
ammonia incorporation and, 157–159
Glutamotransferase,
polyglutamate synthesis and, 399
γ-Glutamyl hydrazide,
glutamine synthetase and, 157
γ-Glutamyl hydroxamate,
formation of, 157, 158
Glutamyl γ-methylamide,
glutamine synthetase and, 157
γ-Glutamyl phosphate,
proline biosynthesis and, 185, 187
γ-Glutamyl transpeptidase,
polyglutamate synthesis and, 399
Glutathione,
γ-glutamyl transfer and, 184–185
protein synthesis and, 536
synthesis of, 451
transferring enzymes and, 514
Glyceraldehyde-3-phosphate,
photosynthesis and, 12, 15
ribose formation and, 297
tryptophan biosynthesis and, 231
Glycerol,
amino acid incorporation and, 516–517
carbon dioxide fixation and, 85
cellulose formation from, 381
fermentation, carbon dioxide and, 44–46
nicotinic acid biosynthesis and, 266
polyglutamate synthesis and, 394
Glycerol dehydrogenase,
enzymic reversion and, 603
Glycerophosphate,
cell wall synthesis and, 419, 457
pyridine nucleotide reduction and, 33
Glycinamide, formylation of, 318
Glycinamide ribotide, 306
formation of, 302
Glycine,
acetate formation from, 99, 104–105, 108
biosynthesis of, 173, 203–206, 242
carbon dioxide fixation in, 99, 104–105
cell walls and, 417–418, 424, 440, 479
conversion to serine, 204
corrin synthesis and, 282
cytochrome formation and, 359
erroneous utilization of, 580
nucleotide accumulation and, 439
permeability to, 205
porphyrin synthesis and, 348–350, 352–357, 367
prodigiosin synthesis and, 366
pyridoxal biosynthesis and, 270
purine fermentation and, 105-107
purine synthesis and, 299, 302, 319
pyrimidine synthesis and, 330
serine deaminase formation and, 612
silk protein and, 524
Glycogen,
dextrandextrinase and, 378
synthesis of, 378–379
Glycolaldehyde,
glycine and, 205
pyridoxal biosynthesis and, 270
Glycolate, glycine and, 205
Glycolipids, cell walls and, 419

Glycolysis, localization of, 495
Glycopeptide,
cell walls and, 415–418
polyol phosphate and, 422–425
Glycyl adenylate,
activating enzymes and, 507
Glycylglycine, activation of, 509
Glycylleucine, activation of, 509
Glycylmethionine, activation of, 509
Glyoxylate,
formation of, 58, 108
glycine and, 205
reductive amination of, 155
Gramicidins, nature of, 479
Green sulfur bacteria,
porphyrin excretion by, 342
Growth,
rate,
ribonucleic acid and, 485, 486
selective advantage and, 615
Guanine,
conversion to adenine, 308
feedback inhibition and, 314
fermentation of, 105–107
histidine synthesis and, 219, 319–320
nucleotide formation from, 308
requirement for, 305, 306
ribosomes and, 483
Guanosine,
phosphorylation of, 311
polysaccharide synthesis and, 414
Guanosine diphosphate,
factor B ester of, 283
formation of, 301
phosphoenolpyruvic carboxykinase and, 64–65, 68
Guanosine diphosphate colitose,
synthesis of, 462–463
Guanosine diphosphate-6-deoxy-D-talose,
isolation of, 464
Guanosine diphosphate fucose,
cell walls and, 462
Guanosine diphosphate-D-glycero-D-mannoheptose,
isolation of, 463
Guanosine diphosphate mannose,
transformations of, 462–463
Guanosine diphosphate-D-rhamnose,
isolation of, 464
Guanosine monophosphate,
ammonia incorporation and, 160–161
feedback control and, 315–316
formation of, 301, 304
hypoxanthine formation and, 308–309, 316
Guanosine triphosphate,
formation of, 301
protein synthesis and, 514, 515, 520, 522–526
soluble ribonucleic acid and, 545

H

Haemophilus influenzae,
hemin requirement of, 339–340
nitrate reductase in, 122
Haemophilus parainfluenzae,
diphosphopyridine nucleotide and, 259, 268
Heart
enzymes in, 592
isocitric decarboxylase of, 57–58
Hematin,
formation of, 360
occurrence of, 338–339
Hematoporphyrin,
accumulation of, 362
growth promotion by, 339, 440
Heme(s),
formation of, 348, 350, 356, 358–362, 367
structure of, 338
Hemin, 338
catalase synthesis and, 362
growth requirement for, 339–341
Hemocyanin, molecular weight of, 473
Hemoglobin,
amino acids of, 473, 474, 477
nitrogen fixation and, 142
sickle cell, 558, 585
synthesis, 492, 515, 521, 522
mutations and, 585
oxygen and, 366–367
Hemoproteins, 338
synthesis of, 358–362
Heteropolymers, synthesis of, 413–415
Heterozygotes, enzyme amounts in, 593
Hexokinase,
β-methylcrotonyl carboxylase and, 76
propionyl carboxylase and, 74

Hexose phosphates,
photosynthesis and, 18, 19, 20
protein synthesis and, 502
Histidase,
enzymic reversion and, 603
glucose and, 611
inducibility of, 616
Histidinal,
histidine synthesis and, 218, 219
Histidine,
acceptor ribonucleic acid for, 512
analogs, feedback and, 175
biosynthesis, 172, 182, 184, 216–221, 241, 242, 309, 310
complementation and, 584
gene mapping and, 584
repression of, 609
flagella synthesis and, 504
glucose and, 205
glutamate synthesis from, 170, 177, 206
photosynthesis and, 18
purine synthesis and, 299, 314–316, 319
requirement, growth rate and, 615
sequential induction and, 616
Histidine deaminase, 156
Histidinol, 240
conversion to histidine, 219
histidine biosynthesis and, 217
Histidinol phosphate,
formation of, 217, 218
Histones, polyglutamates and, 408
Homobiotin, carbon dioxide and, 92
Homocysteine,
formation of, 195-197
methylation, vitamin B_{12} and, 283
Homoserine,
cystathionine formation from, 195-197
formation of, 193, 195–196
function as intermediate, 193–195
isoleucine biosynthesis and, 212
methionine formation from, 195–198
threonine formation from, 195, 196
Homoserine kinase, requirements of, 196
Homozygotes, enzyme amounts in, 593
Hughes press,
nitrogen fixing preparations and, 128
Hydrazine,
glutamine synthetase and, 157, 183
nitrogen fixation and, 138–140, 144, 149–150
nitrogen metabolism and, 120
Hydrazine reductase,
occurrence of, 123, 140
Hydrogen,
carbon dioxide fixation and, 6, 7, 8, 21–22, 26–27, 32
nitrogen fixation and, 133, 144–146, 148–150
thermodynamic efficiency and, 3, 4
Hydrogenase,
diphosphopyridine nucleotide and, 31–32, 35
induction of, 604
nature of, 147
nitrogen fixation and, 126, 131, 133–135, 140, 144–146, 149, 151
Hydrogenomonas, coenzyme Q of, 286
Hydrogenomonas facilis,
carbon dioxide fixation by, 4, 22, 23
photophosphorylation in, 32
Hydrogenomonas ruhlandii,
carbon dioxide fixation by, 4, 22
hydrogenase of, 31
photophosphorylation in, 32
Hydrogen peroxide,
hemin requirement and, 341
Hydrogen sulfide,
carbon dioxide fixation and, 5–6
cysteine synthesis and, 207
3-Hydroxyanthranilic acid,
nicotinic acid biosynthesis and, 265–267
Hydroxyaspartic acid,
pantothenate and, 255
p-Hydroxybenzoate,
requirement for, 194, 222, 223
Hydroxy-β,β-dimethylbutyric acid(s),
pantothenate synthesis and, 255
Hydroxyethylthiamine,
formate exchange and, 89
Hydroxyethylthiamine pyrophosphate,
nitrogen fixation and, 131
carbon dioxide exchange and, 91
β-Hydroxyisovaleryl coenzyme A,
leucine degradation and, 75
3-Hydroxykynurenine,
nicotinic acid biosynthesis and, 265
Hydroxylamine,
amino acid activation and, 505, 509
glutamine synthetase and, 157, 183
nitrogen fixation and, 136, 138–140, 144
nitrogen metabolism and, 120, 121

Hydroxylamine reductase,
occurrence of, 123, 140
6-Hydroxy-2-methylaminopurine, 296
5-Hydroxymethylcytidylate,
bacteriophage and, 323, 325, 326
5-Hydroxymethylcytosine,
occurrence of, 296, 480, 481, 550
β-Hydroxy-β-methylglutaryl coenzyme A,
leucine degradation and, 75
Hydroxymethyltetrahydrofolic acid,
methionine synthesis and, 198
p-Hydroxyphenylpyruvic acid,
formation of, 228, 229
Hydroxyproline, synthesis of, 182
Hydroxypyruvate,
serine formation from, 205, 206
5-Hydroxyuridine,
enzyme synthesis and, 487
Hyphomicrobium vulgare,
carbon dioxide fixation by, 24
Hyponitrite,
assimilation, organisms performing, 122
nitrogen fixation and, 141
nitrogen metabolism and, 120, 121
Hypoxanthine,
acetate formation from 98, 105–108
nucleotide formation from, 308
precursor of, 299
requirement for, 305

I

Imidazole acetol phosphate,
histidine synthesis and, 217, 218
Imidazole glycerol dehydrogenase,
complementation and, 584
Imidazoleglycerol phosphate, 240
formation of, 184, 314–315
histidine synthesis and, 217, 218, 220–221, 241, 242
Imidazole ring, origin of, 219–221
4-(Imidazolidone-2-)caproic acid,
biotin and, 275
Iminoglutaric acid, formation of, 153
Iminopropionate,
alanine dehydrogenase and, 154–155
Incorporation enzyme,
characteristics of, 519
Incorporation factor,
nucleic acid synthesis and, 544
protein synthesis and, 544
Indole, 170, 240, 242
nicotinic acid synthesis and, 266
requirement, transformation and, 586
tryptophan biosynthesis and, 229–230, 233–234
Indole-3-acrylic acid,
nicotinic acid biosynthesis and, 256, 266
Indole-3-glycerol phosphate,
formation of, 230–233
tryptophan formation from, 233–235, 241, 242
Indoleglycerol phosphate hydrolase,
tryptophan auxotrophs and, 233
Indole-serine carboligase, 233
Induction,
function of, 616–618
model of, 620–621
sequential, 616–617
Infection, polyglutamate and, 407–408
Inosine, guanine formation from, 311
Inosine diphosphate,
formation of, 301
phosphoenolpyruvic carboxykinase and, 65
Inosine monophosphate,
adenine formation and, 308
formation, 300–303
guanine and, 309
guanine formation from, 308, 316
xanthine formation from, 308
Inosine monophosphate dehydrogenase,
mutants and, 306
repression of, 317
Inosine monophosphate pyrophosphorylase,
guanine formation and, 309–310
Inosine triphosphate, formation of, 301
Inosinicase, repression of, 317
Inositol, occurrence of, 286
Inositol dehydrogenase, 579
inducer removal and, 603, 611
Insulin,
amino acids of, 473–476
species variations in, 475
strepogenin and, 532
Iodouracil, incorporation of, 541

Iron,
arginine decarboxylase and, 612
hemoprotein formation and, 358
hyponitrite assimilation and, 122
insertion into tetrapyrrole, 358
nitric oxide reductase and, 122
nitrite assimilation and, 122
nitrite reduction and, 122
nitrogen fixation and, 122, 125, 126, 129, 131–132, 147–148, 150
porphyrin excretion and, 344, 345, 354–356
porphyrin interconversion and, 355
Iron-binding compounds,
growth promotion by, 342
Isocitratase, 58
glycine and, 205
heterotrophic growth and, 23
Isocitrate,
carbon dioxide fixation and, 51, 52
glutamate synthesis and, 176
Isocitric dehydrogenase,
carbon dioxide fixation and, 31, 52, 57–58
citrate accumulation and, 617
function of, 58
Isoguanine, utilization of, 311–312
Isoleucine,
alanine formation and, 203
biosynthesis, 172, 182, 192, 212–215, 240, 241, 243
feedback and, 618
polyglutamate synthesis and, 396
thiamine synthesis and, 261
threonine and, 205
transamination of, 180–181
valine utilization and, 580
Isomaltose, dextransucrase and, 377
Isomerase, photosynthesis and, 12
Isonicotinic acid hydrazide,
resistance, hemin requirement and, 339–341
Isopropanol,
carbon dioxide fixation and, 8
Isopropyl-β-D-thiogalactoside,
β-galactosidase and, 602
Isotopes,
biosynthetic pathways and, 172–173
protein assay and, 490–492
Isovaleryl coenzyme A,
leucine degradation and, 75

K

α-Keto acids,
amino acid biosynthesis and, 171, 172
α-Ketobutyrate,
cystathionine cleavage and, 197
formation of, 213
homocysteine and, 197
homoserine and, 197
isoleucine biosynthesis and, 209, 212, 213
nitrogen fixation and, 128–129, 142, 146
threonine and, 243
4-Keto-6-deoxy-D-mannose,
fucose formation and, 462–463
α-Keto-β,β-dimethyl-γ-hydroxybutyric acid, *see* ketopantoic acid
Ketogluconic acids,
cellulose formation from, 381
2-Ketoglucuronoreductase,
induction of, 604
α-Ketoglutaramic acid,
formation of, 185
α-Ketoglutarate, 48
azine formation from, 140
biosynthetic pathways and, 169
carbon dioxide fixation and, 47, 51, 52, 57
glutamate synthesis and, 176, 177
D-glutamate formation and, 397
nitrogen fixation and, 136–138
polyglutamate formation from, 236–237
porphyrin biosynthesis and, 352–354, 357
transamination of, 180–181
α-Ketoglutarate dehydrogenase,
porphyrin synthesis and, 353, 357
β-Ketoglutarate, 48
α-Ketoglutaric oxidase,
carbon dioxide exchange by, 88
α-Ketoglutaric oxime,
nitrogen fixation and, 140, 144
α-Keto-β-hydroxyisovalerate,
valine synthesis and, 214
4-(2′-Keto-3′-hydroxypropyl)-imidazole, *see* Imidazole acetol
α-Ketoisocaproate,
formation of, 215–216
leucine biosynthesis and, 209, 215–216

leucine degradation and, 75
α-Ketoisovalerate,
leucine biosynthesis and, 209, 215
one-carbon units and, 215
pantothenate synthesis and, 258, 259, 271
salicylic acid and, 255, 256
valine biosynthesis and, 209
α-Keto-β-methylvalerate,
isoleucine biosynthesis and, 209
Ketopantoic acid,
pantothenate synthesis and, 258, 271
salicylic acid and, 255, 256
3-Keto-6-phosphogluconate,
phosphogluconate dehydrogenase and, 59
α-Ketosuccinamic acid,
formation of, 185
Kinetosomes,
glutamate incorporation by, 518
Krebs cycle,
enzymes, induction of, 617
Kynurenine,
nicotinic acid biosynthesis and, 265

L

Lactase, genes affecting, 591
Lactate,
acetate formation from, 99
carbon dioxide fixation and, 9–11
carboxydismutase and, 22
citrate transport and, 176
formation,
acetate and, 103
formaldehyde and, 104
isoleucine biosynthesis and, 212
malic enzyme and, 53, 55–56
nitrogen fixation and, 141
photophosphorylation and, 30
valine synthesis from, 208, 210
xylose fermentation and, 88–89
Lactic dehydrogenase,
malic enzyme and, 53, 56
β-methylcrotonyl carboxylase and, 76
methylmalonyl - oxalacetic transcarboxylase and, 83
propionyl carboxylase and, 72
Lactobacilli,
biotin requirement of, 89
hemes in, 357
orotic acid and, 320
protein synthesis by, 503
pyrimidine requirements of, 327, 328
strepogenin and, 532
trans-N-deoxyribosylase of, 318
Lactobacillus acidophilus,
biocytin and, 276
cell walls of, 463–464
deoxyribonucleic acid synthesis in, 550
Lactobacillus arabinosus,
amino acid activation in, 509
D-amino acids in, 236
biocytin and, 276
biotin assay by, 74
biotin deficient, 518
biotin sulfoxide and, 276
carbon dioxide assimilation by, 89
cell walls of, 423, 444, 479
coenzyme A synthesis by, 272, 274
coenzyme F synthesis by, 279
desthiobiotin and, 275
folic acid synthesis by, 278
glutamine and, 183
malic enzyme in, 53, 55–56
oxalacetate decarboxylation by, 51
ribonucleic acid synthesis by, 547, 549
Lactobacillus bifidus,
cell walls of, 461
Lactobacillus bulgaricus,
coenzyme A synthesis by, 274
orotic acid requirement of, 320
pantetheine formation by, 273
Lactobacillus casei,
biotin sulfoxide and, 276
biotin requirement of, 257
cell walls of, 478
desthiobiotin and, 275
histidine and, 219
lipoic acid and, 284
peptide requirement of, 533
purine requirement of, 312, 319
riboflavin requirement of, 262
Lactobacillus delbrueckii,
biocytin and, 276
Lactobacillus fermenti,
thiamine and, 259
thiamine pyrophosphate and, 262
Lactobacillus helveticus,
coenzyme A synthesis by, 274
nucleotide accumulation by, 443
pantothenate requirement of, 257

Lactobacillus helveticus—Continued
pantetheine formation by, 273
Lactobacillus lactis,
vitamin B_6 requirement of, 270
Lactobacillus leichmanii,
deoxyribosides and, 298, 330
purine utilization by, 312
Lactobacillus pentosus,
biocytin and, 276
oxybiotin and, 275
Lactobacillus plantarum,
cell walls of, 478
riboflavin synthesis by, 264
p-aminobenzoylglutamate and, 277
coenzyme F synthesis by, 279
Lactose,
dextransucrase and, 377
enzyme induction and, 594–595, 612
β-galactosidase induction by, 488
Lactsucrose, dextransucrase and, 376
Leguminosae, nitrogen fixation in, 122
Leucine,
acceptor ribonucleic acid for, 512, 513
activation of, 506, 511
alanine formation and, 203
biosynthesis of, 168, 182, 209, 215–216, 242
degradation of, 75
flagella synthesis and, 504
serine deaminase induction by, 600, 612
transamination of, 180–181
Leuconostoc mesenteroides,
biocytin and, 276
coenzyme F synthesis by, 279
dextran of, 374
dextransucrase of, 375, 377-378
phosphogluconate metabolism by, 61
phosphogluconic dehydrogenase in, 60
Leucyltyrosine, activation of, 509
Levan,
structure and occurrence of, 382–383
synthesis of, 373
Levanpolyase, levan structure and, 383
Levansucrase,
induction of, 604
localization of, 383
transfructosylation by, 383–386
Light,
adenosine triphosphate and, 34–35
bacteriochlorophyll synthesis and, 364–365
carboxydismutase and, 28
tetrapyrrole synthesis and, 367
Lipase, protoplasts and, 493
Lipid,
amino acid incorporation and, 517
cell walls and, 418
microsomes and, 495, 499
Lipoic acid, 276
biosynthesis of, 284
requirement for, 283–284
sulfoxide, utilization of, 284
Lipopolysaccharides,
cell walls and, 419, 423, 425
rough strains and, 464
Lipoxidase,
amino acid-lipid complexes and, 517
Lipoyl adenylate,
lipoate activation and, 284
Liver,
amino acid activation in, 507–508
amino acid-lipid complexes in, 517
carbamyl phosphate synthetase of, 159–160
chloramphenicol and, 557
enzymes in, 592
glutamic dehydrogenase of, 154
malic enzyme of, 53
phosphoenolpyruvic carboxykinase of, 64
polyglutamate degradation by, 402
unusual nucleotides in, 482, 513
Luekin, polyglutamate and, 407
Lysine,
aspartokinase and, 194
biosynthesis of, 168, 169, 172, 176, 182, 192, 193, 199, 200, 238–242
biotin binding and, 92
cell walls and, 415, 416, 423, 424, 428, 442, 447-448, 452, 459, 478, 479, 538–540
cross linkages and, 477
deprivation, nucleotide accumulation and, 434–437
diaminopimelate and, 199
excretion of, 237
p-hydroxybenzoate requirement and, 194
lipoate activation and, 284
transamination of, 181
Lysine decarboxylase,

induction, 600
inducer removal and, 603
ultraviolet and, 614
x-rays and, 614
Lysozyme,
cell walls and, 416–417, 424, 459
nitrogen fixing preparations and, 128
polyglutamates and, 400, 408
protein assay and, 490
synthesis, 8-azaguanine and, 487

M

Magnesium,
glutamine synthetase and, 158, 398
nitrogen fixation and, 129
ribosomes and, 483, 495, 497–498, 526–527
Magnesium protoporphyrin,
accumulation of, 362, 363
excretion of, 344, 345
Magnesium vinylpheoporphyrin a_5,
accumulation of, 362
structure of, 363
Maize, chloramphenicol and, 557
Malate,
carbon dioxide fixation and, 9–11, 22, 23, 27, 47
carboxydismutase and, 28
photogas evolution and, 145
porphyrin excretion and, 342
synthesis of, 55
Malic dehydrogenase,
methylmalonyl isomerase and, 85
methylmalonyl-oxalacetic transcarboxylase and, 83
oxalacetic decarboxylase and, 48–50
phosphoenolypruvic carboxylase and, 63–64
pyruvic carboxylase and, 81
Malic enzyme,
affinity for substrates, 63
biotin and, 91, 518
carbon dioxide and, 10, 49–50, 51, 53–57
distribution of, 53
intracellular location of, 68
phosphoenolpyruvic carboxykinase and, 66–68
Malonyl coenzyme A,
carbon dioxide and, 52, 77
fatty acid synthesis and, 78–79
pyruvic carboxylase and, 80
Maltase, induction of, 601
Maltose,
amylose formation and, 379
dextrandextrinase and, 378
dextransucrase and, 375, 377
β-galactosidase induction and, 601
Maltozymase,
induction of, 600
loss of, 603
x-rays and, 614
Mammalian cells,
protein synthesis in, 499–500
Manganese,
glutamine synthetase and, 158, 398
hydroxylamine reductase and, 123
nitrogen fixation and, 126, 129
polyglutamate synthesis and, 395–396, 400, 403
Mannitol dehydrogenase,
transformation and, 586
Mannose, dextransucrase and, 377
Mannosidostreptomycin,
production of, 464
Mastigocladus laminoseus,
nitrogen fixation by, 127
Mating factors, episomes and, 582
Melanophore-stimulating hormones,
amino acids of, 477
Melibiose,
β-galactosidase and, 597, 600
Menadione, *see also* Vitamin K
growth stimulation by, 285
β-Mercaptoethanol,
polyglutamate synthesis and, 400
Mercaptoethylamine,
transferring enzymes and, 514
6-Mercaptopurine,
resistance to, 308–311
Mesaconate,
glutamate fermentation and, 156
Mesoporphyrin,
growth promotion by, 339, 440
Metals,
oxalacetic decarboxylase and, 50–51
Methane,
formation, carbon dioxide and, 108–109
growth on, 23–24
thermodynamic efficiency and, 3
Methanobacterium,
nitrogen fixation by, 122

Methanobacterium omelianskii,
carbon dioxide reduction by, 108–109
Methanobacterium propionicum,
carbon dioxide reduction by, 108–109
Methanobacterium sulfoxidans,
carbon dioxide reduction by, 108–109
Methanol,
growth on, 23–24, 70
methane formation from, 109
Methionine,
activation of, 506
biosynthesis of, 169, 182, 192–198, 239, 242, 283
corrin synthesis and, 282
cysteine formation from, 197, 208
degradation of, 239
flagella synthesis and, 504
p-hydroxybenzoate requirement and, 194
requirement, growth rate and, 615
spermine and, 192
thymidylate synthesis and, 330
transamination of, 180
vitamin B_{12} and, 258, 283
D-Methionine, 397
cell wall synthesis and, 439
1-Methyladenine, occurrence of, 481
2-Methyladenine, 296
occurrence of, 481, 482
Methylamine,
glutamine synthetase and, 157–158
growth on, 23–24
2-Methyl-4-amino-5-hydroxymethylpyrimidine,
thiamine synthesis from, 260
2-Methyl-4-amino-5-hydroxymethyl pyrimidine monophosphate,
pyridoxal phosphate and, 262
2-Methylamino-6-hydroxypurine,
occurrence of, 482
6-Methylaminopurine, 296
occurrence of, 481, 482
2-Methyl-3-amyl pyrrole,
prodigiosin synthesis and, 365–366
4-Methylanthranilate,
indole formation from, 231–232
β-Methylaspartase,
ammonia incorporation and, 156–157
β-Methylcrotonyl carboxylase,
biotin and, 90, 92–95, 97
carbon dioxide fixation and, 52, 75–77
β-Methylcrotonyl coenzyme A,
leucine degradation and, 75
5-Methylcytosine, 296
incorporation of, 552
occurrence of, 481, 482
Methylene blue, hydrogenase and, 146
5,10-Methylene tetrahydrofolic acid,
thymidylate synthetase and, 324, 330
Methyl fumaric acid, aspartase and, 156
Methyl-α-D-galactoside,
β-galactosidase and, 597
Methyl-β-D-galactoside,
β-galactosidase and, 597
α-Methyl glucoside,
dextransucrase and, 377
α-glucosidase induction and, 599
Methyl-β-glucoside, induction by, 600
β-Methylglutaconyl coenzyme A,
carbon dioxide and, 52, 75
1-Methylguanine, 296
occurrence of, 482
4-Methyl-5-β-hydroxyethylthiazole,
thiamine synthesis from, 260
6-Methyl-7-hydroxy-8-N-ribityllumazine,
formation of, 265
6-Methylindole, formation of, 231–232
Methylmalonyl coenzyme A,
carbon dioxide and, 52, 71–75
pyruvic carboxylase and, 80
Methylmalonyl isomerase,
assay of, 85
propionate formation and, 82–83
sources of, 72
Methylmalonyl-oxalacetic transcarboxylase,
biotin and, 90, 97
function of, 98, 111
propionate formation and, 81–86
2-Methyl-1,4-naphthoquinone, *see* Vitamin K
O-Methylribose, occurrence of, 481
Methyl-β-thiogalactoside,
β-galactosidase induction and, 602
5-Methyltryptophan,
activation of, 506
deoxyribonucleic acid synthesis and, 550
mutant selection and, 606
protein synthesis and, 505

Methyl viologen,
nitrogen fixation and, 133
Mevalonic acid,
coenzyme Q and, 286
vitamin K synthesis and, 285
Microbacterium lacticum,
growth factors for, 342
Micrococcus,
amino acid activation in, 507
Micrococcus aerogenes,
tetrahydrofolate formylase of, 279
Micrococcus cinnabareus,
cell walls of, 478
Micrococcus denitrificans,
carbon dioxide fixation by, 22, 23
cytochromes in, 360
denitrification by, 122
enzyme changes in, 617
nitrate reductase in, 122
Micrococcus glutamicus,
N-acetylglutamic semialdehyde formation by, 188
arginine auxotroph of, 238
arginine synthesis by, 192
ornithine formation by, 189
Micrococcus lactilyticus,
hydrazine reductase in, 123, 140, 150
succinate decarboxylation by, 85
Micrococcus lysodeikticus,
carbon dioxide exchange in, 47–48, 51
cell walls of, 416, 417, 418, 419, 452
copolymer synthesis by, 409
coproheme formation by, 358
porphyrin excretion by, 345, 346
porphyrin synthesis by, 354
protein assay in, 490
ribonucleic acid synthesis by, 547
Micrococcus pyogenes,
cell walls of, 479
hemin requirement of, 339, 341
hemoprotein synthesis by, 361
Micrococcus sodonensis,
biocytin and, 276
Microsomes,
protein synthesis and, 495, 499–500, 519–527
Mitochondria,
amino acid incorporation by, 524
carbon dioxide fixation by, 68
chloramphenicol and, 557
enzymes of, 495
Mitomycin c, enzyme synthesis and, 589
Molybdate,
nitrate reductase and, 122
nitrogen fixation and, 122, 125, 126, 129, 147, 151
Monomethylguanine, occurrence of, 481
Muramic acid,
cell walls and, 440–441
chemical synthesis of, 427
isolation of, 428
Mustard gas, enzyme formation and, 614
Mutants,
biosynthetic pathways and, 171–172
constitutive, selection of, 605–606
enzyme synthesis and, 582–584
growth rates of, 615
inducible enzymes and, 608
purine requiring, classes of, 305–307
suppressor, 591
vitamin synthesis by, 257–259
Mycobacteria,
coenzyme Q and, 286
inositol in, 286
Mycobacterium avium,
fatty acid synthesis by, 78
folic acid synthesis by, 277–278
glutamic racemase in, 236
Mycobacterium karlinski,
porphyrin excretion by, 346
Mycobacterium paratuberculosis,
menadione and, 285
Mycobacterium phlei, vitamin K in, 285
Mycobacterium smegmatis,
propionyl carboxylase in, 73
riboflavin production by, 254
Mycobacterium tuberculosis,
alanine dehydrogenase of, 154–155
biotin and, 275
hemin requiring, 339, 440–441
riboflavin synthesis by, 262
vitamin K in, 285

N

Neisseria gonorrhoeae,
glutathione requirement of, 536
thiamine phosphates and, 262
Neisseria perflava,
polysaccharide synthesis by, 378–379
Neolactose,
β-galactosidase and, 597
mutant selection and, 605

Neuraminic acid, cell walls and, 442
Neurospora,
acetate mutant of, 591
adenylosuccinase of, 584, 591
arginine biosynthesis by, 239
aromatic ring biosynthesis by, 222
complementation in, 583–584
cystathionine cleavage by, 197
cystathionine formation by, 195–197
cysteine auxotrophs of, 197
enzyme synthesis in, 612
histidine auxotrophs of, 217
homoserine in, 195
isoleucine synthesis by, 212, 214
lactase of, 591
lysine formation by, 199
methionine auxotroph of, 196
mutant enzymes of, 585
mutants of, 581, 582
nicotinic acid synthesis by, 265
ornithine synthesis in, 189
orotic acid and, 320
pantothenate requirement of, 591
phenylalanine synthesis by, 239
pyrimidine synthesis by, 329
pyrroline-5-carboxylate reductase of, 186
respiratory system, inheritance of, 582
ribonucleic acid synthesis in, 589
serine/glycine auxotrophs of, 205
sulfanilamide requiring, 616
sulfur auxotroph of, 207
supressor mutations in, 591–592
thiamine synthesis by, 260
tryptophan auxotrophs of, 229, 233–234, 617
tryptophan synthetase of, 591–592
tyrosinase of, 585, 591, 608
tyrosine biosynthesis by, 239
valine biosynthesis by, 210, 214
Neurospora crassa,
amino acid biosynthesis in, 176
biotin sulfoxide and, 276
coenzyme A synthesis by, 273–274
deoxyribonucleotide formation by, 318
glutamate synthesis by, 178
glutamic dehydrogenase of, 153
hydroxylamine reductase in, 123
leucine auxotrophs of, 216
nitrate reductase in, 122
nitrite assimilation by, 122
riboflavin synthesis by, 264
thiazole auxotrophs of, 260
Nicotinamide,
purine synthesis and, 319
pyrimidine synthesis and, 330
Nicotinic acid,
biosynthesis of, 253, 256, 265–269
porphyrin synthesis and, 357
requirement, transformation and, 586
sequential induction and, 617
Nicotinic acid mononucleotide,
isolation of, 268
Nitramide, nitrogen fixation and, 141
Nitratase,
basal amount of, 604
enzymic reversion and, 603
oxygen and, 361
Nitrate,
carbon dioxide fixation and, 19, 22
energy metabolism and, 121
metabolism of, 119–121, 127
Nitrate reductase,
occurrence of, 122
synthesis of, 612
Nitric oxide,
nitrogen fixation and, 142
Nitric oxide reductase,
nitrogen fixation and, 151
occurrence of, 122
Nitride(s),
nitrogen fixation and, 150–151
Nitrification,
carbon dioxide fixation and, 2
Nitrite,
assimilation, organisms performing, 122
energy metabolism and, 121
metabolism of, 119, 120
oxidation of, 31
reduction, organisms performing, 122
thermodynamic efficiency and, 3
Nitrite oxidase, occurrence of, 123
Nitrite reductase,
nitrogen fixation and, 151
Nitrobacter, 2
cytochromes in, 31
nitrogen metabolism in, 121, 123
Nitrobacter agilis,
oxidative phosphorylation in, 31

Nitrogen,
fixation, 119–120, 122
biological agents of, 124–125
existing problems of, 151–152
hydrogenase and, 144–146
proposed mechanism of, 147–151
reaction steps in, 135–144
requirements for, 125–127
scope of, 121–124
soluble systems, 127–135
Nitrogenase,
action mechanism of, 148–149
isolation of, 134
nature of, 147–148
Nitrosomonas, 2
efficiency of, 3
nitrogen metabolism in, 121, 123
Nitrous acid, mutagenesis and, 587, 619
Nitrous oxide,
nitrogen fixation and, 139, 145–146, 150
Nitrous oxide reductase,
occurrence of, 122
Nocardia,
nitrogen fixation in, 122, 125
Nocardia rugosa,
factor B esters in, 383
Nodules,
nitrous oxide reductase in, 122
Nostoc, nitrogen fixation by, 122
Notatin, amylomaltase and, 379
Novobiocin,
nucleotide accumulation and, 434, 439
Nuclei, chloramphenicol and, 557
Nucleic acids, *see also* Deoxyribo- and Ribonucleic acids
bacterial content of, 295
bacteriochlorophyll synthesis and, 364
chemical nature of, 479–485
hemoprotein formation and, 359
protein synthesis and, 485–489, 509–513, 614
synthesis, problems of, 485
Nucleoside diphosphates,
nucleic acid synthesis and, 548
Nucleoside diphosphokinase,
cell wall synthesis and, 445
phosphoenolpyruvic carboxykinase and, 65
Nucleoside monophosphates,
polymer synthesis and, 414
Nucleoside phosphotransferase,
cytidine utilization and, 327–328
Nucleoside triphosphates,
interpolynucleotide incorporation of, 546–547
nucleic acid synthesis and, 545, 552
Nucleotides,
accumulation of, 429, 430–439
isolation of, 429–431
peptide complexes of, 536–537
terminal addition of, 545–546
vitamin B_{12} and, 280
Nucleotide pyrophosphorylases,
purine analogs and, 308
Nutrition, enzyme production and, 593

O

Oleic acid, biotin requirement and, 257
Oligerces paradoxus,
chromosome elimination in, 592
Operator gene, repressor and, 609
Ophiostoma multiannulatum,
purine synthesis by, 300
Ornithine,
biosynthesis, 176, 239–241
glutamate and, 186–189
carbamyl phosphate synthetase and, 159
citrulline formation from, 189–191
excretion of, 238
proline and, 206
transaminases and, 181
Ornithine δ-transaminase,
occurrence of, 189
Ornithine transcarbamylase,
constitutive, 593
enzyme repression and, 191, 606, 609
synthesis of, 498
uracil and, 612
Orotic acid,
bacterial nutrition and, 320
formation of, 321
Orotidine, excretion of, 322
Orthophosphate, *see* Phosphate
Oviduct, protein synthesis in, 517
Oxalacetate, 58
acetate formation from, 108
aspartate formation from, 192
biosynthetic pathways and, 169
carbon dioxide fixation and, 47–51, 52

Oxalacetate—*Continued*
glutamate synthesis and, 176
malic enzyme and, 50, 51, 53–55
propionate formation and, 81–83
Oxalacetic decarboxylase,
carbon dioxide fixation and, 48–51
specificity of, 48
Oxalacetic oxime,
nitrogen fixation and, 140, 144
Oxalacetylphosphate,
phosphoenolpyruvic carboxykinase and, 66
Oxalate,
acetate formation and, 108
carbon dioxide fixation and, 25
Oxalosuccinate, 48
isocitric dehydrogenase and, 57–58
Oxamycin,
cell walls and, 415, 453–457, 459, 540
nucleotide accumulation and, 431, 434
Oximes, nitrogen fixation and, 139–140
α-Oxo-, *see* α-Keto-
Oxybiotin, biological activity of, 275
Oxygen,
bacteriochlorophyll synthesis and, 364–365
carbon dioxide fixation and, 3–5
carboxydismutase and, 28
enzyme formation and, 612
hemoprotein synthesis and, 359–361
nitrogen fixation and, 131, 142–143, 147
tetrapyrrole synthesis and, 366–367
Oxytocin, streptogenin and, 532–533

P

Palmitic acid, synthesis of, 78
Pancreas,
acceptor ribonucleic acid of, 512
amino acid activation in, 507–508
chloramphenicol and, 557
polyglutamate hydrolysis by, 402
protein synthesis in, 485
Pantetheine,
formation of, 272, 273
occurrence of, 274
Pantoic acid,
biosynthesis of, 271
coupling with β-alanine, 272
pantothenate requirement and, 257, 258
salicylic acid and, 255, 256
Pantothenate,
biosynthesis, 208, 255–259, 270–274
mutations and, 585
porphyrin synthesis and, 356
Pantothenate synthetase,
reactions catalyzed, 272
Pantothenylcysteine,
decarboxylation of, 272
growth response to, 272
Papain,
polyglutamate hydrolysis by, 402
Pasteurella,
nicotinic acid utilization by, 267–268
Pasteurella pestis,
enzyme induction in, 617
hemin requirement of, 339, 440–441
hemoprotein formation by, 359–360
Pasteurella pseudotuberculosis,
polysaccharide of, 425
rough, lipopolysaccharide of, 464
Peas,
acceptor ribonucleic acid of, 512, 523
amino acid activation in, 507–508
amino acyl ribonucleic acid and, 525
chloramphenicol and, 557
ribosomes, protein release from, 515, 521, 522
Penicillamine, biosynthesis of, 260
Penicillin,
auxotroph selection and, 258
binding of, 598
cell wall synthesis and, 426–429
spheroplasts and, 459–461
Penicillinase,
canavanine and, 581
constitutive, 593, 604, 605
induction of, 595, 597–598, 603, 615
properties of, 597
ultraviolet and, 613–614
Penicillium chrysogenum,
desthiobiotin and, 275
pyridine nucleotide synthesis in, 268
Pentose phosphate,
photosynthesis and, 14, 19
Pentose phosphate isomerase,
photosynthesis and, 17, 20
Pepsin, polyglutamate and, 402
Peptidase, ribosomes and, 497
Peptides,
activation of, 509

capsular, 478
synthesis of, 537
cell walls and, 416–418, 428–429, 442, 537–540
mutants and, 532–534
protein synthesis and, 531–537
synthesis of, 447–452
Perfluorooctanoate, 524
incorporation enzyme and, 519
Periodate,
acceptor ribonucleic acid and, 512, 513
Permease(s),
enzyme formation and, 582, 600-602
glucose repression and, 611
induction of, 601, 603
loss of, 603
operator gene and, 609
ultraviolet and, 613, 614
pH, nitrogen fixation and, 130
Phaseolus vulgaris,
roots, nitrogen fixation in, 125
Phenol,
cell wall extraction and, 423, 425
protein synthesis and, 494
Phenylacetylglucosaminide(s),
antigenicity and, 421
Phenylalanine,
biosynthesis of, 172, 182, 222–229, 241, 242
excretion of, 237
incorporation, 563
lipid and, 517
polyglutamate synthesis and, 396
thiamine synthesis and, 261
transamination of, 180–181
D-Phenylalanine, transamination of, 398
Phenyl-β-D-galactoside,
β-galactosidase and, 597
Phenylisothiocyanate,
cell wall peptide and, 428
Phenylpyruvic acid,
formation of, 227–228
Pheophorbide *a*,
excreted porphyrins and, 345
structure of, 364
Phosphatase, photosynthesis and, 12, 17
Phosphate,
nitrogen fixation and, 126, 129
polyglutamate formation and, 395
3′-Phosphoadenosine-5′-phosphosulfate,
sulfur assimilation and, 207
Phosphoenolypruvate, 239
aromatic biosynthesis and, 225–227
biosynthetic pathways and, 169
carbon dioxide fixation and, 20
cell wall synthesis and, 446
formation of, 66–68, 80
Phosphoenolpyruvic carboxykinase, 51
affinity for substrates, 63
biotin and, 91
carbon dioxide fixation and, 52, 64–68
distribution of, 65
intracellular location of, 67–68
Phosphoenolpyruvic carboxylase,
affinity for substrates, 63
carbon dioxide fixation and, 52, 61–64
Phosphoenolpyruvic carboxytransphosphorylase,
biotin and, 91
carbon dioxide fixation and, 52, 69
Phosphofructokinase, 13
Phosphoglucomutase, synthesis of, 515
6-Phosphogluconate,
carbon dioxide fixation and, 52
metabolism of, 61
ribose formation from, 297
Phosphogluconic dehydrogenase,
carbon dioxide fixation and, 52, 59–61
distribution of, 60
3-Phosphoglycerate,
carbon dioxide fixation and, 52
formate utilization and, 25, 70
photosynthesis and, 12–14, 16–23, 26–27, 29, 34, 36
serine formation from, 205, 206
Phosphoglycerokinase,
photosynthesis and, 12, 21, 29
synthesis of, 28
O-Phosphohomoserine,
formation of, 195, 196
Phosphohydroxypyruvate,
serine formation and, 205, 206
Phospholipid, synthesis of, 414
4′-Phosphopantetheine,
formation of, 272, 273
occurrence of, 274
4′-Phosphopantothenate,
coenzyme A synthesis and, 274
formation of, 273
4′-Phosphopantothenylcysteine,
coenzyme A synthesis and, 274
formation of, 273

Phosphoriboisomerase,
formate utilization and, 25
photosynthesis and, 17, 21
synthesis of, 28
Phosphoribomutase, occurrence of, 298
1-(5′-Phosphoribosyl)-adenosine-5′-phosphate,
histidine biosynthesis and, 221
5-Phosphoribosylamine,
formation, 184, 300–301, 302
inhibition of, 314
5-Phosphoribosyl pyrophosphate, 239
biosynthesis of, 297
diphosphopyridine nucleotide synthesis and, 268
glutamine and, 184
histidine synthesis and, 220–221, 310, 314–315
nucleotide formation and, 308
purine synthesis and, 301, 302
pyrimidine nucleotide synthesis and, 321, 322
tryptophan biosynthesis and, 230, 232, 233
uracil utilization and, 327
Phosphoribulokinase,
formate utilization and, 25
heterotrophic growth and, 23
oxalate utilization and, 25
photosynthesis and, 12–13, 17–21, 23, 29
Phosphorus,
radioactive, enzyme formation and, 614–615
Phosphorylation,
oxidative, reduced pyridine nucleotide and, 32–33
Phosphoserine phosphatase,
serine synthesis and, 206
Phosphotransacetylase,
nitrogen fixation and, 131
occurrence of, 87
Photolithotrophism, definition of, 1–2
Photophosphorylation,
cyclic, 29
noncyclic, 29
Photosynthesis,
cyanide and, 10–11
general equation for, 7
Phthiocol, origin of, 285
Phycomyces blakesleeanus,
nicotinic acid synthesis by, 265
Phytol,
chlorophyll synthesis and, 362, 364
3-Phytylmenadione, occurrence of, 285
Pichia membranaefaciens,
folic acid synthesis by, 278
Picolinic acid,
nicotinic acid biosynthesis and, 267
Pilobolus, growth factors for, 342
Pimelic acid, biotin and, 274
Plants, photosynthesis in, 11–19
Plasmagene theory,
enzyme induction and, 601
Plastoquinone, occurrence of, 286
Pleuropneumonia-like organisms,
deoxyribonucleic acid of, 588
Pneumococcus, transformation in, 586
Polyadenylate, breakdown of, 549
Polycytidylate, priming effect of, 548
Polyglutamate(s),
biological activity of, 407–409
capsules and, 391–393
discovery of, 390
formation of, 236–237, 537
genetic code and, 409
hydrolysis of, 400–402, 405
immunological reactivity of, 405, 408
isomers, precipitation of, 403
occurrence of, 478
physicochemical properties of, 405–407
plasma volume extension by, 401-402, 408-409
structure of, 402–405
synthesis,
enzymes and, 397–400
nutritional requirements for, 394–396
tissue extracts and, 401–402
Polyglutamic acid synthetases,
properties of, 400, 409
Polyglycine, cell walls and, 417–418
Polyguanylic acid,
phosphorylase and, 549
Polyhedral viruses, nucleotides of, 481
Poly-β-hydroxybutyric acid,
acetate metabolism and, 7, 35-36
Polylysine, hydrolysis of, 402
Polymixin, nature of, 479
Polynucleotide phosphorylases,
copolymer synthesis and, 409

Polyol phosphate,
glycopeptide and, 422–425
isolation and structure of, 419–421
Polyphenylalanine,
synthesis of, 409, 563–564
Polyribonucleotide phosphorylase,
distribution of, 548–549
nucleic acid synthesis and, 547–549, 563
precipitation of, 546
role of, 560
soluble ribonucleic acid and, 549
Polysaccharides, 36
cell walls and, 418, 425
photosynthesis and, 18
synthesis, 414
general formulation of, 374
Polyuridylic acid,
amino acid incorporation and, 563
homopolymer synthesis and, 409
Porphobilinogen,
corrin synthesis and, 282
formation of, 353–354
porphyrin synthesis and, 348–351, 352
Porphobilinogen deaminase,
sources of, 351
Porphyrias, 336
porphobilinogen and, 349
Porphyrin(s),
chemistry of, 336–339
excretion of, 342–347
spectral characteristics of, 336–337
synthesis, 367
factors influencing, 347
Porphyrinogens,
iron binding by, 358
interconversion of, 351–352
porphyrin synthesis and, 350
Potassium, polyglutamate and, 407
Prephenic acid, 172
aromatization of, 228, 229, 242
formation of, 227–229
Prephenic aromatase,
action of, 228, 229
Prephenic dehydrogenase,
function of, 229
Primer,
nucleic acid synthesis and, 547, 552–553, 563
Prodigiosin,
chromatography of, 365
distribution of, 365
structure of, 365, 366
Proline,
biosynthesis of, 169, 172, 185–186, 240, 241
incorporation, 5-fluorouracil and, 488
ornithine and, 189, 206
prodigiosin synthesis and, 366
Prolylglycine,
peptidase, ribosomes and, 497
Propionate,
β-alanine formation and, 272
carbon dioxide fixation and, 9–11, 47
fermentation, methane and, 108–109
formation of, 73, 81–86
glycerol fermentation and, 44–46
pyrimidine biosynthesis and, 329
Propionibacteria,
carbon dioxide fixation by, 44–47, 69
hemes in, 357
oxybiotin and, 275
propionate formation by, 82
Propionibacterium pentosaceum,
carbon dioxide utilization by, 46
desthiobiotin and, 274–275
thiamine pyrophosphate and, 261
Propionibacterium shermanii,
cobamide coenzyme of, 281, 282, 283
porphyrin excretion by, 345, 346
succinate formation by, 84
vitamin B_{12} production by, 254
Propionyl carboxylase,
biotin and, 90, 95–96
carbon dioxide fixation and, 52, 71–75
methylmalonyl-oxalacetic transcarboxylase and, 84
reverse reaction of, 73
transcarboxylase function of, 97
Propionyl coenzyme A,
metabolism of, 157
methylmalonyl-oxalacetic transcar boxylase and, 84
pyruvic carboxylase and, 80
Protein(s),
cell walls and, 418
chemical nature of, 472–479
deoxyribonucleic acid and, 586
estimation of, 489–492
nitrogen fixation and, 129–130
ribosomes and, 495, 497
synthesis, 414, 519–531

Protein(s)—*Continued*
cell fragments and, 494
chlorophyll and, 364
contaminated preparations and, 498–499
cytochrome formation and, 359
disrupted cells and, 494
intact cells and, 492–493
mechanism components of, 502–519
nucleic acids and, 485–489
peptides and, 531–537
problems of, 477–478
protoplasts and, 493–494
ribosomes and, 494–498
site of, 499–502
ultraviolet and, 613–614
total possible number of, 578–579
true, 473–475
amino acid sequences and, 475–477
cross linkage and, 477
turnover of, 503, 543–544
Proteus, amino acid activation in, 507
Proteus morganii,
coenzyme A synthesis by, 272, 274
oxalacetate formation by, 51
Proteus vulgaris,
flagella, synthesis of, 493
glutamine synthetase in, 158, 183
γ-glutamyl transfer in, 536
inositol and, 286
thiamine synthesis by, 259
Protocatechuic oxidase,
induction of, 617
Protochlorophyll,
accumulation of, 345
chlorophyll formation and, 362
Protoplasts,
formation of, 459
protein synthesis by, 493–494
Protoporphyrin,
chlorophyll synthesis and, 362
formation of, 348, 351, 354–356
growth requirement for, 339, 340
isomers of, 336
Protoporphyrin IX,
excretion of, 345, 346–347, 362
structure of, 337
Protoporphyrinogen,
formation of, 348, 352
Pseudomonads,
cytochromes of, 359
purine synthesis by, 319
Pseudomonas,
denitrification by, 122
levan of, 383
nitrogen fixation in, 122, 125
tartrate utilization by, 603
Pseudomonas aeruginosa,
β-alanine formation by, 272
cytochromes in, 360
hydroxylamine reductase in, 123
nitrate reductase in, 122
nitric oxide reductase in, 122
nitrite reduction by, 122
nitrous oxide reductase in, 122
Pseudomonas citronellolis,
pyruvic carboxylase of, 80–81
Pseudomonas fluorescens,
aspartase of, 156
coenzyme Q of, 286
cytochrome formation in, 360
enzyme induction in, 594, 600
inositol and, 286
phosphogluconic dehydrogenase in, 60
transaminases in, 180
Pseudomonas methanica,
carbon dioxide fixation by, 24
Pseudomonas oleovorans,
β-methylcrotonyl carboxylase in, 77
Pseudomonas ovalis,
enzyme changes in, 617
Pseudomonas oxalaticus,
carbon dioxide fixation by, 24–25, 70
glyoxylate formation by, 108
phosphoenolpyruvic carboxykinase in, 65
Pseudomonas saccharophila,
α-amylase of, 596, 612
Pseudomonas stutzeri,
cytochromes in, 360
nitric oxide reductase in, 122
nitrite reduction by, 122
nitrogen formation by, 120
Pseudouridine, 296
occurrence of, 481, 482, 513
Pseudovitamin B_{12}, coenzyme form of, 281
Pteridines,
biosynthesis of, 279
folic acid synthesis and, 278

Pteroic acid,
folic acid synthesis and, 277, 278
Pteroylglutamic acid, *see also* Folic acid
peptides, hydrolysis of, 401
Pullalaria,
nitrogen fixation in, 122, 125
Purine(s),
carbon dioxide fixation and, 70–71
fermentation of, 105–107
histidine synthesis and, 219–221
protein synthesis and, 509, 510
pteridine synthesis and, 279
riboflavin synthesis and, 255, 262–263
serine formation and, 204
synthesis,
repression of, 610, 618
suppressor mutations and, 591
Purine nucleotides,
biosynthesis,
ammonia incorporation and, 160
deoxyribonucleotides and, 317–318
Enterobacteriaceae and, 305–307
enzymic reactions and, 299–305
interconversion and, 307–312
precursors and, 298–299
regulation of, 313–317
utilization pattern of, 312–313
vitamins and amino acids and, 318–320
Puromycin, protein synthesis and, 562
Putrescine,
formation of, 186
ribosomes and, 495
Pyridine-2,6-dicarboxylic acid, *see* Dipicolinic acid
Pyridine nucleotides, *see also* Di- and Triphosphopyridine nucleotides
aspartic semialdehyde dehydrogenase and, 195–196
carbon dioxide fixation and, 51–61
folic acid reduction and, 279
iron-oxidizing bacteria and, 32
photosynthesis and, 30
Pyridoxal,
biosynthesis of, 269–270
porphyrin synthesis and, 356
transamination and, 181
Pyridoxal kinase, source of, 270
Pyridoxal phosphate,
cystathionine cleavage and, 197
cystathionine formation and, 195, 196
diaminopimelic acid decarboxylase and, 200
formation of, 270
porphyrin synthesis and, 348, 353
Pyridoxamine phosphate,
transamination and, 270
Pyridoxine,
thiamine and, 262
utilization of, 270
Pyrimidine(s),
biosynthesis, 190, 191, 192, 242, 320–325
ammonia incorporation and, 160–161
alternate pathways and, 328–329
regulation of, 325–327
vitamins and amino acids and, 329–330
feedback and, 618
repression of, 609, 610
interconversions of, 327–328
protein synthesis and, 509, 510
Pyrimidine deoxyribosephosphorylase,
deoxycytidine utilization and, 328
Pyrophosphatase,
carbon dioxide fixation and, 69
Pyrophosphate,
amino acid activation and, 505–506
pyridine nucleotide synthesis and, 269
Pyrophosphorylases,
cell wall synthesis and, 445–446
Pyrrolidone carboxylate,
glutamine synthetase and, 158–159
Δ^1-Pyrroline-5-carboxylic acid,
ornithine synthesis and, 188–189
proline biosynthesis and, 186, 187
Pyruvate,
acetate formation and, 107, 108
alanine formation from, 202–203
biosynthetic pathways and, 168
carbon dioxide fixation and, 9, 47–51, 110–111
cellulose formation from, 381
cell walls and, 442
cystathionine cleavage and, 197
cysteine formation and, 207
diaminopimelate formation and, 200, 201, 453
folic acid reduction and, 279
formate exchange and, 86, 89
isoleucine biosynthesis and, 209
leucine biosynthesis and, 209, 215
methylmalonyl-oxalacetic transcar-

Pyruvate—*Continued*
boxylase and, 84
malic enzyme and, 53–54
nicotinic acid biosynthesis and, 266
nitrogen fixation and, 128–129, 131–134, 141, 142, 146
photophosphorylation and, 30
photosynthesis and, 18
propionate formation and, 81–83, 86
pyridoxal biosynthesis and, 270
pyridoxamine phosphate and, 270
riboflavin synthesis and, 264–265
serine formation and, 205
transamination of, 181, 397–398
valine synthesis from, 208–212
Pyruvic carboxylase,
biotin and, 90, 97
carbon dioxide fixation and, 52, 79–91
Pyruvic kinase,
β-methylcrotonyl carboxylase and, 76
propionyl carboxylase and, 72
synthesis of, 515
Pyruvic oxidase,
carbon dioxide exchange by, 88
Pyruvic oxime,
nitrogen fixation and, 140, 144

Q

Quinic acid, 171, 242
aromatic ring biosynthesis and, 223
oxidation of, 240
Quinic dehydrogenase,
distribution of, 223
Quinolinic acid,
nicotinic acid biosynthesis and, 267

R

Rabbits, atropine esterase in, 593
Racemase(s)
diaminopimelate isomers and, 202
polyglutamate synthesis and, 398
Raffinose,
dextransucrase and, 376, 377
levansucrase and, 385
levan synthesis from, 382
utilization, genes and, 591
Red clover, nitrogen fixation by, 144
Repression,
coordinate, 609–610
function of, 616–618
nucleic acid and, 488–489
pyrimidine biosynthesis and, 326–327
tetrapyrrole synthesis and, 368
Repressor,
combining site of, 609
formation of, 608
identity of, 610, 621
inducible enzymes and, 608, 621
Respiratory systems,
inheritance of, 582, 593
Reticulocytes,
ribosomes of, 515, 521, 522
Rhamnose, cell walls and, 425
Rhizobia,
nitrogen fixation by, 125, 127
Rhizobium trifolii,
β-alanine formation by, 272
oxybiotin and, 275
Rhizopterin,
folic acid synthesis and, 277
Rhodomicrobium vannielii,
nitrogen fixation by, 122
Rhodopseudomonas capsulatus,
carbon dioxide fixation by, 22–23, 27
porphyrin biosynthesis by, 353
porphyrin excretion by, 343, 344
Rhodopseudomonas gelatinosa,
isopropanol utilization by, 8
porphyrin excretion by, 343
Rhodopseudomonas palustris,
carboxydismutase in, 27–28
porphyrin excretion by, 343
Rhodopseudomonas spheroides,
acetate metabolism of, 10
bacteriochlorophyll formation by, 354–356, 363–365
carboxydismutase in, 27–28
enzyme changes in, 617
hematin formation by, 358
hemoprotein synthesis by, 360
photometabolism, cyanide and, 10–11
porphyrin biosynthesis by, 352–354, 357, 367
porphyrin excretion by, 343–346, 367
triose phosphate dehydrogenase of, 30
Rhodospirilla,
amino acid activation in, 507
Rhodospirillum rubrum,
acetate metabolism of, 8–10, 26, 35
alanine formation by, 89
bacteriochlorophyll synthesis by, 364–365

carbon dioxide exchange by, 10
carbon dioxide fixation by, 26–27
carboxydismutase in, 27
coenzyme Q of, 286
fatty acid metabolism of, 7
nitrogen fixation by, 122, 127, 145
organic substrates and, 9, 36
photometabolism, cyanide and, 10–11
photophosphorylation in, 30
porphyrin excretion by, 342, 343
propionyl carboxylase in, 73
Ribitol phosphate,
cell walls and, 419–421, 424, 479
polymer, 415, 418
accumulation of, 438
4-Ribitylamino-5-aminouracil,
riboflavin synthesis and, 263
Riboflavin, 222
accumulation of, 254
aminopterin and, 279
Riboflavin,
biosynthesis of, 255, 262–265
porphyrin formation and, 347, 357
pyrimidine synthesis and, 330
requirement for, 262
Riboflavin-5′-phosphate,
formation of, 265
Ribonuclease,
amino acid-lipid complexes and, 517
amino acids of, 477
cell wall synthesis and, 451
molecular weight of, 473
nucleic acid synthesis and, 545, 546–547
polyglutamate synthesis and, 409
protein synthesis and, 486, 493, 509, 511, 516, 526, 527, 529, 562, 589
protoplasts and, 493–494
ribosomes and, 497
Ribonucleic acid, *see also* Nucleic acids
coding and, 559, 561, 564, 588
components of, 296
composition of, 480–481
deoxyribonucleic acid synthesis and, 551–552
distribution of, 483–484
hereditary specificity and, 582
incorporation factor and, 516
messenger, 561–562, 564
protein synthesis and, 510, 530, 531, 587, 589, 590, 615, 620
repressor and, 610
ribosomal,
composition of, 484
protein synthesis and, 562–563
ribosomes and, 495
soluble,
composition of, 484
extraction of, 482
protein synthesis and, 510–513, 519–527
turnover of, 503–504
role of, 524–525
terminal nucleotides of, 545–546
polyribonucleotide phosphorylase and, 549
synthesis,
amino acids and, 541–544
analogs and, 540–541
cell-free systems and, 545–549
chloramphenicol and, 554
disrupted cells and, 544
intact cells and, 540–544
membrane fragments and, 545
polyribonucleotide pyrophosphorylase and, 547–549
ribonucleotide triphosphates and, 546–547
terminal nucleotides and, 545–546
turnover of, 543–544, 551, 561
Ribonuleotides,
deoxyribonucleic acid synthesis and, 553
protein synthesis and, 530, 531
Ribose,
histidine and, 219
nicotinic acid biosynthesis and, 266
Ribose phosphate(s),
formation of, 296–298
Ribose-1-phosphate,
nucleoside formation and, 298
uracil utilization and, 327
Ribose-5-phosphate,
biosynthetic pathways and, 169
formate utilization and, 25
formation of, 102
phosphoribosylamine formation from, 301, 302
photosynthesis and, 12, 15, 17, 18, 20, 21, 22, 27

Ribosomes,
assembly of, 564
chloramphenicol and, 556–557
composition of, 495, 497, 521, 561, 589
contamination of, 498–499
nature of, 483
protein synthesis and, 494–498, 515, 557, 560–564, 590, 620
protein release from, 515, 521
structure of, 495, 496
turnover in, 503
Ribulose, histidine and, 219
Ribulose-1,5-diphosphate,
formate utilization and, 25
photosynthesis and 12–14, 16, 17, 19, 20, 21, 27, 29
Ribulose diphosphate carboxylase, 111
affinity for substrate, 63
carbon dioxide fixation and, 52, 69–70
Ribulose-5-phosphate,
carbon dioxide fixation and, 52, 59–60
photosynthesis and, 12, 17, 18, 21, 29
ribose formation and, 297
Ruminants,
flora, vitamins and, 254

S

Saccharomyces, *see also* Yeast
tryptophan auxotrophs of, 233
Saccharomyces anamensis,
porphyrin formation by, 345, 346, 347
Saccharomyces carlsbergensis,
coenzyme A synthesis by, 274
Saccharomyces cerevisiae,
biotin sulfoxide and, 276
coenzyme Q in, 286
cytochrome formation by, 359
desthiobiotin and, 275
mutant, heme synthesis by, 354
porphyrin excretion by, 346, 347
pyridine nucleotide synthesis in, 268
thiamine formation by, 262
Salicylic acid,
pantothenic acid and, 255
Salmonella, 579
polysaccharides of, 425
rough, lipopolysaccharide of, 464
Salmonella gallinarum,
cell wall synthesis in, 452
Salmonella paratyphi,
inheritance in, 582
Salmonella typhimurium,
adenine-guanine conversion in, 309–310
alanine formation by, 203
complementation in, 584
flagella, synthesis of, 493, 504
flagellin of, 473, 474
histidine synthesis by, 217–218, 220–221
leucine auxotrophs of, 215, 216
mutant, glucose utilization by, 611
5-phosphoribosylamine formation by, 301
Sarcina lutea,
cell wall synthesis in, 452, 478
vitamin K synthesis by, 285
Saturation amination,
ammonia incorporation and, 155–157
Schardinger dextrins,
Bacillus macerans amylase and, 379–380
dextrandextrinase and, 378
Sedoheptulose,
formate utilization and, 25
Sedoheptulose-1,6-diphosphate,
dehydroshikimate formation from, 225
photosynthesis and, 21
Sedoheptulose-7-phosphate,
photosynthesis and, 12–14, 16, 19, 20, 21
ribose formation from, 297
Sedormid, porphyrin excretion and, 336
Selenomethionine,
enzyme activity and, 581
Serine,
acetate formation and, 107, 108
activation of, 506
biosynthesis of, 203–206, 242
conversion to glycine, 204
cystathionine cleavage and, 197
cysteine formation and, 207
deamination of, 213
methanol utilization and, 70
methionine biosynthesis and, 283
purine synthesis and, 299, 319
pyridoxal synthesis and, 270
pyrimidine synthesis and, 330
tryptophan biosynthesis and, 231, 241, 242
Serine adenylate, formation of, 506
Serine aldolase,
homoserine methylation and, 198
requirements of, 204

Serine deaminase,
glucose and, 611
induction of, 600, 602–603, 612
D-Serine, 397
D-Serine deaminase, 610
induction of, 600, 602–603
ultraviolet and, 613
Serine dehydrase,
5-fluorouracil and, 487
Serine hydroxymethylase, 204
methanol utilization and, 70
Serine sulfhydrase,
requirements of, 207
Serratia, amino acid activation in, 507
Serratia marcescens,
prodigiosin synthesis by, 365
Shikimic acid,
aromatic ring synthesis and, 222–223
formation of, 240
isotope content of, 172
reactions following, 226–227
riboflavin synthesis and, 263
Shikimic acid-5-phosphate,
p-aminobenzoate synthesis and, 276
anthranilate formation from, 235–236
aromatic biosynthesis and, 226–227
tryptophan biosynthesis and, 230
Shope papilloma virus,
deoxyribonucleic acid of, 588
Silkworm, protein synthesis by, 524
Sodium, polyglutamate and, 406–407
Soybean,
root nodules, nitrogen fixation by, 136, 139, 140, 142, 145, 149
Sparine effects, significance of, 205
Spectrophotometry,
nitrogen fixation and, 142–143
Spermidine, 186
β-alanine formation and, 272
ribosomes and, 495
Spermine, 186
β-alanine formation and, 272
nucleic acid synthesis and, 545
ribosomes and, 495
Spheroplasts,
protein synthesis by, 493–494
Spinach,
phosphoenolpyruvic carboxylase of, 61
photosynthesis by, 16
Spirillum itersonii,
denitrification by, 122
Sporulation, protein trunover and, 503
Staphylococci,
disrupted, nucleic acid synthesis in, 544
nutritional requirements of, 434, 435
sulfonamide resistant, 256
Staphylococcus,
amino acid activation in, 507, 508, 524
protein synthesis in, 503, 555
Staphylococcus albus,
cell wall synthesis in, 452
nucleotide accumulation and, 443
Staphylococcus aureus,
amino acid activation in, 509
cell wall of, 415–424, 426, 428, 429, 437–439, 452, 459, 478, 535, 537–538
chloramphenicol and, 557
folic acid synthesis by, 257, 277, 278
glutamine synthetase in, 158
nucleic acid synthesis by, 541–543
nucleotide accumulation by, 431–440, 443–444
peptide formation by, 535
protein synthesis in, 487, 494, 498, 509, 510, 516, 527–529
purine utilization by, 312
strain 209P, cell wall of, 437, 439–440
vitamin K synthesis by, 285
Starch,
formation of, 379
photosynthesis and, 18
Strepogenin, requirement for, 532, 534
Streptobacterium plantarum,
coenzyme A synthesis by, 273–274
folic acid synthesis by, 277
Streptococci,
cell walls of, 418, 421
orotic acid and, 320
polysaccharide synthesis by, 378–379
Streptococcus,
amino acid activation in, 507
biotin requirement of, 89
protein synthesis by, 503
Streptococcus bovis,
dextran of, 374, 377
dextransucrase of, 377–378
Streptococcus cremoris,
lipoic acid and, 284
Streptococcus faecalis,
biocytin and, 276
carbamylaspartate synthesis by, 320
carbamyl phosphate synthetase in, 159

Streptococcus faecalis—*Continued*
cell wall synthesis in, 452, 478
citrulline degradation by, 191
coenzyme F synthesis by, 279
dihydrofolate reduction by, 324
folic acid synthesis by, 277
halogenated uracils and, 541
lipoate liberation by, 284
lysine deprivation in, 435–437, 440
nucleotide accumulation by, 442–443
nucleotide-peptides in, 537
peptide utilization by, 533–534
pyridoxal in, 270
Streptococcus hemolyticus,
amino acid activation in, 507
nucleotide accumulation by, 443
Streptococcus lactis,
citrulline synthesis in, 518
Streptococcus pyogenes,
cell walls of, 478
streptolysin formation by, 548
Streptococcus salivarius,
nucleotide accumulation by, 443
Streptolysin S, formation of, 548
Streptomyces, lytic enzymes of, 415
Streptomyces griseus,
thymidine diphosphate sugars in, 464
vitamin B_{12} synthesis by, 282
Streptomyces olivaceus,
vitamin B_{12} synthesis by, 282
Streptomycetes, vitamin B_{12} in, 282
Streptomycin,
resistance,
hemin requirement and, 341–342
transformation and, 588
Succinate, 58
acetate condensation and, 43
β-alanine formation from, 329
aspartase and, 156, 179
carbon dioxide fixation and, 9–11, 46, 72
decarboxylation of, 85–86
diaminopimelate formation and, 200
glutamate synthesis and, 176
glycerol fermentation and, 44–46
nicotinic acid biosynthesis and, 266
porphyrin synthesis and, 352, 354, 355, 367
photophosphorylation and, 30
propionate formation and, 82–83
pyridine nucleotide reduction and, 33
utilization of, 36
Succinic dehydrogenase,
5-fluorouracil and, 487
induction,
inducer removal and, 603
kinetics of, 600–601
Succinic thiokinase,
porphyrin synthesis and, 353
N-Succinyl-α-amino-ϵ-ketopimelate,
formation of, 201
Succinyl coenzyme A,
biosynthetic pathways and, 169
porphyrin synthesis and, 348–350, 353, 357
N Succinyl-L-diaminopimelate,
deacylation of, 202
formation of, 201–202
N-Succinylglutamic acid,
occurrence of, 189
Sucrose,
dextrandextrinase and, 378
dextran synthesis from, 375–377
levansucrase and, 383–385
levan synthesis from, 373, 382
nitrogen fixation and, 126, 141
photosynthesis and, 18
"starch" formation from, 379
Sulfanilamide,
purine synthesis and, 299
requirement, basis of, 616
Sulfate,
activation of, 207
carbon dioxide fixation and, 6, 19
nitrogen fixation and, 126
Sulfide, sulfur assimilation and, 207
Sulfite, sulfur assimilation and, 207
Sulfonamides, 453
p-aminobenzoylglutamate and, 277–278
folic acid synthesis and, 279
methionine and, 197
resistance to, 256
Sulfur,
assimilation of, 207
carbon dioxide fixation and, 2, 19
thermodynamic efficiency and, 3, 4
Syntrophism,
biosynthetic pathways and, 172

T

Tartrate, utilization, 603
Teichoic acid(s),

alanine and, 423
antigenicity of, 420–421
cell walls and, 479
isolation of, 419
Temperature, enzyme formation and, 613
Template(s),
deoxyribonucleic acid and, 587–588
enzyme induction and, 603
Terregens factor, source of, 342
Tetrahydrofolate formylase,
sources of, 279
Tetrahydrofolic acid,
purine fermentation and, 106–107
hydroxymethyl derivative of, 280
purine synthesis and, 318
serine-glycine interconversion and, 204
Tetrahydropteroyltriglutamic acid,
methionine formation and, 198
Tetrahymena pyriformis (*geleii*),
7-azatryptophan and, 505
chloramphenicol and, 556–557
ethionine and, 505
glutamate incorporation by, 518
porphyrin formation by, 345
purine utilization by, 312–313
Tetrahymena vorax,
porphyrin formation by, 345, 346, 354–357
Tetrapyrroles,
biosynthesis, 347–357
regulation of, 366–368
general distribution of, 335–336
growth requirement for, 339–342, 361–362
Tetrathionase, synthesis of, 612, 613
Tetrathionate,
carbon dioxide fixation and, 6
Thiamine,
biosynthesis of, 259–262, 306
cysteine and, 208
α-hydroxyethyl derivative of, 262
porphyrin synthesis and, 357
purine synthesis and, 319
pyridoxine and, 262
Thiamine monophosphate,
cocarboxylase formation and, 260, 261
Thiamine pyrophosphate,
carbon dioxide exchange and, 91
formation of, 260, 261
lipoic acid and, 284
nitrogen fixation and, 129, 131–132
phosphoroclastic reaction and, 87, 89
ribose synthesis and, 297
Thiaminokinase,
nucleoside triphosphates and, 261
Thiazoles, synthesis of, 260–261
Thiazolidine carboxylic acid,
thiazole auxotrophs and, 260
β-2-Thienylalanine,
deoxyribonucleic acid and, 610
Thiobacillus, cytochromes of, 31
Thiobacillus denitrificans,
carbon dioxide fixation by, 4, 19–20
cytochromes in, 31
denitrification by, 122
photosynthesis by, 16, 19–20, 30
Thiobacillus thiooxidans,
carbon dioxide fixation by, 4, 21, 30
phosphoenolpyruvic carboxykinase in, 65
phosphoenolpyruvic carboxylase of, 63
Thiobacillus thioparus,
carbon dioxide fixation by, 4, 21
denitrification by, 122
Thiogalactosides,
β-galactosidase and, 599
β-Thioglucuronides,
β-glucuronidase induction and, 600, 601
Thiomethylgalactoside,
β-galactosidase induction by, 488
Thiorhodaceae,
carbon dioxide fixation by, 6–7
nitrogen fixation by, 144–145
porphyrin excretion by, 342
Thiosulfate,
carbon dioxide fixation and, 19–20
cytochrome reduction and, 31
sulfur assimilation and, 207
thermodynamic efficiency and, 3, 4
2-Thiouracil,
β-galactosidase formation and, 590
Threonine,
acceptor ribonucleic acid for, 512
activation of, 506
aspartokinase and, 194
biosynthesis, 169, 182, 192–196, 242
feedback and, 618
α-ketobutyrate and, 213, 243
isoleucine biosynthesis and, 205, 209, 212, 213
metabolism of, 173
vitamin B_{12} and, 282

D-Threonine,
D-serine deaminase and, 600
L-Threonine deaminase,
function of, 213
L-Threonine-L-serine deaminase,
adaptive, 213
Thrombin, amino acids of, 477
Thymidine,
polysaccharide synthesis and, 414
utilization of, 328, 329
vitamin B_{12} and, 298
Thymidine diphosphate-6-deoxy-D-glucose,
isolation of, 464
Thymidine diphosphate fucose,
isolation of, 464
Thymidine diphosphate galactose,
formation of, 464
Thymidine diphosphate glucose,
transformations of, 463–464
Thymidine diphosphate-4-keto-6-deoxy-D-glucose,
isolation of, 464
Thymidine diphosphate mannose,
occurrence of, 464
Thymidine diphosphate-L-rhamnose,
occurrence of, 464
synthesis of, 463–464
Thymidine phosphates,
biosynthesis of, 323–325
Thymidine phosphorylase,
induction of, 604
Thymidylate, bacteriophage and, 325
Thymidylate synthetase,
cofactor of, 324
Thymine,
requirement, supressors and, 591
ribonucleic acid and, 296, 481, 482
synthesis of, 330
uracil analogs and, 541
utilization of, 328, 329
Thymus, chloramphenicol and, 557
Tobacco mosaic virus,
8-azaguanine in, 540
molecular weight of, 473
nucleic acid, phosphorylase and, 549
nucleotides of, 481
protein synthesis by, 559
ribonucleic acid of, 588
Tobacco necrosis virus,
nucleotides of, 481
Toluene, aspartase and, 165
Tomato bushy stunt virus,
nucleotides of, 481
Torula cremoris,
biotin requirement of, 89
Torulopsis utilis,
purine utilization by, 312
Toxin, virulence and, 408
Toxopyrimidine phosphate,
pyridoxal phosphate and, 262
Transaldolase,
photosynthesis and 12, 15, 16, 17, 20, 21
xylose fermentation and, 88
Transamidation,
polyglutamate synthesis and, 399
Transaminase(s),
D-amino acids and, 237, 397–398
ω-amino acids and, 181
aspartate formation by, 192
diaminopimelate formation and, 201–202
distribution of, 180–181
glutamate formation by, 177–178, 180–182
histidinol phosphate formation and, 217
leucine synthesis and, 216
phenylalanine formation and, 229
polyglutamate synthesis and, 397–398
pyruvate and, 202
specificity of, 180–181
tyrosine and, 229
valine synthesis and, 215
Transamination,
mechanism of, 181–182
significance of, 181
Trans-N-deoxyribosylase,
function of, 318
thymine utilization and, 328
Transduction,
cistron size and, 588
enzyme formation and, 583, 586–587
Transferring enzymes,
protein synthesis and, 514
Transformation,
deoxyribonucleic acid and, 586
enzyme formation and, 583, 589
molecular weight and, 588
Transketolase,
photosynthesis and, 12, 15, 17, 20, 21

ribose synthesis and, 297
xylose fermentation and, 88
Transpeptidation,
polyglutamate synthesis and, 399–400
Tricarboxylic acid cycle,
biosynthetic pathways and, 169
oxygen and, 359, 360
porphyrin synthesis and, 357, 367
Trichloroacetic acid,
cell wall extraction and, 419, 422
Trichloromethylsulfenyl benzoate,
nitrogen fixation and, 126
Trichophyton equinum,
nicotinic acid synthesis by, 265
Triglycine, activation of, 509
4-(Trihydroxypropyl)-imidazole, *see* Imidazole glycerol
Triose phosphate,
photosynthesis and, 13–14, 29, 36
Triose phosphate dehydrogenase,
photosynthesis and, 12, 17, 20, 21, 30
synthesis of, 28
Triose phosphate isomerase,
photosynthesis and, 12, 17
Triphosphopyridine nucleotide,
N-acetylglutamic semialdehyde formation and, 188
alanine dehydrogenase and, 154
amino acid biosynthesis and, 239–240
biosynthesis of, 269
glutamate synthesis and, 177, 179
glutamic dehydrogenase and, 153, 154
isocitric dehydrogenase and, 57–58
malic enzyme and, 53–55
nitrate reductase and, 122
nitric oxide reductase and, 122
nitrite assimilation and, 122
nitrite reduction and, 122
phosphogluconic dehydrogenase and, 59–60
photosynthesis and, 12, 13, 17, 18, 29, 30
proline synthesis and, 186
ribose formation and, 297
shikimate formation and, 223
Tritium,
decay, protein synthesis and, 615
Trypanosomes,
glucose dissimilation by, 44
porphyrin requirement of, 340
Trypsin,
amino acids of, 477
polyglutamate and, 402
protein assay and, 490
protoplasts and, 493
Tryptazan,
activation of, 506
protein synthesis and, 504–505
Tryptophan,
analogs, reversal of, 175
biosynthesis, 182, 222–227, 229–236, 241, 242
complementation and, 584
feedback and, 618
repression of, 608, 610
degradation of, 177
flagella synthesis and, 504
mutants, 583
nicotinic acid biosynthesis from, 265, 266
photosynthesis and, 18
requirement, transformation and, 586
sequential induction and, 616–617
silk protein and, 524
transamination of, 180
Tryptophan adenylate,
formation of, 506, 507
Tryptophanase, 239, 579
formation of, 612
inducibility of, 602–604, 616
products of, 235
ultraviolet and, 612, 614
Tryptophan desmolase, 233
Tryptophan synthetase, 239
antibody to, 234
cistrons and, 584
components of, 234–235
constitutive, 593
gene mapping of, 584
mutations and, 585
supressor mutations and, 591–592
tryptophan auxotrophs and, 233
Tyrocidin, nature of, 479
Tyrosinase, genes affecting, 591
mutations and, 585
repression of, 608
Tyrosine,
acceptor ribonucleic acid for, 512
activation of, 506
biosynthesis of, 172, 182, 222–229, 241, 242

Tyrosine—*Continued*
cross linkages and, 477
excretion of, 237
incorporation, 5-fluorouracil and, 488
transamination of, 180
Tyrosine decarboxylase,
peptide requirement and, 534
Tyrothrycin, amino acids of, 473

U

Ubiquinone, *see* Coenzyme Q
Ultraviolet,
enzyme synthesis and, 613–614
Uracil,
β-alanine formation from, 272
degradation of, 329
enzyme repression and, 607
halogenated, incorporation of, 541
orotic acid excretion and, 325–326
thymine synthesis and, 323
utilization of, 327
Urea, formation of, 186
4-Ureido-5-imidazolecarboxylic acid,
purine fermentation and, 107
β-Ureidopropionic acid,
β-alanine formation from, 272
Uric acid,
biosynthesis of, 298–299
fermentation of, 105–107
Uridine,
polysaccharide synthesis and, 414, 417
utilization of, 327
Uridine diphosphate,
phosphorylation of, 445
Uridine diphosphate acetylgalactosamine, 427
Uridine diphosphate acetylglucosamine, 427
synthesis of, 445–446
transglycosylation and, 457, 458
Uridine diphosphate acetylglucosamine lactate,
formation of, 446
Uridine diphosphate acetylglucosamine-pyruvate transferase,
cell wall synthesis and, 446
Uridine diphosphate acetylmuramic acid,
formation of, 446
transglycosylation and, 457
Uridine diphosphate acetylmuramyl peptide(s),
cell wall synthesis and, 426–429, 431–433, 435, 442–443, 453, 538–540
Uridine diphosphate arabinose, 427
Uridine diphosphate galactose, 426
Uridine diphosphate galacturonic acid, 427
Uridine diphosphate glucosamine-6-phospho-1-galactose, 427
Uridine diphosphate glucose, 426
cellulose formation and, 381–382
transformations of, 414–415
Uridine diphosphate glucuronic acid, 427
formation of, 415
Uridine diphosphate muramic acid,
accumulation of, 536, 538
Uridine diphosphate xylose, 427
Uridine monophosphate pyrophosphorylase,
uracil utilization and, 327
Uridine nucleotides,
biosynthesis of, 320–322, 444–453
Uridine triphosphate,
amination of, 160–161
cell wall synthesis and, 445–446
cytidine triphosphate formation from, 322
nucleic acid synthesis and, 546
soluble ribonucleic acid and, 545
Urocanase, glucose and, 611
Uronic acid,
enzymes, induction of, 604
Uroporphyrin,
catalase synthesis and, 362
isomers of, 336
Uroporphyrin I,
excretion of, 344, 346
formation of, 336, 348, 354
structure of, 337
Uroporphyrinogen I,
formation of, 348, 351
Uroporphyrin III,
excretion of, 344, 346, 352
formation of, 348, 355
structure of, 337
Uroporphyrinogen III,
formation of, 348, 351, 354
heme formation and, 350

Uroporphyrinogen decarboxylase,
purification of, 354
Ustilago sphaerogena,
ferrichrome and, 342

V

Valeric acid,
carbon dioxide exchange and, 78
Valine,
activation of, 506
biosynthesis of, 168, 169, 172, 174, 182, 208–212, 214–215, 240, 241, 243
alanine formation and, 202–203
erroneous utilization of, 580
excretion of, 238
incorporation of, 505
lipid and, 517
pantothenate synthesis and, 258, 271
salicylic acid and, 255, 256
sickle cell hemoglobin and, 558
thiazole synthesis and, 260, 261
transamination of, 180
transfer ribonucleic acid for, 512
Vasopressin, strepogenin and, 532
Viruses, deoxyribonucleic acid of, 588
Visacosaccharose,
levan synthesis and, 373
Vitamin(s),
excretion of, 254
synthesis,
cell-free preparations and, 257, 259
growing cultures and, 254–256, 257–258
washed suspensions and, 256–257, 259
Vitamin A, bacteria and, 253
Vitamin(s) B_6,
biosynthesis of, 269–270
peptide requirement and, 533
Vitamin B_{12}, 275
biosynthesis of, 280–283
deoxyriboside formation and, 298, 328, 330
methionine and, 198, 258, 283
methylmalonyl isomerase and, 85
nitrogen fixation and, 129
production of, 254
protein synthesis and, 518–519
Vitamin B_{12} anilide,
methionine biosynthesis and, 283
Vitamin D, bacteria and, 253
Vitamin K,
bacterial, 284–285
biosynthesis of, 285

W

Water, photosynthesis and, 34
Waxes, cell walls and, 418, 419

X

Xanthine,
conversion to adenine, 308
fermentation of, 105–107
nucleotide formation from, 308
requirement for, 305, 306
riboflavin synthesis and, 263, 264
Xanthomonas pruni,
nicotinic acid synthesis by, 265, 267
Xanthopterin,
folic acid synthesis and, 278
Xanthosine,
excretion of, 306
riboflavin biosynthesis and, 265
utilization of, 311
Xanthosine-5′-phosphate,
amination of, 160–161
guanine formation from, 308
Xanthosine-5′-phosphate aminase,
adenosine triphosphate and, 317
bacterial, 301
mutants and, 306
X-rays, enzyme formation and, 614
Xylose,
fermentation, carbon dioxide and, 88
Xylsucrose,
dextransucrase and, 376
levansucrase and, 385
Xylulose-5-phosphate,
photosynthesis and, 12, 15, 17
ribose formation from, 297

Y

Yeast, *see also Saccharomyces*
amino acid activation in, 508
aspartic semialdehyde dehydrogenase of, 195–196
aspartokinase of, 194
catalase of, 600
coenzyme A synthesis by, 274
cysteine synthesis by, 207

Yeast—*Continued*
enzyme induction in, 594, 596, 599-601, 614
enzyme synthesis in, 487
β-glucosidase of, 593, 600, 603, 605
glutamic dehydrogenase of, 153
histidine synthesis in, 219
isocitric dehydrogenases of, 57
isoleucine synthesis by, 212, 214
lipoate liberation by, 284
maltozymase of, 600, 603, 614
oxybiotin and, 275
peptide synthesis by, 535–536
phosphoenolpyruvic carboxykinase of, 65
phosphogluconic dehydrogenase of, 59
polyploid, enzymes of, 592
porphyrin synthesis in, 354
purine synthesis by, 319
pyridoxal kinase of, 270
respiratory system, inheritance of, 582, 593
ribonucleic acids of, 513, 561
ribosomes of, 498
thiamine formation by, 260
threonine synthesis in, 195–196
uridine monophosphate formation by, 322
valine synthesis by, 208, 214

Z

Zinc,
enzyme formation and, 612
nitrogen fixation and, 126
polyglutamate synthesis and, 395–396
tryptophan synthetase and, 592
Zygorrhynchus moelleri,
carbon dioxide fixation by, 60–61
Zymobacterium oroticum,
orotic acid formation by, 320–321